Intelligent Information Systems and Knowledge Management for Energy: Applications for Decision Support, Usage, and Environmental Protection

Kostas Metaxiotis
National Technical University of Athens, Greece

INFORMATION SCIENCE REFERENCE
Hershey · New York

Director of Editorial Content: Kristin Klinger
Senior Managing Editor: Jamie Snavely
Assistant Managing Editor: Michael Brehm
Publishing Assistant: Sean Woznicki
Typesetter: Sean Woznicki
Cover Design: Lisa Tosheff
Printed at: Yurchak Printing Inc.

Published in the United States of America by
Information Science Reference (an imprint of IGI Global)
701 E. Chocolate Avenue
Hershey PA 17033
Tel: 717-533-8845
Fax: 717-533-8661
E-mail: cust@igi-global.com
Web site: http://www.igi-global.com/reference

Library of Congress Cataloging-in-Publication Data

Intelligent information systems and knowledge management for energy :
applications for decision support, usage, and environmental protection /
Kostas Metaxiotis, editor.
p. cm.
Includes bibliographical references and index.
Summary: "This book analyzes the need for a holistic approach for the construction and engineering of cities and societies"--Provided by publisher.
ISBN 978-1-60566-737-9 (hardcover) -- ISBN 978-1-60566-738-6 (ebook) 1.
Power resources--Environmental aspects. 2. Power electronics. I. Metaxiotis,
Kostas.
HD9502.A2I554 2010
333.793'20684--dc22

2009004401

British Cataloguing in Publication Data
A Cataloguing in Publication record for this book is available from the British Library.

Dedication

"This book is dedicated to my parents who are talented energetic persons and provided me with their valuable knowledge and energy for many years".

Kostas Metaxiotis

Table of Contents

Foreword xvi

Preface xviii

Section 1
Theories and Concepts

Chapter 1
Artificial Intelligence Techniques for Modern Energy Applications 1
Soteris Kalogirou, University of Technology, Cyprus
Kostas Metaxiotis, University of Piraeus, Greece
Adel Mellit, Jijel University, Algeria

Chapter 2
Intelligent Control of the Energy Generation Systems 40
Nicu Bizon, University of Pitesti, Romania

Chapter 3
Knowledge Management for Electric Power Utility Companies 97
Campbell Booth, University of Strathclyde, UK

Chapter 4
Solving Environmental/Economic Dispatch Problem: The Use of Multiobjective Particle Swarm Optimization 123
M. A. Abido, King Fahd University of Petroleum & Minerals, Saudi Arabia

Chapter 5
Doubly Fed Induction Generators: Overview and Intelligent Control Strategies for Wind Energy Conversion Systems 147
Vinod Kumar, College of Technology and Engineering, India
Steven Kong, University of Queensland, Australia
Yateendra Mishra, University of Queensland, Australia
Z. Y. Dong, The Hong Kong Polytechnic University, Hong Kong
Ramesh C. Bansal, University of Queensland, Australia

Section 2
Models and Tools

Chapter 6
Use of Neural Networks for Modeling Energy Consumption in the Residential Sector 180
Merih Aydinalp Koksal, Hacettepe University, Turkey

Chapter 7
Setting Technology Transfer Priorities with CDM-SET3: Development of Sustainable Energy Technology Transfer Tool 205
Alexandros Flamos, National Technical University of Athens, Greece
Charikleia Karakosta, National Technical University of Athens, Greece
Haris Doukas, National Technical University of Athens, Greece
John Psarras, National Technical University of Athens, Greece

Chapter 8
Formulating Modern Energy Policy through a Collaborative Expert Model 223
Kostas Patlitzianas, REMACO S.A., Greece
Kostas Metaxiotis, University of Piraeus, Greece

Chapter 9
Fuel Reduction Effect of the Solar Cell and Diesel Engine Hybrid System with a Prediction Algorithm of Solar Power Generation 242
Shinya Obara, Kitami Institute of Technology, Japan

Chapter 10
Comprehensive Energy Systems Analysis Support Tools for Decision Making 272
C. Cosmi, National Research Council, Italy
S. Di Leo, University of Basilicata, Italy
S. Loperte, National Research Council, Italy
F. Pietrapertosa, National Research Council, Italy
M. Salvia, National Research Council, Italy
M. Macchiato, National Research Council, Italy
V. Cuomo, National Research Council, Italy

Chapter 11
Poly-Generation Planning: Useful Lessons from Models and Decision Support Tools 296
Aiying Rong, Technical University of Denmark, Denmark
Risto Lahdelma, Helsinki University of Technology, Finland
Martin Grunow, Technical University of Denmark, Denmark

Section 3
Systems and Applications

Chapter 12
Developing an Energy Security Risk Assessment System 337
Alexandros Flamos, National Technical University of Athens, Greece
Christos V. Roupas, National Technical University of Athens, Greece
John Psarras, National Technical University of Athens, Greece

Chapter 13
Formulating National Action Plans for Energy Business Environment: An Intelligent Information System 356
Kostas D. Patlitzianas, REMACO S.A.,—Research Department, Greece

Chapter 14
Information Technology in Power System Planning and Operation under Deregulated Markets: Case Studies and Lessons Learnt 377
Fawwaz Elkarmi, Amman University, Jordan

Chapter 15
An Intelligent Motor-Pump System 400
P. Giridhar Kini, Manipal University, India
Ramesh C. Bansal, The University of Queensland, Australia

Chapter 16
Intelligent Information Systems for Strengthening the Quality of Energy Services in the EU: Case Study in the Greek Energy Sector 423
Alexandra G. Papadopoulou, National Technical University of Athens, Greece
Andreas Botsikas, National Technical University of Athens, Greece
Charikleia Karakosta, National Technical University of Athens, Greece
Haris Doukas, National Technical University of Athens, Greece
John Psarras, National Technical University of Athens, Greece

Compilation of References 438

About the Contributors 491

Index 498

Detailed Table of Contents

Foreword xvi

Preface xviii

Section 1
Theories and Concepts

Chapter 1
Artificial Intelligence Techniques for Modern Energy Applications 1
Soteris Kalogirou, University of Technology, Cyprus
Kostas Metaxiotis, University of Piraeus, Greece
Adel Mellit, Jijel University, Algeria

Artificial intelligence (AI) techniques are becoming useful as alternate approaches to conventional techniques or as components of integrated systems. They have been used to solve complicated practical problems in various areas and nowadays are very popular. They are widely accepted as a technology offering an alternative way to tackle complex and ill-defined problems. They can learn from examples, are fault tolerant in the sense that they are able to handle noisy and incomplete data, are able to deal with non-linear problems and once trained can perform prediction and generalization at very high speed. AI-based systems are being developed and deployed worldwide in a wide variety of applications, mainly because of their symbolic reasoning, flexibility and explanation capabilities. They have been used in diverse applications in control, robotics, pattern recognition, forecasting, medicine, power systems, manufacturing, optimization, signal processing and social/psychological sciences. They are particularly useful in system modeling such as in implementing complex mappings and system identification. This chapter presents a review of the main AI techniques such as expert systems, artificial neural networks, genetic algorithms, fuzzy logic and hybrid systems, which combine two or more techniques. It also outlines some applications in the energy sector.

Chapter 2
Intelligent Control of the Energy Generation Systems 40
Nicu Bizon, University of Pitesti, Romania

In this book chapter are analyzed the Energy Generation System (EGS) topologies, used in automotive systems, and the grid inverter systems, with intelligent control algorithms (fuzzy logic controller, genetic algorithm, etc.). The EGS blocks are modelled using Matlab & Simulink ® program. A necessary block is the EGS power interface between the fuel cell stack and the batteries stack, usually a boost converter that uses a Peak Current Controller (PCC) with a Boundary Control with Current Taper (BCCT). The control law is a function of fuel cell current and battery voltage, which prevents the "boiling" of the batteries. The control objective for this power interface is also the fuel cell current ripple minimization, used in order to improve the fuel cell stack life cycle. Clocked and non-clocked control methods are tested in order to obtain a small fuel cell current ripple, better a dynamic response, and robustness against system uncertainty disturbances. The EGS behaviour is tested by bifurcation diagrams. It is shown that performances increase if the control law is a function that depends by the fuel cell current ripple and battery voltage. The clocked PCC using the BCCT 2-D law is implemented by a fuzzy logic controller. The power load dynamic is compensated using an ultracapacitors stack as a dynamic energy compensator, connected by a bi-directional converter to the batteries stack bus.

Chapter 3
Knowledge Management for Electric Power Utility Companies 97
Campbell Booth, University of Strathclyde, UK

This chapter will present an overview of the challenges presented to modern power utility companies and how many organizations are facing particularly pressing problems with regards to an ageing workforce and a general shortage of skills; a situation that is anticipated to worsen in the future. It is proposed that knowledge management (KM) and decision support (DS) may contribute to a solution to these challenges. The chapter describes the end-to-end processes associated with KM and DS in a power utility context and attempts to provide guidance on effective practices for each stage of the described processes. An overview of one particular power utility company that has embraced KM is presented, and it is proposed that the function of asset management within power utilities in particular may benefit from KM. The chapter focuses not only on KM techniques and implementation, but, equally, if not more importantly, on the various cultural and behavioural aspects that are critical to the success of any KM/DS initiative.

Chapter 4
Solving Environmental/Economic Dispatch Problem: The Use of Multiobjective Particle
Swarm Optimization 123
M. A. Abido, King Fahd University of Petroleum & Minerals, Saudi Arabia

Multiobjective particle swarm optimization (MOPSO) technique for environmental/economic dispatch (EED) problem is proposed and presented in this work. The proposed MOPSO technique evolves a multiobjective version of PSO by proposing redefinition of global best and local best individuals in multiobjective optimization domain. The proposed MOPSO technique has been implemented to solve the EED problem with competing and non-commensurable cost and emission objectives. Several optimization runs of the proposed approach have been carried out on a standard test system. The results demonstrate the capabilities of the proposed MOPSO technique to generate a set of well-distributed Pareto-optimal solutions in one single run. The comparison with the different reported techniques demonstrates the superiority of the proposed MOPSO in terms of the diversity of the Pareto optimal solutions obtained. In addition, a quality measure to Pareto optimal solutions has been implemented where the results confirm the potential of the proposed MOPSO technique to solve the multiobjective EED problem and produce high quality nondominated solutions.

Chapter 5
Doubly Fed Induction Generators: Overview and Intelligent Control Strategies for Wind Energy Conversion Systems 147
Vinod Kumar, College of Technology and Engineering, India
Steven Kong, University of Queensland, Australia
Yateendra Mishra, University of Queensland, Australia
Z. Y. Dong, The Hong Kong Polytechnic University, Hong Kong
Ramesh C. Bansal, University of Queensland, Australia

Adjustable speed induction generators, especially the Doubly-Fed Induction Generators (DFIG) are becoming increasingly popular due to its various advantages over fixed speed generator systems. A DFIG in a wind turbine has ability to generate maximum power with varying rotational speed, ability to control active and reactive by integration of electronic power converters such as the back-to-back converter, low rotor power rating resulting in low cost converter components, etc, DFIG have become very popular in large wind power conversion systems. This chapter presents an extensive literature survey over past 25 years on the different aspects of DFIG. Application of $H\infty$ Controller for enhanced DFIG-WT performance in terms of robust stability and reference tracking to reduce mechanical stress and vibrations is also demonstrated in the chapter.

Section 2
Models and Tools

Chapter 6
Use of Neural Networks for Modeling Energy Consumption in the Residential Sector 180
Merih Aydinalp Koksal, Hacettepe University, Turkey

This chapter investigates the use of neural networks (NN) for modeling of residential energy consumption. Currently, engineering and conditional demand analysis (CDA) approaches are mainly used for residential energy modeling. The studies on the use of NN for residential energy consumption modeling are limited to estimating the energy use of individual or a group of buildings. Development of a national residential end-use energy consumption model using NN approach is presented in this chapter. The comparative evaluation of the results of the model shows NN approach can be used to accurately predict and categorize the energy consumption in the residential sector as well as the other two approaches. Based on the specific advantages and disadvantages of three models, developing a hybrid model consisting of NN and engineering models is suggested.

Chapter 7
Setting Technology Transfer Priorities with CDM-SET3: Development of Sustainable Energy Technology Transfer Tool 205
Alexandros Flamos, National Technical University of Athens, Greece
Charikleia Karakosta, National Technical University of Athens, Greece
Haris Doukas, National Technical University of Athens, Greece
John Psarras, National Technical University of Athens, Greece

There is no much meaning in separating "good" and "bad" technologies. A definitely more critical issue is to identify "good" and "bad" technological options for a specific country & region based on its specific needs and special characteristics. In this framework, aim of this chapter is the presentation of the CDM-SET^3 tool that incorporates the potential host country's priority areas in terms of energy services and the suitable sustainable energy technologies to fulfil these needs and priorities, taking into consideration several criteria that examine the benefits in the economic, environmental and social dimension and through a MCDA approach facilitates the identification of the most proper technology alternatives to be implemented under the umbrella of CDM to a specific host country. The application of CDM-SET^3 in representative case study countries is also presented and the results are discussed. Finally, in the last section are the conclusions, which summarize the main points, arisen in this chapter.

Chapter 8
Formulating Modern Energy Policy through a Collaborative Expert Model 223
Kostas Patlitzianas, REMACO S.A., Greece
Kostas Metaxiotis, University of Piraeus, Greece

Nowadays, a comprehensive and modern energy policy making, which will be characterized by clarity and transparency, is necessary. Indeed, there exists a number of energy policy and planning systems, but there are no decision support systems investigating the energy policy making in an integrated way. In this context, the main aim of this chapter is to present an expert system based on a "multidimensional" approach for the energy policy making, which also incorporates the three objectives (security of supply, competitiveness of energy market and environmental protection) and takes into consideration all the related economical, social and technological parameters. This model was successfully applied in order to support the decisions towards the development of the energy policy priorities in the developing Mediterranean Countries as well as the countries of Gulf Cooperation Council – GCC.

Chapter 9
Fuel Reduction Effect of the Solar Cell and Diesel Engine Hybrid System with a Prediction Algorithm of Solar Power Generation 242
Shinya Obara, Kitami Institute of Technology, Japan

Green energy utilization technology is an effective means of reducing greenhouse gas emissions. The author developed the production-of-electricity prediction algorithm (PAS) of the solar cell. In this algorithm, a layered neural network is made to learn based on past weather data and the operation plan of the hybrid system (proposed system) of a solar cell and a diesel engine generator was examined using this prediction algorithm. In addition, system operation without a electricity-storage facility, and the system with the engine generator operating at 25% or less of battery residual quantity was investigated, and the fuel consumption of each system was measured. Numerical simulation showed that the fuel consumption of the proposed system was modest compared with other operating methods. However, there was a significant difference in the prediction error of the electricity production of the solar cell and the actual value, and the proposed system was shown to be not always superior to others. Moreover, although there are errors in the predicted and actual values using PAS, there is no significant influence in the operation plan of the proposed system in almost all cases.

Chapter 10
Comprehensive Energy Systems Analysis Support Tools for Decision Making 272
C. Cosmi, National Research Council, Italy
S. Di Leo, University of Basilicata, Italy
S. Loperte, National Research Council, Italy
F. Pietrapertosa, National Research Council, Italy
M. Salvia, National Research Council, Italy
M. Macchiato, National Research Council, Italy
V. Cuomo, National Research Council, Italy

Sustainability of energy systems is a common priority that involves key issues such as security of energy supply, mitigation of environmental impacts - the energy sector is currently responsible for 80% of all EU greenhouse gas emissions (European Environment Agency, 2007), contributing heavily to the overall emissions of local air pollutants - and energy affordability. In this framework, energy planning and decision making processes can be supported at different stages and spatial scales (regional, national, pan-European, etc.) by the use of comprehensive models in order to manage the large complexity of energy systems and to define multi-objective strategies on the medium-long term. This chapter is aimed to outline the value of model-based decision support systems in addressing current challenges aimed to carry out sustainable energy systems and to diffuse the use of strategic energy-environmental planning methods based on the use of partial equilibrium models. The proposed methodology, aimed to derive cost-effective strategies for a sustainable resource management, is based on the experiences gathered in the framework of the IEA-ETSAP program and under several national and international projects.

Chapter 11
Poly-Generation Planning: Useful Lessons from Models and Decision Support Tools 296
Aiying Rong, Technical University of Denmark, Denmark
Risto Lahdelma, Helsinki University of Technology, Finland
Martin Grunow, Technical University of Denmark, Denmark

Increasing environmental concerns and the trends towards deregulation of energy markets have become an integral part of energy policy planning. Consequently, the requirement for environmentally sound energy production technologies has gained much ground in the energy business. The development of energy-efficient production technologies has experienced cogeneration and tri-generation and now is moving towards poly-generation. All these aspects have added new dimension in energy planning. The liberalized energy market requires techniques for planning under uncertainty. The growing environmental awareness calls for explicit handling of the impacts of energy generation on environment. Advanced production technologies require more sophisticated models for planning. The energy sector is one of the core application areas for operations research, decision sciences and intelligent techniques. The scientific community is addressing the analysis and planning of poly-generation systems with different approaches, taking into account technical, environment, economic and social issues. This chapter presents a survey on the models and decision support tools for cogeneration, tri-generation and poly-generation planning. This survey tries to reflect the influence of deregulated energy market and environmental concerns on decision support tasks at utility level. Diverse modelling techniques and solution methods for planning problems will co-exist for a long time. Undoubtedly, the application of intelligent techniques is one of the main trends.

Section 3
Systems and Applications

Chapter 12
Developing an Energy Security Risk Assessment System... 337
Alexandros Flamos, National Technical University of Athens, Greece
Christos V. Roupas, National Technical University of Athens, Greece
John Psarras, National Technical University of Athens, Greece

Throughout the last two decades many attempts took place in order policy makers and researchers to be able to measure the energy security of supply of a particular country, region and corridor. This chapter is providing an overview presentation of the Energy Security Risk Assessment System (E.S.R.A.S.) which comprises the Module of Robust Decision Making (RDM) and the Module of Energy Security Indices Calculation (ESIC). Module 1 & 2 are briefly presented throughout section 2 and the application of Module 2 in nine case study countries is discussed at section 3. Finally, in the last section are the conclusions, which summarize the main points, arisen in this chapter.

Chapter 13
Formulating National Action Plans for Energy Business Environment: An Intelligent
Information System.. 356
Kostas D. Patlitzianas, REMACO S.A.,—Research Department, Greece

The penetration of the Renewable Energy Sources (RES) and the development of the Energy Efficiency (EE) is related to the synthesis of an appropriate action plan by each state for its energy business environment (companies such as "clean" energy producers, energy services companies etc.). The aim of this paper is to present an information intelligent system which consists of an expert subsystem, as well as a Multi Criteria subsystem. The system supports the state towards the formulation of a modern business environment, since it incorporates the increasing needs for energy reform, successful energy planning, rational use of energy as well as climate change. The system was successfully applied to the thirteen "new" member states of the EU.

Chapter 14
Information Technology in Power System Planning and Operation under Deregulated Markets: Case Studies and Lessons Learnt 377
Fawwaz Elkarmi, Amman University, Jordan

Power systems have grown recently in size and complexity to unprecedented levels. This means that planning and operation of power systems can not be made possible without the aid of information technology tools and instruments. Even small systems need such aid because of the complexity factor. On the other hand, new trends have recently emerged to solve the problems arising from increased size of power systems. These trends are related to the market structure, legal, and business issues. Other trends also cover technological developments, and environmental issues. Moreover, power systems have special characteristics and features that are not duplicated in other infrastructures. All these issues confirm the need for special information technology tools and instruments which aid in planning and operation of power systems.

Chapter 15
An Intelligent Motor-Pump System 400
P. Giridhar Kini, Manipal University, India
Ramesh C. Bansal, The University of Queensland, Australia

Process industries are energy intensive in nature and are one of the largest consumers of electrical energy that is commercially generated for utilization. Motor driven systems consume more than two-thirds of the total energy consumed by the industrial sector; among which, centrifugal pumps are the most widely used equipment mainly for the purpose of fluid transportation. The efficiency of pumping units is around 40 to 50 %, hence they offer tremendous opportunities of not only improving the efficiency of the process, but also ensure effective energy utilisation and management. With the increasing use of power electronics equipment, power quality (PQ) has become a very serious issue of consideration. On account of the random switching of single-phase loads in addition to time varying operations of industrial loads, PQ problem of voltage variation and unbalance is inevitable across three-phase systems. Application of varying or unbalanced voltages across the three-phase motor terminals results in performance variations leading to inefficient operation. For the purpose of study, the performance of a motor-pump system can be separately analyzed from the motor and pump points of view. The motor efficiency may vary in a very narrow band, pump efficiency depends upon the system head and flow rate but the system efficiency is a combination of the two; hence, necessary to analyze separately. As centrifugal pumps are classified under variable torque-variable speed load category, variation on the input side has a significant effect on the output side. Therefore the system efficiency now becomes an important index for ensuring efficient energy utilisation and efficiency. The main objective of the chapter is to put forward a methodology to analyze the working performance of a three-phase induction motor driven centrifugal pump under conditions of voltage and load variations by, defining additional factors for correct interpretation about the nature and extent of voltage unbalance that can exist in a power system network; define induction motor derating factors for safe and efficient operation based on operational requirements and devise energy management strategies for efficient utilization of electrical energy by the motor-pump system considering the voltage and load conditions.

Chapter 16
Intelligent Information Systems for Strengthening the Quality of Energy Services in the EU:
Case Study in the Greek Energy Sector 423
Alexandra G. Papadopoulou, National Technical University of Athens, Greece
Andreas Botsikas, National Technical University of Athens, Greece
Charikleia Karakosta, National Technical University of Athens, Greece
Haris Doukas, National Technical University of Athens, Greece
John Psarras, National Technical University of Athens, Greece

Nowadays, taking into consideration the prevailing situation of price fluctuations, the rapid population increase and the technology's evolution, the energy efficiency unexploited potential is considered to be extremely significant as a means of partly tackling energy dependence and climate change. This potential can be utilised through the provision of energy services, with the support of intelligent information systems. In particular, up to date several researchers, have proposed energy management tools and methodologies that provide specialized energy management services. However, the majority of the known energy tools are limited to a single equipment type, fuel, or locality. The present chapter introduces an intelligent information decision support system, addressed to Energy Service Companies (ESCOs) for assessing an operational unit's (building or industrial sector) energy behaviour and suggesting the appropriate interventions. Its overall scope is to facilitate the ESCO in reaching a decision quickly and accurately, by simulating the whole unit's energy behaviour.

Compilation of References 438

About the Contributors 491

Index 498

Foreword

Energy is one of the most important basic inputs of economical and industrial development. Nowadays there is an ever increasing concern and interest in important issues like sustainable development, rational use of energy, effective use of renewable energy sources, successful energy planning, energy efficiency, optimization of energy consumption, environmental protection. Since the late 1990s, the European Commission (EC), as well as other international donors (World Bank, Asian Development Bank), has become increasingly active in the areas of sustainable energy and ICT in order to address the challenge of fighting climate changes.

The EC – in its proposal dated June 2005 for a Specific Community Research Programme during the period 2007-2013 – has identified ICT as a major field of progress in terms of research and knowledge addressing and supporting social, economic, environmental and industrial challenges and contributing to sustainable development. Recently, in May 2008 the EC adopted the Communication *"Addressing the Challenge of Energy Efficiency Through Information and Communication Technologies"* which focuses on ICT as an enabler to improve energy efficiency across the economy and the whole of society. Indeed, energy efficiency as a whole is a key challenge in our future world, leading to a twofold bet, both energetic and environmental and ICT can play an important role in this area.

Energy consumption is a key issue nowadays. In Europe, for instance, between 40% and 50% of the energy we generate goes into heating and powering our buildings, homes account for around 30% of the carbon emissions. There is no doubt that a significant reduction of energy consumption is possible with the development and application of new and intelligent ICT-based products to be used in new and existing houses as well as in public buildings such as schools, hospitals, governmental buildings. Renewable energy can play a strong role in contributing to energy provision, not only as a choice for consumers but also to contribute indigenous, non-fossil energy resources into the market.

For this reason, Governments and nongovernmental organizations, including development agencies and businesses are increasingly concerned with the elaboration of a shared vision and roadmap towards ICT-based innovation and intelligent solutions in the energy sector. Decision support systems, intelligent information systems, agent technologies, artificial intelligence-based technologies, knowledge management can heavily contribute to sustainable development and environmental protection, while maintaining growth.

It is then of considerable importance that we not only better understand the processes underlying the innovation and adoption of new technologies – especially ICT – and the broader processes of creativity generating knowledge-based activities, but also that we know more about the resultant outcomes, both economically and socially. In this context, a publication on intelligent information systems and knowledge management in the energy sector deserves primary importance among the academia, researchers, development practitioners, policy initiators and individuals.

This book, edited by Kostas Metaxiotis, represents an ambitious attempt to address the multiple issues concerning how ICT and intelligent information systems can be used with the aim at solving the above presented key issues in the energy sector. The book brings a diverse set of perspectives and provides a rich set of tools, systems, applications and case studies in the energy sector.

Importantly the book provides concepts, approaches, applications and case studies from both developed and developing nations. Some of the 16 chapters focus attention on innovative concepts and theories. Other chapters include descriptions of advanced systems and applications in fields such as energy generation, energy efficiency, energy consumption, renewable energy, poly-generation planning, energy security risk assessment, power system planning and operation, knowledge management in electric power utility companies.

It is a welcome contribution to the growing literature on how to use and apply ICT in the energy sector. The major asset of this book is the accumulation of several theoretical researches, case analysis, and practical implementation processes accompanying profound discussions and techniques for accomplishing tasks that one could easily adopt even in a non-technical environment. In fact, this book will act, not only as a research guide but also as an implementation guide in the longer run.

Prof. Emeritus Emmanuel Samouilidis
National Technical University of Athens, Greece

Preface

"The ICT sector is a key driver to a low carbon society and it is fundamental to fight Climate Change".

In November 2008 three industry representatives of the European Commission's Ad-hoc Advisory Group on ICT for Energy Efficiency expressed the need for a better understanding of the importance of the ICT sector and a more fundamental integration of ICT solutions in EU policies in order to address the challenge of fighting climate change. The World Bank announced in October 2008 an 87% increase in funding for renewable energy (RE) and energy efficiency (EE) projects and programs in developing countries in the past fiscal year. Indeed, the important role that ICT plays in enabling intelligent energy, cleaner energy, greater energy efficiency and carbon reductions is considered as central to energy policy making worldwide. Without any doubt, the role of ICT is crucial in transforming the knowledge-based society to a low carbon society. Several studies published within the last 12 months conclude that the application of digital technology solutions can reduce world energy consumption from 10% to 25% by 2020.

The energy sector is a complex environment, incorporating a variety of organisations, operational frameworks, and internal and external pressures. Application of intelligent systems and knowledge management in the energy sector has been an active area of research for about two decades and significant successes have been achieved. Perhaps unsurprisingly, therefore, some of the biggest players in the sector were among the earliest adopters of knowledge management, notably oil and gas giants Shell and BP. Energy is not only a significant industry in any economy but also a field which needs effective means to manage data as well as information and knowledge. The energy industry is nowadays trying to become a knowledge-based community that is connected to factories, electricians, managers, consultants, researchers and customers for sharing knowledge, reducing administrative costs and improving the quality of service.

The continuous discussion about how ICT can be used in the energy sector has large potential for opening up new areas of opportunities, both social and business. For instance, the Intelligent Energy – EU-funded programme has lent its support so far to more than 400 international projects involving more than 3,000 European organisations; this programme aims at promoting intelligent solutions in energy and creating better conditions for a more energy-wise future, securing environmental protection. *Intelligent Information Systems and Knowledge Management for Energy: Applications for Decision Support, Usage and Environment Protection* is a book aimed at enlightening the above concepts and challenges and therefore at providing understanding as to how ICT can contribute to the energy sector. In particular, its specific purposes are:

The purposes of this publication are as follows:

To create a big knowledge base for scholars, introducing them to all aspects of intelligent computer applications in the energy sector and indicating other areas fertile for research.

To develop scholars' capacity in the design, implementation and application of intelligent information systems and knowledge management in the energy sector

To increase the awareness of the role of intelligent information systems and knowledge management in the energy sector, as well as of the challenges and opportunities for future research.

The book presents insights gained by leading professionals from the practice, research, academic, and consulting side in the field. This is why it should be useful to a variety of target groups, which are interested in the interrelationships between information and communication technologies and energy. The Foreword is written by a senior respected academic researcher and energy policy maker Emmanuel Samouilidis of the National Technical University of Athens, Greece. The book is divided into three sections, each one dealing with selected aspects of information and communication technologies in the energy sector.

Section 1: Theories and Concepts

The five chapters in Section 1 present advanced theories and modern ICT and knowledge management concepts in several fields of energy. Chapter 1 introduces the basic theory of Artificial Intelligence (AI) and its application to the energy sector. AI-based systems are being developed and deployed worldwide in a wide variety of applications, mainly because of their symbolic reasoning, flexibility and explanation capabilities. This chapter presents a review of the main AI techniques such as expert systems, artificial neural networks, genetic algorithms, fuzzy logic and hybrid systems, and describes a wide range of modern AI-based energy applications.

Chapter 2 analyzes the key issue of the control of energy generation systems and presents an innovative concept based on the use of intelligent control algorithms (fuzzy logic controller, genetic algorithm, etc.) in order to reduce the PEMFC (Proton Exchange Membrane Fuel Cells) stress concerning the output load power dynamic.

Chapter 3 presents an overview of the challenges presented to modern power utility companies and how many organizations are facing particularly pressing problems with regards to an ageing workforce and a general shortage of skills; a situation that is anticipated to worsen in the future. It is proposed that knowledge management (KM) and decision support (DS) may contribute to a solution to these challenges. The chapter describes the end-to-end processes associated with KM and DS in a power utility context and attempts to provide guidance on effective practices for each stage of the described processes.

Chapter 4 introduces the multiobjective particle swarm optimization (MOPSO) technique to solve environmental/economic dispatch (EED) problem. The proposed MOPSO technique evolves a multiobjective version of PSO by proposing redefinition of global best and local best individuals in multiobjective optimization domain. The proposed MOPSO technique has been implemented to solve the EED problem with competing and non-commensurable cost and emission objectives.

Adjustable speed induction generators, especially the Doubly-Fed Induction Generators (DFIG) are becoming increasingly popular due to its various advantages over fixed speed generator systems. A DFIG in a wind turbine has ability to generate maximum power with varying rotational speed, ability to control active and reactive by integration of electronic power converters such as the back-to-back converter, low rotor power rating resulting in low cost converter components, etc, DFIG have become very popular in large wind power conversion systems. Chapter 5 presents an extensive literature survey over past 25 years on the different aspects of DFIG.

Section 2: Models and Tools

The second section of this book moves from a more theoretical focus to consider practical models and tools that have evolved to support the solving of specific problems in the energy sector.

The first chapter in this section, Chapter 6, investigates the use of neural networks (NN) for modeling of residential energy consumption. Currently, engineering and conditional demand analysis (CDA) approaches are mainly used for residential energy modeling. The studies on the use of NN for residential energy consumption modeling are limited to estimating the energy use of individual or a group of buildings. This chapter presents a national residential end-use energy consumption model, using NN approach. The comparative evaluation of the results of the model shows that the NN approach can be used to accurately predict and categorize the energy consumption in the residential sector. Based on the specific advantages and disadvantages of three models, developing a hybrid model consisting of NN and engineering models is suggested.

The Kyoto Protocol contains market mechanisms that enable industrialized countries to invest in greenhouse gas emission (GHG) reduction projects on the territory of other countries, either developing, or industrialised, such as the Clean Development Mechanism (CDM). Chapter 7 presents the Clean Development Mechanism Sustainable Energy Technology Transfer Tool (CDM-SET3) which was developed and applied to a selected set of representative developing countries from Asia, Latin America, sub-Sahara Africa and the Mediterranean. The CDM-SET3 steers towards a successful technology transfer through the CDM, taking also into consideration the overall medium to long-term energy and environmental strategy of the host country.

Chapter 8 presents an expert system model, based on a "multidimensional" approach, for the formulation of modern energy policies, which also incorporates the three objectives (security of supply, competitiveness of energy market and environmental protection) and takes into consideration all the related economical, social and technological parameters. This model was successfully applied in order to support the decisions towards the development of the energy policy priorities in the developing Mediterranean Countries as well as the countries of Gulf Cooperation Council – GCC.

Green energy utilization technology is an effective means of reducing greenhouse gas emissions. Chapter 9 presents the development of the production-of-electricity prediction algorithm (PAS) of the solar cell. In this algorithm, a layered neural network is made to learn based on past weather data and the operation plan of the hybrid system (proposed system) of a solar cell, while a diesel engine generator was examined using this prediction algorithm. Numerical simulation showed that the fuel consumption of the proposed system was modest compared with other operating methods.

Chapter 10 is aimed to outline the value of model-based decision support systems in addressing current challenges aimed to carry out sustainable energy systems and to diffuse the use of strategic energy-environmental planning methods based on the use of partial equilibrium models. The proposed methodology, aimed to derive cost-effective strategies for a sustainable resource management, is based on the experiences gathered in the framework of the Energy Technology Systems Analysis Programme of the International Energy Agency (IEA-ETSAP) and under several national and international projects.

Chapter 11 presents a survey on the models and decision support tools for cogeneration, tri-generation and poly-generation planning. This survey tries to reflect the influence of deregulated energy market and environmental concerns on decision support tasks at utility level. Diverse modelling techniques and solution methods for planning problems will co-exist for a long time. Undoubtedly, the application of intelligent techniques is one of the main trends.

Section 3: Systems and Applications

The final section of this book considers some applications and case studies of ICTs in the energy sector.

Throughout the last two decades many attempts took place in order for energy policy makers and researchers to be able to measure the energy security of supply of a particular country, region and corridor. Chapter 12 presents the Energy Security Risk Assessment System (E.S.R.A.S.) which comprises the Module of Robust Decision Making (RDM) and the Module of Energy Security Indices Calculation (ESIC). The application of the system in nine case study countries is also discussed.

Chapter 13 presents an intelligent information system which consists of an Expert subsystem and a Multi Criteria subsystem. The system supports the state towards the formulation of a modern business environment, since it incorporates the increasing needs for energy reform, successful energy planning, rational use of energy as well as climate change. The pilot system was successfully applied to the thirteen "new" member states of the EU.

Chapter 14 presents case studies and lessons learnt concerning the use of ICT in power system planning and operation under de-regulated markets. This chapter aims at discussing the new emerging trends and critical factors which have shaped and continue to influence decisions of power system planners and operators. Some of these are technical issues while others are economic and financial.

Chapter 15 puts forward a methodology to analyze the working performance of a three-phase induction motor driven centrifugal pump under conditions of voltage and load variations by, defining additional factors for correct interpretation about the nature and extent of voltage unbalance that can exist in a power system network. It defines induction motor derating factors for safe and efficient operation based on operational requirements and devise energy management strategies for efficient utilization of electrical energy by the motor-pump system, considering the voltage and load conditions.

The final chapter 16 presents an intelligent decision support system, addressed to Energy Service Companies (ESCOs) for assessing an operational unit's (building or industrial sector) energy behaviour and suggesting the appropriate interventions. Its overall scope is to facilitate the ESCO in reaching a decision quickly and accurately, by simulating the whole unit's energy behaviour.

The work presented in this book has been made possible through the hard work of the contributors who kept the deadlines and were always enthusiastic. The editor would like to thank all the contributors and hopes that this book will encourage the reader to keep strengthening the way ICT can contribute to greater energy efficiency, cleaner energy, environment protection and finally to transforming our knowledge-based society into a low carbon society and economy.

Section 1
Theories and Concepts

Chapter 1
Artificial Intelligence Techniques for Modern Energy Applications

Soteris Kalogirou
University of Technology, Cyprus

Kostas Metaxiotis
University of Piraeus, Greece

Adel Mellit
Jijel University, Algeria

ABSTRACT

Artificial intelligence (AI) techniques are becoming useful as alternate approaches to conventional techniques or as components of integrated systems. They have been used to solve complicated practical problems in various areas and nowadays are very popular. They are widely accepted as a technology offering an alternative way to tackle complex and ill-defined problems. They can learn from examples, are fault tolerant in the sense that they are able to handle noisy and incomplete data, are able to deal with non-linear problems and once trained can perform prediction and generalization at very high speed. AI-based systems are being developed and deployed worldwide in a wide variety of applications, mainly because of their symbolic reasoning, flexibility and explanation capabilities. They have been used in diverse applications in control, robotics, pattern recognition, forecasting, medicine, power systems, manufacturing, optimization, signal processing and social/psychological sciences. They are particularly useful in system modeling such as in implementing complex mappings and system identification. This chapter presents a review of the main AI techniques such as expert systems, artificial neural networks, genetic algorithms, fuzzy logic and hybrid systems, which combine two or more techniques. It also outlines some applications in the energy sector.

DOI: 10.4018/978-1-60566-737-9.ch001

1. INTRODUCTION

Artificial Intelligence (AI) is a term that in its broadest sense would indicate the ability of a machine or artifact to perform the same kind of functions that characterize human thought. The term AI has also been applied to computer systems and programs capable of performing tasks more complex than straightforward programming, although still far from the realm of actual thought. According to Barr and Feigenbaum (1981) AI is the part of computer science concerned with the design of intelligent computer systems, i.e., systems that exhibit the characteristics associated with intelligence in human behavior-understanding, language, learning, reasoning, solving problems and so on (Kalogirou, 2003; 2007). Several intelligent computing technologies are becoming useful as alternate approaches to conventional techniques or as components of integrated systems (Medsker, 1996).

In the early 1950s Herbert Simon, Allen Newell and Cliff Shaw conducted experiments in writing programs to imitate human thought processes (Krishnamoorthy and Rajeev, 1996). The experiments resulted in a program called Logic Theorist, which consisted of rules of already proved axioms. When a new logical expression was given to it, it would search through all possible operations to discover a proof of the new expression, using heuristics. This was a major step in the development of AI. The Logic Theorist was capable of solving quickly thirty-eight out of fifty-two problems with proofs that Whitehead and Russell had devised (Newell *et al.*, 1963). At the same time, Shannon came out with a paper on the possibility of computers playing chess (Shannon, 1950). Though the works of Newell *et al.* (1963) and Shannon (1950) demonstrated the concept of intelligent computer programs, the year 1956 is considered the start of Artificial Intelligence.

One representative definition is pivoted around the comparison of intelligence of computing machines with human beings (McCarthy, 1980; Haugeland, 1985). Another definition is concerned with the performance of machines, which "historically have been judged to lie within the domain of intelligence" (Kurzweil, 1990; Newell and Simon, 1972). None of these definitions or the like has been universally accepted, perhaps because of their references to the word "intelligence", which at present is an abstract and immeasurable quantity. Therefore, a better definition of AI calls for formalization of the term "intelligence". Psychologist and cognitive theorists are of the opinion that intelligence helps in identifying the right piece of knowledge at the appropriate instances of decision-making (Rich and Knight, 1966). The phrase "AI" thus can be defined as the simulation of human intelligence on a machine, to make the machine efficient to identify and use the right piece of "Knowledge" at a given step for solving a problem. A system capable of planning and executing the right task at the right time is generally called rational (Russel and Norvig, 1995). Thus, AI alternatively may be stated as a subject dealing with computational models that can think and act rationally (Luger and Stubblefield, 1993; Winston, 1994; Schalkoff *et al.*, 1992). A common question then naturally arises. Does rational thinking and acting include all possible characteristics of an intelligent system? If so, how does it represent behavioral intelligence such as machine learning, perception and planning? A little thinking, however, reveals that a system that can reason well must be a successful planner, as planning in many circumstances is part of a reasoning process. Further, a system can act rationally only after acquiring adequate knowledge from the real world. Therefore, perception that stands for building up of knowledge from real world information is a prerequisite feature for rational actions. One-step further thinking envisages that a machine without learning capability cannot possess perception. The rational action of an agent (actor) calls for possession of all the elementary characteristics of intelligence. Therefore, relating AI with the computational models capable of thinking and acting rationally has a pragmatic significance (Konar, 1999). AI has been used in several applications, especially for resolving complex problems (Charniak and McDermott, 1985; Chen, 2000; Nilsson, 1998, Zimmermann *et al.*, 2001).

2. ARTIFICIAL INTELLIGENCE TECHNIQUES

AI consists of several branches, namely; Expert Systems (ES), Artificial Neural Networks (ANNs), Genetic Algorithms (GA), Fuzzy Logic (FL), Problem Solving and Planning (PSP), Non-Monotonic Reasoning, (NMR), Logic Programming (LP), Natural Language Processing (NLP), Computer Vision (CV), Robotics, Learning, Planning (Krishnamoorthy and Rajeev, 1996) and various Hybrid intelligent Systems (HIS), which are combinations of two or more of the branches mentioned previously (Kalogirou, 2003; 2007).

Expert systems allow computers to 'make decisions' by interpreting data and selecting from a list of alternatives. Expert systems take computers a step beyond straightforward programming, being based on a technique called rule-based inference, in which pre-established rule systems are used to process the data. Despite their sophistication, systems still do not approach the complexity of true intelligent thought (Kalogirou, 2003; 2007). In addition, expert systems deal with knowledge processing and complex decision-making problems.

A genetic algorithm is a stochastic algorithm that mimics the natural process of biological evolution (Rich and Knight, 1996). GA's are inspired by the way living organisms are adapted to the harsh realities of life in a hostile world, i.e., by evolution and inheritance. The algorithm imitates in the process, the evolution of population by selecting only fit individuals for reproduction. Therefore, a GA is an optimum search technique based on the concepts of natural selection and survival of the fittest. It works with a fixed-size population of possible solutions of a problem, called individuals, which are evolving in time. GA's find extensive applications in intelligent search, machine learning and optimization problems. Problem states in a GA are denoted by chromosomes, which are usually represented by binary strings. A GA utilizes three principal genetic operators; selection, crossover and mutation (Kalogirou, 2003; Konar, 1999; Deyi and Li, 2007).

Artificial neural networks are electrical analogues of the biological neural organs. Biological nerve cells, called neurons, receive signals from neighboring neurons or receptors through dendrites, process the received electrical pulses at the cell body and transmit signals through a large and thick nerve fiber, called an axon. In a similar way, the electrical model of a typical biological neuron consists of a linear activator, followed by a non-linear inhibiting function. The linear activation function yields the sum of the weighted input excitation, while the non-linear inhibiting function attempts to capture the signal levels of the sum. The resulting signal, produced by the electrical neuron, is thus bounded (amplitude limited) (Konar, 1999). An artificial neural network is a collection of such electrical neurons connected in different topologies. The most common application of an artificial neural network is in machine learning (Kodratoff and Michalski, 1990). In a learning problem, the weights and/or non-linearities in an artificial neural network undergo an adaptation (or learning) cycle. The adaptation cycle is required for updating these parameters of the network, until a state of equilibrium is reached. The ANN support both supervised and unsupervised types of machine learning (Konar, 1999).

The logic of fuzzy sets was proposed by Zadeh (1965), who introduced the concept in systems theory and later extended it for approximate reasoning in expert systems. Fuzzy logic deals with fuzzy sets and logical statements for modeling human-like reasoning problems of the real world. A fuzzy set, unlike conventional sets, includes all elements of the universal set of the domain but with varying membership values in the interval [0,1]. It should be noted that a conventional set contains its members with a value of membership equal to one and disregards other elements of the universal set, as they have zero membership. The most common operators applied to fuzzy sets are 'AND' (minimum), 'OR' (maximum) and negation (complementation), where 'AND' and 'OR' have binary arguments, while negation has unary argument (Konar, 1999; George and Williams, 1998). Fuzzy logic is used mainly in control engineering. It is based on fuzzy logic reasoning which employs linguistic rules in the form of IF-THEN statements.

Problem solving and planning (PSP) deals with systematic refinement of goal hierarchy, plan revision mechanisms and a focused search of important goals (Hewitt and Planner, 1971). The need for general-purpose problem-solving techniques to attack problems for which there are no experts has long been recognized. One way of viewing intelligent behavior is as a problem-solver. Many AI tasks can naturally be viewed this way and most AI programmers draw much of their strength from their problem-solving components. Applications of AI in areas such as image analysis, natural language processing and expert systems have significant problem-solving components. Approaches to problem solving such as backward chaining, problem reduction and difference reduction are all investigated under the general umbrella of AI research (Greer, 1986).

A non-monotonic reasoning system allows addition of a new piece of knowledge or fact in the context that can cause a previously believed tentative truth to become false. The belief revision mechanism propagates the effect of any change in belief using dependency-directed backtracking. Planning and design problems use a large number of tentative assumptions based on inexact or partial information. The non-monotonic reasoning system, which has a built-in mechanism through the dependency-directed backtracking, provides increased power and flexibility to solve such problems in a very effective manner. However, its implementation requires a large amount of memory to store dependency information and a large amount of processing time to propagate changes in belief. Doyle presented a TMS (Truth Maintenance System), which is an implementation of the non-monotonic reasoning system (Krishnamoorthy and Rajeev, 1996). It maintains consistency in the knowledge base by calling the reasoning system when a new truth-value is generated. It should be noted that the role of TMS is passive, since it never initiates generation of inferences. When the TMS discovers an inconsistency in the current set of beliefs arising out of a newly added fact, it invokes dependency-directed backtracking to restore the consistency (Krishnamoorthy and Rajeev, 1996). Non-monotonic reasoning is preferred under the following circumstances (Seddon and Brereton, 1996):

- when there is incomplete information (requiring the use of defaults),
- when changing situations occur (reasoning with 'out of date' knowledge),
- when assumptions need to be generated during the solution of complex problems.

Logic programming is in its broadest sense, the use of mathematical logic for computer programming. In this view of logic programming which can be traced at least as far back as John McCarthy's (1958) advice-taker proposal. Logic is used as a purely declarative representation language and a theorem-prover or model-generator is used as the problem-solver. The problem-solving task is split between the programmers, who are responsible only for ensuring the truth of programs expressed in logical form and the theorem-prover or model-generator, which is responsible for solving problems efficiently. The programmer is responsible, not only for ensuring the truth of programs, but also for ensuring their efficiency. In many cases, to achieve efficiency, the programmer needs to be aware of and to exploit the problem-solving behavior of the theorem-prover. In this respect, logic programming is like conventional imperative programming, using programs to control the behavior of a program executor. However, unlike imperative programs, which have only a procedural interpretation, logic programs also have a declarative, logical interpretation, which helps to ensure their correctness. Moreover, such programs, being declarative, are at a higher conceptual level than purely imperative programs and their program executers being theorem-provers, operate at a higher conceptual level than conventional compilers and interpreters.

Natural language processing areas, such as automatic text generation, text processing, machine translation, speech synthesis and analysis, grammar and style analysis of text, graphics, database creation and querying, facts and rules specification, etc., are being used by an overgrowing spectrum of the population (Krishnamoorthy and Rajeev, 1996). All of these topics are included under the general umbrella of artificial intelligence. If this goal of allowing virtually anyone to use computers and specialized software via natural language communication is to be achieved, it is necessary to have a thorough understanding of how humans communicate. Only then will it be possible to create machines with human-like communication skills. Research in this area is quite active and the rapidly accelerating interest in artificial intelligence is generating and promoting more and more research into language comprehension. More detailed information in this area is available from Barr and Feigenbaum (1982) and Winograd (1972). Computational linguistics, a science that incorporates elements of linguistics, psychology and philosophy is devoted to the study of how humans communicate and is an area of research central to artificial intelligence.

Computer vision, deals with intelligent visualization, scene analysis, image understanding and processing and motion derivation (Minsky, 1975). Visual perception employing computers is another active current area of AI research. Often combined with research into robot applications, computer vision involves the sensing and interpretation of images by machines. It has been estimated that in a few years, one-quarter of all industrial robots will be equipped with some form of vision system. Scores of applications of computer vision systems are available, some currently in use and others waiting for the appropriate technology to be developed (Greer, 1986).

Robotics deal with the control of robots to manipulate or grasp objects using information from sensors to guide their actions (Engelberger, 1980; Dean *et al.*, 1995). Mobile robots, sometimes called Automated Guided Vehicles (AGV), are a challenging area of research, where AI finds extensive applications. A mobile robot generally has one or more cameras or ultrasonic sensors, which help in identifying the obstacles on its trajectory. The navigational planning problem persists in both static and dynamic environments. In a static environment, the position of obstacles is fixed, while in a dynamic environment the obstacles may move at arbitrary directions with varying speeds, lower than the maximum speed of the robot. In the near future, mobile robots will find extensive applications in fire fighting, mine clearing and factory automation. In accident-prone industrial environments, mobile robots may be exploited for automatic diagnosis and replacement of defective parts of instruments (Krishnamoorthy and Rajeev, 1996).

Learning deals with research and development in different forms of machine learning (Winston, 1994; Smith *et al.*, 1977). Learning is an inherent characteristic of human beings. By virtue of learning, people, while executing similar tasks, acquire the ability to improve their performance. Machine learning can be broadly classified into three categories: supervised learning, unsupervised learning and reinforcement learning. Supervised learning requires a trainer, who supplies the input-output training instances. The learning system adapts its parameters by using algorithms that generate the desired output patterns from a given input pattern. In the absence of trainers, the desired output for a given input instance is not known and consequently the learner has to adapt its parameters autonomously. Such type of learning is termed 'unsupervised learning'. The third type, called the reinforcement learning, bridges a gap between supervised and unsupervised categories. In reinforcement learning, the learner does not explicitly know the input-output instances, but it receives some form of feedback from its environment. The feedback signals help the learner to decide whether its action on the environment is rewarding or punishable. Thus, the learner adapts its parameters based on the states (rewarding/punishable) of its actions (Krishnamoorthy and Rajeev, 1996 ; Fogel, 1995).

Planning is another significant area of AI. The problems of reasoning and planning share many common issues, but have a basic difference that originates from their definitions (Krishnamoorthy and Rajeev, 1996). The reasoning problem is mainly concerned with the testing of the ability to satisfy a goal from a given set of data and knowledge. The planning problem, on the other hand, deals with the determination of the methodology by which a successful goal can be achieved from the known initial states (Fogel, 1995; Bender, 1996). Automated planning finds extensive applications in robotics and navigational problems.

Hybrid systems combine more than one of the technologies introduced above, either as part of an integrated method of problem solution, or to perform a particular task that is followed by a second technique, which performs some other task. For example, neuro-fuzzy controllers use neural networks and fuzzy logic for the same task, i.e., to control a process, whereas in another hybrid system a neural network may be used to derive some parameters and a GA may be used subsequently to find an optimum solution to a problem (Kalogirou, 2003; 2007; Medsker, 1995).

Almost every branch of science and engineering currently shares the tools and techniques available in the domain of AI. Some important reviews of various applications of AI in combustion (Kalogirou, 2003) renewable energy systems (Kalogirou, 2001), energy (Kalogirou, 2000) and photovoltaic (PV) systems (Mellit and Kalogirou, 2008), are presented where it is shown that AI plays a significant and decisive role in the engineering design process. In this chapter, only five of the major fields of AI, which have been mostly used in modeling and prediction of engineering systems, are presented. These are the artificial neural networks, fuzzy logic, genetic algorithm, expert systems and hybrid systems. Some representative applications in energy systems are also presented.

3. EXPERT SYSTEMS

Feigenbaum (1982) defined an expert system as 'an intelligent computer program that uses knowledge and inference procedures to solve problems that are difficult enough and require significant human expertise for their solution'. Expert systems perform reasoning using previously established rules for well-defined and narrow domains (Turban, 1992). The terms expert system and knowledge-based expert system can therefore be used interchangeably. The Knowledge-Based Expert System (KBES) is the first realization of research in the field of AI, in the form of software. For developers of application software, particularly in medical and engineering disciplines, it was a benefit, as it addressed the decision-making process with the use of symbols rather than numbers. An expert system is not called a program, but a system, because it encompasses several different components such as knowledge base, inference mechanisms, explanation facilities, etc. (Krishnamoorthy and Rajeev, 1996). All these different components interact together in simulating the problem-solving process by an acknowledged expert of a domain. A close examination of any decision-making process by an expert reveals that he/she uses facts and heuristics to arrive at decisions. If the decision to be made is based on simple established facts using a heuristic such as a rule of thumb, then it may be a trivial process. The heuristic type of knowledge and other rules of thumb or guidelines specified in design codes can very well be represented in the form of IF-THEN constructs called production rules. Not only these are easy to understand, but also their representation is very simple thus resulting in an easy implementation in a computer (Krishnamoorthy and Rajeev, 1996). The two main components of an expert system can be identified as:

- *Knowledge Base:* Collection of knowledge required for problem solving.
- *Control Mechanism:* Checks the available facts, selects appropriate knowledge source from knowledge base, matches the facts with the knowledge and generates additional facts.

A knowledge base and an inference engine consisting of one or more inference mechanisms form the major components of an expert system. When an expert system starts the process of inference, it is required to store the facts established for further use. The set of established facts represents the context, i.e., the present state of the problem to be solved. Hence, this component is often called context or working memory.

Whenever an expert gives a decision, one is curious to know how the expert arrived at that decision. In addition, when the expert prompts for information or data, one would like to know why that piece of information is required. The expert uses the knowledge he/she has and the context of the problem to answer the queries such as how a decision is arrived at, or why particular data are needed. A separate module called the explanation facility simulates the process of answering the why and how queries. This module also forms an integral part of any expert system.

The process of collecting, organizing and compiling the knowledge and implementing it in the form of a knowledge base is a laborious task. It does not end with the development of the system. The knowledge base has to be continuously updated and/or appended depending on the growth of knowledge in the domain. A knowledge acquisition facility, which will act as an interface between the expert/knowledge engineer and the knowledge base, can form an integral component of an expert system. As it is not an on-line component, it can be implemented in many ways. A user of the expert system has to interact with it for giving data, defining facts and monitoring the status of problem solving. Conveying the information (textual or graphical), to the user should also be done in a very effective manner. Thus, a user-interface module with the capability to handle textual and graphical information forms, is another component of the expert system. Figure 1 shows the architecture of a KBES with its components and the way the components interact with each other.

Expert systems have been built to solve a range of problems in domains such as medicine, mathematics, engineering, chemistry, geography, computer science, business, law, defense and education. These programs have addressed a wide range of problem types; the following list adapted from Waterman (Kusiak *et al.*, 1989) is a useful summary of general expert system problem categories:

- Interpretation: forming high-level conclusions or deception from collections of raw data
- Prediction: projection probable consequences of given situations
- Diagnosis: determining the cause of malfunctions in complex situations based on observable symptoms
- Design: finding a configuration of system components that meet performance goals while satisfying

Figure 1. Architecture of a KBES with its components and the way the components interact with each other (Kusiak et al., 1989)

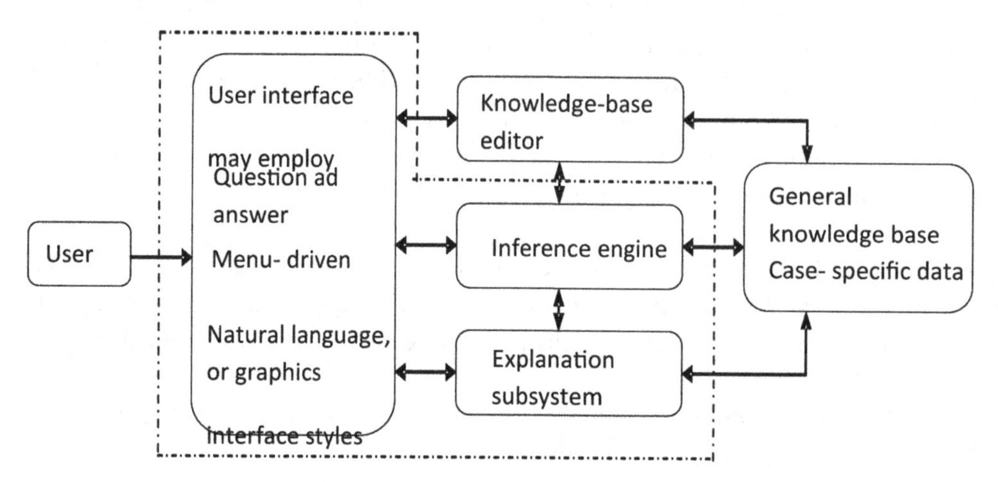

a set design constraints

- Planning: devising a sequence of actions that will achieve a set goal giving certain starting conditions and run-time constraints
- Monitoring: comparing a system's observed behavior to its expected behavior
- Debugging and repair-prescribing and implementing remedies for malfunctions
- Instruction detecting and correcting deficiencies in students understanding of a subject domain
- Control: governing the behavior of a complex environment

Based on the operational mode of the expert systems two classes of expert systems have been identified by Kusiak (1987), the stand-alone and the tandem expert systems.

3.1 Stand-Alone Expert System

An expert system in the stand-alone mode uses data and constraints pertinent to the problem and solves it using rather simple procedures. It does not use the operations research approach. The operations research approach involves modeling the given problem and solving the model using an optimal or a "good" heuristic algorithm. Many existing expert systems fall into the stand-alone class of expert systems.

3.2 Tandem Expert System

The tandem expert system combines the operations research approach with the expert system approach in order to solve a problem. It can be thought as an expert system linked to a database of models and algorithms (see Figure 2). The basic approach utilized in a tandem expert system is the following. A suitable model is either selected or constructed for the given problem. To solve the model, an optimal or heuristic algorithm (available in the model and algorithm base) is selected. The solution generated by the algorithm is modified if necessary, so as to incorporate qualitative aspects and to obtain a solution which can be implemented. There are three variants of the tandem expert system, namely:

Figure 2. Tandem expert system (Kusiak, 1987)

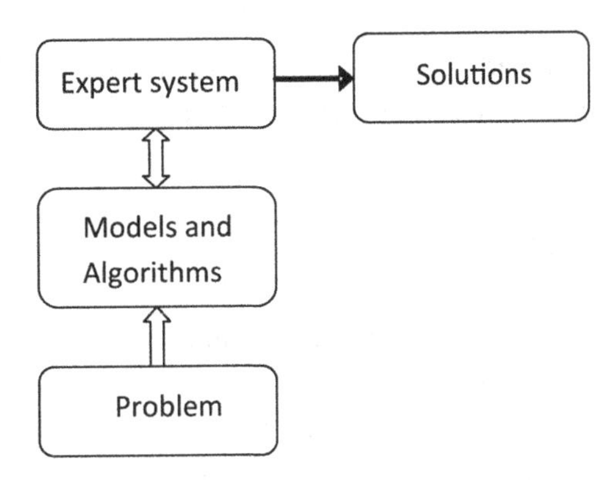

- Data modifying expert system
- Model based expert system
- Model modifying expert system

The basic approach in all variants of the tandem expert system is as mentioned above. However, the emphasis in each of the variants is different. In a data modifying expert system, the main function is to modify, i.e., generate or reduce data (from among available data) as required by the model. It should be noted that the model itself is selected by the expert system. In a model-based expert system, a suitable model and algorithm is selected for the problem considered. The model modifying expert system attempts to modify the selected model, i.e., adds or deletes constraints depending on the nature of the problem considered, or constructs a suitable model. It then selects a suitable algorithm for the model. In each of the above mentioned variants of the tandem expert system, the solution produced by the algorithm is modified, if necessary, in order to incorporate qualitative aspects or to obtain an implementable solution.

3.3. ES Applications in Energy Systems

Many researchers have dealt with the application of expert systems in the enrgy sector (Godart et al., 1991 ; Adapa, 1994 ; Le et al., 1995 ; Azzam et al., 1996 ; Lai, 1998 ; Kandil, 2001). The technology of expert systems and its applications in the energy sector are categorized in Table 1.

Table 1. Categories of experts systems

Experts systems/applications	References
Energy planning	Hung et al. (1998)
Energy efficiency	Lara-Rosano et al. (1998)
Alarm processing, system diagnosis, control and security	Vale et al. (1998)
Electric load forecasting	Metaxiotis et al. (2003)
Power system restoration	Vale et al. (1997)
Reactive power/voltage control	Vajpai et al. (2001)
Unit commitment	Padhy (2000)

In addition, several surveys of ES applications have been made (Ypsilantis et al., 1990 ; Bretthauer et al, 1992 ; Liu et al., 1994). The main advantages of ES are: 1) it is permanent and consistent; 2) it can be easily transferred or reproduced; and 3) it can be easily documented. The main disadvantage of ES is that it suffers from a knowledge bottleneck because it is unable to learn or adapt to new situations.

4. ARTIFICIAL NEURAL NETWORKS

A neural network is a general mathematical computing paradigm that models the operations of biological neural systems. In 1943, McCulloch, a neurobiologist, and Pitts, a statistician, published a seminal paper titled "A logical calculus of ideas imminent in nervous activity" in Bulletin of Mathematical Biophysics (McCulloch and Pitts, 1943) and later in Hebb's famous Organization of Behavior (Hebb, 1949). The early work in artificial intelligence was separated between those who believed that intelligent systems could best be built on computers modeled after brains, and those like Minsky and Papert (1969) who believed that intelligence was fundamentally a symbol processing of the kind readily modeled on the von Neumann computer. For a variety of reasons, the symbol-processing approach became the dominant theme in AI in the 1970s. However, the 1980s showed a rebirth in interest in neural computing.

Hopfield (1982) provided the mathematical foundation for understanding the dynamics of an important class of networks, Kohonen (1984), developed unsupervised learning networks for feature mapping into regular arrays of neurons, whereas Rumelhart and McClelland (1986), introduced the back-propagation learning algorithm for complex, multilayer networks. In mid 1980's, many neural networks research programs were initiated. The list of applications that can be solved by neural networks has expanded from small test-size examples to large practical tasks and large-scale integrated neural network chips have been fabricated (Robert, 1995).

A neural network is a collection of small individually interconnected processing units. Information is passed through these units along interconnections. An incoming connection has two values associated with it, an input value and a weight. The output of the unit is a function of the summed value. ANNs while implemented on computers are not programmed to perform specific tasks. Instead, they are trained with respect to data sets until they learn patterns used as inputs. Once they are trained, new patterns may be presented to them for prediction or classification. ANNs can automatically learn to recognize patterns in data from real systems or from physical models, computer programs, or other sources. An ANN can handle many inputs and produce answers that are in a form suitable for designers (Kalogirou, 2003). Artificial neural networks can be considered as simplified mathematical models of brain-like systems and they function as parallel-distributed computing networks. However, in contrast to conventional computers, which are programmed to perform specific task, most neural networks must be taught, or trained. They can learn new associations, new functional dependencies and new patterns. Neural networks obviate the need to use complex mathematically explicit formulas, computer models and impractical and costly physical models. Some of the characteristics that support the success of ANNs and distinguish them from the conventional computational techniques are (Nannariello and Frike, 2001):

- The direct manner in which ANNs acquire information and knowledge about a given problem domain (learning interesting and possibly non-linear relationships) through the 'training' phase
- Neural networks can work with numerical or analogue data that would be difficult to deal with by other means because of the form of the data or because there are many variables

Figure 3. The basic neuron

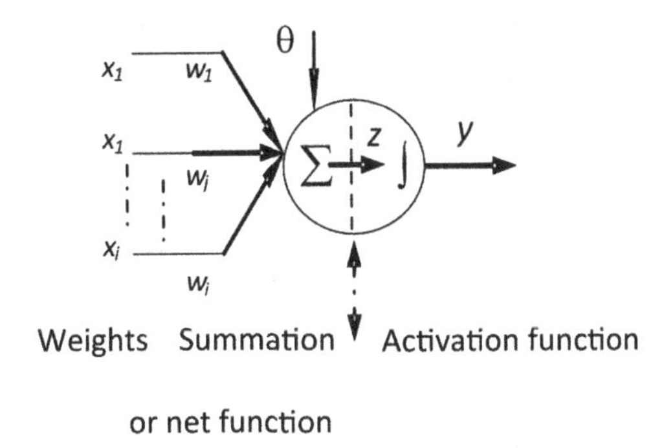

- Neural network analysis can be conceived of as a 'black box' approach and the user does not require sophisticated mathematical knowledge
- The compact form in which the acquired information and knowledge is stored within the trained network and the ease with which it can be accessed and used
- Neural network solutions can be robust even in the presence of 'noise' in the input data
- The high degree of accuracy reported when ANNs are used to generalize over a set of previously unseen data (not used in the 'training' process) from the problem domain

All neural network models that have been proposed over the years, share a common building block, known as a neuron and a networked interconnection structure (Yu and Jenq-Neng, 2001). The most widely used neuron model is based on McCulloch and Pitts' work and is illustrated in Figure 3.

According to Figure 3, the neuron consists of two parts; the net function and the activation function. The net function determines how the network inputs $\{x_i: 1 \leq i \leq N\}$ are combined inside the neuron. In this figure, a weighted linear combination is adopted:

$$z = \sum_{i=1}^{N} w_i x_i + \theta \qquad (1)$$

Table 2. Summary of net functions (Yu and Jenq-Neng, 2001)

Net functions	Formula	Comments
Linear	$z = \sum_{i=1}^{N} w_i x_i + \theta$	Most commonly used
Higher order (2nd order formula exhibition)	$z = \sum_{i=1}^{N} \sum_{j=1}^{N} w_{ij} x_i x_k + \theta$	y_i is a weighted linear combination of their order polynomial terms
Delta (Σ, Π)	$z = \prod_{i=1}^{N} w_i x_i$	Seldom used for the input variables. The number of input terms equals N^d, where d is the order of the polynomial

Table 3. Neuron activation functions (Yu and Jenq-Neng, 2001)

Activation functions	Formula	Derivatives	Comments
Sigmoid	$f(z)=\dfrac{1}{1+e^{-z}}$	$f(z)(1-f(z))$	Commonly used, derivative can be computed from *f(z)*, directly
Hyperbolic tangent	$f(z)=\tanh(z)$	$(1-\lvert f(z)^2\rvert)$	
Inverse tangent	$f(z)=\dfrac{2}{\pi}\tan^{-1}(u)$	$\dfrac{2}{\pi}.\dfrac{1}{1+u^2}$	Less frequently used
Threshold	$f(z)=\begin{cases}1 & x\rangle 0\\ -1 & x\langle 0\end{cases}$	Derivation does not exist	
Gaussian radial basis	$f(z)=\exp\left(-\lVert x-m\rVert^2/\sigma^2\right)$	$-2(x-m).f(z)/\sigma^2$	Used for radial basis network: m and σ
Linear	$f(z)=ax+b$	α	

Parameters $\{w_i: 1 \leq i \leq N\}$ are known as synaptic weights. The quantity θ is called the bias (or threshold) and is used to model the threshold. In literature, many other types of network input combination methods have been proposed. These are summarized in Table 2.

The output of the neuron, denoted by y in Figure 3, is related to the network input x_i via a linear or nonlinear transformation called the activation function; $y = f(z)$. In various neural network models, different activation functions have been proposed. The most commonly used activation functions are summarized in Table 3, which lists both the activation functions as well as their derivatives (provided they exist). In both sigmoid and hyperbolic tangent activation functions, derivatives can be computed directly from the knowledge of $f(z)$.

In a neural network, multiple neurons are interconnected to form a network to facilitate distributed computing. The configuration of the interconnections can be described efficiently with a directed graph. A directed graph consists of nodes (in the case of a neural network consists of neurons as well as external inputs) and directed arcs (in the case of a neural network, synaptic links). Several architectures and algorithms have been developed in literature for solving different problems (Haykin, 1999). The main ones are described in the following sub-sections.

4.1 Multilayer Perceptron (MLP)

Multilayer Perceptron (feed-forward) networks consist of units arranged in layers with only forward connections to units in subsequent layers (Yu and Jenq-Neng, 2001). The connections have weights associated with them. Each signal traveling along the link is multiplied by a connection weight. The first layer is the input layer, and the input units distribute the inputs to units in subsequent layers. In subsequent layers, each unit sums its inputs, adds a bias or threshold term to the sum and nonlinearly transforms the sum to

Figure 4. Feed-forward neural network

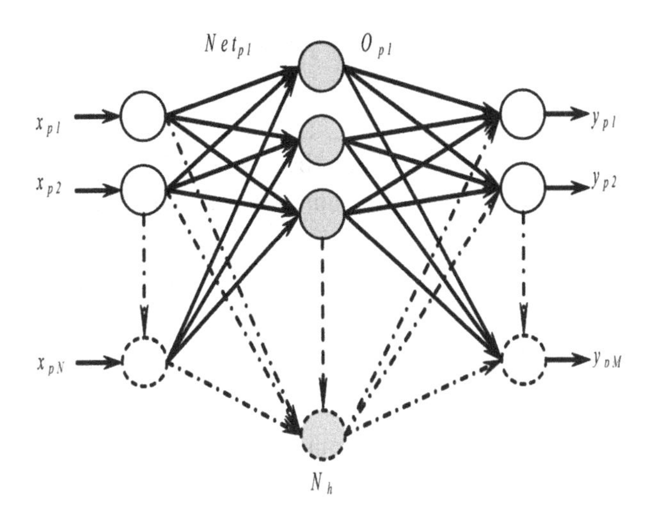

produce an output. This nonlinear transformation is called the activation function of the unit. The output layer units often have linear activations. In the remainder of this section, linear output layer activations are assumed. The layers sandwiched between the input layer and output layer are called hidden layers and units in hidden layers are called hidden units. Such a network is shown in Figure 4.

The training data set consists of N training patterns $\{(x_p, t_p)\}$, where p is the pattern number. The input vector x_p and desired output vector t_p have dimensions N and M, respectively; y_p is the network output vector for the pth pattern. The thresholds are handled by augmenting the input vector with an element $x_p(N + 1)$ and setting it equal to one.

For the jth hidden unit, the net input $net_p(j)$ and the output activation $O_p(j)$ for the pth training pattern are:

$$net_p(j) = \sum_{i=1}^{N+1} w(j,i).xp(i), \text{ where } 1 \le j \le N_h \tag{2}$$

$$O_p(j) = f(net_p(j)) \tag{3}$$

where $w(j, i)$ denotes the weight connecting the ith input unit to the jth hidden unit. For MLP networks, a typical activation function f is the sigmoid, given by:

$$f(net_p(j)) = \frac{1}{1 + \exp(-net_p(j))} \tag{4}$$

For trigonometric networks, the activations can be the sine and cosine functions. The kth output for the pth training pattern is y_{pk} and is given by:

$$y_{pk} = \sum_{i=1}^{N+1} w_{io}(k,i).x_p(i) + \sum_{j=1}^{N_h} w_{ho}(k,j).O_p(j), \quad \text{where } 1 \le k \le M \tag{5}$$

where $w_{io}(k, i)$ denotes the output weight connecting the ith input unit to the kth output unit and $w_{ho}(k, j)$ denotes the output weight connecting the jth hidden unit to the kth output unit. The mapping error for the pth pattern is:

$$E_p = \sum_{k=1}^{Np} \left(t_{pk} - y_{pk}\right)^2 \tag{6}$$

where t_{pk} denotes the kth element of the pth desired output vector. In order to train a neural network in batch mode, the mapping error for the kth output unit is defined as:

$$E(k) = \frac{1}{N_v} \sum_{p=1}^{Nv} \left(t_{pk} - y_{pk}\right)^2 \tag{7}$$

The overall performance of an MLP neural network, measured as mean square error (MSE), can be written as:

$$E = \sum_{k=1}^{M} E(k) = \frac{1}{Np} \sum_{p=1}^{Nv} E_p \tag{8}$$

The key distinguishing characteristic of a MLP with the back-propagation learning algorithm is that it forms a nonlinear mapping from a set of input stimuli to a set of outputs using features extracted from the input patterns. The neural network can be designed and trained to accomplish a wide variety of nonlinear mappings, some of which are very complex. This is because the neural units in the neural network learn to respond to features found in the input. By applying the set of formulations of the backpropagation (BP) algorithm, presented above, the calculation procedure of the learning process is summarized as follow (Gupta *et al.*, 2003):

$$w_{aj}^{(i)} \quad w_{aj}^{(i)} \quad w_{aj}^{(i)}(0) \begin{cases} s_j^{(i)} & = & \left(w_{aj}^{(i)}\right)^T x_a^{(i-1)} \\ x_j^{(i)} & = & \sigma\left(s_j^{(i)}\right) \end{cases} \quad e_j = d_j - \mathrm{x}_j^{(\mathrm{M})} \quad \delta_j^{(M)} = e_j \sigma'\left(s_j^{(M)}\right) \quad e_j^{(i)} = \sum_{l=1}^{n_i+1} \delta_l^{(i+1)} w_{lj}^{(i+1)}$$

$$\delta_l^{(i)} = e_l^{(i)} \sigma'\left(s_l^{(i)}\right) \quad w_{aj}^{(i)} = w_{aj}^{(i)} + \eta \delta_j^{(i)} x_a^{(i-1)} \quad E = E + \frac{1}{k} \sum_{j}^{m} e_j^2$$

In the procedure listed above, several learning factors such as the initial weights, the learning rate, the number of the hidden neural layers and the number of neurons in each layer, may be reselected if the iterative learning process does not converge quickly to the desired point. Although, the BP learning

Figure 5. The radial basis network

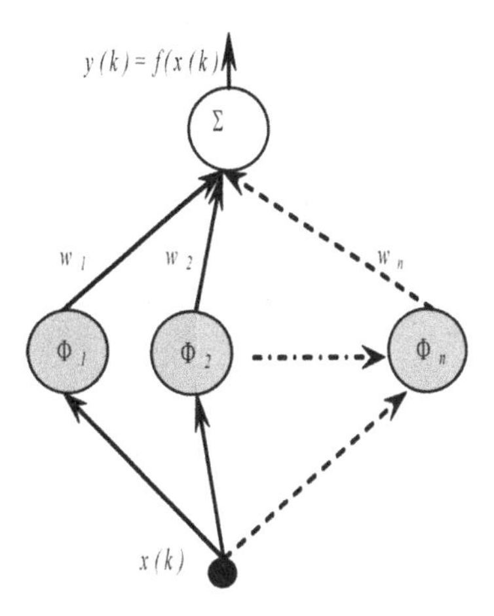

algorithm provides a method for training MLPs to accomplish a specified task in terms of the internal nonlinear mapping representations, it is not free from problems. Many factors affect the learning performance and must be dealt with in order to have a successful learning process. Mainly, these factors include the initial parameters, learning rate, network size and learning database. A good choice of these items may greatly speed up the learning process to reach the target, although there is no universal answer for these issues (Gupta *et al.*, 2003). Advanced methods for learning and adaptation in MLPs are presented in (Haykin, 1999; Lakhmi and Martin, 1998).

4.2 Radial Basis Function Network (RBFNN)

The radial basis function network is considered as a 2-layer network, as shown in Figure 5, because the learning process is done in two different stages, referred to as layers (Gupta *et al.*, 2003; Lakhmi and Martin, 1998). A key aspect is the distinction between the first and second layers of weights. In the first stage, the input data set x_n alone is used to determine the parameters of the basis functions, i.e., the first-layer weights.

As only the input data are used, the training method is called unsupervised. The first layer weights are then kept fixed while the second layer weights are found in the second phase. The second stage is supervised as both input and target data are required. Optimization is done by a classic least squares approach. The basic form of the RBFNN mapping is:

$$y_k(x) = \sum_{j=1}^{M} w_{kj}\Phi_j(x) + w_{k0} \tag{9}$$

where w_{k0} is the bias term which can be absorbed into the summation by including an extra basis function Φ_0 whose activation is set to 1. For the case of Gaussian basis functions, we have:

$$\Phi_j(x) = \exp\left(-\frac{\left\|x - \mu_j\right\|^2}{2\sigma_j^2}\right) \qquad (10)$$

Here x is the d-dimensional input vector with elements x_i and μ_j being the vector determining the centre of basis function Φ_j and has elements μ_{ji}. This Gaussian radial basis functions can be generalized to allow for arbitrary covariance matrices Σ_j. The basis function is therefore taken to have the form:

$$\Phi_j(x) = \exp\left(-\frac{1}{2}\left(x - \mu_j\right)^T \sum_j^{-1}\left(x - \mu_j\right)\right) \qquad (11)$$

Considering the RBFNN mapping defined above (and absorbing the bias parameter into the weights) we obtain:

$$y_k(x) = \sum_{j=1}^{M} w_{kj}\Phi_j(x) \qquad (12)$$

where Φ_0 is an extra "basis function" with activation value fixed at 1. Writing this in matrix notation:

$$y(x) = W\Phi, \qquad (13)$$

where $W = (w_{kj})$ and $\Phi = (\Phi_j)$. The weights can now be optimized by minimization of a suitable error function, e.g. the sum-of-squares error function:

$$E = \frac{1}{2}\sum_n \sum_k \left\{y_k(X^n) - t_k^n\right\}^2 \qquad (14)$$

where t_k^n is the target value for output unit k when the network is presented with the input vector x_n. The weights are then determined by the linear equations:

$$\Phi^T \Phi W^T = \Phi^T T \qquad (15)$$

where $(T)_{nk} = t_k^n$ and $(\Phi)_{nk} = \Phi_j(X^n)$.

This can be solved by:

$$WT = \Phi^* T \qquad (16)$$

Figure 6. (a) Jordan's model (Jordan, 1988), (b) Elman's model (Elman, 1990), (c) Fully recurrent neural network.

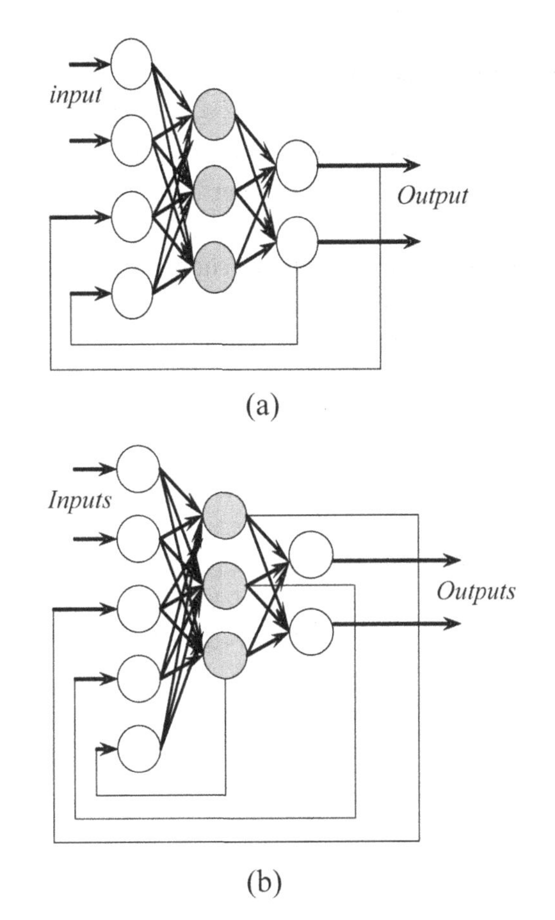

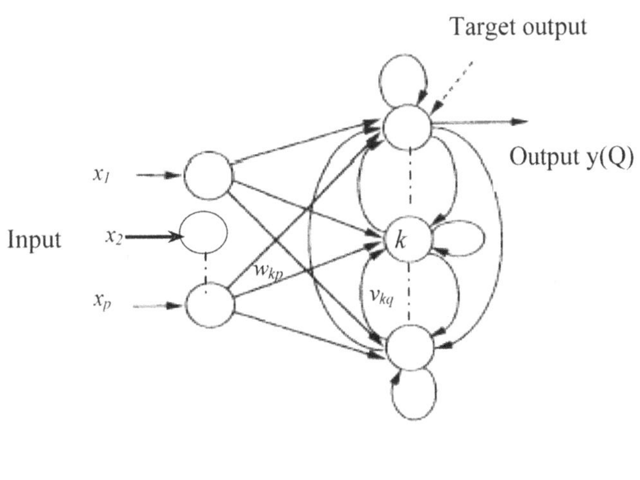

(c)

where the notation Φ^* denotes the pseudo-inverse of Φ. Thus, the second layer weights can be found by fast, linear matrix inversion techniques.

4.3 Recurrent Neural Network (RNN)

Recurrent neural networks (RNN) are shown to have powerful capabilities for modeling many computational structures. For a comprehensive discussion of recurrent neural network model (Herts *et al.*, 1991). Several RNN architectures have been proposed in literature; Jordan (1988) proposed a model that has feedback connections from output units to input units, shown in FigureFigure 6(a), and Elman (1990) proposed a model that has feedback connections from hidden units to input units (see FigureFigure 6(b)). A fully recurrent neural network is illustrated in Figure 6(c).

A real time recurrent learning network (RTRL) comprises input nodes, processing nodes, feed-forward and recurrent connections. Some of the processing nodes are assigned as output nodes. The output of every processing node is connected to all processing nodes including itself, i.e., is a fully connected recurrent network. The output of a processing node at the current time step depends on the input signals and feedback signals in the previous time step.

William and Zipser (1989) showed that this type of recurrent networks can be trained by updating their weights in every processing cycle, i.e., real-time learning.

Since the activation of a processing node is the weighted sum of the current input and feedback signals, we have:

$$s_k(t) = \sum_{p=1}^{p+1}\left(w_{ik}x_p(t)\right) + \sum_{q=1}^{Q}\left(v_{kq}y_q(t)\right) \tag{17}$$

Where $s_k(t)$ is the activation of processing node k at time step t, v_{kq} is the weight connecting processing node q to processing node k, $w_{k,p+1}$ is the bias and $x_{p+1}=1$. The output of the processing node k at the next time step is given by:

$$y_k(t+1) = f_k(s_k(t)) \tag{18}$$

where f is a sigmoid function.

If we denote $T(t)$ as the set of indices such that the processing nodes have target values (teachers) at time step t, and let $e_k(t)$ be the instantaneous error between the target output and the actual output of processing node k, we get:

$$e_k(t) = \begin{cases} d_k(t) - y_k(t) & k \in (t) \\ 0 & otherwise \end{cases} \tag{19}$$

Hence, the total instantaneous squared error at time step t is given by:

Figure 7. (a) A standard global recurrent network (Giles et al., 1997), Figure(b) GR network with embedded memory with order of two (Giles et al., 1997).

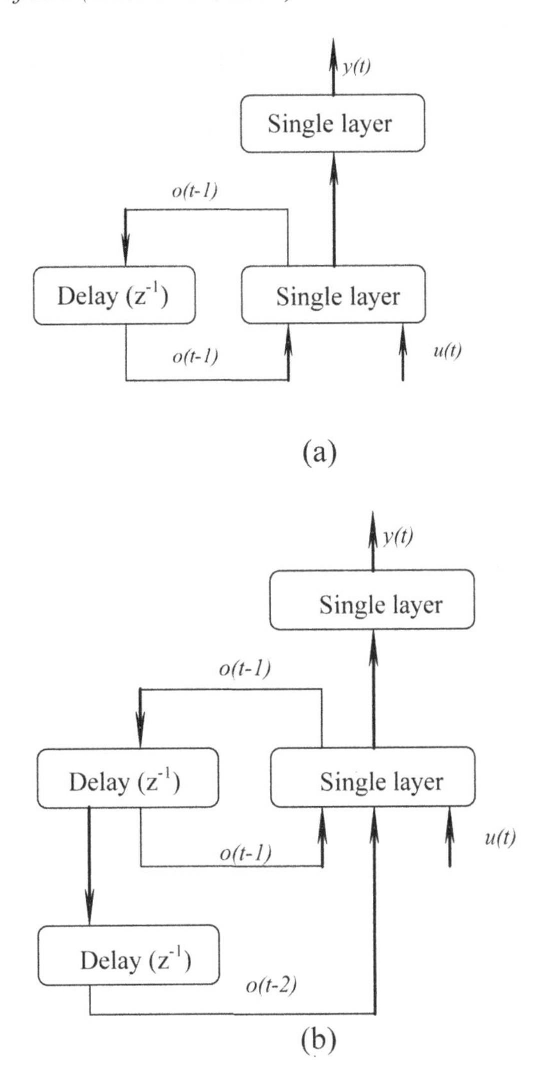

$$E(t) = \frac{1}{2} \sum_{k \in T(t)} \left\{ d_k(t) - y_k(t) \right\}^2 \tag{20}$$

The weights are updated according to the gradient descent rules:

$$\Delta w_{ij}(t) = -\alpha \frac{\partial E(t)}{\partial w_{ij}} = \alpha \sum_{k \in T(t)} e_k(t) \frac{\partial y_k(t)}{\partial w_{ij}} \quad 1 \le i \le Q,\ 1 \le j \le P+1 \tag{21}$$

and

Figure 8. (a) Standard LR network (Giles et al., 1997), Figure(b) A LR network with embedded memory with order of two (Giles et al., 1997).

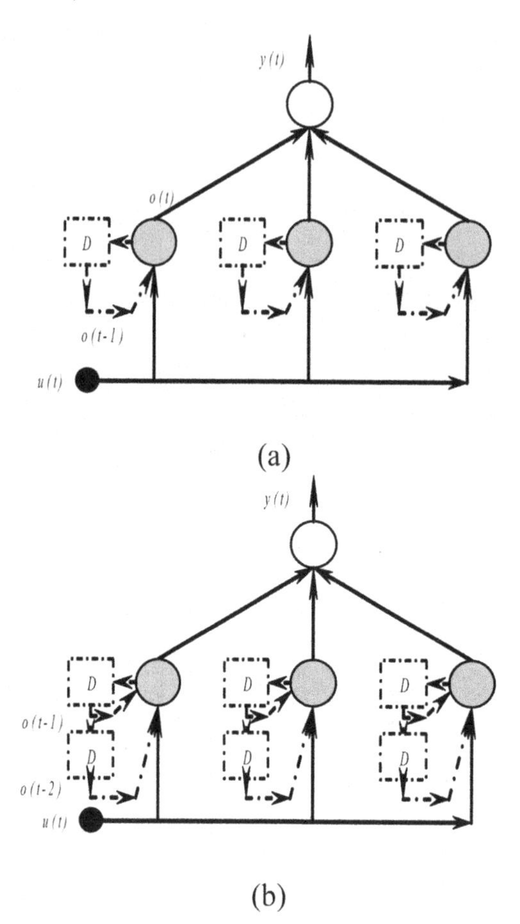

Figure 9. NARX network with memory with order of 3 (Giles et al., 1997)

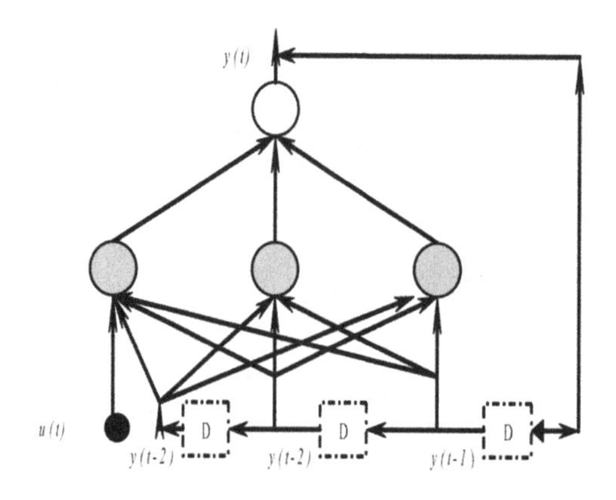

Figure 10. Flow diagram of fuzzy inference system (Lakhmi and Martin, 1998)

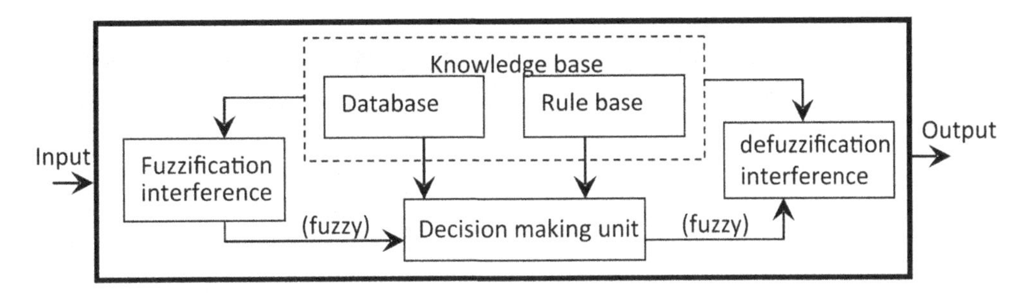

$$\Delta v_{ij}(t) = -\alpha \frac{\partial E(t)}{\partial v_{ij}} = \alpha \sum_{k \in T(t)} e_k(t) \frac{\partial y_k(t)}{\partial v_{ij}} \quad 1 \le i, j \le Q \tag{22}$$

Where α is the learning rate.
By using Eqs. (17) and (18)

$$\frac{\partial y_k(t+1)}{\partial w_{ij}} = \dot{f}_k(s_k(t)) \left\{ x_j(t)\delta_{kj} + \sum_q^Q v_{kq} \frac{\partial y_k(t)}{\partial w_{ij}} \right\} \tag{23}$$

and

$$\frac{\partial y_k(t+1)}{\partial v_{ij}} = \dot{f}_k(s_k(t)) \left\{ y_j(t)\delta_{kj} + \sum_q^Q v_{kq} \frac{\partial y_k(t)}{\partial v_{ij}} \right\} \tag{24}$$

With

$$\frac{\partial y_k(0)}{\partial w_{ij}} = \frac{\partial y_k(0)}{\partial v_{ij}} = 0 \quad \text{and} \quad \frac{\partial y_k(0)}{\partial w_{ij}} = \frac{\partial y_k(0)}{\partial v_{ij}} = 0 \ \delta_{ki} = \begin{cases} 1 & k = i \\ 0 & k \ne i \end{cases} \quad \text{and} \quad \delta_{ki} = \begin{cases} 1 & k = i \\ 0 & k \ne i \end{cases} \tag{25}$$

Equations (17) to (25) completely define the RTRL algorithm. Three classes of networks exists; globally recurrent networks (GR) locally recurrent (LR) networks and Nonlinear Auto-Regressive with eXogeneous inputs (NARX) networks, which are used in several applications such as modeling of dynamical systems, time series prediction and forecasting. NARX networks are a typical model of networks with observable states. Global recurrent networks are a popular class of networks with globally connected hidden states and locally recurrent networks belong to locally recurrent network architecture class with hidden states.

4.3.1 Globally Connected RNNs

These networks are a class of recurrent networks in which the feedback connections come from the state vector to the hidden layer, as illustrated in Figure 7(a). These hidden states are sometimes called context units in literature (Giles *et al.*, 1997).

Suppose such a network with n_u input nodes, n_h hidden nodes and n_y output nodes, then the dynamic equation can be described by:

$$o_i(t) = f\left(\sum_{j=1}^{n_h} w_{ij}^h(t-1) + \sum_{k=1}^{nu} w_{ik}^u u_k(t) + w_i^b\right) \quad (26)$$

$$y_i(t) = f\left(\sum_{j=1}^{n_h} w_{ij}^h o_j(t) + w_i^b\right) \quad (27)$$

where $o(t)$ and $y(t)$ denotes the real valued outputs of the hidden and output neurons at time t, and f is the nonlinear function. This network with a high order of embedded memory differs from the standard globally connected recurrent network in that they have more than one state vector per feedback loop.

For a globally recurrent network with embedded memory of order m, shown in Figure 7(b), the dynamic equations of hidden nodes become:

$$o_i(t) = f\left(\sum_{k=1}^{m}\sum_{j=1}^{n_h} w_{ijm}^h o_j(t-k) + \sum_{k=1}^{nu} w_{ik}^u u_k(t) + w_i^b\right) \quad (28)$$

4.3.2 Locally Recurrent Networks

In this class of networks, the feedback connections are only allowed from neurons to themselves and the nodes are connected together in a feed forward architecture (Giles *et al.*, 1997), as shown in Figure 8(a). The dynamic neurons of locally recurrent networks can be described by:

$$o_i(t) = f\left(w_{ij}^h o_i(t-1) + \sum_{j} w_{ij}^u u_j(t-1) + w_i^b\right) \quad (29)$$

where $o_i(t)$ denotes the output of the ith node at time t, and f is the nonlinearity. For a network with embedded memory of order m, the output of the dynamic neurons becomes:

$$o_i(t) = f\left(\sum_{j=1}^{m} w_{ii}^h o_i(t-n) + \sum_{j} w_{ik}^u u_k(t) + w_i^b\right) \quad (30)$$

Figure 11. (a) Life-cycle of populations, (b) Recombination and mutation

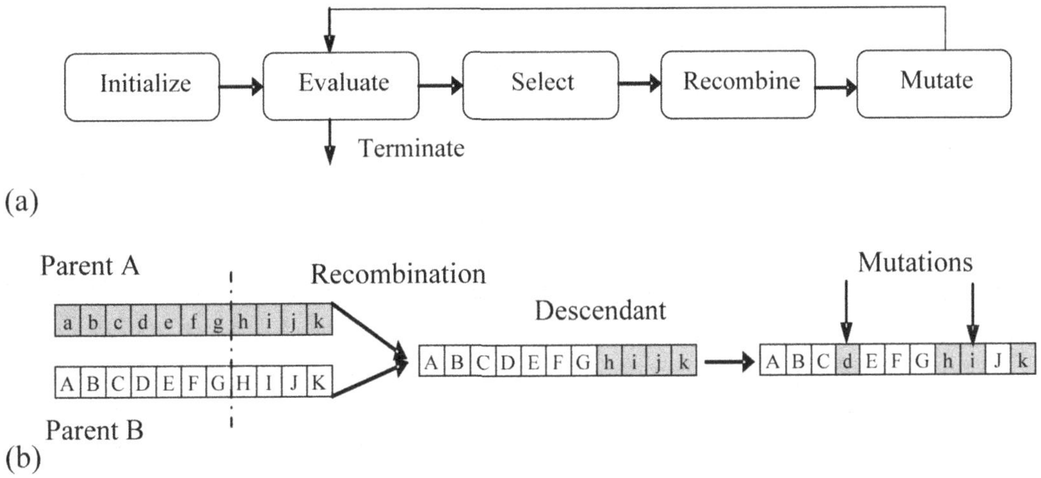

Figure 8(b) shows a LR network with embedded memory with order two. Locally recurrent models usually differ with respect to where and how much output feedback is permitted (Giles *et al.*, 1997).

4.3.3 NARX Recurrent Neural Networks

An important class of discrete-time nonlinear systems is the NARX model (Giles *et al.*, 1997):

$$y(t) = f\left(u(t - D_u), \ldots\ldots, u(t), y(t - D_y), \ldots.. y(t - 1)\right) \tag{31}$$

where $u(t)$ and $y(t)$ represent input and output of the network at time t, D_u and D_y, are the input-memory and output-memory order and the function f is a nonlinear function. When the function f can be approximated by a Multilayer Perceptron, the resulting system is called a NARX recurrent neural network (see Figure 9).

4.4 ANN APPLICATIONS IN ENERGY SYSTEMS

Artificial neural networks have been used by the authors in the field of solar energy for modeling the heat-up response of a solar steam generating plant (Kalogirou *et al.*, 1998), for the estimation of a parabolic trough collector intercept factor (Kalogirou *et al.*, 1996), for the estimation of a parabolic trough collector local concentration ratio (Kalogirou, 1996a), for the design of a solar steam generation system (Kalogirou, 1996b), for the performance prediction of a thermosyphon solar water heater (Kalogirou *et al.*, 1999a), for the modeling of solar domestic water heating systems (Kalogirou *et al.*, 1999b), for the long-term performance prediction of forced circulation solar domestic water heating systems (Kalogirou, 2000) and for the thermosyphon solar domestic water heating systems long-term performance prediction (Kalogirou and Panteliou, 2000). ANNs were also used for the development of a model for the estimation of the sizing parameters of stand-alone PV-systems (Mellit *et al.*, 2003). A review of these models

Figure 12. Intelligent technologies used in hybrid systems (Medsker, 1996)

Hybrid intelligent systems		
Expert system	Fuzzy logic	Neural networks
	Genetic algorithm	
	Case-based reasoning	
Hardware and software		

together with other applications in the field of renewable energy is given in (Kalogirou, 2001). ANNs have also be used in other energy systems, like building thermal load (Kalogirou *et al.*, 1997; Kalogirou and Bojic, 1999), naturally ventilated rooms (Kalogirou *et al.*, 1999c), energy prediction of commercial buildings (Kreider and Wang, 1995), energy consumption optimisation of a large HVAC system (Curtiss *et al.*, 1995), room storage heater model (Roberge *et al.*, 1977), daily solar radiation (Mellit *et al.*, 2005), long-term wind speed and power forecasting (Barbounis *et al.* 2006; 2007) and many other.

Errors reported are well within acceptable limits, which clearly suggest that artificial neural networks can be used for modeling and prediction in other fields of solar energy engineering. What is required is to have a set of data (preferably experimental) representing the past history of a system so as a suitable neural network can be trained to learn the dependence of expected output on the input parameters.

5. FUZZY LOGIC

Fuzzy set (FS) theory is a generalization of conventional set theory and was introduced by Zadeh in 1965 (1972; 1973). It provides a mathematical tool for dealing with linguistic variables associated with natural languages. Systematic descriptions of these topics can be found in several texts (Bellman and Zadeh, 1977; Dubois and Prade, 1980; Kaufmann and Gupta, 1985). A central notion of fuzzy set theory, as described in the following sections, is that it is permissible for elements to be only partial elements of a set rather than full membership. Figure 10 shows the flowchart of fuzzy inference system. The development of fuzzy logic was motivated by the need for a conceptual framework which can address the issue of uncertainty and lexical imprecision. Some of the essential characteristics of fuzzy logic relate to the following (Robert, 1995; Machado and Rocha, 1992):

- In fuzzy logic, exact reasoning is viewed as a limiting case of approximate reasoning
- In fuzzy logic, everything is a matter of degree
- In fuzzy logic, knowledge is interpreted as a collection of elastic or, equivalently, fuzzy constraint on a collection of variables
- Inference is viewed as a process of propagation of elastic constraints
- Any logical system can be fuzzified

There are two main characteristics of fuzzy systems that give them better performance for specific applications:

- Fuzzy systems are suitable for uncertain or approximate reasoning, especially for the system with a mathematical model that is difficult to derive

- Fuzzy logic allows decision making with estimated values under incomplete or uncertain information

Zadeh stated that the attempts to automate various types of activities from assembling hardware to medical diagnosis have been impeded by the gap between the way human beings reason and the way computers are programmed. Fuzzy logic uses graded statements rather than ones that are strictly true or false. It attempts to incorporate the "rule of thumb" approach generally used by human beings for decision-making. Thus, fuzzy logic provides an approximate but effective way of describing the behavior of systems that are not easy to describe precisely.

Fuzzy logic controllers, for example, are extensions of the common expert systems that use production rules like "if-then" statements. With fuzzy controllers, however, linguistic variables like "tall" and "very tall" might be incorporated in a traditional expert system. The result is that fuzzy logic can be used in controllers that are capable of making intelligent control decisions in sometimes volatile and rapidly changing problem environments. Fuzzy logic techniques have been successfully applied in a number of applications like, computer vision, decision-making and system design including ANN training. The most extensive use of fuzzy logic is in the area of control, where examples include controllers for cement kilns, braking systems, elevators, washing machines, hot water heaters, air-conditioners, video cameras, rice cookers and photocopiers (Lakhmi and Martin, 1998). Fuzzy logic has been used for the solar radiation prediction (Sen, 1998) and for the development of a solar tracking mechanisms (Kalogirou, 2002).

6. GENETIC ALGORITHMS

Genetic Algorithms (GA) were envisaged by Holland (1975) in the 1970s as an algorithmic concept based on a Darwinian-type survival-of-the-fittest strategy with sexual reproduction, where stronger individuals in the population have a higher chance of creating an offspring. A genetic algorithm is implemented as a computerized search and optimization procedure that uses the principles of natural genetics and natural selection. The basic approach is to model the possible solutions to the search problem as strings of ones and zeros. Various portions of these bit-strings represent parameters in the search problem. If a problem-solving mechanism can be represented in a reasonably compact form, then GA techniques can be applied using procedures to maintain a population of knowledge structure that represent candidate solutions and then let that population evolve over time through competition (survival of the fittest and controlled variation).

The practicality of using a GA to solve complex problems was demonstrated by (Michelewicz, 1992; Dejong, 1975; Goldberg, 1989; Colin and Jonathan, 2002). Under this paradigm, a population of chromosomes evolves over a number of generations through the application of genetic operators, like crossover and mutation that mimic those found in nature. The evolution process allows the best chromosomes to survive and mate from one generation to the next. Actually, the GA is an iterative procedure that maintains a population of P candidate members over many simulated generations. The GA will generally include three fundamental genetic operations: selection, crossover and mutation. These operations are used to modify the chosen solutions and select the most appropriate offspring to pass on to succeeding generations. The life cycle of such populations and the recombination of the parental and mutation are illustrated in Figure 11(a) and (b).

Figure 13. (a) The first model of fuzzy neural system (Medsker, 1996), Figure(b) The second model of fuzzy neural system (Medsker, 1996)

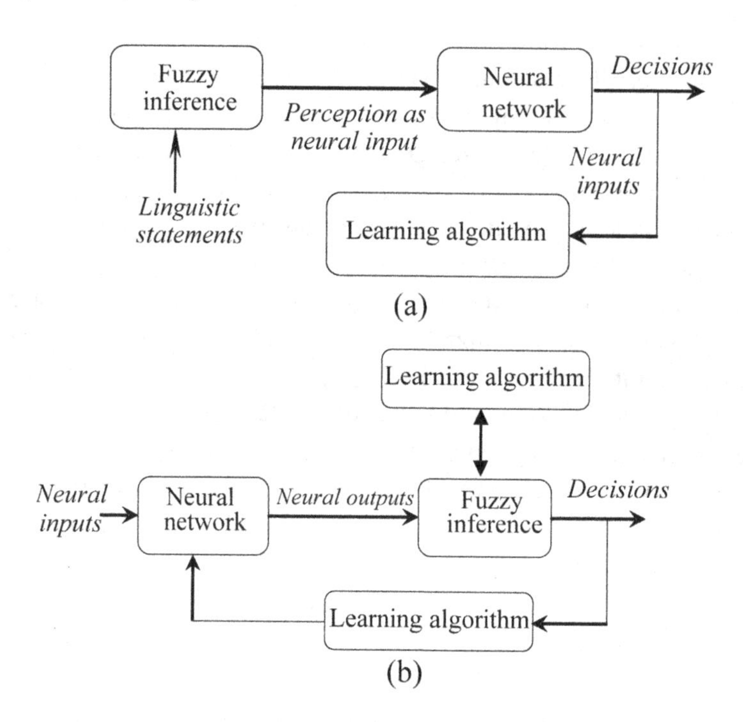

GAs consider many points in the search space simultaneously and have been found to provide a rapid convergence to a near optimum solution in many types of problems; they usually exhibit a reduced chance of converging to local minima. GAs show promise but suffer from the problem of excessive complexity if used on problems that are too large. Genetic algorithms are an iterative procedure that consists of a constant-sized population of individuals, each one represented by a finite linear string of symbols, known as the genome, encoding a possible solution in a given problem space. This space, referred to as the search space, comprises all possible solutions to the optimization problem at hand. In standard genetic algorithms the initial population of individuals is generated at random. At every evolutionary step, also known as a generation, the individuals in the current population are decoded and evaluated according to a fitness function set for a given problem. The expected number of times an individual is

Figure 14. Organization of the Neuro-GENSYS method for optimizing networks with genetic algorithms (Medsker, 1996).

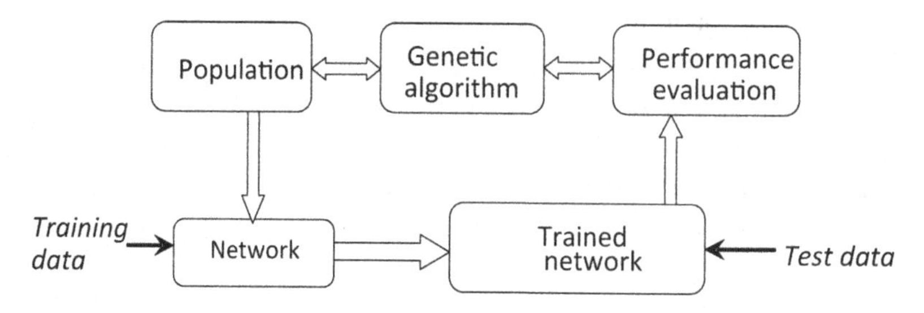

Figure 15. Use of a genetic algorithm to improve the performance of a fuzzy system (Medsker, 1996).

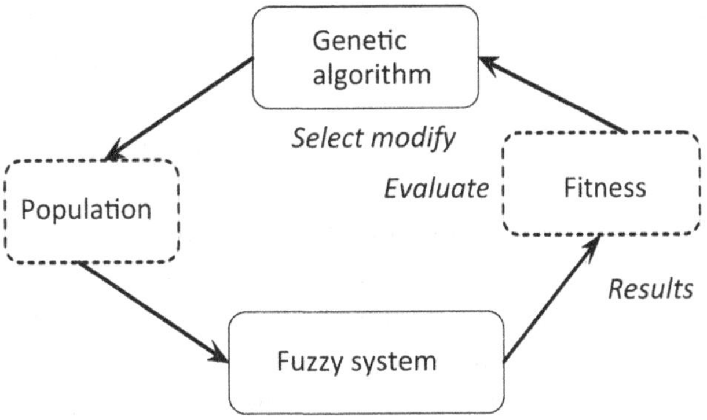

chosen is approximately proportional to its relative performance in the population.

Crossover is performed between two selected individuals by exchanging part of their genomes to form new individuals. The mutation operator is introduced to prevent premature convergence. Every member of a population has a certain fitness value associated with it, which represents the degree of correctness of that particular solution or the quality of solution it represents. The initial population of strings is randomly chosen. The strings are manipulated by the GA using genetic operators, to finally arrive at a quality solution to the given problem (Kalogirou, 2003; 2007).

A basic version of a GA operation is shown in the following procedure (Holland, 1975):

This is a genetic algorithm template, which is a general formulation, accommodating many different forms of selection, crossover and mutation.

6.1 GAs Applications in Energy Systems

Genetic algorithms were used for the development of an optimal configuration of power generating systems in isolated islands with renewable energy (Senjyua *et al.*, 2007) and for the optimal allocation and sizing of photovoltaic grid-connected systems in feeders that provide the best overall impact onto the feeder (Hernadeza *et al.*, 2007). They have also been used for the optimization of other energy systems, such as to estimate the optimal fenestration area (Kalogirou, 2007).

7. HYBRID SYSTEMS

According to Medsker (1995), hybrid intelligent systems, depicted in Figure 12, are usually implemented by means of traditional computing systems, often with microcomputer hardware and software techniques. Expert systems and neural networks are well established as useful technologies that can complement each other in powerful hybrid systems. Other technologies that have more recently been exploited are fuzzy logic, genetic algorithms and case-based reasoning. Developers are finding niches for each of these and the various combinations of the technologies are explored and used. The increased popularity of hybrid intelligent systems (HIS) in recent years lies in the extensive success of these systems in many real-world complex problems. The main reason for this success seems to be the synergy derived by the computational intelligent components, such as machine learning, fuzzy logic, neural networks and genetic algorithms. Each of these methodologies provides hybrid systems with complementary reasoning and searching methods that allow the use of domain knowledge and empirical data to solve complex problems (Tsakonas and Dounias, 2002).

Hybrid systems combining fuzzy logic, neural networks, genetic algorithms and expert systems have proved their effectiveness in a wide variety of real-world problems (Tsakonas and Dounias, 2002). Every intelligent technique has particular computational properties (e.g. ability to learn, explanation of decisions) that make them suitable for particular problems and not for others. For example, while neural networks are good at recognizing patterns, they are not good at explaining how they reach their decisions. Fuzzy logic systems, which can reason with imprecise information, are good at explaining their decisions but they cannot automatically acquire the rules they use to make those decisions. These limitations have been a central driving force behind the creation of hybrid intelligent systems where two or more techniques are combined in a manner that overcomes the limitations of individual techniques. Hybrid systems are also important when considering the varied nature of application domains. Many complex domains have many different component problems, each of which may require different type of processing. If there is a complex application, which has two distinct sub-problems, say a signal processing task and a serial reasoning task, then a neural network and an expert system can be used for solving these separate tasks. The use of intelligent hybrid systems is growing rapidly with successful applications in many areas including process control, engineering design, financial trading, credit evaluation, medical diagnosis and cognitive simulation (Medsker, 1996). While fuzzy logic provides an inference mechanism under cognitive uncertainty, computational neural networks offer exciting advantages, such as learning, adaptation, fault-tolerance, parallelism and generalization. To enable a system to deal with cognitive uncertainties in a manner more like humans, one may incorporate the concept of fuzzy logic into the neural networks (Medsker, 1996). An extended review of hybrid computational intelligence schemes is presented in (Tsakonas and Dounias, 2002).

7.1 Fuzzy Neural Networks

Neural networks can be modified to incorporate fuzzy techniques and produce a neural network with improved performance. One approach is to allow the fuzzy neural network to receive and process fuzzy input. Another option is to add layers on the front end of the network to fuzzify crisp input data to the fuzzy neural processing (Medsker, 1996). The fuzzy neuron is a fundamental concept used in many approaches to integrate fuzzy and neural technologies. In networks that map fuzzy input to crisp output, nodes in every layer of the network can have modified neurons. The input vector consists of a set of fuzzy values and the weights connecting the node with nodes in the previous layer also have fuzzy values. Input values and the weights are each represented by membership functions. A modified summation process is used to find the product of the membership functions of the fuzzy inputs and weights and then add the resulting membership functions to obtain another one that represents the integration of weighted fuzzy inputs to the node. A centroid operation on the resultant can then be used to find a crisp value for the output of the node (Medsker, 1996). The computational process envisioned for fuzzy neural systems is as follows. It starts with the development of a "fuzzy neuron" based on the understanding of biological neuronal morphologies, followed by learning mechanisms. This leads to the following three steps in a fuzzy neural computational process (Medsker, 1996):

- Development of fuzzy neural models motivated by biological neurons,
- Models of synaptic connections which incorporates fuzziness into neural network,
- Development of learning algorithms (i.e., the method of adjusting the synaptic weights).

Two possible models of fuzzy neural systems are:

- In response to linguistic statements, the fuzzy interface block provides an input vector to a multi-layer neural network. The neural network can be adapted (trained) to yield the desired command outputs or decisions (see Figure 13(a)).
- A multi-layered neural network drives the fuzzy inference mechanism (see Figure 13(b)).

7.2 Genetic Algorithms and Neural Networks

Research and development on hybrid genetic and neural systems has grown dramatically since the late 1980's. Most of the activity has been focused on the exploitation of the advantages of genetic algorithms to improve the design and use of neural networks (Kadaba *et al.*, 1991; Harp and Samad, 1991; Scha Ver, 1994). Most of the published works use the ability of genetic algorithms to search large complex spaces to prepare data for neural networks, find initial sets of parameters for training networks and use genetic and the newer evolutionary techniques to evolve neural network topologies. Another application looks at creative ways to couple neural network modules with genetic algorithms and other problem-solving techniques. Research and development efforts also focus on ways to represent neural networks for easier tuning and design by genetic algorithms, ways to substitute genetic algorithms for neural learning algorithms and opportunities for incorporating new techniques from evolutionary computing (Medsker, 1996). Figure 14 illustrates the organization of the NeuroGENSYS method for optimizing networks with genetic algorithms.

7.3 Genetic Algorithms and Fuzzy Logic

Fuzzy and genetic systems operate well in similar environments, including situations involving nonlinearities and requiring high levels of performance. Thus, these two technologies are sometimes alternative and can be used in standalone or transformational modes (Medsker, 1996). They also work well as modules in coupled systems. In hybrid systems, fuzzy components provide a clear representation of knowledge as rules or mathematical expressions. The use of membership functions and associated parameters provide flexibility and abstraction that simplifies the design of highly complex systems. Genetic algorithms facilitate the optimization of fuzzy system performance. In fuzzy system design, genetic algorithms can be used to tune membership values, prune membership functions and derive fuzzy rules. Fuzzy logic control can be applied to the operation of genetic systems and perform the evaluation function required in genetic algorithms. Figure 15 shows the process for use of a genetic algorithm to improve the performance of a fuzzy system. A less-explored area of integration is the use of the two technologies to produce systems with more general learning abilities such as the semantic interpretation of symbols and the understanding of system behavior from their input and output data (Medsker, 1996).

7.4 Hybrid Systems Applications in Energy Systems

A hybrid system which combines combines ANFIS and GA has been used for the sizing of a stand alone PV system (Mellit, 2006). For the same purpose (Mellit, 2006) used a hybrid system which combines neural network and fuzzy logic, whereas in a different work Mellit *et al.* (2004) developed a suitable approach, which combines the ANN with wavelet analysis for the same purpose. Elshatter *et al.* (1997) used fuzzy logic for modeling and simulation of PV-system. Balzani and Reatti (2005) presented an application of artificial neural network for modeling a photovoltaic (PV) module. Moreno *et al.* (2000) proposed an algorithm based on fuzzy logic for the control of stand-alone PV-system. Zhang and Bai (2005) proposed an algorithm based on RBFN and GA for predicting the maximum power point of a PV-panel.

8. CONCLUSION

A number of AI techniques have been reviewed in this chapter, which are suitable for engineering applications. These include expert systems, artificial neural networks, genetic algorithms, fuzzy logic and hybrid systems. AI techniques have been applied in a wide range of fields for modeling, prediction, simulation, optimization and control of engineering systems. The advantages of AI based simulation techniques are that they offer an alternative approach to conventional physical modeling techniques, they do not require the knowledge of internal system parameters, involve less computational effort and offer a compact solution for multi-variable problems. They can also be used for control, supervision, fault diagnosis and monitoring of actual systems. As was seen in this paper, AI techniques have been used successfully in a large variety of energy problems, including solar systems (thermal and PV), buildings and load prediction. The lesson learned from these applications is that to produce a sound model, data that represent the system behavior are required, preferably covering all possible combinations the system will work, which can be used in combination with a suitable AI model. The successful applications presented in this chapter are testimony of the potential of AI techniques in this field. The fact that these techniques are suitable for engineering applications does not mean that they are not suitable for other applications and indeed, they have been used in a large variety of these applications extending in a large number or areas like science, medicine, economics and many others.

ACKNOWLEDGMENT

The contents of this chapter including figures and tables are reprinted from a paper of the same authors presented in Mellit, A. & Kalogirou, S.A., Artificial intelligence techniques for photovoltaic applications: A review, *Progress in Energy and Combustion Science*, Vol. 34, No. 5, pp. 574-632, Copyright (2008), with permission from Elsevier.

NOMENCLATURE

b: Bias
e: Error
E: Mapping error, mean square error
f: activation function, sigmoid function
net: Network
o(t): Real value output of hidden neurons
w_{ij}: Weight of the neuron
y(t): Real value output of output neurons

GREEK

θ: Bias or threshold
Φ: Basis function
Φ^*: Pseudo-inverse of Φ

ABBREVIATIONS

AGV: Automated Guided Vehicles
AI: Artificial Intelligence
ANN: Artificial Neural Network
CV: Computer Vision
ES: Expert System
FL: Fuzzy Logic
GA: Genetic Algorithm
GR: Globally Recurrent
HIS: Hybrid Intelligent System
KBES: Knowledge-Based Expert System
LR: Locally Recurrent
MLP: Multi-Layer Perceptron
NARX: Nonlinear AutoRegressive with eXogeneous
NLP: Natural Language Processing
NMR: Non-Monotonic Reasoning
PSP: Problem Solving and Planning
PV: Photovoltaic
RBFNN: Radial Basis Function Neural Network
RNN: Recurrent Neural Network
RTRL: Real Time Recurrent Learning Network
TMS: Truth Maintenance Systems

REFERENCES

Adapa, R. (1994). Expert system applications in power system planning and operations. *IEEE Power Engineering Review*, *14*(2), 15–18. doi:10.1109/MPER.1994.262142

Azzam, M., & Nour, M. (1996). An expert system for voltage control of a large-scale power system. *Energy Conversion and Management*, *37*, 81–86. doi:10.1016/0196-8904(95)00014-5

Balzani, M., & Reatti, A. (2005). Neural network based model of a PV array for the optimum performance of PV system. *Research in Microelectronics and Electronics*, *2*, 123–126. doi:10.1109/RME.2005.1542952

Barbounis, T. G., & Theocharis, J. B. (2007). Locally recurrent neural networks for wind speed prediction using spatial correlation. *Information Sciences*, *177*, 5775–5797. doi:10.1016/j.ins.2007.05.024

Barbounis, T. G., Theocharis, J. B., Alexiadis, M. C., & Dokopoulos, P. S. (2006). Long-term wind speed and Power Forecasting Using Local Recurrent Neural Network Models. *IEEE Transactions on Energy Conversion*, *21*, 273–284. doi:10.1109/TEC.2005.847954

Barr, A., & Feigenbaum, E. (1982). *The handbook of artificial intelligence* (Vol. 1). Los Altos, CA: Kaufman Publishing Co.

Barr, A., & Feigenbaum, E. A. (1981). *The handbook of artificial intelligence* (Vol. 1). Los Altos, CA: Morgan Kaufmann.

Bellman, R. E., & Zadeh, L. A. (1977). *Local and fuzzy logics in modern uses of multiple-valued logic.* J. M. Dunn & G. Epstein, (Eds.), (pp. 103-165). Dordrecht, the Netherlands: Reidel.

Bender, E. A. (1996). *Mathematical methods in artificial intelligence*. Los Alamitos, CA: IEEE Computer Society Press.

Bretthauer, G., Handschin, E., & Hoffmann, W. (1992). Expert systems application to power systems – state-of-the-art and future trends. *IFAC Symposium on Control of Power Plants and Power Systems*, Munich, Germany, (pp. 463-468).

Charniak, E., & McDermott, D. (1985). *Introduction to artificial intelligence.* Reading, MA: Addison-Wesley.

Chen, Z. (2000). *Computational intelligence for decision support*. Boca Raton, FL: CRC Press.

Colin, R. R., & Jonathan, E. R. (2002). *Genetic algorithms- Principles and perspectives, A guide to GA Theory*. Dordrecht, the Netherlands: Kluwer Academic Publishers.

Curtiss, P. S., Brandemuehl, M. J., & Kreider, J. F. (1995). Energy Management in Central HVAC Plants using Neural Networks. In J.S. Haberl, R.M. Nelson & C.C. Culp, (Eds.). *The use of Artificial Intelligence in Building Systems, ASHRAE*, (pp.199-216).

Dean, T., Allen, J., & Aloimonds, Y. (1995). Artificial intelligence: Theory and practice. Addison-Wesley, Reading, MA.

DeJong, K. A. (1975). An Analysis of the behavior of a class of genetic adaptive systems. Ph.D Dissertation, University of Michigan.

Deyi, L., & Yi, D. (2007). Artificial intelligence with uncertainty. Chapman & Hall/CRC, First edition.

Dubois, D., & Prade, H. (1980.) Fuzzy Sets Systems: Theory and applications. Academic Press, Orlando, FL.

Elman, J. L. (1990). Finding structure in time. *Cognitive Science*, *14*, 179–211.

Elshatter, T. F., Elhagree, M. T., Aboueldahab, M. E., & Elkousry, A. A. (1997). Fuzzy modeling and simulation of photovoltaic system. In 14th European photovoltaic Solar Energy Conference on CD ROM.

Engelberger, J. F. (1980). Robotics in practice. Kogan Page, London.

Feigenbaum, E. A. (1982). Knowledge engineering for the 1980. Department of Computer Science, Stanford University, Stanford, CA.

Fogel, D. B. (1995). Evolutionary computation: Toward a new philosophy of machine intelligence, IEEE Press, Chapter 3.

George, F. L., & William, A. S. (1998). Artificial intelligence structures and strategies for complex problem solving. George Addison Wesley Longman, INC.

Giles, C.L., Tsungnan, L., Bill, G. (1997). Horne remembering the past: The role of embedded memory in recurrent neural network architectures. IEEE Proceedings, 34-43.

Godart, T., & Puttgen, H. (1991). A reactive path concept applied within a voltage control expert system. *IEEE Transactions on Power Systems*, *6*(2), 787–793. doi:10.1109/59.76726

Goldberg, D. E. (1989). Genetic algorithms in search, optimization and machine learning. Addison-Wesley, Reading, MA.

Greer, L. R. (1986). Artificial intelligence and simulation: An introduction. Proceedings of the 1986 Winter Simulation Conference, Wilson, J., Henriksen, J., Roberts, S. (Eds.), 448-452.

Gupta, M.M., Liang, J., Noriyasu, H. (2003). Static and dynamic neural networks, from fundamentals to advanced theory. Forward by Lotfi A. Zadeh, IEEE press John Wiley.

Harp, S. A., & Samad, T. (1991). Optimizing neural networks with genetic algorithms. Proceedings of the American Power Conference, Chicago, 1138–1143.

Haugeland, J. (1985). (Ed.) Artificial intelligence: The very idea. MIT Press, Cambridge.

Haykin, S. (1999). Neural networks: A comprehensive foundation. New York: Macmillan; second edition.

Hebb, D. O. (1949). The Organization of behavior. Wiley, New York.

Hernádeza, J. C. N., Medinaa, A., & Juradob, F. (2007). Optimal allocation and sizing for profitability and voltage enhancement of PV systems on feeders. *Renewable Energy*, *32*, 1768–1789. doi:10.1016/j.renene.2006.11.003

Herts, J., Krogh, A., & Palmer, R. G. (1991). Introduction to the theory of neural computation. Addision-Wesley, Redwood City CA, 163.

Hewitt, C., & Planner, A. (1971). Language for proving theorems in robots, Proceedings of IJCAI, 2.

Holland, J. H. (1975). Adaptation in natural and artificial systems. University of Michigan Press, Ann Arbor.

Hopfield, J. J. (1982). Neural networks and to physical systems with emergent collective computational abilities. *Proceedings of the National Academy of Sciences of the United States of America*, *79*, 2554–2558. doi:10.1073/pnas.79.8.2554

Hung, C., Batanov, D., & Lefevre, T. (1998). KBS and macro-level systems: support of energy demand forecasting. *Computers in Industry*, *37*, 87–95. doi:10.1016/S0166-3615(98)00092-X

Jordan, M. (1988). Serial order: A parallel distributed processing approach, in Elman, J.L. & Rumelhart, D.E. (Eds.) Advanced in connectionist theory: speech Erlbaum.

Kadaba, N., Nygard, K. E., & Juell, P. J. (1991). Integration of adaptive machine learning and knowledge-cased systems for routing and scheduling applications. *Expert Systems with Applications*, *2*, 15–27. doi:10.1016/0957-4174(91)90131-W

Kalogirou, S. (1996a). Artificial neural networks for estimating the local concentration ratio of parabolic trough collectors, Proceedings of the EuroSun'96 Conference, Freiburg, Germany, 1, 470-475.

Kalogirou, S. (1996b). Design of a solar low temperature steam generation system, Proceedings of the EuroSun'96 Conference, Freiburg, Germany, 1, 224-229.

Kalogirou, S. (2000). Long-term performance prediction of forced circulation solar domestic water heating systems using artificial neural networks . *Applied Energy*, *66*, 63–74. doi:10.1016/S0306-2619(99)00042-2

Kalogirou, S. (2000). Applications of artificial neural networks for energy systems. Special issue of Applied Energy journal on Energy systems: Adaptive complexity, 67(1-2), 17-35.

Kalogirou, S. (2001). Artificial neural networks in renewable energy systems: A review . *Renewable & Sustainable Energy Reviews*, *5*, 373–401. doi:10.1016/S1364-0321(01)00006-5

Kalogirou, S. (2001). Artificial neural networks in renewable energy systems: A review. *Renewable & Sustainable Energy Reviews*, *5*(4), 373–401. doi:10.1016/S1364-0321(01)00006-5

Kalogirou, S. (2002). Design of a fuzzy single-axis sun tracking controller . *International Journal of Renewable Energy Engineering*, *4*, 451–458.

Kalogirou, S. (2007). Use of genetic algorithms for the optimum selection of the fenestration openings in buildings, Proceedings of the 2nd PALENC Conference and 28th AIVC Conference on Building Low Energy Cooling and Advanced Ventilation Technologies in the 21st Century, Crete island, Greece, 483-486.

Kalogirou, S., & Bojic, M. (2000). Artificial neural networks for the prediction of the energy consumption of a passive solar building . *Energy*, *25*, 479–491. doi:10.1016/S0360-5442(99)00086-9

Kalogirou, S., Eftekhari, M., & Pinnock, D. (1999c). Prediction of air flow in a single-sided naturally ventilated test room using artificial neural networks. Proceedings of Indoor Air'99, The 8th International Conference on Indoor Air Quality and Climate, Edinburgh, Scotland, Vol. 2, pp 975-980.

Kalogirou, S., Neocleous, C., & Schizas, C. (1996). A comparative study of methods for estimating intercept factor of parabolic trough collectors, Proceedings of the Engineering Applications of Neural Networks (EANN'96) Conference, London, UK, 5-8.

Kalogirou, S., Neocleous, C., & Schizas, C. (1998). Artificial neural networks for modelling the starting-up of a solar steam generator . *Applied Energy*, *60*, 89–100. doi:10.1016/S0306-2619(98)00019-1

Kalogirou, S., & Panteliou, S., Dentsoras. (1999a). A. Artificial neural networks used for the performance prediction of a thermosyphon solar water heater . *Renewable Energy*, *18*, 87–99. doi:10.1016/S0960-1481(98)00787-3

Kalogirou, S., & Panteliou, S. (2000). Thermosyphon solar domestic water heating systems long-term performance prediction using artificial neural networks . *Solar Energy*, *69*, 163–174. doi:10.1016/S0038-092X(00)00058-X

Kalogirou, S., Panteliou, S., & Dentsoras, A. (1999b). Modelling of solar domestic water heating systems using artificial neural networks . *Solar Energy*, *65*, 335–342. doi:10.1016/S0038-092X(99)00013-4

Kalogirou, S. A. (2003). Artificial intelligence for the modeling and control of combustion processes: a review. *Progress in Energy and Combustion Science*, *29*, 515–566. doi:10.1016/S0360-1285(03)00058-3

Kalogirou, S. A. (2007). Artificial Intelligence in energy and renewable energy systems. Nova Publishers, Chapter 2 and Chapter 5.

Kalogirou, S. A., Neocleous, C. C., & Schizas, C. N. (1997). Heating Load Estimation Using Artificial Neural Networks. Proc. of CLIMA 2000 Conf., Brussels, Belgium.

Kandil, M. (2001). The implementation of long-term forecasting strategies using a knowledge-based expert system: part II. *Electric Power Systems Research*, *58*(1), 19–25. doi:10.1016/S0378-7796(01)00098-0

Kaufmann, A., & Gupta, M. M. (1985). Introduction to fuzzy arithmetic, Theory and applications. 2nd ed., Van Nostrand Reinhold, New York (Japanese translation by Atsuka M., Ohmsha Ltd., Tokyo, 1991.
Kodratoff, Y. & Michalski, R.S. (1990). Machine learning: An artificial intelligence approach. Morgan Kaufmann, Vol. 3.

Kohonen, T. (1984). Self-organization and associative memory. Springer-Verlag, New York.

Konar, A. (1999). Artificial intelligence and soft computing behavioral and cognitive modeling of the human brain. CRC Press, Chapter 1.

Kreider, J. F., & Wang, X. A. (1995). Artificial Neural Network Demonstration for Automated Generation of Energy Use Predictors for Commercial Buildings. In Haberl, J.S., Nelson, R.M. and Culp, C.C. (Eds.). The use of Artificial Intelligence in Building Systems. ASHRAE, 193-198.

Krishnamoorthy, C. S., & Rajeev, S. (1996). Artificial intelligence and expert systems for engineers. CRC Press, LLC.

Kurzweil, R. (1990). The Age of intelligent machines. MIT Press, Cambridge.

Kusiak, A. (1987). Artificial intelligence and operations research in flexible manufacturing systems. *Information Processing*, *25*(1), 2–12.

Kusiak, A. Y., Sunderesh, D., & Heragu, S. (1989). Expert systems and optimization. *IEEE Transactions on Software Engineering*, *15*, 1012–1017. doi:10.1109/32.31358

Lai, L. (1998). Intelligent system applications in power engineering. John Wiley & Sons, London, UK.

Lakhmi, C. J., & Martin, N. M. (1998). Fusion of neural networks, fuzzy systems and genetic algorithms: Industrial Applications. CRC Press, LLC.

Lara-Rosano, F., & Valverde, N. (1998). Knowledge-based systems for the energy conservation programs. *Expert Systems with Applications*, *14*, 25–35. doi:10.1016/S0957-4174(97)00069-9

Le, T., Negnevitsky, M., & Piekutowski, M. (1995). Expert system application for voltage control and VAR compensation. *International Journal of Engineering Intelligent Systems for Electrical Engineering and Communications*, *3*(2), 79–85.

Liu, C., Ma, T., Liou, K., & Tsai, M. (1994). Practical use of expert systems in power systems. *International Journal of Engineering Intelligent Systems for Electrical Engineering and Communications*, *2*(1), 11–22.

Luger, G. F., & Stubblefield, W. A. (1993). Artificial intelligence: Structures and strategies for complex problem solving. Benjamin/Cummings, Menlo Park, CA.

Machado, R. J., & Rocha, A. F. (1992). A hybrid architecture for fuzzy connectionist expert systems. In Kandel, A. & Langholz, G. (Eds.), Hybrid architectures for intelligent systems. CRC Press, Boca Raton, FL.

McCarthy, J. (1958). Programs with common sense. Symposium on Mechanization of Thought Processes. National Physical Laboratory, Teddington, England.

McCarthy, J. (1980). Circumscription - A form of non-monotonic reasoning. *Artificial Intelligence*, *13*, 27–39. doi:10.1016/0004-3702(80)90011-9

McCulloch, W. S., & Pitts, W. A. (1943). A logical calculus of the ideas imminent in nervous activity . *The Bulletin of Mathematical Biophysics*, *5*, 115–133. doi:10.1007/BF02478259

Medsker, L. R. (1995). Hybrid intelligent systems. Boston: Kluwer Academic Publishers.

Medsker, L. R. (1996). Microcomputer applications of hybrid intelligent systems. *Journal of Network and Computer Applications*, *19*, 213–234. doi:10.1006/jnca.1996.0015

Mellit, A. (2006). Artificial intelligence based-modeling for sizing of a stand-alone photovoltaic power system: Proposition for a new model using Neuro-Fuzzy system (ANFIS). In Proceedings of the 3rd International IEEE Conference on Intelligent Systems, UK, 1, 605-611.

Mellit, A., & Benghanem, M. Hadj, Arab, A., Guessoum, A. (2003). Modeling of sizing the photovoltaic system parameters using artificial neural network. In Procee. of IEEE, Conference on Control Application, Istanbul, 1, 353-357.

Mellit, A., Benghanem, M., & Bendekhis, M. (2005). Artificial neural network model for prediction solar radiation data: application for sizing stand-alone photovoltaic power system. In proceedings of IEEE Power Engineering Society, General Meeting 2005;1:40-44.

Mellit, A., Benghanem, M., Hadj Arab, A., Guessoum, A., & Moulai, K. (2004). Neural network adaptive wavelets for sizing of stand-Alone photovoltaic systems. Second IEEE International Conference on Intelligent Systems, 1, 365-370.

Mellit, A., & Kalogirou, S. A. (2006). Application of neural networks and genetic algorithms for predicting the optimal sizing coefficient of photovoltaic supply (PVS) systems In Proceedings of World Renewable Energy Congress IX and Exhibition, Florence, Italy on CD-ROM.

Mellit, A., & Kalogirou, S. A. (2008). Artificial intelligence techniques for photovoltaic applications: A review . *Progress in Energy and Combustion Science*, *34*(5), 574–632. doi:10.1016/j.pecs.2008.01.001

Metaxiotis, K., Kagiannas, A., Askounis, D., & Psarras, J. (2003). Artificial intelligence in short term electric load forecasting: a state-of-the-art survey for the researcher. *Energy Conversion and Management*, *44*(9), 1525–1534. doi:10.1016/S0196-8904(02)00148-6

Michalewicz, Z. (1992). Genetic algorithms + Data structures = Evolution programs. Springer-Verlag, Berlin.

Minsky, M. (1975). A framework for representing knowledge. In The psychology of computer vision, Winston, P.H. (Ed.). McGraw Hill, New York.

Minsky, M., & Papert, S. (1969). Perceptrons. MIT Press, Cambridge, Massachusetts.

Moreno, A., Julve, J., Silvestre, S., & Castaer, L. (2000). A fuzzy logic controller for stand alone PV systems, IEEE, 1618-1621.

Nannariello, J., & Frike, F. R. (2001). Introduction to neural network analysis and its applications to building services engineering. *Building Services Engineering Research and Technology*, *22*(1), 58–68. doi:10.1191/014362401701524127

Newell, A., Shaw, J. C., & Simon, H. A. (1963). Empirical explorations with the logic theory machine: A case study in heuristics. In Computers and Thought, Feigenbaum, E. A. & Feldman, J. (Eds.), McGraw Hill, New York.

Newell, A., & Simon, H. A. (1972). Human problem solving. Prentice-Hall, Englewood Cliffs, NJ.

Nilsson, N. (1998). Artificial intelligence: A new synthesis. Morgan Kaufmann.

Padhy, N. (2000). Unit commitment using hybrid models: a comparative study for dynamic programming, expert system, fuzzy system and genetic algorithms. *Electrical Power & Energy Systems*, *23*, 827–836. doi:10.1016/S0142-0615(00)00090-9

Rich, E., & Knight, K. (1996). Artificial intelligence. McGraw-Hill, New York.

Roberge, M. A., Lamarche, L., Karjl, S., & Moreau, A. (1997). Model of Room Storage Heater and System Identification Using Neural Networks, Proc. of CLIMA 2000 Conf., Brussels, Belgium, 265.

Robert, F. (1995). Neural fuzzy systems. Abo Akademi University.

Rumelhart, D. E., & McClelland, J. L. (1986). Parallel distributed processing: Explorations in the microstructure of cognition. The PDP Research Group, MIT Press/Bradford Books, Cambridge, Massachusetts.

Russel, S., & Norvig, P. (1995). Artificial intelligence: A modern approach. Prentice-Hall, Englewood Cliffs, NJ.

Schalkoff, J., Culberson, J., Treloar, N., & Knight, B. (1992). A world championship caliber checkers program. *Artificial Intelligence*, *53*(2-3), 273–289. doi:10.1016/0004-3702(92)90074-8

SchaVer. J.D. (1994). Combinations of genetic algorithms with neural networks or fuzzy systems. In Computational Intelligence: Imitating Life, Zurada, J.M., Marks, R.J., Robinson, C.J. (Eds). New York, IEEE Press, 371–382.

Seddon, A. P., & Brereton, P. (1996). Component selection using non-monotonic reasoning . *Artificial Intelligence in Engineering*, *1*, 235–241. doi:10.1016/0954-1810(95)00034-8

Sen, Z. (1998). Fuzzy algorithm for estimation of solar irradiation from sunshine duration. *Solar Energy*, *63*, 39–49. doi:10.1016/S0038-092X(98)00043-7

Senjyua, T., Hayashia, D., Yonaa, A., Urasakia, N., & Funabashib, T. (2007). Optimal configuration of power generating systems in isolated island with renewable energy. *Renewable Energy*, *32*, 1917–1933. doi:10.1016/j.renene.2006.09.003

Shannon, C. E. (1950). Programming a computer for playing chess. Philosophical Magazine . *Series*, *7*(41), 256–275.

Smith, R. G., Mitchell, T. M., Chestek, R. A., & Buchanan, B. G. (1977). A model for learning systems. Proceedings of IJCAI, 5.

Tsakonas, A., & Dounias, G. (2002). Hybrid computational intelligence schemes in complex domains: An extended review. Vlahavas, I.P. & Spyropoulos, C.D. (Eds.). SETN 2002, LNAI, 2308, 494–511.

Turban, E. (1992). Expert systems and applied artificial intelligence. New York: Macmillan Publishing Company.

Vajpai, J., Singhal, S., & Naizi, K. (2001). Expert system based on line optimal control of reactive power. Proc. All India Seminar, Power Systems: Recent Advances and Prospects in 21st Century, Jaipur, India, 194-208.

Vale, A., Santos, J., & Ramos, C. (1997). SPARSE - A Prolog Based Application for the Portuguese Transmission Network: Verification and Validation. PAP'97 – Practical Application in Prolog, London, U.K.

Vale, Z., Ramos, C., Silva, A., Faria, L., Santos, J., Fernandes, M., et al. (1998). SOCRATES – An Integrated Intelligent System for Power System Control Center Operator Assistance and Training. International Conference on Artificial Intelligence and Soft Computing (IASTED), Cancun, Mexico.

Williams, R. J., & Zipser, D. (1989). Experimental analysis of the real-time recurrent learning algorithm . *Connection Science*, *1*(1), 17–11. doi:10.1080/09540098908915631

Winograd, S. (1972). Understanding natural language, Academic Press, New York.

Winston, P. H. (1994). Artificial intelligence, Addison-Wesley, 2nd ed., Reading, MA.

Ypsilantis, J., & Yee, H. (1990). Survey of expert systems for SCADA-based power applications. 4th Conference in Control Engineering, Gold Coast, Australia, 177-183.

Yu, H. H., & Jenq-Neng, H. (2001). Handbook of neural network signal processing. CRC press.

Zadeh, L.A. (1965). Fuzzy sets . *Information and Control*, *8*, 338–353. doi:10.1016/S0019-9958(65)90241-X

Zadeh, L.A. (1972). A Fuzzy-set-theoretic interpretation of linguistic hedges. Cyber net, 2, 4–34.

Zadeh, L. A. (1973). Outline of a new approach to the analysis of complex systems and decision processes. IEEE Transactions on Syst. *Man Cybernet.*, *3*, 28–44.

Zhang, L. and. Bai, Y. (2005). Genetic algorithm-trained radial basis functions neural networks for modelling photovoltaic panels. Engineering application of artificial intelligence, 18, 833-844.

Zimmermann, H.-J., Tselentis, G., Van Someren, M., & Dounias, G. (2001). Advances in computational intelligence and learning: Methods and applications. Kluwer Academic Publications.

Chapter 2
Intelligent Control of the Energy Generation Systems

Nicu Bizon
University of Pitesti, Romania

ABSTRACT

In this book chapter are analyzed the Energy Generation System (EGS) topologies, used in automotive systems, and the grid inverter systems, with intelligent control algorithms (fuzzy logic controller, genetic algorithm, etc.). The EGS blocks are modelled using Matlab & Simulink ® program. A necessary block is the EGS power interface between the fuel cell stack and the batteries stack, usually a boost converter that uses a Peak Current Controller (PCC) with a Boundary Control with Current Taper (BCCT). The control law is a function of fuel cell current and battery voltage, which prevents the "boiling" of the batteries. The control objective for this power interface is also the fuel cell current ripple minimization, used in order to improve the fuel cell stack life cycle. Clocked and non-clocked control methods are tested in order to obtain a small fuel cell current ripple, better a dynamic response, and robustness against system uncertainty disturbances. The EGS behaviour is tested by bifurcation diagrams. It is shown that performances increase if the control law is a function that depends by the fuel cell current ripple and battery voltage. The clocked PCC using the BCCT 2-D law is implemented by a fuzzy logic controller. The power load dynamic is compensated using an ultracapacitors stack as a dynamic energy compensator, connected by a bi-directional converter to the batteries stack bus. Small fuel cell current ripple using compact batteries and ultracapacitors stacks will be obtained by the appropriate design of the control surface, using an Integrated Fuzzy Control (IFC) for both power interfaces.

INTRODUCTION

Power electronics is an *interdisciplinary* "green" technology (electronics, electrical engineering, automatic control and, obviously, mathematics and physics), with three main aims:

DOI: 10.4018/978-1-60566-737-9.ch002

Figure 1. A typical energy generation system topology with energy storage devices

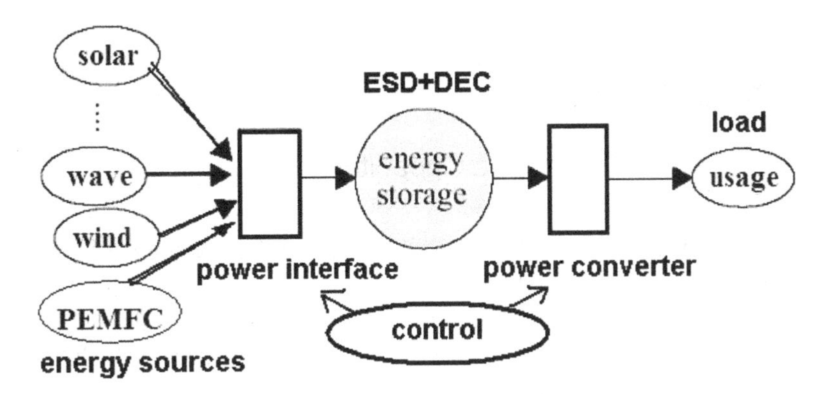

- *To convert electrical energy from one of the other energy forms, facilitating its regulation and control* (main objective of this chapter);
- To achieve high conversion efficiency;
- To minimize the mass of power converters.

The specific objective of this book chapter will be to promote new control techniques for Energy Generation System (EGS) based on intelligent concepts. The typical Energy Generation System topology with Energy Storage Device (ESD) is presented in figure 1. The energy storage technologies, usually used in an EGS, are such as presented below.

Electrochemical Energy Storage

Batteries

The lead-acid battery has a high specific energy as compared to other energy storage technologies, but the specific power is lower due to the high internal impedance (contact resistance between the electrodes and the electrolyte). Researches into this battery type are mainly focused on its construction, so that, after more than 150 years, the chemical reactions remain the same. Efforts to increase the power delivery capabilities of lead-acid batteries are reported in:

- Thin Metal Film (TMF) battery by Bolder Technologies Corporation (USA). The 2V/1,5Ah Bolder TMF cell can be fully discharged in 1s at over 1kA, and then fully recharged in a few minutes. It has a specific power, of about 5kW/kg, and could now reach over 15kW/kg.
- The bipolar lead-acid battery. It achieves a specific power of up to 1kW/kg, and a power density of 35 MW/m^3. For reasons of price, this kind of battery is usually used in an EGS.
- High power Li-ion. These batteries achieve over 1kW/kg specific power, at a higher specific energy than the bipolar lead-acid battery, but lower than the Bolder TMF batteries.

To conclude, these technologies have high-energy storage capabilities, with a lesser performance in power delivery. As far as the charge/discharge time constants are concerned, the batteries belong to the time transfer range of one minute up to several hundreds of hours, which makes them long-term energy storage devices.

Electric Field Energy Storage

Metal Film and Aluminium Electrolytic Capacitors

The aluminium electrolytic capacitors specific power capability is around 150kW/kg, and the power densities range between 500 and 1000MW/m^3. This results into a specific energy (0,1Wh/kg), and energy densities (100Wh/m3), higher than that of the metal film capacitors. The aluminium electrolytic capacitors cost around $500/kW, which is not very expensive considering the high specific power. In an EGS, the capacitors are usually used for filter the high DC voltage bus.

Ultracapacitors (or Supercapacitors)

Ultracapacitors have specific powers (»5kW/kg) lower than electrolytic capacitors, and their specific energies are between those of the electrochemical storage devices and the electrolytic capacitors. Ultracapacitors are fully bi-directional and have a very long life expectancy, up to 200000 cycles, with a cost of approximately $400/kW. These features make ultracapacitors stacks a very attractive option for burst power applications, where they are used as a Dynamic Energy Compensator (DEC). As an EGS example, figure 2 show the Matlab diagram of an Automotive Hybrid Electrical System (AHES), with Proton Exchange Membrane Fuel Cells (PEMFC) as a primary energy source, battery stack as Energy Storage Device (ESD), and ultracapacitors stack as Dynamic Energy Compensator (DEC) (Thounthong, Rael & Davat, 2005; Larminie &Dicks, 2000; Andersen, Christensen, & Korsgaard, 2002). The boost converter is the most widely used power interface between PEMFC and ESD (Bizon, Sofron, & Oproescu, 2005; Xue, Chang, & Kjær, 2004). A Peak Current Controller (PCC) of the boost inductor current (equal to PEMFC current), with a specific control law, can be used to control the ESD voltage bus (V_Low_DC) (Bizon, Sofron, Oproescu, & Raducu, 2007). Two transducers are used: a PEMFC current sensor and a voltage sensor for the V_Low_DC bus (used as input for the buck-boost controller, too). The PCC control low is of boundary type, with current limitation when the V_Low_DC voltage is over to V_knee and up to V_Max (voltage levels correlated with the battery boiling effect). The power load dynamic is compensated using an ultracapacitor stack connected, by a bi-directional buck-boost converter, to the V_Low_DC bus. A simple control strategy may use only two cheap voltage transducers (for the V_low_DC bus voltage – already used, and for the V_UCap – the voltage over the ultracapacitors stack). The control law of the buck-boost controller is of a boundary hysteresis type. Based on the nominal load conditions, the boost power stage can be controlled to draw a specific amount of current (I_peak_PEMFC) from the fuel cell with a limited current ripple, well defined by the frequency, the size of the boost inductor, the duty ratio, and the hysteresis control parameters. If the load power dynamic is high, the frequency of the PCC command signal (PWM_PCC), which is highly dependent of the load power level, may have large variations from the rated frequency (obtained for a nominal load power). Two problems arise. At low frequency (high load power) the PEMFC current ripple can find itself over the imposed value (first made by National Energy Technology Laboratory, 2005) and, if the load power is over a specific limit, the peak current control cannot catch the control low surface (bifurcation and chaos appear; see: Tse, 2003; Parui, 2003; Banerjee, & Verghese, 2001; Bizon, Sofron, & Oproescu,

Figure 2. Automotive Hybrid Electrical System topology © 2007 MEDJMC. Used by permission

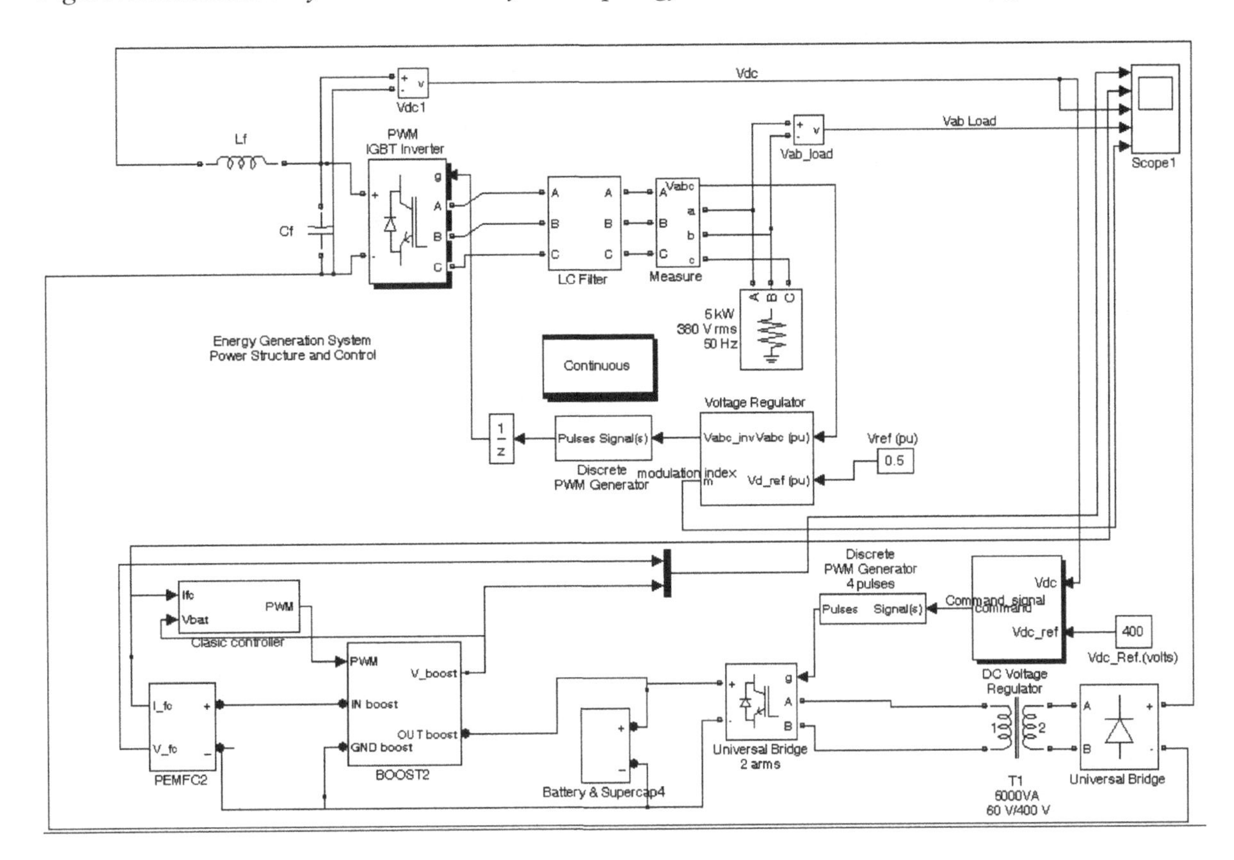

2006; Bizon, & Oproescu, 2007). At high frequency (light load power) the HF PEMFC current ripple is usually up to the imposed limit, but the EGS energy efficiency is dramatically reduced because of the commutation power losses in all power stages (Bizon, & Oproescu, 2005a). So, for a large load power range, a clocked PCC variant is recommended (Bizon, & Oproescu, 2005b). The dynamic of the load power is compensated with an ultracapacitor stack, usually connected to the V_Low_DC bus (which is obligatory when a battery stack is not used in the EGS structure), and in some proposed EGS topologies it is connected to the high DC voltage bus (V_High_DC) together with the ESD (Xue, Chang, & Kjær, 2004). The charge mode for battery stack appears if PEMFC power flow is bigger than the required load power flow. At the same time, the ultracapacitor stack is also charged and the buck-boost converter operates in the buck mode. Alternatively with buck mode, in the rest of time, buck-boost converter operates in the boost mode, when appears an inverted power flow (from V_UCap bus to V_low_DC bus) (Ortuzar, Dixon, & Moreno, 2003). If an appropriate control with grid synchronization is implemented for PWM sine inverter command, then this kind of the EGS topology can be also used for mobile/ stationary grid connected inverter system units (Bizon, & Oproescu, 2007; Bizon, Laurentiu, Mazare, & Oproescu, 2007). The control techniques of the PEMFC power interface and following EGS power converters (see figure 3) will be analyzed in this book chapter. For the following EGS power stages the proposed power converter topologies are designed to ensure high energy efficiency (with minimum power processing stages), easy control, and low EGS price (with minimum switches number) (Raducu, & Bizon, 2007; Bizon, Oproescu, & Raducu, 2006; Balog, &. Krein, 2005; Batarseh, 2003; Rashid,

Figure 3. The analyzed EGS power stages © 2007 MEDJMC. Used by permission

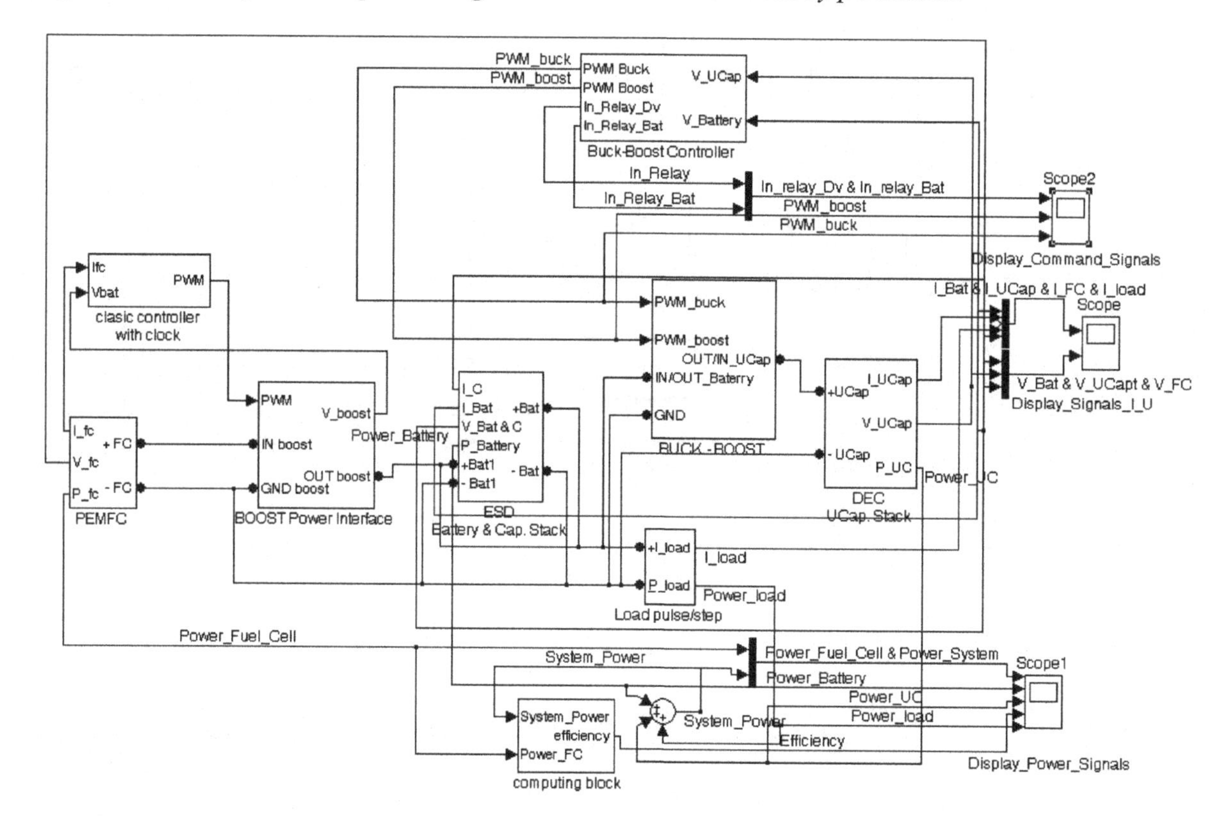

2003; Mohan, Undeland, & Robbins, 2002; Agrawal, 2001).

The Energy Generation System must have respond to a specific load power dynamic, but the chemistry of the fuel cell cannot instantaneously respond to these dynamics of the load power (Larminie, & Dicks, 2003; National Energy Technology Laboratory, & U.S. Department of Energy, 2004; Stimming, de Haart, & Meeusinger, 2005). So, the battery stack must provide momentary load power flow, and the ultracapacitor stack operates as DEC, providing instantaneous load power flow when high load power pulses appear (Bergveld, Wanda, Kruijt, & Notten, 2002; Bizon, 2004). Over a short period of time, and for a constant load up to the nominal one, the state of charge for the battery stack has no net change: during a slow load transient, any energy supplied from the battery to the load must be replenished from the PEMFC. Therefore, the command for the buck-boost converter must include the ultracapacitor stack charging current in order to limit their values.

The advantages and drawbacks of each control methods are seen into the fast- and slow-scale instabilities, with effects in the EGS blocks modelling: the power flow transfer from PEMFC to the Energy Storage Device and load appears as a slow process and the power flow transfer from the ultracapacitor stack to the ESD, and load appears as a fast process. We consider that the bidirectional buck-boost controller, which gives the fast-scale instabilities of the system, must be optimized to ensure the EGS dynamic load power flows.

In order to understand how the proposed EGS control operates, in the next section all EGG blocks are presented and briefly explained, where necessary. The following sections are dedicated to the design methodology of the EGS fuzzy controllers. The aim of the chapter is to show the advantages of the intel-

Figure 4. The EDS model

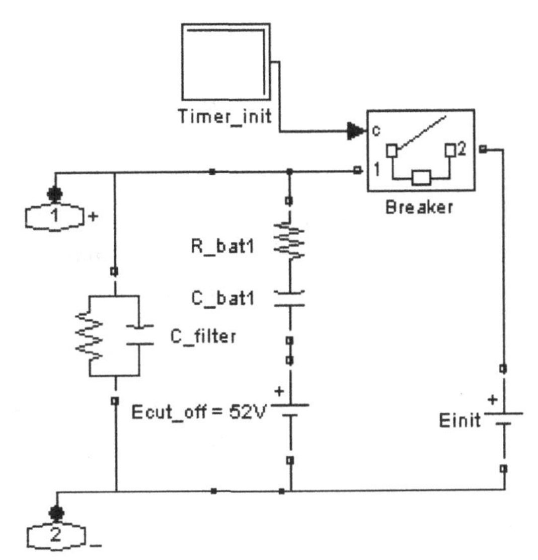

a) EDS Simulink model

Parameters

Cbattery [F]

1e0

Rbattery serie [ohm]

2.4e0

Csupercap [F]

1e-2

Rsupercap paralel [ohm]

1e9

Einitial [V]

60

initial time [s]

1e-6

b) EDS model parameters

ligent EGS control regarding easy design and better control performances. For each fuzzy controller of the used EGS power converters, the simulation results are presented at the end of sections, where some conclusions are given, too. The last section concludes the chapter.

EGS MODELING AND DESIGNING

Battery Model

A lead-acid battery for the ESD is usually a low price solution. Generally, a battery model is complex because the storage device has many model parameters such as capacity, dead-cell voltage, discharge impedance, self-discharge impedance, and shunt capacitance. In order to simplify the simulations, a simple model for a sealed lead acid (SLA) battery is used (Crompton, 2003; Gansky, 2002). The battery is modelled as a capacitor for energy storage, $C_{storage}$, a DC offset voltage, V_{offset}, and a series resistance, R_s, to limit the short-circuit current (figure 4).

For example, if a 60V/7Ah battery stack structure (5 batteries, 6 cells/battery and 2,45V max/cell) is used, and the value of the series resistance is taken 80mΩ/cell, as suggested in Crompton (2003), then equivalent series resistance of the pack is R_S=5·80mΩ=0,4Ω. The typical "dead cell" voltage for SLA battery technology is about 1.75V. Therefore, the total offset voltage is $V_{offset}=5\cdot 6\cdot 1,75\ V=52,5\ V$. Finally, the stored energy in the capacitor can be calculated. First we calculate the maximum voltage over the battery stack: V_{max}=5·6·2,45 V=73,5 V. So, the maximum storage capacitor voltage must be the difference between the maximum expected battery voltage and the dead-cell voltage: $V_{C\_storage}=V_{max} - V_{offset}$=21V. For a 7Ah batteries we obtain

Q=7Ah·3600 sec/hour = 25200 C,and the value for the storage capacitance is $C_{storage}=Q/\ V_{C\_storage}$ =1200 F. The EDS Simulink model is presented in figure 4.

Obviously, by adding the batteries stack to the boost converter output port, the control characteristic is changed, so we must consider the required battery charging parameters in the control law generation (see figure 5). The solid line in figure 5.a represents the boundary control law (named *boundary control with current taper-BCCT)*. This is a function of both current limits and battery voltage. The inductor peak current will back off to prevent the "boiling" of the batteries, when the batteries stack operate in the charging mode and the voltage increases (over knee voltage, V_{knee}, to the maximum value, V_{mas}). In order to minimize the fuel cells current ripple a 2-D control surface is proposed as a control law. A fuzzy system is used to control surface generation as a function of both PEMFC current ripple and batteries stack voltage (figure 5.b).

PEMFC Model

The fuel cell equivalent electrical model has a rather large capacitance that shunts the device. The equivalent circuit for the PEMFC electrical model is shown in figure 6.a, where R_e represents the electrolyte and contact resistance (R_{ohm}), R_{ct} is the charge transfer resistance (represents the activation loss), C_d is the double charge layer capacitance, and Z_d is the diffusion impedance (also called the concentration loss or Warburg impedance).

The proposed model is based on a well-known and simple analytical description (Larminie & Dicks, 2003). The irreversible losses will be determined in terms of three main groups: activation losses, ohmic losses and concentration losses (figure 6). By adding, the cell voltage is determined in terms of the drawn current (i_L) and PEM fuel cell area (A_{PEM}).

The J. Larminie relation used for PEMFC voltage modelling is:

$$E = E^0 - \left(i_l + i_n\right)R_{ohmic} - A_{act}\ln\left(\frac{i_l + i_n}{i_0}\right) + \\ + B_{conc}\ln\left(1 - \frac{i_l + i_n}{i_{\max}}\right), \quad i_l = i_L / A_{PEM} \tag{1}$$

Figure 5. BCCT control law © 2007 EPE. Used by permission.

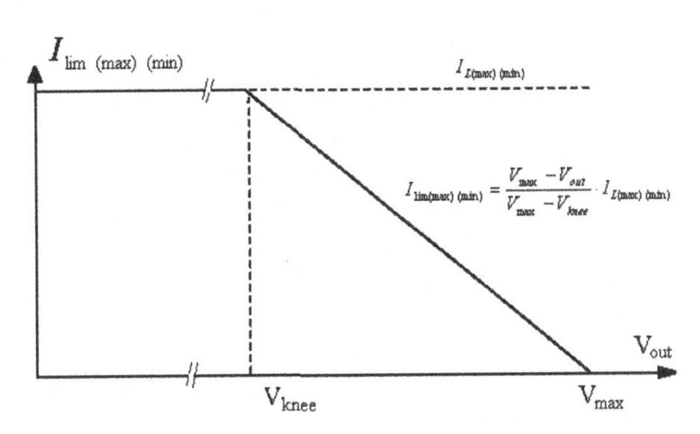

a) BCCT 1-D control law

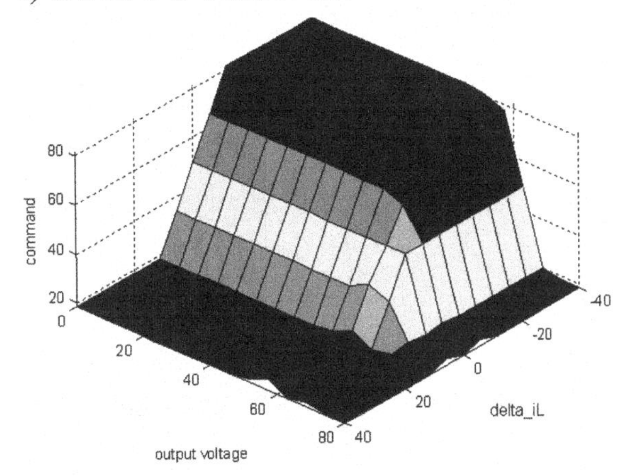

b) BCCT 2-D control law

The proposed relation used for PEMFC modelling is:

$$E = E^0 - \left(i_l + i_n\right) R_{ohmic} - A_{act} \ln\left(\frac{i_l + i_n}{i_0}\right) + \\ + B_{conc} \ln\left(\lambda^{k_1} P_{sys}^{k_2}\left(1 - \frac{i_l + i_n}{i_{\max} \lambda^{k_3} P_{sys}^{k_4}}\right)\right) \tag{2}$$

The k1÷k4 correction coefficients of the PEMFC model can be obtained by different methods (see table 1 and 2). The PEMFC parameters identification algorithm was presented and tested extensively for different real and simulated situations in Bizon, Zafiu, & Oproescu (2005). A short presentation of the identification methodology using genetic algorithm will be following presented.

The genetic algorithm follows above relation (relation 2) and considers:

- 7 constants: E_0, I_n, I_0, $I_{\max}$, R_{ohmic}, A_{act}, and B_{conc}. These values are physical constants and their values will be established by the user from physical specifications.

Figure 6. The PEMFC model and characteristic

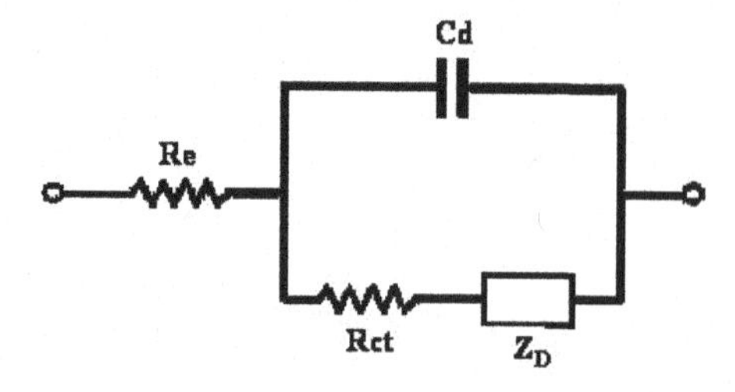

a) The PEMFC equivalent circuit

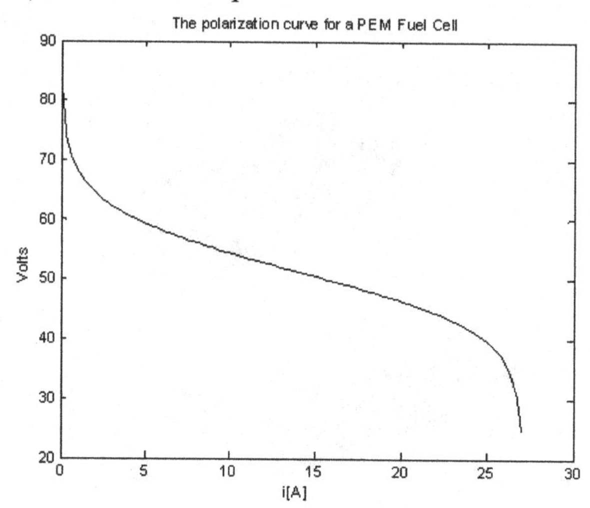

b) The PEMFC u-i characteristic

Table 1. PEMFC model parameters for $\lambda = 1$ and $P_{sys} = 2$

Parameter	Value using "least squares" method	Value using J. Larminie relation	Value using genetic algorithm method
E^o / [Volts]	0.97	1.2	0.967
i_n / [mA/cm^2]	2.7	2	2.755
i_0 / [mA/cm^2]	0.074	0.067	0.074
i_{max} / [A/cm^2]	0.6	0.9	0.624
R_{Ohmic} / [Ω]	0.02	0.03	0.021
A_{act} / [Volts]	0.034	0.06	0.035
B_{conc} / [Volts]	0.16	0.05	0.162
k_2	0.5	N/A	0.531
k_3	0.5	N/A	0.515
k_1	0	N/A	0.012
k_4	0	N/A	0.018

Table 2. RMS errors

Value using genetic algorithm method	Value using "least squares" method	Value using fuzzy method
0,988	0,982	0,973

- 6 parameters λ, P_{sys}, k_1, k_2, k_3, and k_4.

If we denote $\alpha = \lambda^{k_1} P^{k_2}$, $\beta = \lambda^{k_3} P^{k_4}$, and $x = i_l + i_n$, these parameters can be grouped, and relation 2 become:

$$E = E^0 - xR_{ohmic} - A_{act} \ln\left(\frac{x}{i_0}\right) + \\ +B_{conc} \ln\left[\alpha\left(1 - \frac{x}{i_{max}\beta}\right)\right] \quad (3)$$

In this case, there are only two variables that must be computed by the genetic algorithm: α and β. A gene will have two chromosomes with values between 0 and 10.

The best performance was obtained for:

- a populations with 100 peoples;
- 100% mutation probability;
- 5% of old genes;
- 80% percent of new entities for the new population from crossovers;
- m=100 as reference points;
- a dynamic size for the mutation, established as difference between entity values and reference values.
- 5% of new random entities.

The crossover is made from two genes and generates one gene. First *n* chromosomes were provided by first gene, where *n* value is a random number between *0* and *N* (total number of chromosomes): N=2 in this case. The parents of the crossover are chosen with a roulette algorithm that ensures for the good

Figure 7. Best entity evolution

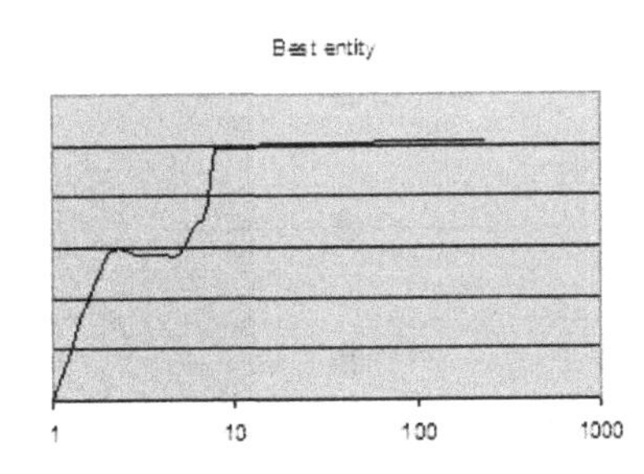

Figure 8. The PEMFC characteristics for different values of the model parameters: λ *and* P_{sys}

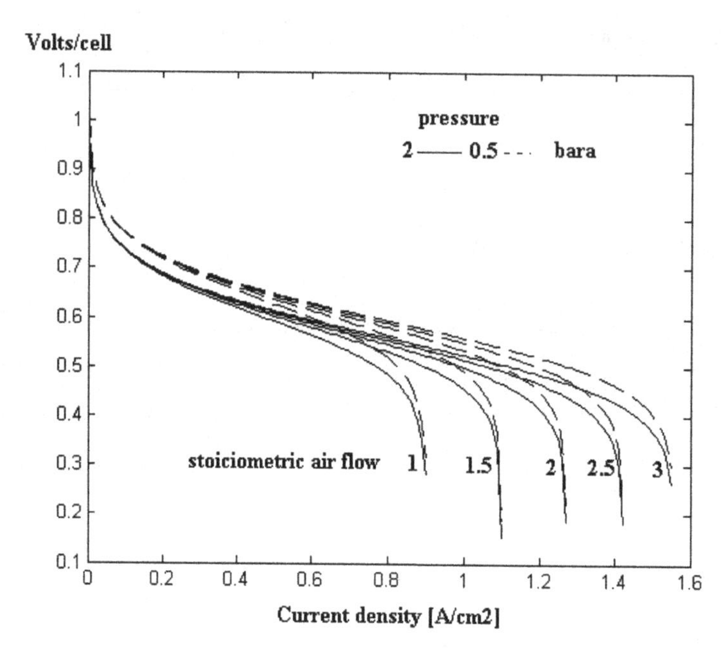

entity a greater probability.

The fitness function F is defined as a sum of absolute differences in all reference points:

$$F = \frac{m}{\sum_{m}\left(abs\left(calculated\ value - reference\ value\right) + 1\right)} \tag{4}$$

Figure 9. ANFIS model topology

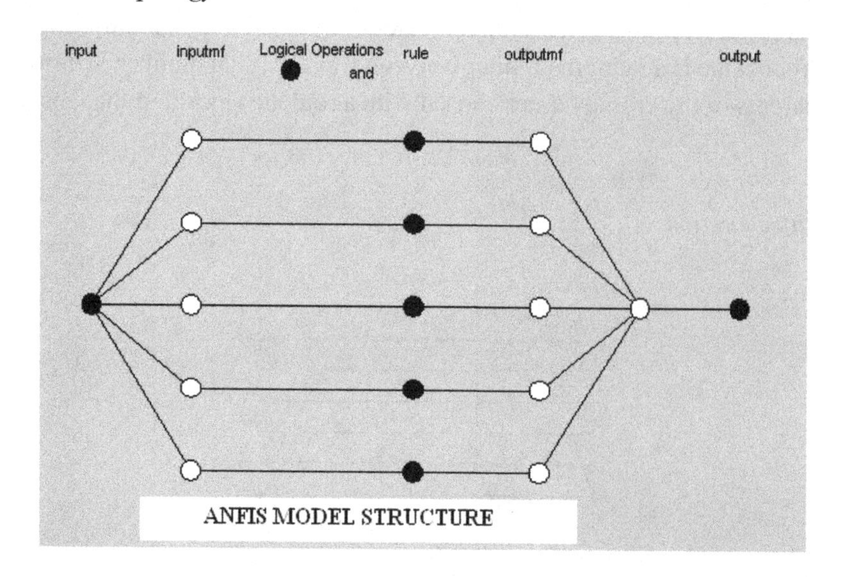

Figure 10. The ANFIS output and training data set

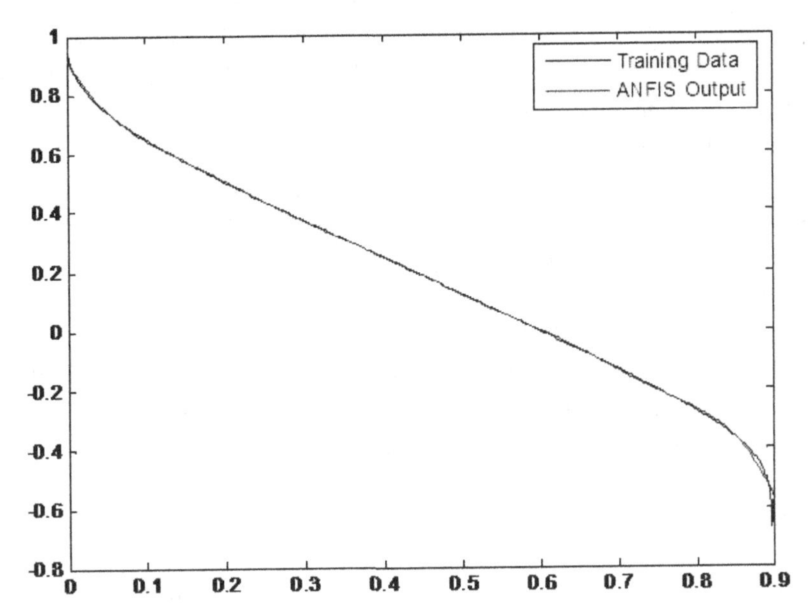

Where m is the number of reference points.

The genetic algorithm computes the sum of fitness values, $SumF$, and then generates a random integer between *0* and $SumF$. Considering that these entities are placed along a line vector (length vector equal to $SumF$), the random integer indicates only one entity (figure 7). In figure 8 some results using the PEMFC model proposed are shown.

The performance criterion used to evaluate the PEMFC models is the identification accuracy given by the RMS error function:

$$RMS\_error = 1 - \frac{\sum_{1}^{m} \sqrt{\left(final\ calculated\ value - reference\ value\right)^2}}{m} \tag{5}$$

The results obtained with a well-known fuzzy identification technique (from the Matlab toolboxes) are given in Figures 9 and 10.

Table 2 compares the RMS errors for three approximation methods. Because it is simpler to use an analytical relation for PEMFC output voltage model, *E*, the relation 3 will be used in the simulation.

A first order system (with death time, τ_{PEMFC}, and time constant, T_{PEMFC}) is used for modelling the Proton Exchange Membrane Fuel Cells dynamic behaviour (where n_{CELL} is the number of the stack cells):

$$\begin{aligned} V_{PEMFC} &= n_{CELL} \cdot E, \\ V_{in}(s) &= \frac{V_{PEMFC} \cdot e^{-t/\tau_{PEMFC}}}{1 + T_{PEMFC} \cdot s} \end{aligned} \tag{6}$$

Using the above data with regard to the batteries stack, the fuel cell voltage is expected to be within the range of 42V to 52V, based on the rated load current and normal operating conditions (see figure 6.b, where n_{CELL}=60 and A_{PEM}=60 cm^2). However, if the preload fails, the open cell voltage can be higher than 72V, and the inverter design must consider this case in order to be protected against it. The PEMFC Simulink model is presented in figure 11.

As an example, figure 12 shows the simulated 30V/4kW PEMFC dynamic characteristics.

The time equivalent systems theory is used to speed up the simulations (Bizon, & Oproescu, 2006a). If the battery and PEMFC time constants are reduced by 10^n, then the values used in simulations are $C/10^n$, $T_{PEMFC}/10^n$, and $\tau_{PEMFC}/10^n$, respectively.

It is possible to model a fuel cell stack in a much more fundamental manner, incorporating electrochemistry, thermal characteristics, and mass transfer (Stamps, & Gatzke, 2005; Pathapati, Xue, & Tang, 2005; Haraldsson, & Wipke, 2004), but such complicated models are not necessary for testing the electrical performances of the PEMFC power interface.

Boost Converter Model

Figure 13 show the Matlam diagram of the boost converter. For a nominal load, the clocked PCC boost converter is designed to operate in the continuous conduction mode (CCM). The converter transfer char-

Figure 11. The PEMFC model

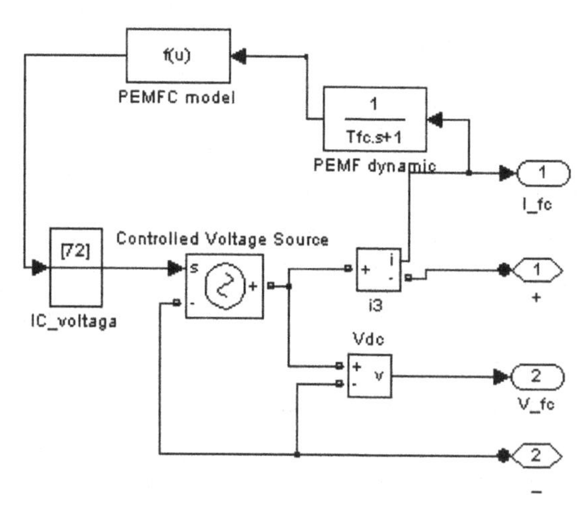

a) The PEMFC Simulink model

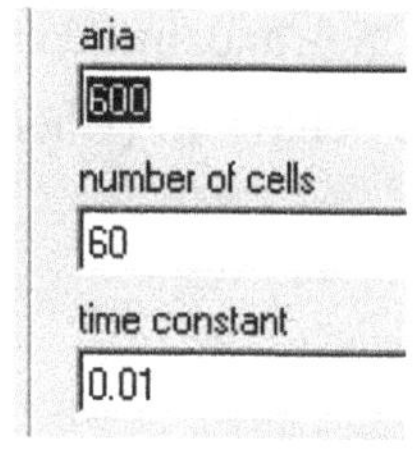

b) The PEMFC parameters

Figure 12. The PEMFC dynamic

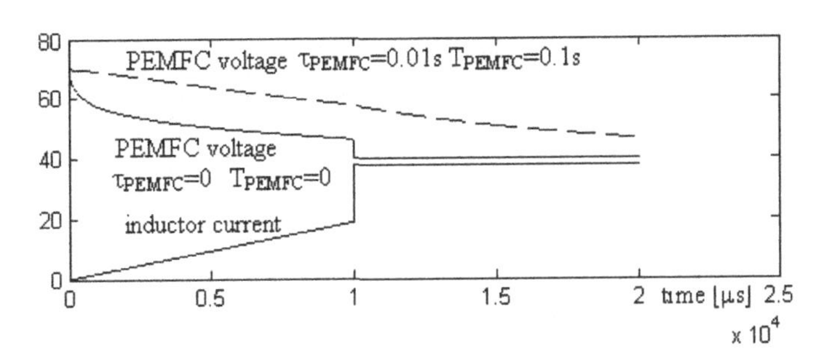

acteristic then operates in CCM is $\frac{V_{out}}{V_{in}} = 1 - \tau$, where $\tau = \frac{t_{on}}{T}$ is the duty ratio. As a designing example, if the rated values for input voltage, output voltage and output power are $V_{in} = 48V$, $V_{out} = 60V$, and $P_{out} = 900W$, respectively, then the duty ratio will be $\tau = 0,2$, and $I_{load} = \frac{P_{out}}{V_{out}} = 15A \Rightarrow R_{load} = 4\Omega$.

The average inductor current is

$$I_L = \frac{I_{out}}{1-\tau} = 18,75A\text{'}$$

and the series resistance (r_L) of the boost inductor can be chosen around 0,05 Ω in the simulations. Obviously, depending on the load power and certain circuit parameters, there are three possible types of operation in any switching period, as illustrated in figure 14, where $i_n = i_L(nT)$ and $I_{ref} = I_{Lmax} = I\_peak\_PEMFC$.

Two types of the PCC controllers are used in the simulations in order to compare the EGS control performances: a clocked variant and a non-clocked variant, with a BCCT control law type 1-D (classic), and 2-D (fuzzy), respectively (see figure 15).

The advantages and drawbacks of each current control method are seen into the fast- and slow-scale instabilities of an EGS, with a PEMFC and a batteries stack as slow processes, and with a boost converter

Figure 13. The boost converter model

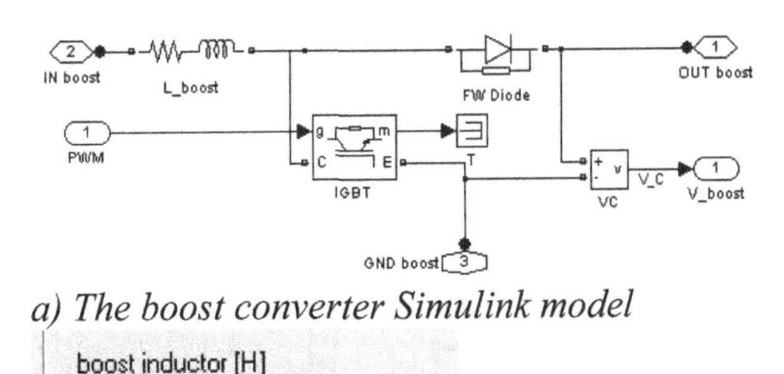

a) The boost converter Simulink model

b) Boost converter parameter

Figure 14. Inductor current waveforms corresponding to three possible types of operation © 2007 EPE. Used by permission.

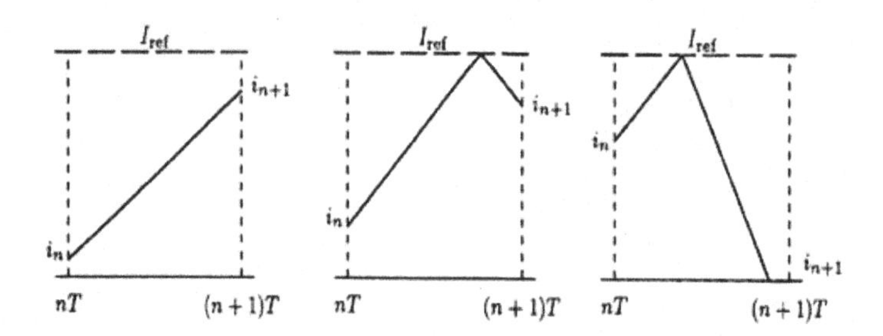

that gives the fast-scale instabilities of the EGS system. The Energy Generation System control must be optimized in order to manage the EGS power flows for any operation mode.

THE EGS CONTROL

In current control mode we must consider the situation where a light load produces a low average inductor current that causes the converter to operate in discontinuous conduction mode (DCM). So, there are

Figure 15. The simulated EGS model

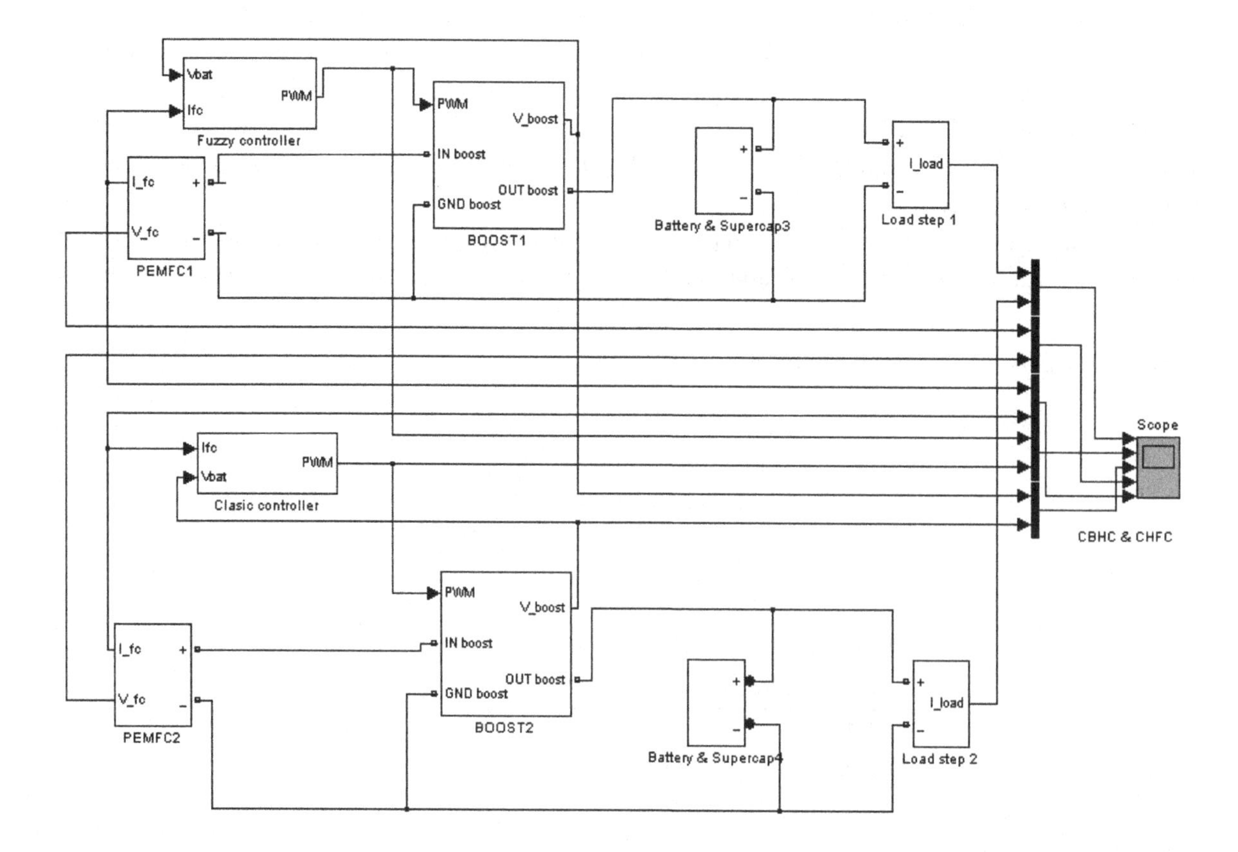

three operation cases which depends of the electronic switch state (*q1*), controlled with PWM command signal, and the diode conduction state (*q2*). Therefore, to study the EGS complex behaviour, a step load model is used (figure 16).

Peak Current Control of the Boost Converter

Several control techniques are available for controlling this kind of nonlinear systems using current control mode (figure 17) (Bizon, 2007c; Batarseh, 2003; Rashid, 2003; Mohan, Undeland, & Robbins, 2002; Agrawal, 2001).

Figure 16. The step load model

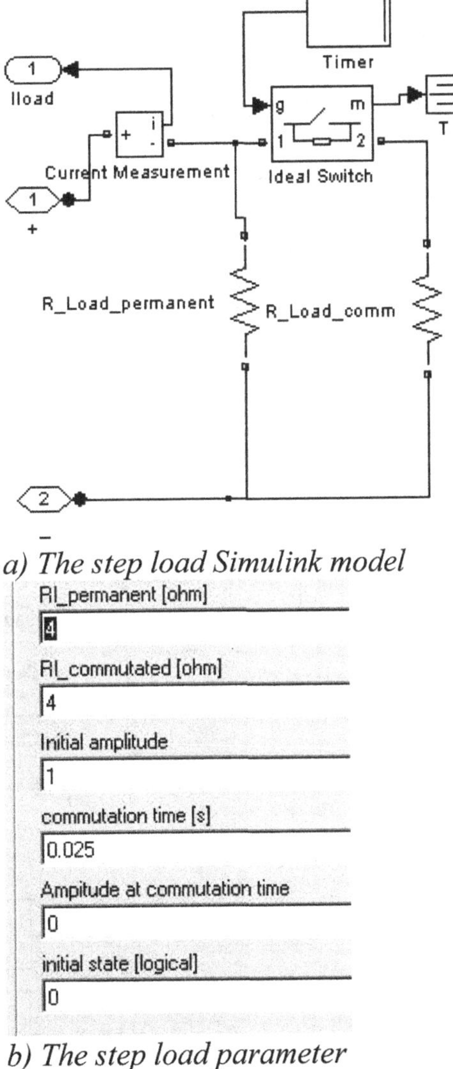

a) The step load Simulink model

b) The step load parameter

In this section the clocked / non clocked peak current control with / without hysteresis, using the BCCT control law (1-D or 2-D), is presented. The controller's models using a

1-D or 2-D boundary control law (Bizon, & Oproescu, 2004) are shown in figures 18, 19, 20, and 21.

In comparison with the non-clocked control methods, the simulation results show that the clocked control methods provide better dynamic response and small PEMFC current ripple, robustness against system uncertainty disturbances, and an implicit stability proof (see next section and Bizon, & Oproescu, M. (2005a&b)). The increased stability and small PEMFC current ripple are obtained by a proper designing of the control surface using a fuzzy system to generate the BCCT 2-D control law (figures 22, 23, and

Figure 17. Control techniques

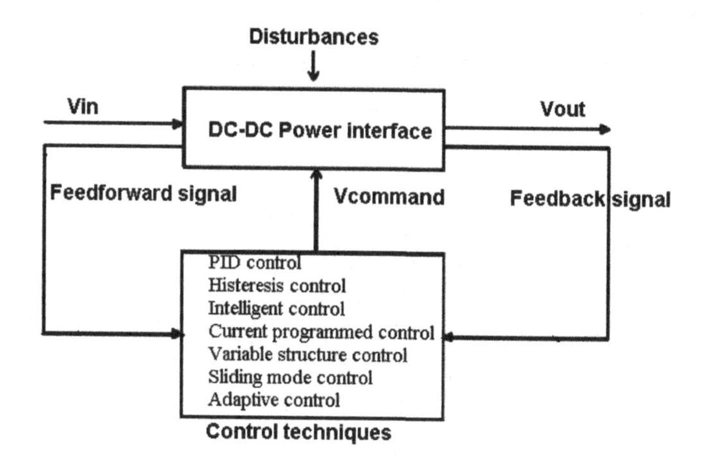

Figure 18. Diagrams of the clocked PCC using the BCCT control law: 1-D (left) and 2-D (right) © 2007 EPE. Used with permission.

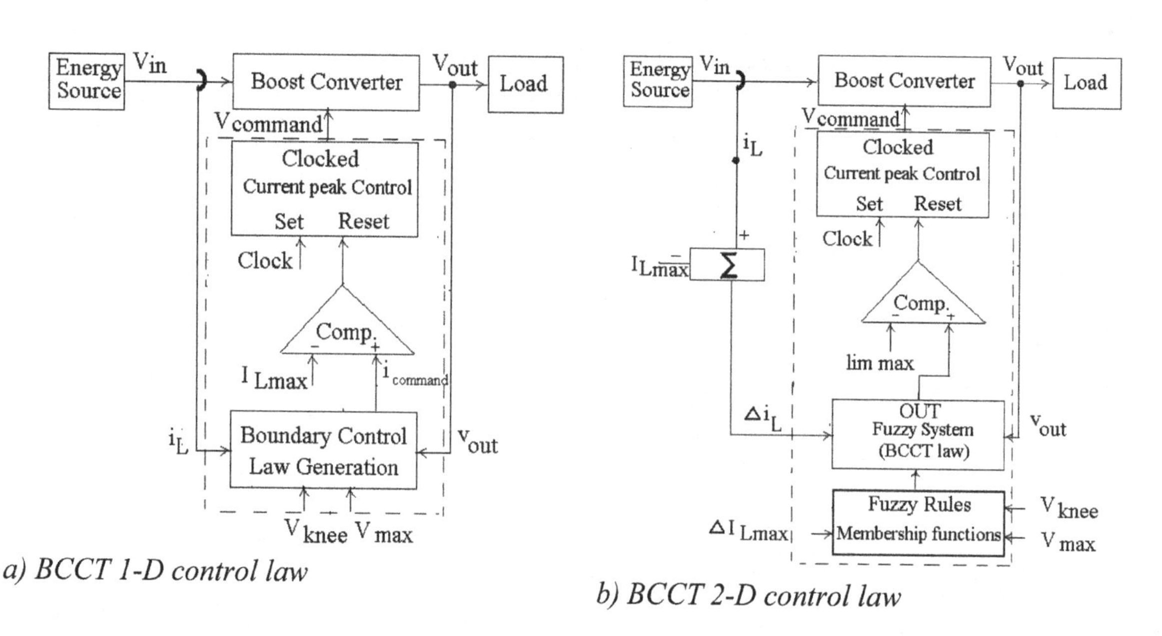

Figure 19. The clocked PCC using the BCCT control law: 1-D (top) and 2-D (bottom)

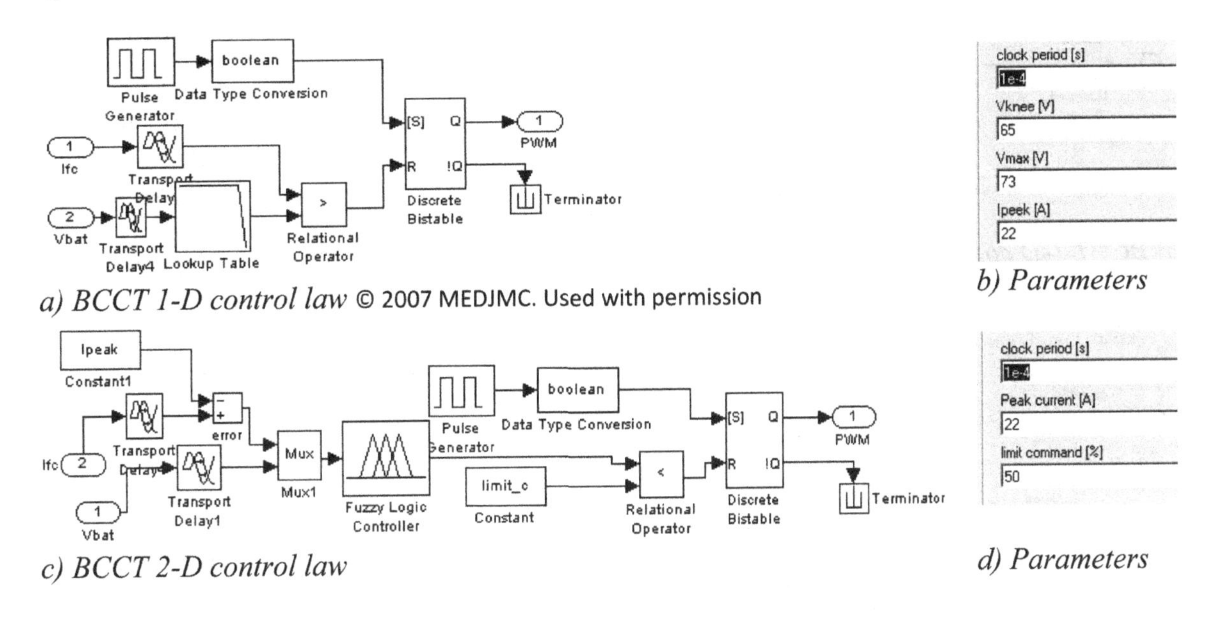

a) BCCT 1-D control law © 2007 MEDJMC. Used with permission

b) Parameters

c) BCCT 2-D control law

d) Parameters

24). In this case the PEMFC current ripple is smaller because the PEMFC current ripple represents one of the control surface variables. At high load steps, the control losing phenomena into an EGS with clocked PCC using the BCCT 1-D control law were extensively presented in Bizon, & Oproescu (2007a).

For a nominal regime of the EGS (with or without ESD) the dynamic of the EGS variables is almost the same for the different control implementations, above presented. For light load, the non-clocked hysteretic control variants can't limit the output voltage of the boost converter because the boost converter

Figure 20. The non-clocked PCC using the BCCT control law: 1-D (top) and 2-D (bottom)

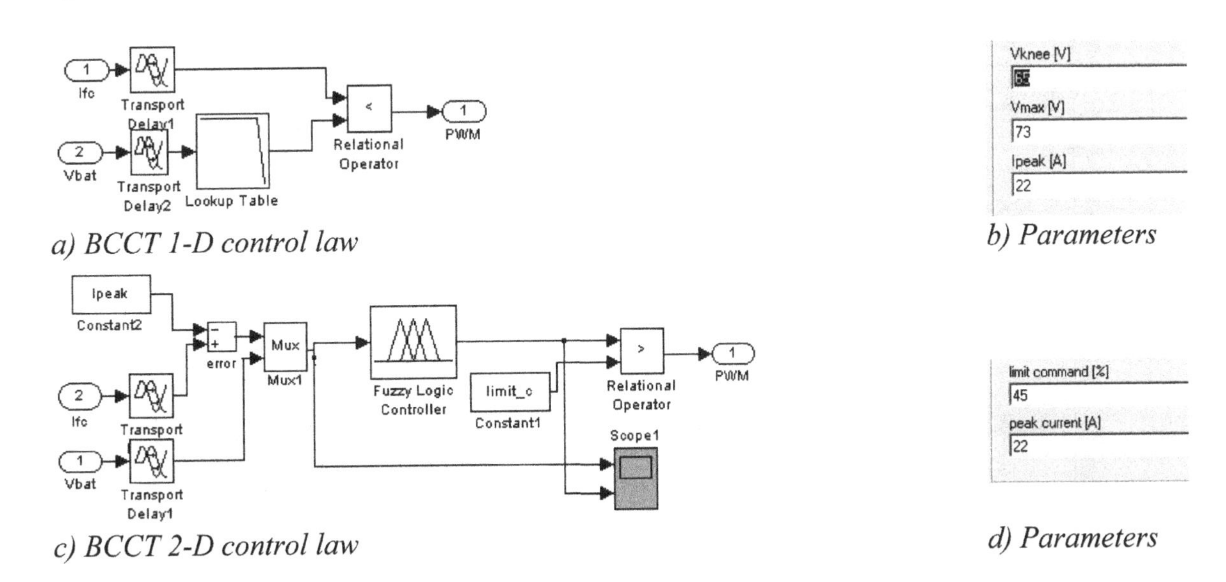

a) BCCT 1-D control law

b) Parameters

c) BCCT 2-D control law

d) Parameters

Figure 21.The non-clocked hysteretic PCC - BCCT control law: 1-D (top) and 2-D (bottom)

Ifc
Vbat
Transport Delay1
Transport Delay2
Lookup Table
Relay
boolean
Data Type Conversion
PWM
Scope

a) BCCT 1-D control law

Vknee [V]
64
Vmax [V]
73
Ipeek [A]
22
histerezis [A]
1

b) Parameters

Ipeak
Constant2
error
Mux
Mux1
Fuzzy Logic Controller
limit_c
Constant1
error1
Scope1
Relay
boolean
Data Type Conversion
PWM
Ifc
Vbat
Transport Delay
Transport Delay1

c) BCCT 2-D control law

peak current [A]
22
limit command [%]
45
Histeresis [%]
5

d) Parameters

operates in DCM. An alternative solution, which can solve this problem, can be the classic clocked PCC variant using a BCCT control law. Unfortunately, the complexity of the controller increases and isn't very robust control method (see next section). The EGS control performances increase if the control law

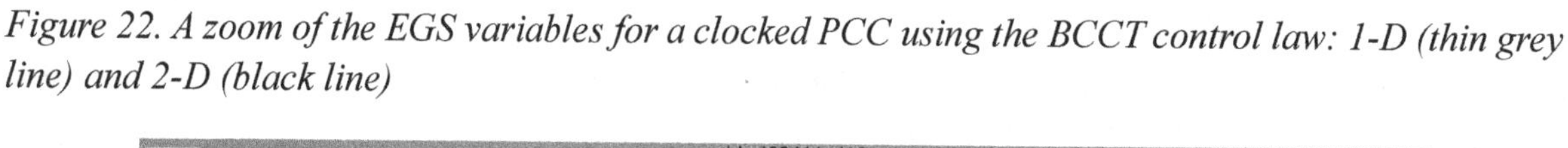

Figure 22. A zoom of the EGS variables for a clocked PCC using the BCCT control law: 1-D (thin grey line) and 2-D (black line)

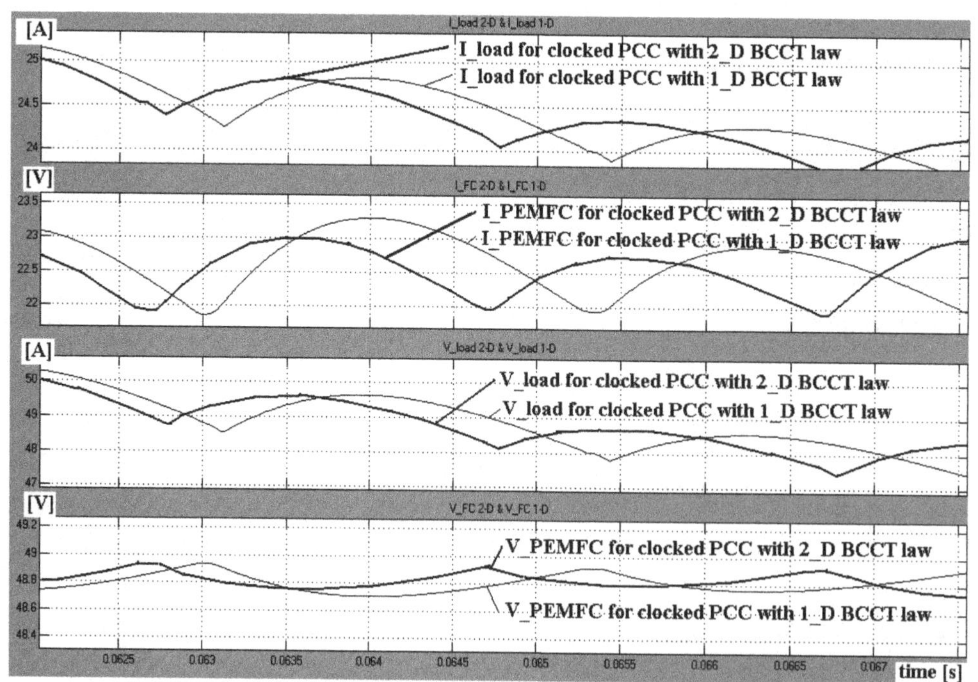

Figure 23. EGS behaviour for a nominal load and a clocked PCC - BCCT control law: 1-D (left) and 2-D (right)

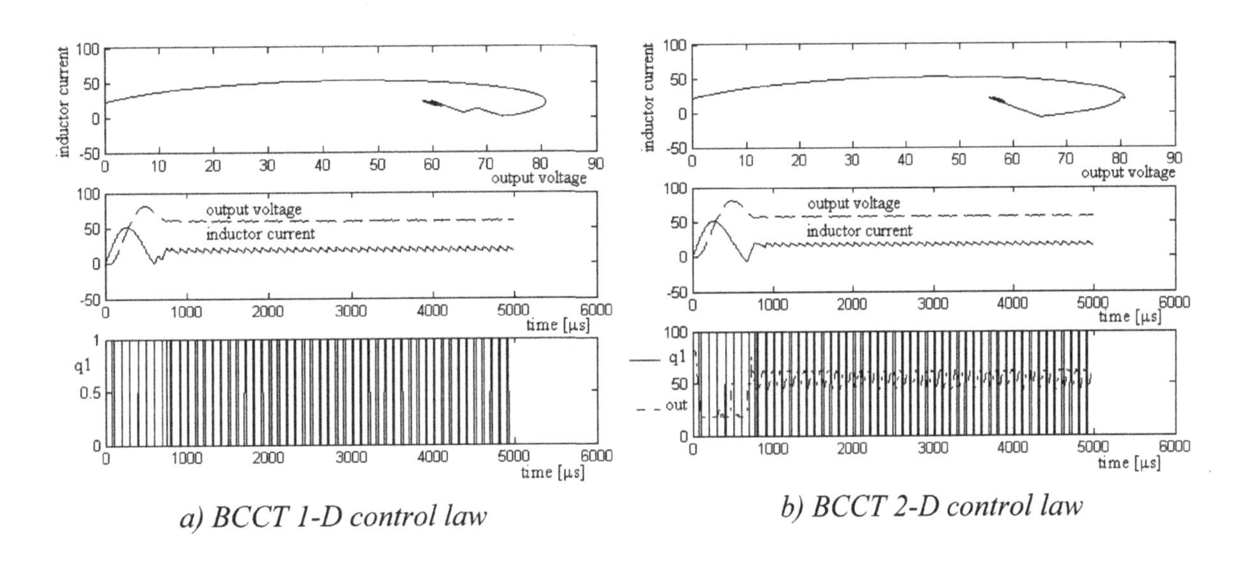

a) BCCT 1-D control law *b) BCCT 2-D control law*

is a function that dependent by the inductor current ripple and voltage over batteries stack, *V_Battery* (equal with voltage on the ESD bus, *V_Low_DC*, and named *Vbat* in figures 19-21). The structure of the Fuzzy Logic Controller (FLC) is following defined (see figure 25):

- The FLC inputs: $\Delta i_L = i_L - I_L$ = *I_FC-I_FCref* (input 1) on [$-\Delta I_{LMM}, \Delta I_{LMM}$] and *V_Battery* (input 2) [0, V_{MM}], respectively, $\Delta I_{LMM} > I_{L\max}$ $V_{MM} > V_{\max}$ (for all presented simulations we choose ΔI_{LMM}=40 and $V_{MM} = 80$);

Figure 24. EGS behaviour for a step-down load and a clocked PCC - BCCT control law: 1-D (left) and 2-D (right)

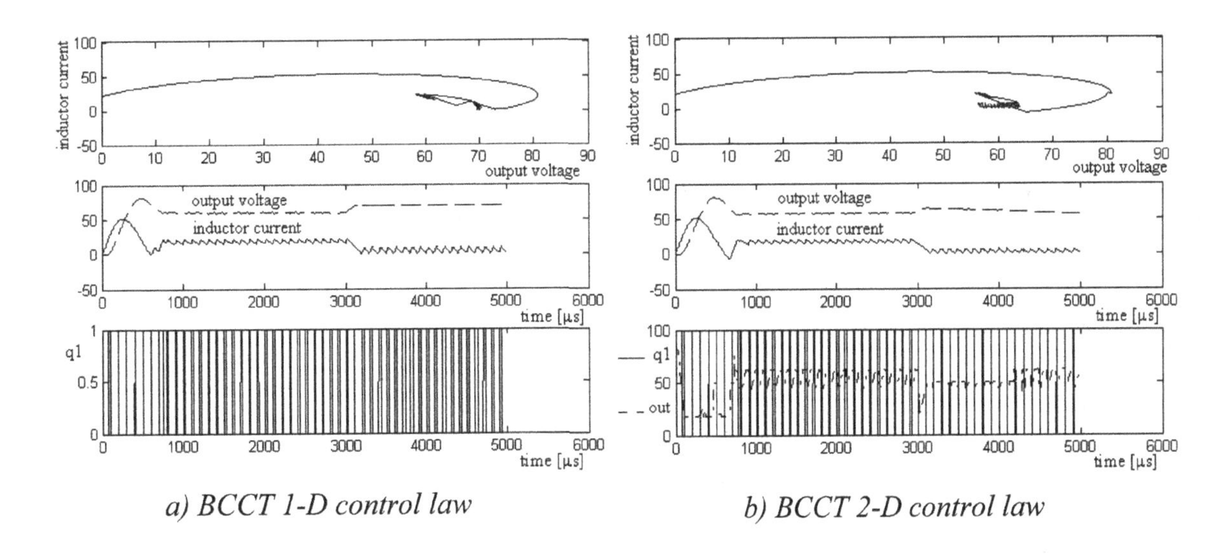

a) BCCT 1-D control law *b) BCCT 2-D control law*

Figure 25. Fuzzy boost controller: FLC FIS (a) and FLC control surface (b) © 2007 MEDJMC. Used with permission

```
fisb=newfis('control-PC');
fisb=addvar(fisb,'input','I FC-I FCref',[-a a]);
fisb=addvar(fisb,'input','V Battery',[0 f]);
fisb=addvar(fisb,'output','PC Fuzzy Command',[0 m]);
fisb=addmf(fisb,'input',1,'Negative','trapmf',[-a -a -c 0]);
fisb=addmf(fisb,'input',1,'Zero Equal','trapmf',[-c 0 0 c]);
fisb=addmf(fisb,'input',1,'Positive','trapmf',[0 c a a]);
fisb=addmf(fisb,'input',2,'Low','trapmf',[0 0 d e1]);
fisb=addmf(fisb,'input',2,'Normal','trapmf',[e1 e1 e2 e2]);
fisb=addmf(fisb,'input',2,'High','trapmf',[e2 e f f]);
fisb=addmf(fisb,'output',1,'Small','trapmf',[0 0 g h]);
fisb=addmf(fisb,'output',1,'Medium','trapmf',[g h h l]);
fisb=addmf(fisb,'output',1,'Big','trapmf',[h l m m]);
ruleListb=[1 1 3 1 1;2 1 2 1 1;3 1 1 1 1;1 2 2 1 1;2 2 2 1 1;
3 2 1 1 1;1 3 1 1 1;2 3 1 1 1;3 3 1 1 1];
fisb=addrule(fisb,ruleListb);
```

a) FLC FIS

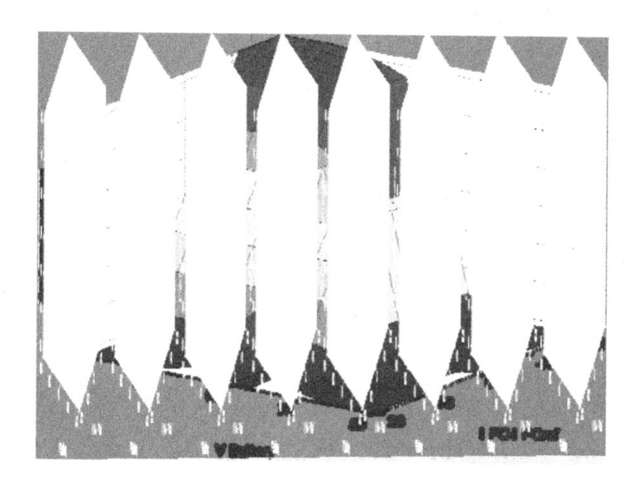

b) FCL control surface for following parameters:

a=20;c=10;d=50;e1=60;e2=65;e=70;f=80;g=20;h=50;l=80;

- The FLC output: PC_Fuzzy_Command on [0, m], $m \in \Re^*$ (for all presented simulation we choose m=100);

Trapezoidal membership functions for the inductor current ripple, Δi_L :

- **N**egative (**N**)=(- ΔI_{LMM}, - ΔI_{LMM}, - ΔI_{Lnom}, 0),
- **Z**ero **E**qual (**ZE**)=(- ΔI_{Lnom}, 0, 0, ΔI_{Lnom}),
- **P**ositive (**P**)=(0, r1, ΔI_{LMM}, ΔI_{LMM}).where ΔI_{Lnom} gives the fixed limits for the nominal inductor

current ripple with the help of limit values, Lim_{max} and Lim_{min} (for all presented simulation we choose ΔI_{Lnom}=10 and the inductor current ripple is minimize by choosing the boundary values).

- Trapezoidal membership functions for voltage over batteries stack, *V_Battery*:
- **Low (L)**=(0, 0, V_{offset}, V_{kneet}),
- **Normal (N)**=(V_{offset}, V_{offset}, V_{kneet}, V_{kneet}),
- **High (H)**=(V_{kneet}, V_{max}, V_{MM}, V_{MM}).
- Trapezoidal membership functions for command signal, *PC_Fuzzy_Command*:
- **Small (S)**=(0, 0, m/4, m/2),
- **Medium (M)**=(m/4, m/2, m/2, 3m/4),
- **Big (B)**=(m/2, 3m/4, m, m).
- The rule list: (N, L, B), (ZE, L, M), (P, L, S), (N, N, M), (ZE, N, M), (P, N, S), (N, H, S), (ZE, H, S), (P, H, S).
- Zadeh fuzzy connectives (max-min) and Mamdani implication are used.

Matlab program and associate control surface of the boost FLC are presented in figure 25.

Using the presented results from this section, and other results from Bizon, & Oproescu (2004) and Bizon, & Oproescu (2007c), we can conclude that clocked PCC with BCCT 2-D control law has the following advantages:

- improves the energy conversion efficiency of the power converter and assures less stress on the switching components. This advantage is obtained for all DC-DC converter topology with fixed switching frequency.
- is a robust controller to parametrical perturbations (and also for designing errors of the boost converter parameters);
- maintains the inductor current ripple in the admitted limits for large dynamic loads (see figure 24);

So, boost converter with clocked PCC using BCCT 2-D control law can be used for a robust and efficient control of the power converter interface between fuel cell (PEMFC) and Energy Storage Device (ESD).

Energy Generation System Behaviour using a clocked Peak Current Control

A simplified EGS model that includes the boost converter and batteries stack is show in figure 27. The boost converter (see figure 13) is described by two differential equations using two state variables, v_{out} and i_L, and two Boolean conduction variables, *q1* and *q2*; *q1*=1 (or 0) when the switch is on (or off, respectively) and *q2*=1 (or 0) when the diode is direct (or reverse, respectively) biased:

$$\frac{di_L}{dt} = q_1 \frac{V_{in} - i_L \cdot R_L}{L} + q_2 \frac{V_{in} - i_L \cdot R_L - v_{out}}{L} = \\ = \frac{V_{in} - i_L \cdot R_L - q_2 \cdot v_{out}}{L} \tag{7}$$

$$\frac{dv_{out}}{dt} = q_1 \frac{-(i_{bat.} + I_{load})}{C} + q_2 \frac{i_L - (i_{bat.} + I_{load})}{C} =$$
$$= \frac{-(i_{bat.} + v_{out} / R_{load}) + q_2 \cdot i_L}{C} \qquad (8)$$

There are three possible types of operation in any switching period (see figure 14 and also figure 26, where $i_n = i_L(nT)$ and $I_{ref} = I_{LMAX}$). In a switching period we can have at least three phase conduction and three time intervals:

- t_1: when the switch is on and the diode is reverse biased;
- t_2: when the switch is off and the diode is direct biased;
- t_3: when the switch is off and the diode is reverse biased.

The battery is modelled as a capacitor for energy storage, $C_{storage}$, a DC offset voltage, V_{offset}, and a series resistance, R_s, which limit the short circuit current (figure 4):

$$i_{bat.} = C_{storage} \frac{dv_{C\_storage}}{dt} =$$
$$= \frac{v_{out} - v_{C\_storage} - V_{offset}}{R_S} \qquad (9)$$

Therefore, the EGS model shown in figure 27 is described by four differential equations that represents a four order system (with four variables: output voltage - v_{out}, PEMFC voltage – V_{in}, voltage over the $C_{storage}$ - $v_{C\_storage}$, and inductor current - i_L). Matlab program that describes the sampled EGS system is following presented:

```
kicel=max(k-dcel,1);
Tstorage=RS*CS;
Ctcel=exp(-Ts/Tcel);
Vcel(k+1)=ncel*(E0-(iL(kicel)/Scel+in)*Rohmic-
-Aact*log((iL(kicel)/Scel+in)/i0)+
+Bconc*log((p^0.5)*(1-iL(kicel)/Scel+in)/((f^0.5)*imax))));
Vin(k+1)=Ctcel*Vin(k)+(1-Ctcel)*Vcel(k);
```

Figure 26. Inductor current waveform in different time intervals

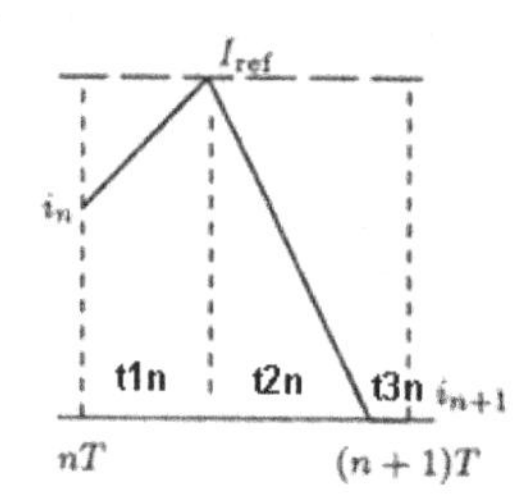

Figure 27. A simplified EGS model

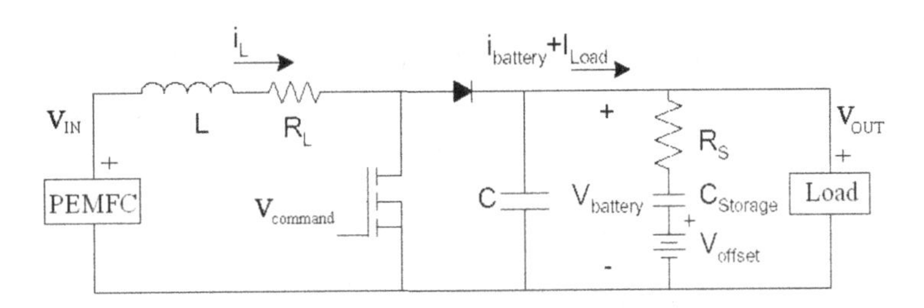

iL(k+1)=iL(k)+(T/L1)*(Vin(k)-iL(k)*RL-q2(k)*vC(k));
vC(k+1)=vC(k)-(T/(C*RS))*(vC(k)-vCS(k)-Vof)-(T/(C*Rout))*vC(k)+(T/C)*q2(k)*iL(k);
vCS(k+1)=vCS(k)+(T/Tstorage)*(vC(k)-vCS(k)-Vof);

Using high sampling frequency, the above quasi-continuous model can be used to analyze the EGS dynamics on the fast scale (Bizon, & Oproescu, in press). Some simulation results are subsequently presented. The used controller and initial conditions are mentioned in every case. Figures 29-31 (EGS phase diagrams) show for I_N =21.75A and L=160μH the effect of the reference current parameter, I_{ref}, when I_{ref} increases linearly from nominal value (I_N) to 4x I_N (see figure 28).

If reference current, I_{ref}, rises over a value that is dependent by the inductor inductance, the clocked PCC with 1-D control surface (see figure 19) can't catch this reference, and the inductor current rise uncontrollable (figure 29).

The clocked peak current control with fuzzy 2-D control surface may operate using a bigger current reference, with penalty in current ripple (figure 30). The fuzzy controller assures the system stabilization for different initial conditions, and for nominal load the PEMFC current ripple value is up to required

Figure 28. Reference current shape

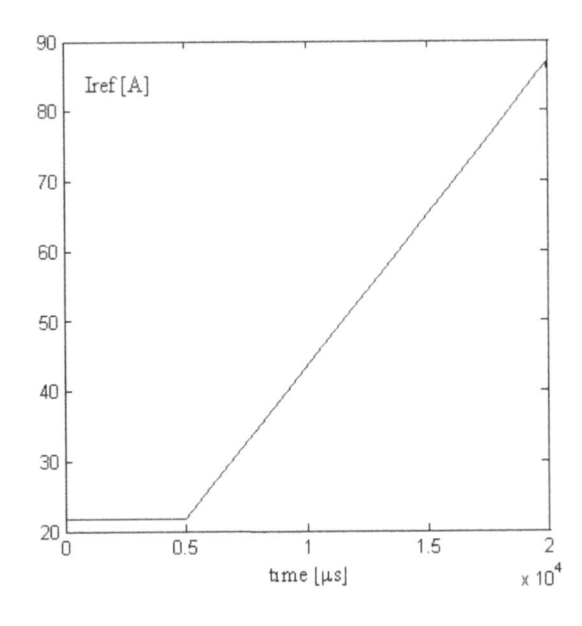

value.

The EGS dynamic with a hysteretic PCC controller using a BCCT 1-D control law (see figure 21) is illustrated in figure 31. The clocked PCC controller using a fuzzy BCCT 2-D control law provides better dynamic response, robustness against system uncertainty disturbances, and an implicit stability proof. These aspects appear more evident in the bifurcation diagrams. The bifurcations diagram is a 2-D representation of an iterative function that expresses the state variables at the end of a switching period $x_{n+1} = f(x_n, p)$, where x_n is a vector consisting of the state variables at $t = nT$, T being the switching period, and p is a control variable or a circuit parameter (like I_{ref} in figures 29-31):

$$x_n = \left(v_{out,\ n} \quad v_{PEMFC,\ n} \quad v_{C\_storage,\ n} \quad i_{L,\ n}\right)^t = \\ = \left(v_{out}(nT) \quad v_{PEMFC}(nT) \quad v_{C\_storage}(nT) \quad i_{L,}(nT)\right)^t \tag{10}$$

The bifurcation diagram relations for an Energy Generation System with/without an ESD, controlled by a clocked PCC controller, are subsequently presented.

Figure 29. The EGS dynamic with clocked PCC controller - BCCT 1-D control law: with (a) /without (b) ESD

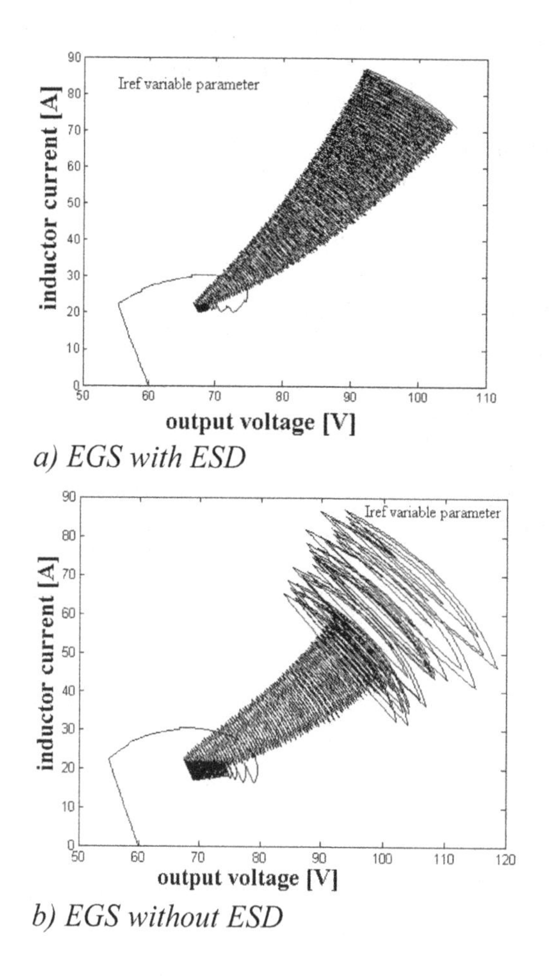

Figure 30. The EGS dynamic with clocked PCC controller - BCCT 2-D control law: with (a) /without (b) ESD. © 2007 EPE. Used with permission.

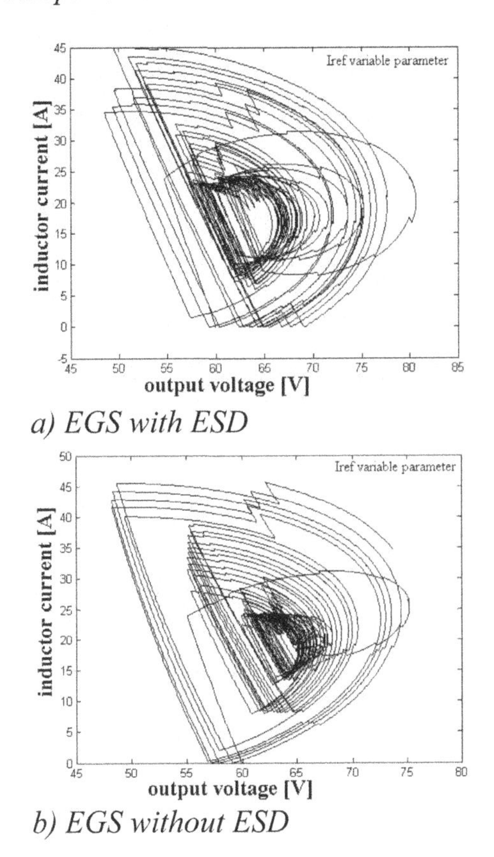

Case a) - t_1 time interval in figure 26: if the inductor current at the beginning of the period is

$$i_{L(n)} < I_{ref} - \frac{V_{PEMFC(n)} \cdot T}{L} \tag{11}$$

then

$$i_L(t) = i_{L(n)} + \frac{V_{PEMFC(n)} \cdot t}{L} \tag{12}$$

The switch will be switched in off state when the inductor current rises over to I_{ref}:

$$\begin{aligned} & i_L(t_{1(n)}) = I_{ref} \Rightarrow \\ & \Rightarrow t_{1(n)} = \frac{L}{V_{PEMFC(n)}}\left(I_{ref} - i_{L(n)}\right) \end{aligned} \tag{13}$$

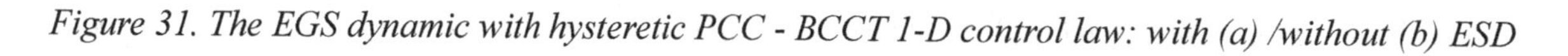

Figure 31. The EGS dynamic with hysteretic PCC - BCCT 1-D control law: with (a) /without (b) ESD

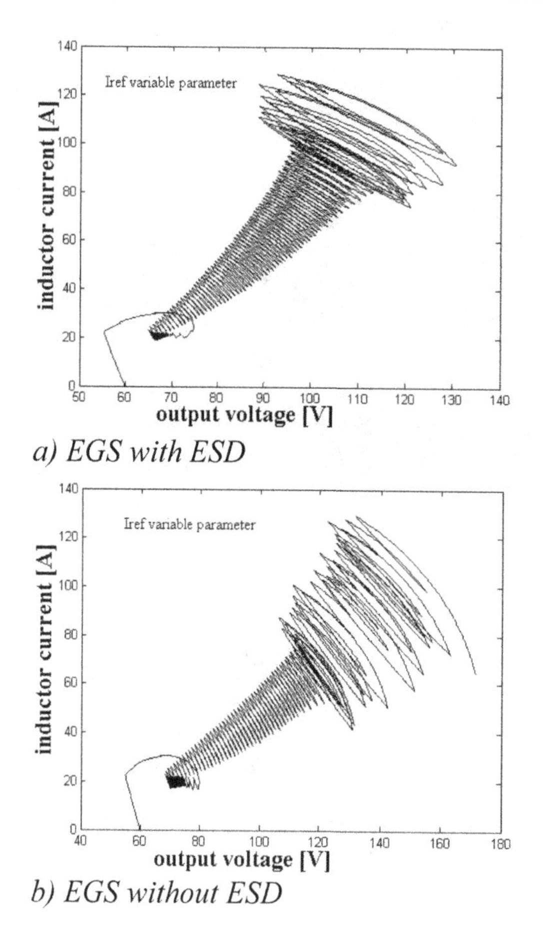

a) EGS with ESD

b) EGS without ESD

So, if $i_{L(n)} < I_{ref} - \frac{V_{PEMFC(n)} \cdot T}{L}$, the bifurcation diagram is:

$$i_{L(n+1)} = i_{L(n)} + \frac{V_{PEMFC(n)} \cdot T}{L} \tag{14}$$

$$v_{C(n+1)} = v_{C(n)} \cdot \exp\left(-\frac{I}{R_{Loud} \cdot C}\right) \tag{15}$$

Case b) - t2 time interval in figure 26: if the inductor current at the beginning of the period is

$$i_{L(n)} > I_{ref} - \frac{V_{PEMFC(n)} \cdot T}{L} \tag{16}$$

then

$$i_L(t) = I_{ref} \cdot \cos \omega_p \left(t - t_{1(n)}\right) + \\ +I_{1(n)} \sin \omega_p \left(t - t_{1(n)}\right) \cdot \exp\left[-\alpha \left(t - t_{1(n)}\right)\right] \tag{17}$$

where

$$\alpha = \frac{1}{2CR_{load}}, \omega_p = \sqrt{\frac{1}{LC} + \frac{1}{4C^2R_{load}^2}}, \tag{18}$$

$$I_{1(n)} = \frac{I_{ref}}{2\omega_p \cdot CR_{load}} + \\ +\frac{1}{\omega_p L} \cdot \left(V_{PEMFC(n)} - v_{C(n)}\right) \cdot \exp\left(-\frac{t_{1(n)}}{CR_{load}}\right) = \\ = \frac{\alpha \cdot I_{ref}}{\omega_p} + \frac{1}{\omega_p L}\left(V_{PEMFC(n)} - v_{C(n)}\right) \cdot \exp\left(-2\alpha \cdot t_{1(n)}\right) \tag{19}$$

If the inductor current falls to zero for time $t = t_{1(n)} + t_{2(n)} > T$, the boost converter maintains the CCM operations, so:

$$i_L\left(t_{1(n)} + t_{2(n)}\right) = 0 \Rightarrow tg\omega_p t_{2(n)} = -\frac{I_{ref}}{I_{1(n)}} \Rightarrow \\ \Rightarrow t_{2(n)} = \frac{1}{\omega_p}\left(\pi + arctg\frac{I_{ref}}{I_{1(n)}}\right) \tag{20}$$

So, if $i_{L(n)} > I_{ref} - \frac{V_{PEMFC(n)} \cdot T}{L}$ and $t_{2(n)} > T - t_{1(n)}$, the bifurcation diagram is:

$$i_{L(n+1)} = I_{ref} \cdot \cos \omega_p \left(T - t_{1(n)}\right) + \\ +I_{1(n)} \sin \omega_p \left(T - t_{1(n)}\right) \cdot \exp\left[-\alpha \left(T - t_{1(n)}\right)\right] \tag{21}$$

$$v_{C(n+1)} = V_{PEMFC(n)} - L \cdot \exp\left[-\alpha\left(T - t_{1(n)}\right)\right] \cdot \\ [\left(-\alpha \cdot I_{ref} + \omega_p \cdot I_{1(n)}\right) \cdot \cos \omega_p \left(T - t_{1(n)}\right) - \\ -\left(\alpha \cdot I_{1(n)} + \omega_p \cdot I_{ref}\right) \cdot \sin \omega_p \left(T - t_{1(n)}\right)] \tag{22}$$

Case c) - t_3 time interval in figure 26: if $i_{L(n)} > I_{ref} - \frac{V_{PEMFC(n)} \cdot T}{L}$ and $t_{2(n)} \leq T - t_{1(n)}$, the bifurcation diagram is:

$$i_{L(n+1)} = 0 \tag{23}$$

$$v_{C(n+1)} = (V_{PEMFC(n)} - \omega_p L \cdot \sqrt{I_{ref}^2 + I_{1(n)}^2}) \cdot \exp\left(-\frac{t_{3(n)}}{CR_{load}}\right) \tag{24}$$

where $t_{3(n)} = T - t_{1(n)} - t_{2(n)}$

To illustrate the complex EGS behaviour (figures 32, 33, 34, 35, 36, and 37, without an ESD and using a clocked PCC controller with/without a BCCT control law) when supplied by a 10 kW PEMFC

Figure 32. The bifurcation diagram for the EGS with L=100μH and clocked PCC without (a) /with (b) BCCT 1-D law

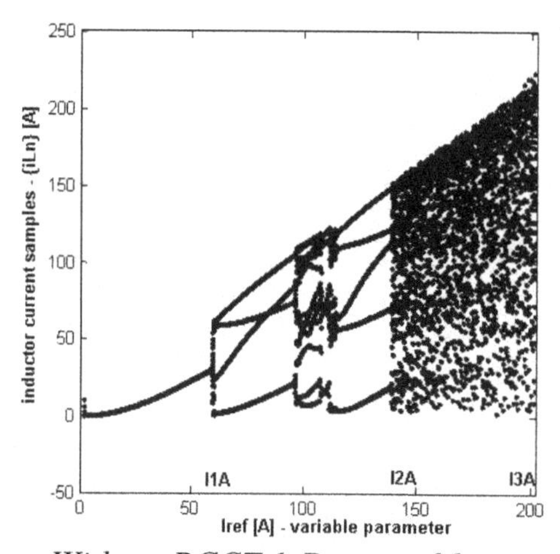

.a. Without BCCT 1-D control law

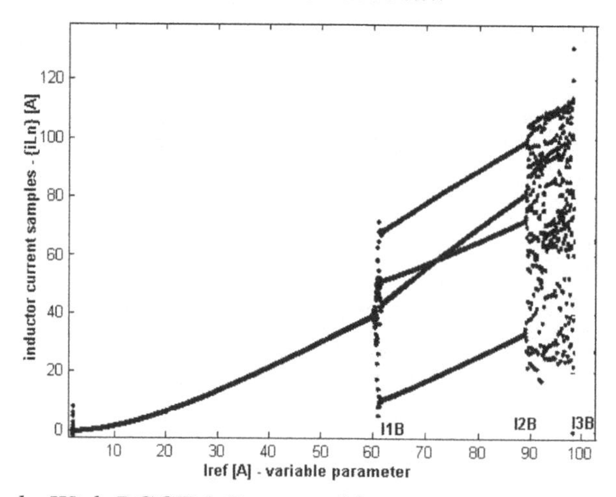

b. With BCCT 1-D control law

Figure 33. The bifurcation diagram for the EGS with $L=160\mu H$ and locked PCC without (a) /with (b) BCCT 1-D law

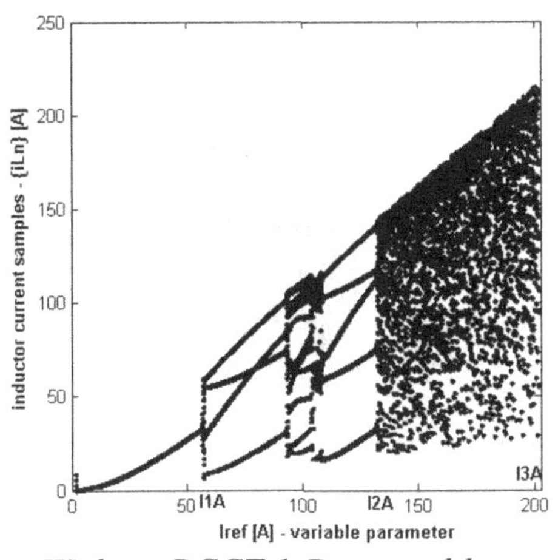

a. Without BCCT 1-D control law

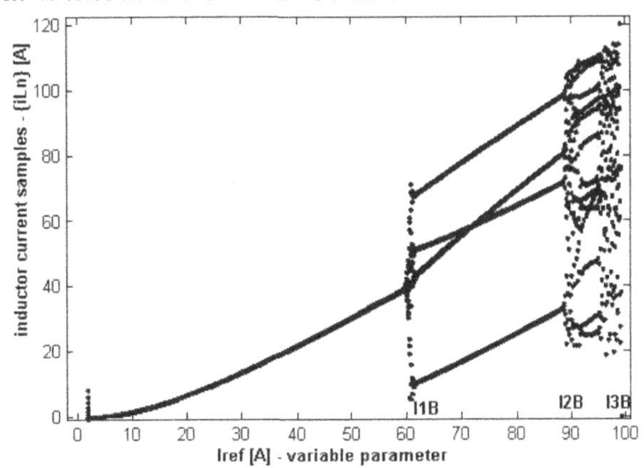

b. With BCCT 1-D control law

(n_{cel}=60 cells; A_{PEM}=600 cm^2), we consider I_{ref} as a variable parameter (I_{ref} increases linearly from $I_N/10$ to $10xI_N$).

The bifurcation diagrams depend on the boost inductor inductance (see figures 32-37). If the inductance value is up to 30 μH, only the first bifurcation appears.

These aspects are more clearly shown in the bifurcation diagram using I_{ref}=21,4A, and L as variable parameter (figure 38).

The bifurcation diagram relations for an EGS with an ESD are obtained in the same manner. To illustrate the EGS behaviour when an ESD is connected at the DC output port, the same rated conditions are considered: a 10 kW PEMFC (n_{cel}=60 cells; A_{PEM}=600 cm^2) a clocked PCC controller with a BCCT 1-D control law, $L=160\mu H$, and I_{ref} as a variable parameter (I_{ref} increases linearly from $I_N/10$ to $3xI_N$). If the EGS have an ESD at the DC output port, the EGS behaviour passes directly from stable operation to chaotic operation, irrespective of what type of the controller is used (figure 39). The usage of an ESD into an EGS introduces a stable pole in the EGS transfer function.

Figure 34. The bifurcation diagram for the EGS with L=200µH and clocked PCC without (a) /with (b) BCCT 1-D law© 2007 EPE. Used by permission.

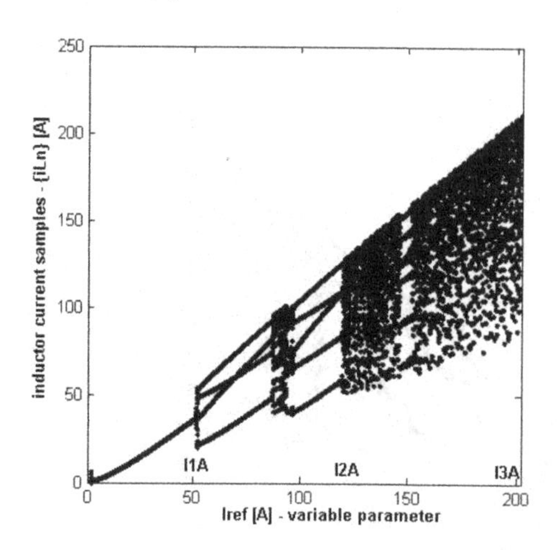

a. Without BCCT 1-D control law

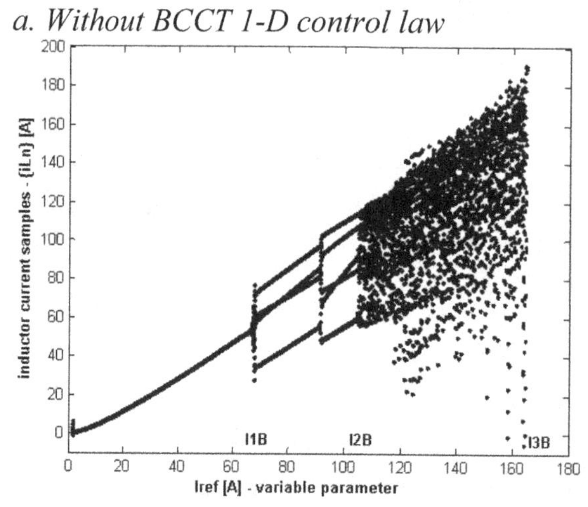

b. With BCCT 1-D control law

The behaviour of the EGS with clocked PCC, without an ESD and without BCCT control law, can be summarized by means of the presented bifurcation diagrams:

- when I_{ref} is up to I_{1A} (see table 3) the EGS operates in stable conditions;
- if the I_{ref} is increased further, the EGS behaviour bifurcates to a period-*n* sub-harmonic oscillation; after 1 up to 5 of such stages, depending on *L* value and controller type, chaotic operation appears if I_{ref} is over to I_{2A}; for different initial conditions the period-*n* sub-harmonic oscillation stage can appear in chaotic operation (as a window);
- chaotic operation is maintained if I_{ref} is up to the maximum reference current:

$I_{3A}= I_{ref(max)} =200A$;

Figure 35. The bifurcation diagram for the EGS with L=300μH and clocked PCC without (a) /with (b) BCCT 1-D law

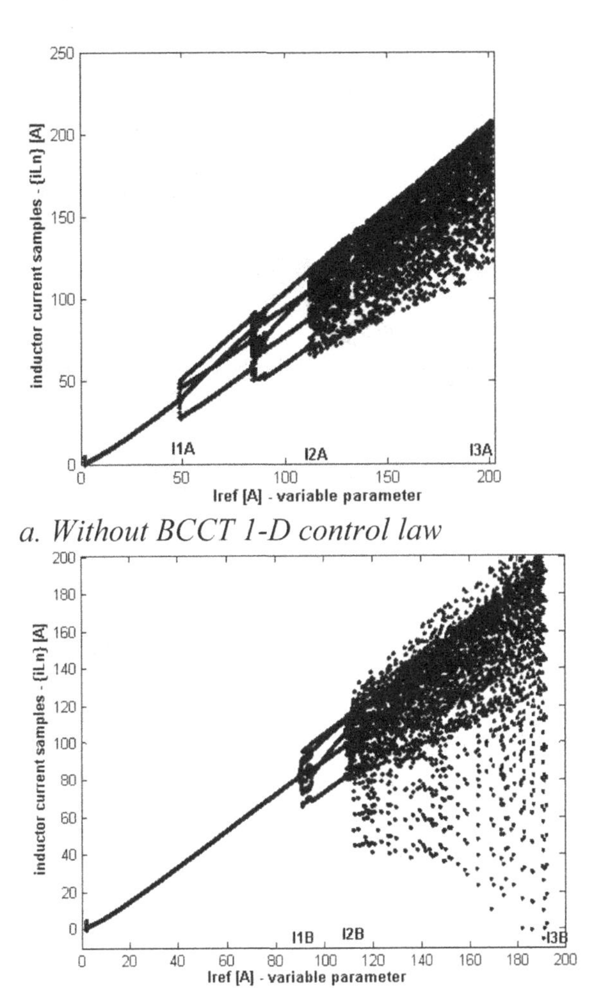

a. Without BCCT 1-D control law

b. With BCCT 1-D control law

Table 3. I_l values as a function of L value

L [μH]	I_{1A} [A]	I_{1B} [A]	L [μH]	I_{1A} [A]	I_{1B} [A]
20	102	102	200	51	66
30	90	90	300	50	87
40	80	80	500	45	103
50	75	75	750	43	67
70	75	75	1000	41	33
80	65	65	1500	40	32
100	60	60	5000	27	30
160	61	60	20000	51	66

Figure 36. The bifurcation diagram for the EGS with L=500μH and clocked PCC Without (a) /with (b) BCCT 1-D law

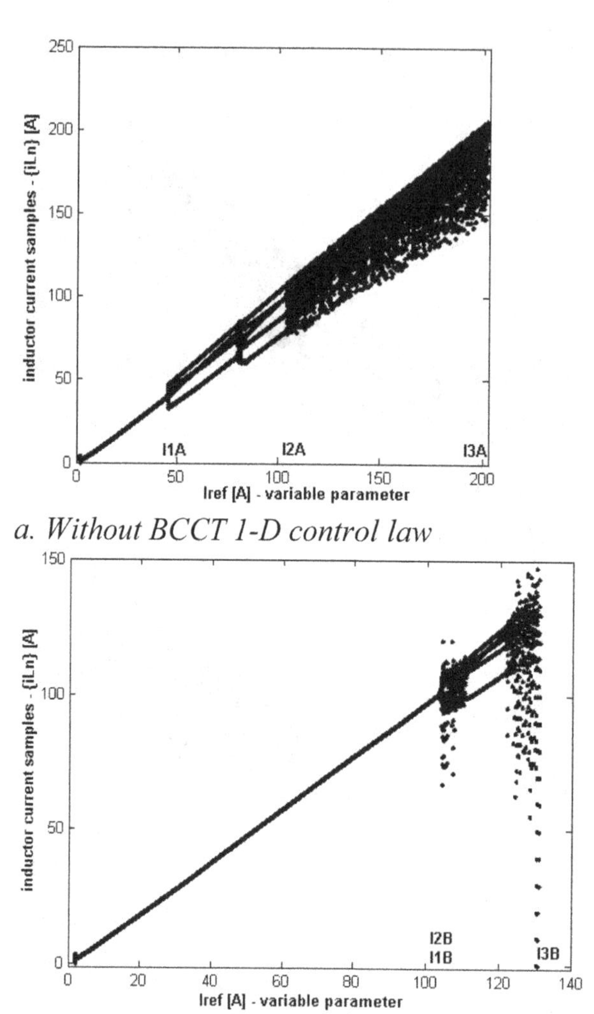

a. Without BCCT 1-D control law

b. With BCCT 1-D control law

The behaviour of the EGS with clocked PCC, without an ESD and with BCCT control law, can be summarized by means of the presented bifurcation diagrams, too:

- when I_{ref} is up to I_{1B} (see table 3) the EGS operates in stable conditions;
- I_{1B} value is lower (in 30A÷33A range) when L value is in 800μH÷5000μH range;
- if L value is up to 100 μH then $I_{1A} \approx I_{1B}$;
- chaotic operation is maintained if I_{ref} is up to $I_{3B} < I_{ref(max)} = 200A$;

The stable operation zone is represented in figure 40. When L value is place within the range 100μH÷800μH, the clocked PCC controller with BCCT control law provides robustness against system uncertainty disturbances, and an implicit stability proof if the 2-D BCCT control law is used.

Figure 37. The bifurcation diagram for the EGS with L=1500µH and clocked PCC without (a) /with (b) BCCT 1-D law

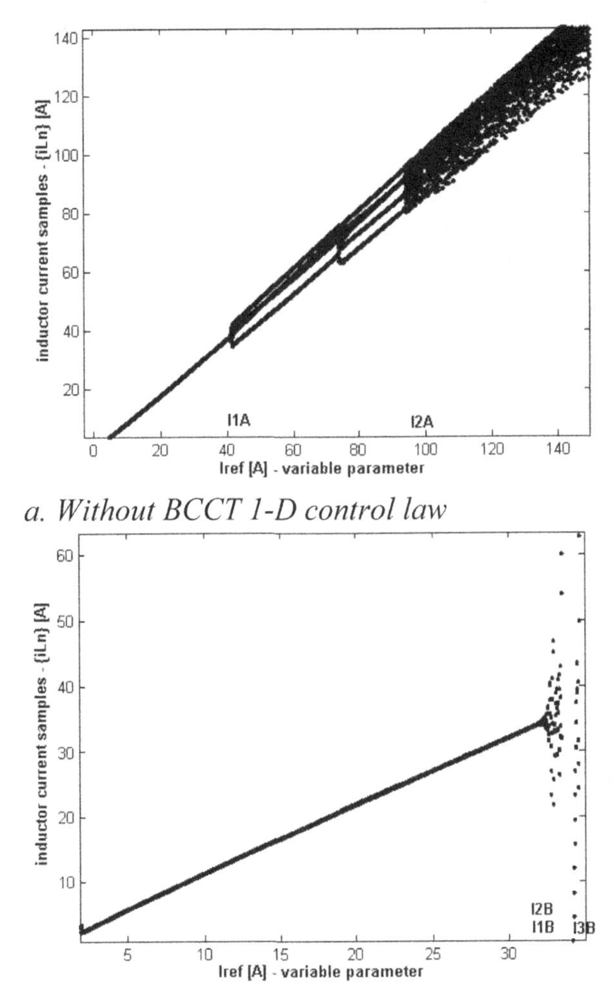

a. Without BCCT 1-D control law

b. With BCCT 1-D control law

Fuzzy Control of the Bidirectional Buck-Boost Converter

The load power dynamic is compensated by using an ultracapacitors stack as Dynamic Energy Compensator (DEC), connected, through a bi-directional buck-boost converter, to the *V_Low_DC* bus (see figure 3). The control law of the buck-boost controller is of a hysteretic type.

Figures 41 show the Matlab diagram of the buck-boost converter. The voltage mode controlled buck-boost converter is designed to operate in continuous conduction mode (CCM).

The converter equation when operate in CCM is $\frac{V_{out}}{V_{in}} = \tau$ (for the buck mode) and $\frac{V_{out}}{V_{in}} = 1 - \tau$ (for the boost mode), respectively, where $\tau = \frac{t_{on}}{T}$ is the duty ratio.

For example, if input voltage, output voltage and output power values are $V_{in} = 24V$, $V_{out} = 48V$ and

Figure 38. The bifurcation diagram for the EGS with: clocked PCC, I_{ref} =21.4A, and L as parameter (different L - zooms):without (a) /with (b) BCCT 1-D law

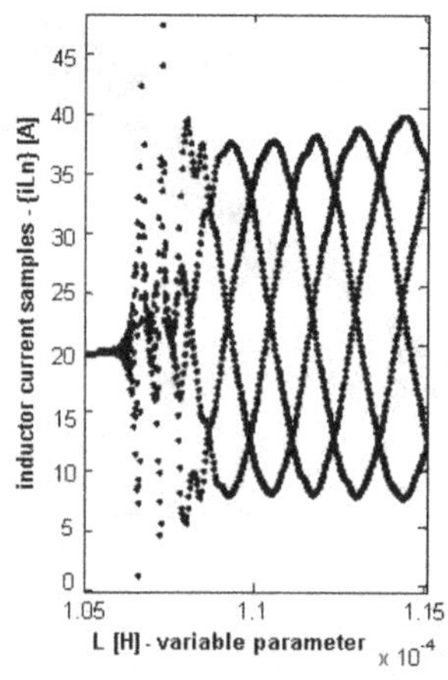

a. *Without BCCT 1-D control law*

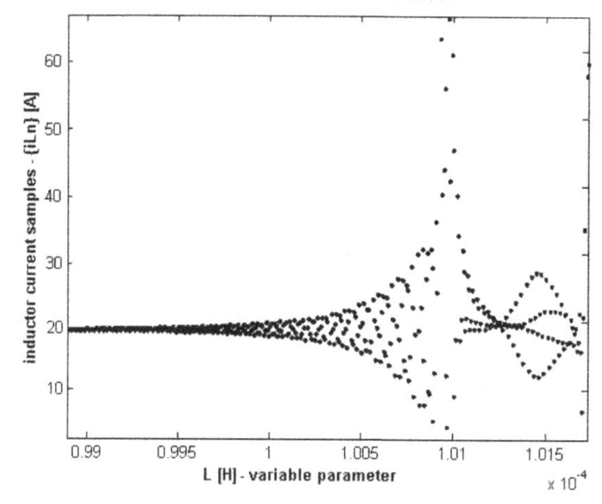

b. *With BCCT 1-D control law*

$P_{out} = 3kW$, respectively, the duty ratio is $\tau = 0,5$, and $I_{load} = \frac{P_{out}}{V_{out}} = 62,5A \Rightarrow R_{load} = 0,768\Omega$.

The maximum average inductor current appears in the boost mode after a high load pulse, $I_L = \frac{I_{out}}{1-\tau} = 130A$.

The series resistance of the boost inductor (r_L) can be chosen around 0,005 Ω.

In order to compare the buck-boost control performances, two types of control surface are used for:

- a hysteretic controller (figure 42);
- a hysteretic fuzzy controller (figure 43).

Figure 39. The bifurcation diagram for the EGS with: ESD, clocked PCC controller with BCCT 1-D, L=160μH, and I_{ref} as parameter© 2007 EPE. Used by permission.

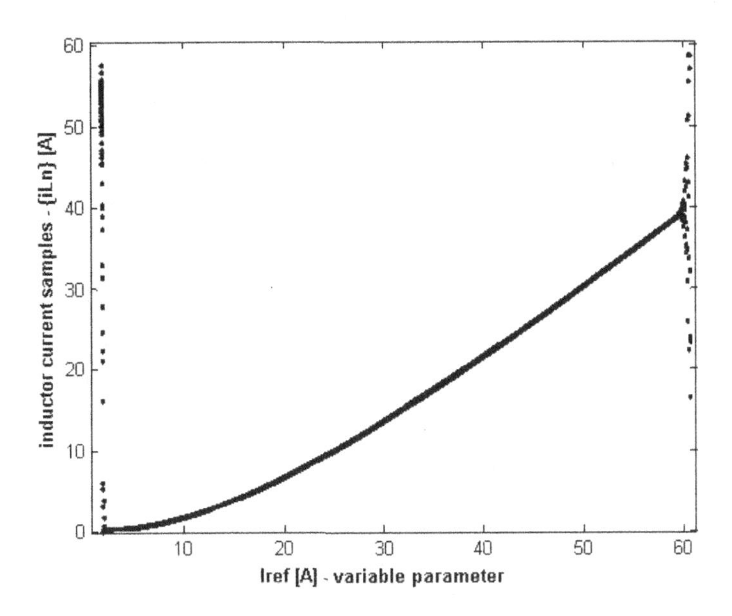

Figure 40. The stable operation zone for the EGS with an ESD, that use a clocked PCC controller with (continuous line) / without (dashed line) BCCT control law © 2007 EPE. Used by permission.

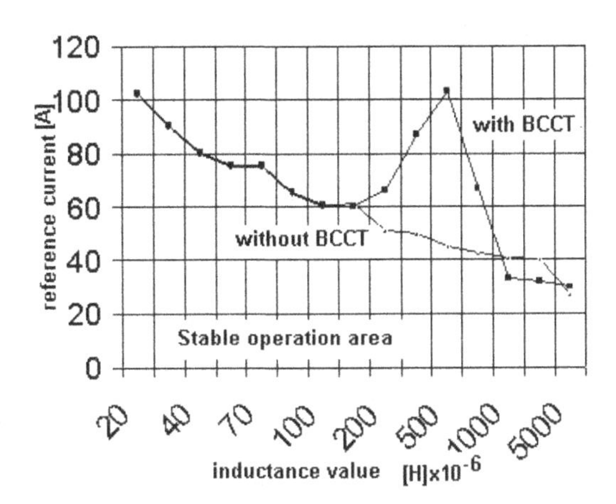

The controller with a voltage hysteretic control law (figure 42) can operate in the buck mode when the voltage difference between DC buses of the batteries stack and ultracapacitors stack (*Dv=V_Battery – V_UCap*) is grreater than high voltage reference (V_H), and in the boost mode when that voltage difference in lower than low voltage reference (V_L). In the buck mode the DEC is charged from the battery DC bus, and in the boost mode the power flow is from DEC to the battery DC bus.

The controller with a voltage hysteretic fuzzy control law (figure 44) has some well-known fuzzy control advantages (Bizon, Sofron, & Oproescu, 2007) and, if implemented, they provide the advantage of an easy protection and an improved control design by using more control variables. For example,

Figure 41. The buck-boost converter

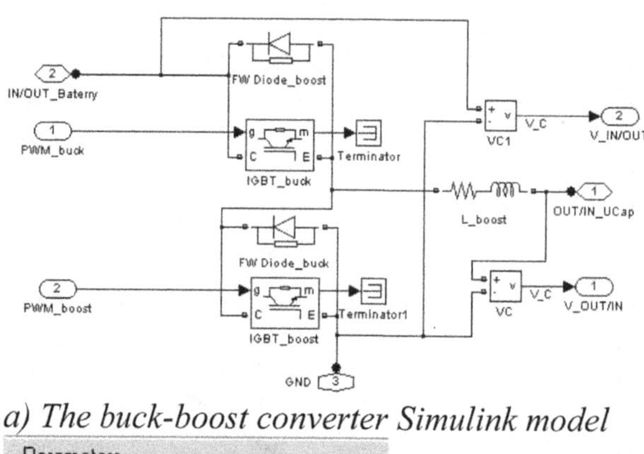

a) The buck-boost converter Simulink model

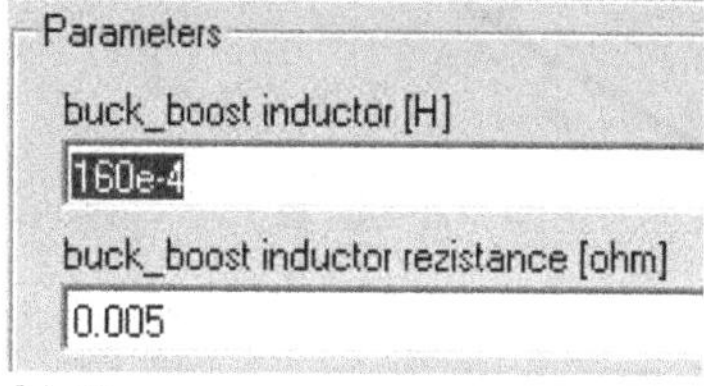

b) Parameters

Figure 42. The buck-boost hysteretic controller © 2007 MEDJMC. Used by permission

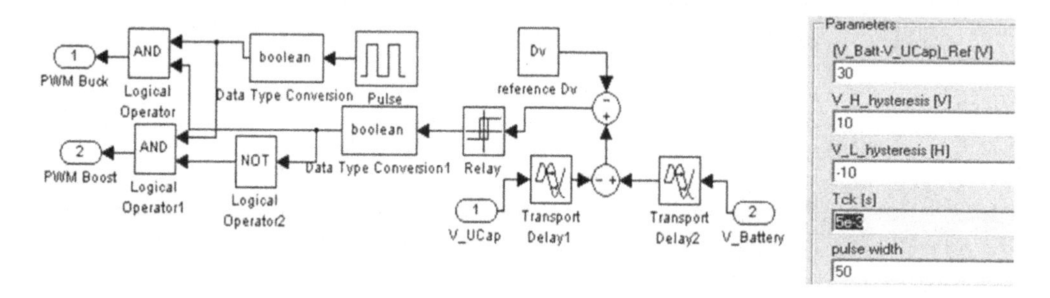

Figure 43. The buck-boost hysteretic fuzzy controller © 2007 MEDJMC. Used by permission

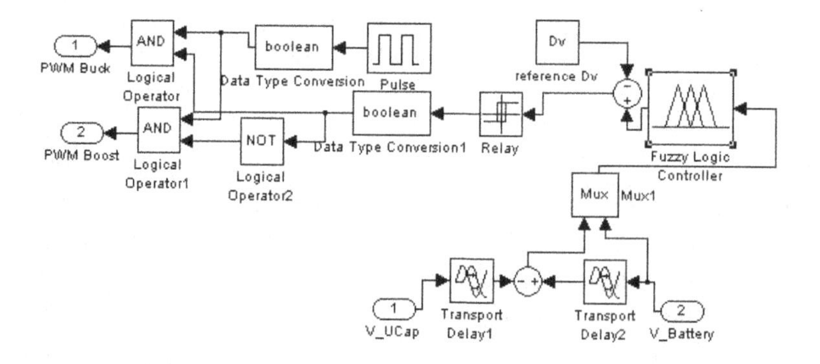

Figure 44. Fuzzy buck-boost controller: FLC FIS (a) and FLC control surface (b) © 2007 MEDJMC. Used by permission

```
fisbb=newfis('control-bb');
fisbb =addvar(fisbb,'input','Vbat-Vucap diff',[20 40]);
fisbb =addvar(fisbb,'input','battery voltage',[40 80]);
fisbb =addvar(fisbb,'output','command bb',[0 m]);
fisbb =addmf(fisbb,'input',1,'Small','trapmf',[0 0 c d]);
fisbb =addmf(fisbb,'input',1,'Nominal','trapmf',[c d d e]);
fisbb =addmf(fisbb,'input',1,'Big','trapmf',[d e f f]);
fisbb =addmf(fisbb,'input',2,'Low','trapmf',[0 0 d e1]);
fisbb =addmf(fisbb,'input',2,'Normal','trapmf',[e1 e1 e2 e2]);
fisbb =addmf(fisbb,'input',2,'High','trapmf',[e2 e3 f f]);
fisbb =addmf(fisbb,'output',1,'Small','trapmf',[0 0 g h]);
fisbb =addmf(fisbb,'output',1,'Medium','trapmf',[g h h l]);
fisbb =addmf(fisbb,'output',1,'Big','trapmf',[h l m m]);
ruleListbb=[1 1 3 1 1;2 1 3 1 1;3 1 1 1 1;1 2 3 1 1;2 2 2 1 1;
3 2 1 1 1;1 3 3 1 1;2 3 1 1 1;3 3 1 1 1];
fisbb =addrule(fisbb,ruleListbb);
Zadeh fuzzy connectives (max-min)
Mamdani implication
```

a. FLC FIS

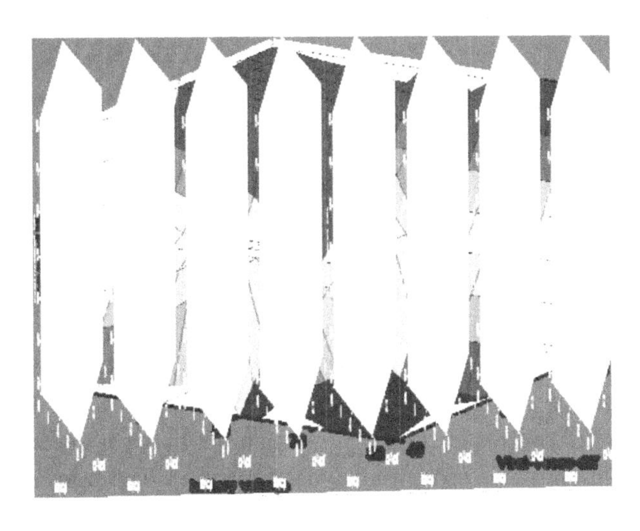

b. FCL control surface for following parameters:

a=60;c=20;d=30;e=40;e1=60;e2=65;e3=70
f=200;g=20;h=50;l=80;

useful variables are: battery current, DEC current, DEC voltage, PEMFC voltage and current, or power flow values for PEMFC, battery, DEC, and load).

The DEC device (ultracapacitors stack) is usually modelled by using a first order model (figure 45).

Figure 45. The DEC model

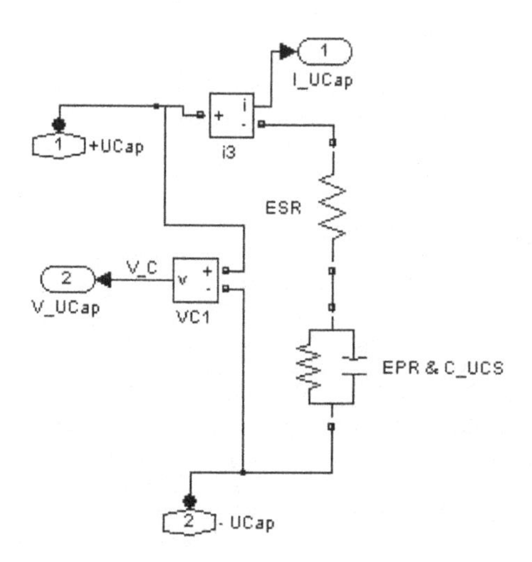

a) The DEC Simulink model

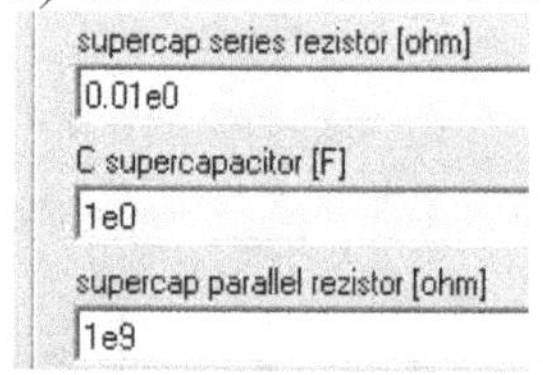

b) Parameters

The EGS dynamic behaviour when using different control laws for the boost and buck-boost converters is presented in figures 46 and 47 for a high load dynamic (EGS topology is shown in figure 3). The load dynamic changes are the most frequent cases in automotive applications (Bizon, Lefter, & Oproescu, 2007), or portable/stationary power units (Bizon, & Oproescu, 2006b), and the controller must be designed to ensure the control of the power flow in any situation. In order to compare the EGS behaviour under different control techniques, the same load dynamic is used. The same state variables are plotted: control and command variables at the top of the plot, currents and voltages for the PEMFC, battery and ultracapacitors in the middle of the plot, and the resulting power flow at the bottom of the plot. Plotted EGS variables are mentioned for every set of figures. Figure 46 shows the simulation results for a classical PCC boost controller (with 1-D BCCT law – figure 19.a), and a classical hysteretic buck-boost controller (figure 42). Figure 10 shows the simulation results for a fuzzy PCC boost controller (with 2-D BCCT fuzzy control surface – figure 19.b) and a fuzzy hysteretic buck-boost controller (figure 43). Observing the signals shape in figure 46, in comparison with the correspondent signals on figure 47, we can say that the signals dynamic are almost the same. Various delays at load changes can be corrected by using a modified boundary surface for the fuzzy buck-boost controller, or by adjusting the decision levels of the fuzzy PCC boost controller.

Obviously, different signals shapes appear in the start-up stage. The fuzzy variants have the advantages mentioned in the section above, and can be integrated into one fuzzy controller (see next section, and Bizon, & Oproescu, 2006c).

Figure 46. Simulation results for a classical PCC boost controller and a classical hysteretic buck-boost controller. © 2007 MEDJMC. Used by permission

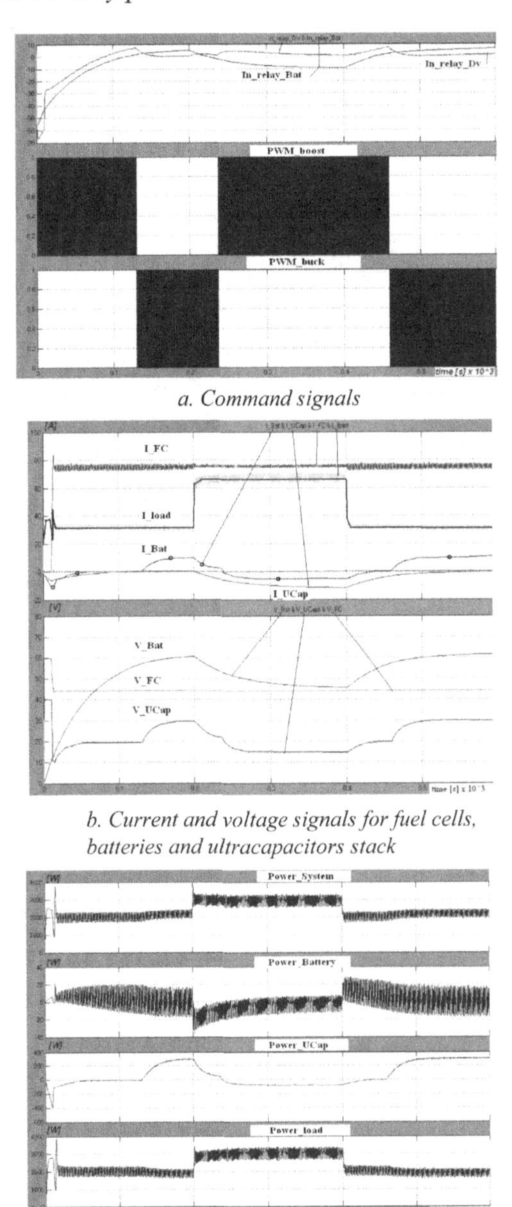

a. Command signals

b. Current and voltage signals for fuel cells, batteries and ultracapacitors stack

c. Output power signals for fuel cells, batteries and ultracapacitors stack

In order to protect the batteries stack, the buck mode (which is the charge mode for the battery stack) appears if the PEMFC power flow is bigger than the required load power flow. At the same time, the ultracapacitors stack is also charged, and the buck-boost converter operates in the buck mode. Alternatively with the buck mode, in the remaining time, the buck-boost converter operates in the boost mode, when the power flow is inverted (from the ultacapacitor stack to batteries stack).

Figure 47. Simulation results for a fuzzy PCC boost controller and a fuzzy hysteretic buck-boost controller © 2007 MEDJMC. Used by permission

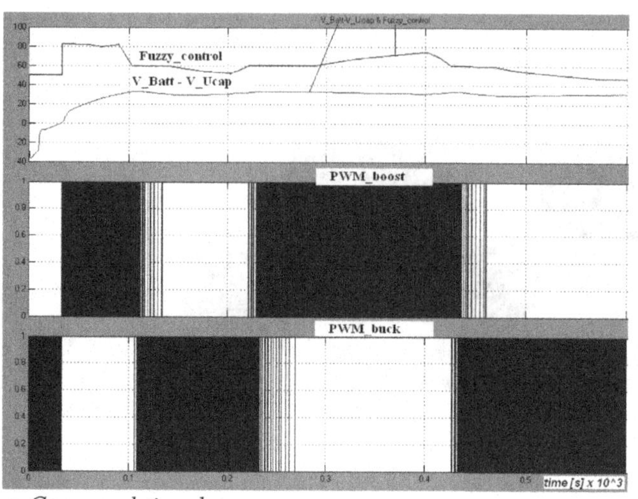

a. Command signals

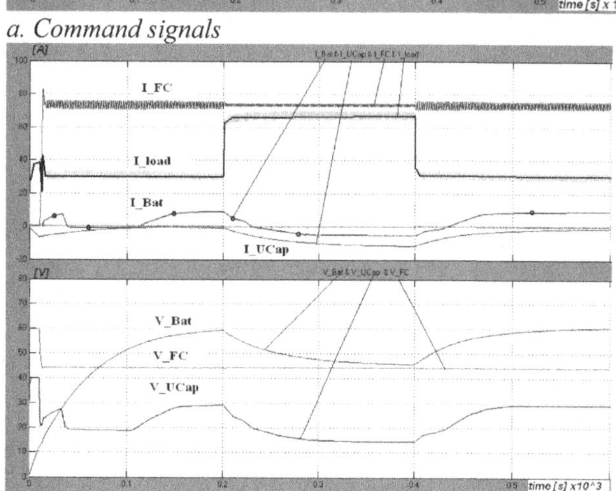

b. Current and voltage signals for fuel cells, batteries and ultracapacitor stack

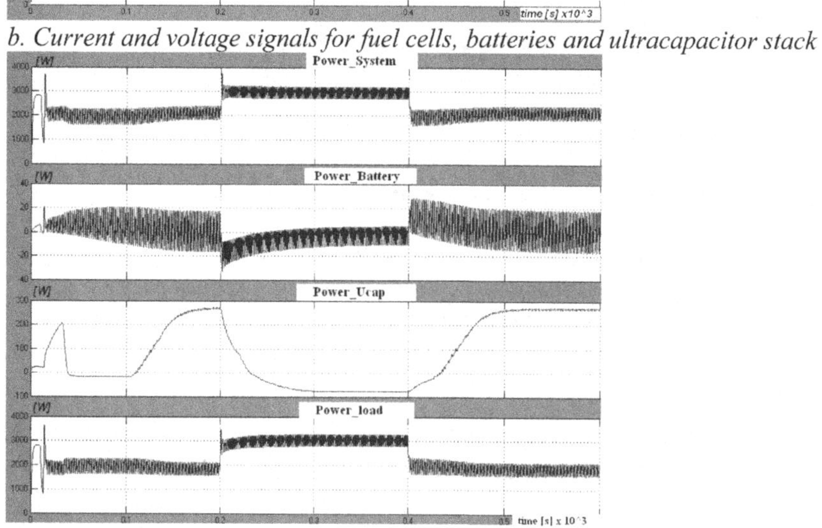

c. Output power signals for fuel cells, batteries and ultracapacitor stack

To speed up the simulations the ultracapacitor, battery and PEMFC time constants are divided by 10^3, so the represented time values must be multiplied by 10^3 to obtain the real values.

Observing the ($V\_Bat - V\_Ucap$) signal (figure 47) we can see that the signal dynamic is small even when a high load pulse appears (from 2 kW to 3 kW at 0.2×10^3 seconds). Consequently, we must use the *V_Bat* signal as a second input variable, which makes the fuzzy buck-boost controller more controllable.

The fuzzy buck-boost controller uses the FIS in figure 44, which combines the hysteretic control strategy for two input variables: first input, $Dv = V\_Bat - V\_Ucap$, and second input, *V_Bat*. For example, if the battery stack is discharged then the boost mode is forced (the fuzzy output, named *control-bb*, has the *big* trapezoidal membership value).

Integrated Control for the EGS Power Flow

Small PEMFC current ripple using compact batteries and ultracapacitor stack will be obtained by a proper designing of the control surface using an Integrate Fuzzy Controller (IFC) for both power interfaces (Bizon, & Oproescu, 2006c). Because the batteries stack voltage appears as a second input variable for both boost PCC fuzzy controller and buck-boost fuzzy controller, the IFC proposed may use only three input control variables: fuel cell current, ultracapacitor stack voltage, and batteries stack voltage. IFC fuzzy rules base is obtained by interblending the independent fuzzy rules bases presented in figures 25 and 44. The fuzzy control proposed integrates two controllers into one, in order to be cheap (only three

Figure 48. EGS topology with IFC. © 2007 MEDJMC. Used by permission

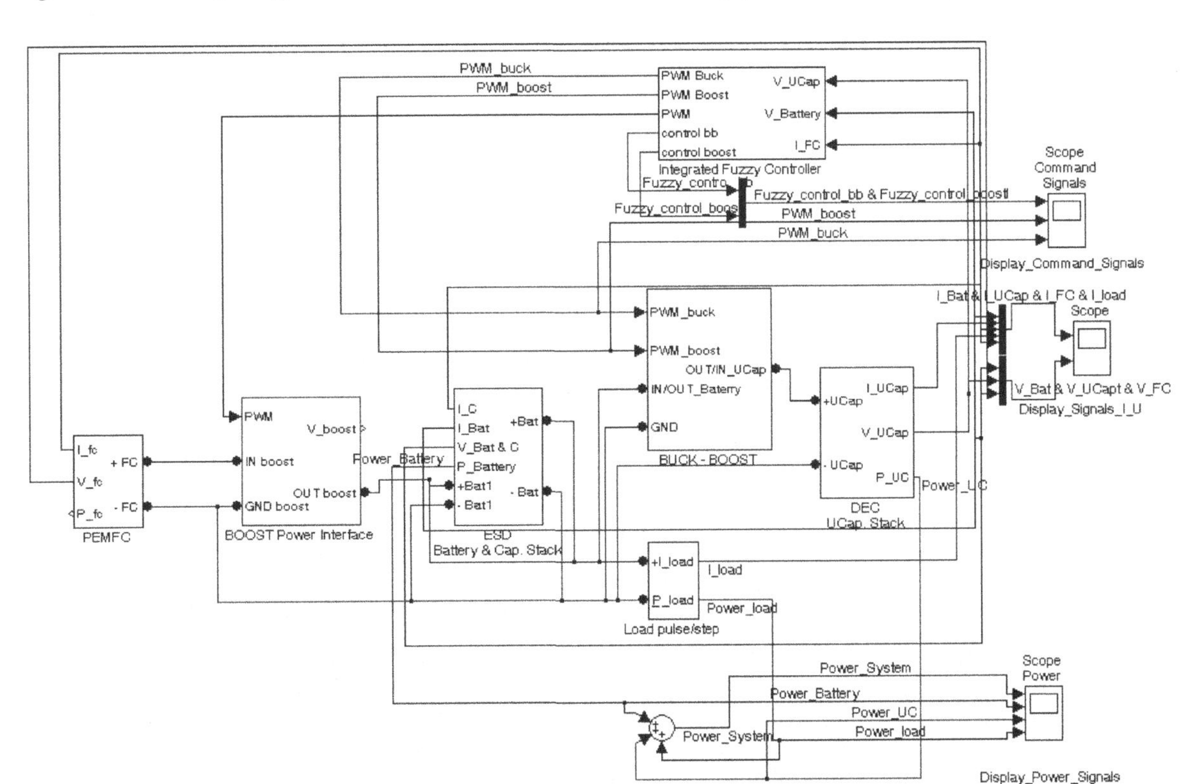

transducers are used) and compact. The EGS topology shown in figure 48 uses the same power converter topologies. The detailed IFC structure is presented in figure 49. The essential IFC block is the fuzzy logic controller, with IFC FIS presented in figure 50. The same input variables (*Dv=Vbat-VUcap, V_Battery* and *I_FC–I_FC_ref*) are used in the IFC FIS. Trapezoidal membership functions for input variables are defined in the same manner as in boost PCC FIS (figure 25), and buck–boost controller FIS (figure 44), respectively. The interblending IFC fuzzy rules base is a mixed fuzzy rules base that maintains the good control characteristics for the IFC controller, too (figure 50). Only minor modifications are made in the IFC fuzzy rules base in order to minimize the PEMFC current ripple.

Obviously, for one output variable the IFC FIS surface is a 4-D map, and if two input variable are used the IFC FIS surface becomes a 3-D map. Some IFC FIS surfaces are shown in figure 51 in order to

Figure 49. IFC structure © 2007 MEDJMC. Used by permission

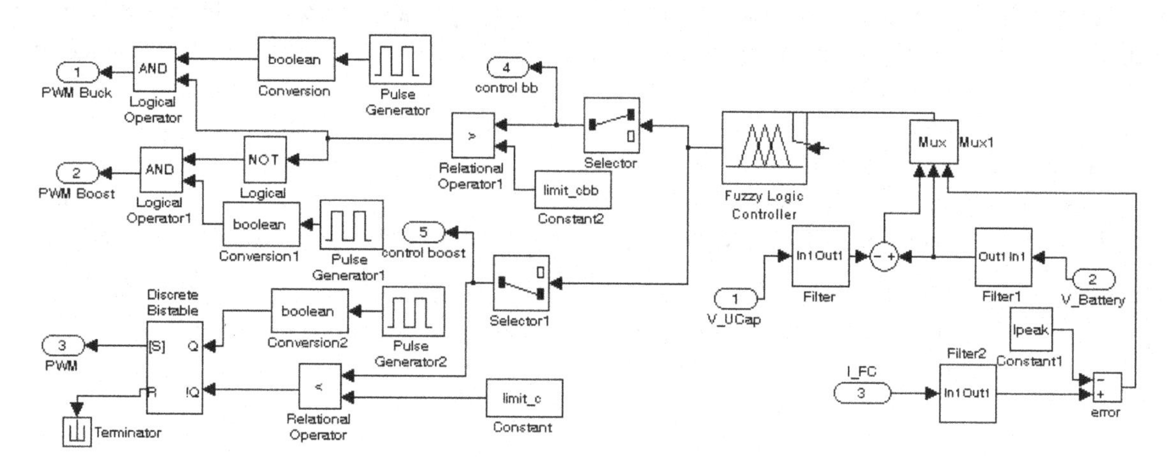

Figure 50. IFC FIS

```
a=30;c=10;mu=20;au=50;cu=25;du=30;eu=40;fu=80;lu=40;
d=50; e1=60;e2=65;e3=70;f=80;g=20;h=50;l=80;m=100;
fisbbb=newfis('control-bbb');
fisbbb=addvar(fisbbb,'input','Vbat-Vucap diff',[mu au]);
fisbbb=addvar(fisbbb,'input','battery voltage',[lu fu]);
fisbbb=addvar(fisbbb,'input','I FC-I FCref',[-a a]);
fisbbb=addvar(fisbbb,'output','command bb',[0 m]);
fisbbb=addvar(fisbbb,'output','PC Fuzzy Command',[0 m]);
fisbbb=addmf(fisbbb,'input',1,'Small','trapmf',[mu mu cu du]);
fisbbb=addmf(fisbbb,'input',1,'Nominal','trapmf',[cu du du eu]);
fisbbb=addmf(fisbbb,'input',1,'Big','trapmf',[du eu au au]);
....

fisbbb=addmf(fisbbb,'output',2,'Small','trapmf',[0 0 g h]);
fisbbb=addmf(fisbbb,'output',2,'Medium','trapmf',[g h h l]);
fisbbb=addmf(fisbbb,'output',2,'Big','trapmf',[h l m m]);
ruleListbbb=[1 1 1 3 3 1 1;1 2 1 3 2 1 1; 1 3 1 3 1 1 1 ;1 1 2 3 2 1
1;1 2 2 3 2 1 1;1 3 2 3 1 1 1;1 1 3 2 1 1 1;1 2 3 2 1 1 1; 1 3 3 2 1 1
1 ;2 1 1 3 3 1 1;2 2 1 3 2 1 1; 2 3 1 2 1 1 1 ;2 1 2 3 2 1 1;2 2 2 2 2
1 1; 2 3 2 1 1 1 1;2 1 3 2 1 1 1;2 2 3 1 1 1 1; 2 3 3 1 1 1 1 ;3 1 1 2
3 1 1;3 2 1 2 2 1 1; 3 3 1 2 1 1 1 ;3 1 2 1 2 1 1;3 2 2 1 2 1 1; 3 3 2
1 1 1 1;3 1 3 3 1 1 1;3 2 3 1 1 1 1; 3 3 3 1 1 1 1 ;];
fisbbb=addrule(fisbbb,ruleListbbb);
```

be compared with the original control FIS surfaces. Output control variables of the IFC FIS (*PC Fuzzy command* and *command bb*) are control signals for the boost and buck-boost converters, respectively.

Obviously, for the same input variables, the IFC FIS surface (figures 51 a and b) is similar to the boost PCC fuzzy controller FIS surface (figure 25), and the buck-boost fuzzy controller FIS surface (figure 44), respectively. These represent the IFC control legacy. Figures 51 (c and d) show that output variable "*command for buck-boost*" is highly dependent on the fuel cell current (this input variable in IFC Mux is named *I_FC–I_FC_ref*), and this represents a new IFC control characteristic. Figure 52 shows the simulation results for the IFC controller proposed for the same load power dynamic and EGS designed parameters, as in the previous section. All mention control requests are well fulfilled by the IFC proposed. The IFC performances are clearly evinced by simulation (extensively presented in Bizon, 2007). The IFC implementation advantages are:

- only two voltage transducer and one current transducer are used;
- the IFC structure is relatively simple, and can be implemented with a cheap microcontroller;
- the IFC control surface can be easily redesigned for other PEMFC, battery, ultracapacitor and load parameters, and the IFC performances remain the same.

Figure 52 shows the simulation results for the IFC controller proposed. The structure of the plot in figure 52 is the following: plot a) Command signals; plot b) Current and voltage signals for fuel cells, batteries and ultracapacitor stack; plot c) Output power signals for fuel cells, batteries and ultracapacitor stack.

Figure 51. IFC control surfaces for different input variables pairs © 2007 MEDJMC. Used by permission

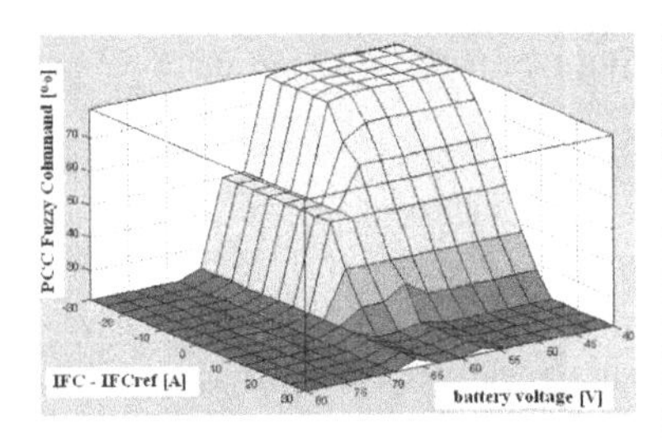

a. PCC Fuzzy command surface as function of "battery_voltage" and "I_FC-I_FC_ref" variables

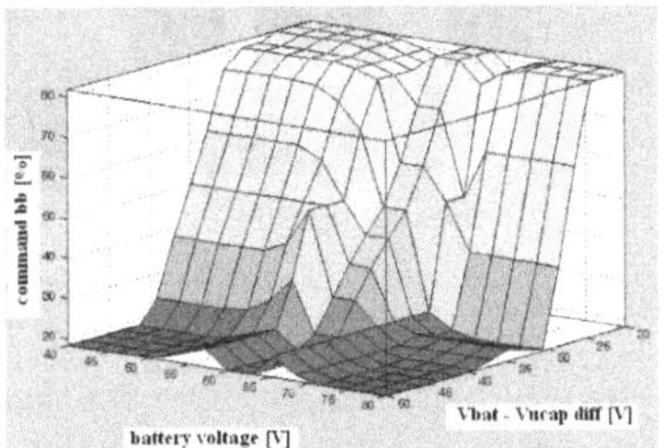

b. Buck-boost command surface as a function of "battery_voltage" and "Vbat-V_Ucap" variables

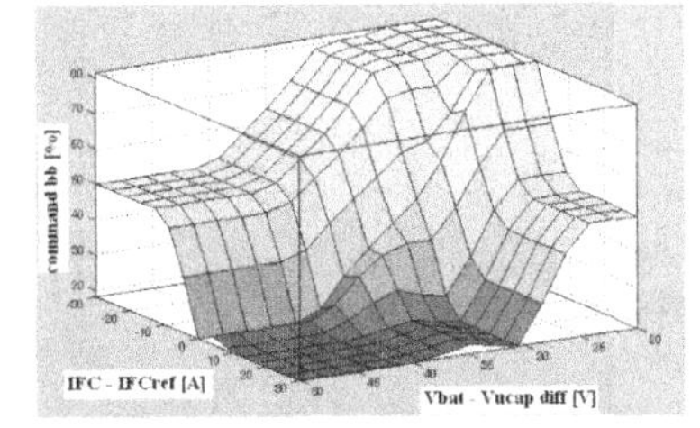

c. Buck-boost command surface as a function of "I_FC-I_FC_ref" and "Vbat-V_Ucap" variables

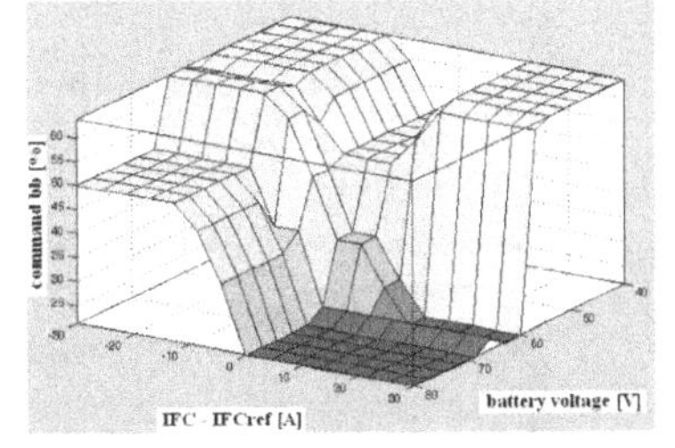

d. Buck-boost command surface as a function of "battery_voltage" and "I_FC-I_FC_ref" variables

For the names of the command signals are used the same names as in figures 48 and 49.

Control of the DC-DC Converter

In this section, a feed-forward fuzzy control for the PWM controller of the DC-DC converter, which directly uses a fuel cell as an energy source, will be presented. For a high load power dynamic the fuel cell output voltage, *V_PEMFC*, varies within a large DC range, and the stabilization of the high DC voltage can be a difficult task. The fuel cell output voltage can be used as feed-forward signal. The advantages of the forward control as far as the stability and robustness of the fuzzy control are compared with the best obtained results by using a well designed proportional-integral control (Bizon, & Oproescu, 2006c; Bizon, Mazare, Laurentiu, Oproescu, & Raducu, 2007).

Proportional-Integral Control

The dynamic behaviour of the DC-DC converter in figure 53 will be analyzed when the DC-DC controller is implemented by a Proportional-Integral (PI) control (figure 54). The input DC source is modelled by a Controllable Voltage Source, with dynamics within the range 25÷45 V given by Repeating Sequences block. The target of the controller is to stabilize the high DC output voltage (*V_High_DC*) to the voltage

Figure 52. Simulation results for the IFC controller © 2007 MEDJMC. Used by permission

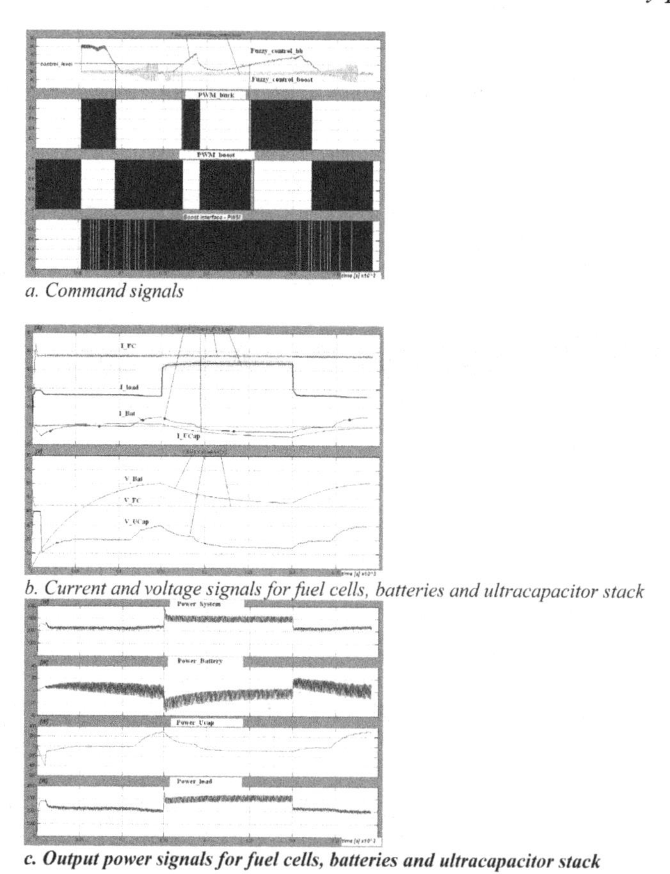

a. Command signals

b. Current and voltage signals for fuel cells, batteries and ultracapacitor stack

c. Output power signals for fuel cells, batteries and ultracapacitor stack

Figure 53. The DC-DC converter Simulink model

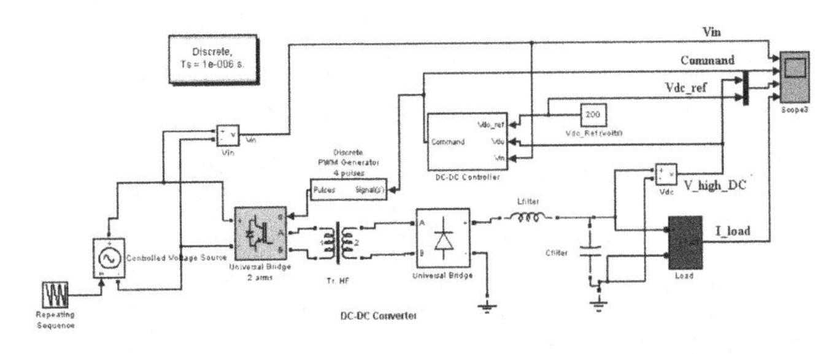

Figure 54. The PI controller for V_High_DC voltage bus

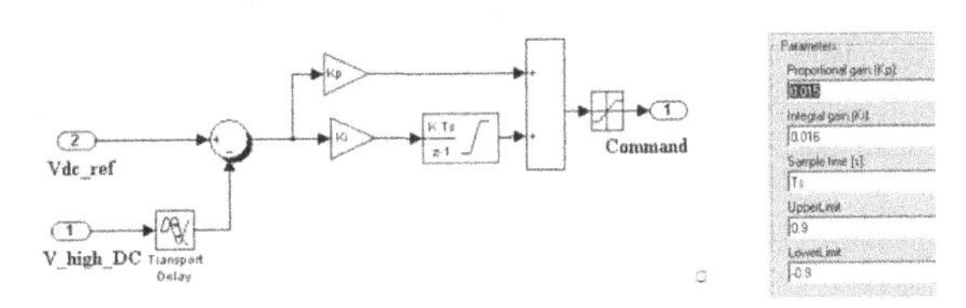

reference value (*Vdc_ref*=200V) when various perturbations are applied (Bizon, & Oproescu, 2006c). The stability of the system was analyzed for the input voltage step, *Vin*, within the range 25V÷45V, and using load values, *R_load*, that makes the converter operate in CCM or DCM. The optimal values for the PI controller parameters (*Kp, Ki*) are designed to ensure the stability of the system for different values within the working range of the perturbation parameters (*Vin, R_load*), when the values of the transformer ratio (*n_tr*) and output filter components (*L_filter, C_filter*) are given. For example, if the design parameters are: *L_filter*=1mH, *C_filter*=200μF, *n_tr*=30:350, *Vin*=25÷45 V, and *R_load*=40÷450 Ω, then the "best values" for DC-DC PI controller parameters are: *Kp*=0.4, *Ki*=0,3 (figure 55.a).

Under the same simulation conditions, the loss of stability for the DC-DC converter on DCM appears when PI controller parameters are *Kp*=0.04 and *Ki*=0,3 (figure 55.b). If we want to decrease the response time, then the gain coefficient, *Kp*, must be increased. But, to the same extent, the load value that ensures the stability of the DC-DC converter must be decreased. For example, when *Kp* is 0.04, the load must have a value up to 100 Ω in order to ensure the stability of the system for various study cases.

Fuzzy Control

For the fuzzy controller the inputs are (figure 56; for details, see Bizon, Mazare, Laurentiu, Oproescu, & Raducu, 2007):

- '*output voltage error*': difference between output voltage, *V_High_DC*, and reference voltage, *Vdc_ref*;
- '*input voltage*': continuous input voltage which may vary from 0 to 50V.

For the fuzzy controller the output variable, '*command*', is the control signal for the PWM command circuit (see figure 54).

The rules base is defined in order to ensure the stabilization of the converter output voltage, and the rules connection is done by min-max Zadeh operators and Mamdani implication.

Figure 55. Behaviour of the DC-DC converter with a PI controller for a input voltage step-down and a pulse load: Best transitory response (left); Loss of stability in DCM (right)

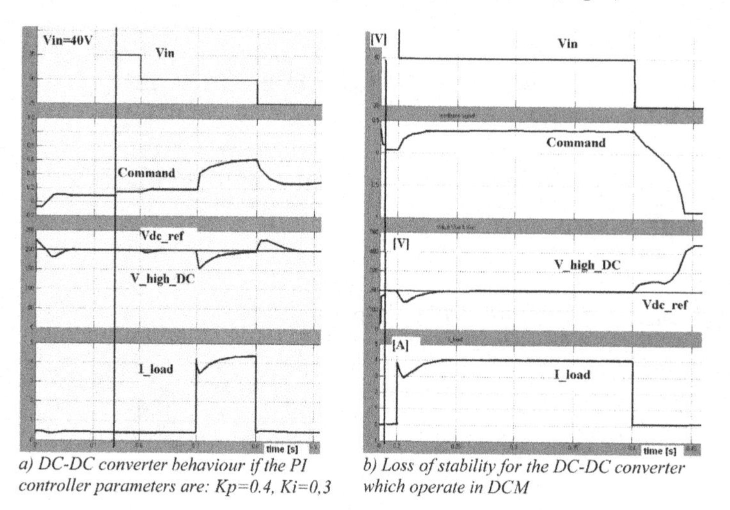

a) DC-DC converter behaviour if the PI controller parameters are: Kp=0.4, Ki=0,3

b) Loss of stability for the DC-DC converter which operate in DCM

Figura 56. Fuzzy controller for the DC-DC converter

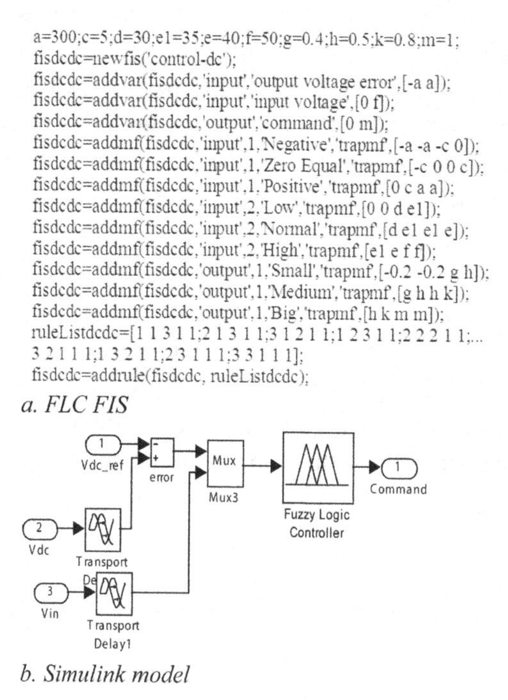

```
a=300;c=5;d=30;e1=35;e=40;f=50;g=0.4;h=0.5;k=0.8;m=1;
fisdcdc=newfis('control-dc');
fisdcdc=addvar(fisdcdc,'input','output voltage error',[-a a]);
fisdcdc=addvar(fisdcdc,'input','input voltage',[0 f]);
fisdcdc=addvar(fisdcdc,'output','command',[0 m]);
fisdcdc=addmf(fisdcdc,'input',1,'Negative','trapmf',[-a -a -c 0]);
fisdcdc=addmf(fisdcdc,'input',1,'Zero Equal','trapmf',[-c 0 0 c]);
fisdcdc=addmf(fisdcdc,'input',1,'Positive','trapmf',[0 c a a]);
fisdcdc=addmf(fisdcdc,'input',2,'Low','trapmf',[0 0 d e1]);
fisdcdc=addmf(fisdcdc,'input',2,'Normal','trapmf',[d e1 e1 e]);
fisdcdc=addmf(fisdcdc,'input',2,'High','trapmf',[e1 e f f]);
fisdcdc=addmf(fisdcdc,'output',1,'Small','trapmf',[-0.2 -0.2 g h]);
fisdcdc=addmf(fisdcdc,'output',1,'Medium','trapmf',[g h h k]);
fisdcdc=addmf(fisdcdc,'output',1,'Big','trapmf',[h k m m]);
ruleListdcdc=[1 1 3 1 1;2 1 3 1 1;3 1 2 1 1;1 2 3 1 1;2 2 2 1 1;...
3 2 1 1 1;1 3 2 1 1;2 3 1 1 1;3 3 1 1 1];
fisdcdc=addrule(fisdcdc, ruleListdcdc);
```

a. FLC FIS

b. Simulink model

The results of the simulation evince better control performance for the fuzzy DC-DC controller (figure 57). The feed-forward control generates quick changes for the command signal, so the voltage variations above the high DC bus are much lower than they are for the PI controller.

The PI control is relatively difficult to design considering the non-linearity of the DC-DC converter system. In general, the PI parameters adjustment is done by mixed techniques of modelling and simu-

Figure 57. DC-DC converter behaviour for a step-down input voltage and a pulse load using a fuzzy logic controller

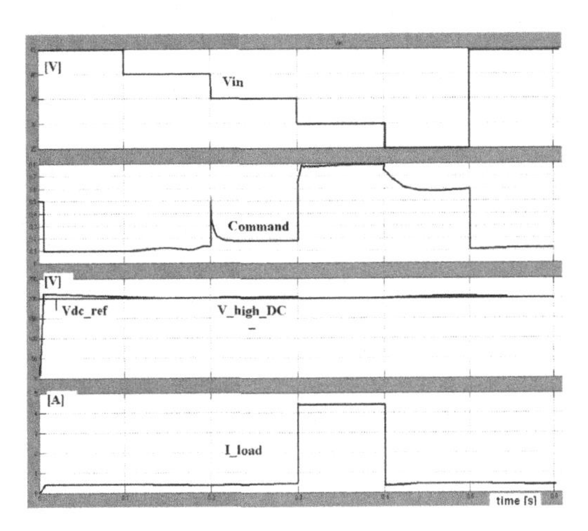

lation. The feed-forward fuzzy control removes the disadvantage of PI adjustment, and it also has the additional advantage of a fast response time to input voltage and load variations. With only nine simple fuzzy rules, the fuzzy control provides high control performances: small stationary error; over-gain up to 5% of the voltage reference value; fast response time; and robustness (output oscillations to step variation of input or load are avoided).

Control of the DC-AC Converter

The effect of the inverter current ripple on the performance of a fuel cell stack (PEMFC life cycle and PEMFC market parameters: reliability, durability and stability) has not been investigated thoroughly, and should be carefully studied (Gemmen, 2003; Woojin, Gyubum, Prasad, & Jo, 2004; Shireen, Kulkarni, & Arefeen, 2006; Fontes, Turpin, Astier, & Meynard, 2007).

The PWM pure sine control is used for medium and high power EGS units, and the PWM modified sine control is habitually used for low power EGS units. The aim of the DC-AC converter control is to reduce the input current harmonics using the on-line control of the PWM modulation indexes; see Kieferndorf, Forster, & Lipo (2004) for tri-phase inverter harmonic reduction with an unbalanced load. Another aim of the control is the EGS grid connection; see Bizon, & Oproescu (2007b) as an example of an intelligent forward control that maintains the output voltage effective value almost constant for a variable supply voltage within a given range.

The experimental PEMFC current ripple is shown in figure 58 for an EGS without a boost power interface. The fuel cell voltage and current are acquired with a USB acquisition board for further signals processing (figure 58 – bottom). For example, if *I_PEMFC* = 100A, then 100 Hz current harmonic normalized value is 14A/100A=14% (without boost power interface) and 10% (with boost power interface). It is demonstrated that LF normalized PEMFC current ripple must be less than 30% for energy efficient operation (see Liu, &Lai (2007); 10% limit is suggested in Gemmen (2003)).

An innovative converter topology as multi-port energy interface and appropriate control is proposed in Fernandez, Sebastian, Hernando, & Martin-Ramos (2006). PEMFC modelling in order to obtain an

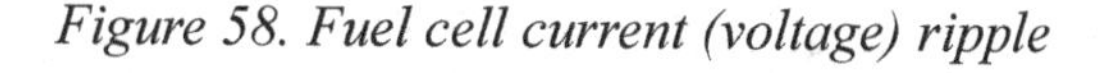

Figure 58. Fuel cell current (voltage) ripple

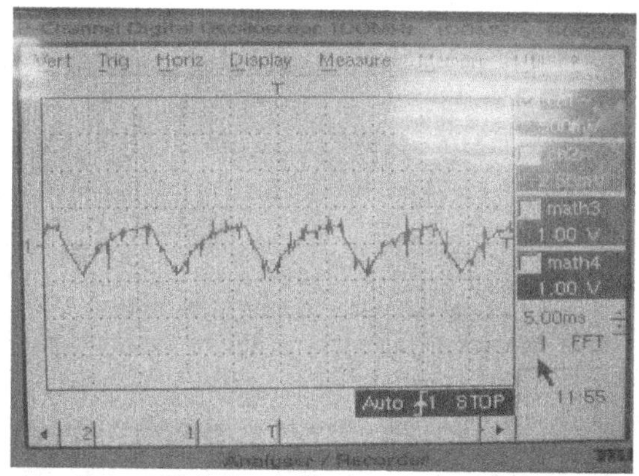

a. using a digital analyser

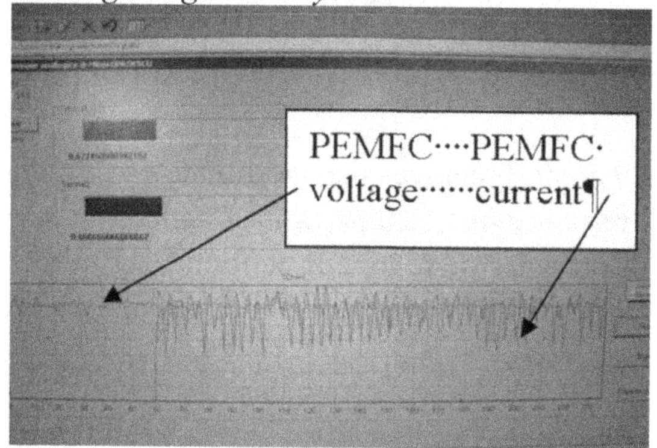

b. using a USB interface

electrical equivalent circuit model of a fuel cell to evaluate the effects of inverter current ripple is presented in Woojin, & Jo (2006).

Various control techniques of minimizing the PEMFC current ripple are proposed: passive/active filter, intelligent control, power spectrum by randomized control, chaotification of control, non-linear control etc. (Yang, 2001; Banerjee, Kastha, & SenGupta, 2002; Wong, Chan, & Ma, 2002; Tse, Ng, Chung, & Hui, 2003; Kuisma, 2003; Morel, Bourcerie, & Chapeau-Blondeau, 2005 ; Mogel, Krupar, & Schwarz, 2005).

Recent researches have been conducted in one possible area for chaos application: *reducing electromagnetic interference (EMI) in switch-mode power supplies*. Switch-mode power supplies, which are notorious generators of both conducted and radiated EMI, owing to high rates of change in voltage and current which, are necessary for efficient operation (Mihali, & Kos, 2006; Li, Qiu, Chen, 2006; Bizon, & Voukalis, 2007; Bizon, in press). However, in order to include this possibility into the practice of engineering, some theoretical issues needed to be solved. First, *design procedures for chaotically operated converters must be mathematically formalized.* Secondly, one needs *a theory to predict the structure of the power spectrum for a chaotically operated power converter using the control methods proposed.* Thirdly, it is well-known that in *most chaotic systems there are periodic windows in the parameters space*, and a slightly inadvertent change of a parameter can bring the system out of chaos, and this aspect must be further analyzed.

CONCLUSIONS AND FUTURE RESEARCH TRENDS

In this book chapter an intelligent management of EGS power flows using fuzzy control concepts is analyzed in order to reduce the PEMFC stress concerning the output load power dynamic. Conclusions are given at the end of each section, and are summarized as follows:

- The boost converter is, ideally, a power interface between PEMFC and DC low voltage bus, and ensures low PEMFC current ripple if this EGS variables are used as control variables.
- The ultracapacitor stack can compensate high load power dynamic at the DC low voltage bus. The stack may be connected directly to the DC low voltage bus, or through a bi-directional converter (a buck-boost converter). The last variant is useful for HEV applications, where an auxiliary converter can be used to charge the ultracapacitor stack from the regenerative power flow. The control variables may be the DC low voltage and voltage above the ultracapacitor stack, and the current protection is done by limitation of the ultracapacitor current.
- The limits for a stable control increase if a batteries stack is connected to the DC low voltage bus. In the battery charge mode, the control law must consider the following control variables: batteries stack voltage and current.
- The integrate fuzzy control is robust, compact and sensorless; EGS behaviour is improved, too.
- The DC high voltage bus must be very well stabilized in order to have a grid-connected EGS. If the PEMFC is directly connected to the DC low voltage bus, the DC-DC converter controller must include DC low voltage as a control variable. A fuzzy control with two control variables (DC low voltage and DC high voltage) can be used, too.

Future research trends concerning EGS are concentrating on the following directions:

- Analysis of stationary fuel cell dynamic ramping capabilities and battery - ultracapacitor power/energy storage capabilities for Hybrid Electrical Vehicle (HEV) (Meacham, Jabbari, Brouwer, Mauzey, & Samuelsen, 2006). A high fuel cell ramp rate is important for the reduction of grid impacts, too.
- Control techniques for maximum power point (MPP) tracking for fuel cell power plants: adaptive control with neuronal network supervising; MPP finding by genetic or chaotic algorithms. The power ripple becomes lower and lower when the operation point gets closer to MPP (Zhong, Huo, Zhu, Cao, & Ren, in press). To ensure an efficient energy operation of the PEMFC stack the PEMFC current ripple must be minimized by means of an active power filter (Mazumder, Burra, & Acharya, 2007; Liu, & Lai, 2007; Bizon, Oproescu, & Raducu, 2008; Bizon, Raducu, & Oproescu, 2008; Auld, Mueller, Smedley, Samuelsen, & Brouwer, 2008), or by using a Z-source inverter topology (Shen, Joseph, Wang, Peng, & Adams, 2007).
- Energy management strategies for HEV EGS power flows (fuel cell / battery / ultracapacitor hybrid system) to improve the operation efficiency, and the response dynamics (Corbo, Corcione, Migliardini, & Veneri, 2006; Williams, Keith, Marcel, Haskew, Shepard, & Todd, 2007; Paladini, Donateo, de Risi, & Laforgia, 2007; Pischinger, Schonfelder, Ogrzewalla, 2007). The fuel cell system requirements regarding transient operation and their dependence on the system structure must be further analyzed for HEV application. HEV modelling is also important in order to analyse the dynamic requirements for fuel cell technology used in automotive applications, and to

understand the relationships between cell structure, physical-chemical properties, operating conditions and EGS performances.

- Designing and developing modular fuel cell stacks for various applications (Rajalakshmi, N., Pandiyan, S., & Dhathathreyan, K. S. (in press). The stability analysis for series/parallel PEMFC modules and associated power interfaces is a challenge for specialists.

A number of future research trends concerning PEMFC stack faults and diagnostic methods (Basu, 2007) are:

- Estimation of the PEMFC's membrane water content using observers (Gorgun, Arcak, & Barbir, 2006);
- In situ diagnostic tools: Electrochemical impedance spectroscopy and PEMFC Power Spectral Density for the AC PEMFC voltage using a sinusoidal signal as small perturbation (Ramschak, Peinecke, Prenninger, Schaffer, & Hacker, 2006);
- Advanced diagnostic tools: methods of identifying the faults that cannot be detected directly by monitoring, or faults that need some type of diagnosis bz the use of observers. Four types of faults in PEMFC's are considered: faults in the air fan, faults in the refrigeration system, growth of the fuel crossover, and faults in the hydrogen pressure. The effects of these faults are included in the mathematical model to analyze the behaviour of the fuel cell system in fault operation conditions in order to find a fault diagnosis model. A neuro-fuzzy network fault diagnostic system for PEMFC using patterns (which record the fault effects) and fuzzy logic reason can be constructed (Riascos, Simoes, & Miyagi, 2007). The neuronal model defines the cause-effect relationship among the variables, and the fuzzy memberships capture the numerical dependence among these variables. Finally, the neuro-fuzzy network is used to conduct the diagnosis of fault causes in the PEMFC model based on the effects observed.

ACKNOWLEDGMENT

The national research projects, CNCSIS Grant #570 and CEEX Project #226, have supported part of the research presented in this book chapter.

REFERENCES

Agrawal, J. P. (2001). *Power electronic systems: theory and design*. Upper Saddle River, N.J.: Prentice Hall.

Andersen, C. A., Christensen, M. O., & Korsgaard, A. R. (2002). *Design and Control of Fuel Cell System for Transport Application* (Tech. Rep.). Aalborg University, Project Group EMSD 10 - 11A, Denmark.

Auld, A., Mueller, F., Smedley, K., Samuelsen, S., & Brouwer, J. (2008). Applications of one-cycle control to improve the interconnection of a solid oxide fuel cell and electric power system with a dynamic load. *Journal of Power Sources*, *179*(1), 155–163. doi:10.1016/j.jpowsour.2007.12.072

Balog, R. S., & Krein, P. T. (2005). Commutation technique for high- frequency link cycloconverter based on state-machine control. *IEEE Power Electronics Letters, 3*(3), 101–104. doi:10.1109/LPEL.2005.858422

Banerjee, S., Kastha, D., & SenGupta, S. (2002). Minimising EMI problems with chaos. *Proceedings of the International Conference on Electromagnetic Interference and Compatibility,* (pp. 162-167). Retrieved December 15, 2002, from http://ieeexplore.ieee.org

Banerjee, S., & Verghese, G. C. (2001). *Nonlinear phenomena in power electronics: attractors, bifurcations, chaos, and nonlinear control.* New York: IEEE Press.

Basu, S. (2007). *Recent Trends in Fuel Cell Science and Technology*. New York: Springer.

Batarseh, I. (2003). *Power electronic circuits*. Hoboken, NJ: John Wiley & Sons Inc.

Bergveld, H. J., Wanda, S. J., Kruijt, S., & Notten, P. H. L. (2002). *Battery Management Systems: Design by Modeling*. Boston: Kluwer Academic Press.

Bizon, N. (2004). Design Consideration for Voltage Regulator Modules used in Automotive Control. In Society of Automotive Engineers of Romania (Ed.) under SAE and FISITA patronage, *International Automotive Congress - CONAT 2004* (pp. 25-29). Brasov, Romania: Transilvania University Press.

Bizon, N. (2007). Intelligent Integrated Control of the Power Flows into an Energy Generation System. *Mediterranean Journal of Measurement and Control, 3*(3), 113–125.

Bizon, N. (in press). Chaotification of the buck converter using a modified Chua's diode. *Fuzzy Systems and AI journal - Reports and Letters*.

Bizon, N., Laurentiu, I., Mazare, A., & Oproescu, M. (2007). Analyze of the Feed-Forward Control for a Pure Sine Inverter. In N. Bizon (Ed.), *International Conference on Electronics, Computers and Artificial Intelligence - ECAI'07* (Vol. 2, pp. 71-79). Bucharest: MatrixROM & University of Pitesti Press.

Bizon, N., Lefter, E., & Oproescu, M. (2007). Modeling and Control of the Energy Sources Power Interface for Automotive Hybrid Electrical System. In Society of Automotive Engineers – SAE of Serbia (Ed.) under FISITA patronage, *21st JUMV International Automotive Conference on Science and Motor Vehicles 2007* (pp. 46-50). Belgrade, Serbia: SAE Press.

Bizon, N., Mazare, A., Laurentiu, I., Oproescu, M., & Raducu, M. (2007). Fuzzy Control of the DC-DC Converter used as Power Interface for a Fuel Cell. In ICSI (Ed.), *International Conference on Progress in Cryogenics and Isotopes Separation* (pp. 169-172). Ramnicu Valcea, Romania: ICSI Press.

Bizon, N., & Oproescu, M. (2004). Hysteretic Fuzzy Control of the Power Interface Converter. In Romanian Academy, (Ed.), *Fuzzy Systems and AI journal - Reports and Letters,* (pp. 139-158). Iasi, Romania: Publishing House of the Romanian Academy.

Bizon, N., & Oproescu, M. (2005a). Hysteretic Fuzzy Control of the Boost Converter. *Electronics, Computers and Artificial Intelligence: University of Pitesti Journal – ECAI'05, 5*(S2), 1-10.

Bizon, N., & Oproescu, M. (2005b). Clocked hysteretic fuzzy control of the boost converter. *Electronics, Computers and Artificial Intelligence - ECAI'05: University of Pitesti Journal – Electronics and Computer Science, 5*(S2), 11-20.

Bizon, N., & Oproescu, M. (2006a). Time Equivalent Systems. *University of Pitesti Journal - Electronics and Computer Science, 6*(1), 12-36.

Bizon, N., & Oproescu, M. (2006b). Modeling and Control of the PEMFC Power Interface. In ICSI (Ed.), *International Conference on Progress in Cryogenics and Isotopes Separation* (pp. 155-168). Ramnicu Valcea, Romania: ICSI Press.

Bizon, N., & Oproescu, M. (2006c). Control of the DC-DC Converter used into Energy Generation System. In ICSI (Ed.), *International Conference on Progress in Cryogenics and Isotopes Separation* (pp. 173-177). Ramnicu Valcea, Romania: ICSI Press.

Bizon, N., & Oproescu, M. (2007a). Energy Generation System Behaviour using a Clocked Fuzzy Peak Current Control. In European Power Electronics Association & University of Aalborg (Eds.), *12th European Conference on Power Electronics and Application - EPE 2007: IEEE Catalog Number 07EX1656C* (pp. 1-8). Denmark: University of Aalborg Press.

Bizon, N., & Oproescu, M. (2007b). Feed-Forward Control of the PWM Sine Inverter. In Romanian Academia (Ed.), *European Conference H2_Fuel_Cells_Millennium _ Convergence* (pp. 35-45). Bucharest: IPA Publishing House.

Bizon, N., & Oproescu, M. (2007c). *Power converters used in Energy Generation Systems (Convertoare de Putere utilizate in Sistemele de Generare a Energiei)*. Bucharest, Romania: MatrixROM.

Bizon, N., & Oproescu, M. (in press). Instabilities Analysis of an Energy Generation System with a Fuzzy Hysteretic Control. *Fuzzy Systems and AI journal - Reports and Letters*.

Bizon, N., Oproescu, M., & Raducu, M. (2006). Cycloconverter Operation in the Low-Cost Energy Generation Systems. *University of Pitesti Journal - Electronics and Computer Science, 6*(1), 24-36.

Bizon, N., Oproescu, M., & Raducu, M. (2008), Fuzzy bang-bang control of a switching voltage regulator. In IEEE Computer Society (Ed.), *16th IEEE International Conference on Automation, Quality and Testing, Robotics – AQTR, IEEE Catalog Number CFP08AQT-CDR, Vol. 2* (pp. 192-197). Cluj-Napoca, Romania: IEEE Press.

Bizon, N., Raducu, M., & Oproescu, M. (2008), Fuel cell current ripple minimization using a bi-buck power interface. In EPE-PEMC (Ed.), *13th International Power Electronics and Motion Control Conference, IEEE Catalog Number CFP0834A-CDR* (pp. 621-628). Poznan, Poland: IEEE Press.

Bizon, N., Sofron, E., & Oproescu, M. (2005). Some Aspects of the PEMFC – Battery Interface Simulation in Automotive Applications. *International Conference for Road Vehicles - CAR2005*. Retrieved December 11, 2005, from http://www.fisita.com/publications/papers

Bizon, N., Sofron, E., & Oproescu, M. (2006). An Investigations into the Fast- and Slow – Scale Instabilities of an Energy Generation System with a Fuzzy Hysteretic Control. In H. N. Teodorescu (Ed.) *European Conference on Intelligent Systems and Technologies - ECIT2006; Advances in Intelligent Systems and Technologies* (pp. 19-36). Iasi, Romania: Performantica.

Bizon, N., Sofron, E., & Oproescu, M. (2007). Intelligent Control of the Power Flows on an Energy Generation System. In International Institute of Informatics and Systemics – IIIS (Ed.), *Fifth Multi-Conference on Systemics, Cybernetics and Informatics – EIC2007, Vol 5* (pp. 314-319). Orlando, FL: IIIS Press.

Bizon, N., Sofron, E., Oproescu, M., & Raducu, M. (2007). Multi-stage Inverter Topologies for an Energy Generation Systems, *13th International Symposium on Modeling, Simulation and System's Identification*. Retrieved May 10, 2007, from http://www.simsis.ugal.ro/simsis13

Bizon, N., & Voukalis, D. C. (2007). *Fundamentals of Electromagnetic Compatibility: Theory and Practice*. Bucharest, Romania: MatrixROM Press.

Bizon, N., Zafiu, A., & Oproescu, M. (2005). PEMFC modeling using a genetics algorithm for parameters tuning, In Bibliotheca Publishing House (Ed.), *International Symposium on Electrical Engineering - ISEE 2005* (pp. 163-169). Targoviste, Romania: Valahia University.

Corbo, P., Corcione, F. E., Migliardini, F., & Veneri, O. (2006). Energy management in fuel cell power trains. *Energy Conversion and Management*, *47*(18-19), 3255–3271. doi:10.1016/j.enconman.2006.02.025

Crompton, T. R. (2003). *Battery reference book*. Oxford, UK: Elsevier Science - Newnes.

Fernandez, A., Sebastian, J., Hernando, M. M., & Martin-Ramos, J. A. (2006). Multiple output AC/DC converter with an internal DC UPS. *IEEE Transactions on Industrial Electronics*, *53*(1), 296–304. doi:10.1109/TIE.2005.862220

Fontes, G., Turpin, C., Astier, S., & Meynard, T. (2007). Interactions between fuel cells and power converters: Influence of current harmonics on a fuel cell stack. *IEEE Transactions on Power Electronics*, *22*(2), 670–678. doi:10.1109/TPEL.2006.890008

Gansky, K. (2002). *Rechargeable batteries applications handbook*. Oxford, UK: Elsevier Science - Butterworth Heinemann.

Gemmen, R. S. (2003). Analysis for the effect of the ripple current on fuel cell operating condition. *Journal of Fluids Engineering*, *125*(3), 576–585. doi:10.1115/1.1567307

Gorgun, H., Arcak, M., & Barbir, F. (2006). An algorithm for estimation of membrane water content in PEM fuel cells. *Journal of Power Sources*, *157*(1), 389–394. doi:10.1016/j.jpowsour.2005.07.053

Haraldsson, K., & Wipke, K. (2004). *Evaluating PEM Fuel Cell System Models* (Tech. Rep. No. 1). Golden, CO: National Renewable Energy Laboratory, DOE Hydrogen and Fuel Cells and Infrastructure Technologies Program.

Kieferndorf, F. D., Forster, M., & Lipo, T. A. (2004). Reduction of DC-bus capacitor ripple current with PAM/PWM converter. *IEEE Transactions on Industry Applications*, *40*(2), 607–614. doi:10.1109/TIA.2004.824495

Kuisma, M. (2003). Variable frequency switching in power supply EMI-control: an overview. *Aerospace and Electronic Systems Magazine*, *18*(12), 18–22. doi:10.1109/MAES.2003.1259021

Larminie, J., & Dicks, A. (2003). *Fuel Cell Systems Explained- Second Edition*. Chichester, UK: John Wiley & Sons Ltd.

Li, Z.-Z., Qiu, S.-S., & Chen, Y.-F. (2006). Experimental Study on the Suppressing EMI Level of DC-DC Converter with Chaotic Map. *Proceedings of the Chinese society for electrical engineering, 26*(5), 76-81. Retrieved December 20, 2006, from http://ieeexplore.ieee.org

Liu, C., & Lai, J. S. (2007). Low Frequency Current Ripple Reduction Technique with Active Control in a Fuel Cell Power System with Inverter Load. *IEEE Transactions on Power Electronics*, *22*(4), 1453–1463. doi:10.1109/TPEL.2007.900505

Mazumder, S. K., Burra, R. K., & Acharya, K. (2007). A Ripple-Mitigating and Energy-Efficient Fuel Cell Power-Conditioning System. *IEEE Transactions on Power Electronics*, *22*(4), 1429–1436. doi:10.1109/TPEL.2007.900598

Meacham, J. R., Jabbari, F., Brouwer, J., Mauzey, J. L., & Samuelsen, G. S. (2006). Analysis of stationary fuel cell dynamic ramping capabilities and ultracapacitor energy storage using high resolution demand data. *Journal of Power Sources*, *156*(2), 472–479. doi:10.1016/j.jpowsour.2005.05.094

Mihali, F., & Kos, D. (2006). Reduced Conductive EMI in Switched-Mode DC–DC Power Converters without EMI Filters: PWM Versus Randomized PWM. *IEEE Transactions on Power Electronics*, *21*(6), 1783–1794. doi:10.1109/TPEL.2006.882910

Mogel, A., Krupar, J., & Schwarz, W. (2005). EMI performance of spread spectrum clock signals with respect to the IF bandwidth of the EMC standard. *Proceedings of the 2005 European Conference on Circuit Theory and Design, 1*, 169-172. Retrieved December 25, 2005, from http://ieeexplore.ieee.org

Mohan, N., Undeland, T. M., & Robbins, W. P. (2002). *Power Electronics: Converters, Applications, and Design*. Hoboken, NJ: John Wiley & Sons Inc.

Morel, C., Bourcerie, M., & Chapeau-Blondeau, F. (2005). Improvement of power supply electromagnetic compatibility by extension of chaos anti-control. *Journal of Circuits, Systems, and Computers*, *14*(4), 757–770. doi:10.1142/S0218126605002556

National Energy Technology Laboratory, & U.S. Department of Energy (2004). *Fuel Cell Handbook*. USA DOE/NETL: EG&G Services Parsons, Inc. Science Applications International Corporation.

National Energy Technology Laboratory. (2005). *NETL published fuel cell specifications for Future Energy Challenge 2005 Competition*. Retrieved June 1, 2005, from http://www.netl.doe.gov.

Ortuzar, M., Dixon, J., & Moreno, J. (2003). Design, Construction and Performance of a Buck-Boost Converter for an Ultracapacitor-Based Auxiliary Energy System for Electric Vehicles. In IEEE Industrial Electronics Society (Ed.), *IECON2003* (Vol. 3, pp. 2889 – 2894), Roanoke, VA.

Paladini, V., Donateo, T., de Risi, A., & Laforgia, D. (2007). Super-capacitors fuel-cell hybrid electric vehicle optimization and control strategy development. *Energy Conversion and Management*, *48*(11), 3001–3008. doi:10.1016/j.enconman.2007.07.014

Parui, S. (2003). *Bifurcation in dc-dc converters: Effects of transition from continuous conduction mode to discontinuous conduction mode and feedback loop delay*. Published doctoral dissertation, IIT Kharagpur, India.

Pathapati, P. R., Xue, & Tang, J. (2005). A new Dynamic Model for Predicting Transient Phenomena in a PEM Fuel Cell System. *Renewable . The Energy Journal (Cambridge, Mass.)*, *30*(1), 1–8.

Pischinger, S., Schonfelder, C., & Ogrzewalla, J. (2007). Analysis of dynamic requirements for fuel cell systems for vehicle applications. *Journal of Power Sources*, *154*(2), 420–427. doi:10.1016/j.jpowsour.2005.10.037

Raducu, M., & Bizon, N. (2007). Efficient Energy Generation System using a Thiristors Inverter Topology. In N. Bizon (Ed.), *International Conference on Electronics, Computers and Artificial Intelligence - ECAI'07* (Vol. 2, pp. 67-70). Bucharest: MatrixROM & University of Pitesti Press.

Rajalakshmi, N., Pandiyan, S., & Dhathathreyan, K. S. (in press). Design and development of modular fuel cell stacks for various applications. *International Journal of Hydrogen Energy*.

Ramschak, E., Peinecke, V., Prenninger, P., Schaffer, T., & Hacker, V. (2006). Detection of fuel cell critical status by stack voltage analysis. *Journal of Power Sources*, *157*(2), 837–840. doi:10.1016/j.jpowsour.2006.01.009

Rashid, M. H. (2003). *Power Electronics: Circuits, Devices and Applications*. Upper Saddle River, NJ: Pearson Education.

Riascos, L. A. M., Simoes, M. G., & Miyagi, P. E. (2007). A Bayesian network fault diagnostic system for proton exchange membrane fuel cells. *Journal of Power Sources*, *165*(1), 267–278. doi:10.1016/j.jpowsour.2006.12.003

Shen, M., Joseph, A., Wang, J., Peng, F. Z., & Adams, D. J. (2007). Comparison of Traditional Inverters and Z-Source Inverter for Fuel Cell Vehicles. *IEEE Transactions on Power Electronics*, *22*(4), 1437–1452. doi:10.1109/TPEL.2007.900505

Shireen, W., Kulkarni, R. A., & Arefeen, M. (2006). Analysis and minimization of input ripple current in PWM inverters for designing reliable fuel cell power systems. *Journal of Power Sources*, *156*(2), 448–454. doi:10.1016/j.jpowsour.2005.06.012

Stamps, A. T., & Gatzke, E. P. (2005). Dynamic Modeling of a Methanol Reformer - PEMFC for Analysis and Design. *Journal of Power Sources*, *161*(1), 356–370. doi:10.1016/j.jpowsour.2006.04.080

Stimming, U., de Haart, L. G. S., & Meeusinger, J. (2005). *Fuel Cell Systems: PEMFC for Mobile and SOFC for Stationary Application*. Berlin: Wiley-VCH Verlag GmbH.

Thounthong, P., Rael, S., & Davat, B. (2005). Fuel Cell and Supercapacitors for Automotive Hybrid Electrical System. *ECTI Transactions on Electrical Engineering, Electronics, and Communications*, *1*(3), 20–30.

Tse, C. K. (2003). *Complex behaviour of switching power converters*. Boca Raton, FL: CRC Press, Taylor & Francis Group.

Tse, K. K., Ng, R. W. M., Chung, H. S. H., & Hui, S. Y. R. (2003). An evaluation of the spectral characteristics of switching converters with chaotic carrier-frequency modulation. *IEEE Transactions on Industrial Electronics*, *50*(1), 171–182. doi:10.1109/TIE.2002.807659

Williams, K. A., Keith, W. T., Marcel, M. J., Haskew, T. A., Shepard, W. S., & Todd, B. A. (2007). Experimental investigation of fuel cell dynamic response and control. *Journal of Power Sources*, *163*(2), 971–985. doi:10.1016/j.jpowsour.2006.10.016

Wong, H., Chan, Y., & Ma, S. W. (2002). Electromagnetic interference of switching mode power regulator with chaotic frequency modulation. *Proceedings of the 23rd Int. Conference on Microelectronics - MIEL 2002*, 2, 577 – 580. Retrieved December 15, 2002, from http://ieeexplore.ieee.org.

Woojin, C., Gyubum, J., Prasad, E. N., & Jo, H. W. (2004). An Experimental Evaluation of the Effects of Ripple Current Generated by the Power Conditioning Stage on a Proton Exchange Membrane Fuel Cell Stack. *Journal of Materials Engineering and Performance*, *13*(3), 257–264. doi:10.1361/10599490419144

Woojin, C., & Jo, H. W. (2006). Development of an equivalent circuit model of a fuel cell to evaluate the effects of inverter ripple current. *Journal of Power Sources*, *158*(2), 1324–1332. doi:10.1016/j.jpowsour.2005.10.038

Xue, Y., Chang, L., & Kjær, S. B. (2004). Topologies of Single-Phase Inverter for Small Distributed Generators: an Overview. *IEEE Transactions on Power Electronics*, *19*(5), 1305–1313. doi:10.1109/TPEL.2004.833460

Yang, K. (2001). Spread-Spectrum DC-DC Converter Combats EMI. *Electronics Design*, (pp. 86-88). Retrieved October 23, 2001, from http://electronicdesign.com/Articles.

Zhong, Z.-D., & Huo, H.-B. (in press). Zhu, -J., Cao, G.-Y., & Ren [Adaptive maximum power point tracking control of fuel cell power plants. *Journal of Power Sources*.]. *Y (Dayton, Ohio).*

Chapter 3
Knowledge Management for Electric Power Utility Companies

Campbell Booth
University of Strathclyde, UK

ABSTRACT

This chapter will present an overview of the challenges presented to modern power utility companies and how many organizations are facing particularly pressing problems with regards to an ageing workforce and a general shortage of skills; a situation that is anticipated to worsen in the future. It is proposed that knowledge management (KM) and decision support (DS) may contribute to a solution to these challenges. The chapter describes the end-to-end processes associated with KM and DS in a power utility context and attempts to provide guidance on effective practices for each stage of the described processes. An overview of one particular power utility company that has embraced KM is presented, and it is proposed that the function of asset management within power utilities in particular may benefit from KM. The chapter focuses not only on KM techniques and implementation, but, equally, if not more importantly, on the various cultural and behavioural aspects that are critical to the success of any KM/DS initiative.

INTRODUCTION

Knowledge management (KM) and decision support (DS) provide opportunities to reduce many forms of risk, to improve business processes, to increase workforce engagement and morale, to enhance training programs and to ultimately deliver financial benefits. Furthermore, there are significant problems presently being experienced throughout the power utility industry associated with the retirement of experienced employees and a shortage of skilled personnel being available to enter the business. These problems are anticipated to worsen in the future. This situation has heightened interest in the potential

DOI: 10.4018/978-1-60566-737-9.ch003

for KM and DS as tools with which such problems can be addressed.

However, it is very important that a measured and systematic approach is taken to KM and DS. There is certainly no single solution that can be used in all circumstances, and there are many examples of KM projects and applications that have resulted in a failure to meet expectations and, in some cases, in the disenfranchisement of personnel. This chapter will provide an overview, and hopefully some guidance, into the various components of KM and will illustrate, through examples and references, the main steps and considerations that must be taken when planning, implementing, maintaining and evolving KM and DS initiatives.

The Problem

Executives in power utility companies are presented with many strategic issues and challenges. These include: privatization and the emergence of new market entrants; an almost continual process of mergers, acquisitions, formation and dissolution of corporate partnerships; changing legislation and government policy, e.g. the UK government's decision to permit future use of nuclear energy, which was previously discounted as a future power generation option (British Broadcasting Corporation, 2007); the drive for green energy and increasing penetration of distributed generation; the demand for increased shareholder return, countered by mounting pressure on wholesale fuel prices and negative publicity and consumer sentiment associated with increasing consumer energy prices; increasing moral and legal obligations with respect to health and safety; and many others.

All of these issues and challenges must be faced in an environment where there is a dwindling skills resource, exacerbated by the fact that there is an increasing shortage of skilled professional engineers. In (American Public Power Association, 2005), it is stated that, with reference to 111 organizations' responses to a survey carried out in 2005 (and a similar survey carried out in 2002):

- *"50 percent of the respondents indicated that more than 20 percent of their workforce would be eligible to retire in the next five years;*
- *63 percent of respondents identified "skilled trades" as being among the utility positions with the most likely retirements over the next five years;*
- *52 percent of respondents indicated that vacancies among the "skilled trade" positions would be among the most difficult to fill;*
- *64 percent of respondents believe that retirements will pose either a moderate or very great challenge to their utility;*
- *Twice as many respondents in 2005 believe that retirements will create a "significant challenge" for their utility than in the 2002 survey; and*
- *The most significant challenges created will be the loss of knowledge due to retirements, the difficulty finding replacements, and the lack of bench strength within the organization."*

It is clear that there will be widespread problems for power utilities in ensuring access to the appropriate levels of skills and expertise to underpin their businesses in the future. Coupled with this are significant changes to both the business environment and to employees' career models. The aforementioned issues of privatization and industry deregulation, an increase in the penetration of distributed generation and the requirement to extensively refurbish and replace generation, transmission and distribution networks, are just some of the issues that will require knowledge and experience to ensure that utility companies react to and effectively manage their businesses through such a changing landscape. Secondly, the concept of a "job for life" is alien to many early and mid-career level employees, with many individuals expecting to occupy several positions, whether with one employer through job rotation, or through frequent changes of employer as their career progresses.

The single most pressing challenge for utilities is therefore to ensure that the knowledge and experience that is required to ensure effective business performance is readily available. Some means of identifying, capturing, sharing and maintaining this knowledge and experience is therefore critical.

Knowledge Management – A Potential Solution?

Knowledge management is a discipline of management science that is often thought of as being a relatively recent development. However, it could be argued that knowledge management has been in existence for centuries, if not longer. There is a biblical quote that states, "there is nothing new under the sun". While the credibility of this statement itself may be the subject of debate, it is difficult to dispute the assertion that the recording of information in an organized fashion, for the subsequent re-use and reference to by others, has been practiced throughout the history of civilization, from cave paintings and carvings, through to hieroglyphics, books, TV and the internet.

The emergence of knowledge management as a stand alone discipline has developed through a recognition that:

- information and data are constantly growing in volume and complexity, and there is a need to sort, filter and organize what is available;
- the power and value of knowledge has been become recognized, particularly in modern, privatized, competitive environments;
- the development of technology and, in particular, internet and intranet-based technologies, can support and enhance the processes of knowledge capture, sharing and exploitation.

With particular reference to electric power utilities, the attractiveness of knowledge management is increased by a number of inter-related factors and drivers. Firstly, as already stated, many utilities face issues with workforce age demographics (American Public Power Association, 2005). The requirement for personnel to rotate within organizations is also becoming increasingly evident. In addition to these HR-related issues, there is an ever-increasing drive for shareholder return through efficient processes and reduction of costs. There is also a requirement to manage and reduce risk in many aspects of a utility's business, and utilities also must always focus on ensuring the health and safety of their employees, other workers and the public.

There is a requirement to make optimal use of experienced personnel, freeing them from "fire fighting" and allowing them to undertake more strategic or intellectual roles. It is also important to ensure that any lessons are learned and acted upon and that error-rates within business processes are reduced, if not eliminated. The requirement to optimize staff training and to permit faster, more efficient induction programmes and training for both new employees and role-changers is also invariably an objective within utility organizations.

All of these factors and drivers can be addressed and/or supported by effective knowledge management programs.

Knowledge Management – Definitions

Before proceeding further, it is necessary to define exactly what is meant by knowledge management in the context of this chapter. The term knowledge management is rather nebulous in nature and many, sometimes conflicting, definitions and interpretations of the term exist. For the purposes of this chapter, knowledge management is defined as:

The identification, capture, structuring and sharing of knowledge and experience in order to provide personnel with access to experience and supporting resources for the purposes of decision support.

In other words, knowledge management and decision support systems should provide access to experience, which may be in the form of lessons learned, frequently asked questions, examples of good and poor practice, risks, recommendations and other experiential-content. This must be organized and provided in the context of the decision making processes to which it applies. It is also very important, in addition to providing access to knowledge and experience, to provide access to any supporting resources, which may be in the form of relevant documents, training material, spreadsheets, software packages, contacts within and external to the organization, websites, etc.

To reiterate, knowledge management and decision support should be completely focused on the requirements of the user and the business; giving users and the business what they need and want in order to make them more effective, while reducing any inefficiencies and frustrations that may be evident.

Knowledge management involves identification, capture, sharing and maintenance of experience and supporting resources, all in a structured and systematic fashion. Technology, which may be in the form of intranets, wikis, forums, document management systems, and other applications, has an important part to play but should be used as a tool, not as a definer of the problem or solution. Any technology should be built around and supportive of the organization's and users' requirements, and any solution should not be technology driven.

KNOWLEDGE MANAGEMENT – CHANGE, CULTURAL AND IMPLEMENTATION ISSUES

Culture and Change

Many businesses have stated that they operate as a collection of 'silos', with an element of isolation and lack of integration and communication between individuals, teams and departments. Consequently, knowledge and experience, which could potentially be used throughout the organization, resides within individuals or is restricted to small, often co-located groups, and is not shared to a great extent.

There are a number of pre-requisites for a successful implementation of any KM program or initiative. It must be ensured that participants (both contributors and consumers of knowledge) are actively included in all stages of defining, implementing and maintaining any KM/DS initiative. A number of guidelines that should be followed in the promotion and management of a KM program are described in the remainder of this section.

Personnel should feel that they have defined the need for and been involved in the specification of the features of the organization's KM solution, not that the solution has been foisted upon them by management, or that KM is a department's or individual's "pet project". Sponsorship and continual engagement from senior management is critical to ensuring the success of any program.

It is important to ensure that the culture within the organization is amenable to a KM initiative. This is not always easy to achieve but is an absolute prerequisite for the success of any KM project; in fact, KM should not be viewed as a project, rather it should be viewed as a permanent change in the way of working and as providing personnel with tools to help them perform their function more effectively. The initiative should not be communicated as providing personnel with more tasks or a higher workload; the opposite should be true if the initiative is well designed and effective. A great deal of thinking must be devoted with respect to how the initiative is presented to personnel, with a clearly stated objective, message and communications strategy.

To encourage employees in all departments to share knowledge and experience with one another, a shift in culture is often required and one of the most effective ways to do this is via a campaign of publicizing both the need to capture and share experience and, possibly more importantly, explaining the benefits of undertaking such a course of action to all participants and inviting questions, comments, suggestions and volunteers.

Promotion and explanation of the benefits and rationale underlying a KM/DS initiative is critical. Participants in a KM initiative should never need to ask, "Why are we doing this?" The reasons for undertaking a decision support program must be explicit and stated in plain language, while at the same time, the benefits both to the business as a whole as well as to each individual employee at various levels within the business must be publicized.

It is also important to identify sponsors, champions and user-representatives within and across the organization in order for the initiative to succeed. Newsletters, road shows, employee meetings, intranet sites, KM charters, and consultation with employees (e.g. asking them what information/knowledge they would have access to in an "ideal world") can all be used to both publicize and encourage engagement with proposed initiatives.

When publicizing or launching such initiatives, the organization and sponsors must be viewed as "listening", and being open to suggestions, from employees' perspectives.

Keeping the program and initiative interesting delivers benefits. It is important that end users *want* to look at any decision support system in order to help them perform their job and in developing and expanding their understanding. This is achieved by ensuring that the decision support system is lean and that the content is supportive of users' needs, which is best achieved by capturing knowledge that:

- is concentrated within individuals;
- is relied upon by many to support critical decisions;
- is related to areas where good practice rewards the business, and poor practice may damage the business;
- is, or may be, at risk of moving within or leaving the business;
- is generated in 'one-off' projects, that may, however, be re-encountered at a later date;
- is in an area where variability in practice is evident.

Implementing and Embedding KM within the Business

KM initiatives and systems should, as far as possible, be connected and fully integrated with existing business processes that personnel are already familiar. This is an important prerequisite for success, and there are various ways of achieving this. For example, a KM/DS system could be:

- an underpinning element of staff induction programs;
- an element of Continuous Professional Development (CPD) and training programs;
- used as a risk (corporate & functional) capture, measurement and management tool;
- used as part of a compliance-management process;
- used as an element of a risk management system;
- used as an element of gathering (and, most importantly, reacting to) employee feedback;
- utilized as an 'At Desk' support tool for specific functions – providing access to decision support and resources (e.g. templates, checklists, etc.) that enable employees to perform their roles effectively;
- used as part of the employee appraisal process;
- part of a regular internal business event, e.g. a system could be used to record the outcomes of lunchtime seminars, which could include the incentive of a free lunch for attendees, but with the requirement that each attendee asks a question and/or shares experience – at such events, an employee presents an area of activity and others discuss and share related experience;
- used to capture and manage output from learning groups and/or communities of practice.

While a KM system may be "nice to look at" and contain valuable information, if it is not embedded or connected with users' normal roles and functions, then there is a significant danger that the system may fall into disuse. Furthermore, careful thought must be devoted to the design, layout, features and content of any system – as already stated, involving users in specifying exactly what type of knowledge, experience and decision support is required, and in what context, prior to the widespread launch of any system, will help to ensure that user uptake and acceptance of the solution is maximized.

Technology – Only an Enabler

A description of the role that technology has to play in any KM initiative has deliberately been reserved to the end of this section. That is not to say that technology is not extremely important and potentially powerful; however it is critical that technology is not permitted to drive the solution. The solution should use technology to achieve its aims. (Marwick, 2001) provides an excellent overview of how technology can be used to support the various forms of knowledge sharing that are typically undertaken. The modes of knowledge sharing covered include:

- ***Socialization (tacit to tacit):****Socialization includes the shared formation and communication of tacit knowledge between people, e.g., in meetings. Knowledge sharing is often done without ever producing explicit knowledge and, to be most effective, should take place between people who have a common culture and can work together effectively*
- ***Externalization (tacit to explicit):****By its nature, tacit knowledge is difficult to convert into explicit knowledge. Through conceptualization, elicitation, and ultimately articulation, typically in collaboration with others, some proportion of a person's tacit knowledge may be captured in explicit form. Typical activities in which the conversion takes place are in dialog among team members, in responding to questions, or through the elicitation of stories.*
- ***Combination: (explicit to explicit):****Explicit knowledge can be shared in meetings, via documents,*

Figure 1. The knowledge management process

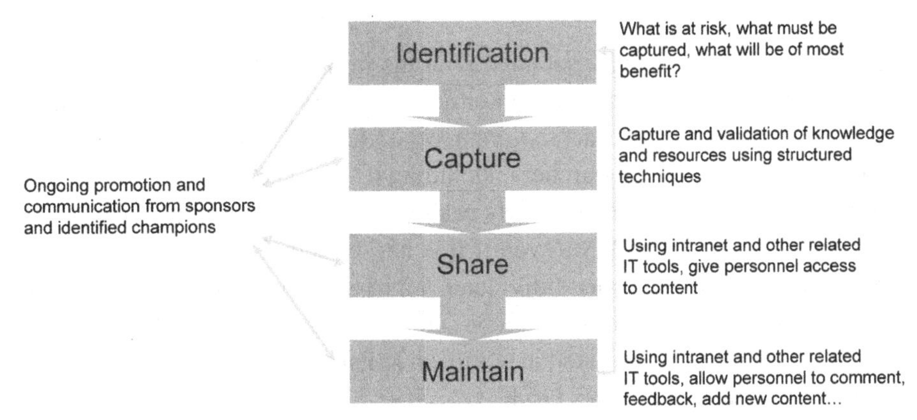

e-mails, etc., or through education and training. The use of technology to manage and search collections of explicit knowledge is well established. However, there is a further opportunity to foster knowledge creation, namely to enrich the collected information in some way, such as by reconfiguring it, so that it is more usable. An example is to use text classification to assign documents automatically to a subject schema. A typical activity here might be to put a document into a shared database.

- ***Internalization (explicit to tacit):****In order to act on information, individuals have to understand and internalize it, which involves creating their own tacit knowledge. By reading documents, they can to some extent re-experience what others previously learned. By reading documents from many sources, they have the opportunity to create new knowledge by combining their existing tacit knowledge with the knowledge of others. However, this process is becoming more challenging because individuals have to deal with ever-larger amounts of information. A typical activity would be to read and study documents from a number of different databases.*

Technology has progressed significantly since 2001 when the text quoted above from (Marwick, 2001) was written. However, many of the technologies described in the paper (such as groupware, intranets, chat software, document management systems) still exist and can provide support to KM. However, as already stated, the focus must remain on KM requirements and users needs, and not on the technology itself.

KNOWLEDGE MANAGEMENT – THE PROCESS

Overview

Knowledge management is a cyclic and continual process. It should not be viewed as a project per se, but rather as an initiative that is intended to introduce a new way of working and thinking across the organization. The end-to-end knowledge management process can be thought of as cycling in nature, and has a number of stages, as presented in Figure 1:

Starting a Knowledge Management Initiative

The phases represented in Figure 1 are associated with the implementation of KM within a business. Prior to this, however, it is important to ascertain or diagnose the level of maturity of the organization with respect to KM and to establish the exact requirements of the organization in this respect. Much work on this has been carried out by other researchers, including (Collison & Parcell, 2001). Diagnosing the organization's present state may involve generally canvassing opinion with respect to KM, evaluating processes to establish where and how KM (even if it is not termed KM) is already being used, either formally or informally. For example, many individuals or groups may produce ad-hoc spreadsheets that contain information for sharing; such activity can be categorized is a form of KM. It is important to not take an overly-theoretical approach. As stated in (Collison & Parcell, 2001), the statement below is often made by members of personnel with respect to KM:

We think this is really good. The only problem we have is that we don't have the time right now on top of everything else we have to do.

An effective response to this statement is proposed:

What if we told you someone else has already done the very task you are about to do. We just need to find out who and what they learned.

This leads to the basic question of "what do we need to learn" and, again in (Collison & Parcell, 2001), this is proposed as a very good starting point and a message that can be used to focus participants' attention, while simultaneously acting to alleviate any concerns that personnel may have over the effort that will be required on the initiative and/or the benefits that may be accrued.

Effectively selling knowledge management to all participants and users is critical to the success of a KM initiative. The approach to this varies depending on the profile of the target party. Directors and business owners will be motivated by the overall protection of (previously) intangible assets, the use of KM to identify and reduce risk and from the improvements in communication, productivity, training and engagement with the workforce that are all facilitated by effective KM. "Experts" will be amenable to KM if it is seen to reduce time spent by them on "firefighting", to reduce time spent on answering questions that have already been asked and answered in the past, to be freeing them to perform more strategic/valuable functions and providing them with increased recognition (and potentially reward) for their efforts. The operational workforce, or users of any KM system, will perceive similar benefits to those offered to "experts", with the additional benefits of having access to a more supportive environment, decreasing the potential for errors, having the opportunity to contribute knowledge to the business and to feel more engaged with the decision making process.

To summarize, initiating a KM program requires an effective communication of the need for KM, the canvassing of opinion as to what KM should offer and publicizing of the benefits of KM to all parties involved at all levels of the organization. Communication is therefore central to the process of preparing for and starting any KM initiative. This communication may be via email, intranet, seminars, articles in company newsletters, etc.

Identifying what Knowledge Should be Captured

Before embarking upon a detailed knowledge capture exercise, it is important to identify exactly what should be captured and to prioritize with respect to areas of the business where knowledge capture will be undertaken. Of course, it is always sensible to try to establish "quick wins" in any project, so selecting an area of the business where benefits will be readily achieved and where these benefits will be apparent to many within the business is the overriding objective in identifying initial areas where KM should be applied.

There are four high level objectives of any identification exercise, and these are to establish areas of the business where:

- Experience may be concentrated.
- Experience may be at risk, for example through retirement, job rotation, moving to another organization.
- There is evidence or a suspicion that practices may be variable.
- There is evidence that good/poor practice directly rewards or damages the business.

A knowledge audit exercise should be an effective mechanism for identifying topics and business functions for subsequent attention. In (De Brún, 2005), it is stated that:

The term 'knowledge audit' is in some ways a bit of a misnomer, since the traditional concept of an audit is to check performance against a standard, as in financial auditing. A knowledge audit, however, is more of a qualitative evaluation. It is essentially a sound investigation into an organization's knowledge 'health'. A typical audit will look at:

- *What are the organization's knowledge needs?*
- *What knowledge assets or resources does it have and where are they?*
- *What gaps exist in its knowledge?*
- *How does knowledge flow around the organization?*
- *What blockages are there to that flow e.g. to what extent do its people, processes and technology currently support or hamper the effective flow of knowledge?*

The knowledge audit provides an evidence-based assessment of where the organization needs to focus its knowledge management efforts. It can reveal the organization's knowledge management needs, strengths, weaknesses, opportunities, threats and risks.

There are many analytical and assessment tools available to support knowledge audits; but the principle activities and objectives of any knowledge audit exercise are generic and include:

- Identification and quantification of an organization's reliance on knowledge and experience in specific areas of their business.
- Identification of areas where knowledge is concentrated and/or is widely used and/or is critically important.

Figure 2. KM involves identification and capture of both knowledge and supporting resources

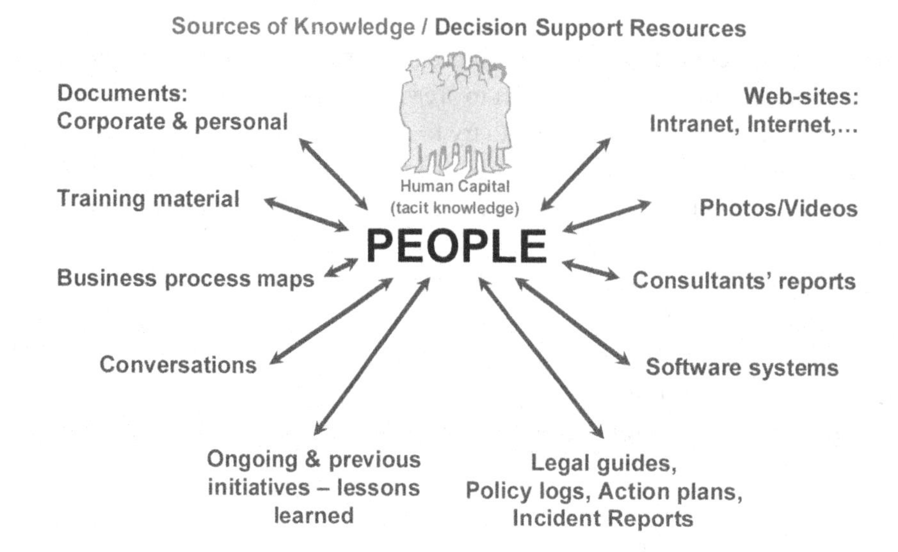

- Identification and evaluation of existing knowledge and experience management methods and tools.
- Identification of areas where there is a known (or suspected) variance in practice.
- Identification of past events within the business that exemplify good and/or poor practice with respect to knowledge and experience management.
- Identification of how knowledge and experience flows around the organization, both intra-departmental and inter-departmental and identification of blockages or other impediments to these flows.
- Identification of how knowledge and experience is interchanged with parties external to the business and any risks associated with these processes.
- Identification of key customers/users of knowledge and experience and their KM requirements.
- Identification of any gaps that may exist in an organization's experience base in order to establish the specific risks faced by the business and to make recommendations on the most appropriate continuing professional development and training activities to mitigate these shortcomings.
- Identification of areas where knowledge is at imminent risk due to staff demographics and/or anticipated movements.
- Quantification of the effects of knowledge loss, deviation from proper practice, improvements to practices through effective knowledge capture and dissemination, business interruptions/accidents due to improper application or non-availability of knowledge and experience.
- Evaluation and quantification of the business benefits of a more comprehensive decision support provision than that which currently exists. This will include identification and prioritization of specific functions that would benefit from the provision of knowledge management and decision support solutions.
- Provision of input to the development of an ongoing and future strategy (including a prioritized list of target topics/individuals) for the capture, retention and dissemination of experience.
- Provision of recommendations relating to improvement of knowledge and experience capture, protection and dissemination.

Figure 3. Knowledge identification and capture process

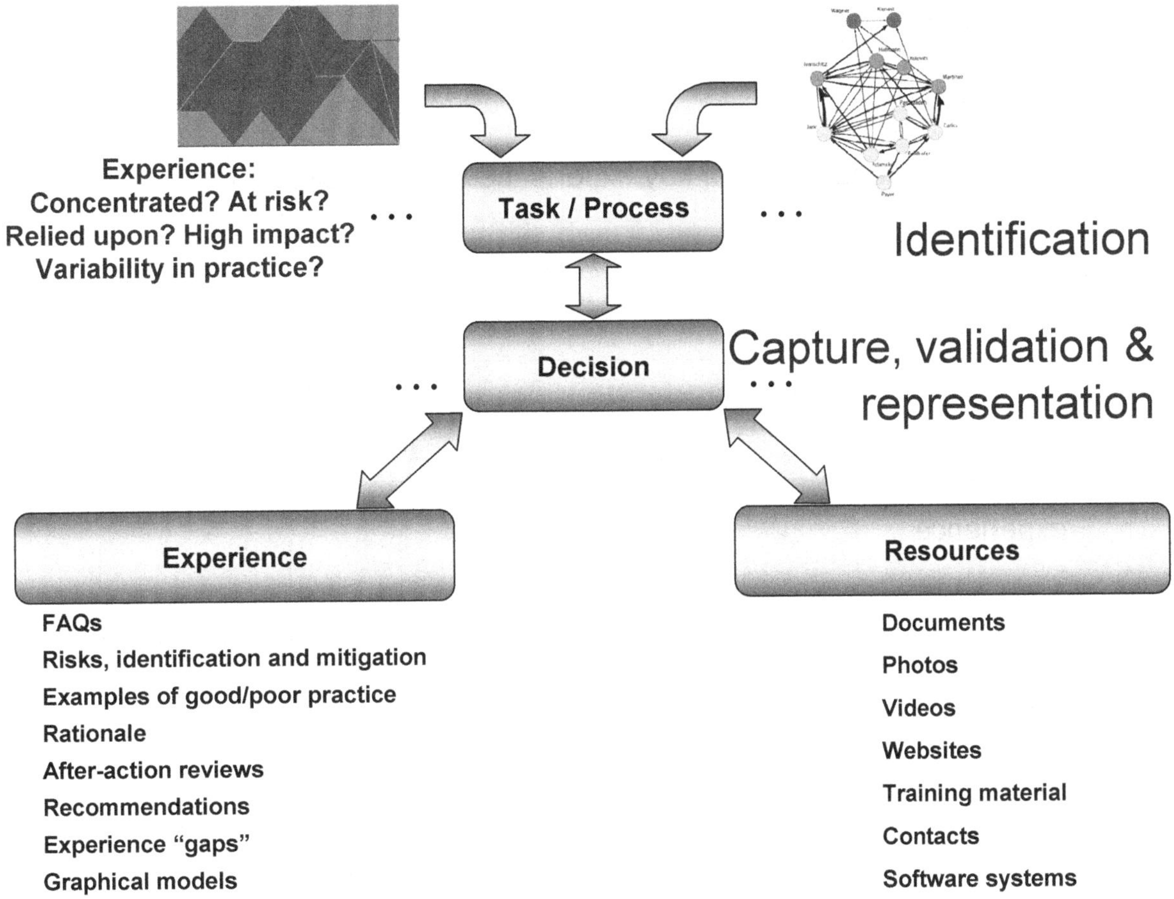

Experience:
Concentrated? At risk?
Relied upon? High impact?
Variability in practice?

Task / Process

Identification

Decision

Capture, validation & representation

Experience

- FAQs
- Risks, identification and mitigation
- Examples of good/poor practice
- Rationale
- After-action reviews
- Recommendations
- Experience "gaps"
- Graphical models

Resources

- Documents
- Photos
- Videos
- Websites
- Training material
- Contacts
- Software systems

It must also be noted that KM does not only involve capture of knowledge from personnel. Individuals, when making decisions, do not only rely on previous experience, but often call upon a wide range of supporting resources in order to make decisions and perform tasks. The identification of these supporting resources must also be included, both during an audit exercise and also during subsequent knowledge capture activity.

To facilitate an audit, carefully designed questionnaires and surveys, seminars, presentations, on-line tools, focused interviews with a cross-section of personnel, and other canvassing techniques, can all be used. For an excellent overview of knowledge audit techniques, refer to (Community Knowledge).

Capturing, Codifying and Disseminating Knowledge

Knowledge capture, or knowledge elicitation, is arguably the most important element of any KM program. There is an adage, "garbage in, garbage out" and this is a very appropriate statement in the context of KM. It is important to capture and share knowledge that is: useful to the organization and individual users; is not necessarily widely known (to avoid the "so what?" reaction); is validated and that all appropriate measures are taken to ensure accuracy; is seen as being open to query, criticism and modification – it is

Figure 4. Structured interview approach

important that captured knowledge is not viewed by others as being absolute and not open to challenge, the last thing that a KM programme should produce is an organization full of automatons.

The processes of knowledge capture (and the preceding step of knowledge identification) are summarized in Figure 3. This process is loosely based on the CommonKADS (CommonKADS) model, where knowledge can be thought of as having multiple "layers", those being: the task layer, which describes the process or function to be carried out; the inference or decision layer, which describes decision that must be taken, with their associated inputs and outputs; and the layer of experience and supporting resources that are used to inform and input to the decision making processes. The model presented in Figure 3 can be used to guide both knowledge capture and the associated process of representing knowledge, either via documents, graphically or via some form of software system.

Knowledge Capture Techniques

The overarching objective in a knowledge capture exercise is to capture and protect knowledge and experience for decision support purposes. It is important to capture only relevant and useful experience. For example, the objective of the exercise not to "re-write" a textbook or manual (unless none exists). The focus of the exercise should be on the end-users of the experience and their specific requirements. For experienced personnel, it may be that only esoteric, unusual and other such experience may be required.

Conversely, when capturing knowledge for consumption by trainees and/or job rotators, a more comprehensive approach, similar to business process mapping, may be required.

There are many techniques that can be used to support knowledge capture. These include

- Observation
- Brainstorming
- Structure problem solving
- Surveying
- Repertory grids
- Graphical techniques/mind mapping
- A number of other techniques

There are also formal frameworks and methodologies that have been developed to assist in many aspects of knowledge capture and structuring, such as the CommonKADS (CommonKADS) framework.

Structured Interviews for Knowledge Capture

Many of the most effective knowledge elicitation methods involve the use of structured interviewing techniques; this therefore warrants further description. Such techniques involve conducting a knowledge elicitation meeting in accordance with a comprehensive pre-prepared and pre-agreed agenda. The interview is recorded, transcribed and the participating expert(s), and ideally a selection of his/her colleagues, subject the transcript document to review and validation.

The approach to conducting structured interviews is summarized in Figure 4.

The elicitation program should be based on areas of expertise identified during the identification/audit phase. The strategy adopted during elicitation should be to initially conduct a scoping meeting, which is based around and builds upon the results of the knowledge audit and communication with the expert(s) prior to the meeting. The aim of the scoping meeting is to obtain a "shallow cut" across the entire area of knowledge, identifying key tasks and processes within the area that require knowledge and experience (and supporting resources) to be employed in their execution. At the end of the scoping meeting, a draft of the agendas for all of the future knowledge elicitation meetings on this area should be drawn up and agreed in principle with the expert(s). Formal, agreed versions of the agendas are produced prior to the beginning of the detailed knowledge elicitation phase.

During the detailed knowledge elicitation meetings, the agenda items should be discussed with the elicitor constantly searching for and attempting to draw out key tasks and sub-tasks, the decisions that require to be made within these tasks and the tacit knowledge and other sources of support (e.g. manuals and other procedural documents) that are required to inform and support these decisions. The elicitor should always attempt to explore the justification for why things are as described by the expert and should never blindly accept explanations offered by the expert. Typical questions that may be asked during elicitation include:

"Scoping" questions (focus on identification, not detail):

- What do you do?
- Can you describe the process and its sub-tasks?
- Which are the most important decisions?
- Which decisions are most reliant on experience?
- Which areas have most risk associated with them?

- Which areas have the most variance in practice?
- Where have mistakes been made in the past?
- Where could improvements be made?
- What are the main sources of support – data, documents, software, other people?

"Detailed elicitation" questions (focus on detailed capture of experience and decision making processes):

- With reference to this particular process, re-confirm the main decisions requiring experience.
- For this decision, can you discuss the background, why it is done the way it is, how it was done in the past?
- For this decision, can you establish the impact of making an error – what are the possible errors, how much do they cost?
- For this decision, could any improvements be made? What would the benefits be?
- What are the FAQs associated with this decision?
- What are the inputs and outputs?
- Are there any risks: how can they be identified and mitigated? What is the impact of these risks – what is the current level of this risk?
- Can you describe examples of good practice?
- Can you describe examples of poor practice?
- Can you describe case studies – these may be structured as concise after-action reviews (what was done, what was good, what was bad, what was learned)?
- Do you have any general or specific recommendations?
- Are there any knowledge gaps around this process?
- For this decision making process, can you detail the supporting resources and how they are used, which are most relevant, which are maybe not so relevant? Should any of these resources be changed/updated? Are there any other recommendations relating to resources?

In addition to these questions, the discussion should be regularly summarized and "teach back", where the interviewer summarizes the main element of the interviewee's description, should be performed periodically throughout the interviews to improve engagement with the interviewee and also to prompt the interviewee to elaborate and extend previous explanations. It is also worthwhile attempting to encourage the interviewee to question their assertions through regular use of the basic question "why?"

A transcript of the meeting should be produced (normally digital audio recorders are used to provide a record of the meeting for transcription purposes). This transcript should be structured according to the agenda, with the key processes, decisions and supporting knowledge being explicitly represented and described within the transcript document – often using a template for the transcript (containing, for example, sub-headings of "background", "FAQ", "Risks", "Recommendations" "After Action Reviews", "Supporting Resources", etc. for each decision making process) can be an effective method of communicating the outcomes of the meeting and aids validation and verification of content. The categorization of content using such a template approach, rather than merely documenting the discussion in uncategorized prose, has been found to be effective in ensuring that interviewees comprehensively engage with the validation process.

Figure 5. Modes of communication, education and utilization (adapted from Carter, 2003).

Communication mode	Average time spent in education in communication mode	Typical proportion of time that communication mode is used
Writing	10-15 years	<10%
Reading	6-10 years	15-20%
Speaking	1-3 years	20-40%
Listening	0-10 hours	40-60%

Following completion of the scoping meeting and the subsequent series of detailed knowledge elicitation meetings, a validated, consolidated transcript document, containing all of the elicited knowledge for the particular area under study, should be available.

Another powerful technique for knowledge elicitation, or to be more accurate, knowledge discovery, is through the use of knowledge discovery in databases (KDD) and data mining techniques (Data mining). This is useful for organizations with large amounts of data relating to their business (e.g. sales and other business performance data, technical process data, etc.). If an expert has a suspicion that inter-relationships and/or patterns exist within such data, KDD can provide a means of investigating whether these hypotheses are in actual fact true. The outcomes from such an exercise could be knowledge, for example in the form of rules, which could subsequently be used in the future to improve aspects of the performance of the business.

The Importance of Listening During Knowledge Capture

Knowledge capture involves extensive periods of listening and a continual attempt to engage with the interviewee on the part of interviewer. Listening is critical to the success of the exercise and effective listening will have the benefit of encouraging the interviewee and fostering a positive relationship between both parties. There are a number of interesting observations that have been made by others (Carter, 2003) relating to the act of listening:

- Individuals listen at a speed of 125-250 words per minute, but can think at a speed of 1000-3000 words per minute.
- 75% of the time, individuals are distracted, preoccupied or forgetful when “listening”.
- 20% of the time, individuals remember the details of what the other party said.

- More than 35% of businesses think listening is a top skill for success.
- Less than 2% of people have had formal education with listening.

Furthermore, the authors of the statistics listed above also summarize the average communications skills education compared to the amount of utilization of each mode of communication:

From Figure 5, it is clear that the mode of communication most used is that for which the least amount of formal training is typically received. It is therefore very important that interviewers in a KM program are aware of the requirements for listening, and they should ideally attend some form of listening/interviewing training course prior to taking part in any KM exercise.

Tacit Knowledge

The capture and codification of tacit knowledge is often discussed as being one of the most difficult challenges to knowledge management practitioners. It is recognized that tacit knowledge, defined in (Wikipedia, 2008) as knowledge that is "*apparently wholly or partly inexplicable*", is difficult, if not impossible to capture. However, the proportion of tacit knowledge within an organization can be reduced through structured sharing and ongoing maintenance of experience within KM systems.

Consider the analogy of riding a bicycle. While explicitly describing (in words) how to ride a bicycle may be viewed as difficult or almost impossible, an experienced cyclist may be able to share valuable knowledge with an inexperienced cyclist (that is capable of basic cycling) relating to technique that can have an immediate positive impact on performance; for example, spinning the pedals at a higher cadence in a certain gear may be more efficient than spinning at a lower cadence in another gear for the same over-ground speed (this could be viewed as an example of good practice). Knowledge management should focus on identifying and capturing such "tricks of the trade", and similarly on capturing "pitfalls of the trade", and making these available so that others may benefit.

It should also be noted that tacit knowledge, while difficult to codify, can often be shared through storytelling and/or demonstration. Many people learn to ride a bicycle through mimicking others and/or by trial and error (often with someone, usually a parent, initially running alongside the embryonic cyclist, holding the bicycle and then letting go). This situation is an example of knowledge sharing where "show-how", through physical demonstration, can be more effective than sharing of "know-how", through words. Technology can often provide effective means of sharing "show-how", for example through video, virtual reality, games/role-playing, etc.

Knowledge management cannot promise to render all tacit knowledge within an organization explicit. However, by publicizing KM and undertaking KM exercises that produce clear benefits to all participants, an environment can be created where knowledge creation, capture and sharing are central to all activities and become second nature to individuals within the organization.

Sharing Knowledge

Knowledge can be shared using a variety of means, from relatively simple, document-based sharing of knowledge, through people-based techniques such as informal exchanges, regular meetings and seminars, to searchable, multi-media systems containing previously captured knowledge and supporting resources. This chapter is deliberately not focused on technology, however, as previously indicated, (Marwick, 2001) describes various technologies (such as groupware, intranets, chats, document management systems) and their applicability to KM. There are also many providers of KM solutions, and information relating to such providers is readily available on the internet.

Encouraging knowledge sharing underpins any successful KM initiative. As previously stated, the benefits of KM must be explicitly stated and continually reinforced. The approach taken will vary depending on the nature of the individual participant in the program. Directors and business owners will be motivated by the overall protection of (previously) intangible assets, the use of KM to identify and reduce risk and from the improvements in communication, productivity, training and engagement with the workforce that are all facilitated by effective KM. "Experts" will be amenable to KM if it is seen to reduce time spent by them on "firefighting", to reduce time spent on answering questions that have already been asked and answered in the past, to be freeing them to perform more strategic/valuable functions and providing them with increased recognition (and potentially reward) for their efforts. The operational workforce, or users of any KM system, will perceive similar benefits to those offered to "experts", with the additional benefits of having access to a more supportive environment, decreasing the potential for errors, having the opportunity to contribute knowledge to the business and to feel more engaged with the decision making process.

The benefits stated above should be used to encourage both the consumption of knowledge and, more importantly, the sharing and contribution of new knowledge, or knowledge that, while not new, was not previously readily available within the organization.

The Role of Reward in Encouraging Knowledge Sharing

Reward and other incentives can be used to encourage knowledge sharing, as discussed in (Bartol, 2002). In this paper, the role of monetary and non-monetary reward in encouraging knowledge sharing is analyzed in detail. It is proposed that, unless potential knowledge sharers obtain a positive answer to the question, "What's in it for me?" then effective knowledge sharing is less likely to happen. The reward itself may be in the form of direct financial bonuses, through non-financial bonuses such as theatre tickets, vacation trips, etc. or through internal or external recognition and praise via company publications, websites, employee meetings, seminars and social events.

The paper also discusses how rewards can be made to individuals, to teams and/or to the entire organization, based on the knowledge shared and the benefit it has delivered. It is also stated that means of measuring the contributions of individuals and teams must be subject to validation, as carried out by Xerox (Brown & Duguid, 2000), and that contributions should be readily quantifiable, both in terms of quantity and quality. The paper describes how Cap Gemini Ernst & Young awards merit pay decisions partly on the basis of the knowledge sharing activities of its employees. On a scale from 2 to 5, employees cannot score more than 3 points if they have not participated in knowledge sharing activities (Stevens, 2000). Similar practices exist at the Lotus Development division of IBM and at Buckman Laboratories and these are also described in (Stevens, 2000). In terms of measuring the benefits arising from contributed knowledge, there are a number of available methods, including measuring the performance increase (if possible) introduced by the shared knowledge. Measurement can be difficult and complex; in the paper, it is stated that Cap Gemini Ernst & Young measures the value of knowledge contributed on the basis of how many people are known to have used that knowledge.

It is proposed that methods of measuring explicit knowledge sharing activities are best facilitated by database/knowledge base systems. However, it must also be recognized that other means of knowledge sharing, both informal and formal, exist and must be encouraged.

Figure 6. CLP KM organizational structure (adapted from Poon, 2005)

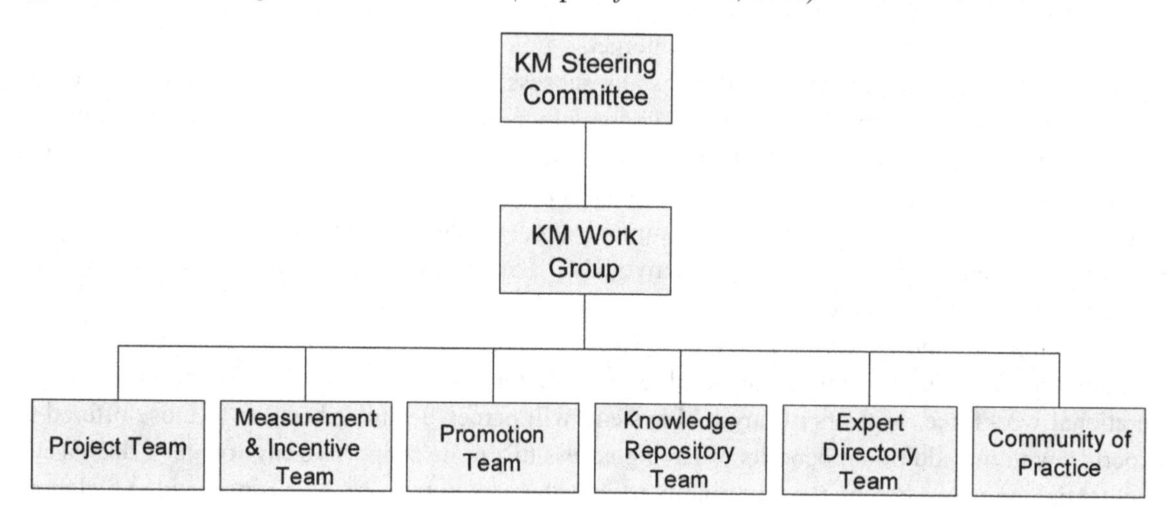

Maintaining Knowledge

At the precise moment when knowledge is created and/or captured; it is effectively "time stamped" and may soon become irrelevant, erroneous or obsolete. Accordingly, it is crucial that all content within any KM/DS system be regularly maintained in order to avoid any problems associated with inaccurate, out of date or lapsed content. There are many ways that this can be achieved; technology can help, but an effective knowledge maintenance program relies largely on effective processes being embedded within the business to ensure that knowledge is regularly reviewed and updated. This can be facilitated through:

- Creation of KM steering groups or similar managerial structures, that are responsible for overseeing and ensuring that KM systems are being used and updated accordingly (and reacting appropriately and sympathetically if this is found to not be happening);
- Regular scheduled knowledge capture/review sessions, where the appropriate material is reviewed and any modifications, additions and/or deletions are made;
- The appointment of "knowledge owners" within the business, with a specific responsibility to review content;
- Introducing regular after-action reviews, where system(s) are updated following completion of tasks by individual(s);
- Creation of communities of practice – these may be supported by technology – chat software, online collaboration tools, blogs and other types of on-line diary/collaboration software;
- Providing (controlled and monitored) access to content to all employees, with a clear message that all staff are encouraged and expected to comment upon/review and/or modify content as necessary;
- Continually publicizing the KM initiative and, in particular, any "success stories", with a reminder and encouragement to all to participate, including reviewing and updating content where relevant.

Maintenance of knowledge relies upon engagement of personnel, and it is important that staff are motivated to participate, are aware of the rationale and benefits of the program and ultimately view knowledge management as a tool, not as a task.

KNOWLEDGE MANAGEMENT IN POWER UTILITIES – APPLICATIONS

China Light and Power

China Light and Power's (CLP) business in Hong Kong is a vertically integrated (generation, transmission and distribution) power utility. CLP, with an installed generation capacity of 6,908 megawatts, and a transmission and distribution network of 12,958 kilometers in length, supplies electricity to approximately 80% of Hong Kong's total population (CLP, 2008).

CLP is one of the leading KM practitioners in the electric power utility industry. This is evidenced by their numerous publications in this area. They are also the first Asian utility to receive PAS 55 accreditation in recognition of their asset management practices.

In (Mak, Choi, 2003), an overview of the application of knowledge management in the project management function associated with power transmission networks is described. In this paper, a diverse range of KM applications, and their attendant benefits, is presented. Applications described in the paper include: knowledge capture and exploitation in the areas of architecture and civil engineering; project engineering in the Chinese mainland, with a particular focus on dealing with process and cultural differences between Hong Kong and mainland China; an application that ensures best practice for cable laying, a general FAQ/ "knowledge quiz" on engineering project management; and finally, and interestingly, an application on the "softer" skills associated with relationship management.

The paper illustrates how a diverse range of business functions, from those that are highly technical and process-oriented, to softer and less procedural functions, can benefit from KM. Another interesting fact to note about this paper is that a diverse range of approaches to the implementation are adopted, showing that, while the high-level KM strategy may be consistent, each application makes use of a number of alternative tools and methods in order to best meet users' needs within the particular application domain. This also emphasizes the requirement, as stated elsewhere in this chapter, to canvass user opinions and requirements and subsequently develop applications to meet these stated needs.

Another paper from CLP (Poon, 2005) discusses more strategic and management-oriented issues associated with KM. It is stated that staff development and growth are facilitated by KM, while also acknowledging the fact that problems associated with an ageing workforce also highlight the need for organized and effective KM within the business.

A valuable insight into the management of KM within CLP is presented in the paper. Figure 6 below highlights how a comprehensive organizational structure has been created to deal with the capture, organization, validation and ongoing maintenance of knowledge within CLP:

The strategic objectives of KM within CLP are to:

- identify and retain critical knowledge;
- enhance infrastructure to facilitate exchange of knowledge;
- promote the value of sharing working knowledge;
- develop new knowledge.

Figure 7. "Listening to those best-placed to know the answers (adapted from Woodhouse, 2006)

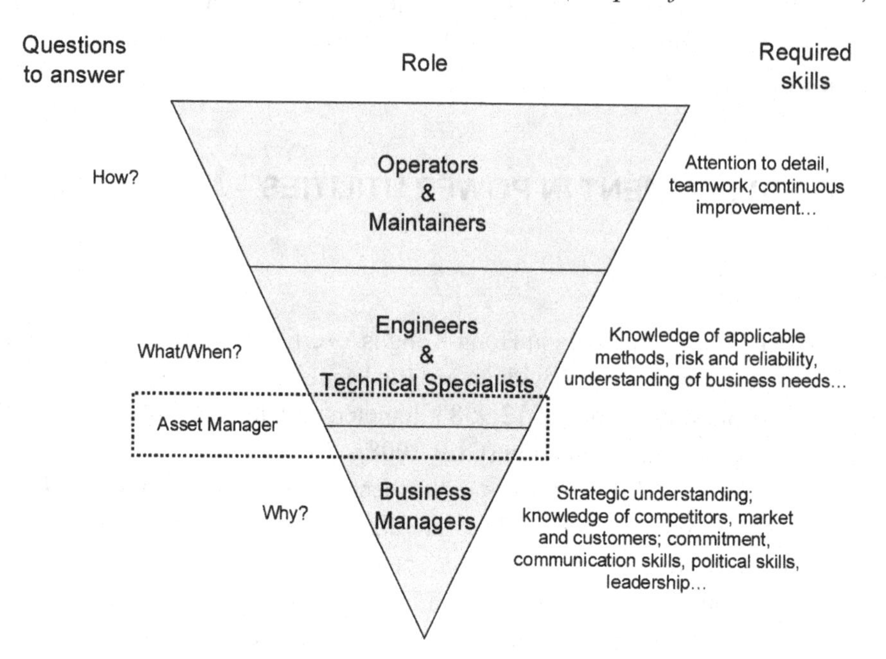

The paper also illustrates how CLP has undertaken a measured approach to identifying and prioritizing knowledge that must be captured and how a structured approach to knowledge capture, sharing and maintenance has been adopted in a number of application areas, using a variety of enabling IT systems.

The topics of promotion and reward and recognition for KM participants are also described in the paper, and it is clear that CLP has a mature and comprehensive approach to KM.

A later publication from CLP (Chan, 2007) discusses four dimensions of KM within CLP: knowledge, process, technology and human. The knowledge dimension is targeted at corralling and categorizing existing documents, presentations and other explicit material that is deemed to contain knowledge and making this available to all via a portal. The process dimension focuses on defining processes that will encourage, and in some cases, mandate the sharing of knowledge within, and as a key component of, business processes. The technology dimension identifies and introduces technology solutions that can facilitate the objectives of KM, and the technology is stated as including: shared workspaces; discussion forums; email alerts to highlight the existence of new or updated knowledge; search engines and content management systems. Finally, the human dimension is discussed in the paper, and an effective "human touch" is highlighted as being a pre-requisite for a successful KM initiative. The need for senior management support and the various means of promoting and recognizing KM activities are described, similar to the description presented in (Poon, 2005).

Other references containing further information on the various aspects of CLP's knowledge management activities include (Lau, Chan, 2000), (Choi, Lau, 2003), (Chan, 2006) and (Chan, Leung, 2006).

Knowledge Management and Asset Management in Power Utilities – PAS 55

Asset management is probably the most important central function that is common to all modern power utilities, and involves the key tasks of: identifying the need for, in some cases designing the application of, and subsequently inviting tenders for and procuring new infrastructure assets; commissioning, operating and maintaining assets; troubleshooting and repairing assets as necessary and refurbishing, replacing or decommissioning obsolete assets. Utilities often outsource elements of their asset management function, but should always retain in-house expertise to effectively monitor and manage the performance of service providers. Increasingly, utilities must demonstrate comprehensive, documented and effective asset management practices to shareholders and external stakeholders such as regulators, health and safety organizations and consumer groups. It is important that these practices include the interfaces with and management of service providers.

PAS 55 is a relatively recent development and is, *"the British Standards Institution's "Publicly Available Specification" for the optimized management of physical assets and infrastructure - it provides clear definitions and a 21-point requirements specification for joined-up, optimized and whole-life asset management systems".* (PAS55)

PAS 55 accreditation has now been recommended for all licensed power utility companies in the UK by Ofgem (the UK regulator for gas and electricity) (AFAQ). PAS 55 documentation is publicly available and provides detailed step-by-step guidance on the processes that are required to enable any organization to gain PAS 55 accreditation. In a paper by (Woodhouse, 2006), it is stated that there is a need for a "joined up approach" to asset management. This approach can be facilitated by clarifying and making explicit all decision-making processes, and knowledge management has an obvious part to play in achieving this.

Interestingly, in (Woodhouse, 2006) there is a section entitled "Human Assets", and in this section it is stated that there is a requirement to create a collaborative environment and that a total of 44 component skills have been identified that are necessary to effectively manage assets. A figure adapted from the paper (Woodhouse, 2006) also outlines how different categories of personnel can provide answers to questions associated with asset management.

It is clear that knowledge management is therefore a fundamental enabler of asset management and can be a facilitator of PAS 55 accreditation. From Figure 7, there are a number of terms that are concordant with KM. "Continuous improvement", "teamwork", "understanding", "communication skills": all of these can be enabled by KM. Additionally, within the PAS 55 documentation itself, knowledge management and decision support systems are both identified as specific components of asset information, risk assessment and planning functions.

Knowledge Management - Benefits and Business Case

At the highest level, KM initiatives can deliver benefits in three main areas: productivity and efficiency improvements; the management of many forms of risk; and in enhancing the efficiency and effectiveness of training processes.

A breakdown of the benefits in each of these areas is presented below, along with an overview of the potential financial benefits of an effective KM initiative.

- **Productivity and Efficiency.** Benefits in this context may include:
    - *Time savings.* Time spent querying others who may be difficult to obtain or unavailable is reduced. Furthermore, answers to questions previously unavailable can be made available on-line via KM initiatives. The incidence of incorrect courses of action that may have been taken in the past can be reduced, directly improving productivity.

  - *Efficiency savings.* Time spent searching for appropriate reference material is reduced. A reduction in the amount of repeated mistakes can be achieved.
  - *Standardization and re-use of existing resources.* Production of duplicates, triplicates, or worse of documents and reports that may already exist can be avoided by providing access to existing documents, which may act as templates, or even ready-made solutions.
  - *Experience-led advice is always available.* Direct access to examples of good and poor practice can be constantly available on-line, therefore acting to reduce the incidence of repeated mistakes.
- **Training.** An obvious use of the experience and resources captured through any KM initiative is for training and e-learning purposes. The benefits of e-learning are well documented and include:
  - *Time savings.* 40-60% time saving in delivery compared to instructor-led programmes (review of 130 case studies) (Hall).
  - *Cost savings.* Reduced attrition costs through the provision of experience-rich training, resulting in new employees reaching full productivity faster. (Costs of attrition can range from 50% to 400% of annual salary depending on the position in question)
  - *Reduced instructor(s) costs.* Reduced travel, accommodation and subsistence costs associated with employee off-site course attendance.
  - *Lost opportunity savings.* Employees training and learning in-situ through accessing at-desk training are still available to respond to events as necessary; this may not be the case if they attend an off-site course or an in-house instructor-led course.
  - *Consistency.* The consistency of delivery is improved through reduced/eliminated use of instructors.
  - *Accessibility.* There is effectively no limitation on the number of trainees and no restrictions on the geographic location of trainees – training can be undertaken at any time and from any location.
  - *Increased retention of information.* It has been established that effective training and e-learning results in a 56% improvement in retention of learning points.
- **Risk Management.** A significant element of any KM initiative should focus upon the identification of risks and how best to mitigate or eliminate them. The benefits of capturing, verifying and disseminating how best to mitigate risks using KM techniques include:
  - *Corporate risk management.* Risks associated with staff turnover and absenteeism (whether through illness, travel, etc) are mitigated if key individuals' knowledge and experience is captured and made available to others via KM initiatives.
  - *Functional risk management.* Risks (and methods for their mitigation) associated with technical procedures, hazards, safety, etc. can be captured and widely disseminated using KM techniques.
  - *Risk identification and ownership.* The visibility, ownership and accountability associated with risks can be captured and widely disseminated using appropriate KM techniques.
  - *Risk mitigation.* The use of multimedia resources (e.g. photographs and video-clips) to illustrate risks, their impact and mitigation techniques are provided through KM portals.
  - *Compliance.* Access to regulations and legislation and an explanation of their impact on individual business processes can be provided via KM portals.
  - *Planning.* Predicting risk occurrence and planning preventative action is greatly enhanced – this minimizes time spent "fire-fighting" through the repeated making of mistakes.

Figure 8. The benefits of knowledge management

Risk Management
- Personnel risk
- Functional risk
- Safeguard reputation
- Improve identification & ownership
- Improve clarity – audits - openness with respect to decision making processes

Training
- Cost savings
- Time savings
- Lost opportunity savings
- Consistency
- Accessibility
- Increased retention of learning

Productivity
- Time/efficiency savings
- Access to good practice
- Reduce mistakes/ re-invention of solutions
- Standardisation or processes and re-use of existing resources
- Experience-led advice always available

Culture
- Learning culture
- Employee engagement
- Improved communication
- Participation encouraged
- Suggestions from all valued – the organisation listens
- "Back to the floor"

    - *Feedback.* Using feedback and communities of practice facilities, feedback on new or emerging risks and ways to mitigate and/or eliminate them can be encouraged. Once captured, the newly identified risks can be made available to the appropriate personnel via some form of KM portal.
- **Financial Benefits.** As already stated, the quantification of any benefits specific to an organisation requires analysis of the individual client's situation. However, considering the following general questions may be useful when attempting to quantify financial benefits, as effective KM initiatives may help to reduce some of these costs or realise a benefit associated with exploiting experience contained within the business. Some of the criteria that will be discussed include[1]:
    - How much do (or could) theoretically avoidable mistakes and errors cost the business in a particular function?
    - How much does variability in practice (e.g. across teams, departments or business groups) cost the business?
    - How much could freeing experts/teams from "fire-fighting" and allowing them to concentrate on their main role save the business?
    - How much could be saved by reducing/eliminating reliance on external consultants in certain areas?
    - How much time is wasted (and what does it cost) waiting for a decision from an experienced individual or in looking for information or resources that should be readily available?
    - How much is spent on training (external and internal) in a particular function? This cost should include both the cost of the training, the cost of trainees being unavailable to their department, the cost of trainees while they are reaching full productivity, the cost of experienced mentors/trainers not being available to their own functions due to training commitments, etc. One utility has publicly stated that it spends over £250,000 on training for each

graduate trainee over a two-year period in their business.

- What opportunities are being missed (and how much is this costing the business) due to insufficient resources/experience being available?
- Are there any commercial opportunities through providing (and charging for) access to the organization's experience in certain areas to third parties?

A summary of the benefits that can be obtained from KM is presented graphically in Figure 8.

SUMMARY AND CONCLUSION

This chapter has presented an overview of the challenges that are presented to modern power utility companies and how many organizations are facing particularly pressing problems with regards to an ageing workforce and a general shortage of skills; a situation that is anticipated to worsen in the future. Knowledge management and decision support can act to address these challenges, but it is important that a number of factors are taken into account when planning, implementing, maintaining and monitoring the performance of a knowledge management initiative.

Firstly, knowledge management should focus on decision support and on the requirements of users. Opinion and needs should be canvassed and taken into account by the sponsors and participants in the initiative. The culture and attitudes towards knowledge sharing, and change in general, within the organization are critical and the objective should be to ensure that KM is viewed as a tool, and not as a task, by all. It is important to establish the benefits of any KM program and to strive to develop metrics by which the performance and/or the contribution of KM to other aspects of the organization's performance may be measured. A number of reward strategies have also been discussed, in addition to the attention that must be paid to ensuring that the content of any KM/DS system is maintained and effectively evolves over time.

An insight into the experience and lessons learned by China Light and Power has been presented in this chapter, and the main message arising from this insight is that this company has clearly invested heavily in KM, has tailored different approaches to KM for different applications within its business and recognizes the importance of the human factors and the need to continually promote and maintain KM within the organization.

Finally, KM must never be viewed as a project, and certainly not as an IT-based project. KM, while not necessarily new, should be seen as a way of encouraging openness, should act to encourage and reward the sharing and documenting of knowledge and the importance of KM must be continually publicized, recognized and valued by sponsors within the organization.

REFERENCES

AFAQ. *PAS 55 Assessment: Ensuring Effective Asset Management.* Retrieved November 6, 2008, from http://www.afnor.co.uk/webai/uk/Site.nsf/0/0B6DE01A86A0C28BC125735400557A6A/$file/PAS55.pdf

American Public Power Association. (2005). *Work Force Planning for Public Power Utilities: Ensuring Resources to Meet Projected Needs.* Retrieved September 3, 2008, from: http://www.appanet.org/files/pdfs/workforceplanningforpublicpowerutilities.pdf

Bartol, K. M. (2002). *Encouraging knowledge sharing: the role of organizational reward systems.* Retrieved November 6, 2008, from http://findarticles.com/p/articles/mi_m0NXD/is_1_9/ai_n25057533

British Broadcasting Corporation. (2006). *Nuclear power plants get go-ahead.* Retrieved September 3, 2008, from http://news.bbc.co.uk/1/hi/uk_politics/5166426.stm

Brown, J. S., & Duguid, P. (2000). Balancing act: How to capture knowledge without killing it. *Harvard Business Review*, *78*(3), 73–80.

Carter, C. (2003). *The power of listening.* Presentation material. Retrieved September 15, 2008, from www.etsu.edu/edc/EDC%20Training%20Handouts/The%20Power%20of%20Listeningpowerpoint2.ppt

Chan, J. K. W. (2006). *From Knowledge to Wisdom: Practical KM Implementation Experience in CLP Power*. Advances in Power System Control, Operation and Management.

Chan, J. K. W. (2007). *Scrutinizing the 4 Key Dimensions in CLP Power's Knowledge Management - The Knowledge Space, Process Space, Technology Space and Human Touch.* International Conference on Engineering Education, 2007. More information and proceedings available from http://icee2007.dei.uc.pt/proceedings/

Chan, J. K. W., & Leung, V. S. Y. (2006). *Integrating KM with Organization Learning - A Case Study on CLP Power*. Knowledge Management Asia Pacific Conference 2006.

Choi Y. H. & Lau, T. W. K. (2003). *Knowledge Management in Transmission Networks and Equipment.* Advances in Power System Control, Operation and Management, 2003. (Conf. Publ. No. 497).

CLP. (2008). *China Light and Power website.* Retrieved September 15, 2008, from https://www.clpgroup.com/Abt/Overview/Pages/default.aspx

Collison, C., & Parcell, G. (2001). *Learning to Fly: Practical Lessons from one of the World's Leading Knowledge Management Companies.* Oxford, UK: Capstone. CommmonKADS. *CommonKADS website.* Retrieved November 6, 2008, from http://www.commonkads.uva.nl/

Community Knowledge. *Introduction to Knowledge Management.* Retrieved September 15, 2008, from http://communityknowledge.co.uk/KMIntro/part_e.html

Data mining. Retrieved November 6, 2008, from http://en.wikipedia.org/wiki/Data_mining

De Brún, C. (2005). *Knowledge audit: conducting a knowledge audit.* Retrieved September 15, 2008, from: http://www.library.nhs.uk/KnowledgeManagement/ViewResource.aspx?resID=93807

Hall, B. *Return on Investment and Multimedia Training: A Research Study.* Multimedia Training Newsletter. Retrieved November 6, 2008, from http://www.brandon-hall.com

Lau, C. H., & Chan, F. C. (2000). *Knowledge Management for Power Systems Application.* Hong Kong Institution of Engineers Annual Symposium 2000. More information available from www.hkie.org.hk

Mak, C. L., & Choi, Y. H. (2003). *Knowledge Management and Project Management for Transmission Network and System*. International Conference on Engineering Education, 2003. More information available from http://www.upv.es/ICEE2003/

PAS55. *PAS 55 website*. Retrieved September 15, 2008, from http://pas55.net.

Poon, P. W. Y. (2005). *Development and Application of Knowledge Management in CLP Power*. Paper presented at IEEE KM Symposium 2005.

Stevens, L. (2000). Incentives for sharing. *Knowledge Management*, *3*(10), 54–60.

Walker, J. (2008). *Engineering Skill Shortage – A Degree of Concern*. Paper presented at New Zealand Electricity Engineers Association (EEA), Conference proceedings, Christchurch, June 2008.

Woodhouse, J. (2006). *Putting the total jigsaw puzzle together: PAS 55 standard for the integrated, optimized management of assets*. Retrieved September 15, 2008, from http://www.twpl.com/confpapers/putting_the_total_jigsaw_together.pdf.

ENDNOTE

[1] This is obviously not an exhaustive list and there are many other questions that may yield costs that may be reduced and/or opportunities that may be exploited via effective KM initiatives. The reduction in costs/increase in opportunities is obviously a subjective issue and this must be addressed and agreed with stakeholders involved in any KM initiative.

Chapter 4
Solving Environmental/ Economic Dispatch Problem:
The Use of Multiobjective Particle Swarm Optimization

M. A. Abido
King Fahd University of Petroleum & Minerals, Saudi Arabia

ABSTRACT

Multiobjective particle swarm optimization (MOPSO) technique for environmental/economic dispatch (EED) problem is proposed and presented in this work. The proposed MOPSO technique evolves a multiobjective version of PSO by proposing redefinition of global best and local best individuals in multiobjective optimization domain. The proposed MOPSO technique has been implemented to solve the EED problem with competing and non-commensurable cost and emission objectives. Several optimization runs of the proposed approach have been carried out on a standard test system. The results demonstrate the capabilities of the proposed MOPSO technique to generate a set of well-distributed Pareto-optimal solutions in one single run. The comparison with the different reported techniques demonstrates the superiority of the proposed MOPSO in terms of the diversity of the Pareto optimal solutions obtained. In addition, a quality measure to Pareto optimal solutions has been implemented where the results confirm the potential of the proposed MOPSO technique to solve the multiobjective EED problem and produce high quality nondominated solutions.

1. INTRODUCTION

Generally, the basic objective of the traditional economic dispatch (ED) of electric power generation is to schedule the committed generating unit outputs so as to meet the load demand at minimum operating cost while satisfying all generator and system equality and inequality constraints. This makes the ED problem a large-scale highly constrained nonlinear optimization problem.

DOI: 10.4018/978-1-60566-737-9.ch004

However, thermal power plants are major causes of atmospheric pollution because of the high concentration of pollutants they cause such as sulpher oxides SO_x and nitrogen oxides NO_x. Nowadays, the pollution minimization problem has attracted a lot of attention due to the public demand for clean air. In addition, the increasing public awareness of the environmental protection and the passage of the U.S. Clean Air Act Amendments of 1990 have forced the power utilities to modify their design or operational strategies to reduce pollution and atmospheric emissions of the thermal power plants (El-Keib et al., 1994; Helsin et al., 1989; Talaq et al., 1994).

Several strategies to reduce the atmospheric emissions have been proposed and discussed in the literature (Talaq et al., 1994). These include

- Installation of pollutant cleaning equipment such as gas scrubbers and electrostatic precipitators;
- Switching to low emission fuels;
- Replacement of the aged fuel-burners and generator units with cleaner and more efficient ones;
- Considering pollution minimization along with cost minimization of economic dispatch problem. This leads to bi-objective environmental/economic dispatch (EED) problem.

The first three options require installation of new equipment and/or modification of the existing ones that involve considerable capital outlay and, hence, they can be considered as long-term options. The emission dispatching option is an attractive short-term alternative in which the emission in addition to the fuel cost objective is to be minimized. In recent years, this option has received much attention (Chang et al., 1995; Dhillon et al., 1993; Farag et al., 1995; Granelli et al., 1992) since it requires only small modification of the basic economic dispatch to include emissions. Thus, the power dispatch problem can be handled as a multiobjective optimization problem with non-commensurable and contradictory objectives since the optimum solution of the economic power dispatch problem is not environmentally the best solution.

Several techniques to handle the environmental/economic dispatch (EED) problem have been reported. Generally speaking, there are three approaches to solve EED problem. The *first* approach treats the emission as a constraint with a permissible limit (Granelli et al., 1992). This formulation, however, has a severe difficulty in getting the trade-off relations between cost and emission.

The *second* approach treats the emission as another objective in addition to usual cost objective (Chang et al., 1995; Dhillon et al., 1993; Farag et al., 1995; Granelli et al., 1992; Yokoyama et al., 1988). However, the EED problem was converted to a single objective problem either by linear combination of both objectives or by considering one objective at a time for optimization. Unfortunately, this approach requires multiple runs as many times as the number of desired Paretooptimal solutions and tends to find weakly nondominated solutions.

The *third* approach handles both fuel cost and emission simultaneously as competing objectives. Stochastic search and fuzzy-based multiobjective optimization techniques have been proposed for the EED problem (Das et al., 1998; Huang et al., 1997; Srinivasan et al., 1994). However, the algorithms do not provide a systematic framework for directing the search towards Pareto-optimal front and the extension of these techniques to include more objectives is a very involved question. In addition, these techniques are computationally involved and time-consuming. Genetic algorithm-based multiobjective techniques have been presented in (Abido, 2003; Abido, 2006; King et al., 2006) where multiple nondominated solutions can be obtained in a single run. However, genetic algorithm-based techniques suffer from premature convergence and the technique presented in (Abido, 2003) is computationally involved due to ranking process during the fitness assignment procedure.

Unlike genetic algorithm and other heuristic techniques, particle swarm optimization (PSO) has a flexible and well-balanced mechanism to enhance and adapt the global and local exploration abilities. It usually results in faster convergence rates than the genetic algorithm (Kennedy et al.; 2001). In recent years, PSO has been successfully implemented to different power system optimization problems including the economic power dispatch problem with impressive success (Al-Rashidi et al., 2007; Bai et al., 2006; Jong-Bae et al., 2005; Selvakumar et al., 2007; Ting et al., 2005; Zwe-Lee, 2003). The potential of PSO to handle nonsmooth and nonconvex economic power dispatch problem was demonstrated and reported (Al-Rashidi et al., 2007; Jong-Bae et al., 2005; Selvakumar et al., 2007; Zwe-Lee, 2003). However, the problem was formulated as a conventional dispatch problem with the fuel cost as the only objective considered for optimization.

In dealing with multiobjective optimization problems, classical search and optimization methods are not efficient for the following drawbacks.

- Most of them cannot find multiple solutions in a single run, thereby requiring them to be applied as many times as the number of desired Paretooptimal solutions.
- Multiple applications of these methods do not guarantee finding widely different Paretooptimal solutions.
- Most of them cannot efficiently handle problems with discrete variables and problems having multiple optimal solutions.
- Some algorithms are sensitive to the shape of the trade-off curve and cannot be used in problems having a nonconvex Paretooptimal front.

On the contrary, the studies on evolutionary algorithms have shown that these methods can be efficiently used to solve multiobjective optimization problems and eliminate most of the above difficulties of classical methods. Since they use a population of solutions in their search, multiple Paretooptimal solutions can be found in one single run.

Generally, changing conventional single objective PSO to a multiobjective PSO requires redefinition of global and local best individuals in order to obtain a front of optimal solutions. In multiobjective particle swarm optimization, there is no absolute global best, but rather a set of nondominated solutions. In addition, there may be no single local best individual for each particle of the swarm. Choosing the global best and local best to guide the swarm particles becomes nontrivial task in multiobjective domain.

A little effort has been recently reported to implement MOPSO for solving power system problems. Wang and Singh (2006, 2007) presented a fuzzified MOPSO to solve EED problem with heat dispatch and with multiple renewable energy sources. The approach presents a fuzzification mechanism for the selection of global best individual with interpreting the global best as an area, not just as a point. On the other hand, only one local best solution is maintained for each particle. This will degrade the search capability and violates the principle of multiobjective optimization. Kitamura et al (2005) presented a modified MOPSO to optimize an energy management system where the problem is solved in three phases by dividing the original optimization problem into partial problems. However, this approach has severe limitation in the case of strong interaction among the constraints in different subproblems. Hazra and Sinha (2007) presented a MOPSO based approach to solve the congestion management problem where the cost and congestion are simultaneously minimized. In this approach the sigma method (Mostaghim et al.; 2003) is adopted to find the best local guide for a particle. However, the use of the sigma values increases the selection pressure of PSO which is already high. This may cause premature convergence in some cases e.g., in multifrontal problems. A vector evaluated PSO (VEPSO) was proposed and examined for determining generator contributions to transmission system (Vlachogiannis et al.; 2005). In VEPSO, fractions of the next generation are selected from the old generation according to each of the objectives separately. However, selection of individuals that excel in one objective without looking to the other objectives implies a problem of killing the middling performance individuals that can be very useful for compromise solutions (Coello; 1999).

In this work, a novel multiobjective particle swarm optimization (MOPSO) technique is proposed and implemented for solving the environmental/economic dispatch problem. The proposed approach extends the single objective PSO by proposing new definitions of the local best and global best individuals in multiobjective optimization problems. Like other multiobjective evolutionary algorithms, a hierarchical clustering technique is implemented to manage Pareto-optimal set size and a fuzzy-based mechanism is employed to extract the best compromise solution. Several runs have been carried out on the standard IEEE 30-bus test system and the results are compared to different techniques reported in literature. The effectiveness and potential of the proposed MOPSO approach to solve the multiobjective EED problem are demonstrated.

2. PROBLEM STATEMENT

The EED problem is to minimize two competing objective functions, fuel cost and emission, while satisfying several equality and inequality constraints. Generally the problem is formulated as follows.

2.1 Problem Objectives

Minimization of Fuel Cost: The total *$/h* fuel cost $F(P_G)$ can be expressed as

$$F(P_G) = \sum_{i=1}^{N} a_i + b_i P_{G_i} + c_i P_{G_i}^2 \tag{1}$$

where N is the number of generators, a_i, b_i, and c_i are the cost coefficients of the ith generator, and P_{Gi} is the real power output of the ith generator. P_G is the vector of real power outputs of generators and defined as

$$P_G = [P_{G_1},\ P_{G2},\ ...,P_{G_N}]^T \tag{2}$$

Minimization of Emission: The atmospheric pollutants such as sulpher oxides SO_x and nitrogen oxides NO_x caused by fossil-fueled thermal units can be modeled separately. However, for comparison purposes, the total *ton/h* emission $E(P_G)$ of these pollutants can be expressed as [12, 31]

$$E(P_G) = \sum_{i=1}^{N} 10^{-2}(\alpha_i + \beta_i P_{G_i} + \gamma_i P_{G_i}^2) + \zeta_i \exp(\lambda_i P_{G_i}) \tag{3}$$

where α_i, β_i, γ_i, ζ_i and λ_i are coefficients of the ith generator emission characteristics.

2.2 Problem Constraints

Generation capacity constraint: For stable operation, real power output of each generator is restricted by lower and upper limits as follows:

$$P_{G_i}^{\min} \leq P_{G_i} \leq P_{G_i}^{\max}, \quad i = 1, \ldots, N \tag{4}$$

Power balance constraints: The total power generation must cover the total demand P_D and the real power loss in transmission lines P_{loss}. Hence,

$$\sum_{i=1}^{N} P_{G_i} - P_D - P_{loss} = 0 \tag{5}$$

Calculation of P_{loss} implies solving the load flow problem which has equality constraints on real and reactive power at each bus as follows.

$$P_{G_i} - P_{D_i} - V_i \sum_{j=1}^{NB} V_j [G_{ij} \cos(\delta_i - \delta_j) + B_{ij} \sin(\delta_i - \delta_j)] = 0 \tag{6}$$

$$Q_{G_i} - Q_{D_i} - V_i \sum_{j=1}^{NB} V_j [G_{ij} \sin(\delta_i - \delta_j) - B_{ij} \cos(\delta_i - \delta_j)] = 0 \tag{7}$$

where $i = 1,2,\ldots,NB$; NB is the number of buses; Q_G and Q_D are the generator and demand reactive power respectively; G_{ij} and B_{ij} are the transfer conductance and susceptance between bus i and bus j respectively. V_i and V_j are the voltage magnitudes at bus i and bus j respectively; δ_i and δ_j are the voltage angles at bus i and bus j respectively.

Then, the real power loss in transmission lines can be calculated as

$$P_{loss} = \sum_{k=1}^{NL} g_k \left[V_i^2 + V_j^2 - 2V_i V_j \cos(\delta_i - \delta_j) \right], \tag{8}$$

where NL is the number of transmission lines; g_k is the conductance of the kth line that connects bus i to bus j.

Security constraints: For secure operation, the transmission line loading S_l is restricted by its upper limit as:

$$S_{l_i} \leq S_{l_i}^{\max}, \quad i = 1, \ldots, NL \tag{9}$$

2.3 Problem Formulation

Aggregating the objectives and constraints, the problem can be mathematically formulated as a nonlinear constrained multiobjective optimization problem as follows.

$$\underset{P_G}{Minimize} \quad [F(P_G), E(P_G)] \tag{10}$$

$$g(P_G) = 0 \tag{11}$$

$$h(P_G) \leq 0 \tag{12}$$

where g is the equality constraints given in Equations (5), (6), and (7) and h is the inequality constraints given in Equations (4) and (7).

3. PRINCIPLE OF MULTIOBJECTIVE OPTIMIZATION

Many real-world problems involve simultaneous optimization of several objective functions. Generally, these functions are non-commensurable and often competing and conflicting objectives. Multiobjective optimization with such conflicting objective functions gives rise to a set of optimal solutions, instead of one optimal solution. The reason for the optimality of many solutions is that no one can be considered to be better than any other with respect to all objective functions. These optimal solutions are known as *Paretooptimal* solutions (Zitzler et al; 1998).

For a multiobjective optimization problem, any two solutions x_1 and x_2 can have one of two possibilities: one dominates the other or none dominates the other. In a minimization problem, without loss of generality, a solution x_1 dominates x_2 if the following two conditions are satisfied:

$$\forall i \in \{1,\ 2,\ \ldots, N_{obj}\} : f_i(x_1) \leq f_i(x_2) \tag{13}$$

$$\exists j \in \{1,\ 2,\ \ldots, N_{obj}\} : f_j(x_1) < f_j(x_2) \tag{14}$$

If any of the above conditions is violated, the solution x_1 does not dominate the solution x_2. If x_1 dominates the solution x_2, x_1 is called the nondominated solution. The solutions that are nondominated within the entire search space are denoted as *Paretooptimal* and constitute the *Paretooptimal set.* This set is also known as *Paretooptimal front.*

4. THE PROPOSED MOPSO TECHNIQUE

4.1 Overview

The recent studies on evolutionary algorithms have shown that the population-based algorithms are potential candidate to solve multiobjective optimization problems and can be efficiently used to eliminate most of the difficulties of classical single objective methods such as the sensitivity to the shape of the Paretooptimal front and the necessity of multiple runs to find multiple Paretooptimal solutions.

In general, the goal of a multiobjective optimization algorithm is not only to guide the search towards the Paretooptimal front but also to maintain population diversity in the set of the Paretooptimal solutions.

In recent years, PSO has been presented as an efficient population-based heuristic technique with a flexible and well-balanced mechanism to enhance and adapt the global and local exploration capabilities. However, changing conventional single objective PSO to a multiobjective PSO requires redefinition of global and local best individuals since, in multiobjective particle swarm optimization, there is no absolute global best, but rather a set of nondominated solutions. In addition, there may be no single local best individual for each particle of the swarm. Choosing the global best and local best to guide the swarm particles becomes nontrivial task in multiobjective domain.

The proposed approach addresses the problem of evolving a multiobjective version of the conventional PSO where the global and local best individuals have been redefined and a mechanism for selection of these individuals has been proposed. It is worth mentioning that the proposed MOPSO technique has been implemented to several nontrivial standard test problems in multiobjective optimization domain with impressive success (Abido et al.; 2007).

4.2 Basic Elements and Definitions

The basic elements of the proposed MOPSO technique are briefly stated and defined as follows:

- ***Particle,*** $X(t)$: It is a candidate solution represented by an m-dimensional vector, where m is the number of optimized parameters. At time t, the jth particle $X_j(t)$ can be described as $X_j(t)=[x_{j,1}(t), \ldots, x_{j,m}(t)]$, where x's are the optimized parameters and $x_{j,k}(t)$ is the position of the jth particle with respect to the kth dimension, *i.e.*, the value of the kth optimized parameter in the jth candidate solution.
- ***Population,*** $pop(t)$: It is a set of n particles at time t, *i.e.*, $pop(t)=[X_1(t), \ldots, X_n(t)]^T$.
- ***Particle velocity,*** $V(t)$: It is the velocity of the moving particles represented by an m-dimensional vector. At time t, the jth particle velocity $V_j(t)$ can be described as $V_j(t)=[v_{j,1}(t), \ldots, v_{j,m}(t)]$, where $v_{j,k}(t)$ is the velocity component of the jth particle with respect to the kth dimension. The particle velocity in the kth dimension is limited by some maximum value, v_k^{max}. This limit enhances the local exploration of the problem space. To ensure uniform velocity through all dimensions, the maximum velocity in the kth dimension is proposed as:

$$v_k^{\max} = (x_k^{\max} - x_k^{\min}) / N \tag{15}$$

where N is a selected number of intervals.

- ***Inertia weight,*** $w(t)$: It is a control parameter that is used to control the impact of the previous velocities on the current velocity. Hence, it influences the trade-off between the global and local exploration abilities of the particles. For initial stages of the search process, large inertia weight to enhance the global exploration is recommended while, for last stages, the inertia weight is reduced for better local exploration. An annealing decrement function (Al-Rashidi et al.; 2007) for decreasing the inertia weight given as $w(t)=\alpha w(t\text{-}1)$, α is a decrement constant smaller than but close to 1, is employed in this study.
- ***Nondominated local set,*** $S_j^*(t)$: It is a set that stores the nondominated solutions obtained by the jth particle up to the current time. As the jth particle moves through the search space, its new position is added to this set and the set is updated to keep only the nondominated solutions. An average linkage based hierarchical clustering algorithm described below is employed to reduce the nondominated local set size if it exceeds a certain prespecified value.
- ***Nondominated global set,*** $S^{**}(t)$: It is a set that stores the nondominated solutions obtained by all particle up to the current time. First, the union of all nondominated local sets is formed. Then, the nondominated solutions out of this union are members in the nondominated global set. The clustering algorithm is employed to reduce the nondominated global set to a prespecified manageable size.
- ***External set***: It is an archive that stores a historical record of the nondominated solutions obtained along the search process. This set is updated continuously after each iteration by applying the dominance conditions to the union of this set and the nondominated global set. Then, the nondominated solutions of this union are members in the updated external set. The clustering algorithm is also used to limit the external set to a prespecified manageable size.
- ***Local best,*** $X_j^*(t)$, and ***Global best,*** $X_j^{**}(t)$: In order to guide the search towards the Paretooptimal front, the global and local best individuals are selected as follows. The individual distances between members in nondominated local set of the jth particle, $S_j^*(t)$, and members in nondominated global set, $S^{**}(t)$, are measured in the objective space. If $X_j^*(t)$ and $X_j^{**}(t)$ are the members of $S_j^*(t)$ and $S^{**}(t)$ respectively that give the minimum distance, they are selected as the local best and the global best of the jth particle respectively.

4.3 Computational Flow

In the proposed MOPSO algorithm, the population has n particles and each particle is an m-dimensional vector, where m is the number of optimized parameters. The computational flow of the proposed MOPSO technique can be described in the following steps.

Step 1 (Initialization): Set the time counter t=0 and generate randomly n particles, $\{X_j(0), j=1, \ldots, n\}$, where $X_j(0)=[x_{j,1}(0), \ldots, x_{j,m}(0)]$. $x_{j,k}(0)$ is generated by randomly selecting a value with uniform probability over the kth optimized parameter search space $[x_k^{min}, x_k^{max}]$. Similarly, generate randomly initial velocities of all particles, $\{V_j(0), j=1, \ldots, n\}$, where $V_j(0)=[v_{j,1}(0), \ldots, v_{j,m}(0)]$. $v_{j,k}(0)$ is generated by randomly selecting a value with uniform probability over the kth dimension $[-v_k^{max}, v_k^{max}]$. Each particle in the initial population is evaluated using the objective functions. For each particle, set $S_j^*(0)=\{X_j(0)\}$ and the local best $X_j^*(0)=X_j(0)$, $j=1, \ldots, n$. Search for the nondominated solutions and form the nondominated global set $S^{**}(0)$. The nearest member in $S^{**}(0)$ to

X_j*(0) is selected as the global best X_j**(0) of the jth particle. Set the external set equal to S**(0). Set the initial value of the inertia weight $w(0)$.

Step 2 (Time updating): Update the time counter $t=t+1$.

Step 3 (Weight updating): Update the inertia weight $w(t)=\alpha\, w(t\text{-}1)$.

Step 4 (Velocity updating): Using the local best X_j*(t) and the global best X_j**(t) of each particle, $j=1, \ldots, n$, the jth particle velocity in the kth dimension is updated according to the following equation:

$$v_{j,k}(t) = w(t)\ v_{j,k}(t-1) + c_1 r_1 (x^{*}_{j,k}(t-1) - x_{j,k}(t-1)) + c_2 r_2 (x^{**}_{j,k}(t-1) - x_{j,k}(t-1)) \tag{16}$$

where c_1 and c_2 are positive constants and r_1 and r_2 are uniformly distributed random numbers in [0,1]. If a particle violates the velocity limits, set its velocity equal to the proper limit.

Step 5 (Position updating): Based on the updated velocities, each particle changes its position according to the following equation

$$x_{j,k}(t) = v_{j,k}(t) + x_{j,k}(t-1) \tag{17}$$

If a particle violates its position limits in any dimension, set its position at the proper limit.

Step 6 (Nondominated local set updating): The updated position of the jth particle is added to S_j*(t). The dominated solutions in S_j*(t) will be truncated and the set will be updated accordingly. If the size of S_j*(t) exceeds a prespecified value, the clustering algorithm will be invoked to reduce the size to its maximum limit.

Step 7 (Nondominated global set updating): The union of all nondominated local sets is formed and the nondominated solutions out of this union are extracted to be members in the nondominated global set S**(t). The size of this set will be reduced by clustering algorithm if it exceeds a prespecified value.

Step 8 (External set updating): The external Pareto-optimal set is updated as follows.

(a) Copy the members of S**(t) to the external Pareto set.
(b) Search the external Pareto set for the nondominated individuals and remove all dominated solutions from the set.
(c) If the number of the individuals externally stored in the Pareto set exceeds the maximum size, reduce the set by means of clustering.

Step 9 (Local best and global best updating): The individual distances between members in S_j*(t), and members in S**(t), are measured in the objective space. If X_j*(t) and X_j**(t) are the members of S_j*(t) and S**(t) respectively that give the minimum distance, they are selected as the local best and the global best of the jth particle respectively.

Step 10 (Stopping criteria): If the number of iterations exceeds its maximum preset limit then go to step 11, else go to step 2.

Step 11 (Best Compromise Solution): Upon having the Pareto-optimal set of nondominated solution, fuzzy-based mechanism to extract the best compromise solution is imposed to present one solution to the decision maker as described below.

4.4 Reducing a Set by Clustering

The Pareto-optimal set can be extremely large or even contain an infinite number of solutions. In this case, reducing the set of nondominated solutions without destroying the characteristics of the trade-off front is desirable from the decision maker's point of view. An average linkage based hierarchical clustering algorithm (Morse, 1980) is employed to reduce the Pareto set to manageable size. It works iteratively by joining the adjacent clusters until the required number of groups is obtained. It can be described as:

Given a set P which its size exceeds the maximum allowable size N, it is required to form a subset P with the size N*

The algorithm is illustrated in the following steps.

Step 1: Initialize cluster set C; each individual $i \in P$ constitutes a distinct cluster.
Step 2: If number of clusters $\leq N$, then go to Step 5, else go to Step 3.
Step 3: Calculate the distance of all possible pairs of clusters. The distance d_c of two clusters c_1 and $c_2 \in C$ is given as the average distance between pairs of individuals across the two clusters

$$d_c = \frac{1}{n_1 . n_2} \sum_{i_1 \in c_1,\ i_2 \in c_2} d(i_1, i_2) \tag{18}$$

where n_1 and n_2 are the number of individuals in the clusters c_1 and c_2 respectively. The function d reflects the distance in the objective space between individuals i_1 and i_2.
Step 4: Determine two clusters with minimal distance d_c. Combine these clusters into a larger one. Go to Step 2.
Step 5: Find the centroid of each cluster. Select the nearest individual in this cluster to the centroid as a representative individual and remove all other individuals from the cluster.
Step 6: Compute the reduced nondominated set P^* by uniting the representatives of the clusters.

4.5 Best Compromise Solution

Fuzzy set theory has been implemented to derive efficiently a candidate Pareto-optimal solution for the decision makers (Dhillon et al.; 1993). Upon having the Pareto-optimal set, the proposed approach presents a fuzzy-based mechanism to extract a Pareto-optimal solution as the best compromise solution. Due to imprecise nature of the decision maker's judgment, the *i*th objective function of a solution in the Pareto-optimal set, F_i, is represented by a membership function μ_i defined as

$$\mu_i = \begin{cases} 1, & F_i \leq F_i^{\min}, \\ \dfrac{F_i^{\max} - F_i}{F_i^{\max} - F_i^{\min}}, & F_i^{\min} < F_i < F_i^{\max}, \\ 0, & F_i \geq F_i^{\max}. \end{cases} \tag{19}$$

where $F_i^{\max}$ and $F_i^{\min}$ are the maximum and minimum values of the ith objective function respectively.

For each nondominated solution k, the normalized membership function μ^k is calculated as

$$\mu^k = \frac{\sum_{i=1}^{N_{obj}} \mu_i^k}{\sum_{j=1}^{M}\sum_{i=1}^{N_{obj}} \mu_i^j}, \tag{20}$$

where M is the number of nondominated solutions. The best compromise solution is the one having the maximum of μ^k. As a matter of fact, arranging all solutions in Pareto-optimal set in descending order according to their membership function will provide the decision maker with a priority list of nondominated solutions. This will guide the decision maker in view of the current operating conditions.

4.6 Settings of the Proposed Approach

In this study, the proposed MOPSO technique has been developed in order to make it suitable for solving nonlinear constrained optimization problems. A procedure is imposed to check the feasibility of the candidate solutions in all stages of the search process. This ensures the feasibility of the nondominated solutions.

The proposed MOPSO technique has been implemented on a 3-GHz PC using FORTRAN language. In all optimization runs, the number of particles and the maximum number of generations were selected as 100 and 1000, respectively. The maximum size of the Pareto-optimal set was selected as 25 solutions while the local best set size is selected as 10 solutions. If the number of nondominated Pareto optimal solutions in global best set and local best set exceeds the respective bound, the clustering technique is used.

The computational flow chart of the proposed MOPSO algorithm is depicted in Fig. 1.

5. RESULTS AND DISCUSSIONS

The proposed MOPSO technique was tested on the standard IEEE 30-bus 6-generator test system as several techniques have been tested on this standard system with reported results in the literature. The single-line diagram of the IEEE test system is shown in Fig. 2 and the detailed data are given in (Farag et al.; 1995). The bus data and line data are given in Appendix. The values of fuel cost and emission coefficients are given in Table 1.

To demonstrate the effectiveness of the proposed approach, three cases with different complexity have been considered as follows:

Case 1: Constraints (4) and (5) with $P_{loss} = 0$ are considered in optimization, i.e., the generation capacity and the power balance constraints with neglecting P_{loss} are considered.

Case 2: Constraints (4) – (8) are considered in optimization, i.e., the generation capacity and the power balance constraints with without neglecting P_{loss} are considered.

Case 3: Constraints (4) – (9) are considered in optimization, i.e., all constraints discussed in the problem formulation are considered.

Figure 1. Computational flow chart of the proposed MOPSO algorithm

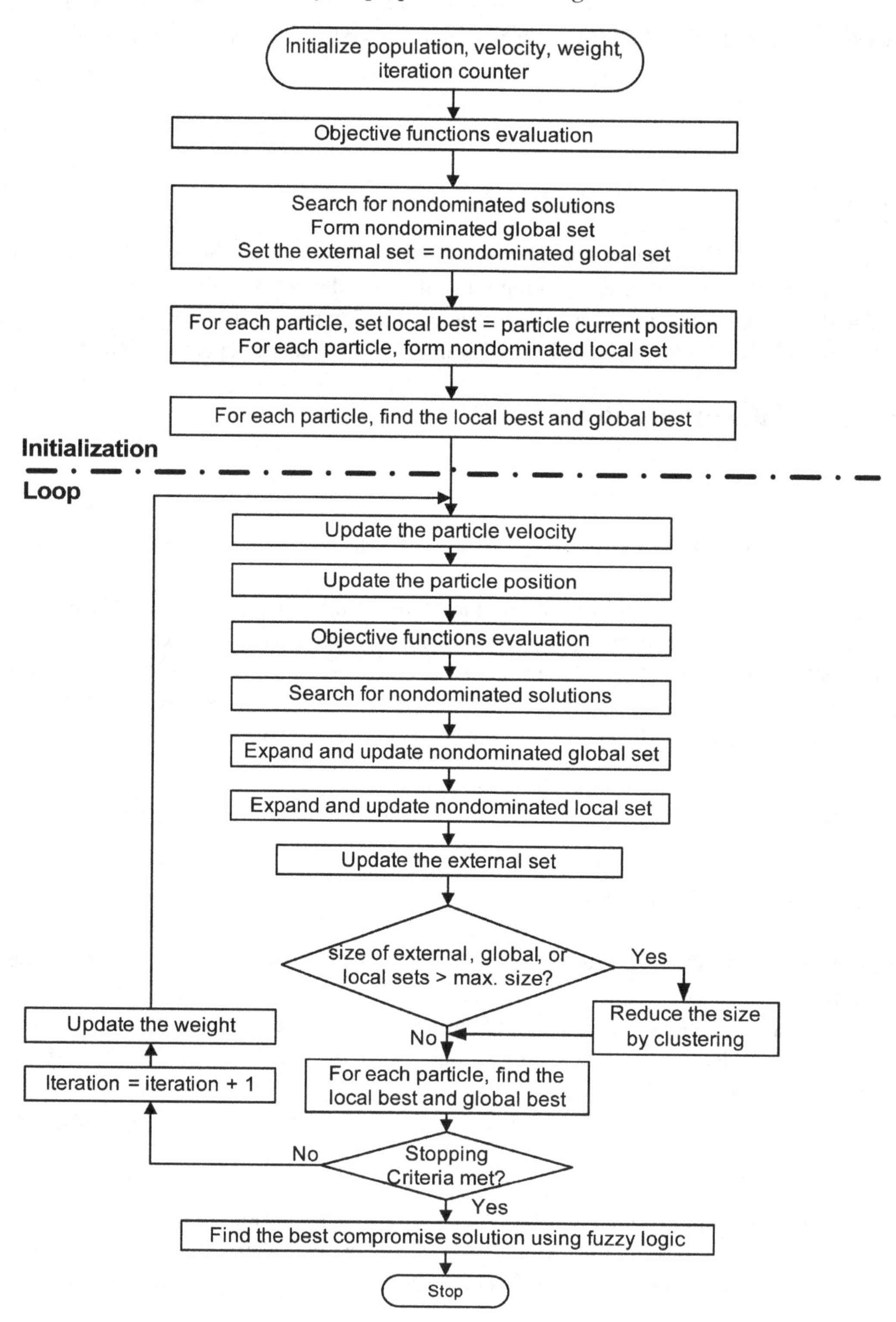

Figure 2. Single-line diagram of IEEE 30-bus test system

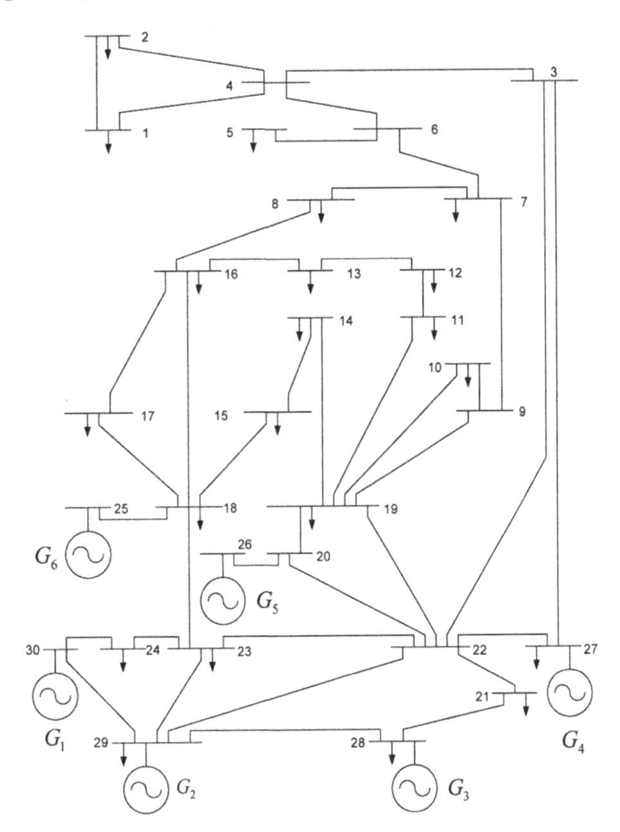

Table 1. Generator fuel cost and emission coefficients

		G_1	G_2	G_3	G_4	G_5	G_6
Cost	*a*	10	10	20	10	20	10
	b	200	150	180	100	180	150
	c	100	120	40	60	40	100
Emission	α	4.091	2.543	4.258	5.326	4.258	6.131
	β	-5.554	-6.047	-5.094	-3.550	-5.094	-5.555
	γ	6.490	5.638	4.586	3.380	4.586	5.151
	ζ	2.0E-4	5.0E-4	1.0E-6	2.0E-3	1.0E-6	1.0E-5
	λ	2.857	3.333	8.000	2.000	8.000	6.667

5.1 Single Objective Optimization

At first, fuel cost and emission objectives are optimized individually using single objective PSO in order to explore the extreme points of the trade-off surface and evaluate the diversity characteristics of the Pareto optimal solutions obtained by the proposed MOPSO technique. The best results of cost and emission functions when optimized individually using single objective PSO are given in Table 2. The PSO convergence with the individual objective optimization in all cases is shown in Fig. 3. The fast

Table 2. The best solutions for cost and emission optimized individually

	Case 1		Case 2		Case 3	
	Cost	Emission	Cost	Emission	Cost	Emission
P_{G1}	0.1098	0.4061	0.1153	0.4104	0.1513	0.4569
P_{G2}	0.2997	0.4590	0.3062	0.4629	0.3369	0.5121
P_{G3}	0.5238	0.5377	0.5962	0.5436	0.7848	0.6518
P_{G4}	1.0164	0.3833	0.9803	0.3896	1.0111	0.4335
P_{G5}	0.5249	0.5379	0.5141	0.5437	0.1050	0.1990
P_{G6}	0.3594	0.5099	0.3550	0.5149	0.4762	0.6142
Cost	**600.11**	638.24	**607.78**	645.23	**618.48**	656.73
Emission	0.2221	**0.1942**	0.2198	**0.1942**	0.2302	**0.2013**

Figure 3. Convergence of cost and emission objective functions in all cases (a) Fuel cost convergence, (b) Emission convergence

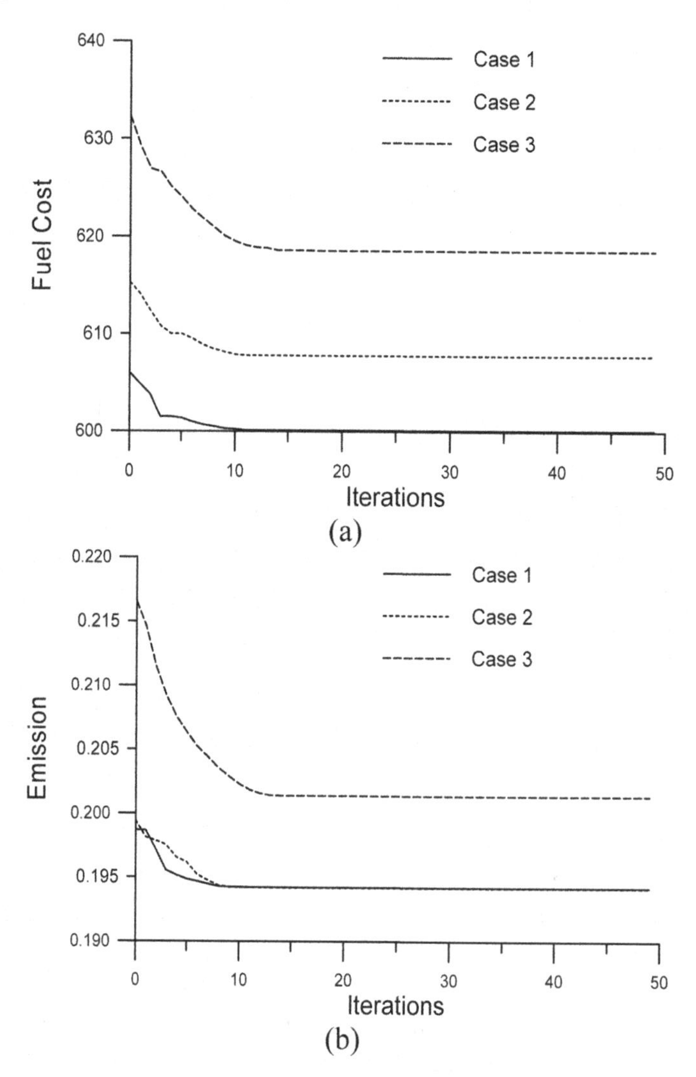

convergence of the PSO technique is quite evident as it takes only few iterations to reach the optimal solution.

5.2 Multiobjective Optimization Using the Proposed MOPSO

The proposed MOPSO approach has been implemented to optimize cost and emission objectives simultaneously considering the three cases stated above. The distribution of the Pareto optimal set over the trade-off surface is shown in Figs. 4, 5, and 6 for the Cases 1, 2, and 3 respectively. It can be seen that the proposed MOPSO technique preserves the diversity of the nondominated solutions over the Pareto-optimal front and solve effectively the problem in all cases considered. It is worth mentioning that, in each case, the Pareto optimal set has 25 nondominated solutions. Out of them, two nondominated solutions that represent the best cost and best emission are given in Table 3.

Comparing the results of single objective PSO given in Table 2 and those of the proposed MOPSO technique given in Table 3, it is clear that the results of the proposed MOPSO in all cases are almost identical with those of single objective PSO. This demonstrates the effectiveness of the proposed MOPSO to span over the entire Pareto-optimal front surface. In addition, the close agreement of the results shows clearly the capability of the proposed MOPSO technique to handle multiobjective optimization problems as the best solution of each objective along with a manageable set of nondominated solutions can be obtained in one single run.

The results of the proposed approach are compared to those reported in the literature with Case 1 using linear programming (LP), multiobjective stochastic search technique (MOSST) (Das et al.; 1998), and the recently developed strength Pareto evolutionary algorithm (SPEA) (Abido; 2003) where the results of these techniques are given in Table 4. Comparing the results of the proposed MOPSO given in Table

Figure 4. Pareto-optimal front of the proposed approach in a single run, Case 1

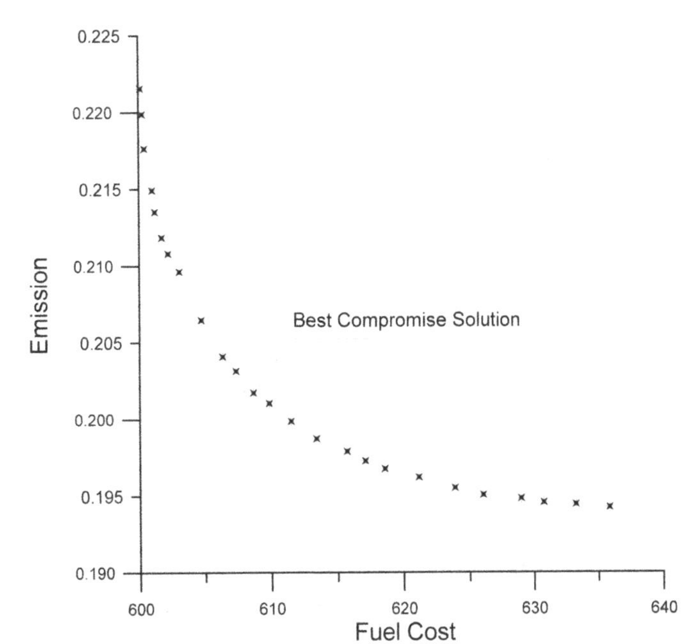

Figure 5. Pareto-optimal front of the proposed approach in a single run, Case 2

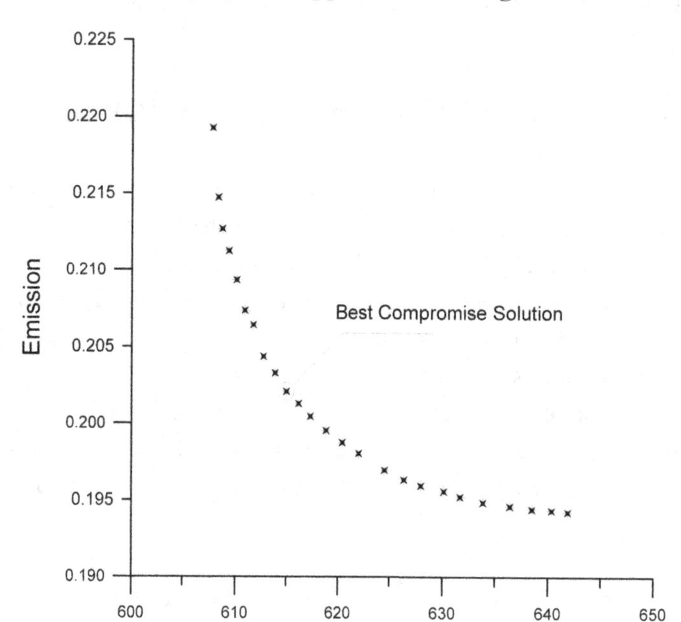

Figure 6. Pareto-optimal front of the proposed approach in a single run, Case 3

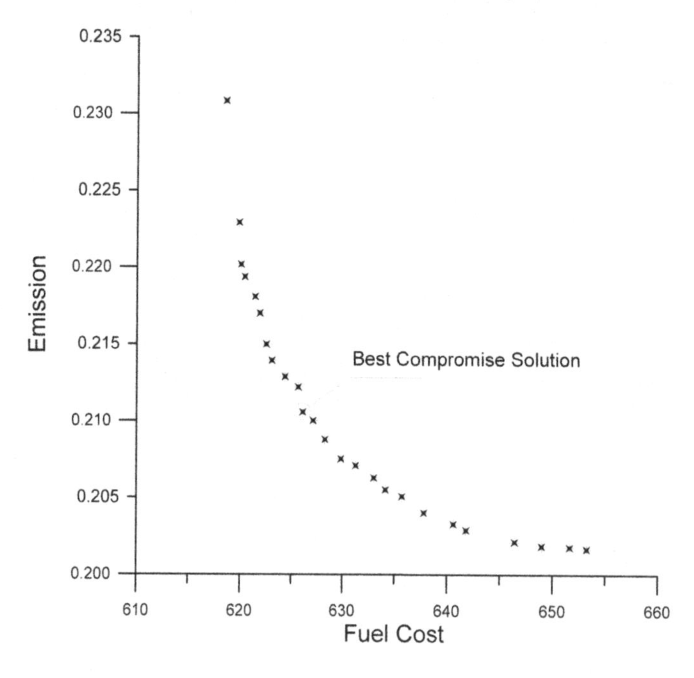

3 and those given in Table 4, it can be seen that the proposed MOPSO technique is superior compared to all reported techniques and gives better results in terms of the fuel cost saving achieved. It can be concluded that the proposed MOPSO is capable of exploring more efficient solutions. This demonstrates the potential and effectiveness of the proposed technique to solve EED problem.

Table 3. Best cost and best emission of the proposed MOPSO technique

	Best Cost			***Best Emission***		
	Case 1	***Case 2***	***Case 3***	***Case 1***	***Case 2***	***Case 3***
P_{G1}	0.1183	01207	0.1524	0.4015	0.4101	0.4589
P_{G2}	0.3019	0.3131	0.3427	0.4590	0.4594	0.5121
P_{G3}	0.5224	0.5907	0.7857	0.5332	0.5511	0.6524
P_{G4}	1.0116	0.9769	1.0180	0.3891	0.3919	0.4331
P_{G5}	0.5254	0.5155	0.0995	0.5456	0.5413	0.1981
P_{G6}	0.3544	0.3504	0.4669	0.5057	0.5111	0.6129
Cost	**600.12**	**607.79**	**618.54**	637.42	644.74	656.87
Emission	0.2216	0.2193	0.2308	**0.1942**	**0.1942**	**0.2014**

Table 4. Best cost and best emission of Case 1 of different techniques

	Best Cost			***Best Emission***		
	LP [12]	*MOSST* [9]	*SPEA* [2]	*LP* [12]	*MOSST* [9]	*SPEA* [2]
Cost	**606.31**	**605.89**	**600.15**	639.60	644.11	638.51
Emission	0.2233	0.2222	0.2215	**0.1942**	**0.1942**	**0.1942**

Table 5. Best compromise solution of the proposed MOPSO technique

	Case 1		***Case 2***		***Case 3***	
	***SPEA* [2]**	**Proposed MOPSO**	***SPEA* [2]**	**Proposed MOPSO**	***SPEA* [2]**	**Proposed MOPSO**
P_{G1}	0.2785	0.2516	0.2594	0.2367	0.2996	0.2882
P_{G2}	0.3764	0.3770	0.3848	0.3616	0.4474	0.3965
P_{G3}	0.5300	0.5283	0.5645	0.5887	0.7327	0.7320
P_{G4}	0.6931	0.7124	0.7030	0.7041	0.7284	0.7520
P_{G5}	0.5406	0.5566	0.5431	0.5635	0.1197	0.1489
P_{G6}	0.4153	0.4081	0.4091	0.4087	0.5364	0.5463
Cost	610.254	608.65	616.069	615.00	629.394	626.10
Emission	0.20055	0.2017	0.20118	0.2021	0.21043	0.2106

The fuzzy-based mechanism (Dhillon et al.; 1993) is used to evaluate each member of the Pareto-optimal set obtained by the proposed MOPSO technique. Then, the best compromise solution that has the maximum value of membership function can be extracted. This procedure has been applied in all cases and the best compromise solutions are given in Table 5. The best compromise solutions are also assigned in Figs. 4, 5, and 6 for Cases 1, 2, and 3 respectively.

5.3 Robustness and Quality Measure of the Proposed MOPSO Technique

To demonstrate the effectiveness and robustness of the proposed MOPSO technique, 20 different optimization runs have been carried out in each case with different initializations. In addition, 20 different runs of SPEA implemented in (Abido; 2003) have been carried out in each case for comparison purposes. The best solutions obtained by each technique in all 20 runs are given in Table 6. It is quite evident that the proposed MOPSO gives better results in all cases compared to SPEA.

In this study, the quality of the Pareto-optimal solutions has been measured as follows. For 20 different optimization runs with 25 nondominated solutions per run, 500 nondominated solutions can be obtained. Combining the solutions obtained by the proposed MOPSO with those obtained by SPEA, a pool of 1000 nondominated solutions can be created. The quality measure presented in (Abido; 2006) has been implemented to the created pool where the dominance conditions have been applied to all solutions in the pool. The nondominated solutions are extracted from the pool to form an elite set of Pareto-optimal solutions obtained by both techniques. Generally, the larger the number of nondominated solutions in the elite set by a certain technique, the better the quality of the solutions obtained by this technique (Abido; 2006).

For each case, this quality measure has been employed. The number of the extracted nondominated solutions in the elite set of each case is 293, 283, and 157 for Cases 1, 2, and 3 respectively. The contributions of SPEA and the proposed MOPSO are given in Table 7. It can be observed that the proposed MOPSO has the majority of the elite set members in all cases considered. It can be concluded that the most of the nondominated solutions obtained by the proposed MOPSO are true Pareto-optimal solutions. This feature of the proposed MOPSO is more pronounced in Case 3 where approximately 72.6% of the elite set size is contributed by the proposed MOPSO technique. This observation is also confirmed in Fig. 7 that shows the Pareto optimal fronts of both techniques where the superiority of the Pareto optimal front of the proposed MOPSO is evident in terms of its diversity and quality.

In addition, the normalized distance between the outer nondominated solutions of each technique represented in the elite set is measured. The normalized distance results are given in Table 7. It can be

Table 6. The best solutions for cost and emission for 20 different optimization runs

	Case 1		*Case 2*		*Case 3*	
	Cost	*Emission*	*Cost*	*Emission*	*Cost*	*Emission*
SPEA	600.15	0.1942	607.81	0.1942	618.70	0.2016
MOPSO	600.12	0.1942	607.79	0.1942	618.54	0.2014

Table 7. Quality Measure Results of nondominated solutions obtained by the proposed MOPSO approach

	# of Pareto Solutions		***Normalized Distance***	
	SPEA	***MOPSO***	***SPEA***	***MOPSO***
Case 1	146	147	0.9821	1.0000
Case 2	130	153	0.9791	1.0000
Case 3	43	114	0.9414	1.0000

seen that the nondominated solutions obtained by the proposed MOPSO span over the entire Pareto-optimal front in all cases as the extreme solutions are related to the proposed MOPSO technique since the normalized distance is 1.0 in all cases. It can be concluded that the proposed MOPSO has better diversity and quality characteristics of the nondominated solutions compared to SPEA for the problem under consideration.

The execution time for the proposed MOPSO is also assessed for the different cases considered. Table 8 shows the average execution time per generation of the proposed MOPSO which is slightly longer than that of SPEA. This can be attributed to the distance measurement between the global nondominated solutions and local nondominated solutions. However, this time can be reduced by reducing the size of the nondominated local set.

Figure 7. Pareto-optimal fronts of the proposed MOPSO and SPEA [2], Case 3

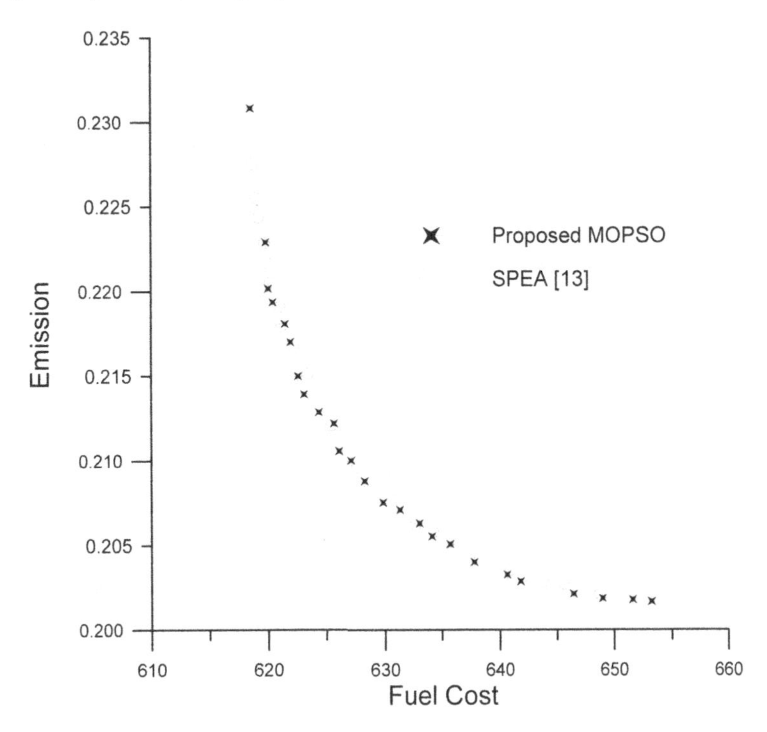

Table 8. The average execution time

	Case 1	*Case 2*	*Case 3*
SPEA	1.210	1.834	5.688
MOPSO	1.224	1.907	5.854

6. CONCLUSION

In this work, a novel multiobjective particle swarm optimization technique has been proposed and applied to environmental/economic dispatch optimization problem. The proposed MOPSO technique presents a multiobjective version of the conventional PSO technique and utilizes its effectiveness to solve the multiobjective optimization problems. The EED problem has been formulated with competing fuel cost and environmental impact objectives. The results show the potential and efficiency of the proposed MOPSO technique to solve multiobjective EED problem and produce multiple Paretooptimal solutions in one simulation run. In addition, the diversity and well-distribution characteristics of the nondominated solutions obtained by the proposed MOPSO technique have been demonstrated. The simulation results also reveal the superiority of the proposed MOPSO technique in terms of the diversity and quality of the obtained Pareto-optimal solutions.

7. ACKNOWLEDGMENT

The author acknowledges the support of King Fahd University of Petroleum & Minerals via Funded Research Project # SAB/2007-01.

8. REFERENCES

Abido, M. A. (2003). A novel multiobjective evolutionary algorithm for environmental/economic power dispatch. *Electric Power Systems Research, 65*, 71–81. doi:10.1016/S0378-7796(02)00221-3

Abido, M. A. (2003). Environmental/Economic power dispatch using multiobjective evolutionary algorithms. *IEEE Transactions on Power Systems, 18*, 1529–1537. doi:10.1109/TPWRS.2003.818693

Abido, M. A. (2006). Multiobjective evolutionary algorithms for electric power dispatch problem. *IEEE Transactions on Evolutionary Computation, 10*, 315–329. doi:10.1109/TEVC.2005.857073

Abido, M. A. (2007, July). Two-level of nondominated solutions approach to multiobjective particle swarm optimization. *Proceedings of the* 2007 *Genetic and Evolutionary Computation Conference GECCO'2007*, (pp. 726-733).

Al-Rashidi, M. R., & El-Hawary, M. E. (2007). Hybrid particle swarm optimization approach for solving the discrete OPF problem considering the valve loading effects. *IEEE Transactions on Power Systems, 22*(4), 2030–2038. doi:10.1109/TPWRS.2007.907375

Bai, H., & Zhao, B. (2006, June). A survey on application of swarm intelligence computation to electric power system. *Proceedings of the 6th World Congress on Intelligent Control and Automation*, (pp. 7587-7591).

Chang, C., Wong, K., & Fan, B. (1995). Security-constrained multiobjective generation dispatch using bicriterion global optimization. *IEE Proceedings. Generation, Transmission and Distribution, 142*, 406–414. doi:10.1049/ip-gtd:19951806

Coello, C. A. C. (1999). A comprehensive survey of evolutionary-based multiobjective optimization techniques. *Knowledge and Information Systems, 1*(3), 269–308.

Das, D. B., & Patvardhan, C. (1998). New multi-objective stochastic search technique for economic load dispatch. *IEE Proceedings. Generation, Transmission and Distribution, 145*, 747–752. doi:10.1049/ip-gtd:19982367

Dhillon, J. S., Parti, S. C., & Kothari, D. P. (1993). Stochastic economic emission load dispatch. *Electric Power Systems Research, 26*, 197–186. doi:10.1016/0378-7796(93)90011-3

El-Keib, A. A., Ma, H., & Hart, J. L. (1994). Economic dispatch in view of the clean air act of 1990. *IEEE Transactions on Power Systems, 9*, 972–978. doi:10.1109/59.317648

Farag, A., Al-Baiyat, S., & Cheng, T. C. (1995). Economic load dispatch multiobjective optimization procedures using linear programming techniques. *IEEE Transactions on Power Systems, 10*, 731–738. doi:10.1109/59.387910

Granelli, G. P., Montagna, M., Pasini, G. L., & Marannino, P. (1992). Emission constrained dynamic dispatch. *Electric Power Systems Research, 24*, 56–64. doi:10.1016/0378-7796(92)90045-3

Hazra, J., & Sinha, A. K. (2007). Congestion management using multiobjective particle swarm optimization. *IEEE Transactions on Power Systems, 22*(4), 1726–1734. doi:10.1109/TPWRS.2007.907532

Helsin, J. S., & Hobbs, B. F. (1989). A multiobjective production costing model for analyzing emission dispatching and fuel switching. *IEEE Transactions on Power Systems, 4*, 836–842. doi:10.1109/59.32569

Huang, C. M., Yang, H. T., & Huang, C. L. (1997). Bi-objective power dispatch using fuzzy satisfaction-maximizing decision approach. *IEEE Transactions on Power Systems, 12*, 1715–1721. doi:10.1109/59.627881

Jong-Bae, P., Ki-Song, L., Joong-Rin, S., & Kwang, Y. L. (2005). A particle swarm optimization for economic dispatch with nonsmooth cost functions. *IEEE Transactions on Power Systems, 20*, 34–42. doi:10.1109/TPWRS.2004.831275

Kennedy, J., & Eberhart, R. (2001) *Swarm Intelligence*. San Francisco: Morgan Kaufmann Publishers.

King, R. T. F., Harry, C. S., & Deb, K. (2006, July). Stochastic evolutionary multi-objective environmental/economic dispatch. *Proceedings of the 2006 IEEE Congress on Evolutionary Computation*, (pp. 3369-3376).

Kitamura, S., Mori, K., Shindo, S., Izui, Y., & Ozaki, Y. (2005). Multiobjective energy management system using modified MOPSO. *IEEE International Conference on Systems, Man and Cybernetics*, (pp. 3497 – 3503).

Morse, J. N. (1980). Reducing the size of nondominated set: Pruning by clustering. *Computers & Operations Research, 7*, 55–66. doi:10.1016/0305-0548(80)90014-3

Mostaghim, S., & Teich, J. (2003). Strategies for finding good local guides in multiobjective particle swarm optimization (MOPSO). *Proceedings of 2003 IEEE Swarm Intelligence Symposium*, Indianapolis, (pp. 26-33).

Selvakumar, A. I., & Thanushkodi, K. (2007). A new particle swarm optimization solution to nonconvex economic dispatch problems. *IEEE Transactions on Power Systems, 22*(1), 42–51. doi:10.1109/TPWRS.2006.889132

Srinivasan, D., Chang, C. S., & Liew, A. C. (1994). Multiobjective generation schedule using fuzzy optimal search technique. *IEE Proceedings. Generation, Transmission and Distribution, 141*, 231–241.

Talaq, J. H., El-Hawary, F., & El-Hawary, M. E. (1994). A summary of environmental/economic dispatch algorithms. *IEEE Transactions on Power Systems, 9*, 1508–1516. doi:10.1109/59.336110

Ting, T. O., Rao, M. V. C., & Loo, C. K. (2005). A novel approach for unit commitment problem via an effective hybrid particle swarm optimization. *IEEE Transactions on Power Systems, 21*(1), 411–418. doi:10.1109/TPWRS.2005.860907

Vlachogiannis, J. G., & Lee, K. Y. (2005). Determining generator contributions to transmission system using parallel vector evaluated particle swarm optimization. *IEEE Transactions on Power Systems, 20*(4), 1765–1774. doi:10.1109/TPWRS.2005.857014

Wang, Lf. & Singh, C. (2007). Environmental/economic power dispatch using fuzzified multi-objective particle swarm optimization algorithm. *Electric Power Systems Research, 77*, 1654–1664. doi:10.1016/j.epsr.2006.11.012

Wang, Lf. & Singh, C. (2006, June) Stochastic combined heat and power dispatch based on multi-objective particle swarm optimization. *Proceedings of 2006 IEEE Power Engineering Society General Meeting.*

Wang, Lf. & Singh, C. (2007) PSO-based multi-criteria optimum design of a grid-connected hybrid power system with multiple renewable sources of energy. *Proceedings of 2007 IEEE Swarm Intelligence Symposium SIS 2007.*

Yokoyama, R., Bae, S. H., Morita, T., & Sasaki, H. (1988). Multiobjective generation dispatch based on probability security criteria. *IEEE Transactions on Power Systems, 3*, 317–324. doi:10.1109/59.43217

Zitzler, E., & Thiele, L. (1998). An evolutionary algorithm for multiobjective optimization: The strength Pareto approach. *TIK-Report, 43.*

Zwe-Lee, G. (2003). Particle swarm optimization to solving the economic dispatch considering the generator constraints. *IEEE Transactions on Power Systems, 18*, 1187–1195. doi:10.1109/TPWRS.2003.814889

APPENDIX

The line and bus data of the IEEE 30-bus 6-generator system are given in Table 9 and Table 10 respectively.

Table 9. IEEE 30-bus test system line data

Line #	From Bus	To Bus	Resistance (pu)	Reactance (pu)	Susceptance (pu)	Rating (MVA)
1	1	2	0.0192	0.0575	0.0264	130
2	1	3	0.0452	0.1852	0.0204	130
3	2	4	0.0570	0.1737	0.0184	65
4	3	4	0.0132	0.0379	0.0042	130
5	2	5	0.0472	0.1983	0.0209	130
6	2	6	0.0581	0.1763	0.0187	65
7	4	6	0.0119	0.0414	0.0045	90
8	5	7	0.0460	0.1160	0.0102	70
9	6	7	0.0267	0.0820	0.0085	130
10	6	8	0.0120	0.0420	0.0045	32
11	6	9	0.0000	0.2080	0.0000	65
12	6	10	0.0000	0.5560	0.0000	32
13	9	11	0.0000	0.2080	0.0000	65
14	9	10	0.0000	0.1100	0.0000	65
15	4	12	0.0000	0.2560	0.0000	65
16	12	13	0.0000	0.1400	0.0000	65
17	12	14	0.1231	0.2559	0.0000	32
18	12	15	0.0662	0.1304	0.0000	32
19	12	16	0.0945	0.1987	0.0000	32
20	14	15	0.2210	0.1997	0.0000	16
21	16	17	0.0824	0.1923	0.0000	16
22	15	18	0.1070	0.2185	0.0000	16
23	18	19	0.0639	0.1292	0.0000	16
24	19	20	0.0340	0.0680	0.0000	32
25	10	20	0.0936	0.2090	0.0000	32
26	10	17	0.0324	0.0845	0.0000	32
27	10	21	0.0348	0.0749	0.0000	32
28	10	22	0.0727	0.1499	0.0000	32
29	21	22	0.0116	0.0236	0.0000	32
30	15	23	0.1000	0.2020	0.0000	16
31	22	24	0.1150	0.1790	0.0000	16
32	23	24	0.1320	0.2700	0.0000	16
33	24	25	0.1885	0.3292	0.0000	16

34	25	26	0.2544	0.3800	0.0000	16
35	25	27	0.1093	0.2087	0.0000	16
36	28	27	0.0000	0.3960	0.0000	65
37	27	29	0.2198	0.4153	0.0000	16
38	27	30	0.3202	0.6027	0.0000	16
39	29	30	0.2399	0.4533	0.0000	16
40	8	28	0.0636	0.2000	0.0214	32
41	6	28	0.0169	0.0599	0.0065	32

Table 10. IEEE 30-bus test system bus data

Bus	P_D (MW)	Q_D (MVAR)
1	0.00	0.00
2	21.70	12.70
3	2.40	1.20
4	7.60	1.60
5	94.20	19.00
6	0.00	0.00
7	22.80	10.90
8	30.00	30.00
9	0.00	0.00
10	5.80	2.00
11	0.00	0.00
12	11.20	7.50
13	0.00	0.00
14	6.20	1.60
15	8.20	2.50
16	3.50	1.80
17	9.00	5.80
18	3.20	0.90
19	9.50	3.40
20	2.20	0.70
21	17.50	11.20
22	0.00	0.00
23	3.20	1.60
24	8.70	6.70
25	0.00	0.00
26	3.50	2.30
27	0.00	0.00
28	0.00	0.00
29	2.40	0.90
30	10.60	1.90

Chapter 5
Doubly Fed Induction Generators:
Overview and Intelligent Control Strategies for Wind Energy Conversion Systems

Vinod Kumar
College of Technology and Engineering, India

Steven Kong
University of Queensland, Australia

Yateendra Mishra
University of Queensland, Australia

Z. Y. Dong
The Hong Kong Polytechnic University, Hong Kong

Ramesh. C. Bansal
University of Queensland, Australia

ABSTRACT

Adjustable speed induction generators, especially the Doubly-Fed Induction Generators (DFIG) are becoming increasingly popular due to its various advantages over fixed speed generator systems. A DFIG in a wind turbine has ability to generate maximum power with varying rotational speed, ability to control active and reactive by integration of electronic power converters such as the back-to-back converter, low rotor power rating resulting in low cost converter components, etc, DFIG have become very popular in large wind power conversion systems. This chapter presents an extensive literature survey over past 25 years on the different aspects of DFIG. Application of H_∞ Controller for enhanced DFIG-WT performance in terms of robust stability and reference tracking to reduce mechanical stress and vibrations is also demonstrated in the chapter.

DOI: 10.4018/978-1-60566-737-9.ch005

Figure 1. Main components of a wind turbine system (adapted from Chen and Blaabjerb, 2006)

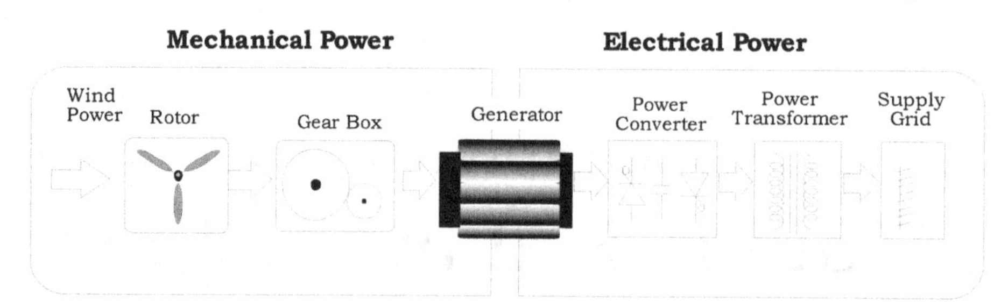

1 INTRODUCTION

Modern power generation relies heavily on conventional power plants, which uses non-renewable sources. The use of non-renewable sources such as coal and Nuclear can have adverse effect of on the environment by increased CO_2 emissions and production and radioactive wastes. To ensure the preservation of the environment and security of future electricity supply, many organizations have shifted their focus towards renewable energy sources such as wind, solar, hydro, tidal wave, biomass, etc. Volatile fuel prices used in conventional power systems are also one of the contributing factors for this change.

Historically, renewable energies have had limited industry investments, however one that has evolved furthest is wind power technology. The Global Wind Energy Council (GWEC) states that wind energy developments has occurred in more than 70 countries around the world, with an installed capacity of 74,223MW in 2006, which is an increase of 26% from 2005 to 2006 (Global Wind Energy Council, 2006).

In recent years, the application of Doubly-Fed Induction Generators (DFIG) in wind turbines has gained significant interests and developments for its high power capacity and network support capabilities. A DFIG has the ability to generate maximum power with varying rotational speed ability to independently control the active and the reactive power by the integration of electronic power converters such as the back-to-back converters, low rotor power rating resulting in low cost converter components and so on. However, the lack of experience and knowledge in grid integration of high levels of wind power remains an issue to date. Considerable amount of research is still required to develop the detailed understanding of the impact of wind power penetration levels and their operational limits on the overall power system stability.

The proposed chapter presents an extensive literature survey over past 25 years on the different aspects of DFIG and control strategies. An advanced controller for the DFIG will also be introduced which aims at improving the operational efficiency of DFIG wind turbine in terms of robust performance and to suppress dynamic loads in terms of mechanical stress and vibrations.

2 GENERATOR OVERVIEW

Wind technology has evolved over the years to today's modern Wind Energy Conversion Systems (WECS) as shown in Fig. 1 (Chen and Blaabjerb, 2006).

All wind turbine systems are comprised of the components as shown in the Fig. 1. It is just a matter of including or excluding the optional components, which is decided by the application and the type of generator used. The generator forms the major link in converting mechanical power to electrical power. Induction generators are commonly used in such applications because of their major advantages such as reduced unit cost, ruggedness, reduced size, ease of maintenance and self-protection against severe overloads and short circuits (Chen and Blaabjerb, 2006; Bansal et al., 2003; Bansal, 2005; Simoes and Farret, 2004; Simoes and Farret, 2004).

The generators used in WECS can be categorized into two types: Fixed Speed Generators (FSGs) and Adjustable Speed Generators (ASGs). A novel detail on these generators is provided in Muller et al. 2002; Datta and Ranganathan, 2002; Holdsworth et al., 2003; Rodriguez et al., 2002; Badrzadeh and Salman, 2006). Overall, FSGs are more expensive in mechanical construction, especially at high- rated power as compared to ASGs, which are widely used in WECS.

2.1 Fixed Speed Generators (FSGs)

FSGs basically operate at constant speed. The performance of fixed speed wind turbines depends a lot on the characteristics of mechanical sub circuits. A fast and strong variation in electrical output power of FSGs is observed when the system experiences strong gusts of winds. The load variations not only require a stiff power grid to enable stable operation, but also require a sturdy mechanical design to absorb high mechanical stresses. This strategy leads to expensive mechanical construction, especially at high-rated power (Muller et al., 2002). A fixed speed system (Datta and Ranganathan, 2002), even though are more simple and reliable, severely limits the energy output of a wind turbine. Since there is no torque control loop, high fluctuations in generated power are observed.

2.2 Adjustable Speed Generators (ASGs)

ASGs are becoming more popular due to their overriding advantages in the WECS. The main advantage of this system is that it is able to operate at variable speed, making it easier to operate at varying wind speeds. Moreover, modern system designs have also incorporated the use of power electronic devices to work together with variable speed generators in order to maximize the generated output power at varying wing speeds, a major improvement compared to fixed speed systems. Two types of wind generators that have become increasingly popular in modern variable speed wind turbines are the direct drive synchronous generator (DDSG) and the DFIG (Chen and Blaabjerb, 2006).

Even though DDSG have been built up to the 1.5 MW range (Muller et al., 2002), there are several disadvantages of using these systems. It requires full 1.0 per unit (p.u.) systems rating of the power converters in the design, which is expensive. Moreover, it also requires the full 1.0 p.u. systems rating of inverter output filters and electro magnetic interference (EMI) filters, making filter design difficult and costly. In contrast, DFIG, which is being analyzed in the present research work, has many advantages associated with it, as discussed in the next section, making it a better alternative for WECS.

3 THE DOUBLY FED INDUCTION GENERATOR

Recently, the overall aim of most of the WECS has been to provide a constant frequency output voltage from a variable speed system. This has given rise to the term Variable Speed Constant Frequency

Figure 2. Doubly fed induction generator wind turbine system (adapted from Muller et al. 2002)

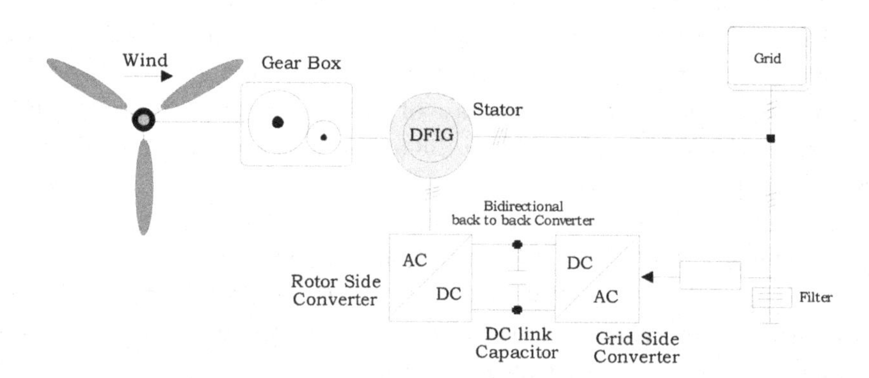

(VSCF) system. A DFIG can supply power at constant voltage and constant frequency while its rotor rotation speed varies (Pena et al., 1996a; Boldea, 2006; Liao et al., 2003; Cadirci and Ermis, 1994; Seman et al., 2004). This represents an asset in providing more flexibility in power conversion and also better stability in frequency and voltage control in the power systems to which such generators are connected. A DFIG consists of a wound rotor induction generator (WRIG) with the stator windings directly connected to the three-phase grid/load and the rotor windings connected to a back-to-back partial scale (20-30% rating) (Boldea, 2003; Hoffman et al., 2002; Ramos et al., 2002; Peterson et al., 2004; Zhi, and Xu, 2007; Xu and Cartwright, 2006; Bozhko et al., 2007; Pena et al., 2007) power converter as shown in Fig. 2 (Muller et al., 2002).

Introduced into the market in 1998, Vestas' OptiSpeed™ wind turbine is among the most widely used DFIGs. OptiSpeed™ includes a variable-speed wind turbine with a DFIG. The system is able to supply constant voltage and frequency meeting electricity supply standards in Australia and Denmark even at different rotor speeds and is able to withstand disturbances from the grid, [www.vestas.com].

The back-to-back converter is a bi-directional power converter consisting of two conventional pulse width modulation (PWM) voltage source converters and a common DC link capacitor. The size of the converter not only relates to the total generator power but also to the selected speed range or the slip power (Hansen et al., 2004; Heier and Waddington, 2006; Leonard, 2001). PWM converters have been widely used and discussed in (Pena et al., 1996 a, b; Helle and Nielsen, 2001; Morel, et al., 1998; Zhan Barker, 2006; Van Wyk and Enslin, 1983; Hudson, et al. 2003; Muller et al., 2000). Due to the bi-directional power flow ability of the converter, the DFIG may operate as a generator or motor both sub-synchronously ($0 < \text{slip} < 1$) and super-synchronously ($\text{slip} < 0$).

Neglecting losses, the rotor power handled by the converter can be represented using slip (s) as (Boldea, 2006; Dufour and Belanger, 2004; Lindholm and Rasmussen, 2003; Ledesma and Usaola, 2005).

$$P_{rotor} \approx -s P_{stator} \tag{1}$$

$$P_{stator} \approx \frac{P_{grid}}{1-s} \tag{2}$$

Table 1. Operation modes and power signs

slip	operation mode	P_{mech}	P_{stator}	P_{rotor}
0 < s < 1 (sub-synchronous)	motor	> 0	> 0	< 0
	generator	< 0	< 0	> 0
s < 0 (super-synchronous)	motor	> 0	> 0	> 0
	generator	< 0	< 0	< 0

and the mechanical power is represented as

$$P_{mech} \approx -P_{rotor}\frac{(1-s)}{s} = P_{stator} + P_{rotor} \tag{3}$$

The higher the slip, the larger is the electrical power, which is either absorbed or delivered through the rotor. At super-synchronous mode, both the stator and rotor powers add up to convert the mechanical power. The operation modes and power signs of a DFIG at sub-synchronous and super-synchronous is given in Table 1 (Boldea, 2006).

Overall, the DFIG has the following advantages over other types of systems (Muller et al. 2002; Ramos et al., 2002; Helle and Nielsen, 2001; Holdsworth et al., 2003):

- Typically, the variable speed range is ±30% around the synchronous speed. The rating of the power electronic converter is only 25–30% of the generator capacity, whereas a synchronous generator design with full-scale converter has a 100% size of the converter, so making DFIG attractive and popular from an economic point of view.
- Reduced cost of inverter filters and EMI filters, because filters are rated at 0.25 p.u. of total system power.
- Improved system efficiency due to the low rating of the converters, which means that the converter losses are less.
- Power factor control can be implemented at lower cost. The DFIG with a four-quadrant converter in the rotor circuit enables decoupled control of active and reactive power of the generator.

The use of a DFIG on a wind turbine not only improves the efficiency of energy transfer from the wind but also provides wind farms with the capability of contributing significantly to network support and operation with respect to voltage control, transient performance and damping (Holdsworth et al., 2003; Hughes et al., 2006). Furthermore, a DFIG can generate an output power greater than its own rating, even as much of twice its rating at a slip value of s = -1, (Eskander and El Hagry, 1993; Eskander et al., 1996). This makes a DFIG wind turbine system very flexible and appropriate for WECS.

Brushless Doubly Fed Induction Machines (BDFM) have also been studied and tested for performance in (Roberts et al., 2004; Runcos et al., 2004; Valenciaga and Puleston, 2007). A BDFM has two stator windings (usually one is called auxiliary winding) and as the name suggests, does not contain a wound rotor but has a cage rotor instead. The major advantage of this machine is that there are no

Figure 3. Equivalent circuit of DFIG (adapted from Peterson, 2003)

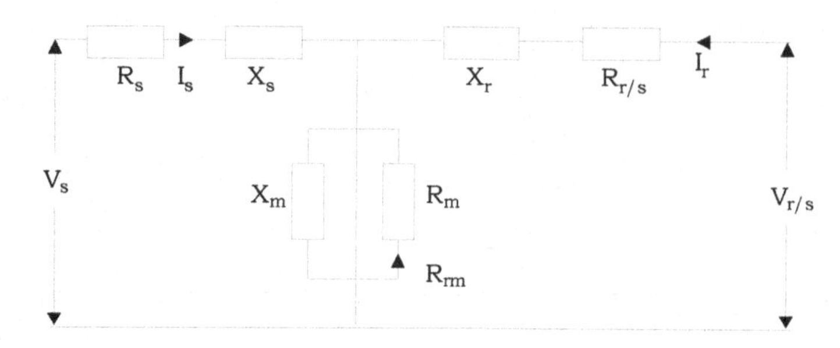

brush gear, making it appropriate for off-shore wind turbine applications where servicing costs are high and it is desirable to avoid brush gear maintenance. The number of rotor bars in the rotor plays an important role in the operation of this machine and as it was experimented in (Runcos et al., 2004), a smaller number of rotor cage bars cause high harmonic content generated in the air gap flux causing high leakage reluctances.

4 EQUIVALENT CIRCUIT AND MODELING OF DFIG

Since a DFIG is an electrical machine, it can be represented electrically using the equivalent circuit shown in Fig. 3 (Peterson, 2003; Cathey, 2001). This equivalent circuit includes the magnetization losses and is valid for one equivalent star (Y) phase and for steady-state calculations. In case of delta (Δ)-connected DFIG machine, it can still be represented by the equivalent Y representation (Fig. 3).

Applying Kirchhoff's voltage law in Fig. 3 and using $X_s = j\omega_1 L_{s\lambda}$ and $X_r = j\omega_1 L_{r\lambda}$ gives the following equations (Peterson, 2003).

$$V_s = R_s I_s + j\omega_1 L_{s\lambda} I_s + j\omega_1 L_m (I_s + I_r + I_{rm}) \tag{4}$$

$$\frac{Vr}{s} = \frac{R_r}{s} I_r + j\omega_1 L_{r\lambda} I_r + j\omega_1 L_m (I_s + I_r + I_{rm}) \tag{5}$$

$$0 = R_m I_{rm} + j\omega_1 L_m (I_s + I_r + I_{rm}) \tag{6}$$

where V_s = stator voltage; R_S = stator resistance; V_r = rotor voltage; R_r = rotor resistance; I_s = stator current; R_m = magnetizing resistance; I_r = rotor current; $L_{s\lambda}$ = stator leakage inductance; I_{rm} = magnetizing resistance current; $L_{m\lambda}$ = rotor leakage inductance; ω_1 = stator frequency; L_m = magnetizing inductance;

$$s = \frac{\omega_1 - \omega_r}{\omega_1} = \frac{\omega_2}{\omega_1} \tag{7}$$

ω_r = rotor speed; ω_2 = slip frequency. Furthermore, if the air-gap, stator and rotor flux are defined as

$$\psi_m = L_m (I_s + I_r + I_{rm}) \tag{8}$$

$$\psi_s = L_{s\lambda} I_s + L_m (I_s + I_r + I_{rm}) = L_{s\lambda} I_s + \psi_m \tag{9}$$

$$\psi_r = L_{r\lambda} I_r + L_m (I_s + I_r + I_{rm}) = L_{r\lambda} I_r + \psi_m \tag{10}$$

Then (4)-(6) become

$$V_s = R_s I_s + j\omega_1 \psi_s \tag{11}$$

$$\frac{Vr}{s} = \frac{R_r}{s} I_r + j\omega_1 \psi_r \tag{12}$$

$$0 = R_m I_{rm} + j\omega_1 \psi_m \tag{13}$$

The modeling of a DFIG has been widely discussed in (Holdsworth et al. 2003; Tapia et al., 2003; Ledesma and Usaola, 2004; Uctug et al., 1994; Ekanayake et al., 2003; Xu and Wang, 2007; Yikang et al., 2005; Meisingset and Ohnstad, 2004; Yazhou et al., 2006; Mei and Pal, 2005; Petersson et al., 2005; Fernandez et al., 2008; Moren et al., 2003; Slootweg et al., 2001; Eskandarzadeh, et al., 1999; Lei et al., 2005; Ekanayake et al., 2003; Feijoo et al., 2000). It is important to note that the three phase asynchronous machine equations are often transformed into direct and quadrature (d-q) axis as well. In order to develop n^{th}-order models for specific applications, whereby higher order models are used for studies requiring high degree of accuracy. Lower order models are used for simplicity and are achieved after certain conditions and assumptions. The transformation into two-phase components and subsequently rotating all variables into a synchronous (d-q) reference frame enables linking of the synchronous frame to stator or rotor flux of an induction machine that is also used in vector control.

Even though the DFIG system has many advantages over other designs, the most important aspect is the control of the DFIG. Wind speed is a highly unpredictable source, which can prove fatal to the DFIG without any control. Likewise, the consumers of electricity can put the electrical system into constant stress by connecting various types of loads, all of which could severely affect the DFIG system if there were no appropriate controls in place.

Through the years, many researchers have attempted to look at various types of DFIG control systems (Boldea, 2003; Ramos et al., 2002; Zhi and Xu, 2007; Bozhko et al., 2007; Hansen et al., 2004; Khatounian et al., 2003; Pena et al., 2002; Salameh and Wang, 1987; Tamura et al., 1989; Brady, 1984; Brady, 1986; Chowdhury and Chellapilla, 2006; Hansen et al., 2006; Hughes et al., 2005; Xiang et al., 2006; Patin et al., 2005; Fernandez et al., 2005; Rongve et al., 2003; Hofmann and Okafor, 2001; Hofmann and Thieme, 1998; Hofmann et al., 1997; Quang et al., 1997; Spee et al., 1995; Papathanassiou and Papadopoulos, 1997; M.P. Burton et al., 2001; Cadirci and Ermis, 1992) for different environment, electrical and mechanical conditions so as to come up with a system which is robust as well as able to handle the unwanted disturbances that the DFIG system can face.

5 CONTROL METHODS USED IN DFIG SYSTEM

Figure 4 shows the general DFIG-WT model which consists of two interconnected controllers, a DFIG controller and a wind turbine controller.

The aim of the control of the grid-side converter is (1) to maintain the DC-link capacitor voltage at a set value, regardless of the magnitude and the direction of the rotor power, and (2) to guarantee converter operation with unity power factor (zero reactive power). This means that the grid-side converter exchanges only active power with the grid.

A rotor side converter (RSC) controls the stator active and reactive power of the DFIG; several control schemes have been proposed for the rotor side DFIG control, which uses vector control techniques to manage stator output power. Direct Torque Control (DTC), Field Oriented Control (FOC) and Flux Magnitude Angle Control (FMAC) are recent research example of DFIG control techniques. The main criterion of these techniques is to enable decoupled stator active and reactive power control of DFIG.

FOC is the most widely used control technique for DFIG; the fundamentals of FOC are presented in where stator-flux or stator-voltage of the DFIG is aligned with the synchronous rotating d-q axis to provide a reference frame for rotor current vector control. Normally Proportional Integral (PI) controllers are implemented for FOC of the RSC. However a DFIG-WT is a nonlinear system with uncertain parameters. The inability of PI control to account for uncertainties means robust performance is not attained for DFIG-WT applications.

Various controller synthesis such as sliding mode control and fuzzy logic control have been investigated for FOC of the RSC to improve their performance over standard PI control. However the disadvantage of these control strategies and alike depends largely on the modeling accuracy of system parameters and dynamics. Discrepancies between mathematical models and real systems are unavoidable, in addition for DFIG-WT the dynamics and parameters can be influenced by many factors such as load variations, disturbances in stator power supply, torn and worn factors, thus the system presents many modeling uncertainties, hence the performance of these control strategies are still not guaranteed.

5.1 Pitch Control for Power Optimization

The pitch control is a mechanical method of controlling the blade angle of the wind turbine. The amount of mechanical power that a turbine can produce in steady state is given by (Rodriguez et al., 2002; Hansen et al., 2004; Helle and Nielsen, 2001).

Figure 4. Control of a DFIG wind turbine system

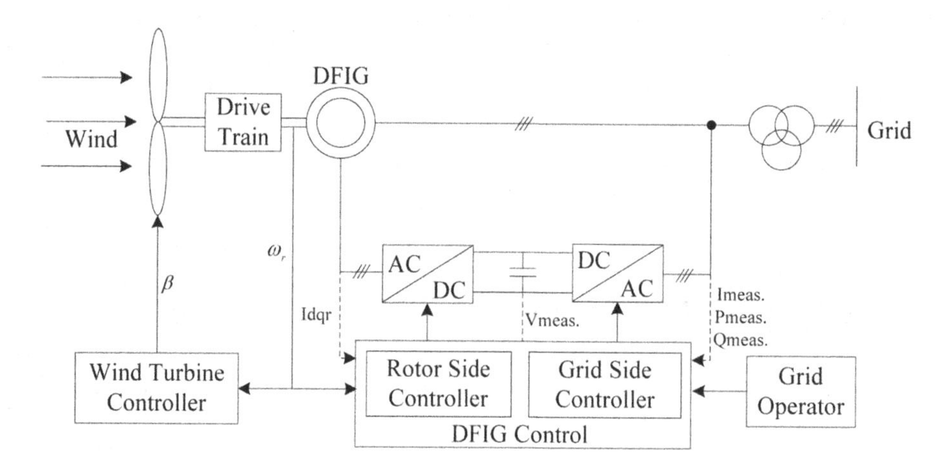

$$P_{mech} = \frac{1}{2}\rho \pi R^2 u^3 C_p(\theta,\lambda) \quad (14)$$

where ρ is the air density (kg/m^3), R the blade radius (m), u the wind speed (m/s), and $C_p(\theta,\lambda)$ the aerodynamic efficiency, which depends on pitch angle, θ and tip speed ratio, λ. The central ambition of wind turbine at low wind speeds is to adjust the rotor speed so that $C_p(\theta,\lambda)$ is always maintained at its maximum value. Fig. 5 shows the optimal power curve at different wind speeds (Holdsworth et al., 2003; Hoffman and Okafor, 2002; Burton et al., 2001; Cadirci and Ermis, 1992).

It is seen that there is a specific generator speed, which yields the maximum possible power at each wind speed. At low wind speeds the pitch angle is usually fixed i.e. it is inactive. But, at high wind speeds whereby the rated generator power is exceeded, at 12 m/s in Fig. 5, pitch control is enabled to limit the maximum output power to be equal to the rated power in order to protect the generator.

5.2 Vector and Decoupling Control

In contrast to the scalar control (Drid et al., 2006), which uses relationships valid in steady state to measure magnitude and frequency (angular speed) of voltage, current and flux linkage vector spaces; vector control uses relationships valid for dynamic states. The vector control not only uses the magnitude and the frequency but also the instantaneous positions of voltage, current and flux space vectors in general (Buja, et al., 2004). Vector control provides the ability to enable independent control of the active and reactive power from the generator. It is a control scheme based on a d-q synchronous reference frame. The stator-flux vector is forced to align with the d-axis of the synchronous frame thus allowing the decoupling of the active and reactive powers. Research work carried out in (Pena et al., 1996a,b; Ramos et al., 2002; Tapia et al., 2003; Ledesma and Usaola, 2004; Pena et al., 2000; Pena et al., 2002; Huang et al., 2006; Marques and Pinheiro, 2005; Kar and Jabr, 2005; Forchetti et al., 2002) show that the vector control is feasible on the DFIG system for large grid as well as stand-alone isolated systems. Fig. 6 shows a typical vector control simulation for controlling a DFIG system.

A different approach was presented in (Khatounian et al., 2003), in which the vector control was

Figure 5. Electrical output power of a generator at different wind speeds

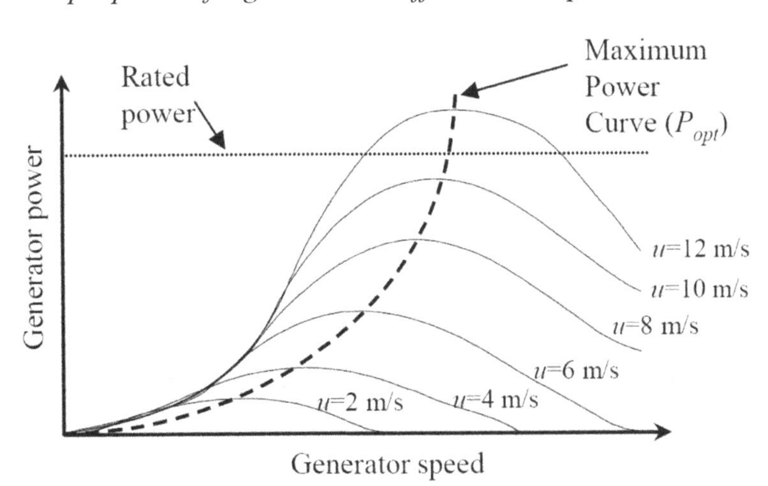

Figure 6. Vector control simulation of DFIG system (adapted from Boldea, 2005)

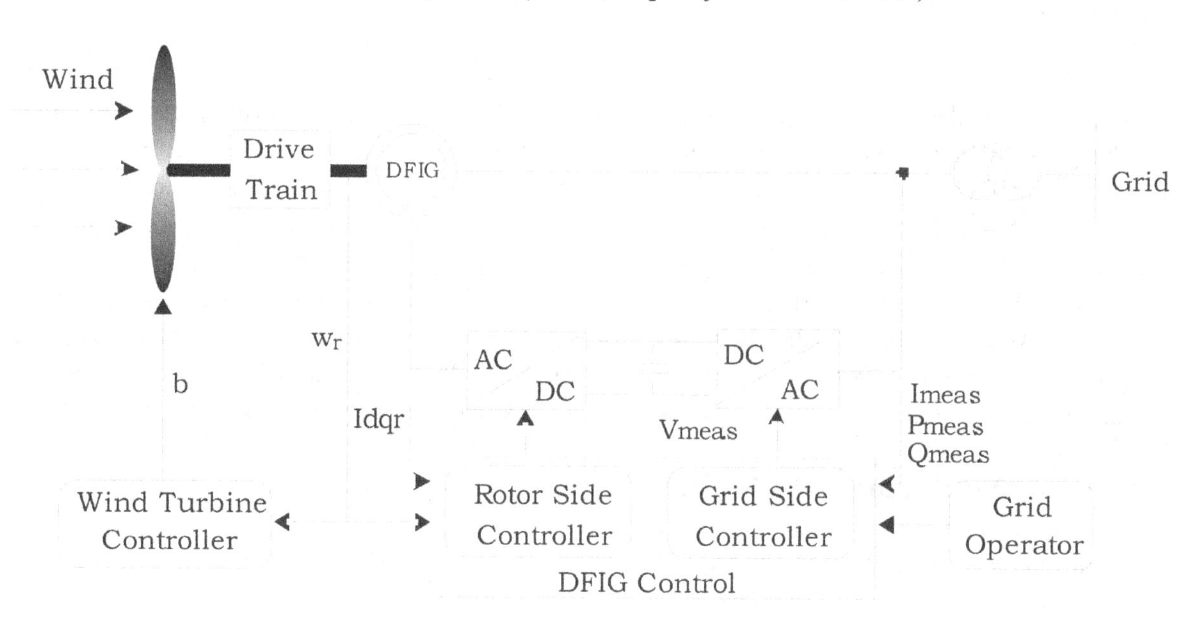

applied to an aircraft power grid, which uses a permanent magnet synchronous machine (PMSM) to provide excitation current to the power converter in the rotor circuit of a DFIG onboard the aircraft. Simulation results were shown for both DFIG as well as PMSM control strategies.

1) Active and Reactive Power Control

Even though active power is usually considered to be more important, reactive power is equally responsible for a systems electrical behavior. An induction generator, such as the DFIG, requires a reasonable amount of reactive power for its operation. In case of grid-connected systems, the generator gets the reactive power from the grid itself. In case of isolated system operation, the reactive power has to be supplied by external sources such as external capacitors (Bansal et al., (2005) and in case of an isolated DFIG system, reactive power is produced exciting the rotor by an external source (i.e. batteries), thus the need for a bank of capacitors is eliminated Caratozzolo et al., (2000). Active and reactive power control has been discussed widely in (Xu and Cartwright, 2006; Peresada et al., 1998; Xu and Cheng, 1995; Rabelo and Hofmann, 2001; Tapia et al., 2001; Tapia et al., 2001; Brekken and Mohan, 2003; Hofmann, 1999; Yamamoto and Motoyoshi, 1991).

In contradiction to (Pena et al., 1996a) which showed position sensorless control of active and reactive power using vector control, ref. (Peresada et al., 1998) showed the development of simple and robust torque tracking and reactive power control algorithm making use of both position and velocity sensors. Moreover, the sensorless control of the DFIG has also been addressed by several researchers (Xu and Cheng, 1995; Cardenas et al., 2005; Radel et al., 2001; Datta and Ranganathan, 2001; Bogalecka and Krzeminski, 2002; Hopfensperger et al., 2000; Ghosn et al., 2003; Ghosn et al., 2002; Shen and Ooi, 2005) for different control methods. The Sensorless schemes are based on estimation techniques. A common way used for estimation of parameters without taking any feedback is the use of model reference adaptive system (MRAS) observer as used in (Cardenas et al., 2005; Ghosn et al., 2003; Ghosn et al., 2002). In (Rabelo and Hofmann, 2001), simulations on active and reactive power controls were made using Laplace transforms (continuous domain) and transfer functions which are easily implemented in Simulink® software.

5.3 Passivity Control Methods

A passivity-based controller (PBC) is used to achieve stabilization via energy balancing; therefore, the regulation of power in a system is automatically achieved. Similar works on passivity control in (Lee and Nam, 2003; Becherif et al., 2003) illustrates the effectiveness of the control scheme for adjusting the mechanical speed, which is connected to the DFIGs stator terminals. This scheme leads to the control of the induction motor torque as well by making use of the rotor voltage of the DFIG as the control variable. However, authors in (Lee and Nam, 2003) went on to further design an equivalent circuit of the DFIG and induction motor control system to provide a better understanding of the performance of the system.

5.4 Flywheel Control

The inclusion of a flywheel in the DFIG control system is not taken as a separate control system of is own. In fact, even though a flywheel is included in the system, the controls are still done by other existing methods. However, in all systems containing flywheels, the sole purpose of the flywheel is to serve as an energy storage device. It is already known that a flywheel is a heavy round mass with a high moment of inertia. Once spun, the flywheel maintains the rotation for a longer period of time. In (Batlle et al., 2004), a flywheel was used as a storage device with passivity based controller as mentioned in the previous section. Whenever the power required by the local loads was less than the generated power, the excess generated energy was used to accelerate the flywheel. As the local load demand increased, the rotation energy of the flywheel was used to rotate the DFIG generator to provide the extra needed power.

Akagi et al. (1999) described a DFIG control system containing a flywheel energy storage system for the purpose of achieving load leveling over the repetitive period of time and also to perform reactive power compensation of the load. In this system, vector and decoupling control method was implemented with the objective of proper reactive power compensation.

5.5 Matrix Converters Control

The most common type of AC-DC-AC power converters used in DFIG rotor circuits is the PWM dc link converter. However, authors in (Zhang and Watthanasarn, 1998; Zhang et al., 1997) have effectively proposed the use of matrix converters to act as direct AC-AC conversion for the control of the rotor-side currents of a DFIG system. The usual PWM AC-DC-AC converters have been known to require low ratings for rotor slip power, low distortion current, enabling of flexible active and reactive power control components, transmission efficiency and voltage stability. However, the matrix converters have been shown to have similar advantages as well.

PWM converters require large DC link capacitors, which are bulky and expensive, whereas, the matrix converter requires no such capacitors. Furthermore, it is claimed that the direct AC-AC matrix conversion scheme is simpler than that used in the two-stage conversion in PWM converters. However, the proposed system together with the DFIG was only simulated and no hardware experimentation were carried out on prototypes to fully verify the practicality of the proposed system, while the use of PWM converters have existed in hardware experimental prototype and have been known to work well, as discussed in the following section.

5.6 Sliding Mode Control

Sliding mode control (De Battista et al., 2000; De Battista and Mantz, 1998) works by providing a compromise between conversion efficiency (maximum power utilization) and torque oscillation smoothing. A wind turbine setup to produce maximum power at various wind speeds is bound to operate near or at maximum rating around high wind speeds. Random wind fluctuations, wind shear and tower shadow (for wind-up turbines), may produce high torque oscillations and ripples and could easily damage the drive train, the power electronics and the generator at this point. The slide mode scheme provides a robust control to protect the system from harmful torque oscillations at the maximum operating point of the turbine by slightly reducing the maximum energy tracking control.

5.7 Direct Torque Control

The main principle of a direct torque control (DTC) (Buja and Kazmierkowski, 2004; Arnalte et al., 2002), is the ability to control directly the rotor flux linkage magnitude and generator torque, made possible by proper selection of the inverter switching states. The resulting switching pattern restricts flux and torque errors within setup flux and torque hysteresis bands. Torque and flux feedback are used for this type of control. The rotor flux is estimated using the rotor and stator current vectors while the final torque is estimated using the estimated rotor flux and the measured rotor currents.

6 SOFTWARE SIMULATIONS AND HARDWARE EXPERIMENTATIONS ON DFIG SYSTEMS

Software simulations have become an important part of research and project work. Even before physical hardware structures and prototypes are constructed, software simulations are very often done so as to predict the actual systems performance and also to provide theoretical explanations on different operational characteristics of components. Majority of the papers reviewed contained software simulations of the proposed systems.

The choice of software is often driven by the appropriateness of the application, its availability and past experience with the software. For studying effects of power generation by wind turbines connected to the grid, the most widely used simulation software's found were Simulink® with SimPower Systems provided as a toolbox in MATLAB®, DIgSILENT® Power Factory, PSCAD®, BLADED® and PSS/E®.

From the review, Simulink® was found to be the most widely used software. In (Holdsworth et al., 2003; Tapia et al., 2003; Khatounian et al., 2003; Pena et al., 2002; Tapia and Tapia, 2005; Morren and de Haan, 2005), Simulink was used for modeling and verification of proposed systems. PSCAD® was used (Sun et al., 2004) for transient analysis of grid connected wind turbines in external short circuit. For the dynamic modeling of variable speed DFIG pitch control, DIgSILENT® was used Hansen et al., 2004; Poller, 2003). Since pitch control is the control by means of mechanical systems, the dynamic modeling tool was most appropriately chosen by the authors. Another simulation software used for dynamic corrections, aerodynamics and structural dynamics is BLADED® (Ramtharan et al., 2007). The least frequently used software for system modeling and simulation purpose was found to be PSS/E® which was used in Norheim et al., 2004).

Hardware experimentations and implementations of proposed systems were found in only a handful of the literatures reviewed (Pena et al., 1996b; Salameh and Wang, 1987; Drid et al., 2006; Caratozzolo et al., 2000; Zhang and Watthanasarn, 1998). A 7.5 kW DFIM was used in experimentations in two very similar work done in Pena et al., 1996b; Zhang and Watthanasarn, 1998). Caratozzolo et al. (2000) used a much smaller 200 W wound rotor induction machine experimental prototype. In this prototype, TMS320C40 Texas Instrument digital signal processor (DSP) was used for the purpose of providing PWM and controlling the PWM converters in the rotor circuit. DSPs are used as an interface between continuous time domain machines and the discrete time domain computers. In these hardware systems, the converter used was the standard PWM AC-DC-AC converter, signifying very good practical performance of this component. In (Drid et al., 2006), experimentations were performed on a 0.8 kW machine supplying isolated loads. For a less complex system control, a microprocessor was used (Salameh and Wang, 1987), in which an INTEL® 8086 microprocessor was used for control purposes.

7 DFIG SYSTEMS CONNECTED TO GRID

There are several papers, which are associated with DFIG operation for a grid-connected system. Major areas of research in grid connected systems found in the review were: control of active and reactive power flow between wind turbine systems and the grid (Pena et al., 1996b); current control of DFIG (Ramos et al., 2002); dynamic interaction of DFIG and fuel cell with grid (Palle et al., 2005); analysis of systems under grid disturbance or fault conditions (Zhan and Barker, 2006; Uctug et al., 1994; Ekanayake et al., 2003; Xu and Wang, 2007; Yikang et al., 2005; Morren and de Haan, 2005; Sun et al., 2004; Thiringer et al., 2003; Seman et al., 2006; Hansen and Michalke, 2007; Xiang et al., 2006; Seman et al., 2006; Douglas et al., 2005); study of wind turbine and power grid synchronization techniques (Gomez and Amenedo, 2002), control comparison of DFIG connected to grid by asymmetric transmission lines (Rubira and McCulloch, 2000), various protection schemes for wind turbine systems under grid fault and stability analysis of wind farms under transient failures and voltage drop conditions Ledesma and Usaola, 2001; Vicatos et al., 2004; Niiranen, 2004; Usaola et al., 2003; Holdsworth et al., 2002).

The research areas, as seen above, are basically attempting to address the difficulties faced by grid connected wind turbine systems. Since a grid can be affected by various disturbances such as lightening, short circuits causing faults or voltage fluctuations, research in such areas were predominant. Furthermore, a grid is known to have a fixed voltage level and frequency; therefore in order to connect a wind turbine to the grid, synchronization has to be performed.

8 DFIG SYSTEMS SUPPLYING ISOLATED LOADS

A relatively small amount of research on DFIG systems supplying isolated or stand-alone systems was found. This could be due to the fact that in large countries, there are more wide grid connected systems hence research are more common in these areas. However, for smaller developing tropical island countries such as Fiji, it becomes fairly impossible to supply isolated settlements using the grid. Hence it is appropriate to develop isolated electricity generation systems to supply the local community with the much-needed electricity.

Unlike the grid connected case, where large consumer loads can be operated without greatly affecting the grid performance, isolated DFIG systems has to be made robust and very good control systems has to be put in place in order to operate the required loads. This is mainly due to the fact that there is usually only one generator supplying the settlement; hence the type of loads that the consumers use greatly affects the performance of the entire system.

The possibility of DFIG supplying an isolated load was proposed in (Vicatos and Tegopoulos, 1989), in which steady state control problem was discussed. In order to achieve constant voltage and constant frequency, the rotor has to be supplied by a voltage phasor having frequency equal to the difference between actual speed and the synchronous speed (Van Wyk and Enslin, 1983; Tamura et al., 1989; Brady, 1984; Brady, 1986; Ioannides, 1992; Salameh and Kazda, 1986; Nakra and Dube, 1988; Pena et al., 1996a). The isolated system in (Khatounian et al., 2003) used a DFIG for aircraft applications. Wind-diesel power system for supplying isolated system was presented in (Aageng et al., 2006). Dynamic modeling of isolated systems containing DFIG was proposed by Caratozzolo et al. in (2000) whereby rotor voltage was controlled in order to control the DFIG characteristics. In addition, (Drid et al., 2006) represented modeling and scalar method of control for DFIG supplying an isolated load.

9 ADVANCED CONTROLLER (H_∞ CONTROL) FORMULATION AND SYNTHESIS

The goal is to synthesize a controller for enhanced DFIG-WT performance in terms of robust stability and reference tracking to reduce mechanical stress and vibrations. Hence a mixed sensitivity H_∞ robust control formulation is applied.

Realization of Generalized Plant for Tracking Control

Figure 7 shows the generalized plant P where we have G which represent the actual mechanics of system consisting of the drive train and DFIG. K is the controller to be synthesized; Wp and Wu are used for the synthesis of the controller. This forms the generalized P which includes exogenous inputs w, control variables u, measured variables v and error signals z which is to be minimized. The input and output signals of P are defined as:

$$w = r = \begin{bmatrix} i_{dr}^{ref} \\ i_{qr}^{ref} \end{bmatrix} \tag{15}$$

$$v = \begin{bmatrix} e_{idr} \\ e_{iqr} \end{bmatrix} \tag{16}$$

$$u = \begin{bmatrix} v_{dr} \\ v_{qr} \end{bmatrix} \tag{17}$$

$$z = \begin{bmatrix} z_1 \\ \cdots \\ z_2 \end{bmatrix} = \begin{bmatrix} W_{p1} e_{idr} \\ W_{p2} e_{iqr} \\ \cdots \\ W_{u1} u_{dr} \\ W_{u2} u_{qr} \end{bmatrix} \tag{18}$$

Where $e_{idr} = i_{dr}^{ref} - i_{dr}$ and $e_{iqr} = i_{qr}^{ref} - i_{qr}$

For a S/KS mixed sensitivity design Fig. 7. *P*, including weights *Wp* and *Wu* can be interpreted as:

$$P = \begin{bmatrix} W_p & -W_p G \\ 0 & W_u \\ \cdots & \cdots \\ I_p & -G \end{bmatrix} = \left[\begin{array}{c|cc} A & 0 & -B \\ \hline C & W_p & 0 \\ 0 & 0 & W_u \\ \cdots & \cdots & \cdots \\ C & I_p & 0 \end{array}\right] \tag{19}$$

Having the objective function:

$$\min \left\| F_l\left(P, K^*_{\text{Kstabilising}}\right) \right\|_\infty = \min \left\| \begin{matrix} W_p \left(I + GK^*_{\text{Kstabilising}}\right)^{-1} \\ W_u K \left(I + GK^*_{\text{Kstabilising}}\right)^{-1} \end{matrix} \right\|_\infty \leq 1 \tag{20}$$

In mixed sensitivity design, the weights *Wp* and *Wu* are used to shape the output sensitivity function and control effort respectively. There are no specific guidelines in the literature with regards to the selection of weights for DFIG wind turbine applications. However based on the general guide lines in (Skogestad Postlethwaite, 2005) it is selected as low pass and high pass filters as:

Figure 7. S/KS Mixed sensitivity minimization in tracking control

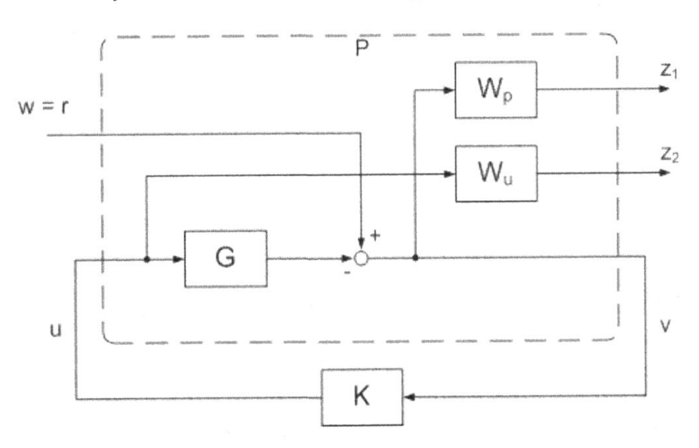

Figure 8. Open loop singular values of system and weight Wp, Wu

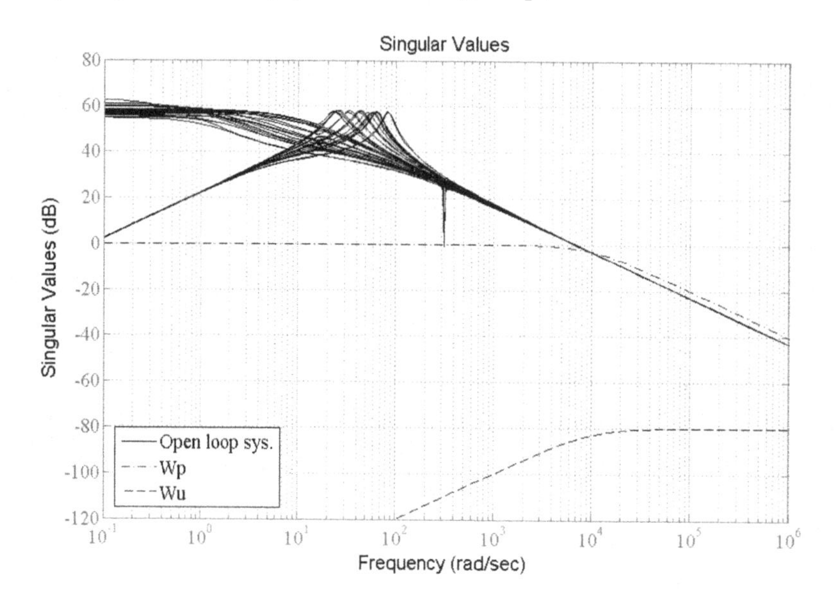

$$W_p(s) = diag(w_p, w_p) \tag{21}$$

$$W_u(s) = diag(w_u, w_u) \tag{22}$$

$$w_p = \frac{k_1}{s + \omega_c} \tag{23}$$

$$w_u = \frac{k_2 s}{s + \omega_c} \tag{24}$$

Where k_1, k_2 are the gains and ω_c is the cut of frequency.

Higher order weights filters can be used, however first order are preferred as the order of the weights adds to the synthesized controller, and a high order controller is not desirable. The parameters of *Wp* and *Wu* are selected based on the relative magnitude of input signals, their frequency dependence, and their relative importance. In our design objective (20), there is a tradeoff between these two, as we can't have very good tracking performance with minimal control effort.

Figure 8 shows the frequency response of the system *G* with uncertainties. The uncertainties considered are the varying operating points of the DFIG-WT. The cut off frequency ω_c is set at 10,000 rad/sec to cover the frequency range of interest and to limit the bandwidth of the controller. The parameters k_1 and k_2 are 10,000 and 0.0001 respectively, it was selected based on trial and error from simulation results. Ideally for tracking performance, we want k_1 to be large. However if it is too large the controller would become unstable. Optimal k_2 would be desirable for vibration suppression and robust stability against uncertainties in the high frequency range due to un-modeled dynamics. From the simulation, a large k_1

Figure 9 (a) Closed loop sensitivity function with full and 5th order controller. (b) zoomed plot showing closed loop response to system uncertainties

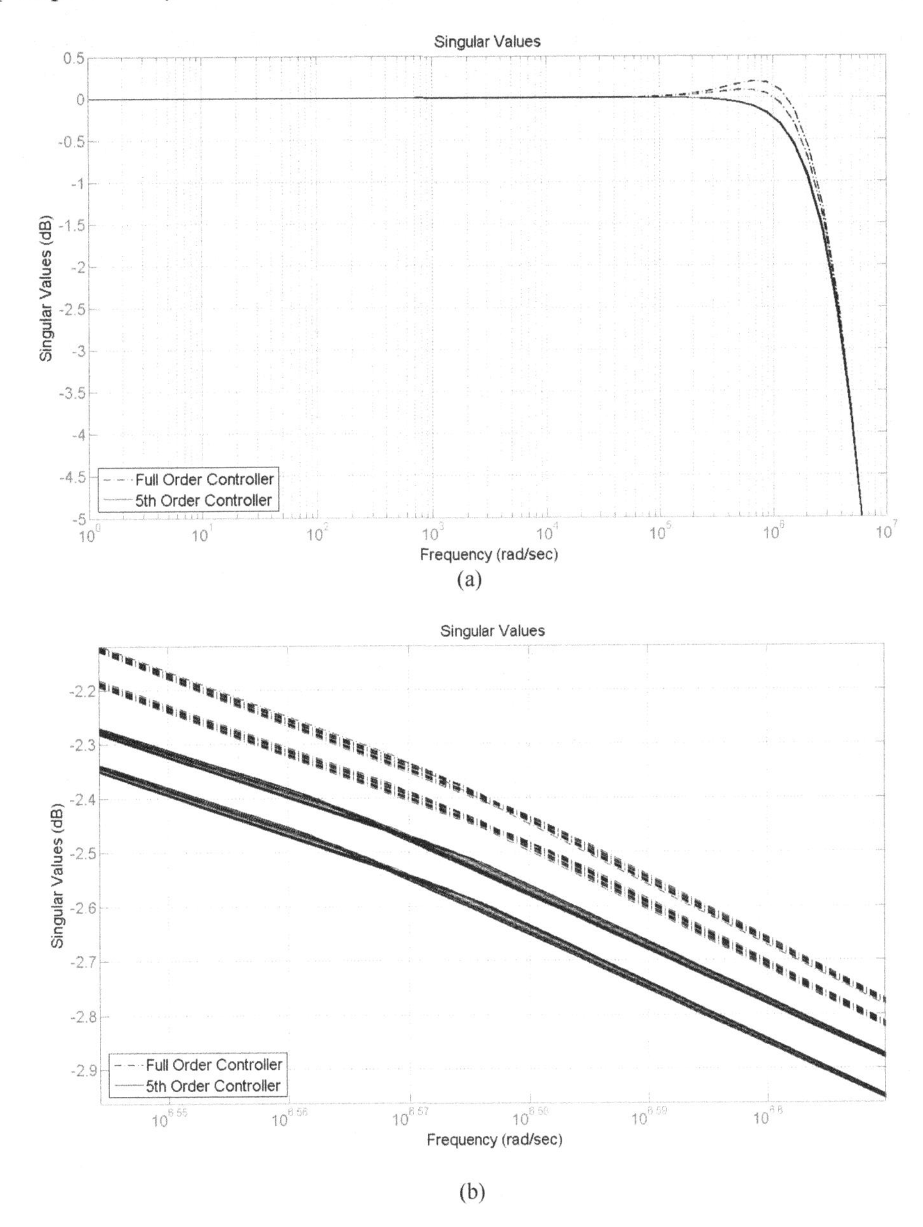

gain produces smooth sensitivity decay above the cut off frequency, in time domain this corresponds to reduced overshoot in the step response of the system. A small k_2 was found to improve steady state tracking error of the controller.

9.1 Model Reduction

The H_∞ infinity control synthesis produced a 9th order controller which was derived and composed of 5 orders from the system G and 2 orders from each of the weights Wp and Wu as it is a MIMO system with 2 inputs and 2 outputs.

Without model reduction a H_∞ controller would be difficult to realize in practice, especially for large systems. Hence Hankel Singular Values (HSV) was calculated for the controller. The HSV provides a measure of control performance to the associated controller states. Based on the HSV, Schur balanced truncation algorithm was used to reduced the control order. The reduced 5th order controller retained similar frequency response characteristics with the full order controller. The result of the perturbed closed loop frequency response of the system is shown in Fig. 9. In spite of modeling uncertainties, the H_∞ controller was able to guarantee robust performance.

The simulations were carried out using Simulink in the MATLAB environment. The results were tested on the full order dynamic system with the reduced 5th order H_∞ controller over a 5 minute (300 seconds) interval. Fig. 10 shows the random wind input which varies between 4 to 15 m/s. The response of the DFIG-WT system is shown in Fig. 11 with a zoomed plot to show in detail of the active power tracking performance. In Fig. 12 the stator reactive power reference was set at zero as current grid codes requires WT to maintain unity power factor. Lastly Fig. 13 shows the comparison between the power generated from PI and proposed controller.

10 CONCLUSION

Asynchronous generators are more common with systems up to 2 MW, beyond which direct-driven permanent magnet synchronous machines are preferred. For a machine of similar rating, energy capture can be significantly enhanced by using DFIG machines. Also, the rated torque is maintained even at super synchronous speeds, therefore it is possible to operate the DFIG system up to higher wind velocities. The voltage rating of the power devices and dc bus capacitor is substantially reduced. The size of the

Figure 10. Random wind speed input

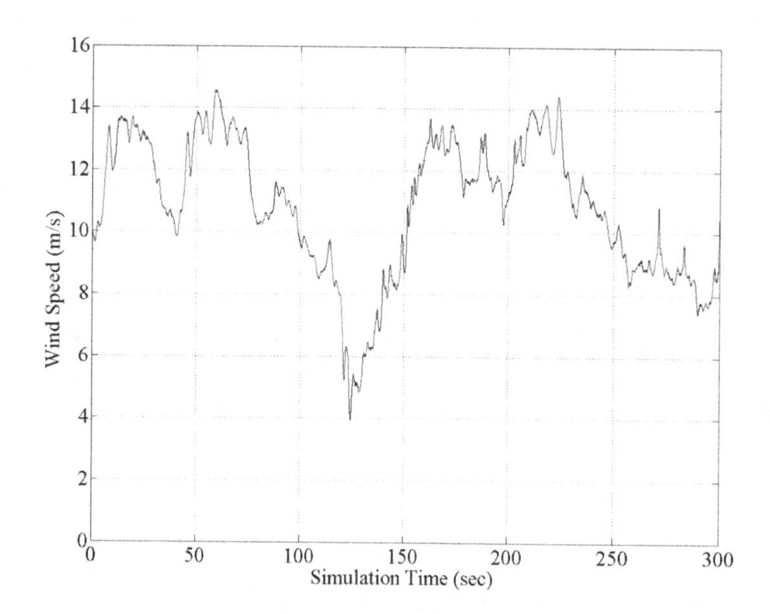

line side inductor also decreases.

The aim of the proposed controller is to provide improved stator active and reactive power generation which in terms can reduce the mechanical stress and vibration experienced by the wind turbine. From the simulation results, it can be seen that the proposed H_∞ controller was able to provide improved performance over standard PI control and at the same time achieve decoupled control of stator active

Figure 11. (a) Stator active power reference and tracking performance over 300 seconds. (b) Zoomed plot of (a)

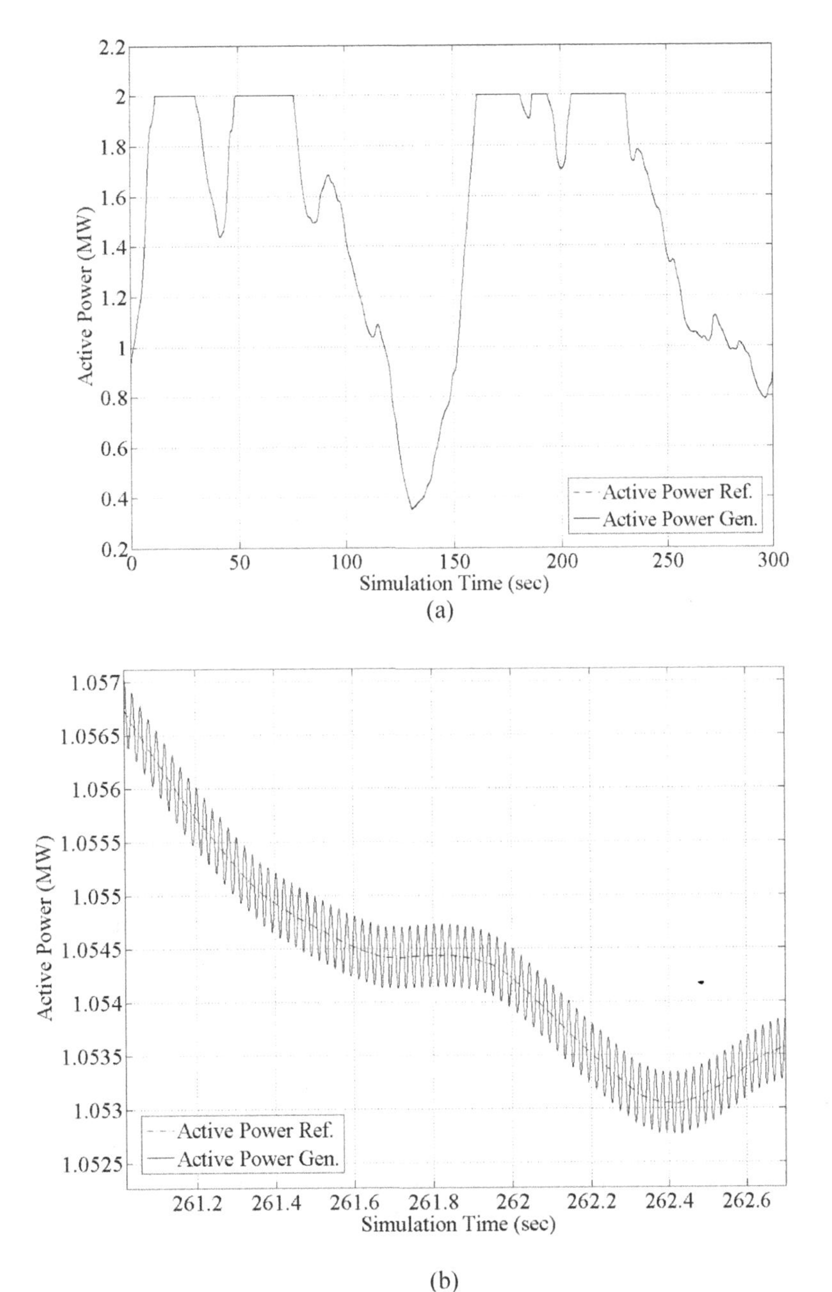

Figure 12. Stator reactive power reference and tracking

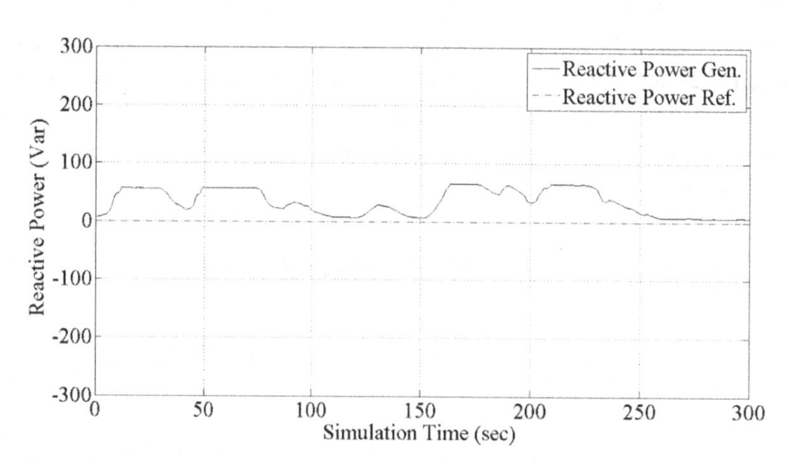

Figure 13. Stator active power output of PI vs H_∞

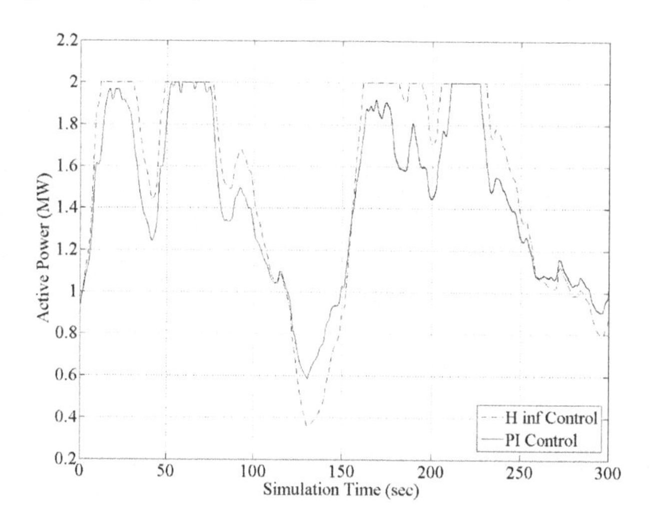

and reactive powers of the DFIG wind turbine system. Robust performance in terms of tracking and stability of the system were also achieved in the dynamic simulation under varying operating conditions. Tracking errors within 600 watts and 100 VAR were maintained, thus the results were very promising for a 2 MW generator. It can, therefore, be concluded from the review that a variable speed systems using doubly fed induction generators is superior because of higher energy output, lower rating and cost of power converters, and better utilization of a generator when compared to other existing systems.

REFERENCES

Aageng, Q., Zongxiang, L., & Yong, M. (2006). Research on isolated diesel-wind power system equipped with doubly fed induction generator, *Proc. 8th IEE International Conf. AC and DC Power Transmission*, March, (pp. 246 – 250).

Akagi, H., & Sato, H. (1999). Control and performance of a flywheel energy storage system based on a doubly fed induction generator-motor for powering conditions, *Proc. 30th Annual IEEE Power Electronics Specialists Conf.*, (p. 1), June – July, 32 – 39.

Arnalte, S., Burgos, J. C., & Amenedo, J. L. R. (2002). Direct torque control of a doubly-fed induction generator for variable speed wind turbines. *Electric Power Components and Systems, 30*(2), 199–216. doi:10.1080/153250002753427851

Badrzadeh, B., & Salman, S. K. (2006). Performance comparison of fixed-speed and doubly fed induction generator under torsional oscillations. *Proc. 8th IEE Int. Conf. AC and DC Power Transmission*, March, (pp. 167 – 171).

Bansal, R. C. (2005). Three-phase self-excited induction generators (SEIG): an overview. *IEEE Transactions on Energy Conversion, 20*(2), 292–299. doi:10.1109/TEC.2004.842395

Bansal, R. C., Bhatti, T. S., & Kothari, D. P. (2003). A bibliographical survey on induction generators for application of non conventional energy systems. *IEEE Transactions on Energy Conversion, 18*(3), 433–439. doi:10.1109/TEC.2003.815856

Bansal, R. C., Zobaa, A. F., & Saket, R. K. (2005). Some issues related to power generation using wind energy conversion systems: an overview. *Int. Journal of Emerging Electric Power Systems, 3*(2), 1070.

Batlle, C., Cerezo, A. D., & Ortega, R. (2004). Power flow control of a doubly-fed induction machine coupled to a flywheel. *Proc. IEEE Int Conf. – Control Applications,* 2, (pp. 1645 – 1650).

Becherif, M., Ortega, R., Mendes, E., & Lee, S. (2003). Passivity-based control of a doubly fed induction generator interconnected with an induction motor. *Proc. 42nd IEEE Conf. Decision and Control*, 6, December, (pp. 2711 – 2716).

Bogalecka, E., & Krzeminski, Z. (2002). Sensorless control of a double-fed machine for wind power generators. *Proc. Eur. Power Electron Conf.-Power Electron., Machines Control*, Dubrovnik and Cavtat, Slovenia.

Boldea, I. (2003). Control of electric generators: a review. *IEEE 26th annual Industrial Electronics Society Conf.*, 1, (pp. 972-980).

Boldea, I. (2005). *The electric generators handbook.* New York: CRC Press.

Boldea, I. (2006). *Variable speed generators: the electrical handbook.* New York: Taylor and Francis.

Bozhko, S. V., Gimnez, R. B., Li, R., Clare, J. C., & Asher, G. M. (2007). Control of offshore DFIG-based wind farm grid with line-commutated HVDC connection. *IEEE Transactions on Energy Conversion, 22*(1), 71–78. doi:10.1109/TEC.2006.889544

Brady, F. J. (1984). A mathematical model for the doubly fed wound rotor generator - Part I. *IEEE Transactions on Power Apparatus and Systems, 103*(4), 798–802. doi:10.1109/TPAS.1984.318356

Brady, F. J. (1986). A mathematical model for the doubly fed wound rotor generator - Part II. *IEEE Transactions on Energy Conversion, 1*(2), 180–183. doi:10.1109/TEC.1986.4765718

Brekken, T., & Mohan, N. (2003). A novel doubly-fed induction wind generator control scheme for reactive power control and torque pulsation compensation under unbalanced grid voltage conditions. *Proc. IEEE 34th Annual Conf. on Power Electronics Specialist*, 2, June, (pp. 760 – 764).

Buja, G. S., & Kazmierkowski, M. P. (2004). Direct torque control of PWM Inverter-fed ac motors – a survey'. *IEEE Transactions on Industrial Electronics*, *51*(4), 744–757. doi:10.1109/TIE.2004.831717

Burton, T., Sharpe, D., Jenkins, N., & Bossanyi, E. (2001). *Wind energy handbook*. Chichester, UK: John Wiley.

Cadirci, I., & Ermis, M. (1992). Double output induction generator operating at subsynchronous and supersynchronous speeds: steady state performance optimization and wind-energy recovery. *IEE Proceedings. Electric Power Applications*, *139*(5), 429–442.

Cadirci, I., & Ermis, M. (1994). Commutation angle analysis of a double output induction generator operating in sub- and super- synchronous modes. *Proc. 7th IEEE Mediterranean Electrotechnical Conf.*, *2*, 793-796.

Caratozzolo, P., Fossas, E., Pedra, J., & Riera, J. (2000 October). Dynamic modeling of an isolated system with DFIG. *Proc. 8th IEEE Int. Power Electronics Congress Conf.*, (pp. 287 – 292).

Cardenas, R., Pena, R., Proboste, J., Asher, G., & Clare, J. (2005). MRAS observer for sensorless control of standalone doubly fed induction generators. *IEEE Transactions on Energy Conversion*, *20*(4), 710–718. doi:10.1109/TEC.2005.847965

Cathey, J. J. (2001) Electric machines: analysis and design applying Matlab®, *McGraw Hill*, Singapore.

Chen, Z., & Blaabjerb, F. (2006). Wind energy – the worlds fastest growing energy source. *IEEE Power Electronics Society Newsletter*, (pp. 15 – 19).

Chowdhury, B. H., & Chellapilla, S. (2006). Double-fed induction generator control for variable speed wind power generation. *Electric Power Systems Research*, *76*(9-10), 786–800. doi:10.1016/j.epsr.2005.10.013

Datta, R., & Ranganathan, V. T. (2001). A simple position sensorless algorithm for rotor side field oriented control of wound rotor induction machine. *IEEE Transactions on Industrial Electronics*, *48*(4), 786–793. doi:10.1109/41.937411

Datta, R., & Ranganathan, V. T. (2002). Variable-speed wind power generation using doubly fed wound rotor induction machine – A comparison with alternative schemes. *IEEE Transactions on Energy Conversion*, *17*(3), 414–421. doi:10.1109/TEC.2002.801993

De Battista, H., & Mantz, R. J. (1998, September). Sliding mode control of torque ripple in wind energy conversion systems with slip power recovery. *Proc. 24th Annual Conf. IEEE Industrial Electronics Society*, *2*, 651 – 656.

De Battista, H., Puleston, P. F., Mantz, R. J., & Christiansen, C. F. (2000). Sliding mode control of wind energy systems with DOIG-power efficiency and torsional dynamics optimization. *IEEE Transactions on Power Systems*, *15*(2), 728–734. doi:10.1109/59.867166

Douglas, H., Pillay, P., & Barendse, P. (2005, October). The detection of interturn stator faults in doubly-fed induction generators. *Proc. 40th Industry Applications Conf.*, 2, 1097 – 1102.

Drid, S., Said, M., Sait, N., Makouf, A., & Tadjine, M. (2006). Doubly fed induction generator modelling and scalar controlled for supplying an isolated site. *J. Electrical Systems*, *2*(2), 103–115.

Dufour, C., & Belanger, J. (2004). Real-time simulation of doubly fed induction generator for wind turbine applications. *IEEE 35th Annual Power Electronics Specialists Conf.*, (pp. 3597-3603).

Ekanayake, J. B., Holdsworth, L., Guang, W. X., & Jenkins, N. (2003). Dynamic modeling of doubly fed induction generator wind turbines. *IEEE Transactions on Power Systems*, *18*(2), 803–809. doi:10.1109/TPWRS.2003.811178

Ekanayake, J. B., Holdsworth, L., & Jenkins, N. (2003). Comparison of 5th order and 3rd order machine models for doubly fed induction generator (DFIG) wind turbines. *Electric Power Systems Research*, *67*(3), 207–215. doi:10.1016/S0378-7796(03)00109-3

Eskandarzadeh, I., Uctug, M. Y., & Demirekler, M. (1999). Refrence frame modeling and steady state analysis of double output induction generator. *Proc. Int. Conf. Environmental Management*, Boston, (pp. 1022-1026).

Eskander, M. N., Abd El Motaleb, M. S., & El Khashab, H. A. (1996, March 1). Optimal control of double output induction generator using optimal regulator theory. *Proc. 4th Int. Workshop on Advanced Motion Control (AMC)*, (pp. 311-315).

Eskander, M. N., & El Hagry, M. T. (1993). Optimal performance of double output induction generator used in WECS. *The Europeans Power Electronics Association*, (pp. 276-281).

Feijoo, A., Cidras, J., & Carrillo, C. (2000). A third order model for the doubly-fed induction machine. *Electric Power Systems Research*, *56*(2), 121–128. doi:10.1016/S0378-7796(00)00103-6

Fernandez, L.M., Garcia, C.A., Jurado, F. & Saenz, J.R. (2005, May). Control system of doubly fed induction generators based wind turbines with production limits. *IEEE Int. Conf. Electric Machines and Drives*, (pp. 1936 – 1941).

Fernandez, L. M., Jurado, F., & Saenz, J. R. (2008). Aggregated dynamic model for wind farms with doubly fed induction generator wind turbines. *Renewable Energy*, *33*(1), 129–140. doi:10.1016/j.renene.2007.01.010

Forchetti, D., Garcia, G., & Valla, M. I. (2002, November 2). Vector control strategy for a doubly-fed stand-alone induction generator. *Proc. IEEE 28th Annual Conf. of the Industrial Electronics Society*, (pp. 991 – 995).

Ghosn, R., Asmar, C., Pietrzak-David, M. & De Fornel, B. (2002, Aug. 26–28). A MRAS-sensorless speed control of doubly fed induction machine, *Proc. Int. Conf. Electrical Machines*, Bruges, Belgium.

Ghosn, R., Asmar, C., Pietrzak-David, M., & De Fornel, B. (2003). A MRAS Luenberger sensorless speed control of doubly fed induction machine. *Proc. Eur. Power Electron. Conf.*, Toulose, France.

Global Wind Energy Council. (2006). Latest News: Global wind energy markets continue to boom – 2006 another record year. Retrieved May 14, 2007 from http://www.gwec.net/index.php?id=30&no_cache=1&tx_ttnews[pointer]=1&tx_ttnews[tt_news]=50&tx_ttnews[backPid]=4&cHash=e42cb8b763

Gomez, S. A., & Amenedo, J. L. R. (2002, November 4). Grid synchronization of doubly fed induction generators using direct torque control. *Proc. IEEE 28th Annual Conf. Industrial Electronics Society,* (pp. 3338 – 3343).

Hansen, A.D., Iov, F., Sorensen, P. & Blaabjerg, F. (2004 March 1 -7). Overall control strategy of variable speed doubly-fed induction generator wind turbine. *Nordic Wind Power Conf.*, Chalmers University of Technology, Goteborg, Sweden.

Hansen, A. D., & Michalke, G. (2007). Fault ride-through capability of DFIG wind turbines. *Renewable Energy, 32*(9), 1594–1610. doi:10.1016/j.renene.2006.10.008

Hansen, A. D., Sorensen, P., Iov, F., & Blaabjerg, F. (2006). Centralized power control of wind farm with doubly fed induction generators. *Renewable Energy, 31*(7), 935–951. doi:10.1016/j.renene.2005.05.011

Heier, S., & Waddington, R. (2006). *Grid integration of wind energy conversion systems*. New York: Wiley.

Helle, L. & Nielsen, S. M. (2001). Comparison of converter efficiency in large variable speed wind turbines. *IEEE 16th Annual Applied Power Electronics Conf. and Exposition, 1*, 628-634.

Hoffman, W., & Okafor, F. (2002). Optimal power utilization with doubly fed full controlled induction generator. *Proc.* 6th*IEEE Africon Conf. Africa*, 693–698.

Hofmann, W. (1999 Sept.). Optimal reactive power splitting in wind power plants controlled by double-fed induction generator. *Proc. IEEE Africon*, Cape Town, South Africa, (pp. 943-948).

Hofmann, W., & Okafor, F. (2001 November - December). Optimal control of doubly-fed full-controlled induction wind generator with high efficiency. *Proc. 27th Annual Conf. IEEE Industrial Electronics Society, 3,* 1213 – 1218.

Hofmann, W. & Thieme, A. (1998 May). Control of a doubly-fed induction generator for wind power plants. *Proc. Power Conversion and Intelligent Motion*, Nurnberg, Germany, (pp. 275-282).

Hofmann, W., Thieme, A., Dietrich, A., & Stoev, A. (1997, September). Design and control of wind power station with double feed induction generator. *Proc. Eupopean Conf. Power Electronics, Brussels, Belgium, 2*, 723–728.

Holdsworth, L., Wu, X. G., Eianayke, J. B., & Jenkins, N. (2002). Steady state and transient behavior of induction generators (including doubly fed machines) connected to distribution networks. In *Proc. IEEE Tutorial, 'Principle and modeling of distributed generators'*, July.

Holdsworth, L., Wu, X. G., Ekanayake, J. B., & Jenkins, N. (2003). Comparison of fixed speed and doubly-fed induction wind turbines during power system disturbances. *Proc. IEE – Gener. Transm. Distrib., 150*(3), 343 – 352.

Holdsworth, L., Wu, X. G., Ekanayake, J. B., & Jenkins, N. (2003). Direct solution method for initializing doubly-fed induction wind turbines in power system dynamic models. *IEEE Proc.- . Gener. Transm. Distrib., 120*(3), 334–342. doi:10.1049/ip-gtd:20030245

Hopfensperger, B., Atkinson, D. J., & Lakin, R. A. (2000). Stator-flux oriented control of a doubly-fed induction machine without position encoder. *Proc. IEE.- Electr. Power Appl., 147*(4), July, 241–250.

Huang, H., Fan, Y., Qiu, R. C., & Jiang, X. D. (2006). Quasi-steady-state rotor emf-oriented vector control of doubly fed winding induction generators for wind-energy generation. *Electric Power Components and Systems, 34*(11), 1201–1211. doi:10.1080/15325000600698597

Hudson, R.M., Stadler, F. & Seehuber, M. (2003). Large developments in power electronic converters for megawatt class wind turbines employing doubly fed generators, *Proc. Int. Power Conversion, Intelligent Motion*, Nuremberg, Germany, June .

Hughes, F. M., Anaya-Lara, O., Jenkins, N., & Strbac, G. (2005). Control of DFIG-based wind generation for power network support. *IEEE Transactions on Power Systems, 20*(4), 1958–1966. doi:10.1109/TPWRS.2005.857275

Hughes, F. M., Lara, O. A., Jenkins, N., & Strbac, G. (2006). A Power system stabilizer for DFIG-based wind generation. *IEEE Transactions on Power Systems, 21*(2), 763–772. doi:10.1109/TPWRS.2006.873037

Ioannides, M. G. (1992). Determination of frequencies in autonomous double output asynchronous generator. *IEEE Transactions on Energy Conversion, 7*(4), 747–753. doi:10.1109/60.182658

Kar, N. C., & Jabr, H. M. (2005, September 2). A novel PI gain scheduler for a vector controlled doubly-fed wind driven induction generator. *Proc. 8th Int. Conf. Electrical Machines and Systems*, (pp. 948 – 953).

Khatounian, F., Monmasson, E., Berthereau, F., Deleleau, E., & Louis, J. P. (2003, November 3). Control of a doubly fed induction generator for aircraft application. *Proc. 29th Annual Conf. IEEE Industrial Electronics Society*, (pp. 2711 – 2716).

Ledesma, P., & Usaola, J. (2001, September 4). Minimum voltage protection in variable speed wind farms. *Proc, IEEE Power Tech Conf.*, Porto, Portugal.

Ledesma, P., & Usaola, J. (2004). Effect of neglecting stator transients in doubly fed induction generators models. *IEEE Transactions on Energy Conversion, 19*(2), 459–461. doi:10.1109/TEC.2004.827045

Ledesma, P., & Usaola, J. (2005). Doubly fed induction generator model for transient stability analysis. *IEEE Transactions on Energy Conversion, 20*(2), 388–397. doi:10.1109/TEC.2005.845523

Lee, S., & Nam, K. (2003, October 3). Dynamic modeling and passivity-based control of an induction motor powered by doubly fed induction generator. *Proc. 38th IEEE Annual Meeting Conf. Industry Applications Conf.*, (pp. 1970 – 1975).

Lei, Y., Mullane, A., Lightbody, G., & Yacamini, R. (2005). Modeling of the wind turbine with a doubly fed induction generator for grid integration studies. *IEEE Transactions on Energy Conversion, 20*(2), 435–441. doi:10.1109/TEC.2005.845526

Leonard, W. (2001). *Control of electric drives*. New York: Springer Verlag.

Liao, Y., Putrus, G. A., & Smith, K. S. (2003). Evaluation of the effects of rotor harmonics in a doubly-fed induction generator with harmonic induced speed ripple. *IEEE Transactions on Energy Conversion, 18*(4), 508–515. doi:10.1109/TEC.2003.816606

Lindholm, M., & Rasmussen, T. W. (2003). Harmonic analysis of doubly fed induction generators. IEEE *Power Electronics and Drive Sys., 5th Int. Conf., 2*, 837–841.

Marques, J., & Pinheiro, H. (2005). Dynamic behavior of the doubly-fed induction generator in stator flux vector reference frame. *IEEE 36th Conf. Power Electronics Specialists*, June, (pp. 2104 – 2110).

Mei, F., & Pal, B. C. (2005, June). Modelling and small-signal analysis of a grid connected doubly fed induction generator. *IEEE Power Engineering Society General Meeting, 3*, 2101-2108.

Meisingset, M., & Ohnstad, T. M. (2004). Field tests and modeling of a wind farm with doubly fed induction generators. *Proc. Nordic Wind Power Conf.*, Goteborg, Sweden, March, 1-6.

Morel, L., Godfroid, H., Mirzaian, A., & Kauffmann, J. M. (1998). Double-fed induction machine: converter optimization and field oriented control without position sensor. *Proc. IEE.- Electr. Power Appl., 145*(4), 360–368.

Moren, J., de Hann, S. W. H., & Bauer, P. (2003). Comparison of complete and reduced models of a wind turbine with doubly fed induction generator. *Proc. 10th Eur. Conf. Power Electronics Applications*, Toulouse, France, September.

Morren, J., & de Haan, S. W. H. (2005). Ridethrough of the wind turbines with doubly-fed induction generator during a voltage dip. *IEEE Transactions on Energy Conversion, 20*(2), 435–441. doi:10.1109/TEC.2005.845526

Muller, S., Deicke, M., & De Doncker, R. W. (2000). Adjustable speed generators for wind turbines based on doubly-fed induction machines and 4-quadrant IGBT converters linked to the rotor. *Proc. IEEE Industry Applications Conf.*, 4, October, 2249 – 2254.

Muller, S., Deicke, M., & De Doncker, R. W. (2002). Doubly fed induction generator systems for wind turbines. *IEEE Industry Applications Magazine, 8*(3), 26–33. doi:10.1109/2943.999610

Nakra, H. L., & Dube, B. (1988). Slip power recovery induction generators for large vertical axis wind turbines. *IEEE Transactions on Energy Conversion, 3*(4), 733–737. doi:10.1109/60.9346

Niiranen, J. (2004). *Voltage dip ride through of a Doubly-fed generator equipped with active crowbar*. Proc. Nordic Wind Power Conf., Chalmers University of Technology, Goteborg, Sweden, March.

Norheim, I., Uhlen, K., Tande, J.O., Toftevaag, T. & Palsson, M. (2004). *Doubly fed induction generator model for power system simulation tools*. Nordic Wind Power Conf, Chalmers University of Technology, Goteborg, Sweden, March.

Palle, B., Simoes, M. G., & Farret, F. A. (2005). Dynamic interaction of an integrated doubly-fed induction generator and a fuel cell connected to grid. *IEEE 36th Conf. Power Electronics Specialists*, June, (pp. 185 – 190).

Papathanassiou, S. A., & Papadopoulos, M. P. (1994). *Simulation and control of a variable speed wind turbine equipped with double output induction generator*. Proc. PEMC, Warsaw, Poland.

Patin, N., Monmasson, E., & Louis, J. P. (2005). Analysis and control of a cascaded doubly-fed induction generator. *IEEE 32nd Annual Conf. Industrial Electronics Society*, Nov., (pp. 2481 – 2486).

Pena, R. Cardenas, R., Clare, J. & Asher, G. (2002). Control strategy of doubly fed induction generators for a wind diesel energy system. *IEEE 28th Annual Conf., 4*, 3297-3302.

Pena, R., Cardenas, R. J., Asher, G. M., & Clare, J. C. (2000, Oct.). Vector controlled induction machines for stand-alone wind energy applications. *Proc. IEEE Industry Applications Conf., 3*, 1409 – 1415.

Pena, R., Cardenas, R. J., Asher, G. M., Clare, J. C., Rodriguez, J., & Cortes, P. (2002, November). Vector control of a diesel driven doubly fed induction machine for a standalone variable speed energy system, *Proc. IEEE 28th Annual Conf. of the Industrial Electronics Society, 2*, 985 – 990.

Pena, R., Clare, J. C., & Asher, G. M. (1996a). A doubly fed induction generator using back-to-back PWM converters supplying an isolated load from a variable speed wind turbine. *IEEE Proc.- . Electric Power Applications, 143*(5), 380–387. doi:10.1049/ip-epa:19960454

Pena, R., Clare, J. C., & Asher, G. M. (1996b). Doubly fed induction generator using back-to-back PWM converters and its application to variable-speed wind-energy generation. *IEE Proceedings. Electric Power Applications, 143*(3), 231–241. doi:10.1049/ip-epa:19960288

Pena, R., Crdenas, R., Escobar, E., Clare, J., & Wheeler, P. (2007). Control system for unbalanced operation of stand-alone doubly fed induction generators. *IEEE Transactions on Energy Conversion, 22*(2), 544–545. doi:10.1109/TEC.2007.895393

Pena, R. S., Asher, G. M., Clare, J. C., & Cardenas, R. (1996). A constant frequency constant voltage variable speed stand alone wound rotor induction generator. *Proc. Int. Conf. Opportunities and Advances in Int. Electric Power Generation, (Conf. Publ. 419)*, March, (pp. 111 – 114).

Peresada, S., Tilli, A., & Tonielli, A. (1998, September). Robust active-reactive power control of a doubly fed induction generator. *Proc. 24th Annual Conf. IEEE Industrial Electronics Society, 2*, 1621 – 1625.

Peterson, A. (2003). *Analysis, modeling and control of doubly fed induction generators for wind turbines*. PhD Thesis, Chalmers University of Technology, Goteborg, Sweden.

Peterson, A., Lundberg, S. & Thiringer, T. (2004, March 1 – 7). *A DFIG wind turbine ride through system influence on the energy production*. Nordic Wind Power Conf., Chalmers University of Technology, Goteborg, Sweden.

Petersson, A., Thiringer, T., Harnefors, L., & Petru, T. (2005). Modeling and experimental verification of grid interaction of a DFIG wind turbine. *IEEE Transactions on Energy Conversion, 20*(4), 878–886. doi:10.1109/TEC.2005.853750

Poller, M. A. (2003, June 23-26). Doubly-fed induction machine model stability assessment of wind farms. *IEEE Power Tech Conf. Proc., 3*, 1-6.

Quang, N. P., Dittrich, J. A., & Thieme, A. (1997). Doubly fed induction machine as generator: control algorithm with decoupling of torque and power factor. *Electrical Engineering, 80*, 325–335. doi:10.1007/BF01370969

Rabelo, B., & Hofmann, W. (2001, October). Optimal active and reactive power control with the doubly fed induction generator in the MW-class wind-turbines. *Proc. 4th IEEE Int. Conf. Power Electronics and Drive Systems, 1*, 53 – 58.

Radel, U., Navarro, D., Berger, G., & Berg, S. (2001). Sensorless field-oriented control of a slipring induction generator for a 2.5 MW wind power plant from Nordex energy GMBH. *Proc. Eur. Power Electron. Conf.*, Graz, Austria.

Ramos, C. J., Martins, A. P., Araujo, A. S. & Carvalho, A. S. (2002). Current control in the grid connection of the double-output induction generator linked to a variable speed wind turbine. *IEEE 28th Annual Conf., 2*, 979-984.

Ramtharan, G., Ekanayake, J. B., & Jenkins, N. (2007). Frequency support from doubly fed induction generator wind turbines. *IET. - . Renew. Power Gener., 1*(1), 3–9. doi:10.1049/iet-rpg:20060019

Roberts, P. C., McMahon, R. A., Tavner, P. J., Maciejowski, J. M., Flack, T. J., & Wang, X. (2004). Performance of rotors in a brushless doubly fed induction machine (BDFM). *Proc. 16th Int. Conf. Electrical Machines*, (pp. 450-455).

Rodriguez, J. M., Fernandez, J. L., Beatu, D., Iturbe, R., Usaola, J., Ledesma, P., & Wilhelmi, J. R. (2002). Incidence on power system dynamics of high penetration of fixed speed and doubly fed wind energy systems: study of the Spanish case. *IEEE Transactions on Power Systems, 17*(4), 1089–1095. doi:10.1109/TPWRS.2002.804971

Rongve, K. S., Naess, B. I., Undeland, T. M., & Gjengedal, T. (2003, June). Overview of torque control of a doubly fed induction generator. *Proc. IEEE Power Tech. Conf.*, Bologna, Italy, 3, 292 – 297.

Rubira, S. D., & McCulloch, M. D. (2000). Control comparison of doubly fed wind generators connected to the grid by asymmetric transmission lines. *IEEE Transactions on Industry Applications, 36*(4), 986–991. doi:10.1109/28.855951

Runcos, F., Carlson, R., Oliveira, A. M., Peng, P. K., & Sadowski, N. (2004). Performance analysis of a brushless double fed cage induction generator. *Proc. Nordic Wind Power Conf.*, Goteborg, Sweden, March, 1-8.

Salameh, Z. M., & Kazda, L. F. (1986). Analysis of the steady state performance of the double output induction generator. *IEEE Transactions on Energy Conversion, 1*(1), 26–32. doi:10.1109/TEC.1986.4765666

Salameh, Z. M., & Wang, S. (1987). Microprocessor control of double output induction generation. I. Inverter firing circuit. *IEEE Transactions on Energy Conversion, 2*(2), 175–181. doi:10.1109/TEC.1987.4765826

Seman, S., Niiranen, J., & Arkkio, A. (2006). Ride-through analysis of doubly fed induction wind-power generator under unsymmetrical network disturbance. *IEEE Transactions on Power Systems*, *21*(4), 1782–1789. doi:10.1109/TPWRS.2006.882471

Seman, S., Niiranen, J., Kanerva, S., & Arkkio, A. (2004). Analysis of a 1.7 MVA doubly fed wind-power induction generator during power system disturbances. *Proc. Nordic Workshop on Power and Industrial Electronics*, Trondheim, Norway, June.

Seman, S., Niiranen, J., Kanerva, S., Arkkio, A., & Saitz, J. (2006). Performance study of doubly fed wind-power induction generator under network disturbances. *IEEE Transactions on Energy Conversion*, *21*(4), 883–890. doi:10.1109/TEC.2005.853741

Shen, B., & Ooi, B. T. (2005). Novel sensorless decoupled p-q control of doubly-fed induction generator (DFIG) based on phase locking to gamma-delta frame. *IEEE 36th Conf. Power Electronics Specialists*, June, (pp. 2670 – 2675).

Simoes, M. G., & Farret, F. A. (2004). *Renewable energy systems: design and analysis with induction generators.* New York: CRC Press.

Skogestad, S., & Postlethwaite, I. (2005). Multivariable Feedback Control: Analysis and Design (2nd Ed.). New York: John Wiley & Sons.

Slootweg, J. G., Polinder, H., & Kling, W. L. (2001, July). Dynamic modelling of a wind turbine with doubly fed induction generator. *IEEE Power Engineering Society Summer Meeting*, Vancouver, Canada, *1*, 644 – 649.

Spee, R., Bhowmik, S., & Eslin, J. H. R. (1995). Novel control strategies for variable speed doubly fed wind power generation systems. *Renewable Energy*, *6*(8), 907–915. doi:10.1016/0960-1481(95)00096-6

Sun, T., Chen, Z. & Blaabjerg, F. (2004). *Transient analysis of grid connected wind turbines with DFIG after an external short circuit fault.* Nordic Wind Power Conf. Chalmers University of Technology, Goteborg, Sweden, March.

Tamura, J., Sasaki, T., Ishikawa, S., & Hasegawa, J. (1989). Analysis of the steady state characteristics of doubly fed synchronous machines. *IEEE Transactions on Energy Conversion*, *4*(2), 250–256. doi:10.1109/60.17919

Tapia, A., Tapia, G., Ostolaza, J. X., & Saenz, J. R. (2003). Modeling and control of a wind turbine driven doubly fed induction generator. *IEEE Transactions on Energy Conversion*, *18*(2), 194–204. doi:10.1109/TEC.2003.811727

Tapia, A., Tapia, G., Ostolaza, J. X., Saenz, J. R., Criado, R., & Berasaregui, J. L. (2001). Reactive power control of a wind farm made up of doubly fed induction generators (I, II). *Proc. IEEE Power Tech Conf.*, Porto, Portugal, 4, Sept.

Tapia, G., & Tapia, A. (2005). Wind generation optimization algorithm for a doubly fed induction generator. *Proc. IEE – Gener. Transm. Distrib.*, *152*(2), 253-263.

Thiringer, T., Petterson, A., & Petru, T. (2003, July). Grid disturbance response of a wind turbine equipped with induction generator and doubly fed induction generator. *IEEE Power Engineering Society General Meeting*, 3, 1542 – 1547.

Uctug, M. Y., Eskandarzadeh, I., & Ince, H. (1994). Modeling and output power optimization of a winf turbine driven double output induction generator. *Proc. IEEE – Electric Power Applications, 141*(2), 33 – 38.

Usaola, J., Ledesma, P., Rodriguez, J. M., Fernandez, J. L., Beato, D., Iturbe, R., & Wilhelmi, J. R. (2003 July). Transient stability studies in grids with great wind power penetration: modelling issues and operation requirements. *IEEE Power Engineering Society General Meeting*, 3, 1541-1544.

Valenciaga, F., & Puleston, P. F. (2007). Variable structure control of a wind energy conversion system based on a brushless doubly fed reluctance generator. *IEEE Transactions on Energy Conversion, 22*(2), 499–506. doi:10.1109/TEC.2006.875447

Van Wyk, J. D., & Enslin, J. H. R. (1983). A study of a wind power converter with microcomputer based maximal power control utilising an oversynchronous electronic Scherbius cascade. *Proc of the lnt. Power Electronics Conf.*, Tokyo, Japan, I, 766-777.

Vernados, P. G., Katiniotis, I. M., & Ioannides, M. G. (2003). Development of an experimental investigation procedure on double fed electric machine-based actuator for wind power. *Sensors and Actuators. A, Physical, 106*(1-3), 302–305. doi:10.1016/S0924-4247(03)00190-0

Vicatos, M. S., & Tegopoulos, J. A. (1989). Steady state analysis of a doubly fed induction generator under synchronous operation. *IEEE Transactions on Energy Conversion, 4*(3), 495–501. doi:10.1109/60.43254

Vicatos, M. S., & Tegopoulos, J. A. (1991). Transient state analysis of a doubly fed induction generator under three phase short circuit. *IEEE Transactions on Energy Conversion, 6*(1), 62–68. doi:10.1109/60.73790

Xiang, D., Ran, L., Bumby, J. R., Tavner, P. J., & Yang, S. (2006). Coordinated control of an HVDC link and doubly fed induction generators in a large offshore wind farm. *IEEE Transactions on Power Delivery, 21*(1), 463–471. doi:10.1109/TPWRD.2005.858785

Xiang, D., Ran, R. L., Tavner, P. J., & Yang, S. (2006). Control of a doubly fed induction generator in a wind turbine during grid fault ride-through. *IEEE Transactions on Energy Conversion, 21*(3), 652–662. doi:10.1109/TEC.2006.875783

Xu, L., & Cartwright, P. (2006). Direct active and reactive power control of DFIG for wind energy generation. *IEEE Transactions on Energy Conversion, 21*(3), 750–758. doi:10.1109/TEC.2006.875472

Xu, L., & Cheng, W. (1995). Torque and reactive power control of a doubly-fed induction machine by position sensorless scheme. *IEEE Transactions on Industry Applications, 31*(3), 636–641. doi:10.1109/28.382126

Xu, L., & Wang, Y. (2007). Dynamic modeling and control of DFIG-based wind turbines under unbalanced network conditions. *IEEE Transactions on Power Systems, 22*(1), 314–323. doi:10.1109/TPWRS.2006.889113

Yamamoto, M., & Motoyoshi, O. (1991). Active and reactive power control for doubly-fed wound rotor induction generator. *IEEE Transactions on Power Electronics, 6*(4), 624–629. doi:10.1109/63.97761

Yazhou, L., Mullane, A., Lightbody, G., & Yacamini, R. (2006). Modeling of the wind turbine with a doubly fed induction generator for grid integration studies. *IEEE Transactions on Energy Conversion, 21*(1), 257–264. doi:10.1109/TEC.2005.847958

Yikang, H., Jiabing, H., & Rende, Z. (2005). Modeling and control of wind-turbine used DFIG under network fault. *Proc. 8th Int. Conf. Electrical Machines and Systems*, 2, Sept., 986 – 991.

Zhan, C., & Barker, C. D. (2006). Fault ride-through capability investigation of a doubly-fed induction generator with an additional series-connected voltage source converter. *Proc. 8th IEE International Conf. AC and DC Power Transmission*, March, 79 – 84.

Zhang, L., Watthanasarn, C. & Shepherd, (1997). Application of a matrix converter for the power control of a variable –speed wind-turbine driving a doubly fed induction generator. *Proc. 23rd Int. Conf. Industrial Electronics, Control and Instrumentation*, 2, Nov., 906 – 911.

Zhang, L., & Watthanasarn, C. (1998). A matrix converter excited doubly-fed induction machine as a wind power generator. *Proc. 7th Int. Conf. on (IEE Conf. Publ. 456) Power Electronics and Variable Speed Drives*, London, Sept., 532 – 537.

Zhi, D., & Xu, L. (2007). Direct power control of DFIG with constant switching frequency and improved transient performance. *IEEE Transactions on Energy Conversion, 22*(1), 110–118. doi:10.1109/TEC.2006.889549

APPENDIX

Nominal Values of the DFIG-WT

Rated Voltage (Y)	**=**	**690 V**
Rated Current	=	1900 A
Rated Frequency	=	50 Hz
Rated Power	=	2 MW

Parameters of the DFIG

R_s	=	**0.0022 Ω**
R_r	=	0.0018 Ω
L_{ss}	=	0.12 mH
L_{rr}	=	0.05 mH
L_m	=	2.9 mH
of Poles	=	2
Moment of Inertia (J_{em})	=	90 kgm^2

Parameters of the Turbine Rotor & Drive Train

Blade Radius	=	40 m
Gearbox Ratio	=	85.97
Moment of Inertia (J_{wt})	=	49.5×10^5 kgm^2

Section 2
Models and Tools

Chapter 6
Use of Neural Networks for Modeling Energy Consumption in the Residential Sector

Merih Aydinalp Koksal
Hacettepe University, Turkey

ABSTRACT

This chapter investigates the use of neural networks (NN) for modeling of residential energy consumption. Currently, engineering and conditional demand analysis (CDA) approaches are mainly used for residential energy modeling. The studies on the use of NN for residential energy consumption modeling are limited to estimating the energy use of individual or a group of buildings. Development of a national residential end-use energy consumption model using NN approach is presented in this chapter. The comparative evaluation of the results of the model shows NN approach can be used to accurately predict and categorize the energy consumption in the residential sector as well as the other two approaches. Based on the specific advantages and disadvantages of three models, developing a hybrid model consisting of NN and engineering models is suggested.

INTRODUCTION

Energy use and associated greenhouse gas (GHG) emissions, and their potential effects on the global climate change have been the worldwide concern especially after the *Kyoto Protocol*. Improving the energy efficiency especially in the residential sector is one of the most effective ways to reduce end-use energy consumption and associated emissions. This is particularly effective for countries with high per capita energy consumptions, such as seen in Canada and USA.

In recent years, the residential energy use and associated GHG emissions in the developed countries account for almost one fifth of the total due to the significant increases in dwelling square footages and the number of appliances used in the households. Figure 1 shows sectoral percentage distributions of

DOI: 10.4018/978-1-60566-737-9.ch006

Figure 1. Sectoral energy consumption and GHG emission distributions in 2005 in Canada and USA (EIA, 2008; OEE, 2008)

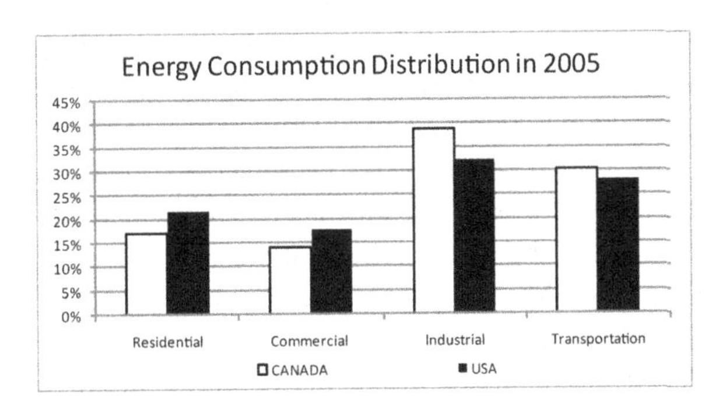

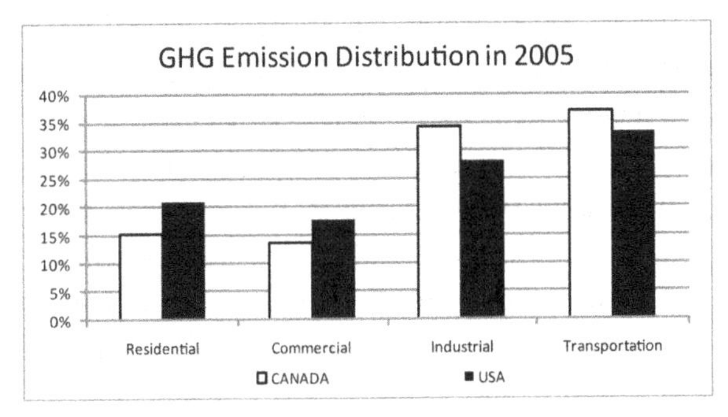

the energy consumption and associated GHG emissions between in 2005 for Canada and USA. Thus, reducing the end-use energy consumption and the associated emissions from the residential sector is one of the effective means of approaching the GHG emission reductions required by the *Kyoto Protocol*.

To reduce the end-use energy consumption and pollutant emissions from the residential sector, a large number of options need to be considered. These include improving the energy efficiency of dwellings through improving envelope characteristics; using higher efficiency heating and cooling equipment, household appliances and lighting; switching to less carbon-intensive fuels for space and domestic hot water heating (DHW); *etc*. Energy efficiency improvements have complex interrelated effects on the end-use energy consumption of households and the associated pollutant emissions (Ugursal & Fung, 1996). As a result, evaluating the effect of various energy efficiency improvement options on residential end-use energy consumption and associated emissions requires detailed mathematical models.

Recently, two types of modes have been used to model residential end-use energy consumption: the Engineering Model (EM) (Farahbakhsh *et al.*, 1998; Larsen & Nesbakken, 2004) and the Conditional Demand Analysis (CDA) Model (Parti & Parti, 1980; Aigner *et al.*, 1984; Caves *et al.*, 1987; Fiebig *et al.*, 1991; Bauwens *et al.*, 1994; Hsiao *et al.*, 1995; Bartels & Fiebig, 2000; Lins *et al.*, 2002; Larsen & Nesbakken, 2004; Aydinalp & Ugursal, 2008).

The EM involves developing a housing database representative of the national housing stock and estimating the end-use energy consumption of the households in the database using a building simulation program. A building simulation program models the end-use energy consumption of a building based on thermodynamic and physics principles taking into consideration such factors as envelope characteristics, internal and solar heat gains, weather conditions, and occupant behavior. Thus, this approach requires a database representative of the housing stock with detailed household description data.

One of the most comprehensive national EM was developed Farahbakhsh *et al.* (1998) using the survey data of 8767 Canadian households and a building simulation software. The main advantage of this model is its capability to evaluate the impact of all types of potential energy saving measures on the residential end-use energy consumption. However, it requires extensive user expertise and lengthy input data preparation time for developing reliable end-use energy estimates. Another national EM was developed by Larsen & Nesbakken (2004) based on the survey data of over 2000 Norwegian households and a building simulation software. As stated by the authors, similar to the Canadian model developed by Farahbakhsh *et al.* (1998), the fundamental weakness of the Norwegian EM was the need of high number of numerical inputs required by the model.

CDA, on the other hand, is a regression-based method. The regression essentially attributes consumption to end-uses on the basis of the household energy consumption. Since CDA does not involve modeling of the energy consumption of each household, it does not require as detailed data on the characteristics of the households as the EM does; however, its results are sometimes unreliable due to multicollinearity problems (Fiebig *et al.*, 1991; Bauwens *et al.*, 1994; Hsiao *et al.*, 1995). The multicollinearity problem makes it difficult to isolate the energy use of appliances with high saturation (*i.e.* appliances owned by a large majority of households), such as the refrigerator. Also, the model requires a very large amount of data due to the high number of independent variables used in the regression equations.

One of the major difficulties associated with the use of the EM to estimate the end-use energy consumption in the residential sector is the inclusion of socio-economic characteristics of the occupants that have a significant effect on the residential energy use. The CDA approach, which is based on regression analysis to decompose household energy consumption into appliance specific levels, can handle socio-economic factors if they are included in the model formulation.

CDA was first introduced by Parti & Parti (1980) in a study in San Diego, CA, USA, that included detailed appliance survey data of over 5000 households. The authors were successfully able to separate the total household electricity consumption to 16 appliances and estimate price and income elasticities. In 1984, Aigner *et al.* used the CDA approach to obtain hourly end-use load profiles by imposing restrictions on the assumption that some appliances were not used at specific hours of the day.

To improve the accuracy of the estimates, prior information on end-use energy consumption in terms of metered data or engineering estimates were integrated into the standard CDA model using random coefficient framework by Fiebig *et al.* (1991) or Bayesian analysis approach by Caves *et al.* (1987), Bauwens *et al.* (1994), and Hsiao *et al.* (1995). In a recent study, Bartels & Fiebig (2000) developed an optimal statistical design technique to select the households and end-uses in the households to be metered. By using these metered data, they were able to successfully estimate the end-use energy consumption of high saturation appliances, such as lighting and refrigerators.

Standard CDA approach was also used to model the national residential end-use energy consumption for Brazil (Lins *et al.*, 2002), Norway (Larsen & Nesbakken, 2004) and Canada (Aydinalp & Ugursal, 2008) using the survey and billing data of about 8000, 2000, and 9000 households, respectively.

In this book chapter, the use of Neural Network (NN) method for modeling residential energy consumption is investigated. Neural Networks are simplified mathematical models of biological neural networks. They are highly suitable for determining causal relationships amongst a large number of parameters such as seen in the energy consumption patterns in the residential sector. The NN approach has been used for prediction problems as a substitute for statistical approaches due to their simplicity of application and accurate estimates.

In the following sections, after a brief background on NN modeling approach, its applications in energy modeling are given followed by a summary table of the specifications of the available residential energy consumption models. Next, an overview of NN modeling approach and information on estimating NN models are presented. Detailed information on the development of a national residential end-use energy model using NN approach and a comparative evaluation of the results of the NN model with those obtained from a CDA and EM are given next. In the following section, future trends in the use of NN for residential energy consumption are discussed. The chapter ends with some conclusion remarks.

BACKGROUND

Neural Networks (NN), also commonly referred to as Artificial Neural Networks, are information-processing models inspired by the way the densely interconnected, parallel structure of the brain processes information. In other words, neural networks are simplified mathematical models of biological neural networks. The key element of the NN is the novel structure of the information processing system. It is composed of a large number of highly interconnected processing elements that are analogous to neurons, and tied together with weighted connections that are analogous to synapses.

NN are capable of finding internal representations of interrelations within raw data. NN are considered to be intuitive because they learn by example rather than by following programmed rules. The ability to learn is one of the key aspects of NN. This typical characteristic, together with the simplicity of building and training NN, has encouraged their application to the task of prediction. Because of their inherent non-linearity, NN are able to identify the complex interactions between independent variables without the need for complex functional models to describe the relationships between dependent and independent variables.

Recently, the NN approach has been proposed as a substitute for statistical approaches for classification and prediction problems. The advantages of NN over statistical methods include the ability to classify in the presence of nonlinear relationships and the ability to perform reasonably well using incomplete databases. The comparison of the results from NN and statistical approaches indicated that neural networks offer an accurate alternative to classical methods such as multiple regression or autoregressive models (Feuston & Thurtell, 1994; AlFuhaid *et al.*, 1997).

Although the NN concept was first introduced in 1943 (McCulloh & Pitts, 1943), it was not used extensively until the mid-1980's owing to the lack of sophisticated algorithms for general applications, and its need for fast computing resources with large storage capacity. Since the 1980's, various NN architectures and algorithms were developed (*e.g.* the multi-layer perceptron (MLP) which is generally trained with the error backpropagation algorithm, Hopfield Network, Kohonen Network, *etc.*). Consequently, NN models have been used extensively as a tool for modeling, control, forecasting, and optimization in many fields of engineering and sciences such as process control, manufacturing, nuclear engineering, and pattern recognition.

Use of NN in Energy Modeling

In the area of energy modeling, the application of NN has been mainly limited to utility load forecasting. There are several hundred papers in the literature on the application of NN for utility hourly load forecasting. These clearly show the superior capability of NN models over conventional methods (such as regression analysis).

Park *et al.* (1991) were among the first group of researchers to use MLP NN for hourly load forecasting. In 1992, Peng *et al.* used an improved NN that used an alternate formulation of the problem in which the input was mapped to the output by both linear and non-linear terms, and an improved method for selecting and scaling the input units. Kiartzis *et al.* (1995) used a MLP NN with 24 output neurons, one for each hour of the day (*i.e.* their model could forecast the next 24-hour load profile on an hourly basis). Chen *et al.* (1996) included humidity in their NN model in addition to ambient temperature to account for the effect of humidity on air-conditioning component of the load. In 1997, AlFuhaid *et al.* tested a cascaded artificial NN (CANN) to capture the sensitivity of the non-linear influence of temperature and humidity on the load for half-hourly load forecasting. In 1999, Al-Shehri developed a MLP type NN to estimate the monthly residential electricity demand in the Eastern Province of Saudi Arabia using the month of the year as the only input parameter. Several other researches who implemented NN to forecast hourly load is mentioned in the work of Metaxiotis *et al.* (2003).

NN models were also used to predict energy consumption of individual buildings since they have a high potential to model nonlinear processes such as building energy loads (Kawashima, 1994). NN applications specifically to building energy analysis were pioneered by the *Joint Centre for Energy Management* at the University of Colorado, Boulder, in the early 1990's. It is reported by Krarti *et al.* (1998) that Kreider & Wang (1991) were the first to apply a NN model to predict the energy consumption of a building. The electricity consumption of a commercial building was predicted and the results showed that the prediction of the NN model was accurate. The authors indicated that NN was easier to use than classical regression methods since they learn from fact patterns, and there was no requirement for a priori statistical analysis. In a later study by the authors, the NN results were compared with statistical results for the same commercial building data (Kreider & Wang, 1992). The regression model attempted to fit all the data globally, but the accuracy at some specific points was not acceptable. The prediction of the NN model was high for those points where regression method completely missed. In 1993, Anstett & Kreider used NN to predict energy use (steam, natural gas, electricity, and water) in a complex institutional building and found the predictive quality of the NN to be satisfactory.

In order to evaluate many of the analytical methods and to asses new methods not widely used in building data studies, an open competition was held in 1993 to identify the most accurate method for making hourly energy use predictions based on limited amount of measured data (Kreider & Haberl, 1994). More than 150 contestants entered the competition and the results of the top six models were presented in the study of Kreider & Haberl (1994), which indicated that NN of various designs and training methods obtained more accurate values than the traditional statistical methods.

Besides predicting commercial building energy consumption, NN was also used to predict energy savings from building retrofits by various researchers, such as Cohen & Krarti (1995), Krarti *et al.*, (1998), and Yalcintas (2008). The results show that NN can be used to determine energy savings from retrofits using available weather, building and/or equipment data.

Yang *et al.* (2005) developed an adaptive NN for on-line prediction of an institutional building in Montreal, Canada. The model is capable of updating the model parameters based on newly available data, thus providing reliable estimates due to unexpected pattern changes in the incoming data. The estimates of the adaptive NN model were found to be in better agreement with the actual data than the estimates of an ordinary "static" NN model. Neto & Fiorelli (2008) compared the estimates of a NN model and a building simulation program based on the energy demand of a university building in Sao Paula, Brazil. Their results show that the estimates of the NN model were closer to the actual values than those of the simulation program.

As for the use of NN to estimate the energy consumption of the residential buildings, the number of studies is very limited. In 1999, Kalogirou *et al.* developed a NN model to estimate the useful energy extracted from a solar domestic water heating system and the stored water temperature rise. Respectively, predictions within 7.1% and 9.7% were obtained. As stated by the authors, the advantages of the NN approach compared to other conventional methods were the speed, simplicity and capacity of the model to learn from examples. Various other modeling applications of NN in energy systems developed by the authors are given in Kalogirou (2000).

Another residential application of NN was conducted by Olofsson & Andersson (2001) to predict annual heating and total energy demands of six Swedish single-family dwellings using data from daily measurements. The authors achieved a deviation of 4% and 5% between the predicted and measured heating and total energy demands, respectively. Also in 2001, Issa *et al.* developed a NN model to estimate the annual energy consumption of the households in central Florida using the *Energy Performance Index* and conditioned living area as inputs.

In Greece, Mihalakakou *et al.* (2002) used time series data in terms of hourly temperature, radiation, and electricity load of a residential building in Athens to develop a NN model to estimate hourly electricity consumption. The results of the model were in good agreement with the actual values with correlation coefficients of 0.96 for summer months and 0.94 for winter months.

Currently, there are many applications of NN to control or model the heating, ventilating and air conditioning (HVAC) systems in commercial or institutional buildings in the literature; however modeling of the residential buildings HVAC systems with NN is limited. El Ferik & Belhadj (2004) are among the first researches to develop a model to investigate the relationship between the energy consumption of a residential building's HVAC system and outdoor weather conditions, mainly temperature and humidity.

Another interesting application of NN to residential building energy consumption was conducted by Entchev & Yang (2007). The authors developed a NN model to predict solid oxide fuel cell (SOFC) performance while supplying both heat and power to an experimental house built by Canadian Centre for Housing Technology in Ottawa, Canada. The SPFC system was connected to the house DHW and space heating system, and modifications were made to the house electrical wiring in order to accept the system. The SOFC performance estimates of the NN model developed by the authors agreed well with the experimental data. The authors concluded that using NN approach the performance of the SOFC system could be modeled with minimum time demand and with a high degree of accuracy.

As this literature review indicates, NN approach has been widely used for load forecasting. Recently, the approach has been used for estimating the energy consumption of various types of commercial, institutional, and some residential buildings; however, NN have been used to model the residential end-use energy consumption at the national level only by Aydinalp (2002a) and Aydinalp *et al.* (2002b; 2004). A brief summary of the NN models developed to estimate residential energy use mentioned above is given Table 1 based on the available information on the model specifications from the published articles. In the following sections, detailed information about the residential end-use NN model developed by Aydinalp (2002a) and Aydinalp *et al.* (2002b; 2004) is presented.

Overview of NN Models

NN use simple processing units, called neurons, to combine data and store relationships between independent and dependent variables. A NN consists of several layers of neurons that are connected to each other. This connection is an information transport link from one sending to one receiving neuron.

A widely used NN model called the multi-layer perceptron (MLP) NN is shown in Figure 2. The MLP type NN consists of one input layer, one or more hidden layers, and one output layer. Each layer employs several neurons, and each neuron in a layer is connected to the neurons in the adjacent layer with different weights.

Signals flow into the input layer, pass through the hidden layer(s), and arrive at the output layer. With the exception of the input layer, each neuron receives signals from the neurons of the previous

Table 1. Specifications of the residential energy consumption NN models

Researcher(s)	Year	NN Type / Configuration	Estimated Parameters	Input Parameters	Activation Functions		Learning Algorithm	Training/ Testing Datasets	Performance	Software
					Hidden Layer	Output Layer				
Kalogirou *et al.*	1999	Feedforward MLP / 8:18:18:18:2	Useful energy extracted from solar DHW heating systems & stored water temp rise in the tank	Collector area, storage-tank heat-loss coefficient, tank type, storage volume, type of system, total daily solar radiation, mean ambient air temperature, & water temperature in the storage-tank	Gaussian, Tanh, Gaussian Complement	Logistic	Standard Back-propagation	264/36	MRE of 7.1-9.7%	
Olofsson & Andersson	2001	Feedforward MLP / 3:12:12:2	Annual total and heating energy demand of six single family dwellings	Daily indoor and outdoor temperature difference, heating demand and demand for electricity and DHW	Tanh	Tanh	Standard Back-propagation		Average Deviation of 4-5%	Professional II Plus
Issa *et al.*	2001	Feedforward MLP / 2:9:1	Annual household energy consumption	Energy performance index and conditioned living area			Standard Back-propagation			
Mihalakakou *et al.*	2002	Feedforward MLP / 2:15:1 to 2:22:1	Hourly electricity consumption of a dwelling	Hourly air temperature and total solar radiation	Logistic	Linear	Standard Back-propagation	5 years / 1 year	R^2 of 0.94-0.96	
Entchev & Yang	2007	Feedforward MLP / 8:10:2	Stack current and stack voltage	Stack fuel flow, stack temperature, stack air flow, stack air inlet temperature, burner fuel flow, burner combustion temperature, burner air flow, burner air inlet temperature	Tanh	Linear	Standard Back-propagation	21000 / 53000	MRE ~ 2%	MATLAB

Figure 2. Architectural graph of an MLP with one hidden layer

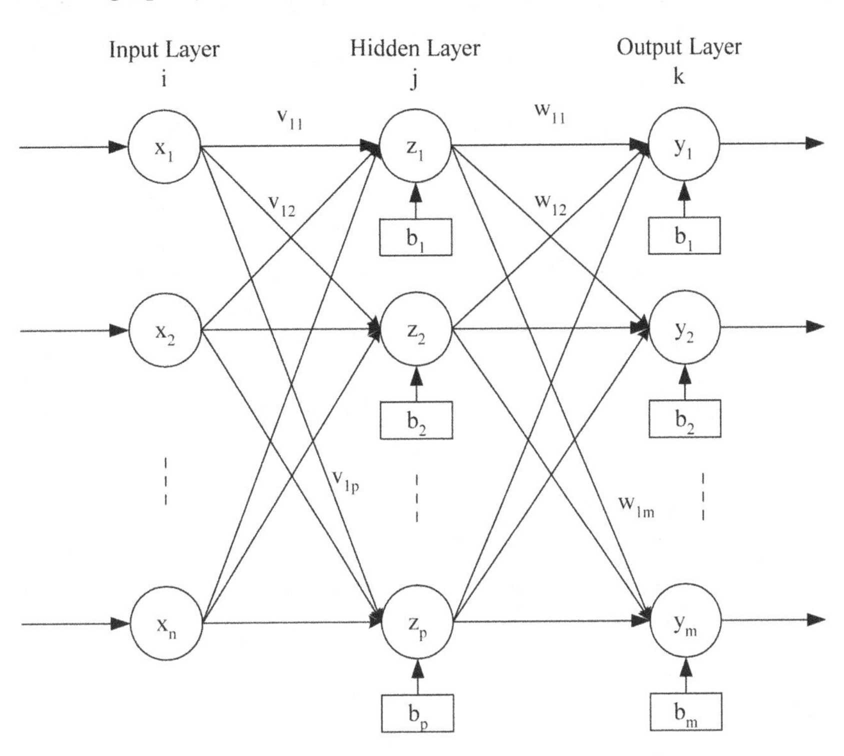

layer. The incoming signals (x_{ij}) are multiplied by the weights (v_{ij}) and summed up with the bias (b_j) contribution.

$$net_j = \sum_{i=1}^{n} x_i v_{ij} + b_j \tag{1}$$

where,

net_j: total input of the hidden layer neuron j
x_i: input to the hidden layer neuron j from input layer neuron i
v_{ij}: weight between the input layer neuron i and hidden layer neuron j
b_j: bias of the hidden layer neuron j
n: number of neurons in the input layer

The output of a neuron is determined by applying an activation function to the total input (net_j) calculated using Equation 1. The bias (b_j) in Equation 1 has the effect of increasing or decreasing the total input to the activation function, depending on whether it has a positive or negative value, respectively, and can be evaluated similar to the intercept term in a linear regression model. The bias avoids the tendency of an activation function to get "stuck" in the saturated, limiting value area of the activation function (Kreider & Wang, 1992). The bias is actually a unit connected to a neuron with a weight of one.

Activation functions for the hidden units are needed to introduce nonlinearity into the network. Without nonlinearity, hidden units would not make MLPs more powerful than just plain networks which do not have any hidden layer units, just input and output units. The sigmoid functions, such as logistic and hyperbolic tangent functions, are the most commonly used activation functions in networks trained by backpropagation (Fausett, 1994). The logistic function with the output amplitude lying inside the range [0.0 to 1.0] is shown as:

$$z_j = f(net_j) = \frac{1}{1+e^{-net_j}} \tag{2}$$

The output amplitude of the hyperbolic tangent function lies inside the range [-1.0 to 1.0], and is shown as:

$$z_j = f(net_j) = \tanh(net_j)\frac{1-e^{-2net_j}}{1+^{e-2net_j}} \tag{3}$$

The outputs of the activation functions become the inputs to the layers downstream. The ultimate output of a NN model, y_k, is the output of the activation function at the output layer. The activation function for the output units is mostly chosen to be logistic, hyperbolic tangent or linear (identity) functions. The identity function can be shown as:

$$z_j = f(net_j) = net_j \tag{4}$$

If the computed outputs do not match the known (*i.e.* target) values, NN model is in error. Then, a portion of this error is propagated backward through the network. This error is used to adjust the weight and bias of each neuron throughout the network so the next iteration error will be less for the same units. The procedure is applied continuously and repetitively for each set of inputs until there are no measurable errors, or the total error is smaller than a specified value.

At this point, the net remembers the patterns for which it was trained and is able to recognize similar patterns in new sets of data. Once the structure and training are completed, predictions from a new set of data may be done, using the already trained network. During the training process, the neural network develops the capability of recognizing different patterns and capturing relevant relationships in the training dataset.

Therefore, the underlying assumption in using the NN model is that the relationships between the input and the output variables in the training dataset, the testing dataset, and prediction dataset are the same. This feature of the NN Models can also be seen in CDA models, but not in EM. The models based on the engineering approach have the capability to estimate a wide range of variables, as long as the detailed dwelling description data are available.

Learning Algorithms

The most commonly used learning algorithm is the standard backpropagation introduced by Rumelhart & McClelland (1986) and its variants, such as resilient propagation (Riedmiller & Braun, 1993) and quickprop (Fahlman, 1988). During training, an associated error is determined for each output layer neuron. Based on this error, the error information term is computed, which is then used to distribute the error of output layer neuron back to all neurons in the hidden layer by updating the weights between the hidden layer and output layer. In a similar way, the error information term is computed for each hidden unit and used to update the weights between the hidden layer and input layer.

The weight correction term for each weight is computed by using the error information term. Also, a learning parameter which is usually taken as a positive number less than one is added to the weight correction formula to reduce the changes in the weights, so that the instability (*i.e.* oscillation) of the network is prevented. The corrections to all weights are done simultaneously by adding the weight correction term to the old weight after each training pattern.

One method to speed up the learning is to use the information about the curvature of the error surface. The quickprop algorithm developed by Fahlman (1988) is based on the assumption that the error curve can be approximated by a quadratic polynomial (*i.e.* parabola), which is concave up. The partial derivative of the error function with respect to the given weight summed over all training patterns is referred as slope and used in the algorithm. The slope term is calculated for all weights.

Resilient propagation algorithm was introduced by Riedmiller & Braun (1993). It performs a direct adaptation of the weight adjustment based on local gradient information. The main difference from other algorithms is that the adaptation is not blurred by gradient behavior. In this algorithm, the size of the weight correction term is changed directly, *i.e.* without considering the size of the partial derivative, whose unforeseeable behavior can disturb the adapted learning rate (Riedmiller & Braun, 1993).

Estimation by NN Models

To develop an NN model, the dataset is first divided into two sets: one to be used for the training of the network, and the other for testing its performance. Approximately 70 or 80 percent of the dataset is used for training and the rest for testing (Anstett & Kreider, 1993).

After deciding which activation function to use for the hidden and output layer units, the datasets are scaled so that each value falls within the range for which the amplitude of the outputs of the chosen activation functions lie. This is done to prevent the simulated neurons from being driven too far into saturation (Highley & Hilmes, 1993), especially when the data span many orders of magnitude. Anstett & Kreider (1993) found that the [0.1 to 0.9] interval provided better results for their dataset, when logistic function and linear function were used as the activation functions for the hidden and output layer units, respectively. Kawashima (1994) scaled the input data in the [0.0 to 1.0] interval and the output data (target values) [0.1 to 0.9] interval for his network, which used logistic function for both hidden and output layer units. Thus, it is not possible to predict which activation and which scaling interval would be best suited for any given network, and these should be chosen after testing various combinations.

There are no rules to establish the number of hidden layers and the number of neurons for each hidden layer for a particular application. Generally, one hidden layer is sufficient for load or building energy prediction (Kawashima, 1994; Stevenson, 1994; Yang *et al.*, 2005). Network architecture is decided basically by trial and error; comparing the performance of the nets with different number of neurons in the hidden layer(s).

The weights and the biases are initialized with distributed random values before the training starts. The learning algorithm resulting in the best performance when compared to those of the other algorithms is chosen for the training of the network. The training is repeated until the sum of the square of errors (SSE) for the entire training data is less than a specified value.

Table 2. Measures used to judge the performance of a NN model

Name	Symbol	Formula	Reference
Multiple Correlation Coefficient	R^2	$1-\frac{\sum_{i=1}^{N}(y_i-t_i)^2}{\sum_{i=1}^{N}t_i^2}$	Anstett & Kreider, 1993
Root Mean Square	RMS	$\sqrt{\frac{\sum_{i=1}^{N}(y_i-t_i)^2}{N}}$	Kreider & Wang, 1992
Coefficient of Variation	CV	$\frac{\sqrt{\sum_{i=1}^{N}(y_i-t_i)^2}}{\frac{N}{\bar{t}}\times 100}$	Kreider & Haberl, 1994
Mean Bias Error	MBE	$\frac{\sum_{i=1}^{N}(y_i-t_i)}{\frac{N}{\bar{t}}\times 100}$	Kawashima, 1994
Sum of Square of Errors	SSE	$\sum_{i=1}^{N}(y_i-t_i)^2$	

A smaller number of data that have never been shown to the network during training, *i.e.* the testing dataset, is used to test the prediction performance of the network after training is complete. The output of the network for the testing dataset is then descaled to get the original units.

Assessing the Prediction Performance of NN Models

To judge the prediction performance of a network, several performance measures are used. Some of these measures are given in Table 2, where

t_i: target value of the ith pattern

$\bar{t}$: mean of the target values

y_i: predicted value of the ith pattern

N: number of patterns

National Residential End-Use NN Model

As the review of the literature presented above indicates, NN have been used to model the residential end-use energy consumption at the national level only by Aydinalp (2002a) and Aydinalp *et al.* (2002b; 2004). In this section of the chapter, this NN model that was developed to model the end-use energy consumption for the Canadian residential sector is presented briefly. Detailed information on this model can be obtained from Aydinalp 2002a and Aydinalp *et al.* (2002b and 2004).

Figure 3. Flowchart diagram depicting the methodology used for the development of the NN Model

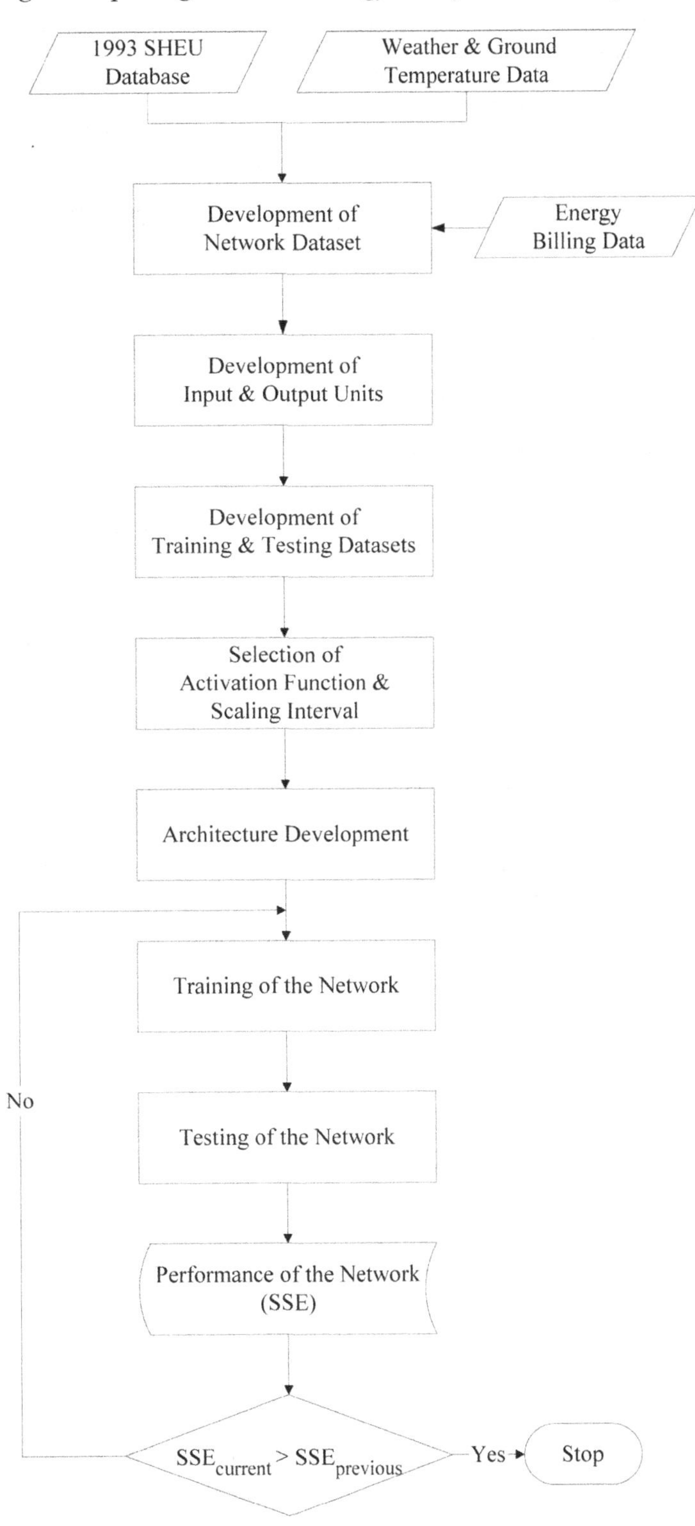

Table 3. Scaling intervals and activation functions chosen for the ALC, DHW, and SH networks

Network	Dataset Scaling Interval	Applied to	Activation Functions	
			Hidden Layer	Output Layer
ALC	-0.5 to 0.5	All data	Logistic	Identity
DHW	0.1 to 0.9	Only continuous data, discrete data left as is	Logistic	Logistic
SH	0.1 to 0.9	All data	Identity	Logistic

This model consists of three networks. Each network is used to predict a single end-use energy consumption: These are:

- Space heating (SH) end-use energy consumption,
- Domestic hot water (DHW) heating end-use energy consumption,
- Appliances, lighting, and cooling (ALC) end-use energy consumption.

The approach used in the development of the networks is shown in Figure 3. The Stuttgart Neural Network Simulator (SNNS) V4.2 software (SNNS, 1998) was used in the development of the NN model.

Sources of Data

Two sources of data were used for the development of the input units of the NN model: the data from the 1993 Survey of Household Energy Use (SHEU) database (Statistics Canada, 1993) which included detailed information on house construction, space heating/cooling and DHW heating equipment, household appliances and some socio-economic characteristics of the occupants for 8767 households in Canada, and the weather and ground temperature data for 1993 (Environment Canada, 1999). The source of data for the output unit of the models was the actual 1993 energy-billing data obtained from fuel suppliers and utility companies for 2,749 households from the 1993 SHEU.

Development of Network Datasets

The selection of households for each network dataset was based on fuel type and energy source information, and availability of energy billing data. The ALC network dataset contains 988 households with electricity bills that do not have electrical SH or DHW heating equipment. The number of households in the DHW dataset that use electricity or natural gas for DHW heating is 563. The SH dataset contains 1,228 households with electricity, natural gas, and oil space heating billing data.

Development of Input and Output Units

The number of input units is different for each network. Input units were selected based on their contribution to the specific end-use energy consumption. For example, the input units of the ALC network were developed using the information available from the 1993 SHEU database on appliances, lighting, and space cooling equipment of the households with electricity bills. The number of input units in the

Table 4. Architectures and learning algorithms of the ALC, DHW, and SH networks

Network	Architecture	Number of Hidden Layer Units			Learning Algorithm	Parameters*
		Layer 1	Layer 2	Layer 3		
ALC	55:09:09:09:1	9	9	9	Quickprop	η: 0.0015, ρ: 2.10, v: 0.000015
DHW	18:29:1	29			Resilient Propagation	β: 1.7, $\phi_{initial}$: 0.02, ϕ_{max}: 30
SH	28:02:1	2			Resilient Poaation	β: 1.1, $\phi_{initial}$: 0.06, ϕ_{max}: 10

*The definitions for the parameters of the learning algorithms are given in Aydinalp (2002a)

ALC, DHW, and SH networks were 55, 18, and 28, respectively. The actual 1993 energy consumption data of the households was used as the output unit of the networks.

Development of Training and Testing Datasets

The network datasets were then divided into two subsets. One of these subsets was used for training (training set) and the other was used for testing (testing set) of the network. The households in each subset were chosen randomly. However, special care was given to include the households with minimum and maximum input and output values into the training dataset. This is done to increase the estimation range of the NN Model. Approximately, testing dataset included 25% of the households which is slightly higher than the percentages used by Kalogirou *et al.* (1999) and Mihalakao *et al.* (2002).

Network Development

Selection of the Activation Functions and Scaling Intervals

Several scaling intervals for the input/output units and activation functions for the hidden and output layers were tested and the ones that resulted in best prediction performances for the ALC, DHW, and SH networks are given in Table 3.

Network Architecture Development and Selection of Learning Algorithm

Since the number of input and output units are decided based on the available data and the desired output, respectively, only the number of units in the hidden layer(s) is left to be determined. Several networks with the number of hidden layer units ranging from one to 30 or 40 were trained with four learning algorithms (quickprop, resilient propagation, enhanced back-propagation, and standard back-propagation). The number of units in the hidden layer of the network and the learning algorithm resulting in the highest prediction performance was chosen as the network architecture for the end-use NN Model.

After determining the number of hidden layer units and the learning algorithm resulting in the highest prediction performance, different networks with the number of hidden layer units in one, two or three layers were trained with the chosen learning algorithm to determine the best network architecture. The performance of the networks was improved by "fine-tuning" the parameters of the chosen learn-

Figure 4. Architectural configuration of the ALC NN Model

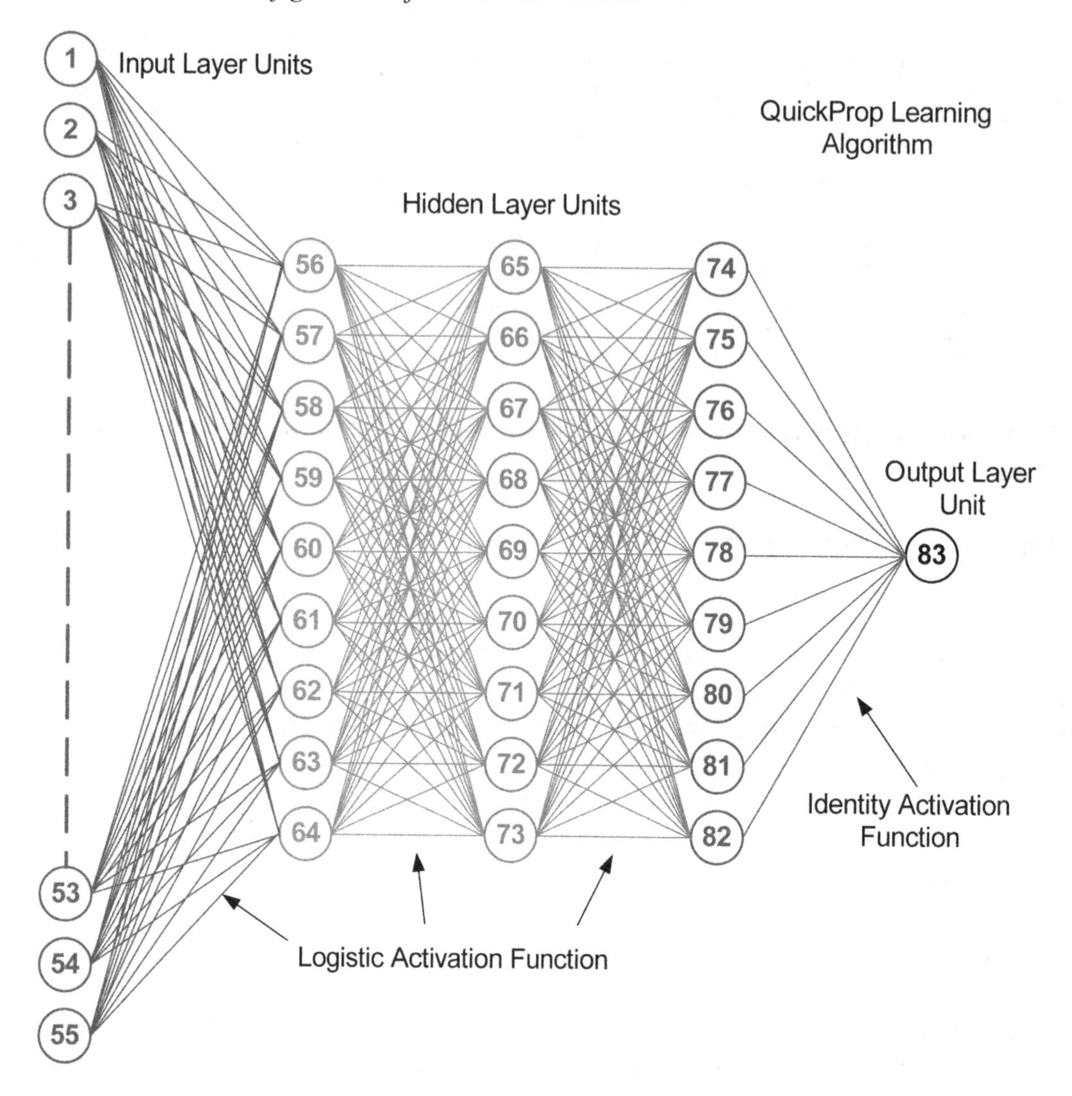

ing algorithm. The "fine-tuning" is done by testing a wide range of values of the learning algorithm parameters.

The learning algorithms and network architecture that produced the best predictions for ALC, DHW, and SH networks are given in Table 4. The architectural configuration of the ALC, DHW, and SH networks are given in Figures 4, 5, and 6. The values of the weights and biases of the end-use networks are given in Aydinalp (2002a).

As seen in Tables 1, 3, and 4, the NN models developed to estimate residential energy consumption in general used identity, logistic and hyperbolic tangent activation functions since they are the most commonly used ones for MLP feedforward network. These models have maximum three hidden layers and are trained using some advance variants of the backpropagation learning algorithm.

Figure 5. Architectural configuration of the DHW NN Model

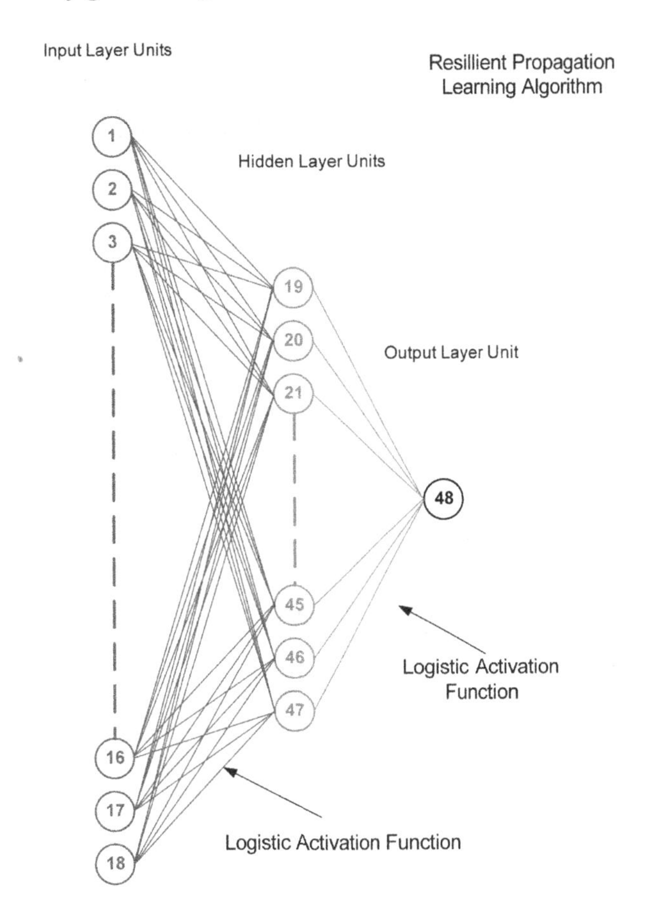

Training and Testing of Networks

The selected network architectures and learning algorithms were used to train and test the network using the training and testing datasets. The training was halted when the testing dataset SSE value stopped decreasing and started to increase, which is an indication of overtraining. The prediction performance in terms of multiple correlation coefficient (R^2) of the ALC, DHW, and SH networks are given in Table 5. As seen here, the network achieved good prediction within the range of R^2 of 0.871 to 0.909. These results showed that NN method was suitable for modeling the three major residential end-use energy consumptions at the national level.

Comparison of the NN, CDA, and Engineering Models

The prediction performance of the NN Model is compared with those of the CDA model developed by Aydinalp & Ugursal (2008) and the EM developed by Farahbakhsh *et al.* (1998) based on the actual energy consumption data of the households in the testing datasets. The results are presented in Table 6. As it can be seen, the CDA and EM have lower multiple correlation coefficient (R^2) and higher coefficient of variation (CV) values than the NN model, which shows that the NN model have a higher prediction

Figure 6. Architectural configuration of the SH NN Model

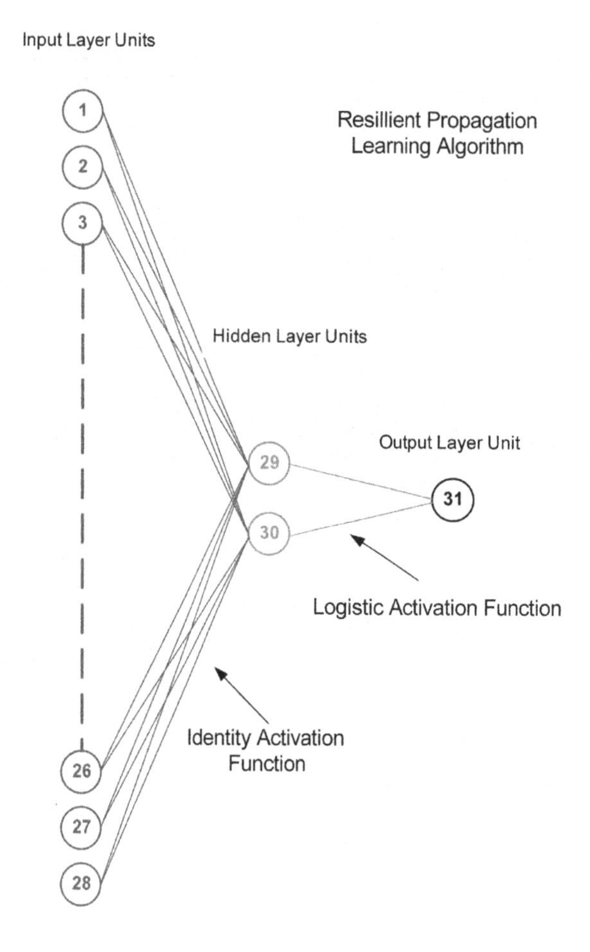

performance than the CDA and EM.

The NN, CDA, and EM were also used to predict the ALC, space and DHW heating energy consumption of households in the survey database that were not used in the training or testing datasets. As seen in Table 6, the average household end-use energy consumption estimates of the three models are close to each other. The estimates of the CDA, NN, and EM on dwelling type, size, and age, as well as the type of fuel used were also compared and were generally in good agreement (Aydinalp, 2002a).

The weighted average household energy consumption was computed by combining the ALC, DHW, and SH end-use energy consumption estimates of the three models. From these estimates, 1993 household energy consumption was estimated as 139 GJ, 132 GJ, and 135 GJ by the NN, CDA, and the EM, re-

Table 5. Prediction performances of the ALC, DHW, SH networks

Network	Architecture	R^2
ALC	**55:09:09:09:1**	0.909
DHW	18:29:1	0.871
SH	28:02:1	0.908

Table 6. Prediction performances and estimates of the ALC, DHW, SH networks

	R^2	CV	Weighted Average End-use Energy Consumption (GJ/yr/household)
ALC Energy Consumption NN	0.909	2.094	32
CDA	0.795	3.343	30
EM	0.780	3.463	32
DHW Heating Energy Consumption NN	0.871	3.337	26
CDA	0.814	4.052	25
EM	0.828	3.898	25
Space Heating Energy Consumption NN	0.908	1.871	80
CDA	0.892	2.007	75
EM	0.778	2.877	77

spectively. These results are close to OEE's estimate of 134 GJ per household for 1993 (OEE, 2008).

The effects of some socioeconomic factors, such as income, dwelling ownership, size of area of residence, on end-use energy consumption were easily estimated using the NN, since these parameters can be incorporated into NN model as input parameters. These socioeconomic factors can also be evaluated by the CDA model if they are not eliminated from the input dataset as a result of statistical significance and multicollinearity problems. On the other hand, to include such factors into the EM, detailed occupancy and preference profiles are required, which are not available in this database.

Various energy saving scenarios on DHW and SH energy consumption were also evaluated by the NN model. The results were then compare with those obtained using EM and other studies. The CDA model did not include any of the variables used in the scenarios, thus it was not possible to evaluate energy saving scenarios using the CDA model.

These results showed that NN Model can estimate the impacts of the energy savings scenarios, as long as the households that undertake the scenarios are well represented in the training datasets. However, the NN Model cannot evaluate the impact of an energy saving scenario on other energy end-uses, since each end-use is predicted separately. On the other hand, the EM has significantly higher level of flexibility in evaluating energy conservation measures, regardless of the number of households in the datasets. It can also evaluate the effects of energy efficiency measures on each end-use, since energy consumption is

Table 7. Qualitative comparison of the models

	NN Model	CDA Model	EM Model
Prediction performance	High	Acceptable	Acceptable
Evaluation of socio-economic factors	Easy	Limited	Difficult
Evaluation of energy saving scenarios	Limited	Poor	Excellent
Ease of use	Moderate	Easy	Requires extensve user expertise

estimated using thermodynamics and heat transfer principles. Based on these information, the specific advantages and disadvantages of the models can be summarized as shown in Table 7.

FUTURE TRENDS IN RESIDENTIAL ENERGY MODELING

There is a growing demand for a versatile residential end-use consumption model that would generate reliable estimates and evaluate several energy saving scenarios, especially from countries aiming to reduce energy consumption and associated CO_2 emissions from the residential sector. By using this national model, these countries would be able to identify the amount of energy reduced due to a specific energy saving scenario and associated polices to be implemented in the residential sector.

The residential energy consumption models presented in Table 1 show that NN can easily be used to estimate the energy consumption in the residential sector. Moreover, the results of the Canadian residential NN model given briefly above show that the NN approach can be used to generate reliable end-use energy consumption estimates at the national level as the other well established approaches such as CDA and EM. All these results show that NN modeling approach is suitable for residential energy modeling on individual building or end-use basis, or the regional/national level.

As shown in Table 7, each model has its specific advantages and disadvantages. The EM is strong in evaluating the impacts of energy saving scenarios, but lacks the ability to evaluate the effects of socio-economic variables. On the other hand, the NN model is flexible in incorporating the socio-economic variables and but has limited capability in evaluating the energy saving scenarios. As stated by Pedersen (2007), the amount of building energy use depends on the attitude and awareness of the energy consumer and this is more noticeable in household energy use than others where many people may have simultaneously influence on energy use. Therefore, incorporating socio-economic variables, especially income, education, age, dwelling ownership, into the residential energy consumption model becomes vital.

Thus, a hybrid model that uses engineering and NN modeling approaches can be developed. The engineering part of the model would use a building energy simulation software for physical and thermodynamic modeling. The NN modeling part would then be used to estimate the end-uses specifically depended on socioeconomic factors such as appliances, lighting, and DHW heating energy consumption. The occupancy and end-use load profiles should be incorporated into the model to achieve reliable estimates.

An example for this type of hybrid residential end-use energy consumption model is given in Swan *et al.* (2008). This new state-of-the-art end-use energy and GHG emission model, named as *Canadian hybrid residential end-use energy and emission model* (CHREM), incorporates a 17,000 house database (CSDDRD) developed from a number of Canadian housing databases and surveys. The appliance, lighting, and DHW annual end-use energy consumption of the households are estimated by a NN model which uses equipment characteristics and usage, weather conditions, dwelling properties, and some socioeconomic and demographic characteristics such as income, employment, dwelling ownership, and population of the city/town the house is located. The annual AL and DHW end-use energy consumptions of each house are then converted to hourly values using representative load profiles which are generated based on experimentally determined probability of use and limiting conditions obtained from various resources. The estimated hourly AL and DHW energy consumptions values are used as inputs to a building energy simulation program, ESP-r (ESRU, 2002) which is an integrated building energy

Figure 7. Structure and flowchart of CHREM. (© 2008, V. Ismet Ugursal. Used with permission.)

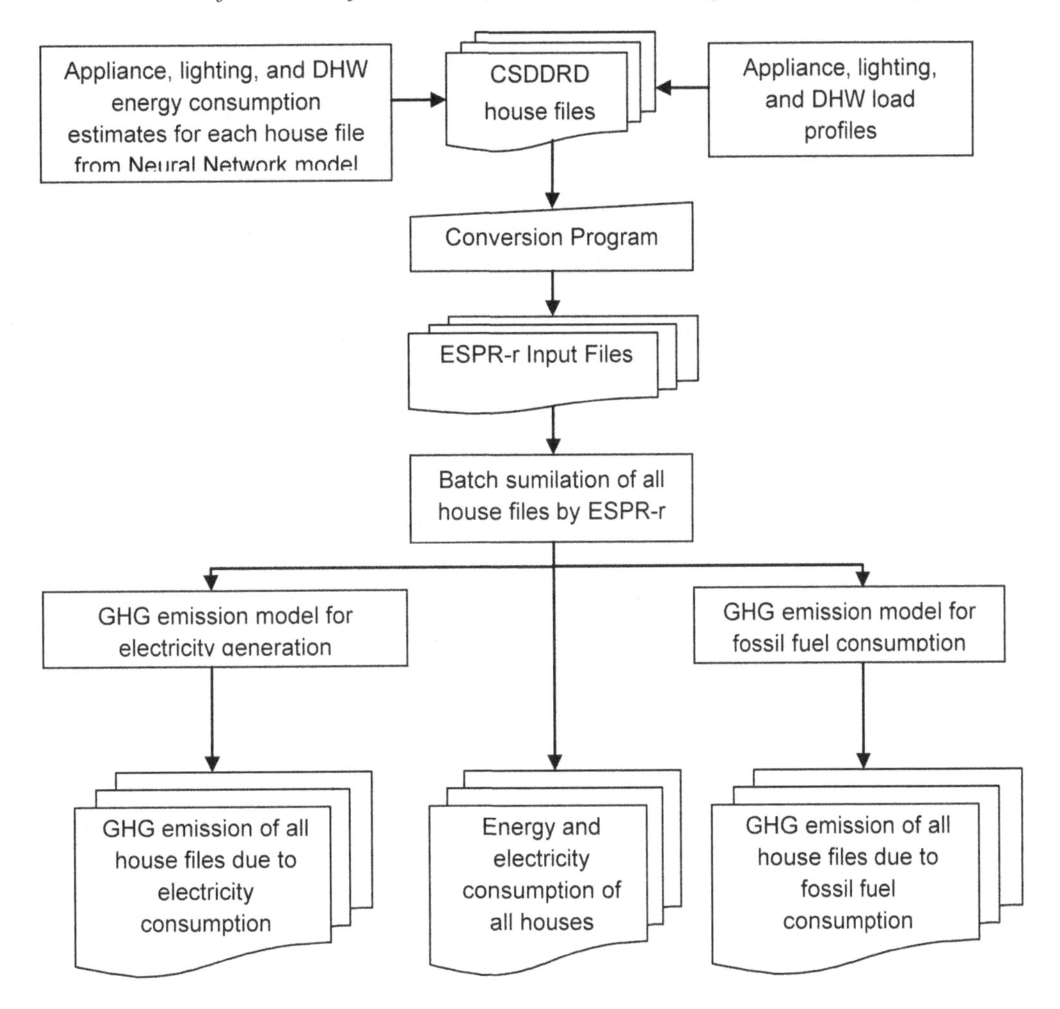

simulation program used to estimate hourly energy and electricity consumption of each house in the database. The GHG emissions due to electricity and fossil fuel consumptions are also calculated based on the energy and electricity estimates of the ESPR-r simulation program. The structure and flowchart of CHREM is given in Figure 7.

The use of this type of hybrid model would allow to easily evaluate various energy saving scenarios on the space heating and cooling end-uses estimated by the EM which utilizes a building energy simulation program. The appliance, lighting, and DHW heating energy consumption end-uses are estimated using a NN model considering the effects of various socio-economic factors. In order to develop a representative and versatile hybrid model, a database with sufficient number of complete household data (*e.g.* >5000), including occupancy and end-use load profiles and energy billing data must be available.

CONCLUSION

Reducing the energy consumption and associated CO_2 emission from the residential sector is one of the effective means of approaching the GHG emission reductions required by the *Kyoto Protocol*, especially for countries with high per capita energy usage. Detailed mathematical models are required to evaluate the effects of various energy efficiency improvement options on the residential energy consumption. Currently, EM, and CDA approaches are mainly used for residential energy consumption modeling. In this chapter, the use of NN approach in residential energy consumption is investigated.

There are numerous studies on the use of NN for load forecasting and quite a lot studies on modeling of building energy consumption. In case of residential energy consumption modeling, use of NN is mostly limited to estimating the total energy consumption of a small number of residential buildings or energy consumption of a specific residential end-use. The results of these studies show that NN is suitable and can easily be used to estimate load and building energy consumption due to its speed, simplicity and high prediction performance.

The NN model developed to estimate the national residential end-use energy consumption for the Canadian residential given briefly as a case study is so far the only application of NN in end-use model in this sector. The results of this model showed that NN can be used to develop reliable end-use estimates, to categorize total and end-use estimates to help the understanding of how energy is used in the residential sector, and to evaluate the effects of some socio-economic factors and energy saving scenarios on end-use energy use.

While NN models have distinct advantages in predicting the energy consumption and the impact of socio-economic factors on energy consumption, they are not flexible in evaluating the impact of energy conservation measures. The EM also provides accurate estimates, has the highest level of flexibility in evaluating the impact of energy saving measures, but has difficulties with the inclusion of the socio-economic factors.

These results suggest that developing a hybrid model that uses the EM for physical and thermodynamic modeling and the NN model for modeling the end-uses those mainly depend on the socio-economic factors would combine the advantages of the NN and engineering in one single model. This model can be used to estimate the total and/or end-use energy consumption and to apply a wide range of energy saving and GHG emission reduction scenarios to a single dwelling or housing stock.

REFERENCES

Aigner, D. J., Sorooshian, C., & Kerwin, P. (1984). Conditional demand analysis for estimating residential end-use load profiles. *The Energy Journal (Cambridge, Mass.)*, *5*(3), 81–97.

Al-Shehri, A. (1999). Artificial neural network for forecasting residential electrical energy. *International Journal of Energy Research*, *23*(8), 649–661. doi:10.1002/(SICI)1099-114X(19990625)23:8<649::AID-ER490>3.0.CO;2-T

AlFuhaid, A. S., El-Sayed, M. A., & Mahmoud, M. S. (1997). Cascaded artificial neural networks for short-term load forecasting. *IEEE Transactions on Power Systems*, *12*(4), 1524–1529. doi:10.1109/59.627852

Anstett, M., & Kreider, J. F. (1993). Application of neural networking models to predict energy use. *ASHRAE Transactions, 99*(1), 505–517.

Aydinalp, M. (2002a). *A new approach for modeling of residential energy consumption.* Unpublished doctoral dissertation, Dalhousie University, Halifax, Nova Scotia, Canada.

Aydinalp, M., Ugursal, V. I., & Fung, A. (2002b). Modeling of the appliance, lighting, and space-cooling energy consumptions in the residential sector using neural networks. *Applied Energy, 72*(2), 87–110. doi:10.1016/S0306-2619(01)00049-6

Aydinalp, M., Ugursal, V. I., & Fung, A. (2004). Modeling of the space and domestic hot-water heating energy-consumption in the residential sector using neural networks. *Applied Energy, 79*(2), 159–178. doi:10.1016/j.apenergy.2003.12.006

Aydinalp Koksal, M., & Ugursal, V. I. (2008). Comparison of neural network, conditional demand analysis, and engineering approaches for modeling end-use energy consumption in the residential sector. *Applied Energy, 85*(4), 271–296. doi:10.1016/j.apenergy.2006.09.012

Bartels, R., & Fiebig, D. G. (2000). Residential end-use electricity demand: results from a designed experiment. *The Energy Journal (Cambridge, Mass.), 21*(2), 51–81.

Bauwens, L., Fiebig, D., & Steel, M. (1994). Estimating end-use demand: a bayesian approach. *Journal of Business & Economic Statistics, 12*(2), 221–231. doi:10.2307/1391485

Caves, D. W., Herriges, J. A., Train, K. E., & Windle, R. J. (1987). A bayesian approach to combining conditional demand and engineering models of electric usage. *The Review of Economics and Statistics, 69*(3), 438–448. doi:10.2307/1925531

Chen, C. S., Tzeng, Y. M., & Hwang, J. C. (1996). The application of artificial neural networks to substation load forecasting. *Electric Power Systems Research, 38*(2), 153–160. doi:10.1016/S0378-7796(96)01077-2

Cohen, D. A., & Krarti, M. (1995). A neural network modeling approach applied to energy conservation retrofits. *Proceedings of the Fourth International Conference on Building Simulation*, (pp. 423 – 430).

Dodier, R., & Henze, G. (1996). Statistical analysis of neural network as applied to building energy prediction. *Proceedings of the ASME ISEC*, San Antonio, TX, (pp. 495 – 506).

EIA. (2008). *State and US Historical Data Overview.* Retrieved July 26, 2008, from http://www.eia.doe.gov/overview_hd.html

El Ferik, S., & Belhadj, C. A. (2004). Neural network modeling of temperature and humidity effects on residential air conditioner load. *Proceedings of the Fourth International Conference on Power and Energy Systems*, # 442-152, 557-562.

Entchev, E., & Yang, L. (2007). Application of adaptive neuro-fuzzy inference system techniques and artificial neural networks to predict solid oxide fuel cell performance in residential microgeneration installation. *Journal of Power Sources, 170*(1), 122–129. doi:10.1016/j.jpowsour.2007.04.015

Environment Canada. (1999). *The National Climate Data and Information Archive*. Retrieved June 29, 1999, from http://climate.weatheroffice.ec.gc.ca/prods_servs/index_e.html

ESRU. (2002). *The ESP-r System for Building Energy Simulation: User Guide Version 10 Series*. Energy Systems Research Unit, University of Strathclyde, Glasgow, Scotland. Retrieved Oct 31, 2008, from http://www.esru.strath.ac.uk/

Fahlman, S. E. (1988). Faster-learning variations on back-propagation: an empirical study. *Proceedings of the 1988 Connectionist Models Summer School*, (pp.38-51). Los Altos, CA: Morgan-Kaufmann.

Farahbakhsh, H., Ugursal, V. I., & Fung, A. S. (1998). A Residential End-use Energy Consumption Model for Canada. *International Journal of Energy Research*, *22*(13), 1133–1143. doi:10.1002/(SICI)1099-114X(19981025)22:13<1133::AID-ER434>3.0.CO;2-E

Fausett, L. (1994). *Fundamentals of Neural Networks*. Englewood Cliffs, NJ: Prentice Hall.

Feuston, B. P., & Thurtell, J. H. (1994). Generalized nonlinear regression with ensemble of neural nets: the great energy predictor shootout. *ASHRAE Transactions*, *100*(2), 1075–1080.

Fiebig, D. G., Bartels, R., & Aigner, D. J. (1991). A random coefficient approach to the estimation of residential end-use load profiles. *Journal of Econometrics*, *50*, 297–327. doi:10.1016/0304-4076(91)90023-7

Highley, D. D., & Hilmes, T. (1993). Load forecasting by ANN. *IEEE Computer Applications in Power*, *6*(3), 10–15. doi:10.1109/67.222735

Hsiao, C., Mountain, D. C., & Illman, K. H. (1995). Bayesian integration of end-use metering and conditional demand analysis. *Journal of Business & Economic Statistics*, *13*(3), 315–326. doi:10.2307/1392191

Issa, R. R. A., Flood, I., & Asmus, M. (2001). Development of a neural network to predict residential energy consumption. *Proceedings of the Sixth International Conference on the Application of Artificial Intelligence to Civil & Structural Engineering Computing*, (pp. 65-66).

Kalogirou, S. A. (2000). Applications of artificial neural-networks for energy systems. *Applied Energy*, *67*(1-2), 17–35. doi:10.1016/S0306-2619(00)00005-2

Kalogirou, S. A., Panteliou, S., & Dentsoras, A. (1999). Modeling of solar domestic water heating systems using artificial neural networks. *Solar Energy*, *65*(6), 335–342. doi:10.1016/S0038-092X(99)00013-4

Kawashima, M. (1994). Artificial neural network backpropagation model with three-phase annealing developed for the building energy predictor shootout. *ASHRAE Transactions*, *100*(2), 1095–1103.

Kiartzis, S. J., Bakirtzis, A. G., & Petridis, V. (1995). Short-term forecasting using NNs. *Electric Power Systems Research*, *33*(1), 1–6. doi:10.1016/0378-7796(95)00920-D

Krarti, M., Kreider, J. F., Cohen, D., & Curtiss, P. (1998). Estimation of energy saving for building retrofits using neural networks. *Journal of Solar Energy Engineering*, *120*, 211–216. doi:10.1115/1.2888071

Kreider, J. F., & Haberl, J. S. (1994). Predicting hourly building energy use: the great energy predictor shootout- overview and discussion of results. *ASHRAE Transactions*, *100*(2), 1104–1118.

Kreider, J. F., & Wang, X. A. (1991). Artificial neural networks demonstrations for automated generation of energy use predictors for commercial buildings. *ASHRAE Transactions*, *97*(1), 775–779.

Kreider, J. F., & Wang, X. A. (1992). Improved artificial neural networks for commercial building energy use prediction. *Solar Engineering*, *1*, 361–366.

Larsen, B. M., & Nesbakken, R. (2004). Household electricity end-use consumption: results from econometric and engineering models. *Energy Economics*, *25*(2), 179–200. doi:10.1016/j.eneco.2004.02.001

Lins, M. P. E., DaSilva, A. C. M., & Rosa, L. P. (2002). Regional variations in energy consumption of appliances: conditional demand analysis applied to Brazilian households. *Annals of Operations Research*, *117*(1), 235–246. doi:10.1023/A:1021533809914

McCulloh, W. S., & Pitts, W. H. (1943). A logical calculus of the ideas immanent in nervous activity. *The Bulletin of Mathematical Biophysics*, *5*, 115–133. doi:10.1007/BF02478259

Metaxiotis, K., Kagiannas, A., Askounis, D., & Psarras, J. (2003). Artificial intelligence in short term electric load forecasting: a state-of-the-art survey for the researches. *Energy Conversion and Management*, *44*(9), 1525–1534. doi:10.1016/S0196-8904(02)00148-6

Neto, A. H., & Fiorelli, F. A. S. (2008). Comparison between detailed model simulation and artificial neural network for forecasting building energy consumption. *Energy and Buildings*.

OEE. (2008). *Energy Use Data Handbook Tables (Canada)*, retrieved July 26, 2008, from http://www.oee.nrcan.gc.ca/corporate/statistics/neud/dpa/handbook_totalsectors_ca.cfm?attr=0

Olofsson, T., & Andersson, S. (2001). Long-term energy demand predictions based on short-term measured data. *Energy and Building*, *33*(2), 85–91. doi:10.1016/S0378-7788(00)00068-2

Park, D. C., El-Sharkawi, M. A., Marks, R. J., Atlas, L. E., & Damborg, M. J. (1991). Electric load forecasting using an ANN. *IEEE Transactions on Power Systems*, *6*(2), 442–449. doi:10.1109/59.76685

Parti, M., & Parti, C. (1980). The total and appliance-specific conditional demand for electricity in the household sector. *The Bell Journal of Economics*, *11*, 309–321. doi:10.2307/3003415

Pedersen, L. (2007). Use of different methodologies for thermal load and energy estimations in buildings including meteorological and sociological input parameters. *Renewable & Sustainable Energy Reviews*, *15*(5), 998–1007. doi:10.1016/j.rser.2005.08.005

Peng, T. M., Hubele, N. F., & Karady, G. G. (1992). Advancement in the application of NN for short-term load forecasting. *IEEE Transactions on Power Systems*, *7*(1), 250–257. doi:10.1109/59.141711

Riedmiller, M., & Braun, H. (1993). A Direct Adaptive Method for Faster Backpropagation Learning: The RPROP Algorithm. *The Proceedings of the IEEE International Conference on Neural Networks*, *1*, 586–591. doi:10.1109/ICNN.1993.298623

Rumelhart, D. E., & McClelland, J. L. (1986). *Parallel Distributed Processing, Vol 1*. Cambridge, MA: The MIT Press.

SNNS. (1998). *SNNS User Manual - Version 4.2.* Computer Architecture Department, University of Tuebingen, Germany. Retrieved Aug 1, 2008, from http://www.ra.cs.uni-tuebingen.de/downloads/SNNS/SNNSv4.2.Manual.pdf

Statistics Canada. (1993). *Microdata User's Guide- The Survey of Household Energy Use*. Ottawa, Canada.

Stevenson, J. S. (1994). Using artificial neural nets to predict building energy parameters. *ASHRAE Transactions*, *100*(2), 1081–1087.

Swan, L., Ugursal, V. I., & Beausoleil-Morrison, I. (2008). A new hybrid end-use energy and emissions model of the Canadian housing stock. *Proceedings of the First International Conference on Building Energy and Environment,* Dalian, China, 1992-1999.

Ugursal, V. I., & Fung, A. S. (1996). Impact of appliance efficiency and fuel substitution on residential end-use energy consumption in Canada. *Energy and Building*, *24*(2), 137–146. doi:10.1016/0378-7788(96)00970-X

Yalcintas, M. (2008). Energy-savings predictions for building-equipment retrofits. *Energy and Buildings*.

Yang, J., Rivard, H., & Zmeureanu, R. (2005). On-line building energy prediction using adaptive artificial neural networks. *Energy and Building*, *37*(12), 1250–1259. doi:10.1016/j.enbuild.2005.02.005

Chapter 7
Setting Technology Transfer Priorities with CDM-SET3: Development of Sustainable Energy Technology Transfer Tool

Alexandros Flamos
National Technical University of Athens, Greece

Charikleia Karakosta
National Technical University of Athens, Greece

Haris Doukas
National Technical University of Athens, Greece

John Psarras
National Technical University of Athens, Greece

ABSTRACT

There is no much meaning in separating "good" and "bad" technologies. A definitely more critical issue is to identify "good" and "bad" technological options for a specific country & region based on its specific needs and special characteristics. In this framework, aim of this chapter is the presentation of the CDM-SET3 tool that incorporates the potential host country's priority areas in terms of energy services and the suitable sustainable energy technologies to fulfil these needs and priorities, taking into consideration several criteria that examine the benefits in the economic, environmental and social dimension and through a MCDA approach facilitates the identification of the most proper technology alternatives to be implemented under the umbrella of CDM to a specific host country. The application of CDM-SET3 in representative case study countries is also presented and the results are discussed. Finally, in the last section are the conclusions, which summarize the main points, arisen in this chapter.

INTRODUCTION

The Kyoto Protocol contains market mechanisms that enable industrialized countries to invest in greenhouse gas emission (GHG) reduction projects on the territory of other countries, either developing, or

DOI: 10.4018/978-1-60566-737-9.ch007

industrialised, such as the Clean Development Mechanism (CDM) and the Joint Implementation (JI) (UNFCCC, 2001).

The CDM has a double objective, helping Annex I countries comply with their emission reduction commitments on the one hand and assisting developing countries in achieving sustainable development (SD). While its primary goal is to save abatement costs, the CDM is also considered as a key means to boost technology transfer and diffusion.

The selection of the most promising CDM technologies as regards their contribution towards the achievement of SD in the host countries is one of the most important elements of the CDM process (Anagnostopoulos *et al*, 2004; Doukas *et al*, 2008). Implementing such sustainable energy technologies in developing countries would stimulate SD on a global scale, which is one of the key objectives of the Kyoto Protocol as well as of the EU. The implementation of a sustainable energy technology generally requires that it passes through a number of research and development phases, such as exploring its technical (Flamos *et al*, 2004a), socio-economic (Flamos *et al*, 2004b) and market potential. In addition, due to the inadequate capacity building and development of the Kyoto mechanisms and the low awareness of the CDM modalities and procedures in developing countries, many projects proposals find great difficulties in order to be appropriately developed under the CDM.

Ideally, a CDM project would therefore be based on a clear assessment of the GHG emission reduction potential and a clear assessment of the technology needs and development priorities in the host country. Actual CDM practice, however, has shown that such projects are largely initiated by the demand for relatively low-cost Certified Emission Reductions (CERs), leading to a series of ad-hoc projects, rather than serving the overall policy objectives of the host countries. Figure 1, shows the amount of CERs issued based on ongoing CDM projects. As it can be seen China, India, South Korea and Brazil have thus far supplied 90% of issued CERs, (actually realised GHG emission reductions which have been verified and issued by the CDM Executive Board). In terms of expected credits (emission reductions expected from projects planned or ongoing), China, India, Brazil, South Korea and Mexico (in that order) presently have a share of 85% of the pipeline; in terms of proposed projects, this percentage is 80% (Fenhann, 2008; Kirkman *et al*, 2006).

In general, a few countries are responsible for the bulk of CDM-activity. This indicates that once the appropriate institutions, such as Designated National Authorities (DNA) are in place, CDM can develop rapidly. Nevertheless, it should be noted that a rapid take-off of CDM activity is usually also the result of genuine investors' interest, which may serve as another indicator for a potential for CDM project development.

When the present pipeline is looked at in terms of division of projects across regions, Asia and the Pacific have 2899 projects in the CDM pipeline, Latin America has 757 projects, Africa has 71 projects, Middle-East has 52 projects, and Non-Annex I Europe and Central Asia have 40 projects (Figure 2) (Fenhann, 2008). For each region, the dominance of one or two countries is clear: China and India dominate the Asian region; Brazil and Mexico have most projects in Latin America; Armenia, Cyprus and Uzbekistan host most projects in the European/Central Asian region, in the Middle-East Israel dominates, while in Africa South Africa.

Indeed, CDM should pass through vigilant approval, monitoring and evaluation procedures that create additional transaction costs unrelated to the physical process of eliminating GHGs. Therefore, the successful implementation of a CDM project in a host country could be hampered by its quite lengthy approval procedures and a huge amount of transaction costs associated with different processes in the CDM life cycle (Chadwick, 2006; Michaelowa & Jotzo, 2005). Consequently, there exist various bar-

Figure 1. Share of largest CDM Host Countries in Issued CERs

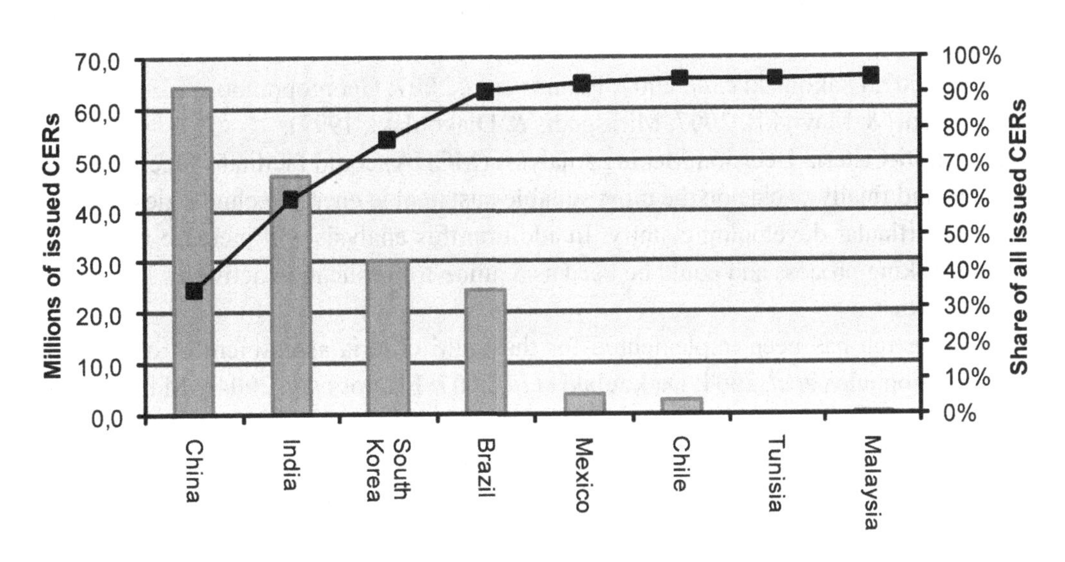

riers which are to be overcome for the success of CDM in host countries.

A Multidimensional Problem

Therefore, a problem that decision makers often face and in which multiple conflicting goals or criteria have to be considered is the identification of the most appropriate environmental friendly technology

Figure 2. Division of CDM Projects across Regions (as of September 2008)

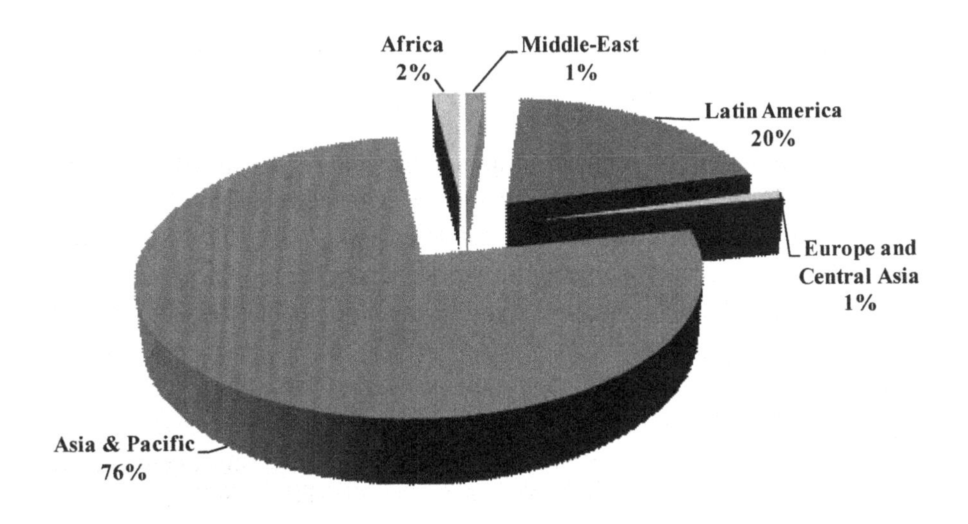

option to be transferred under CDM based on a specific host country's sustainable development needs (Flamos *et al*, 2005). Multicriteria analysis forms a very useful tool in order to take into account simultaneously all the basic aspects (technoeconomic, social etc.) of the problem during the formulation of priorities for the implementation of various technology alternatives and represents a sound methodology applied internationally during the last decade in several problems of environmental and energy planning (Buchholz *et al*, 2007; Diakoulaki *et al*, 2007; Flamos *et al*, 2007; Georgopoulou *et al*, 1998; Karakosta *et al*, 2008; Konidari & Mavrakis, 2007; Mirasgedis & Diakoulaki, 1997).

The use of Multi-Criteria Decision Making Analysis (MCDA) could facilitate "decision makers" to define, evaluate and finally decide on the most suitable sustainable energy technologies to transfer and implement in a particular developing country. In addition this analysis will increase the transparency of the decision making process and could be used as a guide for replication activities in countries with similar characteristics.

Significant research has been implemented for the multi criteria assessment of alternative CDM projects (Anagnostopoulos *et al*, 2004; Diakoulaki *et al*, 2007; Flamos *et al*, 2004b; Manfred, 2007) and the selection of CDM host countries for a specific CDM technology (Georgiou *et al*, 2008). However, it can be observed that the existing literature is characterized by the absence of a system which could facilitate the decision making process towards the selection of the most appropriate technologies to be transferred under the umbrella of CDM in selected countries.

In this regard, the Clean Development Mechanism Sustainable Energy Technology Transfer Tool (CDM-SET^3) has been developed and applied to a selected set of representative developing countries from Asia, Latin America, sub-Sahara Africa and the Mediterranean. The CDM-SET^3 steers towards a successful technology transfer through the CDM taking also into consideration the overall medium to long-term energy and environmental strategy of the host country. It is based on the Electre Tri Method and has the objective to categorize the available sustainable energy technology alternatives, according to their contribution in sustainable development, by examining their economic, environmental and social benefits:

- **Economic benefits:** energy supply diversification and reliability, replicability potential in the country, lower dependency on imported fuels, grid stability, energy price stability, etc.
- **Environmental benefits:** improvement of local air quality, GHG emission reduction, land protection, improved water quality, solid waste management, ecological conservation, etc.
- **Social benefits:** increased socio-economic welfare, poverty alleviation, health improvement, better education, empowerment through training, etc.

Apart from the introduction and problem setting, the chapter is structured along three sections. The third section is devoted to the presentation of the CDM- SET^3 which consists of the MCDA method modeling for the specific problem characteristics and the description of the CDM-SET^3 main characteristics and components. The fourth section presents the application of the tool in Israel and the last section is the discussion of the main outcomes, summarizing the main points that have arisen in this chapter.

The CDM-SET^3

The tool incorporates the SD benefits that a particular technology alternative could deliver to a developing country under the umbrella of CDM with the energy services needs and priorities of a potential host

Figure 3.CDM-SET³ Architecture

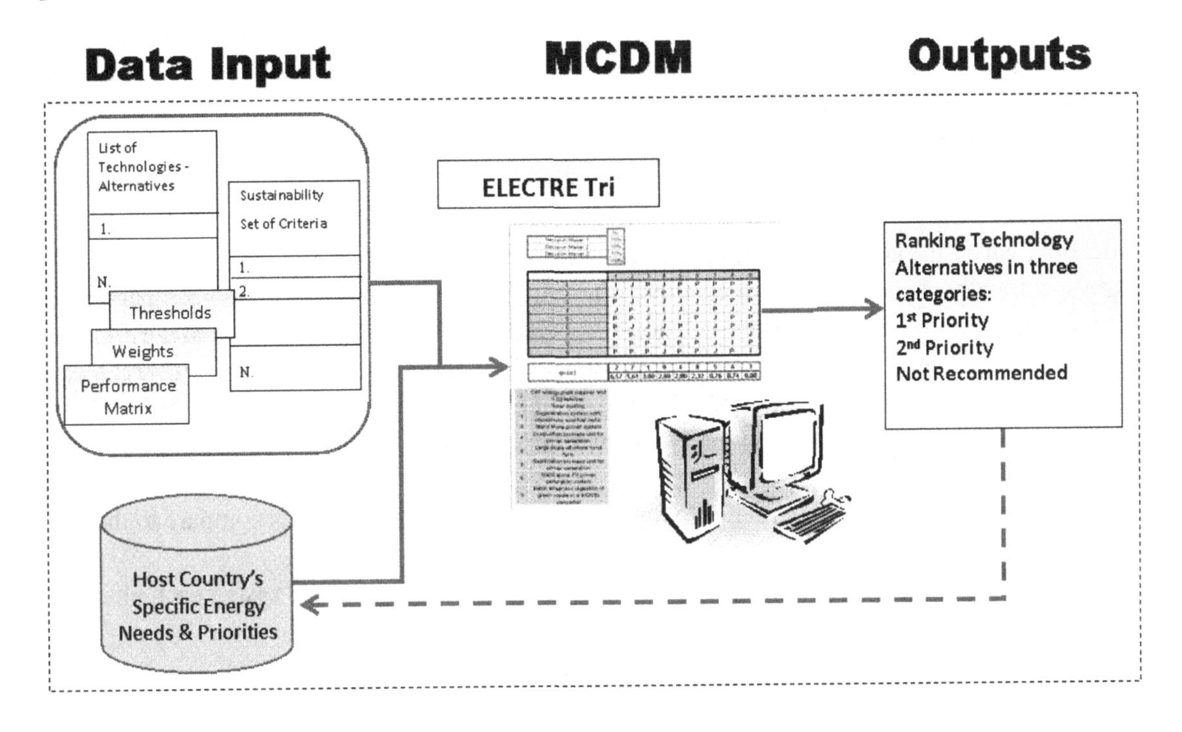

country in order to identify the most suitable sustainable energy technologies to transfer and implement through the specific Kyoto mechanism. The tool's architecture is presented briefly in Figure 3.

The MCDA approach was selected due to the fact that it provides the flexibility and capacity to support the views of many decision makers or stakeholders (Cherni *et al*, 2007) and support the evaluation of the most suitable sustainable energy technologies to transfer and implement in a developing country through the CDM. In particular, these methods have been applied to a wide range of electricity sector issues, such as for environmental assessment (Geldermann *et al*, 2000), for the prioritization of environmental projects (Al-Rashdan *et al*, 1999) and for electricity planning (Beccali *et al*, 2003).

Out of the various outranking multiple criteria decision-aid methods available (e.g. ELECTRE, PROMETHEE, PRAGMA/MACCAP, etc.), the ELECTRE Tri method was selected. It is the most recent out of the methods of the ELECTRE family – was developed in 1992 – and has been applied in problems related to environmental planning, business risk, etc (Arondel & Girardin, 1998; Diakoulaki *et al*, 2000; Montano-Guzman, 2000; Zopounidis & Dimitras, 1998).

ELECTRE Tri is characterised by its ability to assign alternatives to pre-defined categories. Furthermore, ELECTRE Tri can better approximate the attitude of decision-makers, which is usually characterised by a gradual transition from the indifference to the preference state (Georgopoulou *et al*, 2003).

The ELECTRE Tri method was selected to be used in this tool as it was considered appropriate for application in this specific problem due to the following reasons: A particular characteristic of the ELECTRE Tri method is that it provides the possibility of assigning potential actions into pre-defined categories and thus it is suitable for exploring which alternative sustainable energy technologies can be considered as very high, high, medium and low priority options. The latter two categories do not necessarily comprise less desired technologies; they can include technologies that require better (market, technology, social etc) conditions for their implementation through CDM than the existing ones.

The ELECTRE Tri method is basically a two stage process (Mousseau & Slowinski, 1998; Yu, 1992). In the first stage, the outranking hypothesis "action a is at least as good as the reference action of a pre-defined category" is tested. In the second stage, the outranking relations established in the previous step are exploited in order to classify potential actions in the various pre-defined categories (the number of categories is defined by the user). This classification is performed through two processes, the "optimistic" classification, where the action is classified to the highest possible category and the "pessimistic" classification, where the action is classified to the lowest possible category.

More specifically, the assignment of an alternative a in Electre Tri results from the comparison of a with the profiles defining the limits of the categories. Let F denote the set of the indices of the criteria g_1, g_2, ..., g_m $(F=\{1, 2, ..., m\})$ and B the set of indices of the profiles defining $p+1$ categories $(B=\{1,2,...,p\})$, b_h being the upper limit of category C_h and the lower limit of category C_{h+1}, $h=1, 2, ...,p$. ELECTRE Tri assigns alternatives to categories following two consecutive steps (Maystre *et al*, 1994):

- Construction of an outranking relation S that characterises how alternatives compare to the limits of categories;
- Exploitation (through assignment procedures) of the relation S in order to assign each alternative to a specific category.

In this framework, ELECTRE Tri builds an outranking relation S, i.e., validates or invalidates the assertion aSb_h, whose meaning is "a is at least as good as b_h". The assertion b_hSa, whose meaning is "b_h is at least as good as a" is also validated or invalidated in the same way.

ELECTRE Tri requires from the user to (Mousseau *et al*, 1999):

- Define the alternatives, the criteria as well as the pre-defined categories for their classification.
- Determine the values for the performances of the alternatives to the criteria and the indifference, strict preference and veto thresholds.

The Alternatives

The first step of the methodological approach that has been materialized by CDM-SET^3 is the identification of the list of the alternative sustainable energy technologies to be prioritized by the CDM-SET^3.

The selection of technologies to be used as alternatives for CDM-SET^3 has been rather broad by identifying from a range of literature sources, *e.g.* (Martinot, 2006; Matysek *et al*, 2006; WADE, 2003; Van der Gaast *et al*, 2008; WB, 2005; WCI, 2004), clean, sustainable energy technologies that is applicable for CDM.

Table 1 illustrates the alternative sustainable energy technologies for CDM implementation. Although the final list of 42 low-carbon, sustainable energy technologies is not exhaustive, it contains those technologies that have been applied under the CDM and have been identified by different literature sources as important technologies towards a low-carbon energy future.

Table 1. The alternative technologies

Energy service	Related Technologies
Electricity Generation	T1: Fuel switch: coal-to-gas; T2: Biomass combustion for electricity and heat; T3: Biomass gasification; T4: Energy towers; T5: Geothermal electricity production; T6: Hydro dams for large-scale electricity supply (existing and new dams); T7: Small-scale hydro energy; T8: Run of river for large scale electricity supply; T9: Hydrogen; T10: Ocean, wave and tidal energy; T11: Solar lanterns; T12: Solar PV; T13: Wind energy; T14: Small-scale CHP production; T15: SC PC steam power plants; T16: IGCC power plants; T17: Oil-based conventional steam power plants; T18: Concentrating solar power; T19: Hybrid technology; T20: Biogas for cooking and electricity (see biomass gasification)
Heating	T21: Biomass combustion for electricity and heat; T22: Small-scale CHP production; T23: Heat pumps for space heating/cooling and water heating; T24: Solar thermal
Cooking	T25: Improved Cook Stoves; T26: Cook Stoves Based on Ethanol/Methanol and Biomass Gasification; T27: LPG and LNG for household and commercial cooking; T28: Solar Cookers; T29: Charcoal production for cooking and heating; T30: Biogas for cooking and electricity
Cooling	T31: Solar cooling and hybrid systems with heating and hot water; T32: Heat pumps for space heating/cooling and water heating
Energy Efficiency	T33: Energy savings in buildings; T34: Compact fluorescent lamps; T35: Energy efficiency and saving in the agrifood industry; T36: Energy efficiency and saving in the cement industry
Municipal Solid Waste	T37: Methane capture at landfills for electricity and heat; T38: Combustion of MSW for district heat or electricity; T39: Gasification of MSW for large-scale electricity/heat
Transport	T40: Liquid bio fuels for transport
Methane Capture	T41: Coal min/bed methane recovery
CO2 Capture	T42: CO_2 capture and storage technologies, including enhanced oil or gas recovery

The Criteria

The aim of the proposed tool is to assign sustainable CDM energy technologies in categories according to their contribution to SD goals of the host country. For the assessment of these technologies a number of parameters had to be incorporated in the analysis, such as cost aspects, social impacts, GHG reductions as well as environmental impacts, based on the specific characteristics, development needs and perspectives of each host country. Table 2 illustrates fundamental points of view that were identified and associated with the implementation of sustainable energy technologies, which also constitute the criteria of the proposed approach. The benefits, economic, environmental and social ones, from the implementation of a certain technology through CDM in a host country are valuated through the six criteria that are used in the tool for the assessment of each one of the sustainable energy technology alternatives.

The tool focuses on the provision of a small but clearly understood set of evaluation criteria, which can form a sound basis for the comparison of the examined energy options in terms of their contribution to sustainable development through the CDM.

Needs and Priorities Identification

In order to be able to assess the potential of a technology in a country, irrespective of the role of the CDM, it is important to find out:

- Whether the technology is suitable for the country, given the resources needed (e.g. sun for solar-based technology, wind for wind energy, oceans for tidal wave energy, etc.), the research and

Table 2. The criteria

Criterion	Description of criterion
K1: Affinity with strategic/developmental planning	Reflects the accordance of particular technologies with the strategic and developmental planning of each country. The higher the accordance with strategic planning in a specific country the higher the performance of a specific technology in this criterion.
K2: Local and regional economic development	Represents the repercussion of a particular technology in the local and regional development. It does not include the impact on the employment, while it incorporates the extent to which the local enterprises bloom due to the investments in the region. The higher the growth achieved the higher the performance.
K3: CO_2 emissions reduction	Represents the estimated reduction of CO_2 emissions that will be achieved via the implementation of each alternative. The choices with the higher possible reduction are evaluated higher than the options with lower CO_2 reduction potential.
K4: Minimization of the negative effects on the natural environment at national - regional level	Reflects the level of repercussion of the alternative in the natural environment, incorporating the noise levels, esthetic interruptions, pressure on land resources and excessive land use. Options with the least possible impact are ranked higher.
K5: Contribution to the employment	Reflects the impact of the examined option in the social dimension as far as the employment rates are concerned. The higher the contribution to net employment generation, the higher the performance in this criterion.
K6: Contribution to the energy sufficiency (independence)	This criterion depicts the extent to which each examined option contributes to the country's energy independence, by substituting certain amounts of the consumed primary energy.

development status of the technology in general and in the country, and related investment and operation and maintenance costs for the technology;

- Whether the technology is in accordance with the sustainable development needs and priorities of the potential CDM host country.

In the context of the Marrakech Accords of 1991, a decision was adopted by the seventh Conference of Parties to the UNFCCC (COP-7) on a framework for meaningful and effective actions to enhance the implementation of UNFCCC Article 4.5.[1] As part of this decision, an expert group on technology transfer was established with the objective to analyse ways to facilitate the transfer of environmentally sound technologies to developing countries. In this decision COP-7 called upon assessments of technology needs in order to determine the mitigation and adaptation technology priorities of developing countries (and countries with economies in transition). These technology needs assessments (TNA) involve, according to the decision, "different stakeholders in a consultative process to identify the barriers to technology transfer and measures to address these barriers through sectoral analyses".[2]

The purpose of a TNA is to identify technologies that are needed or prioritised in meeting a country's sustainable development needs. One example of a TNA exercise within developing countries has been shown in the "Synthesis Report on Technology Needs identified by Parties not Included in Annex I to the Convention" by the UNFCCC Subsidiary Body for Scientific and Technological Advice (SBSTA), presented at its twenty-fourth session in 2006 (UNFCCC, 2006). The report was based on TNAs undertaken by 23 non-Annex I Parties (Africa: 8; Asia and the Pacific: 6; Latin America and the Caribbean: 6; and Europe: 3) and 26 initial National Communications to the UNFCCC by developing country Parties. Most of the Parties (92%) indicated that their technology priorities are in the energy sector (mainly renewable energy, combined heat and power and demand-side management); 79% of the countries (also) included industry sector technologies in their priority list; about half of the countries included transport sector technologies in their assessments.

Therefore, in order to determine whether a technology is in accordance with the sustainable development needs and priorities of a potential CDM host country, the appropriate information from relevant TNAs should be examined and taken into consideration as well as information from the related DNAs. A broad range of energy services have been identified and embodied in CDM-SET[3].

APPLICATION

The tool has been applied (within the framework of the EC FP7 project ENTTRANS) in five case study countries: Chile, China, Israel, Kenya and Thailand. These countries have been selected because of their different profiles in terms of geographical location, energy profile, population size, and economic profile. Chile has been an active CDM host country since the year 2000 and with an efficiently operating DNA. China is presently the CDM frontrunner among host countries and with an extensive CDM policy since 2005. Moreover, China could and in some cases has already become an important producer of low-carbon technologies to other countries in the world. Israel is the most active country as regards CDM projects implementation in the Mediterranean and it has been active in developing technologies that are particularly suitable for warm regions, e.g., energy towers in deserts. Kenya has recently become a CDM host country in the sub-Sahara African region with a large potential for hydropower and which has recently discovered coal. Finally, Thailand is an example of a host country with an active government policy for supporting clean energy production, but which has long been reluctant to host CDM projects.

This section presents, as an illustration of the CDM-SET[3] in different developing countries the results of research for Israel (Sharan & Vaturi, 2007) and a cross country comparison of prioritised energy service needs.

The stage 1 analysis for the case of Israel resulted in an overview of Israel's CDM portfolio and institutional framework, as well as the identification of some technologies appropriate to be implemented under a CDM project.

The Israeli DNA is under the ministerial responsibility of the Ministry of the Environment and was established in 2004. It includes representatives of governmental and public bodies. Israel, which is classified as a non-Annex I country under the Climate Change Convention, provides an especially attractive option for CDM projects for a wide variety of reasons. These include, among others:

- Technological and scientific expertise, including wide experience in the field of "clean" technologies;
- Open access to a wide range of data, including air pollution monitoring results;
- Availability of local professionals, including scientists, engineers and lawyers.

Israel potential of CDM project opportunities for Annex I countries cover several sectors, such as energy sector, transport, waste, industrial and public buildings, industry and land-use. Within the elaborated desk study analysis, the following areas/market niches where renewable and new energy efficient services could be successfully launched, possibly through the CDM, can be identified:

- Large public buildings (such as universities);
- Isolated areas such as unrecognized Bedouin villages;
- Energy conservation in arid areas;

- Solar collectors for water heating for residences, hotels, and medical centres, and
- Solar collectors for water heating at high temperature for industrial use.

The stage 2 analysis on energy needs and priorities resulted to a ranking of each need and priority by assigning values from 0 - not relevant for Israel to 5 - very high priority (5 - very high, 4 - high, 3 - medium, 2 - low, 1 – very low, 0 - not relevant for Israel, n.a. – no opinion). The data required for stage 2 has been collected trough a participatory workshop and interviews with key experts and representatives from various sectors of the economy. The assessment of Israel's needs and priorities indicated the following:

Electricity for households (N7) was considered very important ('5'), since consumption of electricity by households is one of the main drivers for the annual increase in energy consumption in Israel. Despite that, some stakeholders assumed that this trend would be more moderate since most of the residential buildings in Israel already use air conditioning systems, so that the electricity demand will become more stable. In general, the interviews did not show a clear distinction between rural and urban communities.

The interviewees also gave a very high importance to the service of *electricity for service sectors (N5)* ('5'), which was mainly based on the argument that recent national economic developments have shown a shift from agriculture and industry towards commercial and health service activities. There was a consensus among stakeholders interviewed that the need of electricity in this sector is high and will remain high in the short to medium term, while there have been no incentives to cut electricity usage in these sectors.

Energy efficiency in industry (N10) was also given a score of '5' (very important energy service for Israel) as improvement of energy consumption systems for more efficient energy use was considered critical for the industrial sector, both from an economic and environmental point of view. Most of the interviewees emphasized the importance of investing in provision of efficient energy technologies as the preferred solution to reduce energy costs and CO_2 emissions. Since the steps taken by the government and the private sector have thus far not been sufficient, additional measures in this direction are needed.

Electricity for industry (N1) received a score of 4.5, which is explained by the high annual rate of industrial growth and consequently the growing need for electricity. This growth is largely accompanied by an increased use of automatic machines such as robots, in particular in high-tech industries. Many of the high-tech industries consume mainly electricity for the customer systems and for air conditioning and heating so that their energy needs profile is comparable to that of service sectors. According to some stakeholders, co-generation systems in industry would have good potential in Israeli industrial sectors.

The opinions on *heat for industry (N6)* differed in the sense that in some industries heat is much needed, whereas in other industries the need for heat is very low. Despite the trend of moving from traditional industries to high-tech and services, there is an essential need for efficient heat technologies in the heavy industries. Some stakeholder mentioned the potential for co-generation systems in industrial sectors, which would combine the need for heat with the need for electricity in industrial sectors. Nonetheless, this service was considered important by Israeli stakeholders ('4').

Energy service needs that were considered by stakeholder as either of medium importance, or important ('3.5' on average) were:

Table 3. The alternative needs and priorities

Needs and Priorities	Score
N1 Electricity for Industry	4,5
N2 Electricity for Agriculture	3,5
N3 Electricity for Households - rural communities	5,0
N4 Electricity for Households - urban communities	5,0
N5 Electricity for Service Sectors	5,0
N6 Heat for Industry	4,0
N7 Heat for Households	2,0
N8 Heat for Service Sectors	2,0
N9 Energy for Cooling for all sectors	3,5
N10 Energy Efficiency in Industry	5,0
N11 Municipal Solid Waste Management	3,5
N12 Other Needs and Priorities	-

- **Electricity for agriculture (N2):** On the one hand, the agriculture sector in Israel has been reduced annually, and plays a minor role today in the Israeli economy. On the other hand, the remaining agricultural branches use more electronic and other sophisticated equipment than before (for example, use of lightning and temperature control systems), so the consumption of electricity per capita is higher. Therefore, the interviewed stakeholders tended to give this energy technology need a medium ranking.
- **Energy for cooling purposes (e.g. medicines) (N9):** Energy for cooling is needed mainly for space acclimatisation. Energy for other cooling purposes such as storage of food or medicines is also essential, but its overall impact on the use of energy technologies is minor.
- **Municipal solid waste management (N11):** There was disagreement among the interviewees regarding the need for energy technologies for solid waste management. While some interviewees emphasised that Israel has developed advanced relevant technologies for this service, and that the country could largely improve waste management, others were more skeptical because of the limited sources of waste in a small country like Israel.

Finally, *heat for households (N7)* and for the *service sectors (N8)* was considered unimportant ('2') by stakeholder as these services are generally well covered in Israel through application of air-conditioning systems and radiators.

The Alternatives list is presented in Table 3.

The suitability of clean energy technologies for fulfilling the services that addressed as essential for Israel has been assessed with CDM-SET[3]. Three categories of technologies were found very suitable and have been ranked as very high: *solar based technologies, energy efficiency improvement in residential dwellings,* and *coal-to-gas fuel switch technologies.*

Solar: *Solar photovoltaic (PV)* and *solar thermal for water heating* were top ranked. This can be explained by the high potential for solar-based technologies in Israel. However, suitability of PV technologies might be threatened by the rather high costs associated with the technology. Solar thermal has been relatively well proven as it is already applied in Israel.

Efficiency Improvement in Buildings: Sustainable building design, improved air conditioning and energy-saving lamps were considered very suitable technologies for Israel with the objective to reduce energy demand in households that resulted as a very important priority. Sustainable building design would fit well in the new Israeli building standards which encourage energy use reduction in buildings. The same is assumed for passive cooling technologies and solar cooling technologies that contribute to energy efficiency and have already been applied in construction of buildings in Israel.

Coal to Gas: Coal to gas fuel switching technologies also achieved a high ranking. The current process of switching from coal to gas in energy production in Israel is a large scale application and will continue according to the energy plan of the government and Israel Electric Corporation (IEC).

Wind energy was ranked medium in terms of suitability. However, only a small amount of wind energy capacity has been installed in Israel. By overcoming technological and political obstacles, wind power can be a more important source of clean energy in Israel.

Although municipal solid waste management as a service was ranked relatively low, the *methane combustion technologies* are ranked at a medium suitability level due to their high potential in terms of ecological and economic gains, in particular under the CDM, as it would also contribute to additional low-carbon energy production.

Solar lanterns were also ranked medium as regards their suitability. The suitability for remote areas in the country that are not connected to the grid, but also for remote bus stations, agricultural infrastructures and selected motorways is very high and their use increases every year, mainly in arid areas in the south of Israel. However, the impact of the technology is very much limited to specific areas and applications and that the impact on overall energy saving in Israel is modest.

As medium suitability technologies are ranked the following: *clean coal technologies*, *coal steam improvement*, *hydropower through dams*, *solar towers*, and *biogas for electricity generation*. Regarding clean coal technologies, the lack of in country knowledge about their specific technoeconomic characteristics is considered a significant barrier. In addition, the implementation potential of clean-coal technologies in Israel is rather small. *Coal steam improvement* technology fulfils mainly the needs of countries in which electricity generation is based on coal. In Israel, the technology could be applied in some of the coal fired plants, as well as in some textile industries and in general the potential is limited.

Since there are almost no rivers with a topographic potential for production of hydro energy except for the north part of the Jordan River, the suitability of hydropower technologies is not high in Israel.

The *solar tower technology* is also characterized as a medium suitability technology. The technology has been developed in Israel and theoretically has a large potential in the country. However, it has not yet been implemented in the country, except in a model at the Weizmann Institute for Science. Although there is a large potential for implementation, there are doubts about its practical applicability.

Biogas for electricity generation technology has a very high potential. The benefits are mainly environmental, rather than economic. This technology is considered more appropriate for the rural rather than the urban society of Israel.

Low suitability scores have been achieved by *geothermal energy*, *combined heat and power production*, and *geothermal heat pumps*. The potential of geothermal energy production is a bit ambiguous (there are doubts as regards its applicability in Israel). While the Israeli company *Ormat* is one of the world leaders in geothermal technology production, geothermal energy production in Israel is very limited. In a similar way, *geothermal heat pumps* are scoring very low in CDM-SET3 due to their limited applicability in Israel. The suitability and appropriateness of sustainable energy technologies in Israel is generally not limited by the level of knowledge and R&D activities in the areas of clean energy technologies in the

Table 4. Suitable sustainable energy technologies

Technology	Priority	Technology	Priority
Solar PV	Very high	Clean coal	Medium
Solar Thermal	Very high	Coal steam improvement	Medium
Sustainable Building Design	Very high	Hydropower through dams	Medium
Energy Saving Lamps	Very high	Solar towers	Medium
Air Conditioning	Very high	Biogas for electricity generation	Medium
Coal to Gas for Power	Very high	Geothermal energy	Low
Wind Power	High	Combined heat and power production	Low
Municipal Solid Waste Methane Combustion	High	Geothermal heat pumps	Low
Solar Lanterns	High		

country. Instead, the opportunities for implementation of most technologies are limited since they are not always suitable for the local geographic, climate, and socio-economic circumstances in Israel.

Table 4 presents the application results as regards the most suitable energy technologies to fulfill the country's needs & priorities.

The results across all the case study countries for the prioritization of technologies are listed in Table 5 below.

The energy service priorities listed in Table 5 are in compliance with the current country context for energy service supply. In all the case study countries energy efficiency in industry and electricity for industry are seen as high priority areas and perhaps reflect a need which has been growing over time due to lack of investment in this area and aging of current technologies. In Israel, Chile and Kenya electricity for households was also a priority while Thailand and China considered that this problem had been addressed already and was no longer an issue. Municipal solid waste management for energy was a priority for China, Thailand and Kenya but not for Chile or Israel. Energy services for the service sector were particularly important for Kenya and Israel. For Israel this was related to growth in this sector at the expense of industry and agriculture while for Kenya tourism is a major economic driver. Electricity for agriculture is seen as a priority in Kenya and Thailand due to the importance of this sector in the economy.

Overall, the emphasis in China was on energy services for industry in terms of efficiency, electricity and heat required to maintain their high economic growth rate. This was also a driver in all the other case study countries as mentioned earlier and in the case of Chile was also concerned with the security of supply of energy imports. Kenya had the broadest range of priorities including the need for cooling services for medicines and perhaps this reflects the broad range of investment required in the country.

CONCLUSION

The CDM-SET[3] incorporates the potential host country's priority areas in terms of energy services and the suitable sustainable energy technologies to fulfil these needs and priorities, taking into consideration several criteria that examine the benefits in the economic, environmental and social dimension, and through

Table 5. Cross-country comparison of Priorities (very high & High) for Energy technologies

Chile	Israel
• Energy Saving lamps • Sustainable design of buildings • Passive cooling technologies • Biomass for electricity • Municipal solid waste Landfill methane capture • Wind power • Coal Mine Methane combustion for electricity • Mini/micro hydro • Energy conservation in the cement, agro, chemical and iron and steel industries • Geothermal • Solar Thermal	• Solar PV • Solar Thermal • Sustainable Building design • Energy saving lamps • Air conditioning • Coal to gas for power • Wind power • Municipal solid waste methane combustion • Solar lanterns
Kenya	**Thailand**
• Mini/micro Hydro • Efficient cement production • Solar thermal for heating • Efficient agricultural industry • Energy saving lamps • Solar Thermal for cooling • Solar PV • Efficient iron and Steel Industry • Mini/micro systems • Improved cook stoves • Efficient chemical industry • Efficient charcoal production • LPG for cooking • Wind power • Municipal solid waste gasification for energy • Municipal solid waste biogas for energy • Biomass CHP	• Biogas for electricity • Biomass for electricity • Energy efficiency in cement industry • Biogas for heating • Energy efficiency in Iron and steel industry • Coal/gas based CHP • Steam boiler upgrades • Energy efficiency in agro industry • Passive cooling by building design • Air conditioning • CFLs • Solar thermal water heating • Clean coal • Micro CHP • Municipal solid waste Biogas • Municipal solid waste Methane capture from landfill
China: Yunnan Province	**China: Shandong Province**
• Energy Saving lamps • Solar Coolers • Clean coal for electricity • Energy efficiency in cement industry • Large scale hydro • Supercritical boilers for power • Energy efficiency in iron and steel industry • Coal to gas for electricity • Energy efficiency in chemical industry	• Clean-coal for large-scale power supply Hydropower through dams • Energy saving lamps • Cement industry energy conservation • Solar coolers • Iron & steel industry energy conservation • Wind power for large-scale power supply • Solar cookers (for households) • Combustion of municipal solid waste

a MCDA approach lead to identify the most proper technology alternatives (very high and high priority technology alternatives) to be implemented under the umbrella of CDM to a specific host country.

The selection of the technology to be transferred under the CDM has to be turned towards the host country specific needs and priorities and related decision support approaches and methods can have a significant contribution in this respect. The tool aims to facilitate technology transfer through CDM by providing decision makers with a clear picture of what low-carbon technologies could support a specific host country's national development needs and priorities. Hence, any possible solution to the problems at hand, are approached from a context specific perspective. This means that there is no much meaning in separating "good" and "bad" technologies and it is better to concentrate on the identification of "good" and "bad" technological options for a specific country, region etc., based on its specific needs and special characteristics.

It should be clearly stated that CDM-SET^3 does not intent to replace the decision maker. The pilot application in the case study countries of the ENTTRANS project has indicated that with appropriate use, it could act as a useful decision support tool by structuring the decision process and assisting the decision makers to pre-assess the most suitable sustainable energy technologies to transfer and implement in a particular developing country through CDM.

REFERENCES

Al-Rashdan, D., Al-Kloub, B., Dean, A., & Al-Shemmeri, T. (1999). Environmental impact assessment and ranking the environmental projects in Jordan. *European Journal of Operational Research, 118*(1), 30–45. doi:10.1016/S0377-2217(97)00079-9

Anagnostopoulos, K., Flamos, A., Kagiannas, A., & Psarras, J. (2004). The impact of clean development mechanism in achieving sustainable development. *International Journal of Environment and Pollution, 21*(1), 1–23.

Arondel, C., & Girardin, P. (1998). *Sorting cropping systems on the basis of their impact on groundwater quality.* Cahiers du LAMSADE No. 158, Universite Paris – Dauphine.

Beccali, M., Cellura, M., & Mistretta, M. (2003). Decision-making in energy planning. Application of the ELECTRE method at regional level for the diffusion of renewable energy technology. *Renewable Energy, 28*(13), 2063–2087. doi:10.1016/S0960-1481(03)00102-2

Buchholz, T. S., Volk, T. A., & Luzadis, V. A. (2007). A participatory systems approach to modeling social, economic, and ecological components of bioenergy. *Energy Policy, 35*(12), 6084–6094. doi:10.1016/j.enpol.2007.08.020

Chadwick, B. P. (2006). Transaction costs and the clean development mechanism. *Natural Resources Forum, 30*, 256–271. doi:10.1111/j.1477-8947.2006.00126.x

Cherni, J. A., Dyner, I., Henao, F., Jaramillo, P., Smith, R., & Font, R. O. (2007). Energy supply for sustainable rural livelihoods. A multi-criteria decision-support system. *Energy Policy, 35*(3), 1493–1504. doi:10.1016/j.enpol.2006.03.026

Diakoulaki, D., Georgiou, P., Tourkolias, C., Georgopoulou, E., Lalas, D., Mirasgedis, S., & Sarafidis, Y. (2007). A multicriteria approach to identify investment opportunities for the exploitation of the clean development mechanism. *Energy Policy, 35*(2), 1088–1099. doi:10.1016/j.enpol.2006.02.009

Diakoulaki, D., Kormentza, Y., & Hontou, V. (2000). *An MCDA approach to burden-sharing among industrial branches for combating climate change*. Paper Presented at the 51st Meeting of the European Working Group on MCDA, Madrid.

Doukas, H., Karakosta, C., & Psarras, J. (in press). RES technology transfer within the new climate regime: A "helicopter" view under the CDM. *Renewable & Sustainable Energy Reviews*. doi:.doi:10.1016/j.rser.2008.05.002

Fenhann, J. (2008). *CDM Pipeline Overview.* UNEP Risoe Centre; 1st September 2008, Available at http://www.cd4cdm.org.

Flamos, A., Anagnostopoulos, K., Askounis, D., & Psarras, J. (2004a). e-Serem – A Web-Based Manual For The Estimation of Emission Reductions From JI and CDM Projects. *International Journal Mitigation and Adaptation Strategies for Global Change, 9*(2), 103–120. doi:10.1023/B:miti.0000017728.80734.fb

Flamos, A., Anagnostopoulos, K., Doukas, H., Goletsis, Y., & Psarras, J. (2004b). Application of the IDEA-AM (Integrated Development and Environmental Additionality - Assessment Methodology) to compare 12 real projects from the Mediterranean region. [ORIJ]. *Operational Research International Journal, 4*(2), 119–145. doi:10.1007/BF02943606

Flamos, A., Doukas, H., Karakosta, C., & Psarras, J. (2007). *MCDM Approach for Assessing CDM Energy Technologies Towards Sustainable Development.* Ninth International Conference on "Energy for a Clean Environment", CleanAir 2007, Povoa de Varzim, Portugal.

Flamos, A., Doukas, H., Patlitzianas, K., & Psarras, J. (2005). CDM – PAT: A Decision Support Tool for the Pre-Assessment of CDM Projects. [IJCAT]. *International Journal of Computer Applications in Technology, 22*(2/3). doi:10.1504/IJCAT.2005.006939

Geldermann, J., Spengler, T., & Rentz, O. (2000). Fuzzy outranking for environmental assessment. Case study: iron and steel making industry. *Fuzzy Sets and Systems, 115*(1), 45–65. doi:10.1016/S0165-0114(99)00021-4

Georgiou, P., Tourkolias, C., & Diakoulaki, D. (2008). A roadmap for selecting host countries of wind energy projects in the framework of the clean development mechanism. *Renewable & Sustainable Energy Reviews, 12*(3), 712–731. doi:10.1016/j.rser.2006.11.001

Georgopoulou, E., Sarafidis, Y., & Diakoulaki, D. (1998). Design and implementation of a Group DSS for sustaining renewable energies exploitation. *European Journal of Operational Research, 109*(2), 483–500. doi:10.1016/S0377-2217(98)00072-1

Georgopoulou, E., Sararidis, Y., Mirasgedis, S., Zaimi, S., & Lalas, D. P. (2003). A multiple criteria decision-aid approach in defining national priorities for greenhouse gases emissions reduction in the energy sector. *European Journal of Operational Research, 146*(1), 199–215. doi:10.1016/S0377-2217(02)00250-3

Karakosta, C., Doukas, H., & Psarras, J. (2008). A Decision Support Approach for the Sustainable Transfer of Energy Technologies under the Kyoto Protocol. *American Journal of Applied Sciences, 5*(12), 1720–1729.

Kirkman, A.-M., Aalders, E., Braine, B., & Gagnier, D. [Eds.] (2006). *Greenhouse Gas Market Report 2006: Financing Response to Climate Change: Moving to Action* (v, 136 p.). International Emission Trading Association - IETA.

Konidari, P., & Mavrakis, D. (2007). A multi-criteria evaluation method for climate change mitigation policy instruments. *Energy Policy, 35*(12), 6235–6257. doi:10.1016/j.enpol.2007.07.007

Lenzen, M., Schaeffer, R., & Matsuhashi, R. (2007). Selecting and assessing sustainable CDM projects using multi-criteria methods. *Climate Policy*, *7*, 121–138.

Martinot, E. (2006). *Renewables: Global Status Report, 2006 Update*. REN21, retrieved from www.ren21.net

Matysek, A., Ford, M., Jakeman, G., Gurney, A., & Fisher, B. S. (2006). *Technology: Its Role in Economic Development and Climate Change*. ABARE Research Report 06.6, prepared for Australian government Department of Industry, Tourism and Resources, Canberra, Australia.

Maystre, L. Y., Pictet, J., & Simos, J. (1994). *Méthodes multicritères ELECTRE*. Presses Polytechniques et Universitaires. Romandes, Collection gérer l'environnement, Lausanne.

Michaelowa, A., & Jotzo, F. (2005). Transaction costs, institutional rigidities and the sized of the clean development mechanism. *Energy Policy*, *33*, 511–523. doi:10.1016/j.enpol.2003.08.016

Mirasgedis, S., & Diakoulaki, D. (1997). Multiple criteria analysis vs. externalities assessment for the comparative evaluation of electricity generation systems. *European Journal of Operational Research*, *102*(2), 364–379. doi:10.1016/S0377-2217(97)00115-X

Montano – Guzman. L. (2000). *Une méthodologie d'aide multicritère a la décision pour la diagnostique de l'entreprise*. Paper Presented at the 51st Meeting of the European Working Group on MCDA, Madrid.

Mousseau, V., & Slowinski, R. (1998). Inferring an ELECTRE TRI model from assignment examples. *Journal of Global Optimization*, (12): 157–174. doi:10.1023/A:1008210427517

Mousseau, V., Slowinski, R., & Zielniewicz, P. (1999). *ELECTRE TRI 2.0a – Methodological guide and user's manual*. Documents du LAMSADE No 111, University Paris – Dauphine.

Sharan, Y., & Vaturi, A. (2007). *Interim Report on the ENTTRANS project in Israel*. ENTTRANS Final Meeting, Brussels, December.

UNFCCC. (2001). *FCCC/CP/2001/13/Add.1, Decision 4/CP.7, Annex, Seventh Conference of the Parties to the UNFCCC*. Marrakech, Morocco.

UNFCCC, (2006). *Synthesis Report on Technology Needs Identified by Parties not included in Annex I to the Convention*. FCCC/SBSTA/2006/INF.1

Van der Gaast, W., Begg, K., Flamos. A, Deng, G., Mithulananthan, N., Theuri, D. *et al* (2008). *Synthesis report on technology descriptions 'Sustainable, low carbon technologies for potential use under the CDM'*, EC FP6 ENTTRANS Deliverable 5&6. Joint Implementation Network.

WADE. (2003). *Guide to Decentralized Energy Technologies*. Edinburgh, UK: www.localpower.org.

World Bank. (WB), (2005). *Technical and Economic Assessment: off-grid, mini-grid and grid electrification technologies, prepared for Energy Unit, Energy and Water Department, the World Bank*. Discussion Paper, Energy Unit, Energy and Water Department.

World Coal Institute (WCI). (2004). *Clean Coal Building a Future Through Technology*. Retrieved from http://www.worldcoal.org/assets_cm/files/pdf/clean_coal_building_a_future_thro_tech.pdf

Yu, W. (1992). *ELECTRE TRI: Aspects méthodologiques et manuel d'utilisation.* Document du LAMSADE No. 74, Université Paris – Dauphine.

Zopounidis, C., & Dimitras, A. I. (1998). *Multicriteria Decision Aid Methods for the Prediction of Business Failure*. Dordrecht, The Netherlands: Kluwer Academic Publishers.

ENDNOTES

[1] FCCC/CP/2001/13/Add.1, Decision 4/CP.7, Annex.

[2] FCCC/CP/2001/13/Add.1, Decision 4/CP.7, Annex, para. 3.

Chapter 8
Formulating Modern Energy Policy through a Collaborative Expert Model

Kostas Patlitzianas
REMACO S.A., Greece

Kostas Metaxiotis
University of Piraeus, Greece

ABSTRACT

Nowadays, a comprehensive and modern energy policy making, which will be characterized by clarity and transparency, is necessary. Indeed, there exists a number of energy policy and planning systems, but there are no decision support systems investigating the energy policy making in an integrated way. In this context, the main aim of this chapter is to present an expert system based on a "multidimensional" approach for the energy policy making, which also incorporates the three objectives (security of supply, competitiveness of energy market and environmental protection) and takes into consideration all the related economical, social and technological parameters. This model was successfully applied in order to support the decisions towards the development of the energy policy priorities in the developing Mediterranean Countries as well as the countries of Gulf Cooperation Council – GCC.

1. INTRODUCTION

The energy policy is directed by the security of supply, the competitiveness of energy market and the environmental protection, based on the European Union (EU) energy objectives. In the last years, extensive discussions took place as regards how a competitive energy market would promote, - indirectly perhaps via forces of market - the other energy policy objectives, shifting at the same time the decisions centre from the state to the rest "players" of the market (energy companies / users) (Patlitzianas & Psarras, 2006). This fact caused the intense interest of energy policy analysts and researchers. In this framework, energy policy should compromise the desirable objectives and encourage the close collabo-

DOI: 10.4018/978-1-60566-737-9.ch008

ration between the "players" towards the confrontation of the various obstacles (Doukas, Patlitzianas, Kagiannas, & Psarras, 2008).

According to the developments in European Union (EU), the possible principles of Energy Policy for Europe were elaborated at the Commission's green paper "A European Strategy for Sustainable, Competitive and Secure Energy" on 8 March 2006. As a result of the decision to develop a common energy policy, the first proposals, "Energy for a Changing World" were published by the European Commission, following a consultation process, on 10 January 2007. It is claimed that they will lead to a 'post-industrial revolution', or a low-carbon economy, in the European Union, as well as increased competition in the energy markets, improved security of supply, and improved employment prospects. Although the proposals have been adopted by the European Commission, they require the approval of the European Parliament but were debated and approved at a meeting of the European Council on March 8 and 9, 2007.

The EU has also promoted electricity market liberalisation and security of supply through the 2003 Internal Market in Electricity Directive, which replaced early directives in this area. The 2004 Gas Security Directive has been intended to improve security of supply in the natural gas sector.

Renewable energy has a long history as a central focus of European energy policy. As early as 1986, the European Council listed the promotion of renewable energy sources among its energy objectives. In 1997, the Commission established a target to increase the overall share of renewable energy to 12 percent by 2010 (Doukas et al, 2006). The Commission's most recent initiative, the energy and climate-change package of January 10, 2007, updated this target: 20 percent of all EU energy consumption is to come from renewable sources by 2020. It also established a "minimum target" of 10 percent of the petrol and diesel market to be represented by biofuels by 2020. At the spring meeting of the European Council in Brussels on March 8–9, 2007, the EU heads of state and government endorsed both targets as binding.

In addition to this, the challenges in living conditions and societies have a strong impact on the energy market, since they are related to the increasing energy consumption (Patlitzianas et al, 2006). As a result, uncertainties and conflicts put on the map the important role of the energy market, in spite of its small contribution in the total economic production, which is estimated about 5-10%. (Helm, 2002).

Beyond the bounds of the European Union, energy policy has included negotiating and developing wider international agreements, such as the Energy Charter Treaty, the Kyoto Protocol, the post-Kyoto regime and a framework agreement on energy efficiency; extension of the EC energy regulatory framework or principles to neighbours (Energy Community, Baku Initiative, Euromed energy cooperation) and the emission trading scheme to global partners; the promotion of research and the use of renewable energy.

To sum up, the current plan (2008) of the EU are: 20% increase in energy efficiency, 20% reduction in greenhouse gas (GHG) emissions, 20% share of renewable in overall EU energy consumption by 2020, 10% biofuel component in vehicle fuel by 2020. However, these plans are very ambitious (e.g. today 8.5% of energy is renewable) (Patlitzianas et al., 2008). In particular, the European Commission announced it will issue before the end of 2007 a Framework Directive covering, amongst other sectors, the electricity sector to support renewable energy's development and reach the target of 20% renewable energy by 2020.

As a result, three key targets on "renewable energy", "energy efficiency" and "greenhouse gases" have been agreed by the Heads of EU State for 2020. Each of them makes sense and stands for itself. Efforts to downplay the renewable energy target by stressing the greater significance of the greenhouse gas reduction target - for instance by including nuclear energy to count towards the renewables target - need to be turned down.

In this context, a new European Energy Policy must be ambitious, effective and long-lasting – and involve everyone. In addition, the role of the modern energy policy should be based on the comparison between the failure of policy and the choice of "do nothing policy", based on economical, social and technological parameters (Bloom, 1982). In particular, the energy policy is related to a number of parameters such as the need for improving the reliability and quality of the provided services of the production, transmission and distribution energy companies, the social acceptance of environmental friendly energy technologies, the development of education, the research efforts and the technology (e.g. co-generation of heat, the promotion of renewable energy as well as the rational use of energy etc.) (David, & Zhao, 1989).

Indeed, a comprehensive and modern energy policy making, which will be characterized by clarity and transparency, is necessary (Huang, 1997). In particular, the modern energy policy making has to counterbalance the three objectives and all the related economical, social and technological parameters, avoiding thus unfavourable consequences, e.g. crises, and assuring energy prices stability and accessibility for the users. Based on the literature survey, there exists a number of energy planning systems that are related to the current problem (Kavrakoglou, 1987; Kagiannas, Patlitzianas, Metaxiotis, Askounis & Psarras, 2006; Jebaraj, & Iniyan, 2006).

The energy policy making could be supported through the supply systems (EFOM, ELFIN, MARKAL, MESSAGE, TESOM, UPLAN-E and WASP), demand systems (IIASA, COMMEND, ENPEP, MAED, HELM, MED-PRO) (Weyant, 1990) or the system of integrated planning of resources (IRP-MANAGER, MARKAL) (D'Sa 2005; Wenying 2005). The above decision support systems examine the energy policy in terms of institutional framework, regulations, financing support schemes and environmental issues. However, there are no decision support systems investigating the energy policy making in an integrated way.

Furthermore, according to the Welbank (1985), the use of the expert systems, can help states assess and develop the energy policy. In particular, during the last years the expert systems are considered to be programs with a wide knowledge base in a limited space which use complex knowledge for the execution of works that an expert human could do.

As a result, the philosophy of the expert systems can be used in the current problem for each state that is the "decision maker" of defining importance, shaping the energy policy. In this context, the current chapter presents a collaborative expert model based on a "multidimensional" approach for the energy policy synthesis, which also incorporates the three objectives: security of supply, competitiveness of energy market and sustainable development (save energy and promote climate-friendly energy sources) and takes into consideration all the related economical, social and technological parameters - Energy Policy for Europe - the need for action (Patlitzianas et al. 2007).

The chapter is structured in fifth sections. Apart from the introduction, the second part presents a background as well as a review for the related expert systems aiming at supporting the innovation of the current system. The third part provides a description of the adopted approach and the fourth part presents the developed information system and its application. The last section presents the conclusions, which summarize the main points that have been drawn up in this chapter.

2. ENERGY POLICY MODELS: BACKGROUND AND REVIEW

The energy models reviewed, fall under the following three main areas:

- Energy demand forecasting models (EDF) can be categorized by the method they use to compute the final results. The most common method is the econometric where the results can be anticipated a period longer than one year. The end-use approach is much more complicated and computes the load for each consumer or group of them. It is the only method that can fully satisfy all the requirements of demand side management analysis.
- Generation Expansion Planning or Supply Side Management models, that are designed to find the economically optimal generation expansion policy and the optimum is evaluated in terms of minimum discounted total costs.
- Integrated Resource Planning (IRP) includes models which offer integrated energy resource analysis and the models structure is based on the idea of flow of energy loads.

According to the Kagiannas et al. (2006), the energy policy making could be supported through the following energy policy models:

- The EC (DG-TREN ex-DG-XVII) over the past three decades has sponsored the development of numerous models. The primary models in use are PRIMES and POLES, both of which have evolved through European Commission sponsored programmes. Other EU models are GEM-E3, HERMES, QUEST, MIDAS, MEDEE and EFOM.
- The International Energy Agency (IEA) model is MARKAL. Over 70 teams in more than 35 counties around the world make use of the MARKAL family of energy/economy/environment models. The primary objective of developing MARKAL was to assess the long-term role of new technologies in the energy systems of the 17 IEA member countries.
- The International Atomic Energy Agency (IAEA) model, which has come to the forefront for energy analysis and planning with ENPEP that has been distributed free of charge to more than 80 countries. Included in the latest version of ENPEP as a module, is a windows based version of WASP. Other IAEA models are GTMAx, MAED and EMCAS.
- Energy Information Administration (EIA) models, which the primary model used in policy decision-making is the NEMS. This is a general equilibrium model of the interactions between the U.S. energy markets and the economy.

3. EXPERT SYSTEMS: BACKGROUND AND REVIEW

3.1 Background

An expert system is the incorporation in the computer of a component based on knowledge, so that the system can be able to provide a logical advice regarding a processing function or to have the ability of decision making.

The expert system is a program with a wide knowledge base in a limited space and uses complex knowledge for the execution of works that an expert human could do. An expert system varies from a conventional system of applications as follows (Patlitzianas et al. 2005):

- A conventional system uses mathematical models for the simulation of the problem, while the expert system simulates the human arguing regarding the problem.
- An expert system applies numeral calculations and in the meantime presents the arguments on the "reconstruction" of the human knowledge, while a conventional system whose knowledge is expressed in a special language, other than the code that presents the arguing.
- An expert system is able to solve problems with approximate methods, meaning that it can solve problems, for which we do not have accurate data.

In addition to this, the differences between an expert system and the other systems of artificial knowledge are summarized as follows:

- The expert systems deal with more complicated practical problems, whose solutions are based on the human knowledge. As a result, the expert systems are decision-making tools.
- The expert systems are able to describe the process that they followed for the decision making in order to convince the user that their justification is well-founded and also to propose directly appropriate solutions, compared to the conventional systems that are not so «flexible».
- The expert systems are user friendly and can be used from a wider public.

The expert systems consist of four basic components, which are described as follows:

- Knowledge Base: The knowledge base is composed of rules and facts of the problem. The knowledge base is the most important of the four components. It includes objects, age-groups, characteristics, prices, rules, network of rules. The knowledge can reconstruct with the use of various methods, conceptual networks, frames and logic of predicates. The most common ways of representation of knowledge are the rules sets ("If... then") and they are used in order to represent the consequences of given affair or the action that should be taken in a given situation.
- Working Memory: It constitutes a quantity of working memory in which they are saved the running results of process of resolution of problem. The knowledge and the list of rules are included in the working memory.
- Inference Engine: It is the means with the knowledge, which is included in the working memory, is coordinated. It contains the techniques of justification aiming at the solution of the problems as well as the improvement of rules used in the all process.
- User Interface: It is a kind of connection between the system and the user. This component gives the possibility to the user to communicate with the system, offering continuously actions and data. It gives the possibility to the system of making questions, and providing possible choices, advice and explanations.

3.2 Review in the Energy Market

The expert systems are used for at least 15 years in the energy market. To the best of our knowledge, one of the first expert systems is a system (ALFA) of short-term forecast of electrical charge that was materialised by Jabbour et al. (1988). Since then, many forecast systems were presented, while the most important of them were reported in the works of Hung et al. (1997) and include the following: SOLEXPERT, ALEX, PANexpert, TRANSEPT and CISELEC.

Furthermore, many systems for short-term forecast load exist, as reported in the review of Metaxiotis et al. (2003). Other models, as described by Markovic & Fraissler (1993) and by Kandil (2003), use the expert system for short-term electrical charge forecast with parallel control of the demand and are based on historical data of the last 5 to 10 years. The system of David & Zhao (1989) simulates the extension of electrical installations via a completed expert system which makes use of dynamic planning. Another similar expert system is the NETMAT which was developed by Zitouni & Irving (1999), while COLOMBO (Melli, & Sciuba, 1999) is a system which proposes the planning of electrical generation, as well as the planning of cogeneration plants. Additionally, expert systems exist for the short-term engagement of units (Ouyang, & Shahidehpour, 1990; Li, Shahidehpour,& Wang, 1993), as well as for various other applications of energy and climate policy (Humphreys et al. 2002; Tsamboulas & Mikroudis, 2005; Huang et al., 2005). Furthermore, expert systems are used mainly for the registration and analysis of errors in stations or substations (Kezunovic et al., 2003; Sherwali and Crossley, 1994; Yongli et al. 1994; Minakawa et al., 1995) in situations of network overload, in the regulation of relays so that they function more precisely and efficiently (Lee, Yoon, Yoon, & Jang, 1990), as well as in an alarm process (Eickhoff et al., 1992; Hasan et al., 1995) for the protection of the energy system. Another expert system is SOCRATES (Vale et al., 1998), which is used for the confrontation of extraordinary incidents in the networks transmission and the distribution of electricity, while a study by Khosla & Dillon (1994) constitutes an approach with a combination of an expert system and some networks.

Moreover, the expert systems have penetrated in the restoration of the network after power-offs (Adibi, Kafka, & Milanicz, 1994). The expert systems are mainly used for data collection and decision making after the collapse of the network, but also for the simulation in computer of the situation at which the system collapses (Brunner, Nejdl, Schwarzjirg, & Sturm, 1993). Most of them are connected “on-line”, as for example the system proposed by Brunner et al. (1993), in order to provide help in the operators. Also, the proposal of Nagata et al. is very interesting (1995), because it examines a model, which, combining expert systems with mathematic planning, analyzes the cost involved in the restoration of the operation network.

Another expert system in the energy sector is SPARSE (Zita, Vale, Santos, & Ramos 1995), which is used in the Portuguese distribution network. The application of expert systems in the restoration of networks occupied also Park and Lee (1995) who tested their system in a local network. In addition, the desirable regulations of voltage are achieved automatically via expert systems (Matsuda et al., 1990; Azzam and Nour, 1995; Chokri et al. 1996; Le et al. 1995). Lastly, important applications of expert systems dealt with the safety of RES production units (Singh et al. 1993; Christie et al. 1990; Chang et al. 1990) while various applications involve the energy efficiency (Fouad et al., 1990; Jaber et al. 2005; Doukas, Patlitzianas, Iatropoulos, Psarras, 2006).

Based on the above, the expert systems have many important applications in various issues regarding the operation of the energy market and as a result they are intensely used in specialised technically energy problems. However, their use in problems of knowledge representation for the support of decision making for the energy policy is inexistent.

Figure 1. The philosophy of the model

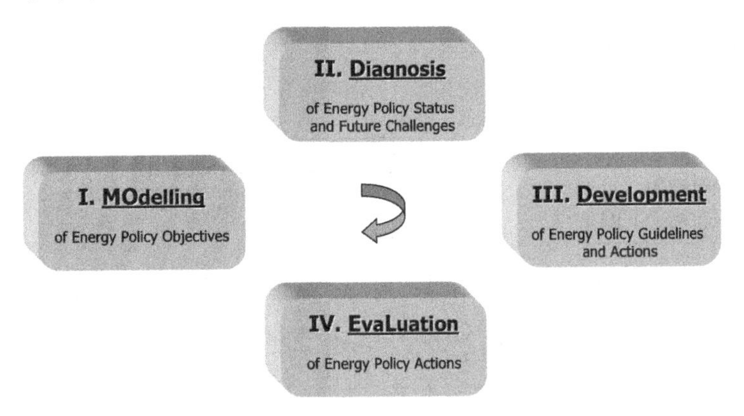

4. SYSTEM APPROACH

The philosophy of the System is presented in the following *Figure 1*:

Two modules are included in the proposed system, as they are presented in the following paragraphs.

4.1 Module A: Diagnosis

The first module is the diagnosis processes of energy policy with the use of an intelligent technique. The approach is able, through the estimation of appropriate indicators (based on the data input), to present to the decision maker the existing situation and future challenges of energy policy based on the "clever" export of conclusions from existing "library" of thresholds. The formulation of the "library" was based on the energy policy review of international organizations (European Commission, International Energy Agency, BP, OPEC, etc.) that concerns the analysis of energy policy for a group of representative countries. In this context, 30 developed countries were analyzed globally and were categorized according to the following characteristics: a) level of economic development; b) level of self-sufficiency in energy resources and c) level of liberalization of energy market.

This step's approach, which describes the diagnosis processes, includes the following steps:

- Estimation of the values' indicators for the years 1990, 1995, 2000,
- Forecast of the values' indicators for the years 2010 and 2020;
- Estimation of the values-objectives (thresholds) for each indicator;
- Determination of possible situation and development of the objectives of energy policy;
- Specification of interactive indicators and parameters;
- Connections of value-fields with the description of the current – future situation and diagnosis of need improvement.

In particular, the procedure of the first step is described as follows:

Figure 2. The procedure of diagnosis (current situation)

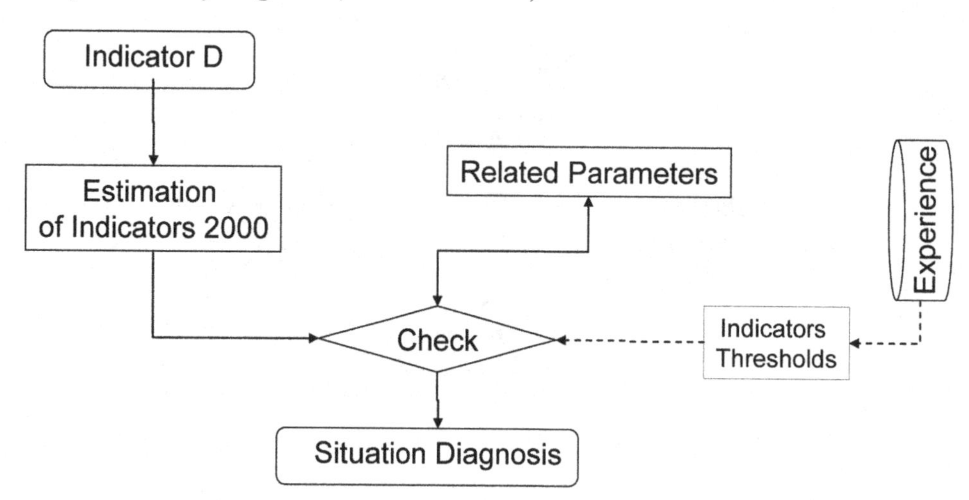

Inputs: The current process constitutes the basic rules for the situation diagnosis as regards the specific indicator. It describes substantially the prevailing situation of a specific parameter of the energy-policy objective that is examined. The used values have the year 2005 as a time base and the typical structure of the procedure is illustrated in the *Figure 2.*

In addition to this, the procedure constitutes the diagnosis of future situation of a concrete parameter of the energy policy objective that is examined. The used value - for the forecast - concern the time-objective 2010. The typical structure of the current process is illustrated in the *Figure 3.*

Outcomes: This is the exporting procedure about the existence or not of dangers in the achievement of the energy policy objectives. It is developed by the connection of processes of more two groups with the parallel examination of processes of related indicators and the individually examination of related parameters' values. *Figure 4* shows the procedure of this process for the existence or not of a need for improvement of the energy policy objective.

Figure 3. The procedure of diagnosis (future situation)

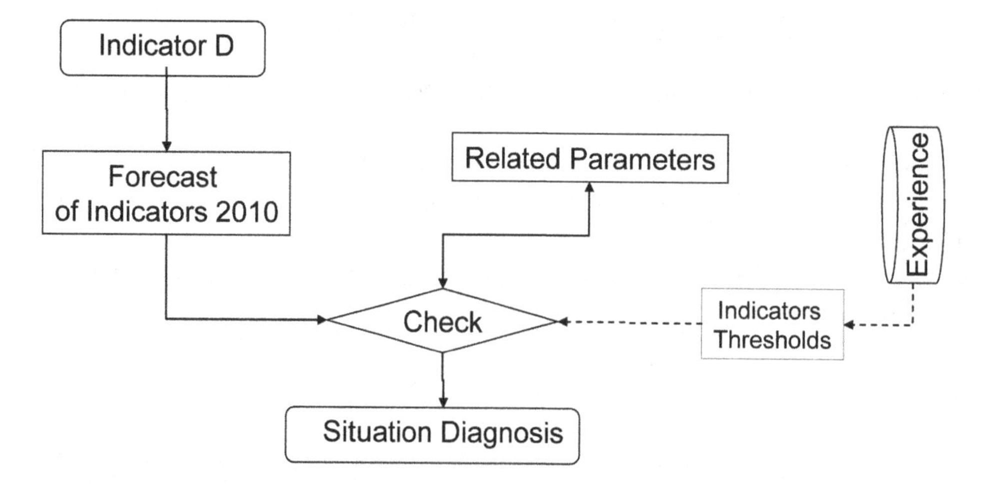

Figure 4. The procedure of diagnosis (overall)

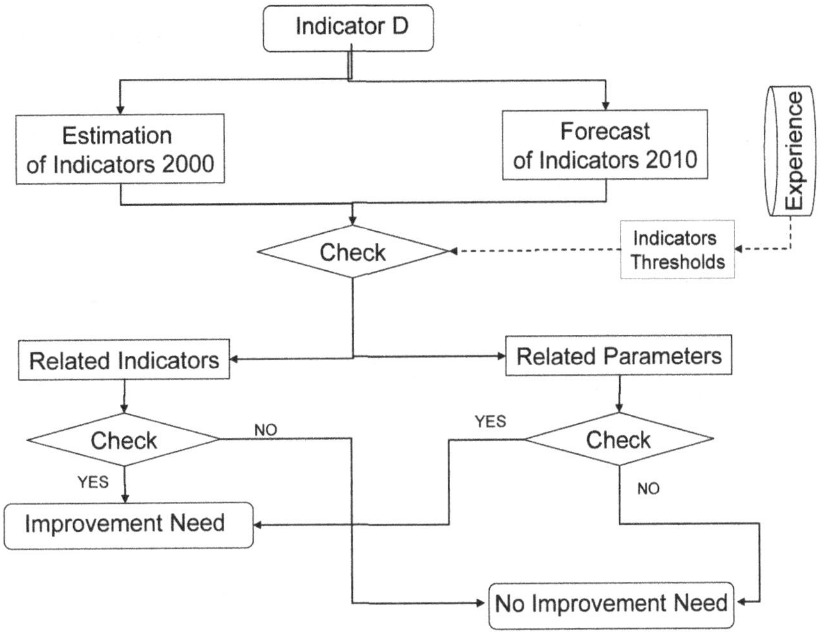

Indicator D
Estimation of Indicators 2000
Forecast of Indicators 2010
Experience
Check
Indicators Thresholds
Related Indicators
Related Parameters
Check
NO
YES
Check
YES
NO
Improvement Need
No Improvement Need

4.2 Module B: Actions' Development

This module's approach aims at the development of energy policy action and is illustrated in the *Figure 5.*

There is direct relation between this step and the previous one. This step was based on the recognition and choice of energy policy actions through an intelligent technique which was based on the suitable description of available experience from international energy organisms as well as institutions of energy

Figure 5. The procedure of actions development

Indicator D
Inputs of the 1st Step
Improvement Needs
Experience
Check
Actions Base
Related Indicators
Related Parameters
Proposed Energy Policy Actions

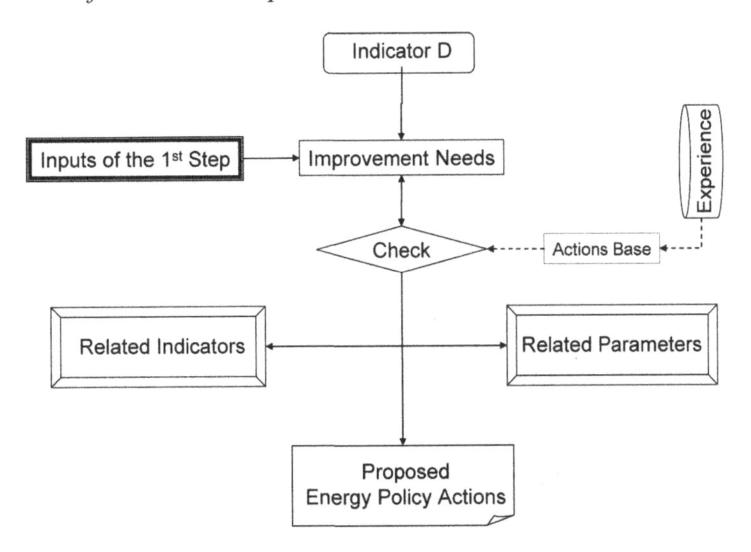

Table 1. The energy policy actions

A.1	**Reduction of the Dependence on Energy Imports**
A.1.1	*Development of the respective mechanisms for the strategic inventories of oil and natural gas (taking into account the characteristics of competitive-liberalized energy)*
A.1.2	*Development of domestic energy sources (taking into account economical, environmental and social parameters)*
A.1.3	*Diversification of energy sources and suppliers*
A.1.4	*Promotion of energy cooperation with neighbouring countries and support of the investments in producing countries*
A.2	Development of Energy Interconnections
A.2.1	*Promotions, enhancement and expansion of energy networks and interconnections*
A.3	Reform the Energy Market
A.3.1	*Independent energy regulator and manager of the transmission system*
A.3.2	*Supporting the Competition (private participation) in the production, distribution and sales.*
A.3.3	*Dividing (natural, book-keeping, and functional) of the utilities*
A.3.4	*Adoption of the institutional framework for the promotion of the competitive energy market.*
A.3.5	*Changes in the energy prices, in order to reflect the real cost (subsidies)*
A.3.6	*Technical support of the process of energy sector reforming.*
A.4	More Efficient Production, Transmission and Distribution of the Energy
A.4.1	*Research and development in the energy technologies.*
A.4.2	*Reduction of the technical dissipation of energy*
A.4.3	*Promotion of modern financing mechanisms for energy investments*
A.5	Promotion of the Renewable Energy Sources and Reduction of CO2 Emissions
A.5.1	*Ratification of the Kyoto Protocol*
A.5.2	*Penetration of the renewable energy sources*
A.5.3	*Promotion of new clean technologies in the energy production*
A.6	Promotion of Energy Efficiency
A.6.1	*Promotion of efficient energy use in the tertiary sector*
A.6.2	*Promotion of efficient use in the transportation*
A.6.3	*Promotion of efficient energy use in the industry sector*
A.6.4	*Adoption of technical energy management (energy audit)*
A.6.5	*Promotion of know-how and technology exchange for the renewable energy sources and energy saving*

policy planning. This step supports the energy policy making through the development of energy policy actions that arise via the connection of framework of indicators with a "library" internationally proposed energy policy action.

In particular, the procedure of the second step is described as follows:

Identification of the Energy Policy Actions: According to the national researches of the international energy organization, over 30 relative reports were analyzed for the total of 30 countries that were mentioned in the previous step. This resulted to the proposal of about 320 energy policy actions. The actions were categorized according to the objectives of energy policy. Afterwards, they were grouped and merged, in order to reflect the main guidelines of energy policy, resulting to the 22 general actions. These actions constitute in fact the "library" of actions of each country and process that is examined.

Figure 6. Selection of the country

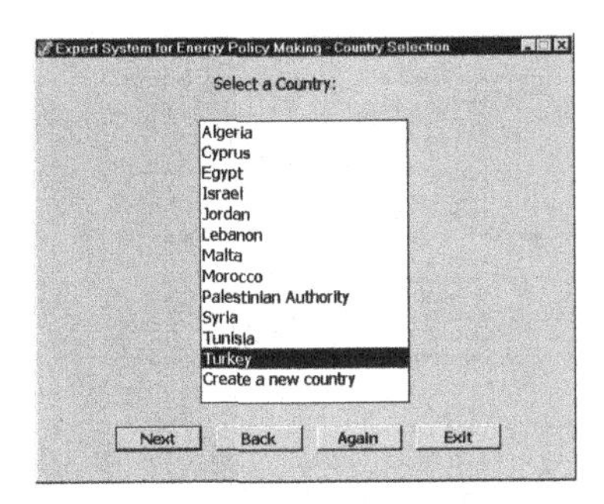

As a result, the energy policy actions are presented in the *Table 1.*

Energy Policy Making: In parallel to the procedures of conclusion-making that were presented in the first step, there were a number of procedures focused on the connection between the value range of the indicators and the "library" of energy policy action. It describes the recognized experience by the reports that were analyzed. In particular, the actions proposed by international organizations were matched with the equivalent values of energy indicators of each country, in order to describe under which circumstances these two are connected.

5. INFORMATION SYSTEM

5.1 Description

The EXPPOL system consists of an expert system which incorporates the abovementioned approach. In particular, its function focuses (a) on the estimations of energy policy indicators, (b) on the description of the situation and recognition of developments and future estimation of the energy policy, as well as (c) the presentation of energy policy actions for each case study country. The system is composed of three typical elements of the expert systems as follows:

Data Base: The data base is categorized in objects, attributes and strings. The data include basic data of an energy balance, economic and social relative data and information, the "library" of conclusion diagnosis and the "library" of energy policy actions. The data base stores simultaneously the list of set

Figure 7. Inputs (initial data)

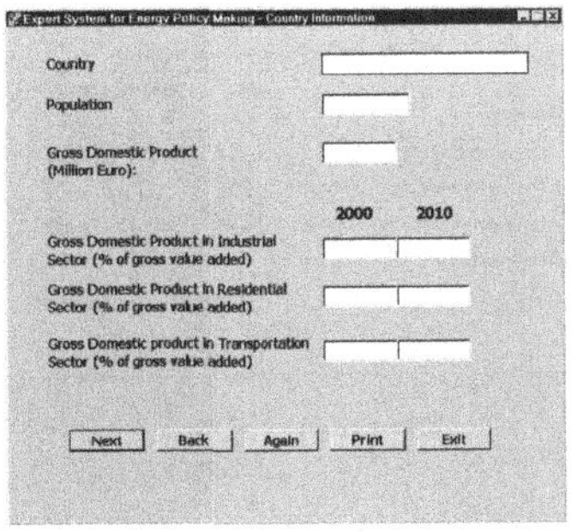

Figure 8. Inputs (energy balance data)

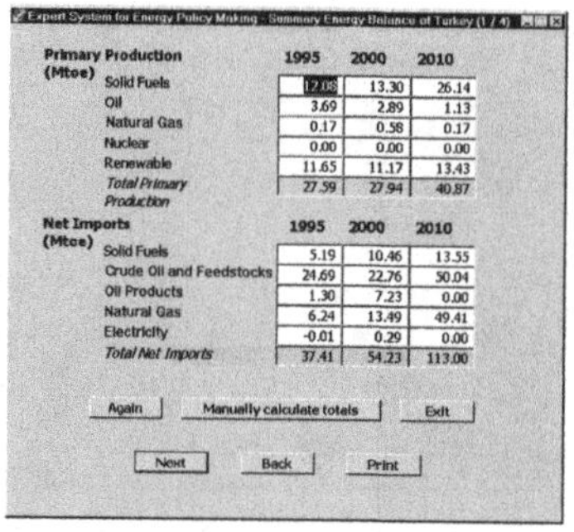

Figure 9. Input (secondary data)

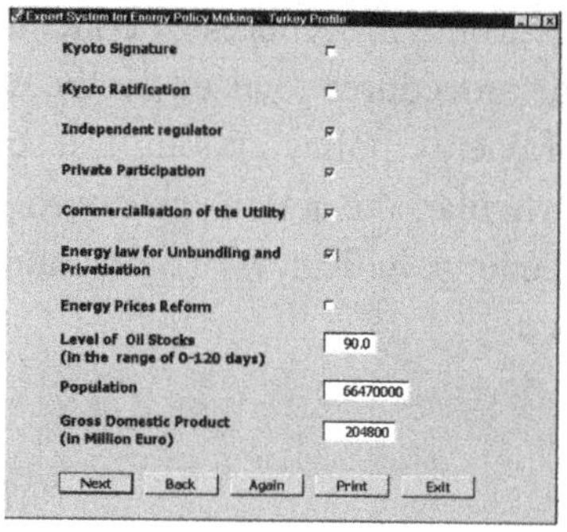

regulations, and therefore maintains a logic sequence and an explanation to the user.

Knowledge Base: It is composed of group regulations that describe the knowledge for the diagnosis of the situation and the expected development, as well as the development of energy policy guidelines. The regulations were based on the analysis of processes. The knowledge base is composed of about 230 regulations, that have been developed with the «if … then …».

Interface with User: It comprises of a variety of "dialogs", which can be categorized in (a) data input and (b) presentation of results. The process of function of the sub-system by the user is described as follows:

Figure 10. Indicators' values

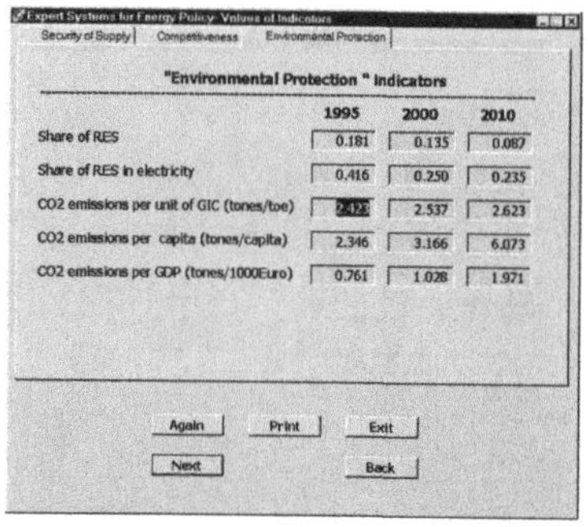

Figure 11. Results

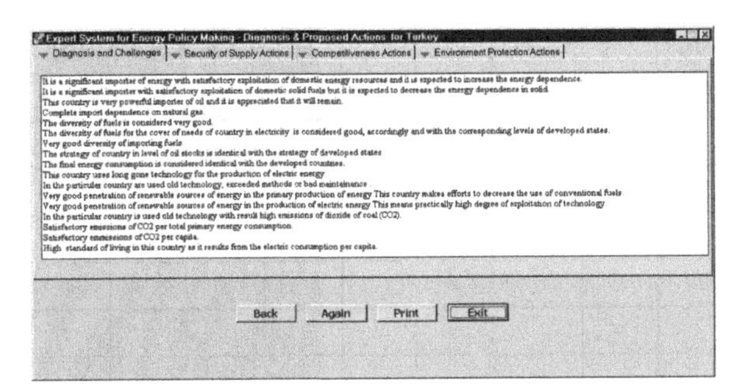

- The initial form (*Figure 6*) of the program gives the possibility to the user to choose one country from the already existing or to input a new country with a specific form (*Figure 7*).
- The following forms (*Figures 8, 9*) are used for the data input of the energy balance and the required data for the calculation of the indicators.
- The form of the *Figures 10* describes a typical form for the environmental protection category. The main results of the sub-system are included in the final form (*Figure 11*) which consists of 4 sub-forms, one for the diagnosis of the situation and the energy policy making and three for the proposed action per objective of the energy policy.

5.2 Application

The application of the information system gave the possibility for the pilot application of the system approach and the applicability of the results that it provides. The application included two groups of countries:

- The developing countries of Northern Africa and Eastern Mediterranean (Algeria-AL, Egypt-EG, Israel-IS, Lebanon-LB, Jordan-JO, Morocco-MO, Palestinian Authority-PL, Syria-SY, Tunisia-TN, Turkey-TU). Among the Mediterranean Countries, considerable disparities are observed, as regards energy production, resulting in the segmentation of the countries into "exporting countries" (Algeria, Egypt and Syria), "self-sufficient countries" (Tunisia), "countries strongly dependent on imports" (Turkey), or "almost totally dependent countries" (the Palestinian Authority, Morocco, Israel, Jordan).
- The six countries of Gulf Cooperation Council – GCC (Saudi Arabia-SA, Kuwait-KU, Oman-OM, Bahrain-BH, Qatar-QT and United Arab Emirates-UAE). In GCC States, a large number of oil and gas accumulations have been found for a long time, including some of the largest and even giant fields in the world. According to the latest issue of BP Statistical Review of World Energy, the six countries contain huge quantities of proved reserves of crude oil and natural gas, estimated in early 2007 at around 478 billion barrels of crude oil and 41.920 billion cubic metres of natural gas, representing about 42 per cent and 24 per cent of the world's total respectively.

The results of the proposed model's application in the above states are presented in the following *Table 2*.

Table 2. The application's results

	EG	AI	JO	IS	LB	MO	PL	SY	TU	TN	SA	BH	QA	OM	KU	UAE
A.1.1			√	√	√	√				√						√
A.1.2	√	√		√				√	√	√	√	√	√		√	
A.1.3			√		√		√							√		√
A.1.4			√	√	√	√			√	√				√		√
A.2.1	√		√	√	√		√			√				√		
A.3.1		√			√	√	√	√		√	√	√	√		√	√
A.3.2		√			√	√	√	√		√	√	√	√	√	√	√
A.3.3	√	√		√	√		√	√		√	√	√	√	√	√	√
A.3.4				√	√		√	√				√	√	√	√	
A.3.5	√	√	√	√	√	√	√	√	√	√	√	√	√		√	√
A.3.6	√	√	√	√	√	√	√	√		√	√	√	√	√	√	
A.4.1	√	√	√		√	√		√		√	√	√	√		√	
A.4.2	√	√	√		√	√		√	√		√	√	√	√	√	√
A.4.3	√		√		√	√				√						√
A.5.1		√	√		√		√	√	√		√	√	√	√	√	
A.5.2	√	√	√		√	√	√	√		√	√	√	√	√	√	
A.5.3			√	√	√	√		√	√			√	√		√	√
A.6.1		√				√	√	√	√		√	√	√	√	√	√
A.6.2	√		√	√	√					√						√
A.6.3		√		√	√			√	√	√	√	√	√		√	
A.6.4	√	√	√	√	√	√	√			√	√			√		
A.6.5	√	√	√	√		√		√		√	√	√	√		√	√

The inputs of indicators were based to a large extent on the context of the two European Commission Projects (EC MEDA project and MEDA Synergy project) and the examination of outputs was based on the conclusions derived from the Euro - Mediterranean ministers meetings (Final Report of the EC MEDA Project, 2004; Final Report of the MEDA Synergy project, 2005; Kagiannas et al. 2002).

In addition, the inputs of indicators for the second group of countries were based to a large extent on the context of the GCC Synergy Project and the examination of outputs was based on the conclusions derived from the project meetings and events (Final Report of the Project, 2007; Patlitzianas et.al. 2006).

As results, most of the information and data presented in this chapter have been derived from the activities carried out within this project and the deliverables produced. Indeed, important advantage of this application was also the availability of the project results (both of projects). This fact makes the application not only realistic but it also gives the possibility of assessment of the results with real data and situations of energy policy making, even in fields where there are no sufficient reliable data or the research experience is limited.

6. CONCLUSION

The main points that derive from the aforementioned analysis have as follows:

- The implementation of an intelligent technique for the decision making is an important innovation in the way of dealing with the matter of energy policy making.
- It is now possible to define the energy policy guidelines in a timeframe much shorter, compared to the one needed without the implementation of the information system.
- The possibility to control the regulations set by the conclusion-making is a fact that attributes significantly to the control of the whole procedure and the correction of any mistakes and wrong choices.
- The use of the information system that integrates the current approach gave the possibility to extract guideline for energy policy for the twelve countries in a very short time.
- The use of an expert system proved significantly important, since it gives the possibility to check the regulations that were "set" for the conclusion making. This fact enables the control as well as the correction of any mistakes of wrong choices.
- Possibility to control the thresholds that were used (in quantitative/ qualitative indicators and in criteria) for exporting the results, fact that can essentially contribute to the control and correction of errors and false choices.

Summing up, the energy policy making should not be a typical scientific area, despite of the need for support in decision-making. The energy policy will always be influenced by external and unpredictable factors, and most times also irrelevant to purely energy matters. Therefore, the current approach is not a solution to energy policy making and it should not substitute the experts of decision making, but it should just be a realistic and consistent "tool" for the decisions support and determination of energy policy guidelines.

The information decision support systems, such as the one presented in this chapter, are necessary in order to identify, diagnose, and develop the appropriate actions in a consistent way, to assist energy policy making. In addition, the model's procedure assisted the specific decision making problem and the outcome might have been quite different if different indicators had been chosen. However, the information system's conceptual can provide a sufficient framework for supporting decisions for other problems.

ACKNOWLEDGMENT

The information and data that correspond to the Mediterranean countries as well as GCC countries have been collected within the framework of the project funded by the European Commission (EC), MEDA programme as well as the EU-GCC SYNERGY programme. The content of the chapter is the sole responsibility of its authors and does not necessarily reflect the views of the EC.

REFERENCES

Adibi, M. M., Kafka, R. J., & Milanicz, D. P. (1994). Expert system requirements for power system restoration. *IEEE Transactions on Power Systems*, *9*(3), 1592–1600. doi:10.1109/59.336099

Azzam, M., & Nour, M. A. S. (1995). Expert system for voltage control of a large-scale power system, Modelling, Measurement Control A: General Physics, Electronics. *Electrical Engineering*, *63*(1-3), 15–24.

Bloom, J. A. (1982). Long-range generation planning using decomposition and probabilistic simulation. *IEEE Transactions on Power Apparatus and Systems*, *101*(4), 797–802. doi:10.1109/TPAS.1982.317144

Brunner, T., Nejdl, W., Schwarzjirg, H., & Sturm, M. (1993). On-line expert system for power system diagnosis and restoration. *Intelligent Systems Engineering*, *2*(1), 5–24.

Chang, C. S., & Chung, T. S. (1990). Expert system for on-line security – Economic load allocation on distribution systems. *IEEE Transactions on Power Delivery*, *5*(1), 467–473. doi:10.1109/61.107314

Chokri, B., Rabei, M., & Dai, D. X. (1996). *An integrated power system global controller using expert system*. Calgary, Canada: Canadian Conference on Electrical and Computer Engineering.

Christie, R. D., Talukdar, S. N., & Nixon, J. C. (1990). A hybrid expert system for security assessment. *IEEE Transactions on Power Systems*, *5*(4), 1503–1509. doi:10.1109/59.99405

D'Sa, A. (2005). Integrated resource planning (IRP) and power sector reform in developing countries. *Energy Policy*, *33*(10), 1271–1285. doi:10.1016/j.enpol.2003.12.003

David, A. K., & Zhao, R. (1989). Integrating expert systems with dynamic programming in generation expansion planning . *IEEE Transactions on Power Systems*, *4*(3), 1095–1101. doi:10.1109/59.32604

David, A. K., & Zhao, R. (1989). Integrating expert systems with dynamic programming in generation expansion planning. *IEEE Transactions on Power Systems*, *4*(3), 1095–1101. doi:10.1109/59.32604

Doukas, H., Patlitzianas, K. D., Iatropoulos, K., & Psarras, J. (2006). Intelligent Building Energy Management System Using Rule Sets. *Building and Environment*, *42*(10), 3562–3569. doi:10.1016/j.buildenv.2006.10.024

Doukas, H., Patlitzianas, K. D., Kagiannas, A. G., & Psarras, J. (2008). Energy Policy Making: An Old Concept or a Modern Challenge? *Energy Sources (Part B)*.

Doukas, H., Patlitzianas, K. D., & Psarras, J. (2006). Supporting the Sustainable Electricity Technologies in Greece Using MCDM. *Resources Policy*, *31*(2), 129–136. doi:10.1016/j.resourpol.2006.09.003

Eickhoff, F., Handschin, E., & Hoffmann, W. (1992). Knowledge based alarm handling and fault location in distribution networks. *IEEE Transactions on Power Systems*, *7*(2), 770–776. doi:10.1109/59.141784

European Commission – EC, (2004). *MEDA programme: Ad-hoc Groups Energy Policy, "Interconnections and Economic Analysis of the Euro-Mediterranean Energy Forum."* Final Report (ME8/B7-4100/IB/98/0480).

European Commission - EC (2005). *Synergy Project "Business Opportunities for CDM Project Development in the Mediterranean."* Final Report (project number 4.1041/D/02-003-507.21086).

European Commission - EC (2007). *Synergy Project - EUROGULF: An EU-GCC Dialogue for Energy Stability and Sustainability.* Final Report (project number: 4.1041/D/02-008-S07 21089).

Fouad, A. A., Vekataraman, S., & Davis, J. A. (1990). An expert system for security trend analysis of a stability-limited power system. *IEEE Transactions on Power Systems, 6*(3), 1077–1084. doi:10.1109/59.119249

Hasan, K., Ramsay, B., & Moyes, I. (1995). Object oriented expert systems for real-time power system alarm processing, Part I, Selection of a toolkit; Part II, Application of a toolkit. *Electrical Power Systems Research, 30*(1), 69-75/77-82.

Helm, D. (2002). Energy policy: security of supply, sustainability and competition. *Energy Policy, 30*, 173–184. doi:10.1016/S0301-4215(01)00141-0

Huang, Y. F., Huang, G. H., Hu, Z. Y., Maqsood, I., & Chakma, A. (2005). Development of an expert system for tackling the public's perception to climate-change impacts on petroleum industry. *Expert Systems with Applications, 1*, 1–13.

Humphreys, P., McIvor, R., & Huang, G. (2002). An expert system for evaluating the make or buy decision . *Computers & Industrial Engineering, 42*, 567–585. doi:10.1016/S0360-8352(02)00052-9

Hung, C. Q., Batanovand, D. N., & Lefevre, T. (1997). *KBS and macro-level systems: support of energy demand forecasting.* Thailand: Asian Institute of Technology.

Hung, C. Q., Batanovand, D. N., & Lefevre, T. (1997). *KBS and macro-level systems: support of energy demand forecasting.* Thailand: Asian Institute of Technology.

Jabbour, K., Riveros, J. F. V., Landsbergen, D., & Meyer, W. (1988). ALFA: automated load forecasting assistant. *IEEE Transactions on Power Systems, 3*(3), 908–914. doi:10.1109/59.14540

Jaber, J. O., Mamlook, R., & Awad, W. (2005). Evaluation of energy conservation programs in residential sector using fuzzy logic methodology. *Energy Policy, 33*, 1329–1338. doi:10.1016/j.enpol.2003.12.009

Jebaraj, S., & Iniyan, S. (2006). A review of energy models. *Renewable & Sustainable Energy Reviews, 10*(4), 281–311. doi:10.1016/j.rser.2004.09.004

Kagiannas, A., Flamos, A., Askounis, D., & Psarras, J. (2002). Energy Policy Indicators for the Assessment of the Euro-Mediterranean Energy Cooperation. *International Journal of Energy Technology and Policy, 2*(4), 301–322. doi:10.1504/IJETP.2004.005738

Kagiannas, A., Patlitzianas, K., Metaxiotis, K., Askounis, D., & Psarras, J. (2006). Energy Models in the Mediterranean Countries: A survey towards a common strategy. International . *Journal of Power and Energy Systems, 26*(3), 260–268.

Kandil, M. S. (2003). The implementation of long-term forecasting strategies using a knowledge-based expert system: part II. *Electric Power Systems Research, 58*(1), 19–25. doi:10.1016/S0378-7796(01)00098-0

Kavrakoglou, I. (1987). Energy models. *European Journal of Operational Research, 16*(2), 2231–2238.

Kezunovic, M., Fromen, C. W., & Sevcik, D. R. (2003). Expert system for transmission substation event analysis. *IEEE Transactions on Power Delivery, 8*(4), 1942–1949. doi:10.1109/61.248306

Khosla, R., & Dillon, T. S. (1994). Neuro-expert system approach to power system problems. *International Journal of Engineering Intelligent Systems for Electrical Engineering and Communications, 2*(1), 71–78.

Le, T. L., Negnevitsky, M., & Piekutowski, M. (1995). Expert system application for voltage control and VAR compensation. *International Journal of Engineering Intelligent Systems for Electrical Engineering and Communications, 3*(2), 79–85.

Lee, S. J., Yoon, S. H., Yoon, M. C., & Jang, J. K. (1990). Expert system for protective relay setting of transmission systems. *IEEE Transactions on Power Delivery, 5*(2), 1202–1208. doi:10.1109/61.53142

Li, S., Shahidehpour, S. M., & Wang, C. (1993). Promoting the application of expert systems in short-term unit commitment. *IEEE Transactions on Power Systems, 8*(1), 286–292. doi:10.1109/59.221229

Markovic, M., Fraissler, W. (1993). Short-term load forecast by plausibility checking of announced demand: an expert-system approach. *European Transactions on Electrical Power Engineering / ETEP, 3*(5), 353-358.

Matsuda, S., Ogi, H., Nishimura, K., Okataku, Y., & Tamura, S. (1990). Power system voltage control by distributed expert systems. *IEEE Transactions on Industrial Electronics, 37*(3), 236–240. doi:10.1109/41.55163

Melli, R., & Sciuba, E. (1999). A prototype expert system for the conceptual synthesis of thermal processes. *Energy Conversion and Management, 38*(15-17), 1737–1749. doi:10.1016/S0196-8904(96)00186-0

Metaxiotis, K., Kagiannas, A., Askounis, D., & Psarras, J. (2003). Artificial intelligence in short term electric load forecasting: a state-of-the-art survey for the researcher. *Energy Conversion and Management, 44*(9), 1525–1534. doi:10.1016/S0196-8904(02)00148-6

Minakawa, T., Ichikawa, Y., Kunugi, M., Shimada, K., Wada, N., & Utsunomiya, M. (1995). Development and implementation of a power system fault diagnosis expert system. *IEEE Transactions on Power Systems, 10*(2), 932–939. doi:10.1109/59.387936

Nagata, T., Sasaki, H., & Yokoyama, R. (1995). Power system restoration by joint usage of expert system and mathematical programming approach. *IEEE Transactions on Power Systems, 10*(3), 1473–1479. doi:10.1109/59.466501

Ouyang, Z., & Shahidehpour, S. M. (1990). Short-term unit commitment expert system. *Electric Power Systems Research, 20*(1), 1–13. doi:10.1016/0378-7796(90)90020-4

Park, Y. M., & Lee, K. H. (1995). Application of expert system to power system restoration in local control center. *International Journal of Electrical Power & Energy Systems, 17*(6), 407–415. doi:10.1016/0142-0615(94)00011-5

Patlitzianas, KD., Doukas, H., & Psarras (2008), Sustainable energy policy indicators: Review and recommendations. *Renewable Energy, 33(5*), 966-973.

Patlitzianas, K. D., Doukas, H., Kagiannas, A., & Askounis, D. (2006). A Reform Strategy of the Energy Sector of the 12 countries of North Africa and Eastern Mediterranean. *Energy Conversion and Management*, *47*(13-14), 1913–1926. doi:10.1016/j.enconman.2005.09.004

Patlitzianas, K. D., Doukas, H., & Psarras, J. (2006). Enhancing Renewable Energy in the Arab States of the Gulf. *Energy Policy*, *34*(18), 3719–3726. doi:10.1016/j.enpol.2005.08.018

Patlitzianas, K. D., Ntotas, K., Doukas, H., & Psarras, J. (2007). Assessing the Renewable Energy Producers' Environment in the EU Accession Member States. *Energy Conservation and Management*, *48*(3), 890–897. doi:10.1016/j.enconman.2006.08.014

Patlitzianas, K. D., Papadopoulou, A., Flamos, J., & Psarras, J. (2005). CMIEM: The Computerized Model for Intelligent Energy Management. *International Journal of Computer Applications in Technology*, *22*(2-3), 120–129. doi:10.1504/IJCAT.2005.006943

Patlitzianas, K. D., & Psarras, J. (2006). Formulating a Modern Energy Companies' Environment in the EU Accession Member States through a Decision Support Methodology. *Energy Policy*, *35*(4), 2231–2238. doi:10.1016/j.enpol.2006.07.010

Sherwali, H., & Crossley, P. (1994). Expert system for fault location on a transmission network. *Proceedings 29th Universities Power Engineering Conference - Part 2,* (pp. 751-754).

Singh, SP., Raju, GS., & Gupta, AK. (1993). Sensitivity based expert system for voltage control in power system. *International Journal on Electrical Power and Energy Systems, IS*(3), 131-136.

Tsamboulas, D. A., & Mikroudis, G. K. (2005). TRANS-POL: A mediator between transportation models and decision makers' policies. *Decision Support Systems*, *42*(2), 879–897. doi:10.1016/j.dss.2005.07.010

Vale, Z., Ramos, C., Silva, A., Faria, L., Santos, J., Fernandes, M., et al. (1998). *SOCRATES – An Integrated Intelligent System for Power System Control Center Operator Assistance and Training*. Cancun, Mexico: International Conference on Artificial Intelligence and Soft Computing (IASTED).

Welbank, M. (1985). *A review of knowledge acquisition techniques for expert systems*. Bristish Telecommunications Research Laboratories Technical Report, Ipswich, UK.

Wenying, Ch. (2005). The costs of mitigating carbon emissions in China: findings from China MARKAL-MACRO modeling. *Energy Policy*, *33*(7), 885–896. doi:10.1016/j.enpol.2003.10.012

Weyant, J. P. (1990). Policy modeling: an overview. *Energy*, *15*, 203–206. doi:10.1016/0360-5442(90)90083-E

Yongli, Z., Yang, Y. H., Hogg, B. W., Zhang, W. Q., & Gao, S. (1994). Expert system for power systems fault analysis. *IEEE Transactions on Power Systems*, *9*(1), 503–509. doi:10.1109/59.317573

Zita, A. Vale, Santos, J., & Ramos, C. (1995). *SPARSE - A Prolog Based Application for the Portuguese Transmission Network: Verification and Validation*. London: PAP'97 – Practical Application in Prolog.

Zitouni, S., & Irving, MR. (1999). *NETMAT: A knowledge-based grid system analysis tool.*

Chapter 9
Fuel Reduction Effect of the Solar Cell and Diesel Engine Hybrid System with a Prediction Algorithm of Solar Power Generation

Shinya Obara
Kitami Institute of Technology, Japan

ABSTRACT

Green energy utilization technology is an effective means of reducing greenhouse gas emissions. We developed the production-of-electricity prediction algorithm (PAS) of the solar cell. In this algorithm, a layered neural network is made to learn based on past weather data and the operation plan of the hybrid system (proposed system) of a solar cell and a diesel engine generator was examined using this prediction algorithm. In addition, system operation without a electricity-storage facility, and the system with the engine generator operating at 25% or less of battery residual quantity was investigated, and the fuel consumption of each system was measured. Numerical simulation showed that the fuel consumption of the proposed system was modest compared with other operating methods. However, there was a significant difference in the prediction error of the electricity production of the solar cell and the actual value, and the proposed system was shown to be not always superior to others. Moreover, although there are errors in the predicted and actual values using PAS, there is no significant influence in the operation plan of the proposed system in almost all cases.

INTRODUCTION

Utilization of the expansion of green energy is an effective method to reduce the amount of greenhouse gas discharge. However, energy supplies using green energy tend to be unstable, hence interconnection with commercial power and operations to link to present generating equipment are required with the

DOI: 10.4018/978-1-60566-737-9.ch009

stabilization of the energy supply in mind. In recent years, there have been remarkable improvements in battery performance and the equipment cost has declined. Therefore, it is considered that the operation optimization method for green energy, including the storage of electricity, will henceforth become important. The objective of this study involves developing an algorithm (**PAS**) to predict the electricity production of a **solar cell**, and to optimize the operation of a diesel-power-plant hybrid system with the green energy of unstable output power (Obara, S. & Tanno, I. (2008(1))). Moreover, the relation between the prediction error of the **PAS** and the energy cost of the proposed system is clarified. To date, there has been considerable research concerning the **operation plan** of a hybrid system, combining a **solar cell** and a diesel power plant (Ashari, M., & Nayar, C. V. (1999), Muselli, M., etc. (1999), Ismail, Y., etc. (2002), Yamamoto, S., etc. (2004)) . These research reports show that a reduction in energy cost can be realized by both the electricity production of a **solar cell** and the power load being predicted. If a large scale computer and considerable weather observation data are used, the **solar radiation** several hours later in arbitrary locations may be predictable (Earth Simulator Center, Japan Agency for Marine-Earth Science and Technology (2008)). In this case, the electricity production of the **solar cell** can be correctly estimated. However, since using this method is costly in terms of communication and information, in this paper, a simpler algorithm for predicting the production-of-electricity of a **solar cell** is developed (**PAS**). In the **PAS** algorithm, a layered neural network (**NN**) is made to learn using data on the amount of **solar radiation** and outside temperature for the past 14 years (NEDO, and the standard weather and the **solar radiation** database in 1990 to 2003 (METPV-3) (NEDO Technical information data base, (2008))). As teaching data in this case, the electricity production of the polycrystalline-silicon-type **solar cell**, as calculated from the amount of slope-face **solar radiation**, is introduced. The weather data (the amount of **solar radiation** and outside air temperature) of each period, ranging from 24 hours earlier to the present, is given to the learned-**NN**, and the **solar cell** output power until several hours have been predicted and the operation of the engine and battery is planned based on this prediction result. Therefore, in this paper, the variables are defined as the **solar cell** area and the battery capacity and the influence of the operation cost is clarified, while an allowance for error is included in the **PAS** prediction result. Furthermore, the fuel consumption of an engine generator is compared with a system where the prediction **operation plan** has not been installed.

System Configurations

Combined Solar Cell and Diesel Engine System

Figure 1 is a schematic diagram of a combined **solar cell** and **diesel engine** system. The power of a **solar cell** can be supplied to the demand side via a DC-AC converter and inverter. Moreover, this power can also be used to charge a battery through a DC-DC converter. The power of an engine generator is supplied to the demand side through an inverter, while the power of the engine generator can also be used to charge a battery through an AC-DC converter. The specification of the engine and the power generator are shown in Tables 1 and 2 of Figure 2, respectively, while the engine output is transmitted to a power generator by a belt.

Data concerning the present weather intelligence (the amount of slope-face **solar radiation** and the outside air temperature) and the extent to which the battery is charged are input into a system controller. The **PAS** is then introduced into the system controller and the predictive value of a **solar cell** output can be analyzed (Obara, S., & Tanno, I. (2008(2))). In this system controller, optimal operation is planned,

Figure 1. System scheme

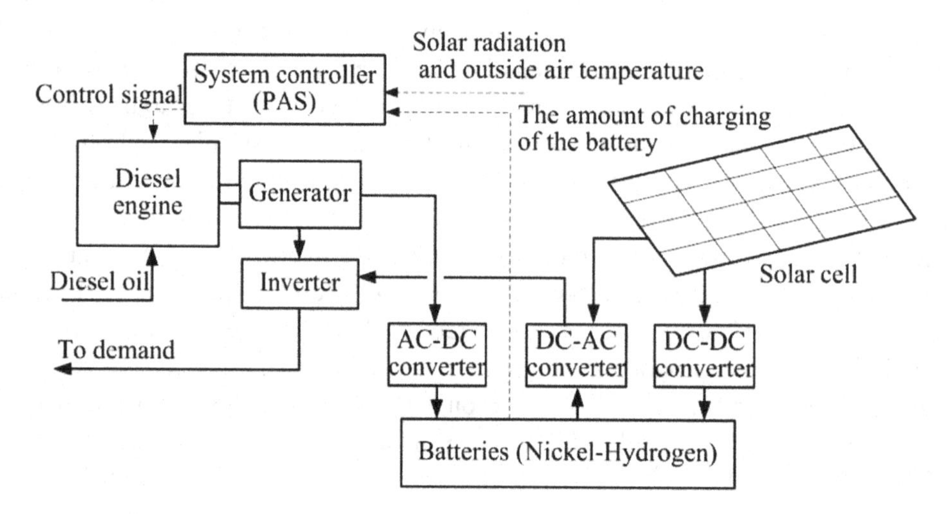

based on the predictive value of **solar cell** electricity production. The operation and rundown of the engine generator are controlled based on this result.

Solar Cell and Battery

The **solar cell** is a monotonous type of polycrystalline silicon, which is installed in the roof tilted at 30 degrees from the southward direction. The battery is of the nickel hydrogen type, with a specification determined from the reference (SANYO Co. Ltd. (2007)). The analytical conditions of a **solar cell**, battery, and each converter and inverter are shown in Table 3 of Figure 2. In the **solar cell**, a rise in the cell temperature will reduce the **generation efficiency**. The temperature coefficient is set to 0.4%/K in this paper, while a loss of charging efficiency, discharge efficiency, and self-discharge is included in the battery efficiency in the table.

Diesel Engine Generator

Figures 3 and 4 show the relation of the fuel calorific value and output characteristics of the **diesel engine** power generator, and the examination result of a power generation output and **generation efficiency** (Obara, S., (2007(1)), Obara, S., (2007(2))). The maximum **generation efficiency** of a test engine generator is about 3kW in output. In the analysis described later, engine generator efficiency is calculated using the approximate expression shown in Figure 4. The system investigated in this paper is a specification capable of supplying power to an apartment house comprising three average households in Sapporo. This assumption is because the output power range of a general-purpose **diesel engine** is several kilowatts.

Figure 2.

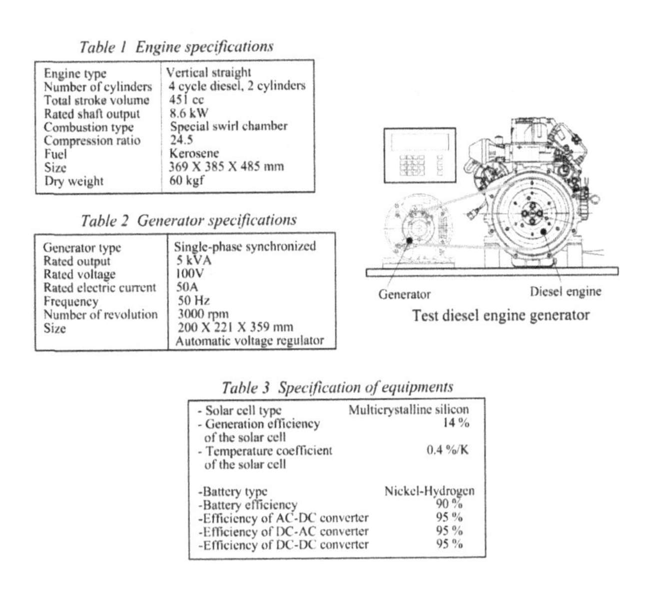

Table 1 Engine specifications

Engine type	Vertical straight
Number of cylinders	4 cycle diesel, 2 cylinders
Total stroke volume	451 cc
Rated shaft output	8.6 kW
Combustion type	Special swirl chamber
Compression ratio	24.5
Fuel	Kerosene
Size	369 X 385 X 485 mm
Dry weight	60 kgf

Table 2 Generator specifications

Generator type	Single-phase synchronized
Rated output	5 kVA
Rated voltage	100V
Rated electric current	50A
Frequency	50 Hz
Number of revolution	3000 rpm
Size	200 X 221 X 359 mm
	Automatic voltage regulator

Test diesel engine generator

Table 3 Specification of equipments

- Solar cell type	Multicrystalline silicon
- Generation efficiency of the solar cell	14 %
- Temperature coefficient of the solar cell	0.4 %/K
-Battery type	Nickel-Hydrogen
-Battery efficiency	90 %
-Efficiency of AC-DC converter	95 %
-Efficiency of DC-AC converter	95 %
-Efficiency of DC-DC converter	95 %

Analysis Method

Load Pattern

In the analysis of this paper, the power load pattern in three common households in Sapporo shown in Figure5 is used (Narita, K., (1996)). The power load consumption by household electric appliances and electric lights is the main focus, and cooling and heating loads are excluded, meaning there is little difference from one month to another in the figure.

Figure 3. Output of the diesel engine generator

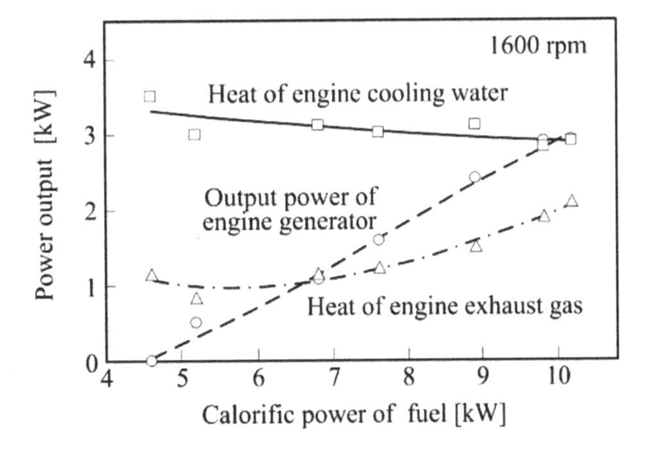

Figure 4. Generator efficiency of the diesel engine

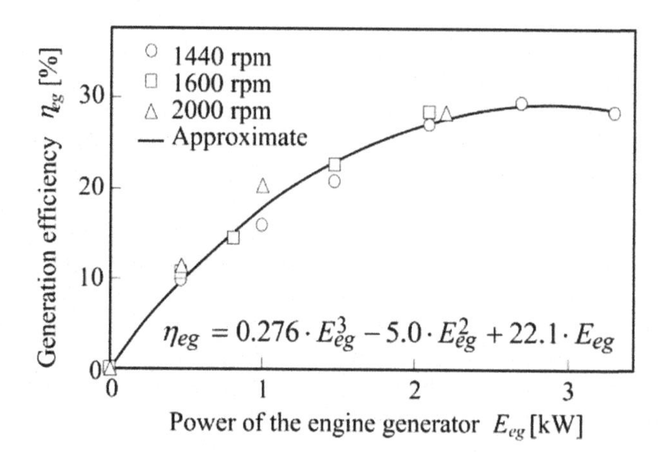

Amount of Slope-Face Solar Radiation

In a monotonous type **solar cell**, direct and diffuse forms of **solar radiation** are used to generate power. The means of calculating the **solar radiation** reinforcement, which enters a solar power generation system, is described in the following (Kosugi, T., etc. (1998), Japan Solar Energy Society, (1985)). Global-**solar-radiation** reinforcement, direct-**solar-radiation** reinforcement, and water surface diffuse-**solar-radiation** reinforcement are expressed with I_H, I_D, and I_M, respectively. When the intensity of direct and diffuse forms of **solar radiation** are expressed with H_D and H_M among the **solar radiation** which enters an acceptance surface, H_D will be calculated by Eq. (1). θ on the right hand side of Eq. (1) expresses the incident angle to the acceptance surface of the sunlight, and is calculated using Eq. (2).

$$H_D = I_D \cdot \cos\theta \tag{1}$$

$$\sin\theta = \cos\phi \cdot \sin\delta - \sin\phi \cdot \cos\omega \cdot \cos\delta \tag{2}$$

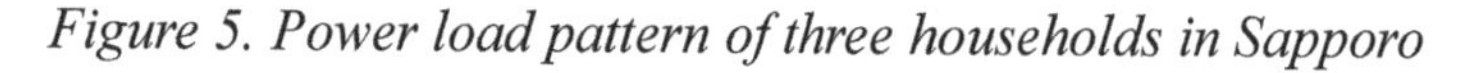
Figure 5. Power load pattern of three households in Sapporo

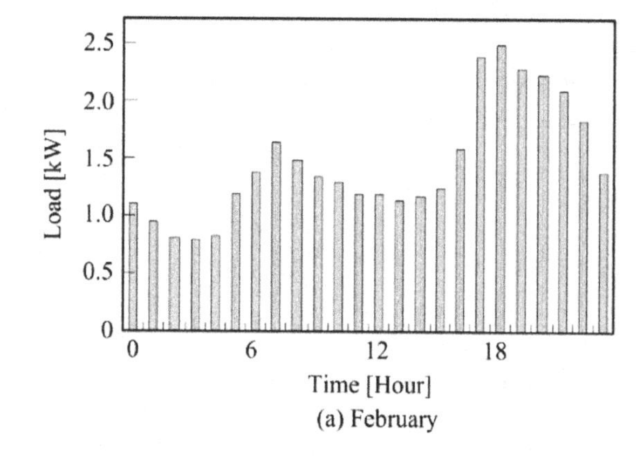

(a) February

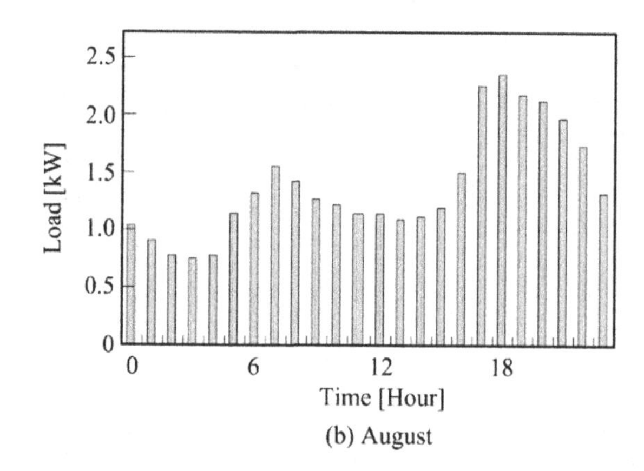

(b) August

Figure 6. Slope-face solar radiation and the production of electricity of a solar cell module in Sapporo

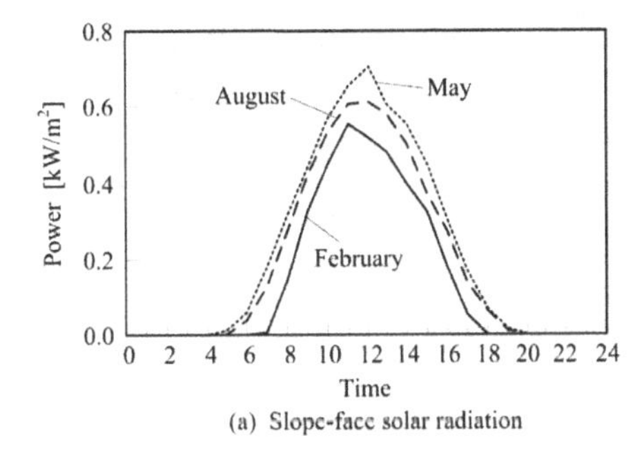

(a) Slope-face solar radiation

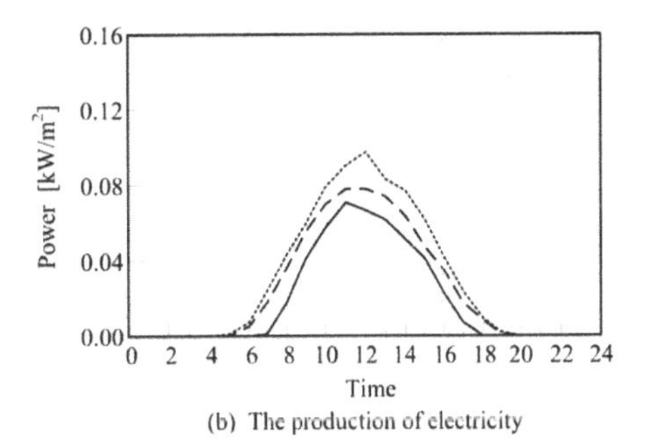

(b) The production of electricity

Here, ϕ, δ, and ω express the latitude, the solar celestial declination, and the hour angle of an establishment place, respectively, while Equation (3) is a formula for calculating the diffuse-**solar-radiation** component H_M. The 1st term of the right hand side of Eq. (3) is a diffuse-sky-radiation component, the 2nd is a reflected-**solar-radiation** component, and β and ρ express the angle of gradient of an acceptance surface and the reflection coefficient of the ground surface as calculated by Eq. (4). In this paper, H_D and H_M are obtained by introducing "Standard weather and the **solar radiation** database of the meteorological government office and AMEDAS (METPV-3, 1990 to 2003) (NEDO Technical information data base, (2008))" and the installation requirements of a **solar cell** into Eqs. (1) to (4). Figure 6 (a) is the monthly average amount of slope-face **solar radiation** in Sapporo installed in roofs tilted at an angle of 30 degrees southward.

$$H_M = I_M \cdot \frac{1+\cos\beta}{2} + \rho \cdot I_H \cdot \frac{1-\cos\beta}{2} \tag{3}$$

$$\cot\beta = \cos\phi \cdot \cot\omega + \sin\phi \cdot \mathrm{cosec}\omega \cdot \tan\delta \tag{4}$$

Production of Solar Cell Electricity

A polycrystal solar module of the area S_s is used and the average production-of-electricity $P_{s,dw,t}$ of Date dw and the solar module for time t to $t+1$ is calculated by Eq. (5), where Section 3.4.3 describes how to express with detailed dw. The rise in temperature $T_{c,dw,t}$ of the **solar cell** will reduce **generation efficiency**. In this paper, as shown in Table 1, the temperature coefficient R_T is set to 0.4%/K, where T_o is the reference temperature and η_s expresses the **generation efficiency** in T_o. In this paper, they are determined at 298K and 14%, respectively. The temperature $T_{c,dw,t}$ of the **solar cell** is calculated from the specific heat of the polycrystalline silicon, and the **solar radiation** in Time t. Figure 6 (b) shows the electricity production per unit area of the **solar cell**. This result calculated the amount of slope-face **solar radiation** shown in Figure 6 (a) via the introduction into Eq. (5). The **solar cell** area of the average individual house in Japan is about 25m² to 40m² (based on a **solar cell** capacity of 3 to 5 kW).

Figure 7. Prediction algorithm of solar power generation (PAS)

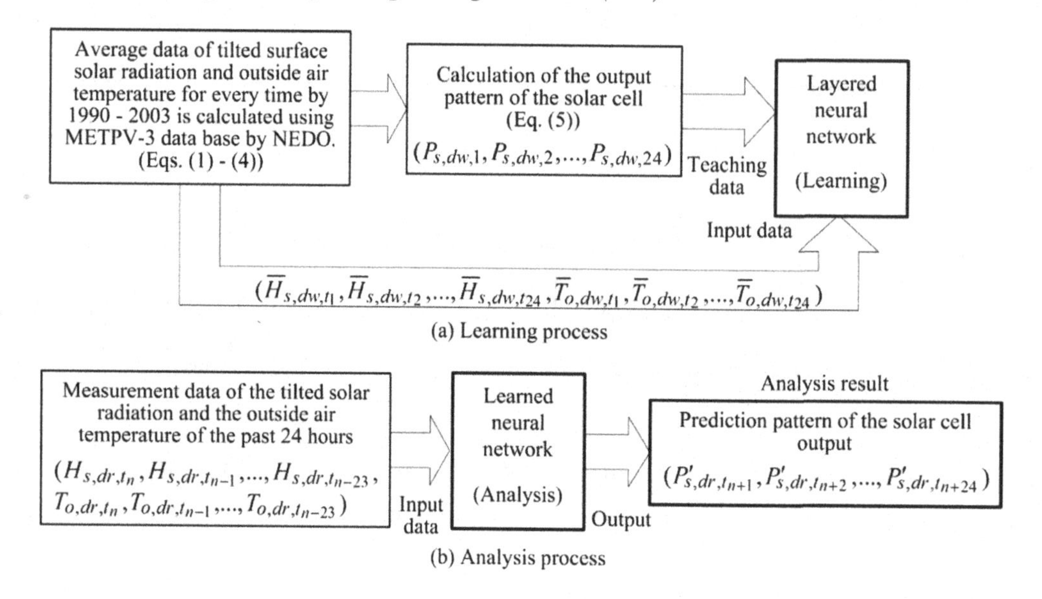

$$P_{s,dw,t} = S_s \cdot \eta_s \cdot \left(H_{D,dw,t} + H_{M,dw,t}\right) \cdot \left\{1 - \left(T_{c,dw,t} - T_o\right) \cdot \left(R_T / 100\right)\right\} \tag{5}$$

Production-of-Electricity Prediction Algorithm of Solar Cell (**PAS**)

In this study, the operation of an engine generator and battery is planned based on the **PAS** prediction result of the electricity production of a **solar cell**. In the **PAS**, a layered **NN** is introduced and the electricity production of a **solar cell** is predicted via the following procedures:

Data of the Amount of Slope-face Solar Radiation, and Outside Air Temperature

The data flow of **NN** used by the **PAS** is shown in Figure 7. In the learning process of the **NN** (Figure 7 (a)), the amount of slope-face **solar radiation** of METPV-3 (1990 to 2003) $\overline{H}_{s,dw,t_1}$ and the time average in t to $t+1$ of the outside air temperature $\overline{T}_{o,dw,t}$ are taken in (NEDO Technical information data base, (2008)), while the input in the learning process of the **NN** is the data aggregate shown in Eq. (6).

$$(\overline{H}_{s,dw,t_1}, \overline{H}_{s,dw,t_2}, \ldots, \overline{H}_{s,dw,t_{24}}, \overline{T}_{o,dw,t_1}, \overline{T}_{o,dw,t_2}, \ldots, \overline{T}_{o,dw,t_{24}}) \tag{6}$$

Where, $\overline{H}_{s,dw,t_i}$ and $\overline{T}_{o,dw,t_i}$ express data concerning the amount of slope-face **solar radiation** and outside air temperature in Date dw and Time t_i, which were respectively obtained from METPV-3. Moreover, it is i=1 to 24. When the data group of Eq. (6) is input into the **NN**, the learning result based on the average of the past 14 years will be obtained. However, the repeatability of the meteorological data within the narrow time range is very low. Therefore, in this paper, the prediction calculation by the **NN** is analyzed for one week so that it may be described in the following section.

Input Data in the **NN** Learning Process

For example, the present time may be 23:00 on August 2 and the electricity production of the **solar cell** is predicted at 2:00 on August 3. The data input by the learning process of the **NN** are prepared as follows:

(a) Integrate the data each time from August 1-7 among the data concerning the past amount of slope-face **solar radiation** (data for every time for the past 14 years from METPV-3). This value is set to X.
(b) Integrate the data on the amount of slope-face **solar radiation** each time by 23:00 during the present period August 1-2. This value is set to Y.
(c) Remove Y from X. The average daily integration slope **solar radiation** is set to Z, and calculated by the following formula. The denominator of this equation is represented by the remaining days for one week (seven days - two days = five days).

$$Z = \frac{X - Y}{5} \tag{7}$$

(d) Allocate at each time of a prediction day, August 3, by dividing Z. This result is the input data group $H_{s,dw,t}$ of the **NN** learning process concerning the amount of slope-face **solar radiation,** where the allocation rate to each time of **solar radiation** is decided with parallelism for the rate distribution of the amount of slope-face **solar radiation** for each time of X.
(e) Determine using the same method as (a) to (d), also for the input data group $T_{o,dw,t}$ concerning the outside air temperature during the **NN** learning process.
(f) Obtain the teaching-data group $P_{s,dw,t}$ of the **NN** by giving input data concerning the amount of slope-face **solar radiation**, and the outside air temperature obtained by (d) and (e) to Eq. (5). The **NN** is made to learn according to the flow shown in Figure 7 (a), using both the input and teaching data during the **NN** learning process obtained in the procedure described in the top. In the following section, details of the **NN** learning calculation are described.

Learning Calculation of the **NN**

The prediction day for the electricity production of a **solar cell** is expressed in dw, while the data obtained by METPV-3 described at the top are expressed with $\bar{H}_{s,i,t}$ and $\bar{T}_{o,i,t}$, where dw expresses "the week in a specific position and a specific month", and "the day of what position of the week". However, the amount of slope-face **solar radiation** and data on the outside air temperature (both of which are actual measurement) are expressed with $H_{s,j,t}$ and $T_{o,j,t}$, respectively. Here, *i* and *j* express the prediction day and measurement date. The daily means of each sum of the amount of slope-face **solar radiation** and the outside air temperature in this case are calculated by Eqs. (8) and (9) in this paper. $H_{day,dw}$ in Eq. (8) is the value whereby "the sum total of the actual measurement before the prediction time (the 2nd term of the numerator)" is subtracted from "the sum total for one week of the slope-face amount of **solar radiation** of the same week of past data (the 1st term of the numerator)", and divided by "the remaining days $8 - dw$ (however, $dw = 1, 2, \ldots, 7$)." Around $T_{day,dw}$ of Eq. (9) is obtained, as well as

$$H_{day,dw} = \frac{\sum_{i=1}^{7}\sum_{t=1}^{24}\overline{H}_{s,i,t} - \sum_{j=1}^{dw-1}\sum_{t=1}^{24}H_{s,j,t}}{8 - dw} \qquad (8)$$

$$T_{day,dw} = \frac{\sum_{i=1}^{7}\sum_{t=1}^{24}\overline{T}_{o,i,t} - \sum_{j=1}^{dw-1}\sum_{t=1}^{24}T_{o,j,t}}{8 - dw} \qquad (9)$$

$$R_{H,dw,t} = \left(\sum_{k=dw}^{7}\overline{H}_{s,k,t} \middle/ \left(\sum_{i=1}^{7}\sum_{t=1}^{24}\overline{H}_{s,i,t} - \sum_{j=1}^{dw-1}\sum_{t=1}^{24}H_{s,j,t}\right)\right) \qquad (10)$$

$$R_{T,dw,t} = \left(\sum_{k=dw}^{7}\overline{T}_{s,k,t} \middle/ \left(\sum_{i=1}^{7}\sum_{t=1}^{24}\overline{T}_{s,i,t} - \sum_{j=1}^{dw-1}\sum_{t=1}^{24}T_{s,j,t}\right)\right) \qquad (11)$$

$$H_{s,dw,t} = R_{H,dw,t} \cdot H_{day,dw} \qquad (12)$$

$$T_{o,dw,t} = R_{T,dw,t} \cdot T_{day,dw} \qquad (13)$$

the method described at the top. Each of $H_{day,dw}$ and $T_{day,dw}$ is multiplied by the rates $R_{H,dw,t}$ and $R_{T,dw,t}$ for each time obtained by Eqs. (10) and (11). The slope-face amount of **solar radiation** and outdoor air temperature on the prediction Day dw and at Time t are also predicted using this result (Eqs. (12) and (13)). Correction is made and it is the value whereby the same amount of slope-face **solar radiation** of a past prediction day is divided by the time sum of the amount of slope-face **solar radiation** in the same week after a prediction day. The same definition applies to $R_{T,dw,t}$ as to $R_{H,dw,t}$.

As $\left(H_{D,dw,t} + H_{M,dw,t}\right) = H_{s,dw,t}$, the electricity production $P_{s,dw,t}$ in a solar module can be obtained by assigning this value and $T_{o,dw,t}$ to Eq. (5). $P_{s,dw,t}$ is calculated for each time of t=1 to 24 in the days and months dw, and these teaching data are given to the **NN**.

Prediction Calculation by the NN

The data flow in the analysis process of the **NN** is shown in Figure 7 (b). The present days, months and time are expressed with dr and t_n, respectively, while the electricity production of the **solar cell** from the present time t_n to 24-hours after (t_{n+24}) is predicted. The meteorological measurement data (actual value of the amount of slope-face **solar radiation** and outside air temperature) for the previous 24 hour period is input into the learned **NN**, thus allowing, as shown in Figure 7 (b), the production-of-electricity

Figure 8. Layered neural network of the proposed system

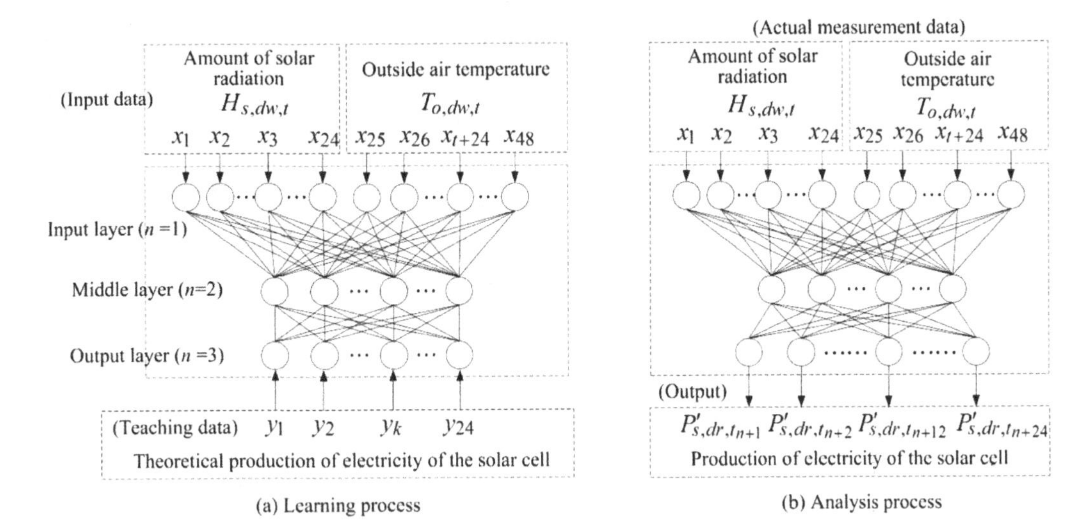

pattern $(P'_{s,dr,t_{n+1}}, P'_{s,dr,t_{n+2}}, \ldots, P'_{s,dr,t_{n+24}})$ of the **solar cell** from t_n to t_{n+24} to be obtained.

Structure of the NN

Figure 8 shows the structure of layered **NN** used by the **PAS**. Figure 8 (a) shows a learning process and Figure 7 (b) shows an analysis process. The **NN** is a three-layer structure. In a learning process, the input data and teaching data, which were described in Sections 3.4.2 and 3.4.3, are input into the 3rd layer (output layer). In this process, the weight of each neuron, which connects the middle layer and each layer, is set up. During the analyzing process, the amount of slope-face **solar radiation** for the previous 24 hours and the data (actual value) of outside air temperature are input into an input layer. Consequently, the predicted electricity production of the **solar cell** is output, from the output layer of the learned **NN** for each time up to 24 hours after it is output.

Analysis Case (1)

Analysis Condition

The **PAS** is introduced into the system shown in Figure 1, and the case of operation concerning the power supply to three houses is investigated. In the analysis example in this paper, the electricity production of the **solar cell** for the 1st weeks in February and August in Sapporo is predicted. Figures 2 and 4 are used for the output characteristics of the engine generator, while the characteristics of each system use the values in Table 3.

Figure 9. Energy balance

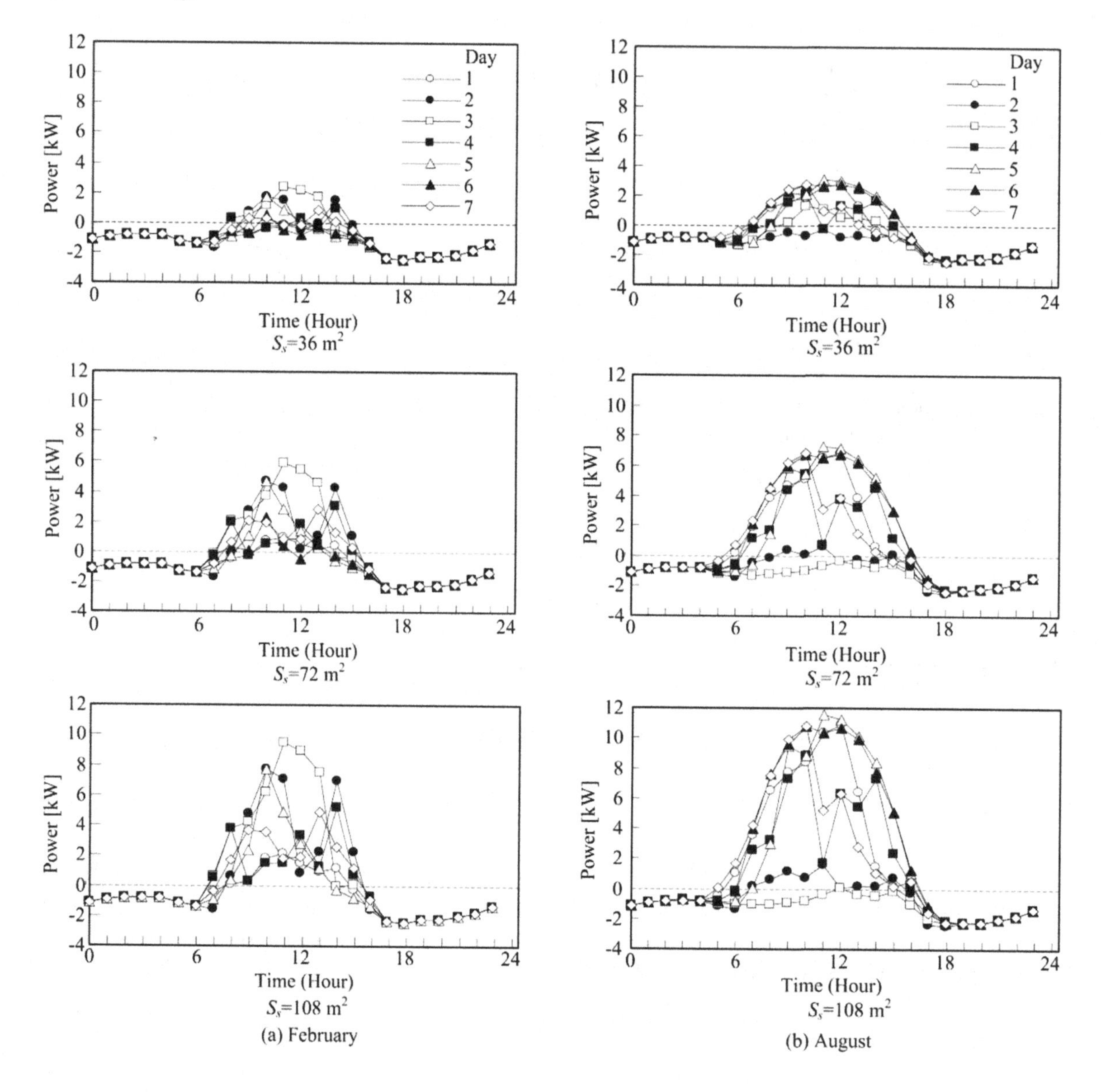

Energy Balance

The output photovoltaic power is calculated by introducing data concerning the average METPV-3 slope-face amount of **solar radiation**, and the average outside temperature into Eq. (5). The difference, excluding the amount of power demand (Figure 5) from these values, is shown in Figure 9. In Figure 9, the surplus daytime power increases, meaning the **solar cell** area does the same. Such surplus power shifts over time using a battery, and the power can be supplied to satisfy the demand side. When the amount of **solar radiation** for the same past week is measured, there is considerable daily variation. In the example of Figure 9, the **solar radiation** on August 2 and 3 is very small compared with other days the same week. Moreover, prediction of change over time for the amount of **solar radiation** is difficult.

Figure 10. Predictive value of the output per unit area of the solar cell in February 1 to 7

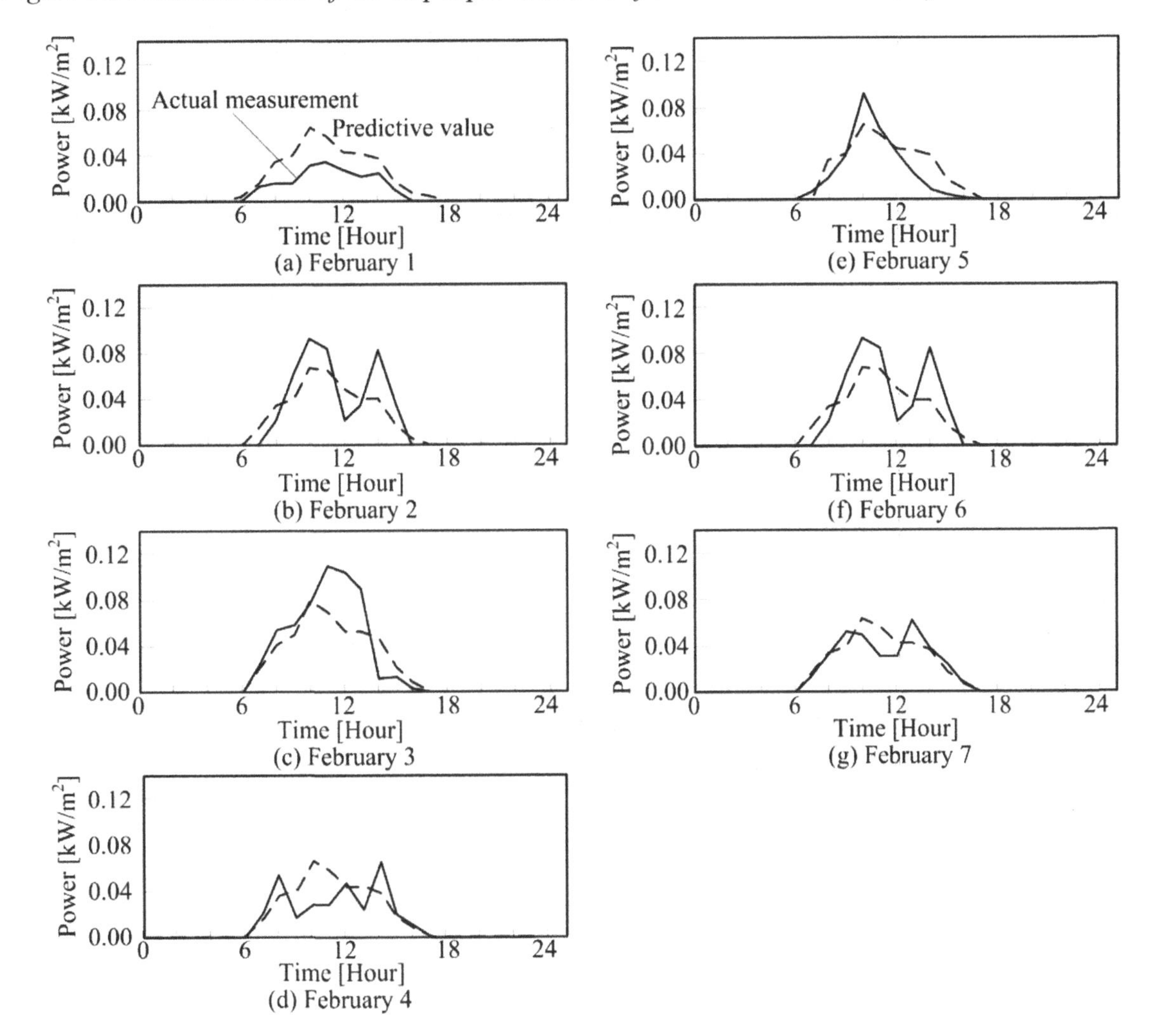

Even if the average data is compared to past data, it is difficult to predict the output power of **solar cells** for equivalent days and months in future.

Production-of-Electricity Prediction of Solar Cell

Figures 10 and 11 show the prediction result in terms of electricity production based on the actual measurement and **PAS** of a **solar cell** in February and August, which were calculated in the above paragraph, where the electricity production of these **solar cells** was analyzed per unit area of the **solar cell**. There was very little **solar radiation** on August 2 and 3 compared with other days the same week. In these cases, the predicted value by the **PAS** shows the characteristic of low electricity production. We should clarify the prediction error of each figure, and the influence of operational planning on the system. So, in the following analysis, the relation between the prediction error and the system fuel consumption is investigated.

Figure 11. Predictive value of the output per unit area of the solar cell in August 1 to 7

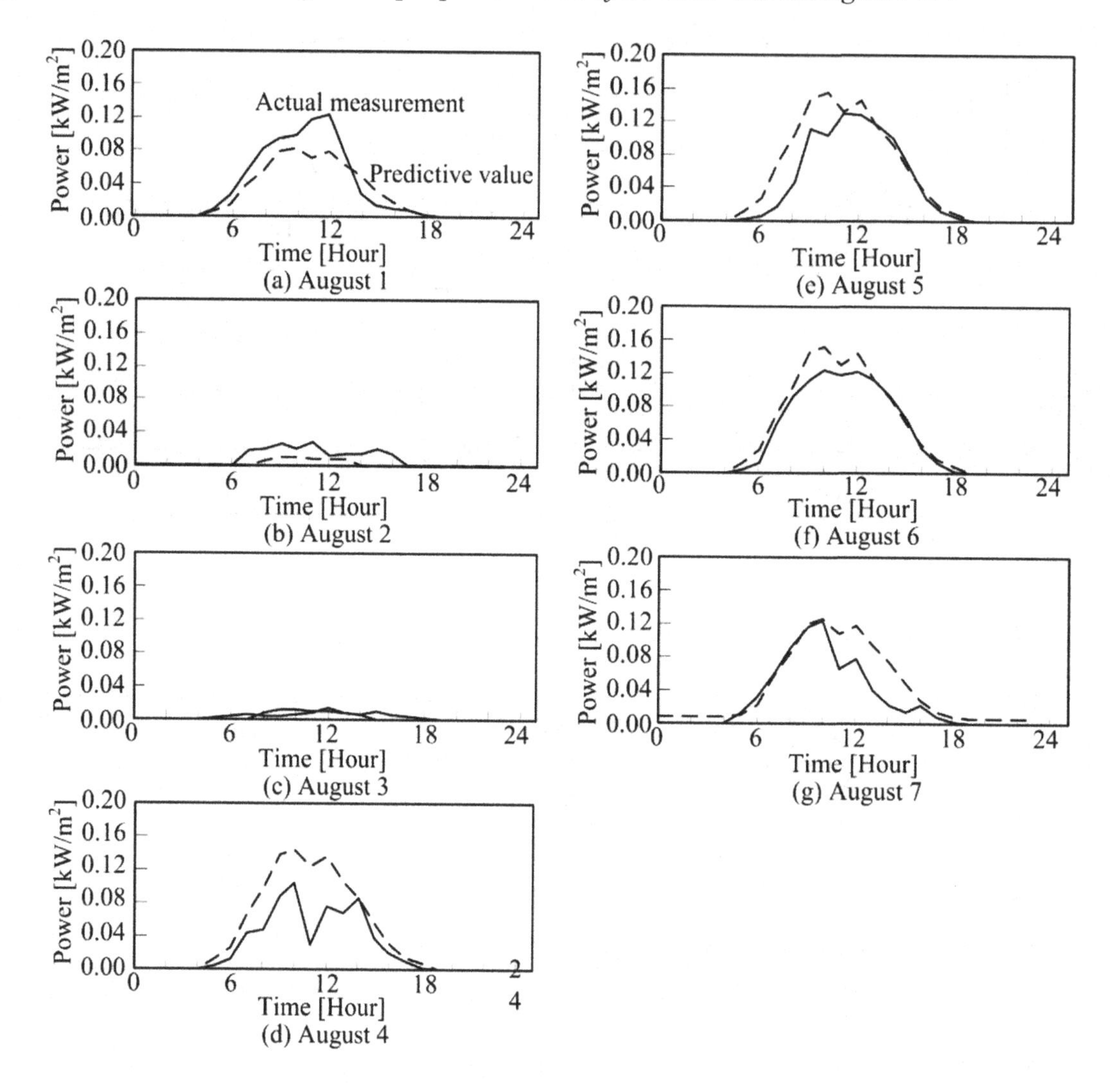

Engine Generator Operation Method

The charge amount of a battery is calculated based on the electricity production of the **solar cell**, as predicted by the **PAS**. The engine generator is operated at the point just before the amount of battery charge is expected to go below 0, and at the point of maximum efficiency (3kW of output power). The power of the engine generator is initially supplied to the demand side, and the surplus is charged by the battery, meaning the operational planning and working hours of the engine generator are strongly dependent on the electricity production and battery capacity of a **solar cell**. It is decided that based on the fuel cost of the engine generator, the energy cost of the proposed system will be the depreciation of each piece of equipment of a **solar cell**, battery, and engine generator. Therefore, the following analyses consider the relation between a **solar cell**, the installed capacity of a battery and the energy cost.

Figure 12. Error of prediction production of electricity and actual measurement power. These errors are the values per unit area of a solar cell.

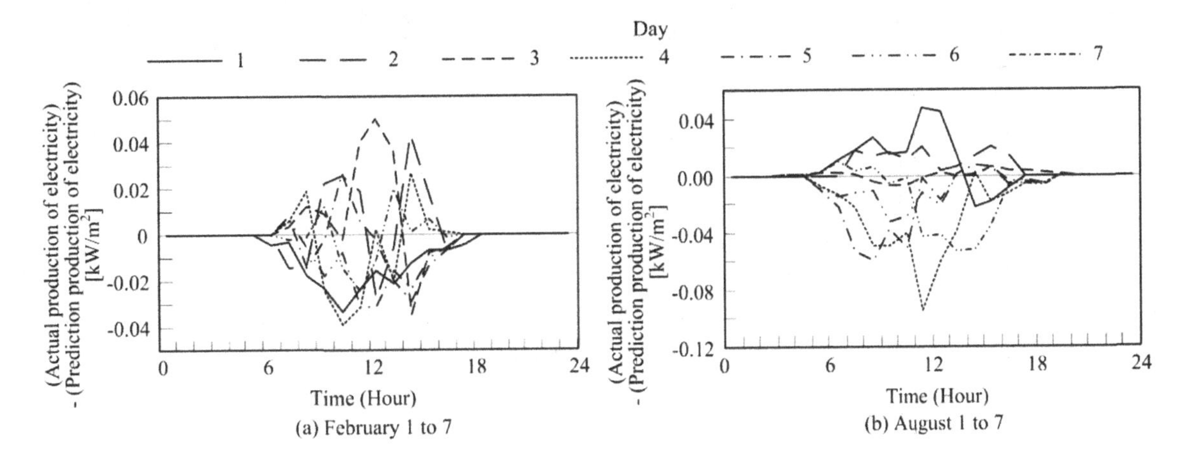

Analysis Result (1)

Prediction Error of the **PAS**

Figure 12 expresses the error of the prediction electricity production in Figure 10 and 11, and the actual electricity production. The error in February represents 4% to 89% of the range, while that in August is

Figure 13. Operating pattern of the engine generator. Area of the solar dell is 36 2 m^2

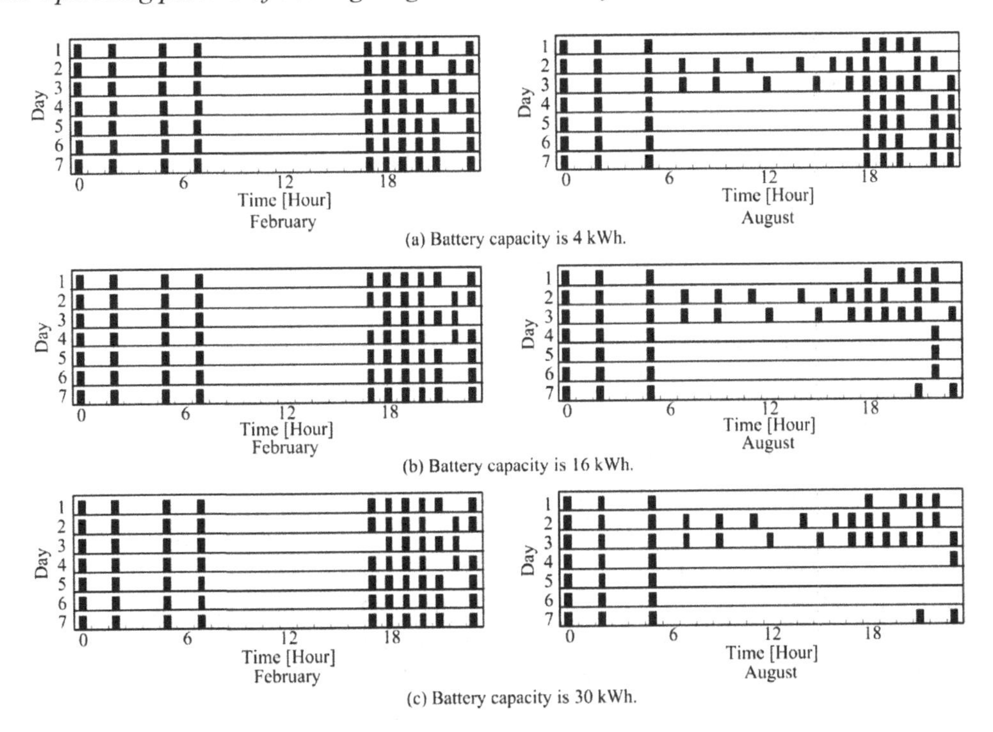

Figure 14. Operating pattern of the engine generator. Area of the solar dell is 108 m²

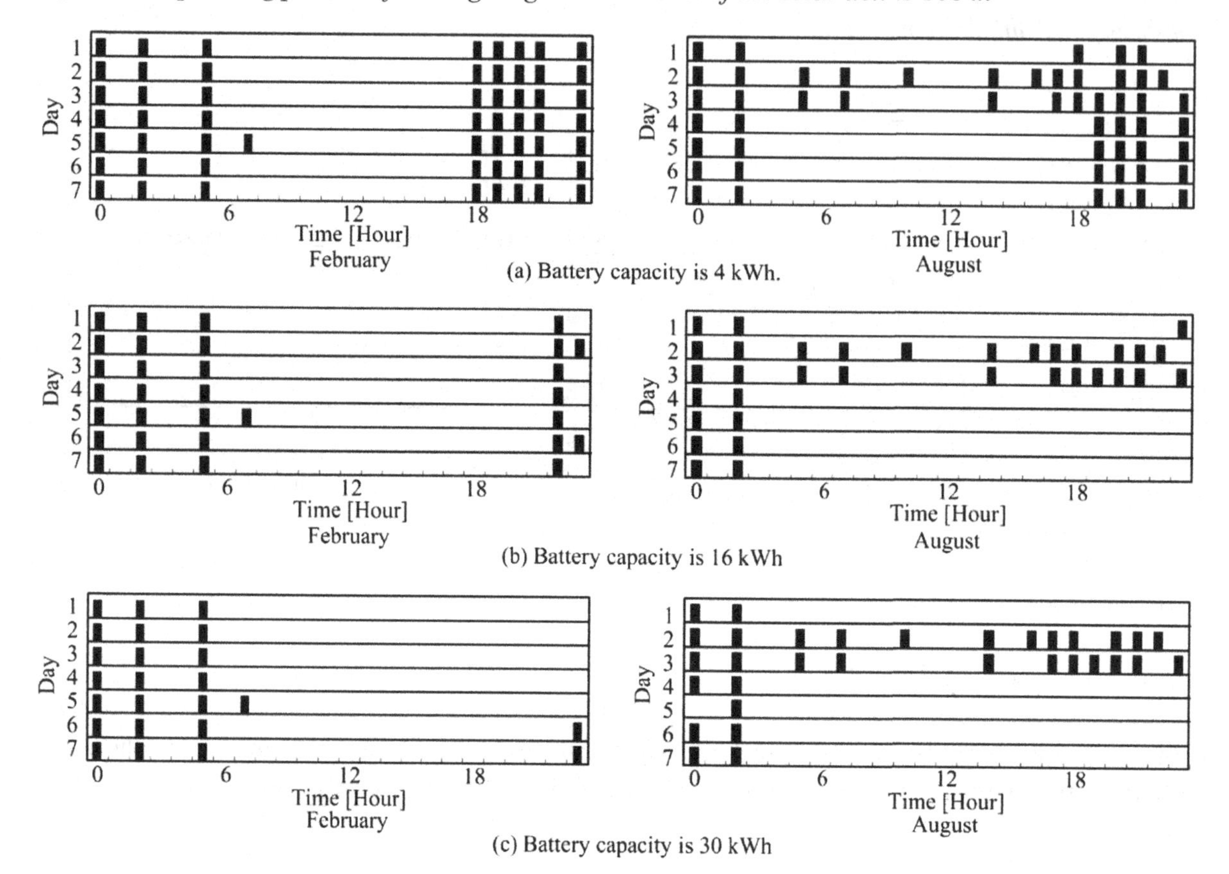

15% to 73% of the range. Although the error is comparatively small in the morning and evening, it is considerable in daytime. Moreover, the predictive value in August exceeds the actual electricity production in many cases. In the **NN**, learning takes place using the average value of the previous slope-face amounts of **solar radiation** and the outdoor air temperature. For this reason, it is thought that correspondence of the **NN** is difficult for rapid changes in weather conditions over time.

Operation Plan of the Engine Generator

Figures 13 and 14 show the analysis result of the **operation plan** of the engine generator. In the case of the **solar cell** of area 36 m², the operating time of the engine generator in August is -6% to -26% to February. On the other hand, in the case of the **solar cell** of area 108 m², the operating time of the engine generator in August is -9% to +30% to February. If the **solar cell** area is small, it will take considerable time to charge a battery. In this case, since the battery is always charged to some extent, the operation time of an engine generator is extended. Because there are many time zones and given the low electricity storage capacity of the battery, the hours of operation of the engine generator will be long in February with the little **solar radiation** available. On the other hand, when the **solar cell** area is considerable, a lot of power is available to supply a battery, resulting in a fall in the operating time of the engine generator, and a considerable battery capacity. However, there is insufficient electricity production within a **solar cell** on August 2 and 3, meaning extended operation hours of an engine generator on these days.

Figure 15. Prediction error of PAS algorithm

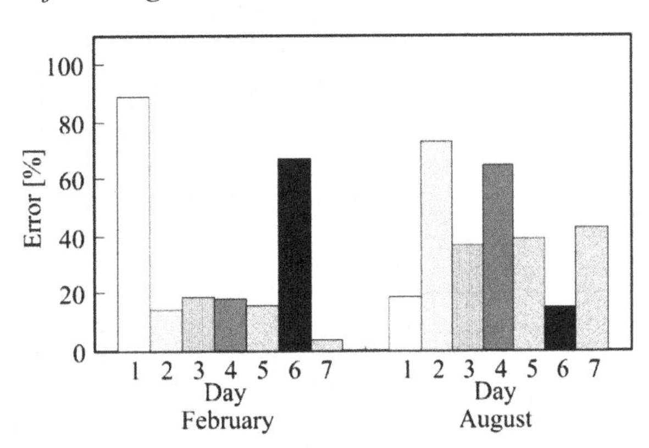

Prediction Error of the **PAS**, and Operation of Engine Generator

Figure 15 shows the prediction error of the analysis result of Figures 10 and 11, showing the error of the integrated value in one day of actual electricity production, and the integrated value of one day of a prediction slope-face amount of **solar radiation**. On the other hand, Figure 16 shows the correction factor of the engine generator hours of operation when adding the prediction error shown in Figure 15 to the **operation plan**ning, based on the prediction result for **solar cell** electricity production by the **PAS**. The correction factor is an extended rate of operation hours to the **operation plan** of the engine

Figure 16. Correction of the engine generator operation hours required according to the prediction error of PAS algorithm

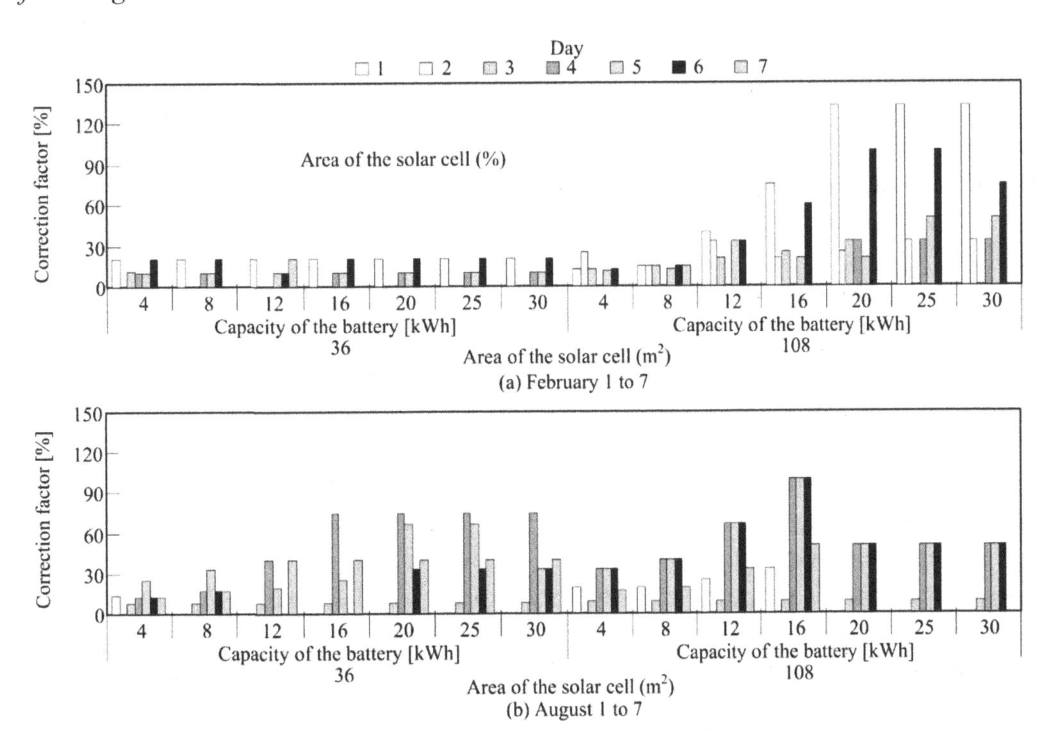

Figure 17. Operation plan of the battery in February

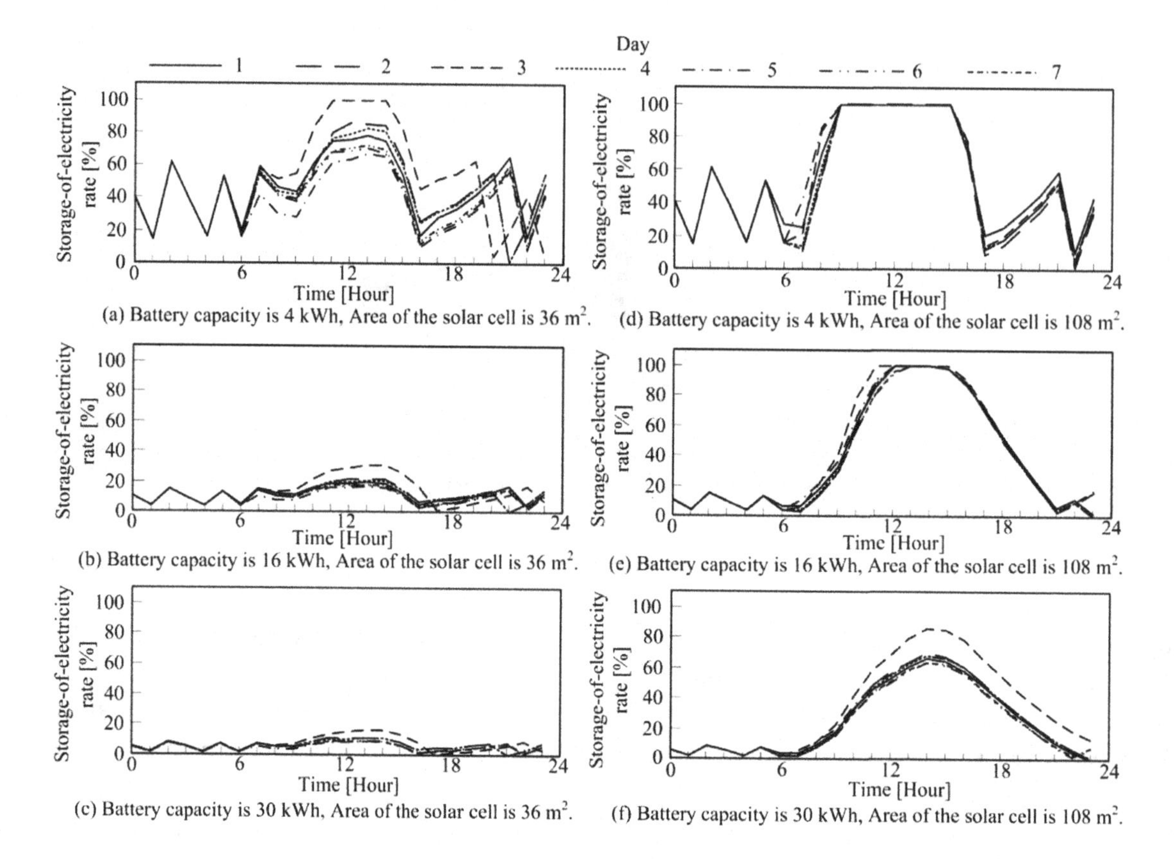

generator, based on the prediction of the **PAS** shown in Figures 13 and 14, while the large day of prediction error shown in Figure 15 includes a trend whereby the correction factor shown in Figure 16 also expands. Moreover, because the **operation plan** of an engine generator is strongly related to the electricity production of a **solar cell**, the correction factor influences the area of the **solar cell**. There is the method of repeating the analysis by **PAS** in order to control the increase in the correction factor, whenever its operation is corrected due to the shortage of **solar radiation** etc.

Operation Plan of a Battery

Figures 17 and 18 show the analysis results of the **operation plan** of a battery with **PAS**. When the engine generator is operated, the charge amount of a battery will increase rapidly, reflecting the characteristic whereby the electricity storage volume increases rapidly in each figure when charged by the engine generator. On many days, a large number of operations of the engine generator are observed during the period from evening to morning. However, regarding the amount of **solar radiation** on August 2 and 3, many engine operations are also observed in daytime (Figures 18 (a) and (d) especially). The advantages of operation via the proposed system are the facts that the installed capacity of the **solar cell** and battery is small, the fuel consumption of the engine generator is low, and there is the ability to plan with the electricity storage volume near 0 at 24:00 (this time is the final point of **operation planning**).

Figure 18. Operation plan of battery in August

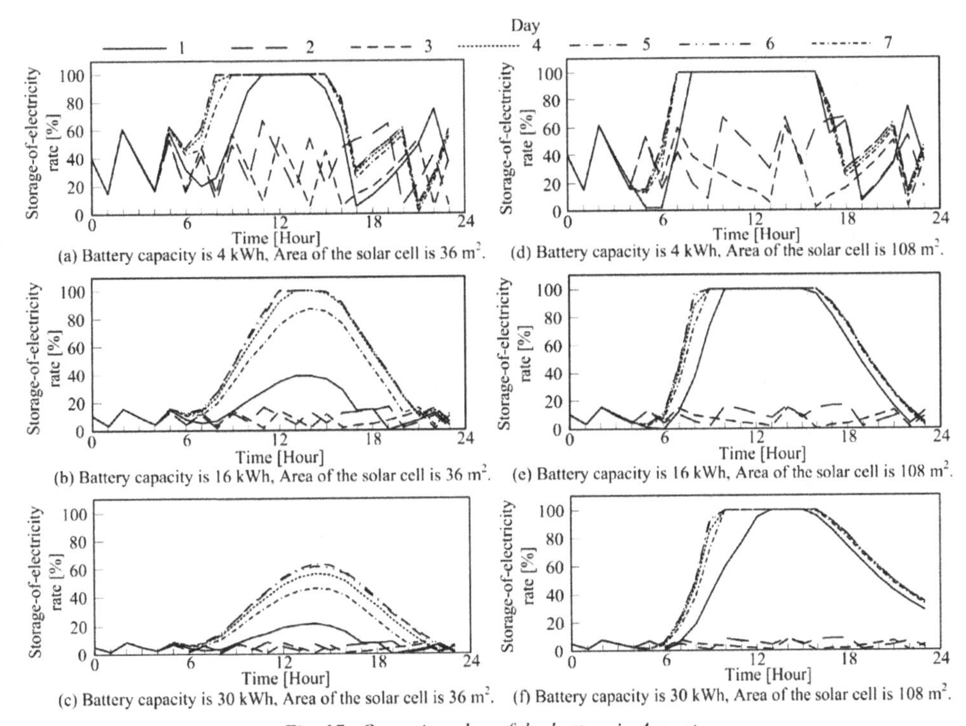

However, if the prediction error of the **PAS** is considerable, **operation planning** of Figures 17 and 18 will adversely affect profit, with considerable stored electricity still remaining at 24:00.

Energy Cost of the System

Figure 19 shows the result of the operation cost for one week, when the proposed system is used within three households in February and August. Each value used to calculate the operational cost is shown in

Figure 19. The first week of operating cost of every month

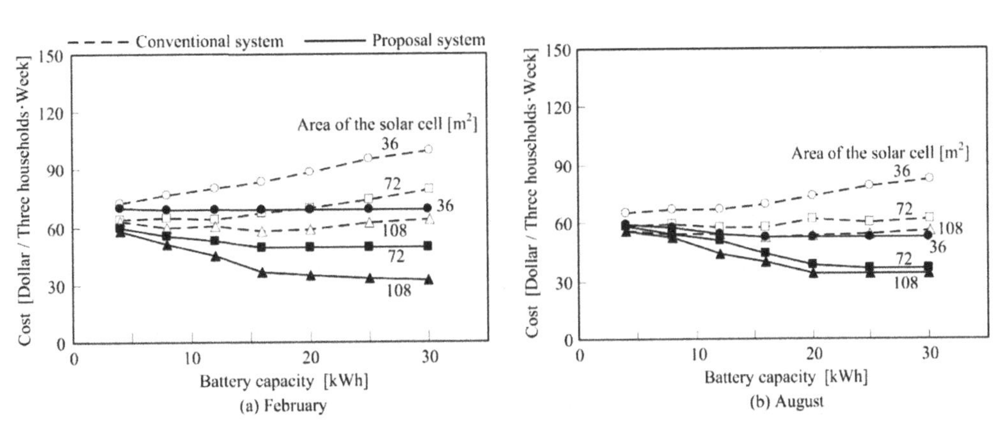

Table 4. Analysis conditions of cost

Exchange rate of U.S. dollar and Japanense yen **Equipment cost of the solar cell** **Equipment cost of the nickel-hydoride batter Kerosene**	**110 Yen/Dollar** **590 Dollar/m²** **890 Dollar/kWh** **900 Dollar/m³**

Table 4, as calculated from the kerosene fuel consumed with the engine generator, and excluding maintenance costs. The conventional system in this figure is the control method whereby the engine generator is operated automatically and charged, when the battery charge amount goes below 25%. The operation cost of the proposed system falls, meaning considerable battery capacity and **solar cell** area. In the conventional system, if the capacity of the battery to be introduced becomes large, the charge amount will also increase, since the operation and shutdown of the engine generator are settled based on the battery charge rate, meaning extending the operation hours of the engine generator will increase battery capacity but also operation cost. A battery capacity of 30 kWh and **solar cell** area of 108 m^2 are introduced, and cost reductions of 50% and 40% are anticipated with the proposed system using **PAS** in February and August compared with the minimum operation cost in the conventional system, respectively.

Figure 20 shows the analysis result of the annual operation cost of the proposed system, and the equipment cost, with the values in Table 4 used to calculate each cost. When the area of the **solar cell** is small, even if this results in an increase in the battery capacity, there are very few effects reducing the system operation cost in comparison to the result of Figure 20 (a). The system specification for significantly reducing operation cost is the range of battery capacity of 15kWh or more and a **solar cell** area of 100 m^2 or more. However, within these ranges, as shown in Figure 20 (b), there is a significant rise in equipment cost.

Because the equipment cost of the proposed system is considerable, compared with the present commercial power, the energy cost is disadvantageous. In order to compete with commercial power, without the sale of electricity to utilities, the equipment cost of a **solar cell** must be 1/3 or less of the present level, while the battery cost must also be further reduced. On the other hand, as shown in Figure 19,

Figure 20. Equipment cost and annual operation cost of the proposal system

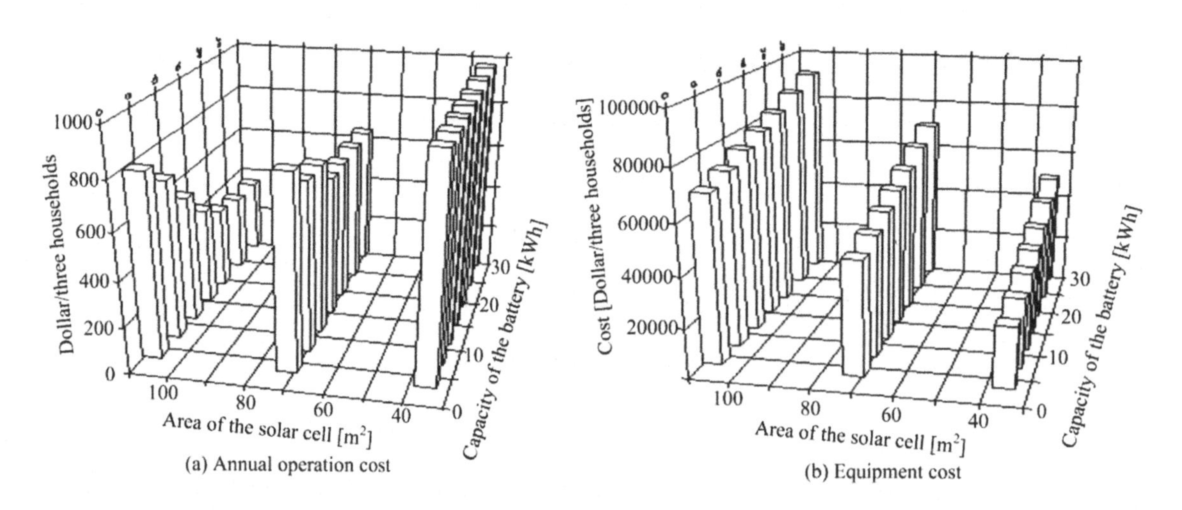

Table 5. Specification of solar cell and battery

Solar cell type	Multicrystalline silicon
Area of the solar cell	72 m^2
Generation efficiency of the solar cell	14%
Temperature coefficient of the solar cell	0.4%/K
Battery type	Nickel-Hydrogen
Battery efficiency	95%
Battery capacity	4kWh

compared with the method of controlling an engine generator solely based on the battery charge volume, the system using **PAS** is more advantageous.

Conclusions from the 1st Case Study

In this paper, details of the **operation plan** and the installed capacity of a combined **solar cell** and **diesel engine** power generator system are studied, with the production-of-electricity **prediction algorithm** of a **solar cell (PAS)**. Moreover, the operation method for the 1st weeks in February and August in Sapporo was analyzed, presuming the effective use of the load management of an engine generator, and the surplus power of a **solar cell** by battery. However, since the operation methods of the proposed system differ significantly based on the electricity production of a **solar cell** (cell area), and the battery capacity, the relation among these values, **operation planning** and operation cost was clarified. Furthermore, although the error was included in the prediction result of the **PAS**, the magnitude of this value and the influence of the system operation method were both investigated and the following conclusions were consequently obtained:

1. The method of the **operation plan** of a system based on **PAS** was discussed. Furthermore, the extended rate of the operational hours of the engine generator (correction factor), which occurred during the **PAS** prediction error, was clarified. Because the operation hours of the engine generator have a significant influence on the electricity production of a **solar cell**, the correction factor changes significantly based on the area and **solar radiation** of the **solar cell**.
2. Compared with the method of controlling the engine generator by the magnitude of the battery charge volume (as in a conventional system), the system using the energy cost **PAS** is more advantageous. The proposed system, which introduces a battery capacity of 30 kWh and **solar cell**

Figure 21. Solar power and power balance in February

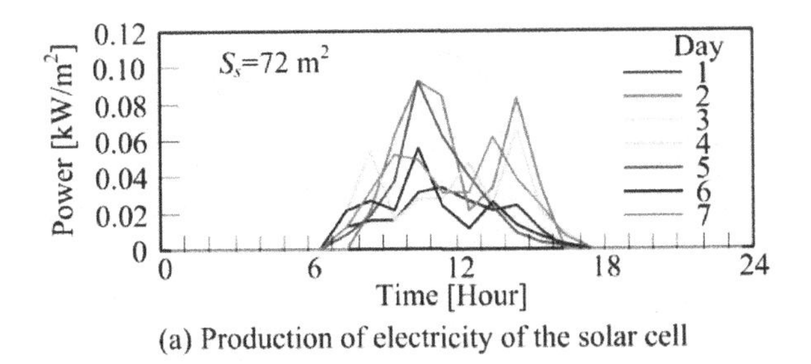

(a) Production of electricity of the solar cell

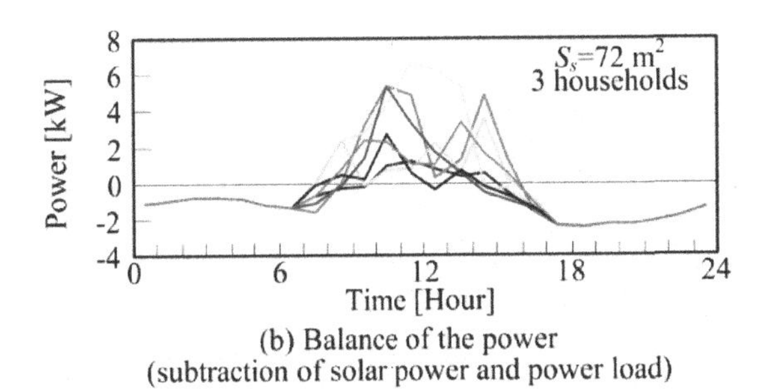

(b) Balance of the power (subtraction of solar power and power load)

area of 108 m², as well as the **PAS**, operates in February and August compared with the minimum operation cost in the conventional system, and cost reductions of 50% and 40% are respectively anticipated.

ANALYSIS CASE (2)

Analysis Condition

Examination System

In this paper, the following three systems (S-1, S-2, S-3) are examined, none of which involve power being supplied from the outside.

S-1 system: The engine generator is operated according to load fluctuation and no battery is installed. The surplus power of the **solar cell** can be sold off. In this system, because the engine is operated by a large area from a low to a high load, partial load operation with low efficiency occurs frequently.

S-2 system: When operating the engine, operate by the fixed load of maximum efficiency (3kW output power). However, the residual quantity of the battery is measured for every sampling period, and the operation or stop status of the engine is judged. Accordingly, if the battery capacity drops to 25% or less, the engine generator will be operated and the battery will be charged. The power of the **solar cell** exceeding the power load charges the battery.

S-3 system: The production-of-electricity **prediction algorithm** (**PAS**) of a **solar cell** is introduced, and the operation method of the system determined. Based on the predicted output characteristic of the **solar cell**, a plan is made concerning the operation or stop of the engine generator, and operation of the charge or discharge of the battery. When operating the engine, it is operated with a fixed load and at maximum efficiency (3kW output power). When the power supplies from the battery to the demand side run short due to a prediction error, the engine generator is operated compulsorily.

Battery Capacity

The cost of a nickel hydrogen type battery is about 100,000 yen/kWh (940 US$/kWh) (SANYO Co. Ltd. (2007)). The battery capacity introduced into the system in this paper shall be 4kWh, which corresponds to about 170 minutes of an average household's power load (0.47kW x 170 minutes x 3 households = 4kWh). The optimal capacity of the **solar cell** and the battery must be discussed and determined for the dynamic **operation plan**. Table 5 shows the battery specifications in this study.

Figure 22. Solar power and power balance in August

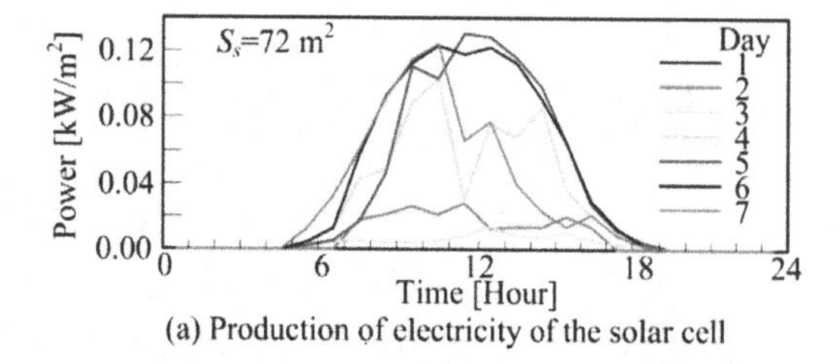

(a) Production of electricity of the solar cell

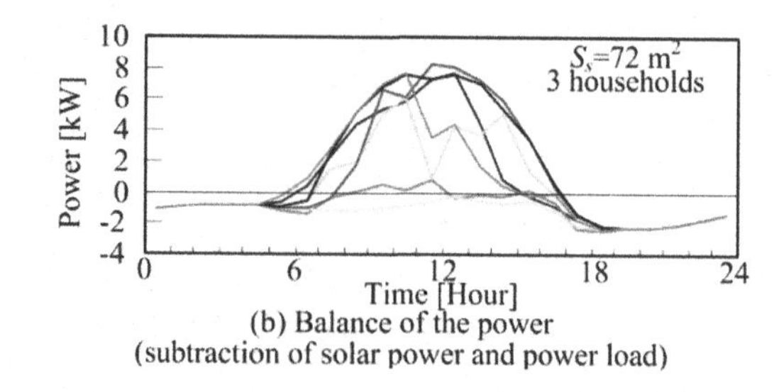

(b) Balance of the power (subtraction of solar power and power load)

Figure 23. Operation result of S-1 system in February

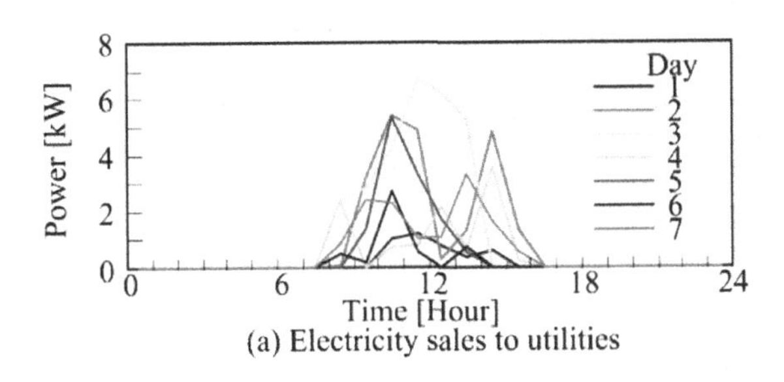

(a) Electricity sales to utilities

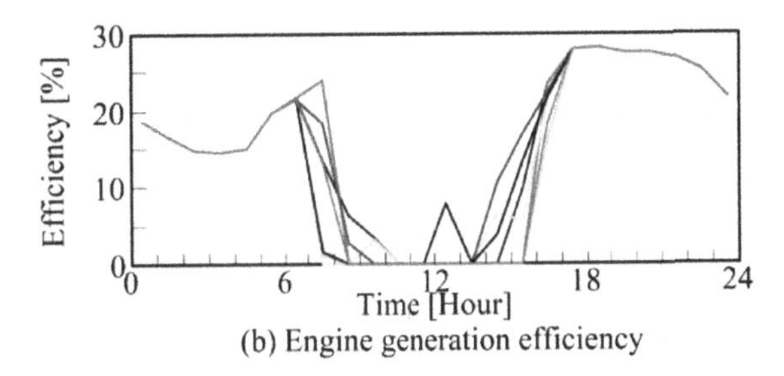

(b) Engine generation efficiency

Figure 24. Operation result of S-1 system in August

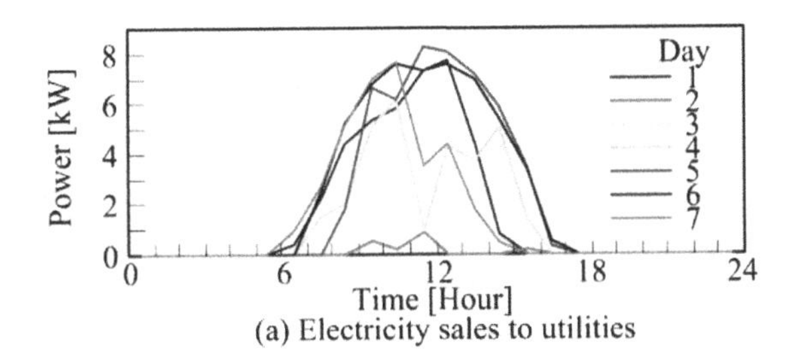

(a) Electricity sales to utilities

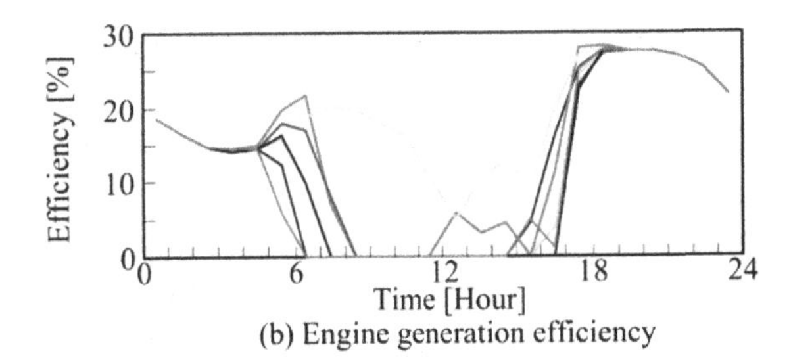

(b) Engine generation efficiency

Demand Pattern

In the analysis in this paper, the power load pattern in the general household in Sapporo shown in Figure 5 is used (Narita, K., (1996)). As for the power load, the electricity consumption of a household appliance and an electric light is mainly included, and air conditioning load is not contained. For this reason, the difference in every month in the figure is small. The system of Figure 1 with the operating method of S-1 to S-3 is introduced into 3 household apartment house with the load pattern shown in Figure 5.

Figure 25. Output characteristics of the diesel engine generator

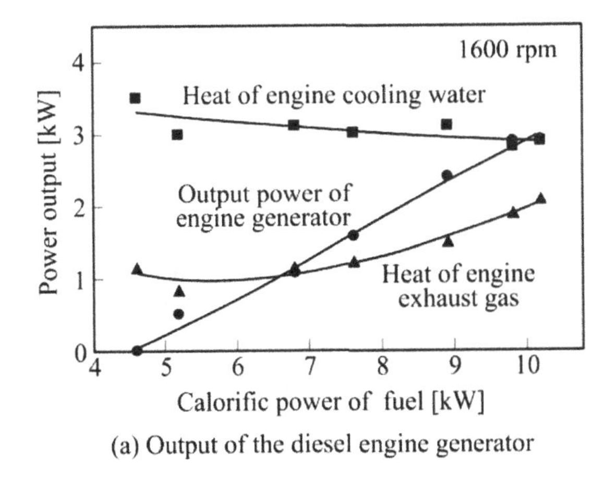

(a) Output of the diesel engine generator

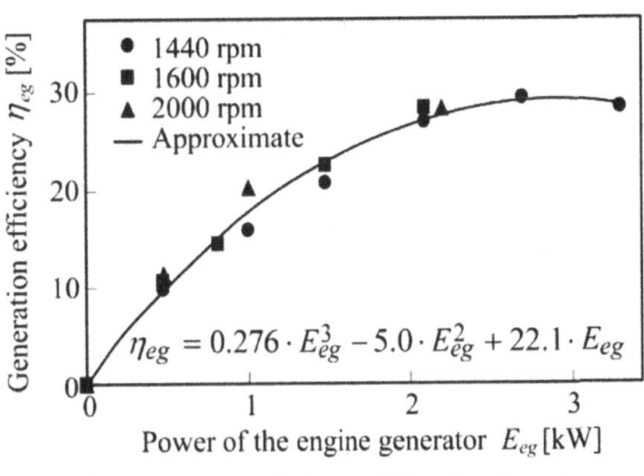

(b) Generator efficiency of the diesel engine

Figure 26. Operation result of the battery (S-2)

Analysis Result (2)

Electricity Production of the Solar Cell, and Power Balance

Figures 21 and 22 show the electricity production of the **solar cell** and power balances (power load is subtracted from the electricity production of the **solar cell**) from the 1st to 7th for every month. Except for August 2 and 3, the majority of the daytime power load can be supplied with the power of the **solar cell**. However, around 7:00 and around 18:00, where there is significant power demand, there is insufficiency in the electricity production of the **solar cell**.

Operation Result of the S-1 System

Figures 23 and 24 show the operation result of the S-1 system, the amount of electricity sales to utilities, and the **generation efficiency** of the engine generator. No battery is used in the S-1 system. During the period when the electricity production of the **solar cell** cannot meet the demand, the engine generator operates. Due to the particularly high power demand during every morning and evening period, in addition to the power of the **solar cell**, the operating load of the engine generator also increases. Consequently, as shown in Figures 23 (b) and 24 (b), partial load operation with low efficiency occurs frequently. The average efficiency for each February day of the engine generator is 14% to 16%, and will be 12% to 19% in August. There is an extreme lack of **solar radiation** on August 2 and 3 compared with other days in the same week, explaining the wide efficiency range of the engine generator in August. As shown in Figure 25, the peak **generation efficiency** of the engine generator is about 30%. In the S-1 system, however, high efficiency cannot be maintained. When consideration is made of electricity sales to utilities, the economic efficiency of the S-1 system is good.

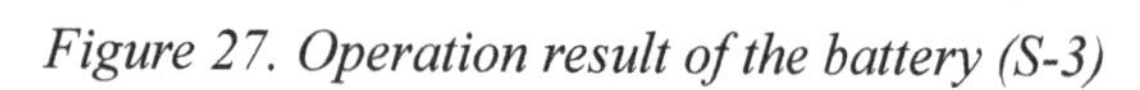
Figure 27. Operation result of the battery (S-3)

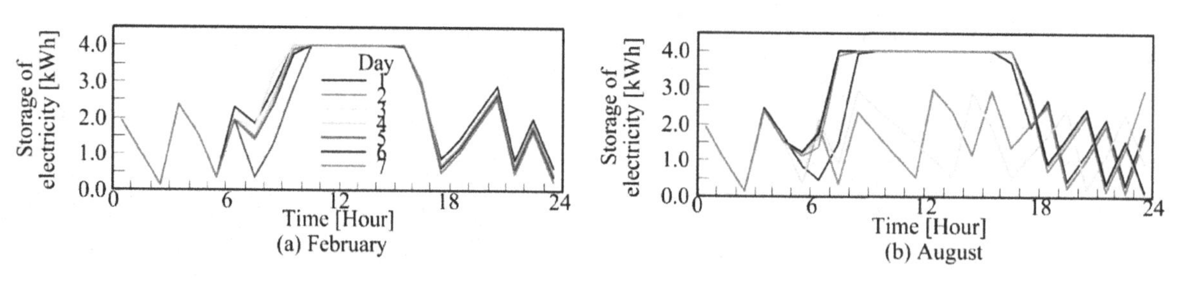

Figure 28. Predictive value of the production of electricity of the solar cell in February 1 to 7

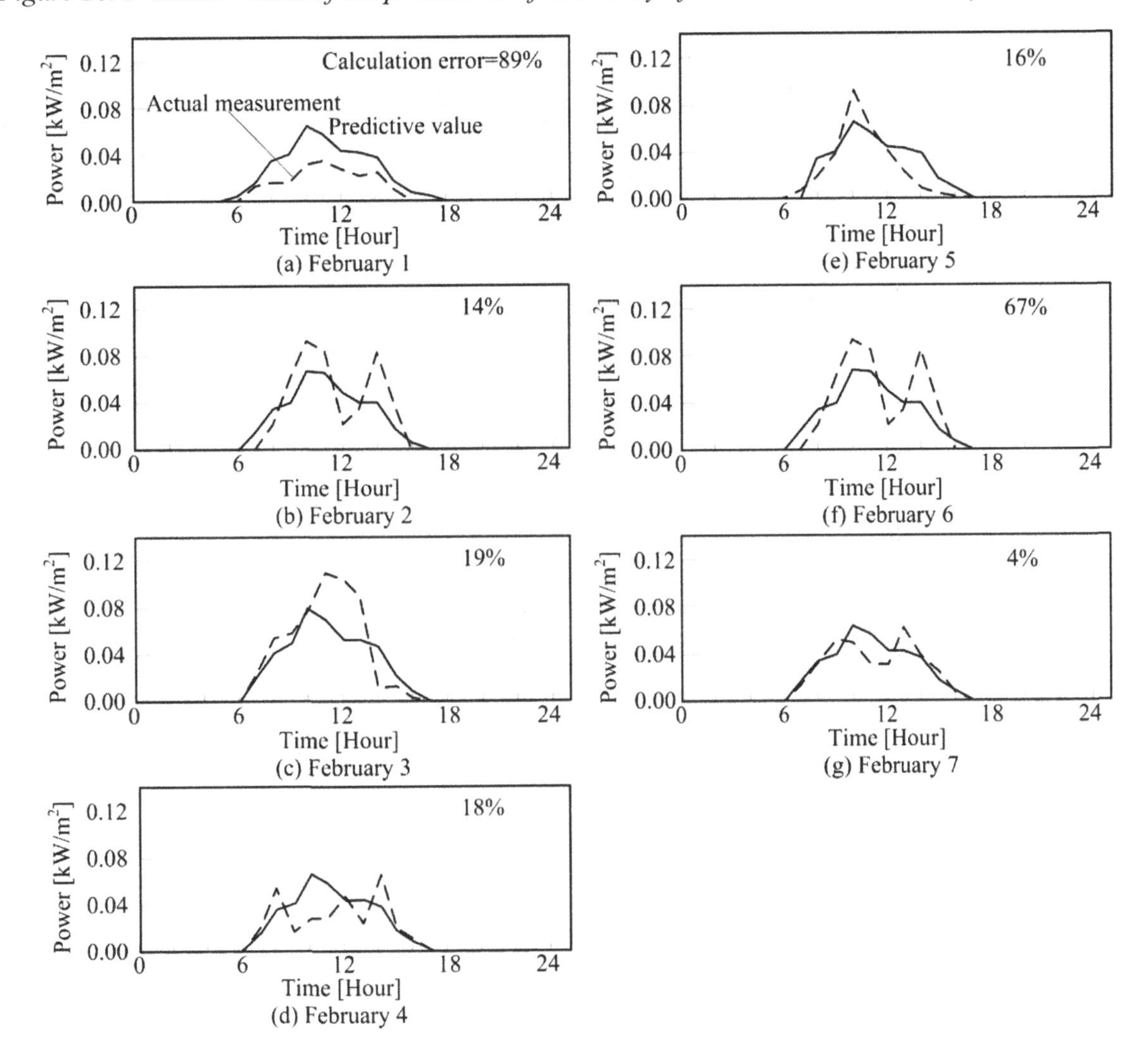

Battery Operation Plan

Figures 26 and 27 show the result of the operation analysis of the battery for the S-2 and S-3 systems. In the battery capacity (4kWh) set up in this paper, none of the surplus power of the daytime **solar cell** can be stored as electricity. Except for August 2 and 3 in Figure 27 (b), when Figures 26 and 27 are compared, the frequency of battery charges and discharges of the battery occurs more in the S-2 systems. The extreme lack of **solar radiation** on August 2 and 3, means the cycle of slight charge and discharge is repeated with the battery. In the S-2 system, when the battery charge amount declines to 25% or less, it will charge with the engine generator. On the other hand, operation of the charge and discharge of the battery can be planned within the S-3 system, meaning here that the frequency of charging by the engine generator can be reduced as much as possible. Consequently, the battery charge-and-discharge loss of the S-3 system and the fuel consumption of the engine generator can be reduced compared with the S-2 system.

Figure 29. Predictive value of the production of electricity of the solar cell in August 1 to 7

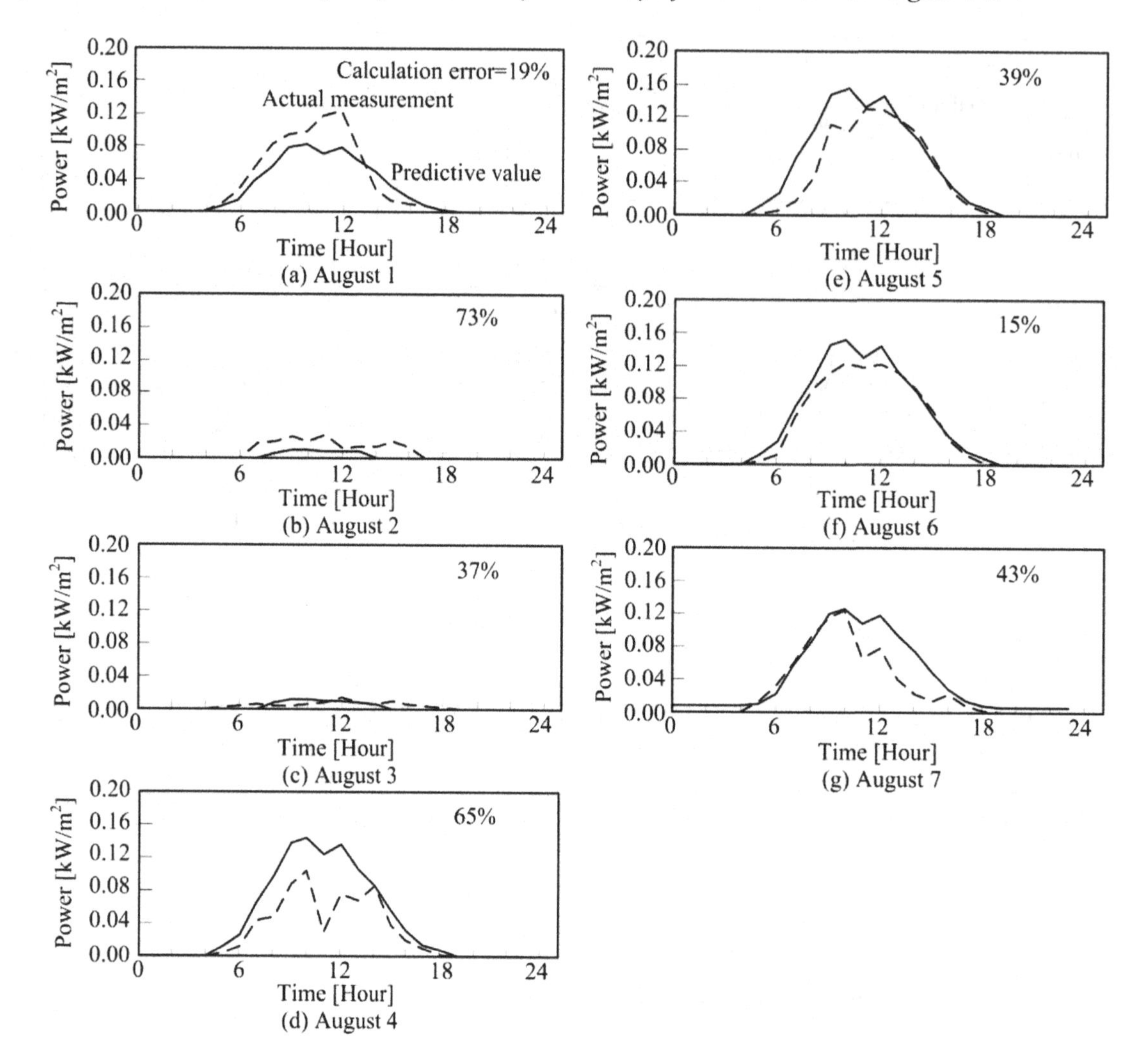

Figure 30. Error of power generation prediction of solar cell (S-3)

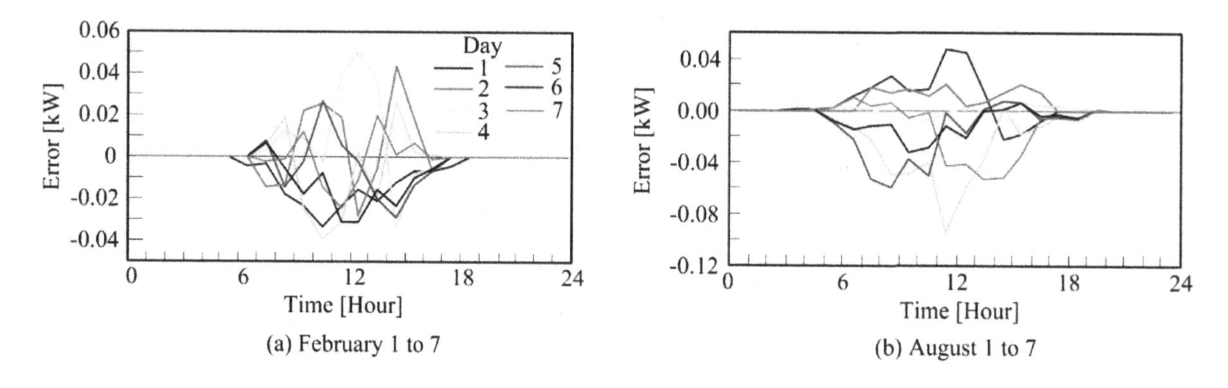

Figure 31. Operating patter of the engine generator in February

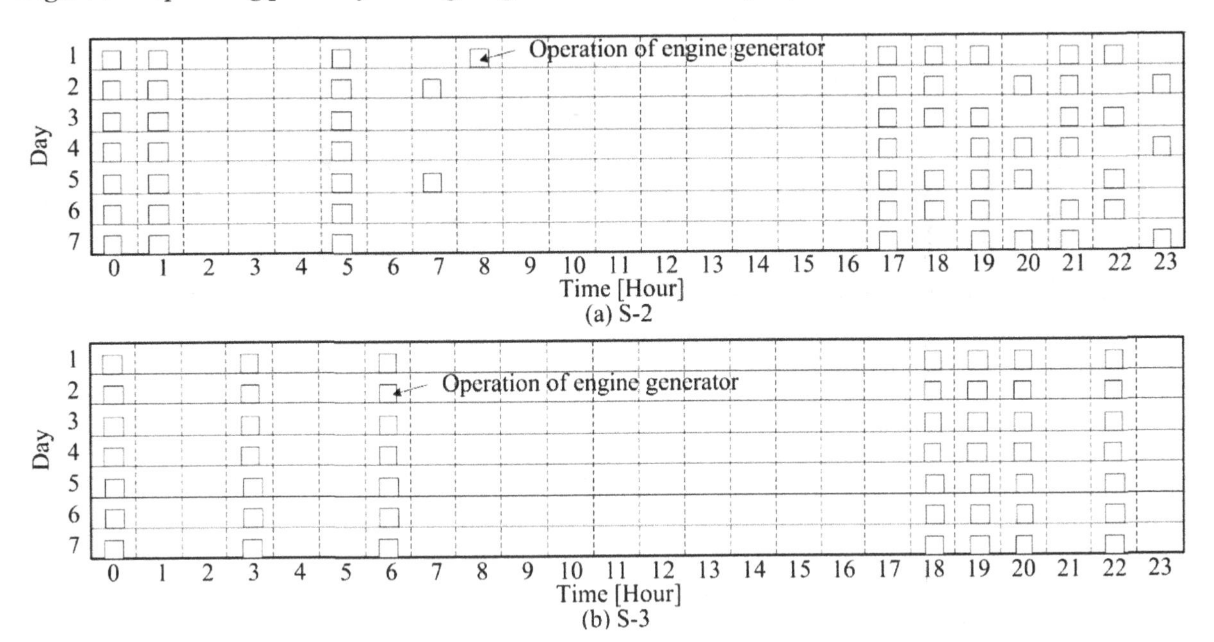

In Figures 26 and 27, power input exceeding battery capacity occurs in daytime, due to the considerable electricity production of the **solar cell** and overall system operation will be influenced when the area of a **solar cell** is changed. Concerning optimization of the area of a **solar cell**, this is a vital topic and separately reported upon.

Figure 32. Operating pattern of the engine generator in August

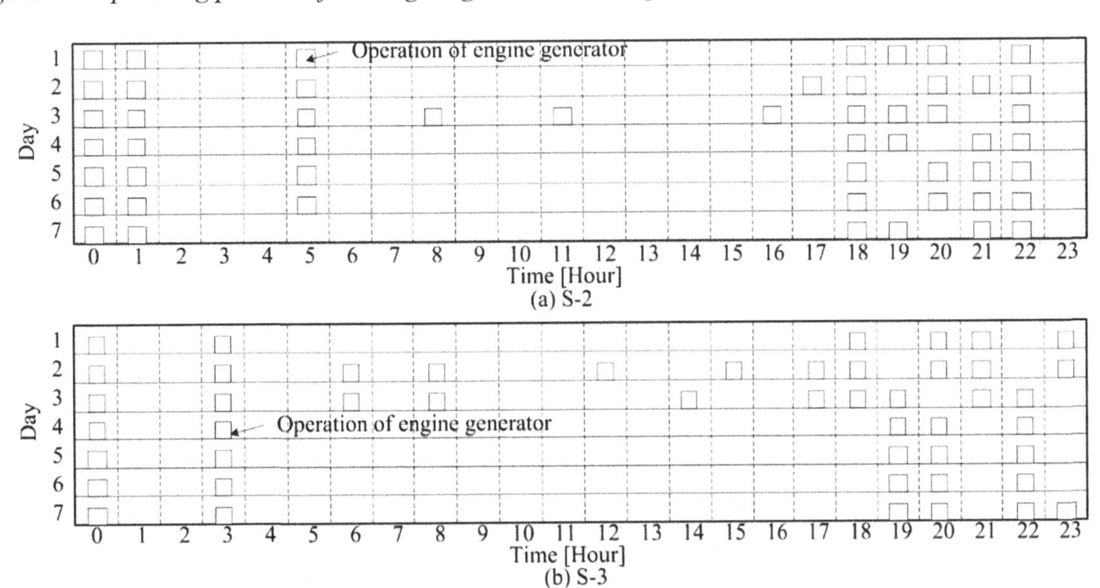

Figure 33. Fuel consumption in each system

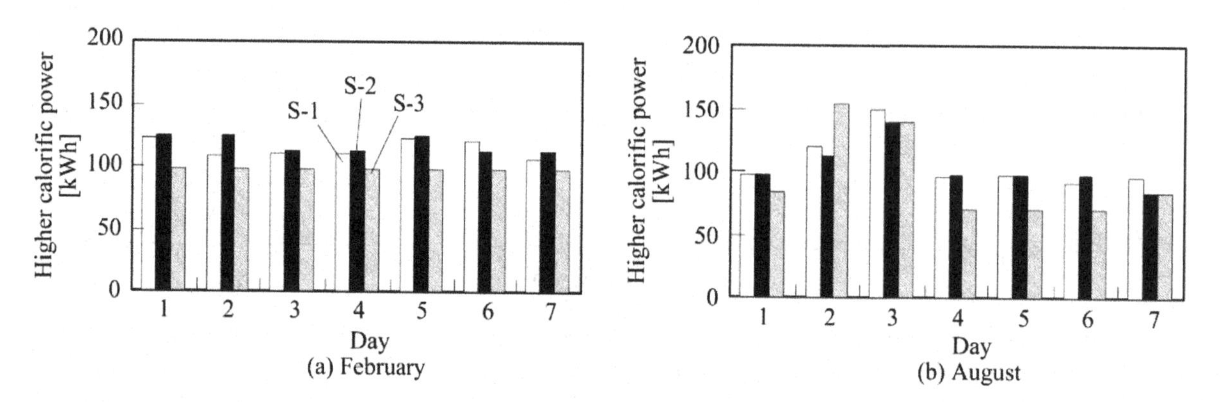

Result of the Power Generation Prediction of Solar Cell

Figures 28 and 29 shows the result of the production-of-electricity predictive value of the **solar cell** and the actual value for each day in February and August. The addition error of the **solar radiation** for one day is described in each figure. As shown in Figures 29 (b) and 29 (c), the predictive value of the **solar radiation** on August 2 and 3, when an extreme lack of **solar radiation** was recorded, shows few characteristics. Figure 30 shows the difference in the predictive value between the amount of **solar radiation** and the actual value for each day and every month. The error for the 1st week in February ranged from 4% to 89%, and between 15% and 73% during the 1st week in August.

The magnitude of the error of the predictive value and the actual value was investigated, namely in terms of its impact on the **operation plan** of the S-3 system. Accordingly, the difference in the prediction electricity production of the **solar cell** and the actual electricity production clarifies details of how this influence affects the fuel consumption of the engine generator. In the following, the analysis result of the **operation plan** of the engine generator in the S-3 system is described.

Operation Result of the Engine Generator

Figures 31 and 32 show the operation result of the engine generator of the S-2 and S-3 systems in February and August, respectively, with the operation of each system planned for every sampling time. In the S-2 system, if the battery residual quantity falls to 25% or less, the engine generator will be operated. In the S-3 system meanwhile, based on the predicted electricity production of the **solar cell,** the operation of the engine generator is planned so that the power balance for every sampling time need not become negative. However, when there is a considerable operation prediction error, the battery power supply may be insufficient, in which case, the engine generator is operated immediately.

Compared with the engine operation hours of the S-2 system, the time for which the S-3 system is used is short in many respects. There is an excessive lack of **solar radiation** on August 2 and 3, hence the S-3 system operation for those days differs compared with the operation method of other days in the same week. However, the operation on August 3 in the S-2 system also differs from other days in the same week, although this is true to a lesser extent on August 2. The engine operation hours of the S-2 system on August 2 are 8 hours, as compared to 11 hours for the S-3 system. Moreover, despite the fact the engine operation hours of the S-2 system on August 3 total 10 hours, the operation hours of the S-3 system remain the same. The prediction errors of the electricity production of the **solar cell** of S-3 system on August 2 and 3 are 73% and 37%, respectively, as shown in Figures 29 (b) and (c). When there is significant prediction error, the operation time of the engine generator of the S-3 system will become long. However, for example, although the prediction error in Figure 28 (a) is 89%, the operation hours of the engine generator on the same day shown in Figure 31 (b) do not differ from other days in the same week. The fuel consumption of the S-3 system is more disadvantageous than that of the S-2 system, when the prediction error of the **PAS** is significant and the extreme lack of electricity production of the **solar cell** becomes a more frequent phenomenon. When the electricity production of the **solar cell** diminishes, the battery charge volume is insufficient. Moreover, when the prediction error of the **PAS** is considerable, the time shift of the power means a plan cannot be appropriately established, thus increasing the frequency of operation of the engine generator.

Fuel Consumption of Each System

Figure 33 shows the result of the higher calorific power of the fuel consumed with the engine generator of each system. Except for August 2 and 3, the S-3 system shows little fuel consumption when clearly compared with other systems. However, it is difficult to compare the benefits of the S-1 and S-2 systems. When the fuel consumption of the S-1 system for the 1st week in February is set to 1.0, the fuel consumption rates of the S-2 and S-3 systems are 1.02 and 0.85, respectively. Moreover, when the fuel consumption of the S-1 system for the 1st week in August is set to 1.0, the fuel consumption rates of the S-2 and S-3 systems are 0.97 and 0.90, respectively.

conclusions from the 2ND Case Study

An algorithm to predict the production of electricity of the **solar cell** was developed, and **operation plans** in the case of its introduction into the hybrid system of a **solar cell** and a **diesel engine** generator were discussed. The following systems were examined in this paper:

- An engine generator operated according to a load fluctuation system (S-1).
- When operating the engine, operate by a fixed load of maximum efficiency (3kW output power). If the battery capacity drops to 25% or less, the engine generator will be operated and the battery will be charged (S-2).
- Introduce the production-of-electricity **prediction algorithm** (**PAS**) of a **solar cell**, and determine the system operation method. Based on the predicted output characteristic of the **solar cell**, a plan is made for the operation or stop of the engine generator, and the operation of the charge or discharge process of the battery (S-3).

The operation of each system was analyzed using the meteorological data for the 1st weeks in February and August, the differences in the operation results in each system were examined, and the following conclusions obtained.

1. Estimation of relative merits is difficult for the fuel consumption of the S-1 and S-2 systems. Although we can state that the fuel consumption of the S-3 system is 10% to 15% lower when compared with other systems, there is significant prediction error of the electricity production of the **solar cell**, and when this latter declines, the fuel consumption of the S-3 system shows no superiority over other systems.
2. The predictive accuracy of the **PAS** developed in this paper ranged from 4% to 89% in the 1st week in February and August. Even if the error of this range is added to the S-3 system, if the electricity production of the **solar cell** is that expected for an average year, there will be no significant difference in the operational planning within this system.
3. The **operation planning** of the system with **PAS** is demonstrated as effective in terms of reducing the fuel consumption. Compared with the fuel consumption of the S-1 system in the 1st week in February, the S-2 and S-3 systems show a 2% increase, and 15% decrease, respectively. Moreover, compared with the S-1 system for the 1st week in August, they will record decreases of 3% and 10% respectively.

OVERALL CONCLUSIONS AND FUTURE RESEARCH

We think that operation with few consumption of fossil fuel can be planned by introducing **PAS** into the complex system of renewable energy and engine generators. The examination of the energy system introduced in **PAS** is planned from 2009 in Kitami Institute of Technology in Japan.

REFERENCES

Ashari, M., & Nayar, C. V. (1999). An optimum dispatch strategy using set points for a photovoltaic (PV)–diesel–battery hybrid power system. *Solar Energy*, *6*(1), 1–9. doi:10.1016/S0038-092X(99)00016-X

Earth Simulator Center. Japan Agency for Marine-Earth Science and Technology, (2008). Retrieved from http://www.jamstec.go.jp/esc/index.en.html

Ismail, Y., etc. (2002). An operating method for fuel savings in a stand-alone wind/diesel/battery system. *Journal of Japan Solar Energy Society*, *28*(2), 31–38.

Japan Solar Energy Society. (1985). *Solar Energy Utilization Handbook* (pp. 10-88).

Kosugi, T., etc. (1998). Evaluation of output and unit cost of power generation systems utilizing solar energy under various solar radiation in the world. *Transaction of the Institute of Electrical Engineers of Japan . Publication of Power and Energy Society*, *118*(3), 246–253.

Muselli, M., etc. (1999). Design of hybrid-photovoltaic power generator, with optimization of energy management. *Solar Energy*, *65*(3), 143–157. doi:10.1016/S0038-092X(98)00139-X

Narita, K. (1996). *Research on unused energy of cold region cities and utilization for district heat and cooling*. Ph.D. thesis, Dep. Socio-Environmental Eng. Faculty of Eng., Hokkaido Univ. Sapporo, Japan.

NEDO Technical information data base, (2008). *Standard meteorology and solar radiation data (MET-PV-3)*. Retrieved from http://www.nedo.go.jp/database/index.html

Obara, S. (2007a). Improvement of power generation efficiency of an independent micro grid composed of distributed engine generators. *Transactions of the ASME . Journal of Energy Resources Technology, 129*(Issue 3), 190–199. doi:10.1115/1.2748812

Obara, S. (2007b). Energy cost of an independent micro-grid with control of power output sharing of a distributed engine generator. *Journal of Thermal Science and Technology, 2*(1), 42–53.

Obara, S., & Tanno, I. (2008a). Operation Prediction of a Bioethanol Solar Reforming System Using a Neural Network. *Journal of Thermal Science and Technology, 2*(2), 256–267. doi:10.1299/jtst.2.256

Obara, S., & Tanno, I. (2008b). Fuel reduction effect of the solar cell and diesel engine hybrid system with a prediction algorithm of solar power generation. *Journal of Power and Energy Systems, 2*(4), 1166–1177. doi:10.1299/jpes.2.1166

SANYO Co. Ltd. (2007). Nickel-Metal Hydride Production Information, Batteries. Retrieved from http://www.sanyo.co.jp/energy/english/product/twicell_2.html

Yamamoto, S., etc. (2004). An operating method using prediction of photovoltaic power for a photovoltaic-diesel hybrid power generation system. *Transactions of the Institute of Electrical Engineers of Japan, B . Power and Energy, 124*(4), 521–530. doi:10.1541/ieejpes.124.521

Chapter 10
Comprehensive Energy Systems Analysis Support Tools for Decision Making

C. Cosmi
National Research Council, Italy

S. Di Leo
University of Basilicata, Italy

S. Loperte
National Research Council, Italy

F. Pietrapertosa
National Research Council, Italy

M. Salvia
National Research Council, Italy

M. Macchiato
National Research Council, Italy

V. Cuomo
National Research Council, Italy

ABSTRACT

Sustainability of energy systems is a common priority that involves key issues such as security of energy supply, mitigation of environmental impacts - the energy sector is currently responsible for 80% of all EU greenhouse gas emissions (European Environment Agency, 2007), contributing heavily to the overall emissions of local air pollutants - and energy affordability. In this framework, energy planning and decision making processes can be supported at different stages and spatial scales (regional, national, pan-European, etc.) by the use of comprehensive models in order to manage the large complexity of energy systems and to define multi-objective strategies on the medium-long term. This Chapter is aimed to outline the value of model-based decision support systems in addressing current challenges aimed to carry out sustainable energy systems and to diffuse the use of strategic energy-environmental planning methods based on the use of partial equilibrium models. The proposed methodology, aimed to derive cost-effective strategies for a sustainable resource management, is based on the experiences gathered in the framework of the IEA-ETSAP program and under several national and international projects.

DOI: 10.4018/978-1-60566-737-9.ch010

INTRODUCTION

Climate change mitigation as well as sustainable, secure and competitive energy supply are high Community priorities and a need for each Member State, as outlined also by the EU common Energy Policy adopted by the European Council on 9 March 2007 that proposes the following targets and objectives (Commission of the European Communities, 2007):

- Reducing greenhouse gas emissions from developed countries by 30% by 2020; the EU has already committed to cutting its own emissions by at least 20% and would increase this reduction under a satisfactory global agreement.
- Improving energy efficiency by 20% by 2020.
- Raising the share of renewable energy to 20% by 2020.
- Increasing the level of biofuels in transport fuel to 10% by 2020.

Interrelations between sustainable development and energy planning are very strong thus a decisive step for translating the EU's political directions into concrete actions is to work in the direction of promoting sustainable energy systems, capable of fulfilling an increasing and more differentiated market demand but guaranteeing, at the same time, reduced impacts of energy production and use.

Pursuing this aim involves an in depth medium long-term analysis of energy systems and public intervention to boost investments in energy efficiency, renewable energy and new technologies, to increase the capacity of existing infrastructures and to limit demand growth.

It is therefore necessary to foster the adoption of energy planning methods based on analytical tools as well as to promote the implementation of best practices at local, national and supra-national scale, fostering consensus building among the stakeholders .

This chapter is aimed to provide a summary of the rationale of strategic energy planning, to discuss the main features of the proposed methodology and the reference analytical tools as well as to describe some exemplificative results of a real case study.

BACKGROUND

Decision making concerning policy issues aimed to address major energy challenges is a complex process, consisting of many steps and often involving different groups of interest, with different backgrounds, roles and ambitions. Usually, a huge amount of data is managed to describe the current situation as well past trends and future constraints/opportunities and a deep understanding of the pathways along which new energy systems can emerge and develop over time is required (International Institute for Applied Systems Analysis, n.d.).

A wide range of software tools and database is available to support different aspects and aims of energy analysis, in particular, addressing the two main aspects: information management and decision-making. Therefore many tools and methods based on geographic information systems (GIS) and linear programming techniques have been developed to respond to the necessities of energy modelers (Wierzbicki *et al.*, 2000). In this framework, energy system models are fundamental to calculate the implications of certain policy strategies on the energy system, the economy and the environment.

International bodies like the World Energy Council (WEC), the International Energy Agency (IEA) or the Intergovernmental Panel on Climate Change (IPCC) investigate the future of energy systems on a multinational or even global level (Messner and Strubegger, 1995a). Many research institutions and scientists focus their research on the long-term future of energy systems. In this framework a leading position in maintaining and expanding databases on technologies and resources and developing new methods and modeling techniques for exploring alternative energy pathways is played by the International Institute for Applied Systems Analysis (IIASA) (International Institute for Applied Systems Analysis, n.d.) and the Energy Technology Systems Analysis Programme of the International Energy Agency (IEA/ETSAP) (Energy Technology Systems Analysis Programme, n.d..). Since the 1970s a number of energy analytical tools have been developed at IIASA's Energy Program, such as the MESSAGE model, a systems engineering optimization model used for medium- to long-term energy system planning, energy policy analysis, and scenario development (Messner and Strubegger, 1995b).

The IEA/ETSAP promotes since it was established, more than two decades ago, within specific Annexes the co-operation among researchers and authorities and the utilization of quantitative methods for policy assessment, aimed to achieve important advances in the representation of local, national and multi-country energy systems as well as the analysis of their medium-long term development under different policy scenarios. A short review of recent studies carried out by the ETSAP international communities in the period 2005-2008 is reported in Annex X "Global Energy Systems and Common Analyses" (Goldstein & Tosato, 2008).

The ETSAP models generators are currently largely used worldwide in many national and international projects at a Pan-European scale, among which, the FP6 NEEDS (New Energy Externalities Development for Sustainability, n.d) Integrated Project, the IEE RES2020 (Monitoring and Evaluation of the RES directives implementation in EU27 and policy recommendations for 2020, n.d.) Project, and the FP7 REACCESS (Risk of Energy Availability: Common Corridors for Europe Supply Security, n.d.) project. At local scale, the IEA – ECBCS Annex 33 on Advanced Local Energy Planning (Jank *et al.,* 2005) was addressed to develop a decision support methodology for strategic local energy planning, based on the integration of MARKAL with sub-system energy models and on a participative involvement of affected groups, and to apply it in several local case studies around Europe with different planning framework and purposes (Cosmi *et al*., 2000). Recent applications are also addressed to develop sustainable energy strategies in the context of the liberalization of the energy markets in Europe and climate change protection policies, taking into account the international dimension for energy and environment policies (Fahl, *et al.,* 2004, Van Regemorter, 2005).

Other models have been developed under the auspices of the European Commission, among which the Energy Flow Optimization Model (EFOM), a linear programming energy model for the assessment of national energy strategies and its extension EFOM-ENV developed to support studies on energy related environmental strategies, and PRIMES, a general purpose model conceived for forecasting, scenario construction and policy impact analysis (Capros, n.d.).

In particular, the EFOM-ENV "bottom-up" models are national dynamic optimization models representing the energy producing and consuming sectors in each Member State. They optimize the development of these sectors under given fuel import prices and useful energy demand over a pre-defined time horizon. The model data base contains a wide range of conversion and end use technologies and power and energy conservation measures in the demand sectors. (Worldbank, n.d.). On the other hand, PRIMES, is a large scale model of the energy systems of European Union member-states linked together through energy markets model providing simulations of the energy system and the decisions of the agents and the markets, covering in detail several sectors, uses, and technologies (Capros et al., 1999).

MESSAGE, a systems engineering optimization model used for medium- to long-term energy system planning, energy policy analysis, and scenario development (Messner and Strubegger, 1995b). In order to facilitate the assessment of salient feedbacks between the energy system and the economy as a whole the MESSAGE model has been linked to MACRO, the macroeconomic module of the top-down macroeconomic model MERGE (Manne et al. 1995).

MESSAGE, EFOM-ENV and MARKAL/TIMES, the latter extensively described in the following section, belong to the class of least-cost optimization models, that share many common features integrated assessment of a national energy system including environmental consequences but vary in the user supporting system and the level of detail (Voogt *et al.,* 2000).

An exhaustive documentation on the most diffused energy system models can be found in literature (Bardouille, 1998; Beeck, van, 1999; Cuomo *et al.*, 2000; WorldBank, n.d.; COMMunity for ENergy environment & Development, n.d.; Electric Power Research Institute, 1998; Institute for Energy Engineering - Technical University of Berlin, 2003; Schlenzig, 1998).

Energy System Modelling

A model, as well known, is a simplified abstraction of the system under investigation that consists of a description of its structure, data and by a set of mathematical equations that can be based on different mathematical methodologies (e.g. simulation, linear or non-linear optimization, etc.) (Cuomo et al., 2000). Energy models are utilized both for representing the whole energy systems and for specific purposes like investment calculation and operation planning for conversion plants and they can be classified according to different criteria, as exhaustively reported in literature (for instance, Worldbank, 1991; Electric Power Research Institute, 1998; Bergman & Henrekson, 2003).

Representing and analyzing energy systems by mathematical models is, in particular, of utmost importance in decision making processes as it allows managing efficiently large amounts of data, providing at the same time a common basis for the discussions, with transparent and easily accessible methods of calculation, input data and assumptions. Thus, comprehensive modeling can help energy analysts and planners in revealing the strengths and weaknesses of the present energy system and in identifying needs, threats and opportunities for the future. Moreover, these models are interactive and support communication since new ideas and questions can be evaluated very quickly varying data input and/or modifying the set of user-defined constraints and comparing the related results (sensitivity analysis).

Furthermore, long-term strategic planning can be usefully combined with subsystem analysis, with additional tools for a more detailed investigation of specific system components, to support comprehensive models in devising a single clear decision between two competing technologies or more specific solutions for subsystems to be embedded in the general framework.

The Main Planning Phases

Energy planning is not only a technical issue. This implies that, it is necessary to take into account also social and political factors in order to achieve the necessary consensus. As concerns the organizational set-up, it should be implemented a cooperative process involving, at the various stages, experts in the subject area, modelers, and decision support experts, leading to an iterative process as the refinement and improvement of strategies is accomplished through continuous communication and discussion of partial results and new ideas between the groups (Jank *et al.*, 2005). As stakeholder involvement is essential for the success of the project, additional efforts should be devoted to promote different initiatives aimed to involve decision makers and to share their view and experience. To this extent different target groups should be identified, including public authorities stakeholders, policy makers, energy planners, market actors and analysts, energy agencies, researchers. All these groups can suggest the policy priorities and discuss the operative implementation of planning strategies, taking advantage from the different points of view and experiences as well as by an improved knowledge of the local systems. The guidelines derived from this information exchanges among the different groups, constitute a sound basis for the decision making processes on medium-long term, promoting energy efficiency and encouraging the application of best practices.

In brief, the main phases involved in energy planning are:

- **Preparation and analysis:** It is aimed to focus on the main objectives of the planning process and to set the boundaries of the modeling framework. The foreseen outputs are a detailed list of the problems/features that will require a more detailed analysis, the definition of the overall planning objectives, budget and duration of the process, the organizational set up and roles, and some preliminary indications on possible development scenarios.
- **Characterization of the case study:** By a detailed description of energy supply and use, with reference to infrastructures, availability of present and future technologies, energy needs by end use, environmental impacts and, if possible, their related externalities. A structured and updateable database is envisaged for cataloguing the gathered information and to allow the users their utilization in further studies.
- **Setting up of the modeling environment:** The large variety of models actually available makes the choice of modeling tools and methods a key step that has to be carefully evaluated before undertaking the energy analysis. Firstly the Reference Energy and Materials System (REMS) is set up, a flow diagram describing the essential energy and material flows through the analyzed technology network, from resource extraction, to demand for useful energy services through energy transformation and end-use devices (Figure 1). In order to prepare the basic model's data input a set of techno-economic and environmental model parameters (e.g. technology's efficiency, capital cost, unit emission factor ect.) is associated to each element of this network. Moreover a number of modeling assumptions and superimposed constraints are introduced to fully describe, in a realistic way, the energy system and its envisaged future development, with regard to energy demand and socio-economic parameters (e.g. population, GDP, money discount rate, etc). After the basic model implementation, an user interface allows to automatically translate the whole data input into model's equations, to then solve them by an optimization routine that select, among all the possible combinations of fuels and technologies, the "best" one for each time period in order to minimize the total system cost over the entire planning horizon. The first step of results analysis deals with the calibration of the model to verify its reliability in terms of model's answer. In this phase, the correspondence among the models results and statistic data has to be verified, with particular regard to the energy and materials balances of the base year (usually the latest year for which reliable data are available) and to the standard demand projections. This is a critical issue both for tuning the model in order to obtain sound results in agreement with the reference database and for identifying the key variables for scenario analysis.
- **Scenario analysis:** Once the comprehensive model is calibrated the next steps are addressed to

the definition of alternative hypotheses for the development of the energy system (scenarios) and the comparison of the alternative sets of results for devising robust policy strategies. The basis of comparison among scenarios is provided by the reference scenario (BASE), that is the most likely representation of the existing system along with the expectations of its future development, following a "business as usual trend". Starting from the basic set of assumptions, the key issues to be examined are identified and the alternative scenarios are specified in terms of quantitative parameters. In this phase, a participative process involving modelers and stakeholders is envisaged to set policy scenarios of general interest and setting measurable targets to be achieved on the medium-long term (e.g. percentages of reduction of CO2 and local air pollutant emissions). Among the full set of possible choices a selected number of representative scenarios can be identified. The scenario assumptions are then translated into exogenous model's constraints for the analyzed energy system and the model runs by scenario are compared to analyze the consequences upon the model results in terms of equilibrium quantities (optimal mix of fuels and materials as well as technologies and demand levels), CO2 avoidances, GHG and pollutant emissions reduction. A critical review of the scenarios results allows individuating among the optimal energy-technology development paths by scenario, the "robust solutions", defined by those elements which are less sensitive to exogenous changes for the formulation of policy recommendations and the individuation of tailored strategies on energy, waste and air quality. Scenarios analysis provides, therefore, a precious basis of knowledge for assessing different normative and tariffs strategies as well as for supporting local actors and energy planners in decisional processes.

- **Energy plan implementation and monitoring:** The latest steps of a planning process deals with the implementation of the devised strategies, identifying policy measures and incentives that allow translating the model results into concrete actions. Moreover being energy planning a dynamic process, the monitoring phase is essential to check the achievement of the planning objectives and to improve/adjust the planning strategies according to the new inputs arisen after their actuation and changing framework conditions (Jank et al., 2005). In this context, it could be useful to select suitable indicators for a qualitative and quantitative evaluation of the effectiveness of the devised strategies and to identify the sectors and/or territorial areas that require ad hoc solutions.

Methods and Models

The ETSAP MARKAL and TIMES Models Generators

The ETSAP MARKAL/TIMES models generators are currently widely used for studies at different spatial scales, as they have proven their flexibility, effectiveness and robusteness of results in very different applications ranging from global multi-region to community level. MARKAL (an acronym for MARket ALlocation) and TIMES (The Integrated MARKAL-EFOM System) were developed in the framework of the Energy Technology Systems Analysis Programme (ETSAP) of the International Energy Agency and constitute powerful comprehensive tools for representing and analyzing the evolution of a specific energy system over a medium-long time horizon (usually a period of 40 to 50 years, split in several time periods of fixed or variable length, typically 3 – 5 years). MARKAL and TIMES are models generators, that is, they generate multi-period, bottom-up, demand-driven optimization models based on mathematical programming (Fishbone & Abilock, 1981; Fishbone et al., 1983; Loulou et al., 2004; Loulou et al., 2005). The energy system evolution is modeled by defining different sets of data

Figure 1. A schematic representation of energy flows from supply sources to demand sectors for the REMS definition

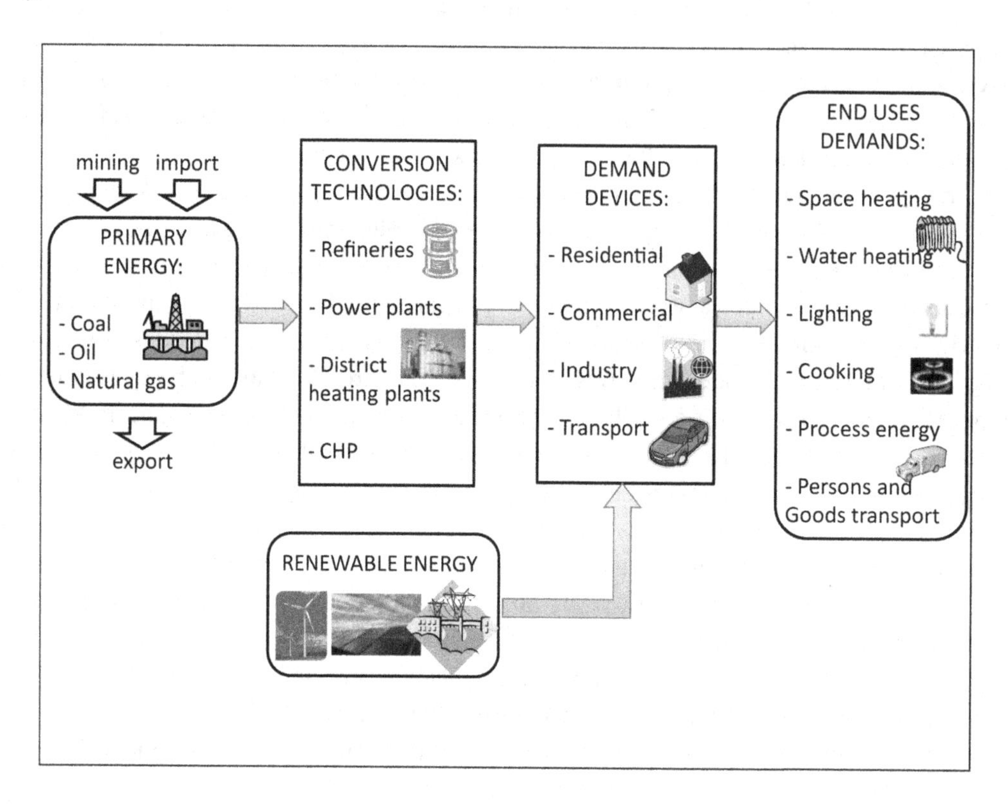

input for each period to take into account the subsequent availability of resources and different boundary conditions. This feature is of fundamental importance to follow technology development and to compare different scenarios.

The general modeling platform for a MARKAL/TIMES based model is actually constituted by the following components (Energy Technology Systems Analysis Programme, n.d.):

- The source code MARKAL or TIMES, which processes the set of data files that fully describes the energy system (technologies, commodities, resources and demands for energy services) generating the matrix that specifies the mathematical programming problem, and post-processing the optimization results.
- The General Algebraic Modeling System (GAMS) the computer programming language in which the source code is written (General Algebraic Modelling System, n.d.)
- A "solver", the software package integrated with GAMS to solve the mathematical programming problem set by the model.
- A "shell", i.e. an user interface that allows creating, managing the data input and running the model generator and analyzing results (e.g. the Versatile Data Analyst-VEDA Front End-FE (KanORS Consulting Inc., 2003a) and Back End-BE (KanORS Consulting Inc., 2003b), as well as the ANSWER shell (Noble, 2005)).

Available resources, economic, technical and environmental characteristics of technologies together with useful energy demands are the main input parameters in a MARKAL/TIMES model. Technologies are modeled as processes and transform one energy carrier into another. They are described by economic parameters, that include investment and maintenance costs, life duration, date of availability of new technologies and the residual capacity of existing technologies; technical parameters, among which, the efficiency of the technology (expressed as input/output ratio); whereas emission factors are introduced in relation with each fossil fuel and/or technology.

Due to new installations and decommissioning of old capacity, the market shares of different technologies vary throughout the periods. The salvage values of existing capacities at the end of the time horizon are taken into account, and all costs are discounted to their present value by a given discount rate fixed by the user (Mäkelä, 2000).

Additional constraints can be set on the atmospheric pollutant emissions as well as on annual energy carrier balances, seasonal district heating balances, diurnal electricity balances, annual availability and demand equations.

Among the numerous model versions actually available (Seebregts *at al.*, 2002), the multi-regional one allows to link multiple country specific models including trade of emission permits (Bahn et al. 1997) and is a valuable tool to evaluate the consequences of liberalization of energy markets analyzing, for instance, the impact of trade on the electricity price, on the choice of technologies and on the environmental conditions (Van Regemorter, 2005).

The principal outputs of the models deal with:

- total and annual energy system costs,
- investments and capacity utilization of technologies,
- primary energy, final energy - by sector and/or by fuel,
- air pollutants emissions - by fuel, sector and end-use sector,
- average and marginal emissions prices,
- electricity generation mix– by fuel and by technology,
- imports, exports & domestic production of fossil & renewable fuels,
- use of processes and energy carriers

MARKAL and TIMES are worldwide utilized for perspective analyses and for the assessment of the effectiveness of policy strategies in response to energy – environmental directives, as they allows a quick comparison of different scenarios identifying, for each of them, the cost-effective responses in terms of policy measures (e.g. optimal energy-technology pathways in compliance with emissions targets, emphasizing role of new technologies and effects on the economy of regulations, taxes, and subsidies, etc.).

Mathematical Background

MARKAL and TIMES generate from each set of data files a matrix with all the coefficients that specify the economic equilibrium model of the energy system as a mathematical programming problem.

The objective function of the primal problem (EQ_OBJ) built up with a bottom-up approach from demand categories and technological options, represents the total discounted cost of the energy system i.e. the sum of the discounted annualized costs minus the annualized revenues of an energy system, as specified in the following equation (Mäkelä, 2000):

EQ_OBJ = (1)

+ Costs for sunk material

+ Investment costs

+ Fixed costs

+ Variable costs

+ Taxes

+ Surveillance costs

+ Decommissioning costs

- Subsidies

- Recuperation of sunk material

- Salvage value

In a vectorial notation, the primal problem can be written as:

$$C = \mathbf{c}\,\mathbf{X} \tag{2}$$

$$\mathbf{X} \geq \mathbf{0} \tag{3}$$

$$\mathbf{AX} \leq \mathbf{B} \tag{4}$$

Where:

- **c** is the vector of the activities' costs per unit of activity, actualized to a base year (usually the middle year of the first time period);
- **X** is the vector of the anthropogenic activities, characterized by the fuels and the technologies utilized as well as by the pollutants emitted
- **A**, a [n, m] matrix, contains all the coefficients necessary to define the constraints of the local energy system environment;
- **B** is the vector of the exogenous constraints.

A least cost solution is then determined by minimizing the total system cost in compliance with the "primal constraints' system" represented by the set of linear inequalities of the equations (3) and (4) (i.e.

the non-negativity of activities, the fulfillment of end-use demands, other relationships among supply and demand, additional exogenous constraints defined by the model user representing the boundary conditions, such as availability of technologies and fuels, maximum levels of emissions allowed, etc.).

The minimum cost optimal solution select for each set of data the best allocation of activities, identifying the fuel mix and the set of technologies which are able to satisfy the given demands at the minimum feasible cost, taking into account the super imposed constraints.

In brief, MARKAL and TIMES share the same basic modeling paradigm, being technology explicit, dynamic partial equilibrium models of energy markets. In both cases the equilibrium is obtained by maximizing the total surplus of consumers and suppliers via Linear Programming. The two models also share the multi-regional feature, which allows the modeler to construct geographically integrated (even global) instances (Loulou *et al.*, 2005).

However, there are also significant differences in the two models, that do not affect the common basic paradigm, but rather some of their technical features and properties as discussed in (Loulou *et al.*, 2005).

The Role of Auxiliary Tools

Auxiliary tools can provide three main contributions in the different planning phases:

1. Data collection and characterization of the analyzed case studies: to allow management, retrieval and update of information
2. Subsystems analysis: to allow an in depth analysis of specific components or subsystems, reducing the number of equations and allowing more homogenous representation of sub-sectors (the main results can be fed in the comprehensive model)
3. Data and results representation: in particular this deals with the REMS preparation, for which specific software can be used to schematize the network of technologies and energy/materials flows, as well as with a geographic visualization of the models results, in which GIS based tools could be helpful.

The main general data on national and local energy systems (by region, by province) as well as other socio-economic aspects of interest can be derived by databases provided by the national statistical offices as well as by the Statistical Office of the European Communities (The Statistical Office of the European Communities, n.d.). Data on operating, maintenance and investment costs, as well as technical information on energy technologies are usefully provided by energy information systems, such as (Schlenzig, 1998):

- CO2DB: a database on carbon mitigation technologies (Strubegger, 2003);
- DECPAC/DECADES - Databases and Methodologies for Comparative Assessment of Different Energy Sources for Electricity Generation (DECADES, 1993);
- IKARUS - Instrumente für Klimagas Reduktionsstrategien (Forschungszentrum Jülich, 1994), a comprehensive techno-economic databases developed for the German Ministry for Technology and Research (Eichhammer, 1999);
- ENIS: Energy Information System on Time Series, Energy Balances and Technology Data, included in MESAP (Baumhögger et al., 1998).

Subsystem analysis, aimed to the design of some parts of energy systems as well as to integrate comprehensive models, can be usefully supported by a large number of tools, with different scope, methodology and platform. As an example, auxiliary modeling tools can be applied to respond to the following requests:

- To optimize sustainable or conventional energy conversion or storage devices according to different criteria (e.g. costs, primary energy use, CO2 production). One of the tools aimed to achieve this goals is PRODESign, the PRogram for Optimisation and Dimensioning of Energy conversion Systems, developed by the Energy Research Centre of the Netherlands (ECN) (Ouden, den, 2000).
- To determine the optimal long-term expansion plan for a power generating system. WASP, the Wien Automatic System Planning Package (International Atomic Energy Agency, 1980), is the most widely used model in developing countries for this scope .
- To analyze and optimize industrial energy supply systems. As an example, TOP-Energy, the Toolkit for Optimization of Industrial Energy Systems (Augenstein *et al.,* 2004), developed by the Engineering Thermodynamics Group (LTT) of Aachen, respond exhaustively to this request.
- To help plan, analyze, and operate district heating and cooling systems, as done by HeatMap, District Energy Analysis Software for Steam, Hot-Water, and Chilled-Water Systems (Washington State University - Extension Energy Program, n.d) developed by the Washington State University Extension Energy Program.
- To optimize local and national waste management systems. This is the main aim pursued by the MIMES/Waste model, developed by Chalmers University of Technology (Sundberg and Wene, 1994).
- To assess the environmental impacts and resulting external costs from electricity generation systems and other industrial activities. This is the main aim of EcoSenseWeb (Institute of Energy Economics and the Rational Use of Energy - University of Stuttgart, n.d.), an integrated computer system developed in the ExternE-Project and based on the Impact Pathway Approach (IPA). It provides relevant data and models required for an integrated impact assessment of emissions to air, soil and water, considering different impact categories (human health, crops yield loss, damage to building materials, loss of biodiversity and climate change, Preiss, 2007).
- To carry out Life Cycle Assessment (LCA) studies (The Society of Environmental Toxicology and Chemistry, n.d.). There are many software support tools available to purpose this aim as reported in (Eco-smes - Services for green products, n.d). One of the most widely used LCA software tool is SimaPro (PRé Consultants., n.d.), used also in the following applicative example that fully integrates also the Ecoinvent database that contains international industrial life cycle inventory data on energy supply, resource extraction, material supply, chemicals, metals, agriculture, waste management services, and transport services (Ecoinvent Centre, n.d.).

Geographic Information Systems (GIS) technologies can be very useful in energy-environmental planning, for their capability to handle large amount of data and to graphically represent and visualize them effectively (Cuomo *et al.,* 2000).

GIS are used, for instance, to provide guidance for planners and developers in relation to the development of renewable energy schemes, outlining the key renewable energy resource opportunities and constraints, as done in North Yorkshire (Land Use Consultants, n.d.), or to help decision-makers in their assessment of public policies and strategies concerning urban air quality policy, aim pursued by the AIDAIR Information and Decision Support System (University of Geneva, 1998).

An interesting example of integration among different data bases and modeling tools in provided by IIASA, where beside the main tool MESSAGE (Model for Energy Supply Strategy Alternatives and their General Environmental Impact), linked to the top-down macroeconomic model (MACRO), a simulation model is used to help formulate scenarios of economic and energy development whereas a climate model helps to estimate aggregate climate impacts of E3 scenarios (The International Institute for Applied Systems Analysis, n.d.). A database on carbon mitigation technologies (CO2DB) provides the data input necessary to derive the main technical, economic and environmental characteristics as well as data on innovation, commercialization and diffusion. A soft link with the 'Regional Air Pollution INformation and Simulation' (RAINS)-model, that describes the pathways of emissions of sulfur dioxide, nitrogen oxides and ammonia and explores their impacts on acidification and eutrophication (Amann et al. 1996), it is useful for an integrated assessment of alternative strategies to reduce acid deposition.

INTEGRATION OF MARKAL, LCA AND EXTERNE IN THE VAL D'AGRI LOCAL SCALE APPLICATION

In order to clarify the potentialities and operating aspects of comprehensive models in energy-environmental planning, a local scale application is herewith presented. This work was aimed to support the definition of sustainable energy-environment strategies focused on the reduction of the whole environmental burdens related to the Val D'Agri (Basilicata region, Southern Italy) energy system. The case study considered is a place of naturalistic interest that has been interested, in the last decades, by a huge development of oil mining activities, representing the largest oilfield of Italy.

In the following, the energy system structure (reference energy system and data input), the main modeling assumptions, and the main results obtained in the scenario analysis are briefly discussed. More details can be found in (Pietrapertosa *et al.*,2003) and in (Pietrapertosa *et al., in press*).

Setting Up of the Modeling Environment

Characterization and Modeling of the Energy System

The first step of this study dealt with the acquisition of an in depth knowledge of the Val d'Agri energy system. The characterization of energy flows is a fundamental step in the analysis of the supply and demand systems. Data collection in a sub-regional area (29 towns located in two provinces, nearly 73,000 inhabitants representing 12% of the total regional population) is not an easy task, being available most of energy statistics only at a higher level (regional or, at least, provincial) and thus requiring data processing based on specific proxy variables (population being the simplest) and cross-checking the final assumptions with *ad hoc* surveys.

In 1997, the base year of the analysis, the total production of electricity was about 163 GWh/year, of which the most part (71%) produced by hydropower plants and only 18% by fossil-fired and 11% by biomass power plants. The estimated future contribution was calculated according to the guidelines of the Regional Energy Plan, which fosters towards the realization of mini-hydroelectric plants, photovoltaic panels and a biomass plant (using forest waste products and grain field residues).

In the same year, energy demand of the main macroeconomic sectors was about 40,16 ktoe, representing about 5% of the regional energy demand. Industry and Residential were the most energy consuming sectors (representing respectively 36% and 34%), followed by Agriculture (18%) and Commercial and Services (12%). More details on the Residential and Industrial energy demands are represented in Figure 2. As concerns the fuels mix, electricity was the most used energy carrier (39%), followed by gaseous fuels (26%), fossil liquids (23%) and solids (13%). This preliminary analysis highlights the important role assumed by fossil fuels to meet the energy demand in the system under study.

Based on the local energy balance and on the development hypotheses outlined by the Regional Energy Plan, an accurate description of the reference technological network for the Val d'Agri, the Reference Energy and Material System (REMS), was carried out (Figure 3). In the adopted REMS representation, the right side shows the main end uses demands, grouped in six demand sectors: Agriculture, Industry, Residential, Commercial and Services, Waste Management and Oil mining activities. Each demand is met by a set of end use devices (DMD), which can be linked to processes (PRC) and conversion technologies (CON). On the left of the network, energy sources (fossil fuels and renewable sources) and wastes (Municipal Solid Waste–MSW, Agriculture and Industry waste) are represented. As concerns waste management, taking into account the national directives, landfills are progressively substituted by an integrated system of technologies aimed to waste valorization and pollution prevention.

Using the ANSWER shell (Noble-Soft Systems, n.d.) each element of the REMS (technology, fuel and material) was characterized by technical, economic and environmental data, with reference to the base year as well as to certain periods of the time horizon for which specific data and assumptions were available.

Integration with LCA and ExternE Data

In order to perform an in-depth analysis of the Val D'Agri energy system (Pietrapertosa *et al.*,2003) and to determine the overall environmental impact due to the different life cycle phases, with particular regard to oil extraction activities and conversion technologies, an innovative approach was applied, based on the integration of MARKAL models generator with Life Cycle Assessment (LCA) and ExternE results.

Figure 2. Residential energy demand, by end-use (a) and sharing out of the Industrial energy demand among sub-sectors (1997)

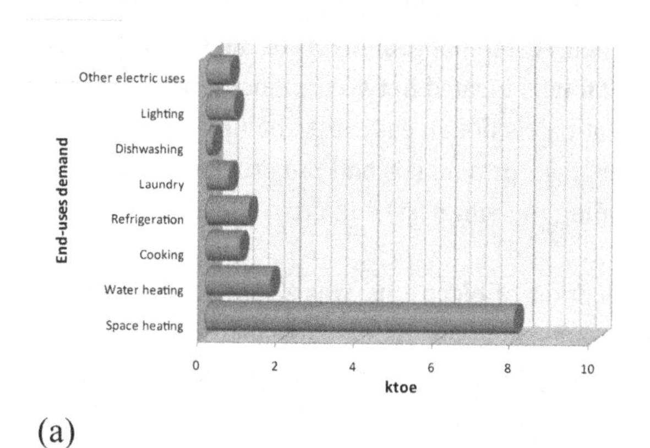

(a)

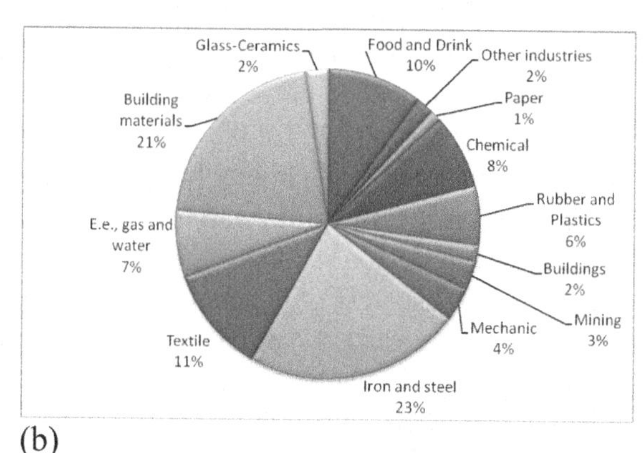

(b)

Figure 3. Schematic representation of the REMS for MARKAL-VdA: the grey boxes represent renewable energy sources and technologies.

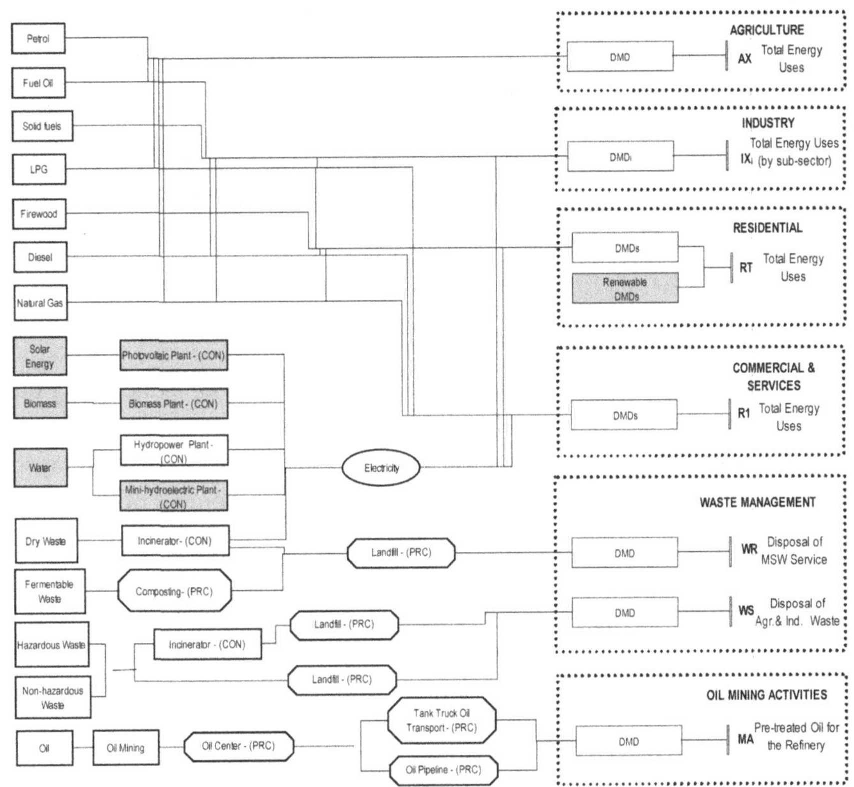

The integration of comprehensive tools such as MARKAL with LCA, a methodology largely utilized by the international scientific community for the assessment of the environmental performances of technologies, can be useful for estimating in much detail the environmental impacts, allowing a "cradle to grave" characterization of sub-systems in the framework of the global energy system. On the other hand, the internalization of external costs of atmospheric emissions can be useful to identify the best paths for implementing technology innovation and strategies for a more sustainable energy supply and use.

In this study, the LCA of oil mining activities, and conversion technologies was performed, estimating their impacts by using respectively the SimaPro (PRé Consultants, n.d) and GEMIS (Öko-Institut, 2008) software. The results were subsequently fed by a soft linking into the MARKAL - Val d'Agri model, inserting the impact parameters as new variables. The global impact of anthropogenic activities was thus assessed in terms of a selected set of indicators, taken by the "Core set of Indicators" of the European Environment Agency – EEA (total energy uses, total energy use per fuel, energy production from renewable sources, total emissions of several pollutants, etc.) in order to identify the key parameters of the analysis (European Environment Agency, 2003). This work was in particular addressed to include in the model the impacts due to the construction and dismantling phases of technologies that could have a fundamental role in assessing their sustainability as well as in defining sound policy strategies on a medium –long term. Moreover in order to take into account the main health and environmental damages due to the analyzed energy system, externalities were introduced as taxes per unit of primary pollutant emitted.

Model's Calibration and Preparation to Scenario Analysis

Processing the data files that characterize the energy model (including also LCA/ExternE data), the MARKAL source code, written in GAMS, generates a matrix that sets the mathematical programming problem and solves it. If a feasible solution is obtained by the optimization routine, results are imported into answer and carefully analyzed in order to report on them. First runs are always aimed to "calibrate" the model, that is to verify the correspondence of results to the base year energy balance and to specific values on the time horizon. Eventual verified anomalies are thus corrected acting on data input and assumptions, in an iterative way, as represented in Figure 4.

Scenario Analysis

Taking into account the hypotheses outlined by the Regional Energy Plan concerning the development of electric power generation technologies (in particular those using renewable energy, e.g. photovoltaic, mini-hydro, wind and biomass), a reference scenario (BAU) was implemented to calibrate the MARKAL-Val d'Agri model and to point out the unconstrained optimized development of the energy system with reference to the standard commodities and to the additional environmental parameters (aggregated impacts indicators) provided by LCA.

Beside the reference scenario, three environmental scenarios were analyzed: CO_2, Impacts and Eco-taxes. The CO_2 scenario includes three cases with increasing constraints on CO_2 emissions (from 1% to 5% of the whole time horizon BAU levels); the Impacts scenario, is made up of four cases that analyze the effects of exogenous constraints on aggregated impacts indicators (respectively, Greenhouse effect, Acidification, Smog, and a combination of these three indicators-Mix), whereas in the Eco-taxes sce-

Figure 4. Schematic representation of the modeling process for the Val d'Agri case study.

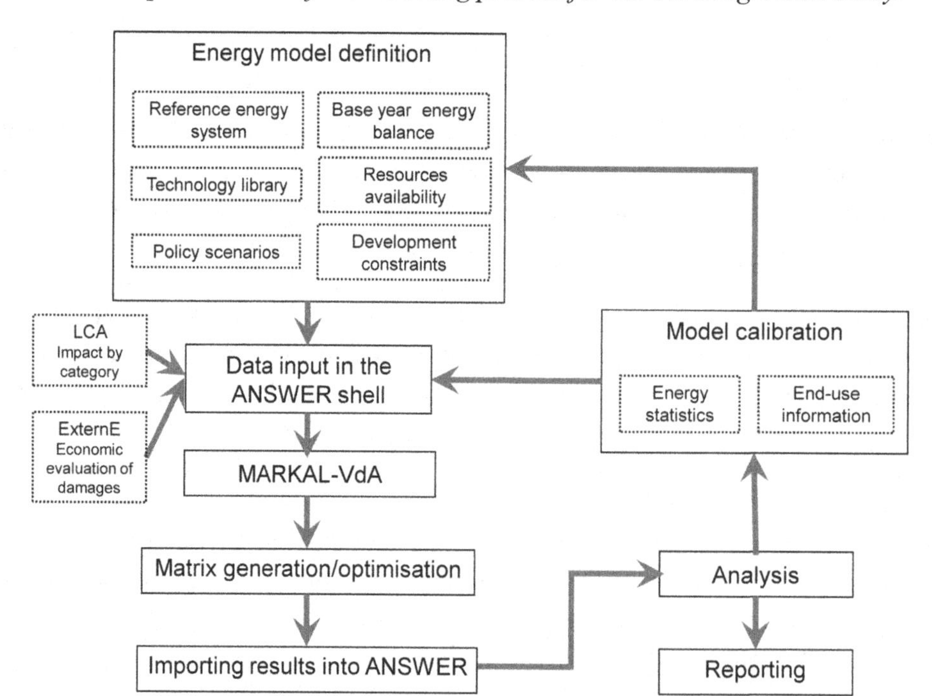

Figure 5. Electric energy production by fuel (EE25- Indicator of EEA): a comparison among the four cases of the Impacts scenario.

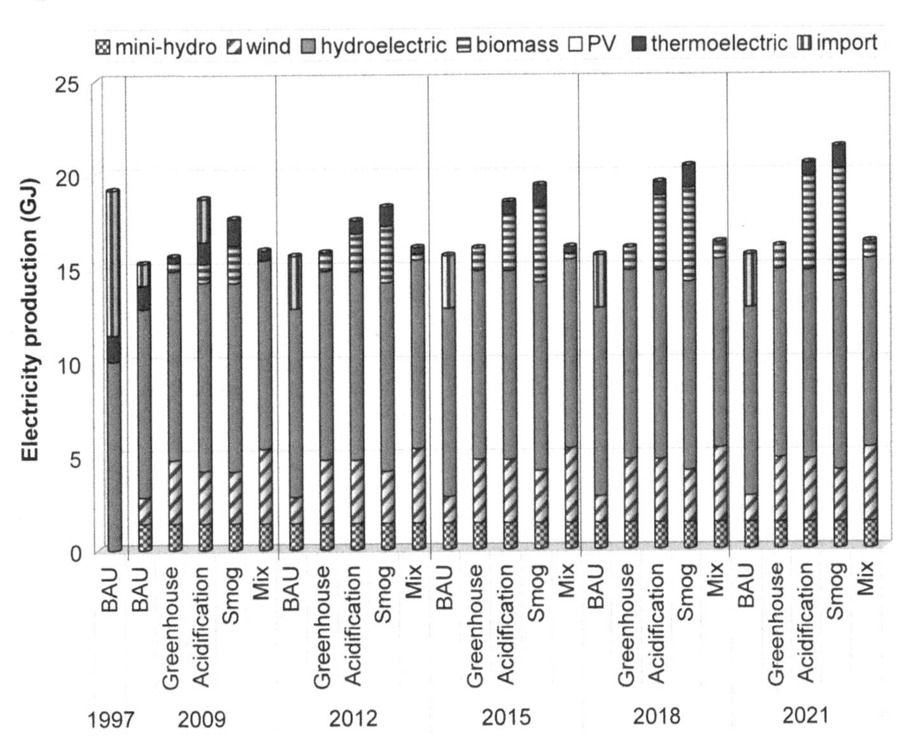

nario, external costs were introduced as taxes on CO_2, NOx, SO_2, TSP e VOC emissions, one at a time and then all together (Tax-TOT case), to evaluate the economic impact of environmental pollution and the effectiveness of environmental taxes in mitigation strategies. None of the analyzed scenarios takes into account the effects of the liberalisation of the energy market, that is still limited in Italy where the final users, at least in the domestic sector, have no choice as to the original supplier of the electricity or gas (Farinelli et al., 2005).

As concerns the final energy consumption, a 10% reduction of the total amount can be observed in the BAU case, due to an optimized use of resources, no regret insulation interventions in Residential and to the increase of efficiency in end-uses devices (boilers, electrical appliances).

The observed energy reduction increases about 1% in the environmental constrained scenarios, in which some changes in resource use can be observed, with particular regard to electricity production. The reduction of anthropogenic environmental impacts is in fact obtained by increasing the endogenous production of electricity by renewable (+17%), in particular wind energy and mini-hydro, which substitute the thermal plants and the imports from neighbour regions (e.g. Figure 5).

The environmental constraints determine obviously an increase of the total discounted energy system cost which ranges from 2 to 15% (in detail, Smog: +1.5%, Acidification: +1.8%, Mix: +8.7%, Greenhouse effect: +15.2%).

Taking into account the total emission levels achieved in the three CO_2 scenario cases (Figure 6), the average cost of CO_2 reduction is about 167 Euro/ton. This value is comparable with the average Italian costs, that is about 60 Euro/ton CO_2 (Ente per le Nuove Tecnologie, l'Energia e l'Ambiente,

Figure 6. Total system cost increase vs. CO_2 abatement percentages

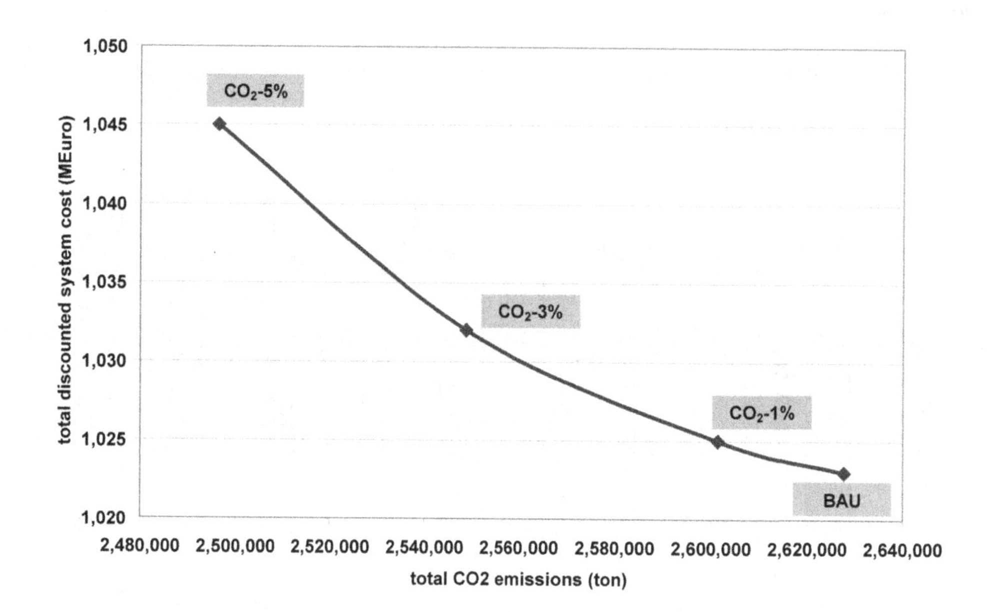

2000), which are estimated imposing a weaker constraint on CO_2 emissions (half of the national Kyoto Protocol target).

The introduction of environmental taxes has consequences on the emissions levels of the related pollutants depending both on the ecotaxes values and on the availability of alternative technologies and fuels. Moreover, the ecotaxes induce also a variation of the emission levels of the untaxed pollutants: as shown in Table 1.

A comparison of the results obtained in the different scenarios allows deriving some general considerations:

- the reduction of energy consumption is a privileged tool for defining sustainable energy-environmental policies and can be obtained by an increase of efficiency and energy saving interventions;
- an increasing use of renewable energy sources is a key issue for improving air quality, allowing a reduction of atmospheric pollution due to combustion processes and reducing energy dependence, which represents one of the main objectives of the European Commission;
- the evaluation of environmental impacts should be based on the overall life cycle of goods and services, taking into account also construction and disposal phases, in order to avoid wrong conclusions related only to the use phase;
- the cost increase due to the necessity to reduce pollutant emissions and the cost gap among traditional and innovative technologies should be evaluated also with reference to external costs, in order to take into account environmental benefits related to avoided emissions;
- Eco-taxes can be an effective tool in environmental and economic terms to reorient consumers and enterprises towards eco-compatible products, processes and services.

A further analysis of the solution is envisaged to individuate the "robust" strategies and to define the priorities in the local energy environmental strategies.

Table 1. Total emissions (ton) in the Ecotax scenario.

Cases	TSP (ton)	VOC (ton)	NOx (ton)	SO_2 (ton)	CO_2 (ton)
BAU	3,345	38,371	155,589	15,256	2,627,645
TAX-CO2	3,350	38,382	156,675	15,237	2,614,431
TAX-NOX	3,339	38,361	154,488	15,260	2,628,419
TAX-SO2	3,346	38,371	155,625	15,208	2,648,299
TAX-TSP	3,345	38,371	155,585	15,255	2,627,645
TAX-VOC	3,345	38,371	155,585	15,255	2,627,645
TAX-TOT	3,339	38,360	154,480	15,236	2,615,256

CONCLUSION

Since 1978, when the International Energy Agency adopted the newborn MARKAL and created the Energy Technology and Systems Analysis Program (ETSAP) to oversee its development, many advances have been made in the ability to analyze local, national and multi-country energy systems and to support energy analysis and planning with multi-criteria evaluations. In the latest years, significant progresses have been made in terms of the size and complexity of the models that can be solved as well as the issues that can be addressed. These tools are also the core of large research projects that involve many institutions (both public and private), Industry, NGOs.

This Chapter was aimed to provide a review of the main modeling tools actually available worldwide, with special focus on the MARKAL/TIMES models generators, and to explain their role in the framework of the overall energy-environmental planning process.

The Val D'Agri case study shows an innovative attempt of overcoming the typical boundaries of energy system modeling in order to extend the analysis to the whole life cycle of energy and waste flows as well as of electric power generation technologies. Moreover the economic evaluation of the energy system takes into account also externalities related to greenhouse gases and local air pollutants emissions. To achieve this goal the MARKAL comprehensive model was integrated, via a soft-link with LCA and ExternE sub-systems results. The scenario analysis shows once more that efficiency increase and energy saving are privileged tools for driving a steady reduction of energy consumption, whereas renewable energy sources have a key role in the supply system but need an in depth characterization of the construction and dismantling phases, that may contribute heavily to environmental damage. This is well provided by the LCA evaluation of environmental impacts whereas the introduction of externalities outlines the importance of eco-taxes in the estimation of fair prices of resources and for promoting the use of eco-compatible technologies and resources.

ACKNOWLEDGMENT

The *Val d'Agri Case study* was performed by the National Institute for Physics of Matter – INFM (from June 2003 merged into the National Research Council - CNR) in the framework of the project "Development of Industrial Districts for Earth Observation - COS (OT)" - research stream OR3: "Implementation of comprehensive models for supporting energy-environmental planners and decision makers in the definition of local scale strategies aimed to the reduction of local air pollutants and GHGs", funded by the Italian Ministry of University and Research (MIUR) under the Italian National Operative Programme (PON) 2002 – 2006.

REFERENCES

Amann, M., Bertok, I., Cofala, J., Gyarfas, F., Heyes, C., Klimont, Z., & Schöpp, W. (1996). *Cost-effective Control of Acidification and Ground-Level Ozone.* International Institute for Applied Systems Analysis, Laxenburg, Austria.

Augenstein, E., Wrobel, M., Kuperjans, I., & Plessow, M. (2004). TOP-Energy computational support for energy system engineering processes. In *First international conference from scientific computing to computational engineering,* Athens.

Bahn, O., Barreto, L., Büeler, B., & Kypreos, S. (1997). *A multi-regional MARKAL-MACRO model to study an international market of CO2 emission permits: a detailed analysis of a burden sharing strategy among the Netherlands, Sweden and Switzerland.* PSI Technical Report, no. 97-09, Villigen, Switzerland.

Bardouille, P. (1998). *Towards the Operationalisation of Sustainable Energy Systems: Examination of Environmental Decision-Support Tools and Models, and Elaboration of a Conceptual Framework for Sustainable Power Sector Investment Analysis and Decision-Support.* Masters Dissertation, International Institute for Industrial Environmental Economics, Lund University, Lund, Sweden.

Baumhögger, F., Baumhögger, J., Baur, J.-M., Kühner, R., Schellmann, U., Schlenzig, C., et al. (1998). *Manual MESAP, Version 3.1.* Institute of Energy Economics and the Rational use of Energy (IER), Germany.

Bergman, L., & Henrekson, M. (2003). *CGE Modeling of Environmental Policy and Resource Management.* Retrieved June, 21, 2008, from http://users.ictp.it/~eee/workshops/smr1533/Bergman%20-%20 Handbook-1.doc.

Capros, P. (n.d.). *The PRIMES Energy System Model. Summary Description.* National Technical University of Athens - European Commission Joule-III Programme. Retrieved June, 25, 2008, from http:// www.e3mlab.ntua.gr/manuals/PRIMsd.pdf.

Capros, P., Mantzos, L., Vouyoukas, L., & Petrellis, D. (1999). European Energy and CO2 Emissions Trends to 2020: PRIMES model v.2. *Bulletin of Science, Technology & Society, 19*(6), 474–492. doi:10.1177/027046769901900604

Commission of the European Communities. (2007). *An Energy Policy for Europe COM(2007)1.* Retrieved April, 9, 2008, from http://ec.europa.eu/energy/energy_policy/doc/01_energy_policy_for_europe_en.pdf.

COMMunity for ENergy environment & Development - COMMEND (n.d.). Retrieved August, 18, 2008, from http://www.energycommunity.org/default.asp?action=71.

Consultants, L. U. (n.d.). Retrieved July 25, 2008, from http://www.landuse.co.uk/Bristol/Downloads/NorthYorkshire/Explanatory_Note.pdf.

Cosmi, C., Cuomo, V., Macchiato, M., Mangiamele, L., & Salvia, M. (2000). Chapter 5.1: Basilicata Regional Environmental Plan. In R. Jank (Ed.), *Annex 33 Advanced Local Energy Planning (ALEP) – A Guidebook* (pp. 171-187). Jülich, Germany: Forschungszentrum Jülich, BEO on behalf of the International Energy Agency, Energy Conservation in Buildings and Community Systems Programme.

Cuomo, V., Salvia, M., & Schlenzig, C. (2000). Appendix A.1: Modelling the Energy System. In R. Jank (Ed.), *Annex 33 Advanced Local Energy Planning (ALEP) – A Guidebook* (pp. 171-187). Jülich, Germany: Forschungszentrum Jülich, BEO on behalf of the International Energy Agency, Energy Conservation in Buildings and Community Systems Programme.

DECADES. Inter-agency Joint Report on Databases and Methodologies for Comparative Assessment of Different Energy Sources for Electricity Generation (1993). *Computerized Tools for Comparative Assessment of Electricity Generation Options and Strategies.* An inter-agency project of CEC, ESCAP, IAEA, IBRD, IIASA, OECD, OPEC, UNIDO, WMH, DECADES-Project, Working Paper No. 5.

den Ouden, A. C. B. (2000). PRODESign: a program for optimisation and dimensioning of energy conversion systems. In GG Hirs (ed). *Proceedings of ECOS 2000: The International Conference on Efficiency, Cost, Optimisation, Simulation and Environmental Aspects of Energy and Process Systems.* 05–07 July 2000, University of Twente, Enschede, The Netherlands. Part 3, p1691–1703. [ECN report number: ECN-RX--00-024].

Eco-smes - Services for green products. (n.d.). Retrieved September, 5, 2008, from http://ex-elca2.bologna.enea.it/cm/navContents?l=IT&navID=info&subNavID=1&pagID=7

Ecoinvent Centre. (n.d.). Retrieved September, 1, 2008, from http://www.ecoinvent.org/.

Eichhammer, W. (1999). Energy Efficiency technologies: the IKARUS Database. In *New Equilibria in the Energy Markets: The Role of new Regions and Areas* (pp.379-389). Cleveland, OH: International Association for Energy Economics -IAEE, Fraunhofer-ISI.

Electric Power Research Institute – EPRI. (1998). *Inventory of available methods and processes for assessing the benefits, costs, and impacts of demand-side options, Volume 3: Description and review of computer tools for integrated planning,* (TR-108506-V3, Final report). G. Heffner, EPRI California. Retrieved June, 8, 2008, from www.epriweb.com/public/TR-108506-V3.pdf.

Energy Technology Systems Analysis Programme - ETSAP (n.d.). Retrieved August, 25, 2008, from http://www.etsap.org.

Ente per le Nuove Tecnologie, l'Energia e l'Ambiente – ENEA. (2000). *Il processo di attuazione del Protocollo di Kyoto in Italia. Metodi, scenari e valutazione di politiche e misure.* [Carrying out the Kyoto Protocol in Italy. Methods, scenarios and valuation of action policies] (*in Italian*).

European Environment Agency - EEA (2003). *EEA core set of indicators* (Tech. Rep.). Revised version April, 2003.

European Environment Agency - EEA. (2007). *Indicator: EN01 Energy and non energy-related greenhouse gas emissions [2007.04]*. Retrieved April, 23, 2008, from http://themes.eea.europa.eu/Sectors_and_activities/energy/indicators/EN01,2007.04

Fahl, U., Rühle, B., & Voss, A. (2004). *Wissenschaftliche Begleitung des Energieprogramm Sachsen, Schlussbericht.* Institute for Energy Economics and the Rationale Use of Energy (IER), University of Stuttgart. Retrieved from www.ier.uni-stuttgart.de/sachsen

Farinelli, U., Johansson, T. B., McCormick, K., Mundaca, L., Oikonomou, V., & Örtenvikc, M. (1981). MARKAL, a linear-programming model for energy system analysis: technical description of the BNL version. *International Journal of Energy Research, 5*, 353–375. doi:10.1002/er.4440050406

Fishbone, L. G., Giesen, G., & Vos, P. (1983). *User's guide for MARKAL (BNL/KFA version 2.0), IEA Technology System analysis Project. BNL 51701*. Julich, Germany: Brookhaven National Laboratory and Kernforschungsanlange.

Forschungszentrum Jülich - KFA. (1994). IKARUS—Instrumente für Klimagas Reduktionsstrategien, Teilprojekt 4: Umwandlungssektor Strom- und Wärmeerzeugende Anlagen auf fossiler und nuklearer Grundlage. Teil 1 u. 2, Jülich.

General Algebraic Modelling System - GAMS (n.d.). Retrieved August, 25, 2008, from http://www.gams.com.

Goldstein, G., & Tosato, G. C. (2008). *International Energy Agency Implementing Agreement for a Programme of Energy Technology Systems Analysis - Annex X: Global Energy Systems and Common Analyses*. Retrieved July, 24, 2008, from http://www.etsap.org/official.asp.

Institute for Energy Engineering - Technical University of Berlin. (2003). *deeco Related models (which embed infrastructure capacity limitations)*. Retrieved August, 18, 2008, from http://www.iet.tu-berlin.de/deeco/related-models.html.

Institute of Energy Economics and the Rational Use of Energy - University of Stuttgart. (n.d.) *EcoSenseWeb*. Retrieved July, 25, 2008, from http://EcoSenseWeb.ier.uni-stuttgart.de/.

International Atomic Energy Agency – IAEA. (1980). *Wien Automatic System Planning Package (WASP), A Computer Code for Power Generating System Expansion Planning*.

International Institute for Applied Systems Analysis – IIASA (n.d.). Retrieved May, 25, 2008, from http://www.iiasa.ac.at/Research/ENE/model/

Jank, R., Johnsson, J., Rath-Nagel, S., Steidle, T., Ryden, B., Skoldberg, H., et al. (2005). Annex 33 information based on the final reports of the project. In R. Burton (Ed.), *Technical Synthesis Report. Annexes 22 & 33. Energy Efficient Communities & Advanced Local Energy Planning (ALEP)* (pp. 15-28). Birmingham, UK: Faber Maunsell Ltd on behalf of the international Energy Agency, Energy Conservation in Building and Community Systems Programme.

KanORS Consulting Inc. (2003a). *VEDA-FE User Guide - Version 1.5.11*. Retrieved May 25, 2006, from http://www.kanors.com/userguidefe.htm

KanORS Consulting Inc. (2003b). *VEDA4 User Guide - Version 4.3.8.* Retrieved May 25, 2006, from http://www.kanors.com/userguidebe.htm

Loulou, R., Goldstein, G., & Noble, K. (2004). *Energy Technology Systems Analysis Programme - Documentation for the MARKAL Family of Models. Part I: Standard MARKAL.* Retrieved January 14, 2006, from http://www.etsap.org/MrklDoc-I_StdMARKAL.pdf.

Loulou, R., Remme, U., Kanudia, A., Lehtila, A., & Goldstein, G. (2005). A comparison of the TIMES and MARKAL models. In *Energy Technology Systems Analysis Programme - Documentation for the TIMES Model - Part I: TIMES concepts and theory.* Retrieved January 14, 2006, from http://www.etsap.org/Docs/TIMESDoc-Intro.pdf.

Mäkelä, J. (2000). *Development of an energy system model of the Nordic electricity production system.* Unpublished master's thesis, Helsinki University of Technology, Finland.

Manne, A., Mendelsohn, R., & Richels, R. (1995). MERGE: a model for evaluating regional and global effects of GHG reduction policies. *Energy Policy, 23*, 7–34.

Messner, S., & Strubegger, M. (1995a). *Model-Based Decision Support in Energy Planning.* WP-95-119, International Institute for Applied Systems Analysis, Laxenburg, Austria.

Messner, S., & Strubegger, M. (1995b). *User's Guide for MESSAGE III.* WP-95-69. International Institute for Applied Systems Analysis, Laxenburg, Austria.

Monitoring and Evaluation of the RES directives implementation in EU27 and policy recommendations for 2020 – RES2020 (n.d.). Retrieved June, 2, 2008, from http://www.res2020.eu/.

New Energy Externalities Developments for Sustainability - NEEDS (n.d.). Retrieved May, 25, 2008, from http://www.needs-project.org.

Noble, K. (2005, November). *ANSWERv6 for TIMES: Status Report.* Noble-Soft Systems, Australia. ETSAP Annex X Meeting. Oxford, UK. Retrieved June, 22, 2008, from http://www.ukerc.ac.uk/Downloads/PDF/E/ETSAP_24Noble_ANSWER.pdf

Öko-Institut. (2008). *Global Emission Model for Integrated Systems (GEMIS) - Version 4.4.* Retrieved September, 2, 2008, from http://www.oeko.de/service/gemis/en/.

Pietrapertosa, F., Cosmi, C., Macchiato, M., Marmo, G., & Salvia, M. (2003). Comprehensive modelling for approaching the Kyoto targets on local scale. *Renewable & Sustainable Energy Reviews, 7*(3), 249–270. doi:10.1016/S1364-0321(03)00041-8

Pietrapertosa, F., Cosmi, C., Macchiato, M., Salvia, M., & Cuomo, V. (in press). Life Cycle Assessment, ExternE and Comprehensive Analysis for an integrated evaluation of the environmental impact of anthropogenic activities. *Renewable & Sustainable Energy Reviews.*

PRé Consultants. (n.d.) Retrieved August, 28, 2008, from http://www.pre.nl/.

Preiss, P. (2007). EcoSenseWeb V1.2. *User's manual & "Description of updated and extended draft tools for the detailed site-dependent assessment of External Costs"* [Draft version 4]. *Risk of Energy Availability: Common Corridors for Europe Supply Security – REACCESS* (n.d.). Retrieved June 7, 2008, from http://reaccess.epu.ntua.gr/.

Schlenzig, C. (1998). *PlaNet: Ein entscheidungs-unterstützendes System für die Energie-und Umweltplanung (in German)*. IER Universität Stuttgart, Germany. Retrieved January, 28, 2008, from http://elib.uni-stuttgart.de/opus/volltexte/2001/742/pdf/diss_cs.pdf.

Seebregts, A. J., Goldstein, G. A., & Smekens, K. (2002). Energy/environmental modelling with the MARKAL family of models. In P. Chamoni, R. Leisten, A. Martin, J. Minnemann, and H. Stadtler (Eds). *Operations Research Proceedings 2001 — Selected Papers of the International Conference on Operations Research (OR 2001)* (pp. 75–82). Duisburg, Germany: Springer.

Strubegger, M. (2003). *CO2DB Software: Carbon Dioxide (Technology) Database. Version 3.0.* International Institute for Applied Systems Analysis, Laxenburg, Austria. Retrieved June, 22, 2008, from http://www.iiasa.ac.at/Research/ECS/docs/CO2DB_manual_v3.pdf

Sundberg, J., & Wene, C.-O. (1994). Integrated Modelling of Material Flows and Energy Systems (MIMES). *International Journal of Energy Research, 18*(3), 359. doi:10.1002/er.4440180303

Systems, N.-S. (n.d.). ANSWER MARKAL Energy modeling Software. Retrieved May, 25, 2008, from http://www.noblesoft.com.au/.

The Society of Environmental Toxicology and Chemistry -SETAC (n.d.). Retrieved July, 23, 2008, from http://www.setac.org/

The Statistical Office of the European Communities – Eurostat. (n.d.). Retrieved May, 14, 2008, from http://epp.eurostat.ec.europa.eu/

University of Geneva. (1998). *AIDAIR-GENEVA: a GIS based Decision Support System for air pollution management in an urban environment.* Retrieved June, 18, 2008, from http://ecolu-info.unige.ch/recherche/eureka/

van Beeck, N. (1999). *Classification of Energy Models.* Communicated by dr.ing. W. van Groenendaal. FEW 777. Tilburg University & Eindhoven University of Technology. Retrieved June, 18, 2008, from http://arno.uvt.nl/show.cgi?fid=3901.

Van Regemorter, D. (2005). *Impact of the liberalisation of the electricity market: an analysis with MARKAL Belgium.* Retrieved August, 25, 2008, from http://www.ukerc.ac.uk/Downloads/PDF/E/ETSAP_1Regemorter.pdf

Voogt, M. Oostvoorn, Frits van, Leeuwen, M. L. Van, & Velthuijsen, J. W. (2000). Energy modelling for economies in transition. In J. W. Maxwell & Jürgen von (Eds.), *Empirical Studies of Environmental Policies in Europe.* Boston: Kluwer Academic Publishers.

Washington State University - Extension Energy Program. (n.d.). *HeatMap, District Energy Analysis Software for Steam, Hot-Water, and Chilled-Water Systems*. Retrieved July, 25, 2008, from http://www.energy.wsu.edu/software/heatmap.cfm.

Wierzbicki, A., Makowski, M., & Wessels, J. (Eds.). (2000). *Model-Based Decision Support Methodology with Environmental Applications.* Dordrecht, The Netherlands: Kluwer Academic Publishers.

Worldbank (1991). *Assessment of personal computer models for energy planning in developing countries. Report Number 17336.* Retrieved July, 7, 2008 from http://go.worldbank.org/LCFRW3DDZ1.

Chapter 11
Poly-Generation Planning: Useful Lessons from Models and Decision Support Tools

Aiying Rong
Technical University of Denmark, Denmark

Risto Lahdelma
Helsinki University of Technology, Finland

Martin Grunow
Technical University of Denmark, Denmark

ABSTRACT

Increasing environmental concerns and the trends towards deregulation of energy markets have become an integral part of energy policy planning. Consequently, the requirement for environmentally sound energy production technologies has gained much ground in the energy business. The development of energy-efficient production technologies has experienced cogeneration and tri-generation and now is moving towards poly-generation. All these aspects have added new dimension in energy planning. The liberalized energy market requires techniques for planning under uncertainty. The growing environmental awareness calls for explicit handling of the impacts of energy generation on environment. Advanced production technologies require more sophisticated models for planning. The energy sector is one of the core application areas for operations research, decision sciences and intelligent techniques. The scientific community is addressing the analysis and planning of poly-generation systems with different approaches, taking into account technical, environment, economic and social issues. This chapter presents a survey on the models and decision support tools for cogeneration, tri-generation and poly-generation planning. This survey tries to reflect the influence of deregulated energy market and environmental concerns on decision support tasks at utility level. Diverse modelling techniques and solution methods for planning problems will co-exist for a long time. Undoubtedly, the application of intelligent techniques is one of the main trends.

DOI: 10.4018/978-1-60566-737-9.ch011

1. INTRODUCTION

Energy is a vital input for social and economic development of any nation. Nowadays, several innovations have been driven by fast evolution of the technologies in the energy sector. Increasing environmental concerns and the trends towards deregulation of energy markets have become an integral part of energy policy planning. Consequently, the requirement for environmentally sound energy production technologies has gained much ground in the energy business. **Poly-generation** has been gradually accepted as one of the most efficient and economical ways of producing energy commodities in the future. **Poly-generation** means that two or more energy commodities (e.g. electricity, heat, cooling, hydrogen or other chemical products) are generated simultaneously in a single integrated production process. It is a leading technology responding to competitive and economic pressures to cut expenses, to increase efficient use of energy, and to reduce emissions of air pollutants and greenhouse gases. Combined heat and power (CHP, also called **cogeneration**) has been a proven and reliable technology with a history of more than 100 years. Thomas (1994) stated that CHP is a global solution to voltage dip, pollution and energy efficiency. Combined cooling, heating and power (CCHP), or **tri-generation**, is booming. **Tri-generation** is directly derived from CHP by making use of the low-value waste heat energy in the process to produce high-value cooling energy. **Cogeneration** and **tri-generation** have found wide applications in district heating, large buildings and different industries such as paper, wood, food and semiconductors (Chicco & Mancarella, 2007). The concept of **poly-generation** can be derived from the concept of **tri-generation**, as referred to the production of electric power, cooling and heat, with two latter ones in case available at different enthalpy (temperature/pressure) levels. Furthermore, **poly-generation** can encompass the provision of additional outputs such as hydrogen, dehumidification, or other chemical substances used in the specific process (Chicco & Mancarella).

The energy sector is one of the core application areas in operations research, decision science and artificial intelligent techniques due to the fact the energy systems require large investments and are technologically challenging to implement. However, in contrast to the well-established planning and decision support tools for power-only generation systems, the tools for **poly-generation** systems are scant because **poly-generation** planning is inherently more complicated than power-only generation planning. The interdependence between the generation of different energy commodities in **poly-generation** plants imposes great challenge in planning. In a **poly-generation** plant, generation of different energy commodities follows a joint characteristic, which means that planning must be done in coordination among different energy commodities. A survey of the relevant publications on scientific journals in the last years (2001-2007) shows that the largest part of these publications is still related to **cogeneration**. However, the number of research papers related to **poly-generation** concepts and applications has been increasing lately. Onovwiona and Ugursal (2006) reviewed the current technology used for residential **cogeneration** systems. Salgado and Pedrero (2008) collected the literature for the short-term operational planning on **cogeneration** systems from 1983 to 2006. But this survey did not include the literature for **tri-generation** and **poly-generation** (more than three energy commodities) planning. In addition, the survey did not cover the literature for decision analysis, long-term planning and strategic decision. Huang, Poh and Ang (1995) and Zhou, Ang and Poh (2006) reviewed the decision analysis in energy and environmental modeling in general but without focus on **cogeneration**, **tri-generation** or **poly-generation**. Pohekar and Ramachandran (2004) reviewed the application of multi-criteria decision making (MCDM) to energy planning in general. Jebaraj and Iniyan (2006) reviewed comprehensive energy modeling at macro and national levels in general. Hiremath, Shikha and Ravindranath (2007) focused on decentralized energy planning in general. Metaxiotis and Kagiannas (2005) reviewed the application of intelligent techniques in the energy sector in general. Wu and Wang (2006) reviewed technology development for **tri-generation** and Chicco and Mancarella (2007) reviewed the application of **cogeneration**, **tri-generation** and **poly-generation** systems.

Here we intend to give a comprehensive survey on the state-of-art of models and decision support tools for energy planning, focusing on, **cogeneration**, **tri-generation** and **poly-generation**. Firstly, we try to capture the changes requested by the deregulation of energy markets and environmental concerns, including planning, optimization and decision support tools. Secondly, we point out the potential usefulness of intelligent techniques in energy planning as the planning problems become more complicated with the emergence of new production technologies and new uncertain planning environment.

The chapter is organized as follows. In Section 2, we describe the development and challenges of **poly-generation** systems. In Section 3, we define the criteria for classifying models and methodologies. In Section 4, we classify the literature based on the criteria definded in Section 3 and summarize the observations from the current literature. In Section 5, we give the research directions and point out the usefulness of intelligent techniques.

2. DEVELOPMENT AND CHALLENGE OF POLY-GENERATION SYSTEMS

Currently electric power market deregulation is going on all over the world, for example, in North America, Australia, New Zealand and Europe. In addition, fuel (e.g. natural gas) markets are also deregulated. At the same time, a growing awareness for the environment impacts of energy consumption and power generation is noticed. Forced by the Climate Protection Protocol in Kyoto, many countries raise their effort to cut greenhouse gas emissions. For example, a European Union (EU) wide emissions trading scheme (ETS) started in 2005 and an emissions allowance market was introduced. Consequently, energy efficient and environmentally friendly production technologies such as **cogeneration**, **tri-generation** and **poly-generation** gain more interest. The EU commission announces to raise the share of electricity by **cogeneration** technology from 9% to 18% during the years 1997-2010 (CEC Commission of the European Communities, 1997). In the following, we discuss some issues about **poly-generation** planning associated with the evolution of the economical energy–efficient technologies, the evolution of the energy markets and the evolution of environmental protection.

2.1 General Background of Poly-Generation Technologies

Compared with the separate generation of heat in boilers and power in condensing plants, **cogeneration** systems offer considerably higher energy efficiency levels (up to 90%) that lead to fuel (emissions) savings of typically between 10-40%, depending on the techniques used and the systems replaced (Madlener & Schmid, 2003). In addition, the related production technologies such as reciprocating engines, micro-turbines, Stirling engines, and fuel cells have ability to utilize sustainable fuels, like regenerative biomass, which makes them environmental friendly (Alanne & Saari, 2004). Afgan, Gobasi, Carvalho and Cumo (1998) also mentioned that **cogeneration** technologies support sustainable energy development. Accordingly, Begic and Afgan (2007) have developed sustainability assessment tools for the decision making in selection of energy systems including **cogeneration** plants. Classical **tri-generation** plant solutions are represented by coupling a **cogeneration** prime mover to an absorption chiller fired by **cogeneration** heat (Maidment & Prosser, 2000). Goodell (2002) showed that **tri-generation** saves about 24.5% of primary energy (fuel) as compared with **cogeneration** plant with electric-driven compression chillers. Chicco and Mancarella (2006) showed that **tri-generation** is a profitable alternative in a competitive market as compared with **cogeneration**. Similarly, **poly-generation** is derived from **cogeneration** and **tri-generation**. Considering the combination of **cogeneration** systems with different technologies for thermal/cooling generation, the enthalpy levels at which thermal/cooling energy can be manifold. For example, sometimes steam itself is produced at several pressure levels (high, medium and low) and the enthalpy levels of steam at different pressure levels are different. If high, medium and low pressure steams are treated as different energy commodities, then it is possible to identify the case of **tri-generation**, quad-generation and so forth. Furthermore, the **poly-generation** can encompass the provision of additional outputs such as hydrogen, dehumidification, or other chemical substances used in the specific process (Chicco & Mancarella, 2007). **Poly-generation** leads to the more efficient use of primary energy (fuel) and less emission impacts on the environment. Wang, Zheng and Jin (2007) showed that the efficiency of a **poly-generation** system integrating a chemical production process is 20.2% higher than for the traditional production process. Undoubtedly, **poly-generation** represents the development directions for future energy systems.

2.2 Impact of New Production Technologies on Planning Problems

Generally speaking, for **poly-generation** planning, it is necessary to model the individual plant characteristics based on different technologies and connection between different plants. New emerging technologies like small scale **cogeneration**, gas engine, fuel cell, adsorption chiller and desiccant dehumidifiers enable an increasing flexibility in energy service systems. This will yield new alternatives and better possibilities to design a sustainable energy system. But modeling such techniques requires more sophisticated models, which make the optimization and planning problems more difficult than before. The reason why the planning problems in **poly-generation** are much more complicated than power-only generation lies in the fact that the complexity of the plant model increases as the number of energy commodities increases because the interaction between different energy commodities becomes complicated. In a **poly-generation** plant, generation of different energy commodities follows a joint characteristic, which means that planning must be done in coordination among different energy commodities. In addition, such technologies will also result in more complex systems to design, plan, operate and maintain as they introduce physical connections between traditionally separate supplies.

2.3 Impact of Competitive Markets on Planning Problems

Similar to power only generation systems, the introduction of the competitive market implied that new planning models needed to be implemented for several previously non-existent tasks such as risk analysis (Al-Mansour & Kozuh, 2007; Makkonen and Lahdelma, 1998; Rong & Lahdelma, 2005a) and bidding strategy (Andersen & Lund, 2007). On the other hand, the deregulation has also changed the traditional planning tasks such as optimal dispatching (Chen, Tsay, & Gow, 2005). Ventosa, Baillo, Ramos and Revier (2005) identified three trends for the electricity market modeling: optimization models, equilibrium models and simulation models. Here we mainly focus on optimization models in the competitive market. It means that the price clearing process is exogenous to the firm's optimization programs, i.e. the power marginal price is an input parameter for the optimization program. This neglects the influence of the firm's decisions on the market clearing price. This category represents a large collection of literature for planning and scheduling problems in the competitive market. In addition, we also collect some literature on the application of the equilibrium models in the economic dispatch and environment impacts.

2.4 Impact of Emissions Trading Scheme on Planning Problems

Emissions trading scheme (ETS) has several effects on planning and decision problems. Firstly, ETS introduces emission allowance market and the price of emission allowance is volatile based on the allowance demand and supply. In this respect, the impact of ETS on planning problems is similar to that imposed by the power market (Rong & Lahdelma, 2007a). Secondly, due to the strict legal regulations on environment protection, emission should be explicitly handled in the decision and optimization problems. The decision and optimization problem may become a conflicting multi-objective problem (minimizing the fuel cost and emission levels simultaneously) (Curti, von Spakovsky, & Fvarat, 2002a; Curti, von Spakovsky, & Fvarat, 2002b). Thirdly, environmental concerns will result in choice of environmentally friendly technologies such as **cogeneration**, **tri-generation** and **poly-generation**. A plan for connecting the **cogeneration** or **tri-generation** or **poly-generation** plants with a utility company can involve various interest groups. These interest groups express different conflicting concerns. Thus, a utility planner must create an acceptable plan by utilizing the planning knowledge intelligently to accommodate the interest of various parties (Wang, Liu, & Luu, 1994). Finally, production technologies cannot be selected only based on capital costs and energy costs. The impact of resource utilization on the environment and society must be considered (Alanne & Saari, 2004; Alanne, Salo, Saari, & Gustafsson, 2007) and the three dimensions (economic, environmental and social impacts) of sustainability justify the multi-criteria approaches.

2.5 Challenge and Complexity of Poly-Generation Planning

Based on the above analysis, deregulated energy market, ETS and advanced production technologies pose new challenges to **poly-generation** planning. The new market situations (energy market and emission allowance market) require that decision support and planning tools must be more versatile, accurate and efficient than before. Versatility is needed in that it must be easy to keep the model and its parameters up to date when market situations change. The planning environments become more heterogeneous than before. Not only do we need to model different new production technologies, as mentioned by Chicco and Mancarella (2007), accurately but also we need to model contract portfolio containing both large and small contracts accurately. Otherwise, the smaller contracts can be optimized incorrectly due to numerical inaccuracies. Optimization must be more efficient than before for at least two reasons. Firstly rapid re-optimization is required when the situations on the market change. Secondly, advanced computations, such as risk analysis through stochastic simulation require solving a large number of deterministic models rapidly (Makkonen & Lahdelma, 1998; Makkonen & Lahdelma, 2001; Rong, Hakonen, Makkonen, Ojanen, & Lahdelma, 2004; Rong & Lahdelma, 2005a ; Rong & Lahdelma, 2007a).

Here we can see that the challenges lie in the complexity of the planning problem itself, modeling techniques and solution methodologies. The complicated problems need to be solved quickly to respond to the practical need of the fast change on the market. The answers can be found in two ways. Firstly, we can resort to the specialized modeling techniques, which can result in the structured model and facilitate the efficient solution of the problem. Secondly, we can find the solution by resorting to intelligent tech-

Figure 1. Decision making levels and time span in energy industry (Makkonen, 2005). min: minute(s); h: hour(s); d: day(s). w: week(s); m: month(s); y: year(s)

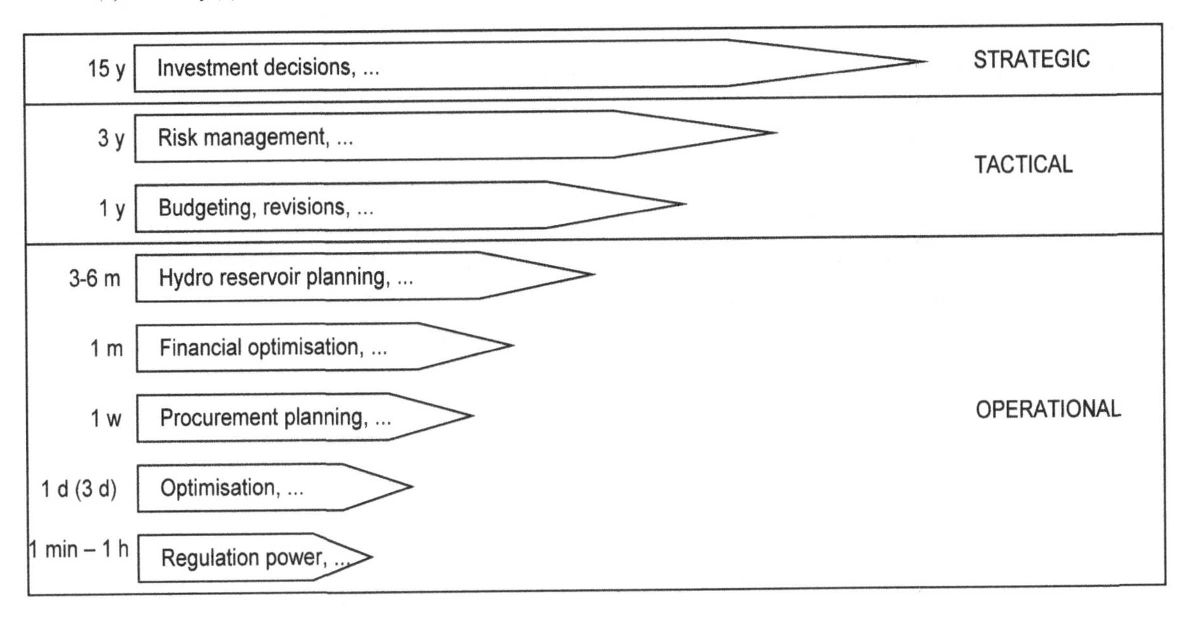

niques and **meta-heuristics** to obtain the satisfactory solution within the acceptable time limits.

3. CLASSIFICATION CRITERIA

We classify the planning and decision problems based on the following seven criteria: (1) system type, (2) problem type, (3) objectives, (4) modeling techniques, (5) solution methods, (6) power or fuel market, and (7) environmental concerns. But it is not necessary for all of the problems to include all of seven criteria.

3.1 System Type

The **poly-generation** systems are classified based on the number of energy commodities in the system, i.e., **cogeneration**, **tri-generation** and **poly-generation** (more than three commodities). This is consistent with the development stage of **poly-generation** systems (Chicco & Mancarella, 2007).

3.2 Problem Type

In terms of planning problems types, we classify them into optimization problems and decision problems with discrete alternatives.

The goal of energy system planning is to determine optimal strategies over a time horizon (day, week, month, year or multi-year) to produce the required energy commodities and satisfy the corresponding (forecast) energy demand so that some performance criteria (investment cost, energy cost, environment impact and so on) can be optimized. Planning generally covers short-, medium- and long-term horizons and these three planning horizons are associated with three levels of decision-making: operational, tactical and strategic. Figure 1 illustrates the time scale of different tasks extending from strategic decision making tasks to operational activities.

The strategic level encompasses the investment and capacity expansion planning. The tactical level considers fuel allocation and emissions allowance as well as risk management. The operational level considers solving the operational planning and optimization problems such as unit commitment (UC) optimization and optimal bidding. However, some planning problems such as economic dispatch (ED) problem can span across different decision levels. ED is the basic problem in energy planning. It determines how to dispatch the committed (switched-on) plants to meet (forecast) demand and other constraints cost-efficiently. Long-term capacity expansion planning problem (strategic) is the combination of investment decision and the long-term ED problem. UC problem is the combination of unit scheduling problem and the short-term ED problem. ED plays an important role in energy planning and it is the underlying problem for many complicated planning problems. Therefore, it is not surprising that a lot of research focuses on the efficient solution of the ED problem.

We further classify ED related problems into general operational planning (GOP) problems (operational/tactical), and combined design and operational planning problems (operational/tactical). GOP covers a wide range of problems including operational planning in normal senses as well as decision related optimization problem & risk analysis related problem using the operational planning as a background. In addition, we also discuss load management problems and energy-economy planning problems, similar to those (Figueira, Greco, & Ehrgott, 2005, p. 859-897).

Finally, in terms of decision problems with discrete alternatives, we classify them into two groups. Group A: Comparative evaluation of energy generation technologies; Group B: Selection among alternative energy plans and policies.

3.3 Objectives

Traditional energy planning mainly focuses on minimization of energy costs or maximization of the benefits. This results in single objective optimization problems. The growing environmental awareness and the apparent conflict between economic and environmental objectives turn energy planning problems into multi-objective (or multi-criteria) problems. We can classify multi-objective energy planning problems into multi-objective optimization problems and decision problems with discrete alternatives (Figueira et al., 2005, p. 859-897).

3.4 Modeling Techniques

Modeling techniques refer to describing the problem based on mathematical programming techniques (Bazaraa & Shetty, 1993; Dantzig, 1963). For decision problems with discrete alternatives, we do not discuss modeling techniques. In this category, the problem refers to the way in which decision analysis is envisaged. The problem deals with the situation where how the decision maker sees himself fit into the decision process to aid in arriving at some kind of results (Figueira et al., 2005). The problems are not formulated based on mathematical programming techniques. It means that the modeling techniques apply only to single and multi-objective optimization problems. Model-based methods have long been the main way to support planning in the energy sector (Dyner & Larsen, 2001; Kagiannas, Askounis, & Psarras, 2004). Cost-efficient operation of a **poly-generation** energy system is generally planned by us-

Figure 2. Relationship between objective, modeling technique and solution methods. LP: linear programming; MILP: mixed integer linear programming; NLP: non-linear programming

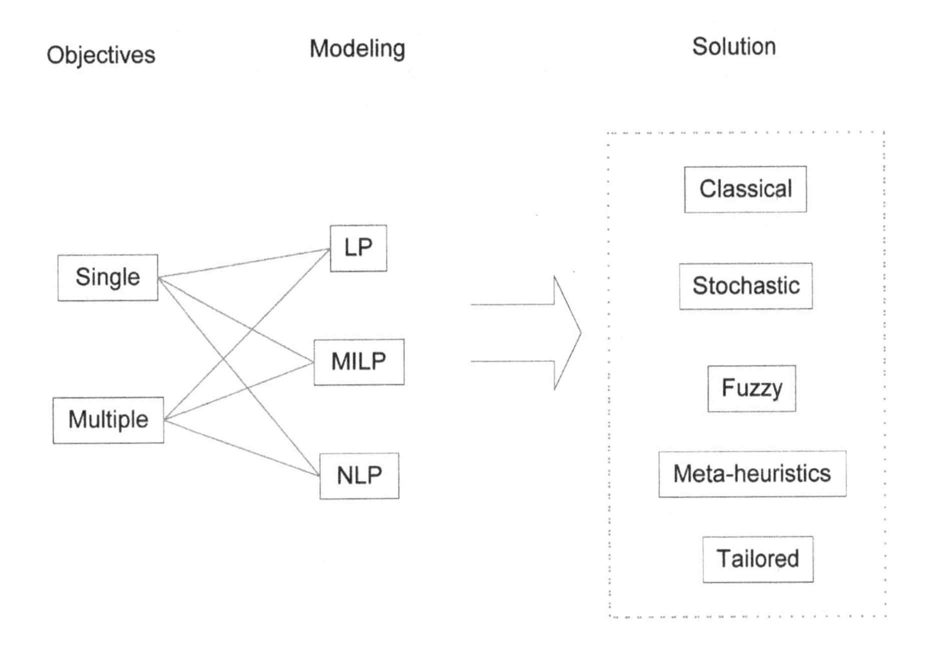

ing an optimization model. The planning problems can be formulated as linear programming (LP)-based models including mixed integer linear programming (MILP) models (Rong, 2006; Rong & Lahdelma, 2005b), and non-linear programming (NLP) (e.g. quadratic programming) models (Dotzauer, 2001; Linkevics & Sauhauts, 2005).

3.5 Solution Methods

In terms of solution techniques, both simulation methods (Spinney & Watkins, 1996) and model-based methods are widely used. Sometimes, simulation techniques are applied in conjunction with model-based methods. For model-based methods, it is worth mentioning that three criteria (objectives, modeling techniques and solution methods) are related to each other, i.e., objectives and modeling techniques together determine the solution methods as shown in Figure 2.

In terms of model-based methods, we classify them into **classical optimization, stochastic optimization, fuzzy optimization, meta-heuristic**s and **tailored heuristics**. **Classical optimization** techniques refer to optimization techniques based on classical operations research optimization theory (Bazaraa & Shetty, 1993; Dantzig, 1963). In this category, typical solution methods for LP-based models are standard LP and MILP solution methods, specialized LP and MILP solution methods based on the specific structure of the problem. Typical solution methods for NLP-based models are general nonlinear optimization (GNLO), sequential nonlinear programming (SNLP), quadratic programming (QP), sequential quadratic programming (SQP), successive approximation (SA) algorithm, Newton method and algorithms based on Karush-Kuhn-Tucker (KKT) conditions. Other methods such as Branch and Bound (BB) algorithm, Lagrangian Relaxation (LR), dynamic programming (DP), DP-based algorithms (some heuristic techniques are used in conjunction with DP to reduce the time complexity of pure DP), and dual programming (DuP) can be applied to both LP- and NLP-based models. **Stochastic optimization** methods are optimization algorithms which include probabilistic elements, either in the problem data (the objective function, the constraints) or in the algorithm itself or both. **Fuzzy optimization** techniques refer to the optimization techniques based on fuzzy set theory (Zadeh, 1978). **Meta-heuristics** are heuristics for solving a general class of computational problems while **tailored heuristics** are heuristics for solving the specific problems. **Meta-heuristics** include a broad range of algorithms such as tabu search (TS), simulated annealing, harmony search (HS) (in analogy with music improvisation process) and evolution computation-based algorithms. Evolution computation-based algorithms further include a wide class of algorithms that mimic biological evolution mechanisms such as genetic algorithm (GA), evolutionary programming (EP), immune algorithm (IA), ant colony (AC) algorithm, and particle swarm optimization (PSO). In addition, artificial neural network (ANN) has also found application in optimization of energy systems.

In terms of decision problems with discrete alternatives, we classify solution methods into **multi-attribute utility theory (MAUT)** approaches, **outranking approaches** and other multi-criteria decision making (OMCDM) approaches that cannot be classified under the heading of **MAUT** and **outranking approaches** such as fuzzy set, NAIADE (The novel approach to imprecise assignment and decision environment) (Munda,1996) and TOPSIS (technique for order preference by similarity to ideal solution) (Hwang & Yoon, 1981) approaches.

Multi-attribute utility theory (MAUT) approach tries to assign utility values to each action (alternative). This utility is a real number representing the preferability of the considered action. Very often the utility is the sum of the partial utilities that each criterion assigns to the considered action (Figueira et al., 2005). **Outranking methods** obtain the values of the criteria by using an outranking relation defined on the set of potential alternatives given what is known about the decision-maker's preferences and given the quality of the performances of alternatives and the nature of the problem. If the criteria values are sufficiently close to each other, they are indifferent to the decision maker. If the difference between the criteria values is sufficiently large, there is no doubt which alternative is better according to the criterion. In between there is an area, in which the decision maker is assumed to hesitate between indifference and strict preference.

The analytical hierarchy process (AHP) method (Figueira et al., 2005) and SMAA-2 (stochastic multi-criteria acceptability analysis) method (Lahdelma & Salminen, 2001) belong to **MAUT**. A family of ELECTRE (ELimation Et Choice Traduisant la REalite (The ELimination and Choice Expressing Reality) I, II, III, and IV methods and PROMETHEE (The Preference Ranking Organization METHods for Enrichment Evaluation) belong to **outranking methods**.

3.6 Power or Fuel Market

This criterion means that the effect of the volatile power or fuel price or both is considered in the planning problem.

3.7 Environmental Concerns

This criterion means that the impact of energy planning on the environment (emissions) is considered in the planning problem.

Table 1. LP-based general operational planning models. P: Power or fuel market; E: Environmental concerns; O: Objective; S: single objective; M: Multi-objective; N: No; Y: Yes; Co: Cogeneration; Tri: Tri-generation

Reference	System	P	E	O	Solution
Gardner and Rogers (1997)					
Lahdelma and Hakonen (2003)	Co	N	N	S	Specialized LP
Lahdelma and Makkonen (1996)					
Rong, Hakonen and Lahdelma (2006)					
Rong and Lahdelma (2007b)	Co	Y	N	S	Specialized LP
Rong and Lahdelma (2005b)					
Rong and Lahdelma (2007a)	Co	Y	Y	S	Specialized LP
Rong, Hakonen, Makkonen and Ojanen (2004)					
Bengiamin (1983)	Co	N	N	S	Standard LP
Ghoudjehbaklou and Puttgen (1988)					
MacGregor and Puttgen (1991)	Co	Y	N	S	Standard LP
Puttgen and MacGregor (1989)					
Grohnheit (1993)	Co	Y	Y	S	Standard LP
Mohanty and Panda (1993)					
Lee, Jung, and Lyu (1999)	Co	N	N	S	Fuzzy
Rong and Lahdelma (2007d)	Co	Y	N	S	Tailored Heuristic
Rong and Lahdelma (2005c)	Tri	N	N	S	Specialized LP
Kong, Wang and Huang (2007)	Tri	N	N	S	Standard LP
Rong, Lahdelma and Luh (2008)	Tri	Y	N	S	LR
Makkonen and Lahdelma (1998)	Co	Y	N	M	Simulation+ LP
Makkonen and Lahdelma (2001)					
Rong and Lahdelma (2005a)	Co	Y	Y	M	Simulation+ LP

Table 2. MILP-based general operational planning models. P: Power or fuel market; E: Environmental concerns; O: Objective; S: single objective; M: Multi-objective; N: No; Y: Yes; Co: Cogeneration; Tri: Tri-generation

Reference	System	P	E	O	Solution
Bojic and Stojanovic (1998)	Co	N	N	S	Specialized MILP
Marechal and Kalitventzeff (1998)	Co	N	N	S	Standard MILP
Dotzauer (2003)	Co	Y	Y	S	Standard MILP
Arivalagan, Raghavendra and Rao (1995)					
Gómez-Villalva and Ramos (2003)					
Gustafsson (1993)	Co	Y	N	S	Standard MILP
Rolfsman (2004a)					
Venkatesh (1995)					
Makkonen and Lahdelma (2006)					
Rong and Lahdelma (2007c)	Co	N	N	S	BB
Seo, Sung, Oh, Oh and Kwak (2008)					
Maifredi, Puzzi, and Beretta (2000)	Co	N	N	S	DP
Carraretto and Lazzaretto (2004)	Co	Y	N	S	DP
Faille, Mondon and Al-Nasrawi (2007)					
Oh, Oh and Kwak (2007)	Tri	N	N	S	BB
Piacentino and Cardona (2008b)	Tri	Y	Y	S	Standard MILP

4. CLASSIFICATION OF PUBLICATIONS AND OBSERVATIONS

Based on the classification criteria introduced in Section 3, the publications are organized mainly based on problem types in Section 3.2. Tables 1, 2, 3, 4, 5, 6, 7, 8, and 9 give different types of optimization problems. Table 10 gives decision problems with discrete alternatives. Table 11 gives the application of artificial neural networks (ANN) in **poly-generation** planning.

In terms of optimization problems, based on Tables 1-9, the following observations are obtained.

- The general operational planning (GOP) problems (Tables 1-4) account for more than 60% of all of the collected publications for optimization problems. This justifies the importance of GOP in energy planning. In addition, we can see that LP-based (including MILP) (Tables 1 and 2) approaches and NLP-based approaches (Tables 3 and 4) are almost equally distributed among the

Table 3. NLP-based single-objective general operational planning models using classical optimization techniques. P: Power or fuel market; E: Environmental concerns; N: No; Y: Yes; Co: Cogeneration; Tri: Tri-generation

Reference	System	P	E	Solution
Dotzauer and Ravn (2000)	Co	Y	N	LR
Pribicevic, Krasenbrik and Haubrich (2002)	Co	Y	N	LR+SQP
Moslehi, Khadem, Bernal and Hernandez (1991)	Co	Y	N	GNLO
Chen and Hong (1996)	Co	Y	N	NT
Meibom, Kiviluoma, Barh, Brand,Weber and Larsen (2007)	Co	Y	N	Stochastic
Weber and Woll (2006)				
Haurie, Loulou and Savard (1992)	Co	Y	N	SA
Casella, Maffezzoni, Piroddi and Pretolani (2001)	Co	Y	N	SQP
Wu and Rosen (1999)				
Rifaat (1998)	Co	N	Y	LR+KKT
Rifaat (1997)	Co	N	N	LR+KKT
Guo, Henwood and van Ooijen (1996)	Co	N	N	LR
Gonzalez Chapa andVega Galaz (2004)	Co	N	N	LR+SQP
Rooijers and van Amerongen (1994)	Co	N	N	DuP
Palsson and Ravn (1994)	Co	N	N	Stochastic
Marik, Schindler and Stluka (2008)	Tri	Y	N	GNLO
Chicco and Mancarella (2008)	Tri	Y	N	GNLO

literature.

- Environmental concerns can be considered either in single-objective or multi-objective optimization problems. LP-based (including MILP) approaches generally resort to single objective optimization methods to handle environmental concerns (emissions).

Table 4. NLP-based general operational planning models using intelligent techniques. P: Power or fuel market; E: Environmental concerns; O: Objective; S: single objective; M: Multi-objective; N: No; Y: Yes; Co: Cogeneration

Reference	System	P	E	O	Solution
Mazur (2007)	Co	N	N	S	Fuzzy
Achayuthakan and Srivastava (1998)					
Huang, Chen, Chu and Lee (2004)					
Lai, Ma and Lee (1997)	Co	N	N	S	GA
Manolas, Frangopoulos, Gialamas and Tsahalis (1997)					
Su and Chiang (2004)					
Sudhakaran and Slochanal (2003)	Co	N	N	S	GA+TS
Algie and Wong (2004)					
Lai, Ma and Lee(1998)					
Lee and Jeong (2001)					
Tsay and Lin (2000)	Co	N	N	S	EP
Tsay, Lin and Lee (2001a)					
Tsay, Lin and Lee (2001b)					
Wong and Algie (2002)					
Sudhakaran, Vimal Raj and Palanivelu (2007)					
Tsukada, Tamura, Kitagawa and Fukuyama (2003)	Co	N	N	S	PSO
Song, Chou and Stonham (1999)	Co	N	N	S	AC
Vasebi, Fesanghary and Bathaee (2007)	Co	N	N	S	HS
Hong and Li (2002)	Co	Y	N	S	GA
Hong and Li (2006)					
Chen, Tsay and Gow (2005)	Co	Y	N	S	IA
Tsay, Chang and Gow (2004)					
Matics and Krost (2007)	Co	N	Y	M	Fuzzy
Chang and Fu (1998)	Co	N	Y	M	GA+fuzzy
Wang and Singh (2007)	Co	N	Y	M	PSO
Wang and Singh (2008)					
Tsay (2003)	Co	N	Y	M	EP

- In most cases, multi-objective optimization takes environmental concerns into account. However, sometimes it may only handle the uncertainty in the energy market (Frangopolous & Dimpoulos, 2004; Makkonen & Lahdelma, 1998; Makkonen & Lahdelma, 2001; Toffolo & Lazzaretto, 2002).
- In most cases, NLP-based single-objective approaches (Table 3: **classical optimization**; Table 4: intelligent techniques) do not handle environmental concerns. They mainly focus on handling the

Table 5. Combined design and operational planning problems. P: Power or fuel market; E: Environmental concerns; O: Objective; S: single objective; M: Multi-objective; N: No; Y: Yes; Co: Cogeneration; Tri: Tri-generation.

Reference	System	P	E	O	Modeling	Solution
Arcuri, Florio and Fragiacomo (2007)	Tri	Y	Y	S	MILP	Standard MILP
Cardona and Piacentino (2007)	Tri	Y	N	S	NLP	GNLO
Oh, Lee, Jung and Kwak (2007)	Tri	N	N	S	MILP	BB
Hirai, Hiroyasu, Miki, Tanaka, Shimosaka, Aoki and Umeda (2004)	Co	N	N	S	NLP	GA
Weber, Marechal, Favrat and Kraines (2006)	Tri	N	Y	S	MILP	GA
Toffolo and Lazzaretto (2002)	Co	Y	N	M	LP	EP
Balestieri and De Borros Correia (1997)	Co	N	Y	M	LP	Interactive GDF
Curti, von Spakovsky and Fvarat (2002a) Curti, von Spakovsky and Fvarat (2002b)	Co	Y	Y	M	NLP	GA
Frangopoulos and Dimopoulos (2004)	Co	Y	N	M	NLP	GA
Li, Song, Favrat, and Marechal (2004)	Co	Y	Y	M	NLP	EP
Mavromatisl and Kokossis (1998)	Co	Y	Y	M	NLP	Logic-based
Burer, Tanaka, Favrat and Yamada (2003)	Tri	Y	Y	M	NLP	EP
Li, Nalim and Haldi (2006)	Tri	Y	Y	M	NLP	GA
Piacentino and Cardona (2008a)	Tri	Y	Y	M	NLP	Tailored Heuristic

complicated plant characteristics. It implies that the joint characteristics of **poly-generation** plant alone are difficult to handle.

- Stochastic problems are not widely studied as compared with deterministic ones.
- UC optimization problems (Table 6) concerned with **cogeneration, tri-generation** and **poly-generation** are not widely studied. There may be two reasons for it. Firstly, the ED problems for **poly-generation** have already been complicated due to the interdependence between different energy commodities. Secondly, in a broader context of energy systems, **cogeneration, tri-generation** or **poly-generation** plants are generally treated as "must-on" plants (Voorspools & D'haeseleer, 2003).
- Capacity expansion planning problems (Table 9) are mainly studied using simulation approaches. These types of problems are also not widely studied.

Based on Table 10, all of decision problems with discrete alternatives consider environmental concerns. This justifies the major motivation for introducing multi-criteria decision making in the energy sector: the need to incorporate environmental concerns in energy planning (Figueria et al., 2005; Pohekar & Ramachandran, 2004).

Based on Table 11, artificial neural networks (ANN) have found application in **poly-generation** problems. Their major functions are to model complicated plant characteristics, predict the performance of the system, automatic generation of power planning schedules and demand forecast.

Table 6. Unit commitment problems. P: Power or fuel market; E: Environmental concerns; O: Objective; S: single objective; N: No; Y: Yes; Co: Cogeneration; Tri: Tri-generation

Reference	System	P	E	O	Modeling	Solution
Rong, Hakonen and Lahdelma (2008)	Co	Y	N	S	MILP	DP-based
Rong, Lahdelma and Grunow (2009)						Heuristic
Bos, Beune and van Amerongen (1996)	Co	N	N	S	MILP	LR+specialized LP
Thorin, Brand and Weber (2005)	Co	Y	N	S	MILP	LR
Illerhaus and Verstege (1999)	Co	Y	N	S	MILP	Standard MILP
Seeger and Verstege (1991)						
Voorspool and D'haeseleer (2003)	Co	Y	N	S	NLP	Tailored Heuristic
Eriksen (2001)	Co	N	Y	S	NLP	Simulation
Ummels, Gibescu, Pelgrum, Kling and Brand (2007)	Co	Y	Y	S	NLP	Simulation
Sakawa, Kato and Ushiro (2002)	Tri	Y	N	S	MILP	GA

Table 7. Energy economy planning problems. P: Power or fuel market; E: Environmental concerns; O: Objective; S: single objective; M: Multi-objective; N: No; Y: Yes; Co: Cogeneration; Tri: Tri-generation; Poly: Poly-generation

Reference	System	P	E	O	Modeling	Solution
Cormio, Dicorato, Minoia and Trovato (2003)						
Lehtilä and Pirilä (1996)	Co	Y	Y	S	LP	Standard LP
Henning, Amiri and Holmgren (2006)						
Rolfsman (2004b)	Co	Y	N	S	MILP	Standard MILP
Martinsen and Krey (2008)	Co	Y	Y	S	LP	Fuzzy
Aki, Oyama andTsuji (2000)						
Cardona, Sannino, Piacentino and Cardona (2006)	Tri	Y	Y	S	LP	Standard LP
Asano, Sagi, Imamura, Ito and Yokoyama (1992)	Co	Y	N	S	NLP	GNLO
Liu, Gerogiorgis and Pistikopoulos (2007)	Poly	Y	Y	S	NLP	GNLO
Ensinas, Nebra, Lozano and Serra (2007)	Co	N	Y	S	NLP	Simulation
Hollmann and Voss (2005)						
Papaefthymiou, Schavemaker, van der Sluis, Kling, Kurowicka and Cooke (2006)	Co	Y	N	S	NLP	Simulation
Calderan, Spiga and Vestrucci (1992)	Co	N	N	S	NLP	Simulation
Gao, Jin, Liu and Zheng (2004)	Poly	N	N	S	NLP	Simulation
Uche, Serra, and Sanz (2004)						
Aguiar, Pinto, Andrea and Noguerira (2007)	Tri	Y	Y	S	NLP	Simulation
Mavrotas, Diakoulaki, Florios and Georgiou (2008)	Tri	Y	Y	M	NLP	Fuzzy

5. DEVELOPMENT DIRECTIONS FOR POLY-GENERATION PLANNING

Based on Section 4, model-based methodologies are the main way to support **cogeneration, tri-generation** and **poly-generation** planning in the energy sector. We have found that LP-based (including MILP) and NLP-based modeling and solution techniques are almost equally divided among the publications. In the

Table 8. Load management problems. P: Power or fuel market; E: Environmental concerns; O: Objective; S: single objective; M: Multi-objective; N: No; Y: Yes; Co: Cogeneration;Poly: Poly-generation

Reference	System	P	E	O	Modeling	Solution
Ashok and Banerjee (2003)	Co	Y	N	S	NLP	Newton
Babu and Ashok (2008)	Co	Y	N	S	NLP	BB
Tsikalakis and Hatziargyriou (2008)	Co	Y	Y	S	NLP	Heuristic + QP
Geidl and Andersson (2007)	Poly	Y	Y	S	NLP	GNLO
Matics and Krost (2008)	Co	Y	Y	M	NLP	Fuzzy

Table 9. Capacity expansion planning problems. P: Power or fuel market; E: Environmental concerns; O: Objective; S: single objective; N: No; Y: Yes; Co: Cogeneration

Reference	System	P	E	O	Modeling	Solution
Bakken, Skjelbred and Wolfgang (2007)	Co	Y	Y	S	MILP	Newton
Larsen, Palsson and Ravn (1998)						
Manhire and Jenkins (1982)	Co	N	N	S	NLP	Simulation
Millan, Campo and Sierra (1998)						
Sødergren and Ravn (1996)						
Laurikka (2006)	Co	Y	Y	S	NLP	Simulation

future, this trend will continue. On the one hand, it is possible to use LP-based formulation to handle the increasing complexity of **poly-generation** planning problems under the deregulated energy market and emissions trading scheme. On the other hand, the planning problems can be formulated as NLP-based models and sophisticated solution techniques are needed to solve the problem. In this branch, intelligent techniques such as evolution computation based algorithms, fuzzy logic and ANN can find their applications. In addition, planning under uncertainty and emission concerns are still the important topics in energy planning. In the following, we discuss the development directions in these aspects.

5.1 Development Direction of LP-Based Approaches

The reason why the planning problems for **tri-generation** and **poly-generation** are much more complex than for **cogeneration** problem lies in the fact that the complexity of the plant model increases as the number of energy commodities increases because the interaction between different energy commodities becomes complicated. In a **poly-generation** plant, generation of different energy commodities follows a joint characteristic, which means that planning must be done in coordination among different energy

Table 10. Decision problems with discrete alternatives. P: Power or fuel market; E: Environmental concerns; N: No; Y:Yes; Co: Cogeneration; Tri: Tirgeneration

Reference	System	Problem	P	E	Solution methods	
Georgopoulou, Sarafidis and Diakoulaki (1998)	Co	A	Y	Y	Outranking	PROMETFEE II
Patlitzianas, Pappa and Psarras (2008)	Co	A	Y	Y	Outranking	ELECTRE III
Tzeng, Shiau and Lin (1992)	Co	A	N	Y	Outranking	PROMETFEE
					MAUT	AHP
Alanne, Salo, Saari and Gustafsson (2007)	Co	A	N	Y	MAUT	PARIS
Begic and Afgan (2007) Sheen (2005)	Co	A	N	Y	OMCDM	Fuzzy
Afgan, Pilavachi and Carvalho (2007) Afgan and Carvalho (2008)	Co	A	N	Y	OMCDM	General Index
Liposcak, Afgan, Duic and Carvalho (2006)	Co	A	Y	Y	OMCDM	General Index
De Paepe and Mertens (2007)						
Pilavachi, Roumpeas, Minett and Afgan (2006)	Co	A	Y	Y	OMCDM	Agglomeration
Dinca, Badea, Rousseaux and Apostol (2007)	Co	A	Y	Y	OMCDM	NAIADE
Wang, Jing, Zhang, Shi and Zhang (2008)	Tri	A	N	Y	MAUT	Fuzzy +AHP
Georgopoulou, Lalas and Papagiannakis (1997)	Co	B	N	Y	Outranking	ELECTRE III
Makkonen, Lahdelma, Asell and Jokinen (2003)						
Lahdelma, Makkonen and Salminen (2006)	Co	B	Y	Y	MAUT	SMAA-2
Lahdelma, Makkonen and Salminen (2009)						
Ojanen, Makkonen and Salo (2005)	Co	B	Y	Y	MAUT	RICH
Wang, Liu and Luu (1994)	Co	B	Y	Y	OMCDM	Negotiation

Table 10. continued

Reference	System	Problem	P	E	Solution methods	
Alfonso, Perez-Navarro, Encinas, Alvarez, Rodrıguez and Alcazar (2007)	Co	B	Y	Y	OMCDM	Segment Ranking

Problem A: Comparative evaluation of energy generation technologies
Problem B: Selection among alternative energy plans and policies
Agglomeration: Agglomeration function based on statistical evaluation of weight factors
General Index: General Index of the resource, economics and social aspects

Table 11. Artificial neural networks. P: Power or fuel market; E: Environmental concerns; N: No; Y: Yes; Co: Cogeneration

Reference	System	P	E	Function
Perryman (1995)	Co	N	N	Control and Monitor
De, Kaiadi, Fast and Assadi (2007)	Co	Y	Y	Plant model
Fast, Assadi and De (2009)	Co	N	Y	Plant model
Wu, Shieh, Jang and Liu (2005)	Co	Y	N	Plant model
DeLong and Speegle (2003)	Co	N	N	Framework for plant analysis
Maiorano, Sbrizzai, Torell and Trovato (1999)	Co	N	N	Load Management
Cerri, Borghetti and Salvini (2006)	Co	Y	Y	Load Management
Entchev and Yang (2007)	Co	N	N	Predicting plant performance
Moghavvemi, Yang and Kashem (1998)	Co	N	N	Automating generation schedules
Celli, Pilo, Pisano and Soma (2005)	Co	Y	Y	Energy Management System
Matsumoto, Tamaki and Murao (2007)	Co	N	N	Demand Forecast
Ogaji, Sampath, Singh and Probert (2002)	Co	N	N	Fault diagnostics

commodities. Modeling **poly-generation** planning problems as generic LP-based problems are always a viable choice. Then we can resort to the generic LP-based solvers to solve the problem. However, the speed of the generic LP-based solvers is limited by the size of the problem. Another alternative is to adopt the specialized modeling techniques, which render the structure for the complicated problem and facilitate the development of the efficient solution algorithms. In the following, we discuss a specialized modeling technique for a **cogeneration** plant, which may be extended to model a **poly-generation** plant. Figure 3 shows the plant characteristic of a convex **cogeneration** plant. Figure 4 shows the plant characteristic of a non-convex **cogeneration** plant.

In Figure 3, the corner points are called extreme points of the plant. An extreme point is represented as (c_j, p_j, q_j) (cost, power, heat). Lahdelma and Hakonen (2003) formulated a convex plant model as an extremal problem (Dantzig, 1963), i.e., the plant characteristic is represented as the convex combination of extreme points. This formulation in fact was the reformulation of a general linear **cogeneration** plant model (Lahdelma & Rong, 2005). Then the specialized efficient algorithms for different **cogeneration** planning problems (Lahdelma & Hakonen, 2003; Rong, Hakonen & Lahdelma, 2006; Rong, Hakonen & Lahdelma, 2008; Rong & Lahdelma, 2005b, Rong & Lahdelma, 2007b; Rong & Lahdelma, 2007d; Rong, Lahdelma & Grunow, 2009) were developed based on this plant formulation. In addition, some of these algorithms have been applied to risk analysis and decision making for **cogeneration** systems (Lahdelma, Makkonen, & Salminen, 2006; Lahdelma, Makkonen, & Salminen, 2009; Makkonen & Lahdelma, 1998; Makkonen & Lahdelma, 2001; Makkonen, Lahdelma, Asell, & Jokinen, 2003; Rong & Lahdelma, 2005a; Rong & Lahdelma, 2007a). Rong and Lahdelma (2005c) and Rong, Lahdelma and Luh (2008) extended this **cogeneration** plant formulation to modeling a **tri-generation** plant and developed specialized efficient algorithms for **tri-generation** planning.

A non-convex plant (Figure 4) can be encoded as multiple convex sub-models based on convex partition (Makkonen & Lahdelma, 2006). Then the specialized efficient BB algorithms (Makkonen & Lahdelma; Rong & Lahdelma, 2007c) for non-convex **cogeneration** planning were developed based on this plant formulation.

The advantages of the above formulation are given below. (1) Physical plants and the purchase and

Figure 3. Plant characteristic of a convex cogeneration plant (Rong and Lahdelma, 2007b).

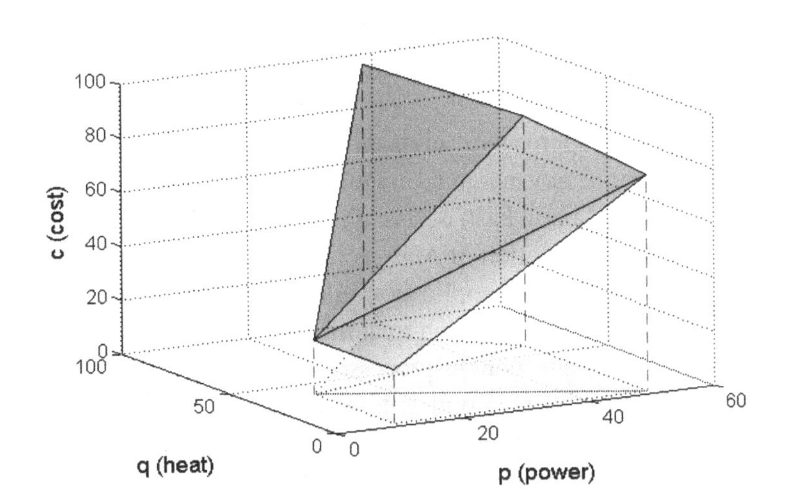

Figure 4. Plant characteristic of a non-convex cogeneration plant (Rong and Lahdelma, 2007c). p= power: q= heat; points 1,...,9 are extreme points of the plant. A1, A2 and A3 are three convex sub-areas included in the non-convex plant

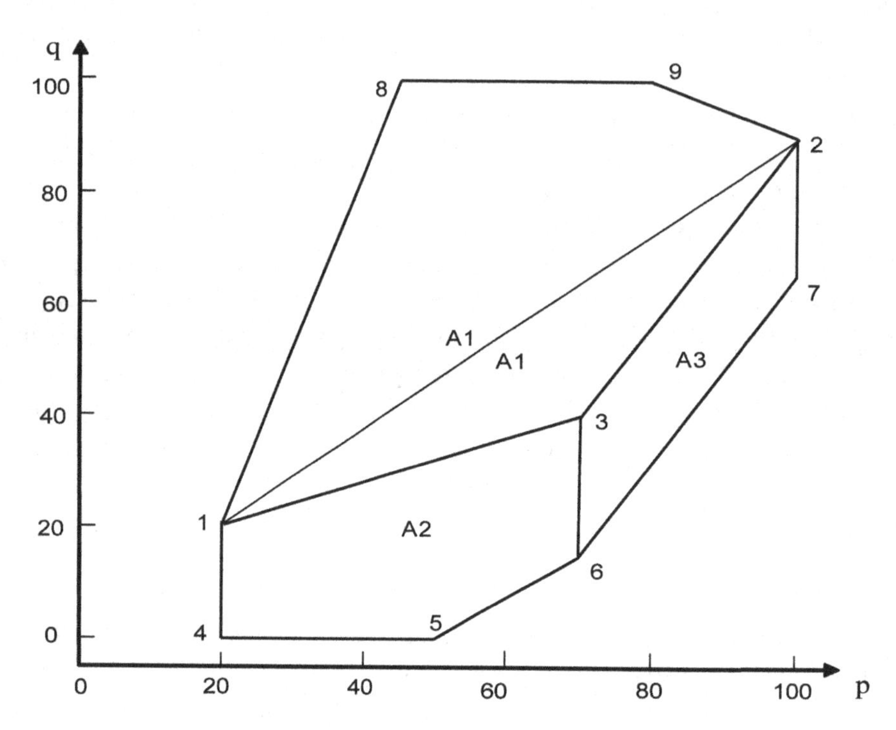

sales contracts for different energy commodities can be modeled as a special case of the plant formulation (Rong, 2006); (2) Power market can be viewed as a pair of open purchase and sales contracts, i.e. this formulation can accommodate power market; (3) The emission cost can be modeled using cost coordinate of the extreme point. Without considering the influence of the emission trading, the cost coordinate of the extreme point is fuel cost for the physical plant while the cost coordinate is penalized fuel cost (fuel cost + specific CO_2 emissions for fuel × hourly allowance price) (Rong) if the allowance price of CO_2 is considered; and (4) It is easy to extend to **poly-generation** systems.

5.2 Development Direction of NLP-Based Approaches

As mentioned in Section 4, for NLP-based approaches, **meta-heuristics**, fuzzy logic and ANN should be the directions. Firstly, ANN has superior advantage to model different complex functions such as non-linear and non-convex functions (De, Kaiadi, Fast, & Assadi, 2007; Fast, Assadi, & De, 2009; Wu, Shieh, Jang, & Liu, 2005). Currently, the application of ANN is limited to **cogeneration** plant model. However, there is no doubt that the ANN can find application in modeling **tri-generation** and **poly-generation** plant. Secondly, fuzzy logic can deal with non-quantifiable concepts and the vagueness of thinking which is inherent in human reasoning and thought-process. It can find application in multi-criteria decision making problems for both multi-objective optimization and decision problems with discrete alternatives. Thirdly, as the robustness of **meta-heuristics** increases, especially, the evolution computation based techniques, have found and will find a way in ploy-generation planning (Wong, 2002).

5.3 Planning under Uncertainty

Planning under uncertainty is a lasting topic under the deregulated energy market emissions trading scheme. The introduction of the competitive market implies that the actors such as producers, traders, distributors and end customers in the market are exposed to substantial risks caused by volatile market situations. Now we summarize the approaches for dealing with uncertainty in literature. In term of optimization approaches, the literature includes (1) **stochastic optimization** (Chang & Fu, 1998; Mavrotas, Diakoulaki, Florios, & Georgiou, 2008; Meibom, Kviluoma, Barh, Brand, Weber, & Larsen, 2007; Palsson & Ravn, 1994; Weber & Woll, 2006); (2) **fuzzy optimization** (Mavrotas, Demertzis, Meintani, & Diakoulaki, 2003); and (3) scenario analysis (Makkonen & Lahdelma, 1998; Makkonen & Lahdelma, 2001; Rong & Lahdelma, 2005a; Rong & Lahdelma, 2007a) (stochastic simulation in conjunction with solving the underlying deterministic planning problems and **stochastic optimization** based on scenarios). These approaches are similar to the optimization approaches for dealing with uncertainty is other areas (Sahinidis, 2006). In addition, we have found that simulation models are commonly used for handling uncertainty. For example, production costing simulation models are one of the commonly used approaches (Campo, & Sierra, 1998; Larsen, Palsson & Ravn, 1998; Laurikka, 2006; Manhire & Jenkins, 1982; Millan, Sødergren & Ravn, 1996) for long-term capacity expansion planning. One of the most important short-term UC problems are also dealt with by simulation approaches (Eriksen, 2001; Ummels, Gibescu, Pelgrum, Kling, & Brand, 2007). Moreover, **meta-heuristics**, especially evolution computation based algorithms (Chang & Fu, 1998; Wang & Singh, 2008) have been developed to adapt to the uncertainty in energy system planning.

5.4 Environmental Concerns

The growing environmental awareness and apparent conflict between economic and environmental objectives was the main impetus that pushed energy planners during early eighties towards the use of multi-criteria decision making methods (Figueira et al., 2005, p. 859-897). Undoubtedly, the rapid changes and the increasing complexity of the energy market and emissions allowance market as well as the development of advanced energy efficient production technologies such as **cogeneration**, **tri-generation** and **poly-generation** gave rise to further methodological development. Increasing competition and risk along with the prerequisite for sustainability have broadened and will broaden the energy application field by bringing out new challenges for the development of integrated multi-criteria approaches by taking uncertainty also into account (Figueira et al.). In the meanwhile, environmental modeling and environmental economic dispatch (ED) solution methods as studied in (Talaq, EI-Hawary, & EI-Hawary, 1994) will experience further development.

5.5 Usefulness of Intelligent Techniques

It is difficult to solve the **poly-generation** problem exactly because it a large, multi-period, combinatorial, non-convex, stochastic problem. The intelligent techniques can find to overcome this difficulty.

(1) ANN can be used to model the non-linear and non-convex feature of **poly-generation** plant and **poly-generation** planning problem as well as the complicated interaction between generation of different energy commodities because ANN can be trained with data from an existing plant and

ANN can learn complex interrelationship between the parameters by simply observing typical examples of the condition.

(2) Fuzzy logic can play a role in both dealing with uncertainties and obtaining the tradeoff between multiple conflicting objectives in the multi-criteria decision making environment. It helps to represent the information that is not readily in quantitative format (e.g. soft goals and constraints, incomplete information and subjective preferences).

(3) ANN and Fuzzy logic together can automate the construction and maintenance of models and optimization systems (symbolic manipulation, automated transformation of models to more easily solved format, automatic classification of model type and choice of solution methods and their parameters).

(4) The robustness of evolution computation techniques is especially useful in solving hard non-linear and non-convex **poly-generation** planning problems as well as dealing with uncertainty.

(5) Finally, it is worth mentioning that expert systems are also useful in **poly-generation** planning, similar to the application in power-only generation planning (Metaxiotis & Kagiannas, 2005; Rahman & Lauby, 1993). There have been some applications in **cogeneration** planning (Patlitzianas, Pappa, & Psarras, 2008). But we did not highlight the expert systems in the literature because the entire system is the integration of multi-criteria decision making techniques and expert systems. However, it is no doubt that expert systems can support decision-making and operations of energy companies with ploy-generation facilities.

6. CONCLUSION

Change of energy markets, growing environmental awareness and novel energy-efficient **poly-generation** production technologies create a need for new kinds of planning and decision support tools in energy industries. The energy sector is of vital importance for satisfaction of societal needs. In this chapter we have provided a comprehensive review of models and decision support tools for **poly-generation** planning. Diverse modeling and solution techniques have been applied in energy planning with **cogeneration**, **tri-generation** and **poly-generation** facilities. The solution methods range from **tailored heuristics**, artificial intelligence and **meta-heuristics** to simulation, mathematical programming and multi-criteria decision making approaches.

The real-life energy system planning problem may neither fit the assumption of a single technique nor be effectively solved by the strength and capabilities of a single technique. Integration of different solution techniques represents the trend for dealing with complicated energy systems. Both object-oriented paradigm (Lahdelma, 1994) and constraint programming (Marriott & Stuckey, 1998) are suitable for implementing large and complex structured systems to be used and maintained for a long time. The following aspects need to be further addressed in the future.

- An effective technique for modeling sophisticated **poly-generation** systems need to be explored.
- Stochastic problems and related solution techniques need to be further investigated.
- Multi-criteria decision making tools including multi-objective optimization in **poly-generation** systems need to be further strengthened.
- The application of fuzzy logic, artificial neural networks and evolution based computation methods need to be further strengthened.

REFERENCES

Achayuthakan, C., & Srivastava, S. C. (1998). A genetic algorithm based economic load dispatch solution for eastern region of EGAT system having combined cycle and cogeneration plants. *Proceedings of International Conference on Energy Management and Power Delivery* (pp. 165-170).

Afgan, N. H., & Carvalho, M. G. (2008). Sustainability assessment of a hybrid energy system. *Energy Policy, 36*(8), 2903–2910. doi:10.1016/j.enpol.2008.03.040

Afgan, N. H., Gobaisi, D. A., Carvalho, M. G., & Cumo, M. (1998). Sustainable energy development. *Renewable & Sustainable Energy Reviews, 2*, 235–286. doi:10.1016/S1364-0321(98)00002-1

Afgan, N. H., Pilavachi, P. A., & Carvalho, M. G. (2007). Multi-criteria evaluation of natural gas resources. *Energy Policy, 35*(1), 704–713. doi:10.1016/j.enpol.2006.01.015

Aguiar, A. B. M., Pinto, J. O. P., Andrea, C. Q., & Noguerira, L. A. (2007). Modeling and Simulation of Natural Gas Microturbine Application for Residential Complex Aiming Technical and Economical Viability Analysis. *2007 IEEE Canada Electrical Power Conference* (pp. 376-381).

Aki, H., Murata, A., Yamamoto, S., Kondoh, J., Maeda, T., Yamaguchi, H., & Ishii, I. (2005). Penetration of residential fuel cells and CO_2 mitigation—case studies in Japan by multi-objective models. *International Journal of Hydrogen Energy, 30*, 943–952. doi:10.1016/j.ijhydene.2004.11.009

Aki, H., Oyama, T., & Tsuji, K. (2000). Analysis on the energy pricing and its reduction effect on environmental impact in urban area including DHC. *2000 IEEE Power Engineering Society Winter Meeting* (pp. 1097-1102).

Aki, H., Oyama, T., & Tsuji, K. (2003). Analysis of energy pricing in urban energy service systems considering a multiobjective problem of environmental and economic impact. *IEEE Transactions on Power Systems, 18*(4), 1275–1282. doi:10.1109/TPWRS.2003.818599

Aki, H., Oyama, T., & Tsuji, K. (2006). Analysis of energy service systems in urban areas and their CO2 mitigations and economic impacts. *Applied Energy, 83*, 1076–1088. doi:10.1016/j.apenergy.2005.11.003

Al-Mansour, F., & Kozuh, M. (2007). Risk analysis for CHP decision making within the conditions of an open electricity market. *Energy, 32*, 1905–1916. doi:10.1016/j.energy.2007.03.009

Alanne, K., & Saari, A. (2004). Sustainable small-scale CHP technologies for buildings: the basis for multi-perspective decision-making. *Renewable & Sustainable Energy Reviews, 8*, 403–431. doi:10.1016/j.rser.2003.12.005

Alanne, K., Salo, A., Saari, A., & Gustafsson, S. I. (2007). Multi-criteria evaluation of residential energy supply systems. *Energy and Building, 39*, 1218–1226. doi:10.1016/j.enbuild.2007.01.009

Alfonso, D., Perez-Navarro, A., Encinas, N., Alvarez, C., Rodrıguez, J., & Alcazar, M. (2007). Methodology for ranking customer segments by their suitability for distributed energy resources applications. *Energy Conversion and Management, 48*(5), 1615–1623. doi:10.1016/j.enconman.2006.11.006

Algie, C., & Wong, K. O. (2004). A test system for combined heat and power economic dispatch problems. *2004 IEEE International Conference on Electric Utility Deregulation, Restructuring and Power Technologies* (vol.1, pp. 96-101).

Andersen, A. N., & Lund, H. (2007). New CHP partnerships offering balancing of fluctuating renewable ectricity productions. *Journal of Cleaner Production, 15*, 288–293. doi:10.1016/j.jclepro.2005.08.017

Arcuri, P., Florio, G., & Fragiacomo, P. (2007). A mixed integer programming model for optimal design of trigeneration in a hospital complex. *Energy, 32*, 1430–1447. doi:10.1016/j.energy.2006.10.023

Arivalagan, A., Raghavendra, B. G., & Rao, A. R. K. (1995). Integrated energy optimization model for a cogeneration based energy supply system in the process industry. *Electrical Power & Energy Systems, 17*, 227–233. doi:10.1016/0142-0615(95)00037-Q

Asano, H., Sagi, S., Imamura, E., Ito, K., & Yokoyama, R. (1992). Impacts of time-of-use rates on the optimal sizing and operation of cogeneration systems. *IEEE Transactions on Power Systems, 7*(4), 1444–1450. doi:10.1109/59.207366

Ashok, S., & Banerjee, R. (2003). Optimal operation of industrial cogeneration for load management. *IEEE Transactions on Power Systems, 18*, 931–937. doi:10.1109/TPWRS.2003.811169

Babu, C. A., & Ashok, S. (2008). Peak Load Management in Electrolytic Process Industries. *IEEE Transactions on Power Systems, 23*(2), 399–405. doi:10.1109/TPWRS.2008.920732

Bakken, B. J., Skjelbred, H. I., & Wolfgang, O. (2007). eTransport: Investment planning in energy supply systems with multiple energy carriers. *Energy, 32*, 1676–1689. doi:10.1016/j.energy.2007.01.003

Balestieri, J. P., & De Borros Correia, P. (1997). Multiobjective linear model for pre-feasibility design of cogeneration systems. *Energy, 22*, 537–548. doi:10.1016/S0360-5442(96)00151-X

Bazaraa, M. S., & Shetty, C. M. (1993). *Nonlinear programming theory and algorithms*. New York: Wiley.

Begic, F., & Afgan, N. H. (2007). Sustainability assessment tool for the decision making in selection of energy system—Bosnian case. *Energy, 32*, 1979–1985. doi:10.1016/j.energy.2007.02.006

Bengiamin, N. N. (1983). Operation of cogeneration plants with power purchase facilities. *IEEE Transactions on Power Appratus Systems PAS, 102*(10), 3467–3472. doi:10.1109/TPAS.1983.317845

Bojic, M., & Stojanovic, B. (1998). MILP optimization of a CHP energy system. *Energy Conversion and Management, 37*, 637–642. doi:10.1016/S0196-8904(97)00042-3

Bos, M. F. J., Beune, R. J. L., & van Amerongen, R. A. M. (1996). On the incorporation of a heat storage device in Lagrangian Relaxation based algorithms for unit commitment. *Electrical Power & Energy Systems, 18*(4), 205–214. doi:10.1016/0142-0615(95)00059-3

Burer, M., Tanaka, K., Favrat, D., & Yamada, K. (2003). Multi-criteria optimization of a district cogeneration plant integrating a solid oxide fuel cell-gas turbine combined cycle, heat pumps and chiller. *Energy, 28*, 497–518. doi:10.1016/S0360-5442(02)00161-5

Calderan, R., Spiga, M., & Vestrucci, P. (1992). Energy modelling of a cogeneration system for a food industry. *Energy, 17*, 609–616. doi:10.1016/0360-5442(92)90096-I

Cardona, E., & Piacentino, A. (2007). Optimal design of CHCP plants in the civil sector by thermoeconomics. *Applied Energy, 84*(7-8), 729–748. doi:10.1016/j.apenergy.2007.01.005

Cardona, E., Sannino, P., Piacentino, A., & Cardona, F. (2006). Energy saving in airports by trigeneration. Part II: Short and long term planning for the Malpensa 2000 CHCP plant. *Applied Thermal Engineering, 26*, 1437–1447. doi:10.1016/j.applthermaleng.2006.01.020

Carraretto, C., & Lazzaretto, A. (2004). A dynamic approach for the optimal electricity dispatch in the deregulated market. *Energy, 29*, 2273–2287. doi:10.1016/j.energy.2004.03.025

Casella, F., Maffezzoni, C., Piroddi, L., & Pretolani, F. (2001). Minimising production costs in generation and cogeneration plants. *Control Engineering Practice, 9*, 283–295. doi:10.1016/S0967-0661(01)00002-8

CEC Commission of the European Communities. (1997). *Communication from the Commission to the Council, the European Parliament, the Economic and Social Committee and the Committee of the Regions—A Community strategy to promote combined heat and power (CHP) and to dismantle barriers to its development,* COM (1997) 514 final, Brussels.

Celli, G., Pilo, F., Pisano, G., & Soma, G. G. (2005). Optimal Participation of a Microgrid to the Energy Market with an Intelligent EMS. *International Power Engineering Conference* (pp. 663-668).

Cerri, G., Borghetti, S., & Salvini, C. (2006). Neural management for heat and power cogeneration plants. *Engineering Applications of Artificial Intelligence, 19*, 721–730. doi:10.1016/j.engappai.2006.05.013

Chang, C. S., & Fu, W. (1998). Stochastic multiobjective generation dispatch of combined heat and power systems. *IEE Proc. Generation . Transmission and Distribution, 145*, 583–591. doi:10.1049/ip-gtd:19981997

Chen, B. K., & Hong, C. C. (1996). Optimum Operation for a back-Pressure cogeneration system under time-of-use rates. *IEEE Transactions on Power Systems, 11*, 1074–1082. doi:10.1109/59.496197

Chen, S., Tsay, M., & Gow, H. (2005). Scheduling of cogeneration plants considering electricity wheeling using enhanced immune algorithm. *Electrical Power & Energy System, 27*, 31–38. doi:10.1016/j.ijepes.2004.07.008

Chicco, G., & Mancarella, P. (2006). From cogeneration to trigeneration: Profitable alternatives in a competitive market. *IEEE Transactions on Energy Conversion, 21*, 265–272. doi:10.1109/TEC.2005.858089

Chicco, G., & Mancarella, P. (2007). (article in press). Distributed multi-generation: A comprehensive view. *Renewable & Sustainable Energy Reviews*. doi:10.1016/j.reser.2007.11.014

Chicco, G., & Mancarella, P. (2008). (article in press). Matrix modelling of small-scale trigeneration systems and application to operational optimization. *Energy*. doi:.doi:10.1016/j.energy.2008.09.011

Cormio, C., Dicorato, M., Minoia, A., & Trovato, M. (2003). A regional energy planning methodology including renewable energy sources and environmental constraints. *Renewable & Sustainable Energy Reviews, 7*, 99–130. doi:10.1016/S1364-0321(03)00004-2

Curti, V., von Spakovsky, M. R., & Fvarat, D. (2002a). An environomic approach for the modeling and optimization of a district heating network based on centralized and decentralized heat pumps, cogeneration and / or gas furnace, Part I: Methodology. *International Journal of Thermal Sciences, 39*, 721–730. doi:10.1016/S1290-0729(00)00226-X

Curti, V., von Spakovsky, M. R., & Fvarat, D. (2002b). An environomic approach for the modeling and optimization of a district heating network based on centralized and decentralized heat pumps, cogeneration and / or gas furnace, Part II: Application. *International Journal of Thermal Sciences, 39*, 731–741. doi:10.1016/S1290-0729(00)00225-8

Dantzig, G. (1963). *Linear programming and extensions*. Princeton, NJ: Princeton University Press.

De, S., Kaiadi, M., Fast, M., & Assadi, M. (2007). Development of an artificial neural network model for the steam process of a coal biomass cofired combined heat and power (CHP) plant in Sweden. *Energy, 32*, 2099–2109. doi:10.1016/j.energy.2007.04.008

De Paepe, M., & Mertens, D. (2007). Combined heat and power in a liberalised energy market. *Energy Conversion and Management, 48*, 2542–2555. doi:10.1016/j.enconman.2007.03.019

DeLong, S. O., & Speegle, C. T. (2003). Applying the systems methodology in the design of the West Point cogeneration power plant. *Proceedings of the 2003 IEEE Systems and Information Engineering Design Symposium*.

Dinca, C., Badea, A., Rousseaux, P., & Apostol, T. (2007). A multi-criteria approach to evaluate the natural gas energy systems. *Energy Policy, 35*, 5754–5765. doi:10.1016/j.enpol.2007.06.024

Dotzauer, E. (2001). *Energy system operation by Lagrangian Relaxation*. Ph.D. Thesis, Department of Mathematics, Linköping University, Sweden.

Dotzauer, E. (2003). Experiences in mid-term planning of district heating system. *Energy, 28*, 1545–1555. doi:10.1016/S0360-5442(03)00151-8

Dotzauer, E., & Ravn, H. F. (2000). Lagrangian relaxation based algorithms for optimal economic dispatch of cogeneration systems with a storage. In *proceedings from the second International Conference on Simulation, Gaming, Training and Business Process Reengineering in Operations* (pp. 51-55). Riga, Latvia.

Dyner, I., & Larsen, E. R. (2001). From planning to strategy in the electricity industry. *Energy Policy, 29*, 1145–1154. doi:10.1016/S0301-4215(01)00040-4

Ensinas, A. V., Nebra, S. A., Lozano, M. A., & Serra, L. M. (2007). Analysis of process steam demand reduction and electricity generation in sugar and ethanol production from sugarcane. *Energy Conversion and Management, 48*, 2978–2987. doi:10.1016/j.enconman.2007.06.038

Entchev, E., & Yang, L. (2007). Application of adaptive neuro-fuzzy inference system techniques and artificial neural networks to predict solid oxide fuel cell performance in residential microgeneration installation. *Journal of Power Sources, 170*(1), 122–129. doi:10.1016/j.jpowsour.2007.04.015

Eriksen, P. B. (2001). Economic and environmental dispatch of power: CHP production systems. *Electric Power Systems Research, 57*, 33–39. doi:10.1016/S0378-7796(00)00116-4

Faille, D., Mondon, C., & AI-Nasrawi, B. (2007). mCHP Optimization by Dynamic Programming and Mixed Integer Linear Programming. *2007 International Conference on Intelligent Systems Applications to Power Systems.*

Fast, M., Assadi, M., & De, S. (2009). Development and multi-utility of an ANN model for an industrial gas turbine. *Applied Energy, 86*(1), 9–17. doi:10.1016/j.apenergy.2008.03.018

Figueira, J., Greco, S., & Ehrgott, M. (Eds.). (2005). *Multiple criteria decision analysis: state of the art surveys.* Berlin: Springer.

Frangopoulos, C. A., & Dimopoulos, G. G. (2004). Effect of reliability considerations on the optimal synthesis, design and operation of a cogeneration system. *Energy, 29*, 309–329. doi:10.1016/S0360-5442(02)00031-2

Gao, L., Jin, H., Liu, Z., & Zheng, D. (2004). Exergy analysis of coal-based polygeneration system for power and chemical production. *Energy, 29*, 2359–2371. doi:10.1016/j.energy.2004.03.046

Gardner, D. T., & Rogers, J. S. (1997). Joint planning of combined heat and power and electric power systems: An efficient model formulation. *European Journal of Operational Research, 102*(1), 58–72. doi:10.1016/S0377-2217(96)00221-4

Geidl, M., & Andersson, G. (2007). Optimal Power Flow of Multiple Energy Carriers. *IEEE Transactions on Power Systems, 22*(1), 145–155. doi:10.1109/TPWRS.2006.888988

Georgopoulou, E., Lalas, D., & Papagiannakis, L. (1997). A multicriteria decision aid approach for energy planning problems: The case of renewable energy option. *European Journal of Operational Research, 103*, 38–54. doi:10.1016/S0377-2217(96)00263-9

Georgopoulou, E., Sarafidis, Y., & Diakoulaki, D. (1998). Design and implementation of a group DSS for sustaining renewable energies exploitation. *European Journal of Operational Research, 109*, 483–500. doi:10.1016/S0377-2217(98)00072-1

Ghoudjehbaklou, H., & Puttgen, H. B. (1988). Optimum electric utility spot price determinations for small power producing facilities operating under PURPA provisions. *IEEE Transactions on Energy Conversion, 3*(3), 575–582. doi:10.1109/60.8070

Gómez-Villalva, E., & Ramos, A. (2003). Optimal Energy Management of an Industrial Consumer in Liberalized Markets. *IEEE Transactions on Power Systems, 18*, 716–723. doi:10.1109/TPWRS.2003.811197

Gonzalez Chapa, M. A., & Vega Galaz, J. R. (2004). An economic dispatch algorithm for cogeneration systems. *2004 IEEE Power Engineering Society General Meeting.*

Goodell, M. (2002). *Trigeneration advantage*. http://www.cogeneration.net/The%20Trigeneration%20Advantage.htm.

Grohnheit, P. E. (1993). Modelling CHP within a national power system. *Energy Policy, 21*, 418–429. doi:10.1016/0301-4215(93)90282-K

Guo, T., Henwood, M. I., & van Ooijen, M. (1996). An algorithm for combined heat and power economic dispatch. *IEEE Transactions on Power Systems, 11*, 1778–1784. doi:10.1109/59.544642

Gustafsson, S. I. (1993). Mathematical modelling of district heating and electricity loads. *Applied Energy, 46*, 149–159. doi:10.1016/0306-2619(93)90064-V

Haurie, A., Loulou, R., & Savard, G. (1992). A two-player game model of power cogeneration in New England. *IEEE Transactions on Automatic Control, 37*(9), 1451–1456. doi:10.1109/9.159591

Henning, D., Amiri, S., & Holmgren, K. (2006). Modelling and optimisation of electricity, steam and district heating production for a local Swedish utility. *European Journal of Operational Research, 175*(2), 1224–1247. doi:10.1016/j.ejor.2005.06.026

Hirai, S., Hiroyasu, T., Miki, M., Tanaka, Y., Shimosaka, H., Aoki, S., & Umeda, Y. (2004). Satisfactory design of cogeneration system using genetic algorithm. *2004 IEEE Conference on Cybernetics and Intelligent Systems*.

Hiremath, R. B., Shikha, S., & Ravindranath, N. H. (2007). Decentralized energy planning; modeling and application—a review. *Renewable & Sustainable Energy Reviews, 11*, 729–752. doi:10.1016/j.rser.2005.07.005

Hollmann, M., & Voss, J. (2005). Modeling of decentralized energy supply structures with "system dynamics." *2005 International Conference on Future Power Systems*.

Hong, Y. Y., & Li, C. H. (2002). Genetic algorithm based economic dispatch for cogeneration units considering multiplant and multibuyer wheeling. *IEEE Transactions on Power Systems, 17*, 134–140. doi:10.1109/59.982204

Hong, Y. Y., & Li, C. Y. (2006). Back-pressure cogeneration economic dispatch for physical bilateral contract using genetic algorithms. *2006 International Conference on Probabilistic Methods Applied to Power Systems*.

Houwing, M., Heijnen, P. W., & Bouwmans, I. (2006). Deciding on micro-CHP: a multi-level decision-making approach. *Proceedings of the 2006 IEEE International Conference on Networking, Sensing and Control* (pp. 302-307).

Huang, J. P., Poh, K. L., & Ang, B. W. (1995). Decision analysis in energy and environmental modelling. *Energy, 20*, 843–855. doi:10.1016/0360-5442(95)00036-G

Huang, S. H., Chen, B. K., Chu, W. C., & Lee, W. J. (2004). Optimal operation strategy for cogeneration power plants. *Conference Record of the 2004 IEEE Industry Applications Conference*.

Hwang, C. L., & Yoon, K. (1981). *Multiple Attribute Decision Making: Methods and Application*. New York: Springer.

Illerhaus, S. W., & Verstege, J. F. (1999). Optimal operation of industrial CHP-based power systems in liberalized energy markets. *IEEE Power Tech'99 Conference,* Budapest, Hungary, Aug 29-Sept 2, 1999.

Jebaraj, S., & Iniyan, S. (2006). A review of energy models. *Renewable & Sustainable Energy Reviews, 10*, 281–311. doi:10.1016/j.rser.2004.09.004

Kagiannas, A. G., Askounis, D. T., & Psarras, J. (2004). Power generation planning: a survey from monopoly to competition. *Electrical Power & Energy Systems, 26*, 413–421. doi:10.1016/j.ijepes.2003.11.003

Kong, X. Q., Wang, R. Z., & Huang, X. H. (2007). Energy optimization model for a CCHP system with available gas turbines. *Applied Thermal Engineering, 25*, 377–391. doi:10.1016/j.applthermaleng.2004.06.014

Lahdelma, R. (1994). *An objected-orientd mathematical modeling system.* Ph.D. thesis, Acta, Polytechnica, Scandinavica Mathematics and Computing in Engineering Series No. 66. Systems Analysis Labortory, Helsinki University of Technology.

Lahdelma, R., & Hakonen, H. (2003). An efficient linear programming algorithm for combined heat and power production. *European Journal of Operational Research, 148*, 141–151. doi:10.1016/S0377-2217(02)00460-5

Lahdelma, R., & Makkonen, S. (1996). Interactive graphical object-oriented Energy modelling and optimisation. In *Proceedings of the International Symposium of ECOS'96, Efficiency, Cost, Optimisation, Simulation and Environmental Aspects of Energy Systems* (pp. 425-431), June 25-27, Stockholm, Sweden.

Lahdelma, R., Makkonen, S., & Salminen, P. (2006). Multivariate Gaussian criteria in SMAA. *European Journal of Operational Research, 170*, 957–970. doi:10.1016/j.ejor.2004.08.022

Lahdelma, R., Makkonen, S., & Salminen, P. (2009). Two ways to handle dependent uncertainties in multi-criteria decision problems. *Omega, 37*, 79–92. doi:10.1016/j.omega.2006.08.005

Lahdelma, R., & Rong, A. (2005). Efficient re-formulation of linear cogeneration planning models. In M.H. Hamza (Ed.) *Proceedings of the 24th IASTED International Conference Modelling, Identification, and Control* (pp. 300-305). February 16 -18, 2005, Innsbruck, Austria.

Lahdelma, R., & Salminen, P. (2001). SMAA-2: Stochastic Multicriteria Acceptability Analysis for Group Decision Making. *Operations Research, 49*(3), 444–454. doi:10.1287/opre.49.3.444.11220

Lai, L. L., Ma, J. T., & Lee, J. B. (1997). Application of genetic algorithms to multi-time interval scheduling for daily operation of a cogeneration system. *1997 International Conference on Advances in Power System Control, Operation and Management* (pp. 327-331).

Lai, L.L., Ma, J.T., & Lee, J.B. (1998). Multitime-interval scheduling for daily operation of a two cogeneration system with evolutionary programming. *Electrical power & Energy Systems, 20*, 305-311.

Larsen, H. V., Palsson, H., & Ravn, H. F. (1998). Probabilistic production simulation including combined heat and power plants. *Electric Power Systems Research, 48*, 45–56. doi:10.1016/S0378-7796(98)00080-7

Laurikka, H. (2006). Option value of gasification technology within an emissions trading scheme. *Energy Policy, 34*, 3916–3928. doi:10.1016/j.enpol.2005.09.002

Lee, J. B., & Jeong, J. H. (2001). A daily optimal operational schedule for cogeneration systems in a paper mill. *2001 Power Engineering Society Summer Meeting* (vol.3, pp. 1357-1362).

Lee, J. B., Jung, C. H., & Lyu, S. H. (1999). A daily operation scheduling of cogeneration systems using fuzzy linear programming. *1999 IEEE Power Engineering Society Summer Meeting* (vol.2, pp. 983-988).

Lehtilä, A., & Pirilä, P. (1996). Reducing energy related emissions using an energy system optimization model to support policy planning in Finland. *Energy Policy, 24*(9), 805–919. doi:10.1016/0301-4215(96)00066-3

Li, H., Nalim, R., & Haldi, P. A. (2006). Thermal-economic optimization of a distributed multi-generation energy system—a case study of Beijing. *Applied Thermal Engineering, 26*(7), 709–719. doi:10.1016/j.applthermaleng.2005.09.005

Li, H., Song, Z., Favrat, D., & Marechal, F. (2004). Green heating system: characteristics and illustration with multi-criteria optimization of an integrated energy system. *Energy, 29*, 225–244. doi:10.1016/j.energy.2003.09.003

Linkevics, O., & Sauhauts, A. (2005). Formulation of the objective function for economic dispatch optimisation of steam cycle CHP plants. *2005 IEEE Russia Power Tech.*

Liposcak, M., Afgan, N. H., Duic, N., & Carvalho, M. (2006). Sustainability assessement of cogeneration sector development in Croatia. *Energy, 31*(13), 2276–2284. doi:10.1016/j.energy.2006.01.024

Liu, P., Gerogiorgis, D. I., & Pistikopoulos, E. N. (2007). Modeling and optimization of polygeneration energy systems. *Catalysis Today, 127*, 347–359. doi:10.1016/j.cattod.2007.05.024

MacGregor, P. R., & Puttgen, H. B. (1991). A spot price based control mechanism for electric utility systems with small power producing facilities. *IEEE Transactions on Power Systems, 6*(2), 683–690. doi:10.1109/59.76713

Madlener, R., & Schmid, C. (2003). Combined heat and power generation in liberalized markets and a carbon-constrained world. *GAIA-Ecological Perspective in Science . Humanities and Economics, 12*, 114–120.

Maidment, G. G., & Prosser, G. (2000). The use of CHP and absorption cooling in cold storage. *Applied Thermal Engineering, 20*, 1059–1073. doi:10.1016/S1359-4311(99)00055-1

Maifredi, C., Puzzi, L., & Beretta, G. P. (2000). Optimal power production scheduling in a complex cogeneration system with heat storage. *Proceedings of 35th Intersociety Energy Conversion Engineering Conference.*

Maiorano, A., Sbrizzai, R., Torell, F., & Trovato, M. (1999). Intelligent load shedding schemes for industrial customers with cogeneration facilities. *1999 IEEE Power Engineering Society (Winter Meeting).*

Makkonen, S. (2005). *Decision modeling tools for utilities in the deregulated energy market.* Ph.D. Thesis, Research Report A93, Systems Analysis Laboratory, Helsinki University of Technology.

Makkonen, S., & Lahdelma, R. (1998). Stochastic simulation in risk analysis of energy trade. Trends in Multicriteria Decision Making: In *Proceedings of 13th International Conference on Multiple Criteria Decision Making* (pp. 146-156). Berlin: Springer.

Makkonen, S., & Lahdelma, R. (2001). Analysis of power pools in the deregulated energy market through simulation. *Decision Support Systems, 30*(3), 289–301. doi:10.1016/S0167-9236(00)00106-8

Makkonen, S., & Lahdelma, R. (2006). Non-Convex power plant modelling in energy optimisation. *European Journal of Operational Research, 171*, 1113–1126. doi:10.1016/j.ejor.2005.01.020

Makkonen, S., Lahdelma, R., Asell, A. M., & Jokinen, A. (2003). Multi-criteria decision support in the liberated energy market. *Journal of Multi-Criteria Decision Analysis, 12*(1), 27–42. doi:10.1002/mcda.341

Manhire, B., & Jenkins, T. B. (1982). A new technique for simulating the operation of multiple assigned-energy generating units suitable for use in generation system expansion planning models. *IEEE Transactions on Power Apparatus and Systems, 10*, 3861–3868. doi:10.1109/TPAS.1982.317036

Manolas, D. A., Frangopoulos, C. A., Gialamas, T. P., & Tsahalis, D. T. (1997). Operation optimization of an industrial cogeneration system by a genetic algorithm. *Energy Conversion and Management, 38*, 1625–1636. doi:10.1016/S0196-8904(96)00203-8

Marechal, F., & Kalitventzeff, B. (1998). Process integration: Selection of the optimal utility system. *Computers & Chemical Engineering (Supplementary), 22*, S149–S156. doi:10.1016/S0098-1354(98)00049-0

Marik, K., Schindler, Z., & Stluka, P. (2008). Decision support tools for advanced energy management. *Energy, 33*(6), 858–873. doi:10.1016/j.energy.2007.12.004

Marriott, K., & Stuckey, P. J. (1998). *Programming with constraints.* Cambridge, MA: MIT Press.

Martinsen, D., & Krey, V. (2008). Compromises in energy policy-using fuzzy optimization in an energy systems model. *Energy Policy, 36*(8), 2973–2984. doi:10.1016/j.enpol.2008.04.005

Matics, J., & Krost, G. (2007). Computational Intelligence Techniques Applied to Flexible and Auto-adaptive Operation of CHP Based Home Power Supply. *2007 International Conference on Intelligent Systems Applications to Power Systems* (pp. 1-7).

Matics, J., & Krost, G. (2008). Micro combined heat and power home supply: Prospective and adaptive management achieved by computational intelligence techniques. *Applied Thermal Engineering, 28*, 2055–2061. doi:10.1016/j.applthermaleng.2008.05.002

Matsumoto, T., Tamaki, H., & Murao, H. (2007). Controlling residential co-generation system based on hierarchical decentralized model. *2007 IEEE Conference on Emerging Technologies&Factory Automation* (pp. 612-618).

Mavromatisl, S. P., & Kokossis, A. C. (1998). A logic based model for the analysis and optimisation of steam turbine networks. *Computers in Industry, 36*, 165–179. doi:10.1016/S0166-3615(98)00070-0

Mavrotas, G., Demertzis, H., Meintani, A., & Diakoulaki, D. (2003). Energy planning in buildings under uncertainty in fuel costs: The case of a hotel unit in Greece. *Energy Conversion and Management, 44*, 1303–1321. doi:10.1016/S0196-8904(02)00119-X

Mavrotas, G., Diakoulaki, D., Florios, K., & Georgiou, P. (2008). A mathematical programming framework for energy planning in services' sector buildings under uncertainty in load demand: The case of a hospital in Athens. *Energy Policy, 36*, 2415–2429. doi:10.1016/j.enpol.2008.01.011

Mazur, V. (2007). Fuzzy thermoeconomic optimization of energy-transforming systems. *Applied Energy, 84*, 749–762. doi:10.1016/j.apenergy.2007.01.006

Meibom, P., Kiviluoma, J., Barh, R., Brand, H., Weber, C., & Larsen, H. V. (2007). Value of electric heat boiler and heat pumps for wind power integration. *Wind Energy (Chichester, England), 10*, 321–327. doi:10.1002/we.224

Metaxiotis, K., & Kagiannas, A. (2005). Intelligent computer applications in the energy sector: a literature review from 1990 to 2003. *International Journal of Computation Applications in Technology, 22*, 53–64. doi:10.1504/IJCAT.2005.006936

Millan, J., Campo, R. A., & Sierra, S. (1998). A modular system for decision-making support in generation expansion planning. *IEEE Transactions on Power Systems, 13*, 667–671. doi:10.1109/59.667398

Moghavvemi, M., Yang, S. S., & Kashem, M. A. (1998). A practical neural network approach for power generation automation. *Proceedings of International Conference on Energy Management and Power Delivery, 1*, 305–310.

Mohanty, B., & Panda, H. (1993). Integrated energy system for industrial complexes. Part I: A linear programming approach. *Applied Energy, 46*, 317–348. doi:10.1016/0306-2619(93)90048-T

Moslehi, K., Khadem, M., Bernal, R., & Hernandez, G. (1991). Optimization of multiplant cogeneration system operation including electric and steam networks. *IEEE Transactions on Power Systems, 6*, 484–490. doi:10.1109/59.76690

Munda, G. (1996). *Naiade. Manual and tutorial.* In Ispra: Joint Research Centre - EC, ISPRA SITE.

Ogaji, S., Sampath, S., Singh, R., & Probert, D. (2002). Novel approach for improving power-plant availability using advanced engine diagnostics. *Applied Energy, 72*, 389–407. doi:10.1016/S0306-2619(02)00018-1

Oh, S. D., Lee, H. J., Jung, J. Y., & Kwak, H. Y. (2007). Optimal planning and economical evaluation of cogeneration system. *Energy, 32*, 760–771. doi:10.1016/j.energy.2006.05.007

Oh, S. D., Oh, H. S., & Kwak, H. Y. (2007). Economic evaluation for adoption of cogeneration system. *Applied Energy, 84*(3), 266–278. doi:10.1016/j.apenergy.2006.08.002

Ojanen, O., Makkonen, S., & Salo, A. (2005). A multi-criteria framework for the selection of risk analysis methods at energy utilities. *International Journal of Risk Assessement and Management, 5*, 16–35. doi:10.1504/IJRAM.2005.006609

Onovwiona, H. I., & Ugursal, V. I. (2006). Residential cogeneration systems: review of the current technology. *Renewable & Sustainable Energy Reviews, 10*, 389–431. doi:10.1016/j.rser.2004.07.005

Palsson, O. P., & Ravn, H. F. (1994). Stochastic heat storage problem -- Solved by the progressive hedging algorithm. *Energy Conversion and Management, 35*(12), 1157–1171. doi:10.1016/0196-8904(94)90019-1

Papaefthymiou, G., Schavemaker, P. H., van der Sluis, L., Kling, W. L., Kurowicka, D., & Cooke, R. M. (2006). Integration of stochastic generation in power systems. *Electrical Power & Energy Systems, 28*, 655–667. doi:10.1016/j.ijepes.2006.03.004

Patlitzianas, K. D., Pappa, A., & Psarras, J. (2008). An information decision support system towards the formulation of a modern energy companies' environment. *Renewable & Sustainable Energy Reviews, 12*, 790–806. doi:10.1016/j.rser.2006.10.014

Perryman, R. (1995). Condition monitoring of combined heat and power systems. *IEEE Colloquium on Condition Monitoring of Electrical Machines* (pp. 1-4).

Piacentino, A., & Cardona, F. (2008a). An original multi-objective criterion for the design of small-scale polygeneration systems based on realistic operating conditions. *Applied Thermal Engineering, 28*, 2391–2404. doi:10.1016/j.applthermaleng.2008.01.017

Piacentino, A., & Cardona, F. (2008b). EABOT – Energetic analysis as a basis for robust optimization of trigeneration systems by linear programming. *Energy Conversion and Management, 49*, 3006–3016. doi:10.1016/j.enconman.2008.06.015

Pilavachi, P. A., Roumpeas, C. P., Minett, S., & Afgan, N. H. (2006). Multi-criteria evaluation for CHP system options. *Energy Conversion and Management, 47*, 3519–3529. doi:10.1016/j.enconman.2006.03.004

Pohekar, S. D., & Ramachandran, M. (2004). Application of multi-criteria decision making to sustainable energy planning. *Renewable & Sustainable Energy Reviews, 8*, 365–381. doi:10.1016/j.rser.2003.12.007

Pribicevic, B., Krasenbrik, B., & Haubrich, H. J. (2002). Co-generation in a competitive market. *IEEE Power Engineering Society Summer Meeting* (vol.1, pp. 422-426).

Puttgen, H. B., & MacGregor, P. R. (1989). Optimum scheduling procedure for cogenerating small power producing facilities. *IEEE Transactions on Power Systems, 4*(3), 957–964. doi:10.1109/59.32585

Rahman, S., & Lauby, M. (1993). Identification of Potential Areas for the Use of Expert Systems in Power System Planning. *Expert Systems with Applications, 6*, 203–212. doi:10.1016/0957-4174(93)90010-4

Retrieved from http://www.cogeneration.net/The%20Trigeneration%20Advantage.htm

Rifaat, R. M. (1997). Practical considerations in applying economical dispatch models to combined cycle cogeneration plants. *IEEE WESCANEX 97 Communications, Power and Computing* (pp. 59-63).

Rifaat, R. M. (1998). Economic dispatch of combined cycle cogeneration plants with environmental constraints. *Proceedings of International Conference on Energy Management and Power Delivery, 1*, 149–153.

Rolfsman, B. (2004a). Combined heat-and-power plants and district heating in a deregulated electricity market. *Applied Energy, 78*, 37–52. doi:10.1016/S0306-2619(03)00098-9

Rolfsman, B. (2004b). Optimal supply and demand investment in municipal energy systems. *Energy Conversion and Management, 45*, 595–611. doi:10.1016/S0196-8904(03)00174-2

Rong, A. (2006). *Cogeneration planning under the deregulated power market and emissions trading scheme*. Ph.D. Thesis, University of Turku, Turku Center for Computer Science, Turku, Finland.

Rong, A., Hakonen, H., & Lahdelma, R. (2006). An efficient linear model and optimisation algorithm for multi-site combined heat and power production. *European Journal of Operational Research, 168*(2), 612–632. doi:10.1016/j.ejor.2004.06.004

Rong, A., Hakonen, H., & Lahdelma, R. (2008). A variant of the dynamic programming algorithm for the unit commitment of combined heat and power systems. *European Journal of Operational Research, 190*, 741–755. doi:10.1016/j.ejor.2007.06.035

Rong, A., Hakonen, H., Makkonen, S., Ojanen, O., & Lahdelma, R. (2004). CO_2 emissions trading optimization in combined heat and power generation. In P. Neittaanmäki, T. Rossi, K. Majava, O. Pironneau (eds.), *Proc. ECCOMAS 2004*.

Rong, A., & Lahdelma, R. (2005a). Risk analysis of expansion planning of combined heat and power energy system under emissions trading scheme. In W. Tayati (eds) *Proceedings of IASTED International Conference on Energy and Power Systems* (pp. 70-75), April 18-20, 2005, Krabi, Thailand

Rong, A., & Lahdelma, R. (2005b). An efficient model and specialized algorithms for cogeneration planning. In W. Tayati (Ed.) *Proceedings of the IASTED International Conference Energy and Power System* (pp. 1-7), April 18-20, 2005, Krabi, Thailand.

Rong, A., & Lahdelma, R. (2005c). An efficient linear programming model and optimization algorithm for trigeneration. *Applied Energy, 82*, 40–63. doi:10.1016/j.apenergy.2004.07.013

Rong, A., & Lahdelma, R. (2007a). CO_2 emissions trading planning in combined heat and power production via multi-period stochastic optimization. *European Journal of Operational Research, 176*, 1874–1895. doi:10.1016/j.ejor.2005.11.003

Rong, A., & Lahdelma, R. (2007b). Efficient algorithms for combined heat and power production planning under the deregulated electricity market. *European Journal of Operational Research, 176*, 1219–1245. doi:10.1016/j.ejor.2005.09.009

Rong, A., & Lahdelma, R. (2007c). An efficient envelope-based Branch and Bound algorithm for non-convex combined heat and power production planning. *European Journal of Operational Research, 183*, 412–431. doi:10.1016/j.ejor.2006.09.072

Rong, A., & Lahdelma, R. (2007d). An effective heuristic for combined heat and power production planning with power ramp constraints. *Applied Energy, 84*, 307–325. doi:10.1016/j.apenergy.2006.07.005

Rong, A., Lahdelma, R., & Grunow, M. (2009). An improved unit decomitment algorithm for combined heat and power systems. *European Journal of Operational Research, 195*, 552–562. doi:10.1016/j.ejor.2008.02.010

Rong, A., Lahdelma, R., & Luh, P. (2008). Lagrangian relaxation based trigeneration planning with storages. *European Journal of Operational Research, 188*, 240–257. doi:10.1016/j.ejor.2007.04.008

Rooijers, F. J., & van Amerongen, R. A. M. (1994). Static economic dispatch for co-generation systems. *IEEE Transactions on Power Systems, 9*, 1392–1398. doi:10.1109/59.336125

Sakawa, M., Kato, K., & Ushiro, S. (2002). Operational planning of district heating and cooling plants through genetic algorithms for mixed 0—1 linear programming. *European Journal of Operational Research, 137*, 677–687. doi:10.1016/S0377-2217(01)00095-9

Salgado, F., & Pedrero, P. (2008). Short-term operation planning on cogeneration systems: A survey. *Electric Power Systems Research, 78*, 835–848. doi:10.1016/j.epsr.2007.06.001

Seeger, T., & Verstege, J. (1991). Short term scheduling in cogeneration systems. *1991 Power Industry Computer Application Conference* (pp. 106-112).

Seo, H., Sung, J., Oh, S. D., Oh, H. S., & Kwak, H. Y. (2008). Economic optimization of a cogeneration system for apartment houses in Korea. *Energy and Building, 40*(6), 961–967. doi:10.1016/j.enbuild.2007.08.002

Sheen, J. N. (2005). Fuzzy evaluation of cogeneration alternatives in a petrochemical industry. *Computers & Mathematics with Applications (Oxford, England), 49*(5-6), 741–755. doi:10.1016/j.camwa.2004.10.035

Sødergren, C., & Ravn, H. F. (1996). A method to perform probabilistic production simulation involving combined heat and power units. *IEEE Transactions on Power Systems, 11*, 1031–1036. doi:10.1109/59.496191

Song, Y. H., Chou, C. S., & Stonham, T. J. (1999). Combined heat and power economic dispatch by improved ant colony search algorithm. *Electric Power Systems Research, 52*, 115–121. doi:10.1016/S0378-7796(99)00011-5

Spinney, P. J., & Watkins, G. C. (1996). Monte Carlo simulation technique and electric utility resource decisions. *Energy Policy, 24*(2), 155–164. doi:10.1016/0301-4215(95)00094-1

Su, C. H., & Chiang, C. L. (2004). An incorporated algorithm for combined heat and power economic dispatch. *Electric Power Systems Research, 67*, 187–195. doi:10.1016/j.epsr.2003.08.006

Sudhakaran, M., & Slochanal, S. M. R. (2003). Integrating genetic algorithms and tabu search for combined heat and power economic dispatch. *Conference on Convergent Technologies for the Asia-Pacific Region.*

Sudhakaran, M., Vimal Raj, P. A. D., & Palanivelu, T. G. (2007). Application of particle swarm optimization for economic load dispatch problems. *2007 International Conference on Intelligent Systems Applications to Power Systems.*

Talaq, J.H., El-Hawary, F., & El-Hawary, M.E. (1994). A summary of environmental/economic dispatch algorithms. *IEEE Transactions on Power Systems, 9*(3), 1508–1516. doi:10.1109/59.336110

Thomas, M. (1994). Combined heat and power, the global solution to voltage dip, pollution, and energy efficiency. *International Conference on Power System Technology, PowerCon 2004* (vol. 2, pp.1975-1980).

Thorin, E., Brand, H., & Weber, C. (2005). Long-term optimization of cogeneration systems in a competitive market environment. *Applied Energy, 81*, 152–169. doi:10.1016/j.apenergy.2004.04.012

Toffolo, A., & Lazzaretto, A. (2002). Evolutionary algorithms for multi-objective energetic and economic optimization in thermal system design. *Energy, 27*, 549–567. doi:10.1016/S0360-5442(02)00009-9

Tsay, M.T. (2003). Applying the multi-objective approach for operation strategy of cogeneration systems under environmental constraints. *Electrical Power & Energy systems, 25*, 219-226.

Tsay, M. T., Chang, C. Y., & Gow, H. J. (2004). The operational strategy of cogeneration plants in a competitive market. *2004 IEEE Region 10 Conference*.

Tsay, M.T., & Lin, W.M. (2000). Application of evolutionary programming to optimal operational strategy cogeneration system under time-of-use rates. *Electrical Power & Energy systems, 22*, 367-373.

Tsay, M. T., Lin, W. M., & Lee, J. L. (2001a). Interactive best-compromise approach for operation dispatch of cogeneration systems. *IEEE proc. Generation, Transmission . Distribution, 148*(4), 326–332. doi:10.1049/ip-gtd:20010163

Tsay, M.T., Lin, W.M., & Lee, J. L. (2001b). Application of evolutionary programming for economical dispatch of cogeneration system under emission constraints. *Electrical Power & Energy systems, 23*, 805-812.

Tsikalakis, A. G., & Hatziargyriou, N. D. (2008). Centralized Control for Optimizing Microgrids Operation. *IEEE Transactions on Energy Conversion, 23*(1), 241–248. doi:10.1109/TEC.2007.914686

Tsukada, T., Tamura, T., Kitagawa, S., & Fukuyama, Y. (2003). Optimal operational planning for cogeneration system using particle swarm optimization. *Proceedings of the 2003 IEEE Swarm Intelligence Symposium* (pp. 138-143).

Tzeng, G. H., Shiau, T. A., & Lin, C. Y. (1992). Application of multicriteria decision making to the evaluation of new energy system development in taiwan. *Energy, 17*, 983–992. doi:10.1016/0360-5442(92)90047-4

Uche, J., Serra, L., & Sanz, A. (2004). Integration of desalination with cold-heat-power production in the agro-food industry. *Desalination, 166*, 379–391. doi:10.1016/j.desal.2004.06.093

Ummels, B. C., Gibescu, M., Pelgrum, E., Kling, W. L., & Brand, A. J. (2007). Impacts of wind power on thermal generation unit commitment and dispatch. *IEEE Transactions on Energy Conversion, 22*, 44–51. doi:10.1109/TEC.2006.889616

Vasebi, A., Fesanghary, M., & Bathaee, S.M.T. (2007). Combined heat and power economic dispatch by harmony search algorithm. *Electrical power & Energy system, 29*, 713-719.

Venkatesh, B. N. (1995). Decision models for management of cogeneration plants. *IEEE Transactions on Power Systems, 10*, 1250–1256. doi:10.1109/59.466530

Ventosa, M., Baillo, A., Ramos, A., & Rivier, M. (2005). Electricity market modeling trends. *Energy Policy, 33*, 897–913. doi:10.1016/j.enpol.2003.10.013

Voorspools, K.R. & D'haeseleer, W.D. (2003). Long-term unit commitment optimization for large power systems: unit decommitment versus advanced priority listing. *Applied Energy, 76*, 157–167. doi:10.1016/S0306-2619(03)00057-6

Wang, J., Jing, Y., Zhang, C., Shi, G., & Zhang, X. (2008). A fuzzy multi-criteria decision-making model for trigeneration system. *Energy Policy, 36*(10), 3823–3832. doi:10.1016/j.enpol.2008.07.002

Wang, L., & Singh, C. (2007). Environmental/economic power dispatch using a fuzzified multi-objective particle swarm optimization algorithm. *Electric Power Systems Research, 77*, 1654–1664. doi:10.1016/j.epsr.2006.11.012

Wang, L., & Singh, C. (2008). Stochastic combined heat and power dispatch based on multi-objective particle swarm optimization. *Electrical power & Energy system, 30*, 226-234.

Wang, S., Liu, C., & Luu, S. (1994). A Negotiation methodology and its application to Cogeneration Planning. *IEEE Transactions on Power Systems, 9*, 869–875. doi:10.1109/59.317661

Wang, Z., Zheng, D., & Jin, H. (2007). A novel polygeneration system integrating the acetylene production process and fuel cell. *International Journal of Hydrogen Energy, 32*(16), 4030–4039. doi:10.1016/j.ijhydene.2007.03.018

Weber, C., Marechal, F., Favrat, D., & Kraines, S. (2006). Optimization of an SOFC-based decentralized polygeneration system for providing energy services in an office-building in Tokyo. *Applied Thermal Engineering, 26*, 1409–1419. doi:10.1016/j.applthermaleng.2005.05.031

Weber, C., & Woll, O. (2006). Valuation of CHP power plant portfolios using recursive stochastic optimization. *9th International Conference on Probabilistic Methods Applied to Power systems,* KTH, Stockholm, Sweden, June 11-15, 2006.

Wong, K. P. (2002). Recent development and application of evolutionary optimisation techniques in power systems. *Proceedings of International Conference on Power System Technology, 1*, 1–5. doi:10.1109/ICPST.2002.1053493

Wong, K. P., & Algie, C. (2002). Evolutionary programming approach for combined heat and power dispatch. *Electric Power Systems Research, 61*, 227–232. doi:10.1016/S0378-7796(02)00028-7

Wu, D. W., & Wang, R. Z. (2006). Combined cooling, heating and power: A review. *Progress in Energy and Combustion Science, 32*, 459–495. doi:10.1016/j.pecs.2006.02.001

Wu, T., Shieh, S., Jang, S., & Liu, C. (2005). Optimal energy management integration for a petrochemical plant under considerations of uncertain power supplies. *IEEE Transactions on Power Systems, 20*(3), 1431–1439. doi:10.1109/TPWRS.2005.852063

Wu, Y. J., & Rosen, W. A. (1999). Assessing and optimizing the economic and environmental impacts of cogeneration/district energy systems using an energy equilibrium model. *Applied Energy*, *62*, 141–154. doi:10.1016/S0306-2619(99)00007-0

Zadeh, L. A. (1978). Fuzzy sets a basis for a theory of possibility. *Fuzzy Sets and Systems*, *1*, 3–28. doi:10.1016/0165-0114(78)90029-5

Zhou, P., Ang, B. W., & Poh, K. L. (2006). Decision analysis in energy and environmental modeling: An update. *Energy*, *31*, 2604–2622. doi:10.1016/j.energy.2005.10.023

LIST OF ABBREVIATIONS

AC: Ant Colony algorithm
AHP: Analytical Hierarchy Process
ANN: Artificial Neural Network
BB: Branch and Bound algorithm
CCHP: Combined Cooling, Heating and Power
CHP: Combined Heat and Power
DP: Dynamic Programming
Dup: Dual Programming
ED: Economic Dispatch
ELECTRE: ELimation Et Choice Traduisant la REalite(Elimination and Choice Expressing Reality)
EP: Evolutionary Programming
ETS: Emissions Trading Scheme
EU: European Union
GA: Genetic Algorithm
GDF: Interactive algorithm for multi-criteria optimization, developed by Geoffrion, Dyer and Feinberg
GNLO: General Non-Linear Optimization
GOP: General Operational Planning
HS: Harmony Search
IA: Immune Algorithm
KKT: Karush-Kuhn-Tucker conditions
LP: Linear Programming
LR: Lagrangian Relaxation algorithm
MAUT: Multi-Attribute Utility Theory
MCDM: Multi-Criteria Decision Making
MILP: Mixed Integer Linear Programming
NAIADE: Novel Approach to Imprecise Assignment and Decision Environment
NLP: Non-Linear Programming
OMCDM: Other Multi-Criteria Decision Making methods

PARIS: Preference Assessment by Imprecise Ratio Statement method
PROMETHEE: Preference Ranking Organization METHods for Enrichment Evaluation
PSO: Particle Swarm Optimization
QP: Quadratic Programming
RICH: Rank Inclusion in Criteria Hierarchies
SA: Successive Approximation
SMAA: Stochastic Multi-criteria Acceptability Analysis)
SNLP: Sequential Non-Linear Programming
SQP: Sequential Quadratic Programming
TOPSIS: Technique for Order Preference by Similarity to Ideal Solution
TS: Tabu Search
UC: Unit Commitment

Section 3
Systems and Applications

Chapter 12
Developing an Energy Security Risk Assessment System

Alexandros Flamos
National Technical University of Athens, Greece

Christos V. Roupas
National Technical University of Athens, Greece

John Psarras
National Technical University of Athens, Greece

ABSTRACT

Throughout the last two decades many attempts took place in order policy makers and researchers to be able to measure the energy security of supply of a particular country, region and corridor. This chapter is providing an overview presentation of the Energy Security Risk Assessment System (E.S.R.A.S.) which comprises the Module of Robust Decision Making (RDM) and the Module of Energy Security Indices Calculation (ESIC). Module 1 & 2 are briefly presented throughout section 2 and the application of Module 2 in nine case study countries is discussed at section 3. Finally, in the last section are the conclusions, which summarize the main points, arisen in this chapter.

INTRODUCTION

Energy planning and energy security of supply became one of the most important topics in the field of global economy. Recent problems that emerged due to the sudden energy supply disruptions pointed in the most emphatic way that further research and development must take place in order to mitigate problems regarding the smooth supply of energy.

Furthermore, the unstable economic environment that most economies are facing and the high dependency that most countries have in foreign imports, made essential for the policy makers to focus on the concept of energy security of supply. Energy security can be defined as the continuous and uninterrupted availability of energy, to a specific country or region. The European Union (EU) has many times highlighted as a key priority, the need to assess the current energy system and the risks of energy

DOI: 10.4018/978-1-60566-737-9.ch012

disruptions in order to design and adopt the required policies. The increased dependence on fossil fuel imports, sometimes from politically unstable regions of the world; the increase in the volatility of Primary Energy Supply fuel (P.E.S) prices, especially oil and the significant environmental impacts from their extensive consumption are considered non sustainable on the long term. In the aforementioned statement the three most important concepts that affect the design of security of supply can be summarized. The dependence that an economy has to a particular energy fuel conducts a very important role in the measurement of security of supply and can be identified as supply risk. Accordingly, the risk that an economy faces due to the volatility of P.E.S. prices can be identified as market risk, whilst the global climate change that is caused due to the usage of fossil fuels can be included under the broad category of environmental impacts risk.

A significant number of researchers have focused on the topic of energy security of supply. However, a common point of the existing research work is that the methodologies that were adopted in order to model the parameters that affect the smooth energy supply in a particular country or region, were limited to the development of particular indicators that measure only specific dimensions of energy security, without integrating in their approach the overall set of technoeconomic, social and environmental aspects of national and international energy systems. Notable research has been conducted by the Dutch Energy Research Centre (E.C.N.) (Jansen *et al*, 2004 and Schaepers *et al*, 2007). Jansen was the first that utilized the Shannon-Wiener diversity index as basic indicator (Gnansounou, 2008). The second study by ECN and the Clingendael International Energy Programme (CIEP) proposed quantitative indicators for quantifying the concept of Security of Energy Supply. Additionally, they were the first who created a weighting and scoring system for the synthesis of the supply/demand Index (S/D Index) taking into account final energy demand, energy conversion and primary energy supply. Apart from the research work that has been conducted by ECN, other research groups measured the concept of vulnerability that the European energy systems exhibit (Costantini *et al*, 2007; Gnansounou, 2008; Gupta, 2007, 2008; World Energy Council, 2008). All the aforementioned research efforts used subjective weights in order to derive a composite index (Jansen *et al*, 2004; Schaepers *et al*, 2007). Grubb et al, 2006 explored the strategic security of electricity in the context of the United Kingdom electricity system. Using the concept of diversity of fuel source mix they measured the impact of source variability on a second dimension of security, the reliability of generation availability. Additionally, Chevalier, 2005 used several indicators in order to measure the dependency that countries within the European Union exhibit to oil and natural gas. They found that most European countries are exposed to energy availability risks and stated the need for the European Union to develop a common energy security policy.

It can be observed that the existing literature is characterized by the absence of a system which through its sub modules will measure the energy security of supply for a particular country, integrating the concept of vulnerability and using the Principal Component statistical Analysis (PCA) for estimating objectively the weights that will be used for the synthesis of the index, as Gupta (2007) points. In addition, taking into account the long term planning that is required for the energy system expansion, it has been considered that the traditional risk management techniques should be further supported by appropriate Robust Decision Making (RDM) models that could facilitate decision making under deep uncertainty.

Based on the above needs, a system has been developed in order to facilitate the decision making process towards the formulation of energy policy strategies.

Figure 1. E.S.R.A.S Overview

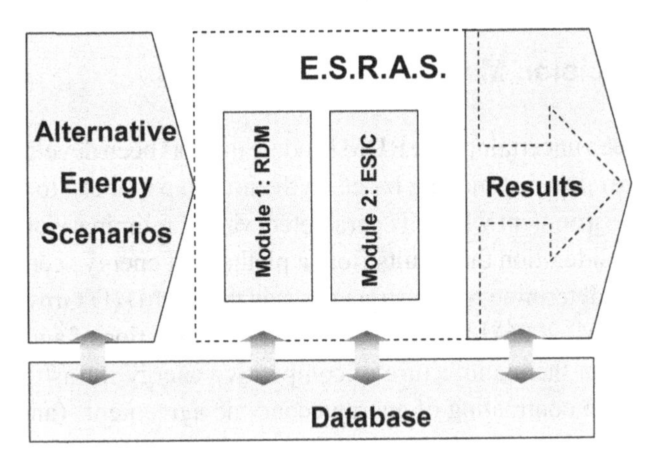

ENERGY SECURITY RISK ASSESSMENT SYSTEM

The name of the system is E.S.R.A.S. which stands for Energy Security Risk Assessment System. E.S.R.A.S. has been developed in order to materialize the methodology that throughout measurable indicators provides quantitative outputs about the level of energy security of a particular country or region. The general objective is the measurement of energy vulnerability that economies face relative to their primary energy fuels measuring the three dimensions of vulnerability: supply risk, market risk & environmental risk. In addition through the calculation of the robustness function, that expresses the largest level of uncertainty in which the failure cannot be present, and of the opportunity function, that expresses the lowest level of uncertainty that permits the possibility of sweeping victory, the decision making under deep uncertainty is facilitated. More specifically, through its sub modules E.S.R.A.S, examines the robustness of energy systems and assess the vulnerability that a country exhibits to oil and natural gas. This is the main advantage of the developed system, as it formulates the basic concepts that affect and define the security of supply of a specific country or region. Contrary to the existing methodologies, E.S.R.A.S. measures both the dependence that an energy system has to the main Primary Energy Supply fuels and the robustness of the system. Additionally, by using partial regression techniques it defines the significance of each aspect that affects the security of supply and provides the reader with aggregate results enabling comparisons among various energy systems, countries, regions and alternative policy scenarios.

Currently, E.S.R.A.S. incorporates two main modules (Fig 1):

Module 1: Robust Decision Making (RDM)
Module 2: Energy Security Indices Calculation (ESIC)

The following sections are describing the two modules of E.S.R.A.S and their sub-modules.
Module 1 incorporates 3 sub-modules:

- Sub-module 1: Robustness Function Calculation given Cost Threshold
- Sub-module 2: Cost Threshold Function Calculation given Robustness

- Sub-module 3: Calculation of Regret and Relative Regret

Module 1: Robust Decision Making

Seeking decisions under deep uncertainty, the RDM Module that has been developed (MATLAB 7.0) tries via the philosophy of robust decision making based on the info-gap models; to determine the robustness of energy systems in an environment where its parameters vary. Preliminary objective of the developed software is to take into consideration the results from a plethora of energy scenarios (via model generator TIMES), secondly, is to determine robust strategies (via the RDM) (D.Groves, 2005) with the use of info-gap theory (Ben-Haim Y., 2006) by receiving as an input scenarios of aggregated expenses. These expenses could be costs from the manufacture of competitive energy infrastructures, from the growth of energy corridors, from the contracting of energy/economic agreements (and dependences) between countries and from the restriction of environmental damage. The output of the model is the mapping of future energy strategies helping the promotion of better long-term choices in the sector of energy.

Even if initially the decision makers map out the energy policy that has decided to be followed, this policy is an estimate or approach of final policy which will be followed over time horizons of twenty or more years. The reasons that create the divergence between initial and finally realised policy are:

1. The environmental impacts are significant and the natural resources are exhausted. This may result to the reception of more intensive and stricter environmental and resource exploitation measures.
2. The energy policy followed by each one of the countries or coalitions is a function which must take under consideration the persons that live and work in these countries. Energy referenda, new governments, more environmental friendly budgets, human tendencies, energy fashions are some of the human concerns that can render an energy system uncertain.
3. The future, energy, technological developments increase the uncertainty with regard to the policy that will finally be followed (even if no intense).
4. The energy corridors, specific countries relations with EU and the EU neighbours, the energy subsidies etc. may influence the initial policy choice.

The above factors could render the energy policy decision as a decision of deep uncertainty with reward function the relative cost (cumulated cost increase).

Methodology

An info-gap model is a without limit nested sets. The structure of sets in an info-gap model emanates from the information on the uncertainty. Generally, the structure of an info-gap model of uncertainty is selected to determine the smallest or the most accurate family of sets which elements are conformed to the previous information.

In any uncertainty range α, the set $U\left(\alpha, \tilde{u}\right)$ contains all the functions not bigger than α. Nevertheless, the uncertainty range is unknown and thus the info-gap model is an unlimited family of sets. Uncertainty could be either pernicious or propitious. That is to say the uncertain variants can be unfavourable or favourable. Adversity expresses the possibility of failure, while the favourability expresses the occasion for the sweeping success. The pernicious and propitious aspects of uncertainty could be expressed by two decision functions:

1. the robustness function which expresses the immunity to the failure,
2. the opportunity function which expresses the immunity to sweeping success.

The robustness function expresses the largest level of uncertainty in which the failure cannot be present while the opportunity function expresses the lowest level of uncertainty that permits the possibility of sweeping victory.

Let consider q be the parameter decision vector. In other words, when the energy policy decision makers of a country decide, they determine this decision vector (estimate of final policy that will be followed). This vector has the form:

$$q = \begin{bmatrix} per\_NoExchg \\ per\_NorGas \\ per\_SwePO \\ per\_GhgLim \end{bmatrix} \quad (1)$$

where, $per\_NoExchg$, per_NorGas, per_SwePO and per_GhgLim are the participation percentages of indicative scenarios: NoExchg, NorGas, SwePO and GhgLim in the policy respectively that is taken into consideration by the decision makers. The robustness function is expressed as the biggest values set of uncertainty α of an info-gap model and is given by:

$$\hat{\alpha}(q) = \max\{\alpha : \text{ minimal requirements are satisfied}\} \quad (2)$$

The robustness function does not include maximization of performance or decision result. The largest bearable uncertainty is found in the decision vector q that satisfies the performance (performance above a critical level of survival or threshold). In this way, the robustness function comes from a decision algorithm that maximizes the immunity against the pernicious uncertainty.

Furthermore, a reward function $R(q,u)$ is taken into consideration. The used reward function $R(q,u)$ in this chapter is the function of cumulated increase cost. The reward function depends on the vector q and the info-gap uncertainty function u. Using (equation 2), the reward function $R(q,u)$ which is the cumulated increase cost should be equal or smaller than a threshold r_c.

The robustness function can now be expressed from the equation:

$$\hat{\alpha}(q,r_c) = \max\left\{\alpha : \left(\max_{u\in U(\alpha,\tilde{u})} R(q,u)\right) \le r_c\right\} \quad (3)$$

where the set $U(\alpha,\tilde{u})$ includes the possible Nordic energy policy decisions that the divergence from (initial) policy is not bigger than α, r_c is the threshold of policy cumulated increase cost and u is the candidate policy that belongs in the set $U(\alpha,\tilde{u})$. Observing the equation (3), $\hat{\alpha}(q,r_c)$ can be determined given r_c or vice-versa r_c can be determined given $\widehat{\alpha}(q, r_c)$ (Fox *et al*, 2007; Krokhmal, 2003; 2004; Regan *et al* 2005).

Sub-module 1: Robustness Function Calculation given Cost Threshold

One of the most important aspects of the problem described above is the robustness function calculation given policy (cumulative increase) cost threshold. Having an initial policy for the energy system of case study countries q and the given cumulative cost increase threshold r_c, the algorithm results finally in the robustness $\widehat{\alpha}(q, r_c)$. This sub-module calculates the column vector $\overline{u}_k$ and in this point, the political uncertainty is imported in the model.

Sub-module 2: Cost Threshold Function Calculation given Robustness

The second important issue is the calculation of the (cumulative increase) cost threshold function given robustness. This is the inverse problem from which that has been presented above. Having initial policy for the energy system of case study countries q and the robustness $\widehat{\alpha}(q, r_c)$, the algorithm which is used is presented giving as an output the cost threshold $r_c(q, \widehat{\alpha})$.

Sub-module 3: Calculation of Regret and Relative Regret

Regret is defined as the difference between the performance of a future policy, given some value function, and that of what would have been the best performing policy in that same future. As Savage (1954) notes, the regret of policy j in future f using values m is given as

$$\mathrm{Re}\,gret_m(j,f) = Performance_m(j,f) - \underset{j'}{Min}\left[Performance_m(j',f)\right] \tag{4}$$

where policy j' indexes through all policies in a search to determine the one best suited to the conditions presented by future f. A regret measure envisions people in the future benchmarking the result of a strategy against the one that would have been chosen by past decision makers if they had had perfect foresight. It also supports the desire of today's decision makers to choose policies that, in retrospect, will not appear foolish compared to available alternatives. It is also sometimes useful to consider the relative regret where the regret is

$$\mathrm{Re}\,lative\_\mathrm{Re}\,gret_m(j,f) = \frac{Performance_m(j,f) - \underset{j'}{Min}\left[Performance_m(j',f)\right]}{\left|\underset{j'}{Min}\left[Performance_m(j',f)\right]\right|} \tag{5}$$

scaled by the best performance attainable among the candidate strategies in a given scenario (Kouvelis *et al* 1997; Lempert *et al* 1996; 1998; 2000a; 2000b; 2003; 2005).

Observing equation (5) in the denominator of fraction the absolute price of minimal price of cumulative cost increase is presented. Generally, the relative regret in this chapter as mathematic depiction determines the ratio of deviation from the best price concerning the absolute distance of best price. The best price, generally, is the lowest value of cumulative cost increase and generally this value is positive.

Module 2: Energy Security Indices Calculation

Having described the first Module, we now proceed to the description of Module 2. The main purpose of Module 2 is to implement the methodology that throughout measurable indicators provides quantitative outputs about the level of energy security of a particular country. The general objective is the measurement of energy vulnerability that economies face relative to their main Primary Energy Fuels measuring the three dimensions of vulnerability: Supply Risk, Market Risk and Environmental Risk.

Module 2 has been applied, in nine European countries: Greece, Cyprus, France, Belgium, Germany, Austria, Spain, Italy and Bulgaria, focusing on the Supply and Market Risk. The selection of the countries occurred in terms of geographical position, since all countries are European and close to each other and therefore face similar conditions of energy imports; while on the other side their energy systems different structure provides useful insights about the effects of various technoeconomic parameters in their energy supply security.

Module 2 integrates selected indices that measure the dependence and vulnerability of the security of supply that a country or region has, relative to Oil & Natural Gas. In this framework, six Energy Security Indicators that measure the energy mix of a particular country and the degree of supply concentration were selected and calculated. Consequently these indicators were combined using the multivariate statistical approach of Principal Component Analysis (P.C.A.) in order to develop two Sub-Modules that will enable us to synthesize two composite indices named Oil Vulnerability Index (O.V.I.) and Natural Gas Vulnerability Index (N.G.V.I.).

Methodology

The steps that were followed in order to develop the indices, were firstly to examine the energy mix of the countries under consideration, secondly to use indicators that measure the two major risks that contribute to the overall P.E.S fuels vulnerability of an economy, which are market risk and supply risk, and thirdly to build a model that will allow us to aggregate the results that were obtained from the indicators and will enable us to construct an index that will measure the vulnerability that the countries under examination have to the P.E.S fuels.

More specifically, the procedure that will be explained in the next section will be performed for oil & natural gas in order to develop separate sub-indices. The first index will measure the vulnerability that the countries under consideration have relative to oil, whereas the second index will exhibit the vulnerability that the aforementioned countries have to natural gas.

Energy Mix of Case Study Countries

As the first step of our approach is to examine the energy mix of the countries that constitute our sample, we used as sources several databases in order to obtain the essential information. The basic intuition behind the identification of the energy mix of each country is to examine the percentage of the P.E.S fuels that each country consumes and to identify their dependency on energy imports. To be more specific, the percentage of total imports from other countries is the first step towards the development of indicators that measure the geopolitical risk. This procedure will be explained in detail in another section of this chapter. The databases or citations that were used in the identification of the sample countries energy mix are presented below: Oil Information 2005 for the Organization for Economic Cooperation and Development (OECD) countries (IEA, 2006a; 2006b; 2006;c), BP Statistics 2006, Integrated Energy Policy 2006 and Energy Statistics Yearbook 2006 (DESA, 2006). The data on the oil reserves, production and consumption has been taken from BP Statistical Review 2006 and the Energy Information Administration (EIA) 2006.

Market Risk and Supply Risk

The second step in our procedure is to measure the countries under consideration vulnerability relative to Market and Supply risk. The first important element that defines the security of supply of a particular country is the dependency that a country has on a particular P.E.S. fuel (Market risk). Market risk of an economy refers to the risks of macroeconomic effects due to erratic price fluctuations in energy markets, whilst Supply risk of an economy refers to the risks of physical disruptions in P.E.S supplies, as Gupta (2007) points. In order to measure and quantify these two risks, several indicators that capture the essence of the two risks for oil and natural gas have been utilized.

For the measurement of market risk for the two primary energy fuels, oil and natural gas two indicators were used: P.E.S. consumed in an economy to its Gross Domestic Product (GDP) and P.E.S consumption in total primary energy consumption. The first indicator measures the P.E.S. fuel that is consumed in the countries' economies under consideration relative to their GDP and it is expressed as tones of oil per unit of GDP, whilst the second indicator is calculated as the percentage of the P.E.S fuels under examination relative to the total P.E.S fuels that each country consumes.

Additionally, for the measurement of supply risk, which refers to the risks of physical disruptions in P.E.S fuel supplies as Gupta (2007) points, four indicators were used.

Each indicator was calculated for the P.E.S fuel under examination. The selection of the supply risk measurement indicators occurred on the basis of four factors: net P.E.S import dependence of an importing country, diversification of P.E.S. imports, political risks in P.E.S. supplying countries, and market liquidity as Gupta (2007) notices.

More specifically, we have selected two indicators that capture the essence of the geopolitical risks: Political Measurement Indicator and the Market Liquidity Indicator, and two indicators that quantify the diversification of the energy mix that the sample countries have: Diversification of Primary Energy Demand (DPED) and Net Energy Import Dependency (NEID).

The first pair of indicators captures and quantifies the impact of the Geopolitical risks that the countries under examination have, whilst the second pair of indicators measures the diversification that each country has relative to the Primary Energy Demand and the Net Import Dependency that the countries under examination have. In the following section the methodology in order to calculate each indicator will be provided.

Principal Component Analysis

Having explained the intuition behind the selection of meaningful indicators that will measure and quantify the two main concepts of security of supply, in this section we will proceed to the description of the model that was used for the synthesis of the two sub indices that will measure the vulnerability that the countries under examination have relative to natural gas and oil.

For the development of the two sub-indices, the PCA has been used. PCA is a multivariate statistical approach that transforms a set of correlated variables into a set of uncorrelated variables called components. These components are linear combinations of the original variables. PCA is used to reduce the dimensionality problems and to transform interdependent coordinates into significant and independent ones as Smith, (2006) notices.

Using the PCA, the vulnerability that each country under examination has relative to natural gas and oil will be examined.

The main reason for employing PCA is that it makes possible to define a synthetic measure that is able to capture interactions and interdependence between the selected set of indicators that were developed in order to constitute the two sub-indices. These indicators are called causal variables; while the two corresponding sub indices are the explained variables.

The main advantage of PCA statistical approach is that while standard regression techniques require the explained/dependent variable to be observed, PCA treats the latter as a latent variable (Basu, 2002; Shukla & Kakar, 2006). Principal component analysis constitutes a canonical form and helps to understand both the individual contribution of each of the indicators to the Index and their aggregate contribution.

Development of Indices

More specifically, the general form of our model, Gas and Oil Vulnerability Index (GOVI), that will be used for the development of the two indices, Oil Vulnerability Index (OVI) and Natural Gas Vulnerability Index (NGVI), is presented below:

$$GOVI = a + \beta_1 X_1 + \dots + \beta_K X_K + e \tag{6}$$

where X_1, X_K, are set of the proposed indicators that were used for capturing and quantifying the two main elements of energy security of supply, market and supply risk. Additionally, β1, βk are the corresponding vectors of parameters in each domain, and e is the error term.

Oil Vulnerability Index and Natural Gas Vulnerability Index

Having presented the general form of our model, we can proceed to the development of the two indices that will measure the vulnerability that our sample countries exhibit relative to oil & natural gas. Equation (7) exhibits the Oil Vulnerability Index (OVI) and equation (8) exhibits the Natural Gas Vulnerability Index (NGVI).

$$OVI = \beta_1 \chi_1 + \beta_2 \chi_2 + \beta_3 \chi_3 + \beta_4 \chi_4 + \beta_5 \chi_5 + \beta_6 \chi_6 + \varepsilon \tag{7}$$

$$NGVI = \beta_1 \chi_1 + \beta_2 \chi_2 + \beta_3 \chi_3 + \beta_4 \chi_4 + \beta_5 \chi_5 + \beta_6 \chi_6 + \varepsilon \tag{8}$$

Table 1. Correlation Matrix of Oil Indicators

Relative Indicators	OIL/GDP	OIL/T.P.E.S	ML	PM	DPED	NEIS
OIL / GDP	1.00000					
OIL / T.P.E.S	0.05621	1.00000				
ML	-0.42560	-0.01200	1.00000			
PM	-0.36540	-0.15800	0.45200	1.00000		
DPED	-0.38100	0.52100	-0.32500	-0.08560	1.00000	
NEID	0.50000	0.75200	0.56150	-0.23650	0.85200	1.00000

The total variation in the two indices is composed of: (i) the variation due to sets of indicators, and (ii) the variation due to error. If the model is well specified, including an adequate number of indicators in each domain, so that the mean of the probability distribution of ε is zero, ($E(\varepsilon) = 0$), and error variance is small relative to the total variance, we can reasonably assume that the total variation in the two indices is largely explained by the variation in the indicator variables in each domain included for the computation of this composite index.

Taking into account that the number of indicators, which are considered as variables in the model, is large and the indicators might be mutually correlated, the replacement of the set of indicators by an adequate number of their principal components (PCs) will be performed (Basu, 2002).

The PCs are considered to be normalized linear functions of the indicator variables and are mutually orthogonal. The first PC accounts for the largest proportion of total variation (trace of the covariance matrix) of all indicator variables. The second PC accounts for the second largest proportion and so on.

Principal Components

Firstly, we transform the indicators into their standardized form, subtracting the mean from each of the data dimensions in order to produce a data set that will have mean equal to zero:

$$X_k = \frac{X_k - \overline{X}_k}{stdX_k} \tag{9}$$

The aforementioned procedure will transform the variables on a 0 to 1 scale and will enable us to calculate the covariance matrix. Since the data is 6 dimensional the covariance matrix will be 6×6.

Secondly, the calculation of the eigenvectors and eigenvalues for the 6×6 matrix will take place. For calculating the eigenvectors and the eigenvalues the statistical program Xlstat is used.

The main intuition behind the calculation of the eigenvalues is the usage of the following detrimental equitation:

$$\left(R - \lambda I\right) = 0 \tag{10}$$

Solving for λ, a sixth degree polynomial equation is obtained and six λ which correspond to the correlation matrix R are calculated. Furthermore, in order to derive the eigenvectors the following matrix equation is solved:

$$\left(R - \lambda_j I\right) F_j = 0 \tag{11}$$

Solving the aforementioned matrix equation we derive six eigenvectors that are presented in table 1.

Taking in account that eigenvectors and eigenvalues have different values, we order them by the highest eigenvalue, to lowest, for obtaining the components in order of significance.

$$\lambda_1 \rangle \lambda_2 \rangle \lambda_3 \rangle \lambda_4 \rangle \lambda_5 \rangle \lambda_6$$

Finally, we develop a matrix of vectors by placing the calculated eigenvectors in the columns. Having formed this matrix we transpose the matrix of vectors and multiply it on the left of the original normalized indicators, transposed (Smith, 2006).

Using the eigennectors the calculation of the PCs will take place.

Taking into account that λ_ι = var Pj and hence $\lambda 1 > \lambda 2 > \lambda 3 > \lambda 4 > \lambda 5 > \lambda 6$ = total variation of the OVI and NGVI. Using therefore the PCs and the eigenvalues the two indices are developed.

Thus, the OVI and the NGVI are developed using the variances of the calculated six principal components as weights:

$$OVI = \frac{\lambda_1 \chi_1 + \lambda_2 \chi_2 + \lambda_3 \chi_3 + \lambda_4 \chi_4 + \lambda_5 \chi_5 + \lambda_6 \chi_6}{\lambda_1 + \lambda_2 + \lambda_3 + \lambda_4 + \lambda_5 + \lambda_6} \tag{12}$$

$$NGVI = \frac{\lambda_1 \chi_1 + \lambda_2 \chi_2 + \lambda_3 \chi_3 + \lambda_4 \chi_4 + \lambda_5 \chi_5 + \lambda_6 \chi_6}{\lambda_1 + \lambda_2 + \lambda_3 + \lambda_4 + \lambda_5 + \lambda_6} \tag{13}$$

APPLICATION OF MODULE 2

Having explained the methodology that is followed in order to obtain two sub indices that will measure the energy security of supply for oil & natural gas in the previous section, we will now proceed to the discussion of the results that were obtained.

To begin with, table 1 displays the correlation matrix of the normalized indicators for the computation of the composite index that measures oil vulnerability for the selected nine countries, whilst table 2 exhibits the correlation matrix for the indicators that are used for the computation of natural gas index.

It must be stated that the normalized indicators are made positively related to both the oil and natural gas vulnerability indices. From the results obtained from table 1 and table 2, it can be observed that the indicators that measure Diversification of Primary Energy Demand (DPED) and Net Energy Import Dependency (NEID) are negatively related to Market Liquidity and Political Measure. This can be explained by the fact that energy import dependency and primary energy demand are affected by the level of political risk that the exporting countries are exhibiting. More specifically, the highest the level of political risk, that is measured using the Political Measure indicator, of a particular exporting country

Table 2. Correlation Matrix of Natural Gas Indicators

Relative Indicators	OIL/GDP	OIL/T.P.E.S	ML	PM	DPED	NEIS
OIL / GDP	1.00000					
OIL / T.P.E.S	0.03865	1.00000				
ML	-0.38970	-0.08500	1.00000			
PM	-0.08213	-0.24950	0.48990	1.00000		
DPED	-0.40260	0.41250	-0.2009	-0.09852	1.00000	
NEID	0.62450	0.68710	0.45620	-0.38710	0.72600	1.00000

the lower the value of Net Energy Import Dependency indicator. Accordingly, it can be noticed that both indicators P.E.S. consumed in an economy to its GDP and P.E.S. consumption in total primary energy consumption are negatively related to the Market Liquidity and Political Measure indicators.

As it was previously mentioned the OVI and NGVI are calculated as the sum of the six individual indicators that were used for the quantification and measurement of the concept of security of supply, using as weights the eigenvalues. Thus, before discussing the results of the aggregate OVI and NGVI, it would be useful to examine the performance of the sample countries relative to each indicator, as useful insights will be derived relative to the overall performance of the countries.

Ranking of Countries According to Individual Indicators

The visualization of the application results depicts the performance of the sample countries relative to the four indicators that measure the supply risk, exhibiting also the ranking of the countries in relation with the two indicators that quantify market risk.

Supply Risk Indicators

From the four supply risk indicators, the DPED indicator is calculated as $DPED = -\sum_i \left(p_i \ln p_i\right)$ and measures the dependency that the sample countries have, in the primary energy fuels.

A lower value implies that a country is dependent on limited energy sources whilst a higher value indicates a diversification of primary energy fuels.

Among the nine case study countries, Austria is the best performing one as regards DPED. This is reasonable, as its economy energy sources are evenly distributed among the main P.E.S. On the contrary, the lowest performing country is Cyprus as it is mainly dependant in two primary energy fuels.

The second supply risk indicator, NEID, is measured similarly to the DPED indicator adding a correction factor c_i that measures the energy import dependency of the sample countries.

$$NEID = -\sum_i \left(p_i \ln p_i\right)$$

$$c_i = 1 - m_i$$

NEID measures the level that a country relies on energy imports in order to meet its Primary Energy Demand. A higher value of this indicator implies that a country is highly dependent on energy imports.

Figure 2a. Oil Supply Risk Indicators; Figure 2b. Natural Gas Supply Risk Indicators

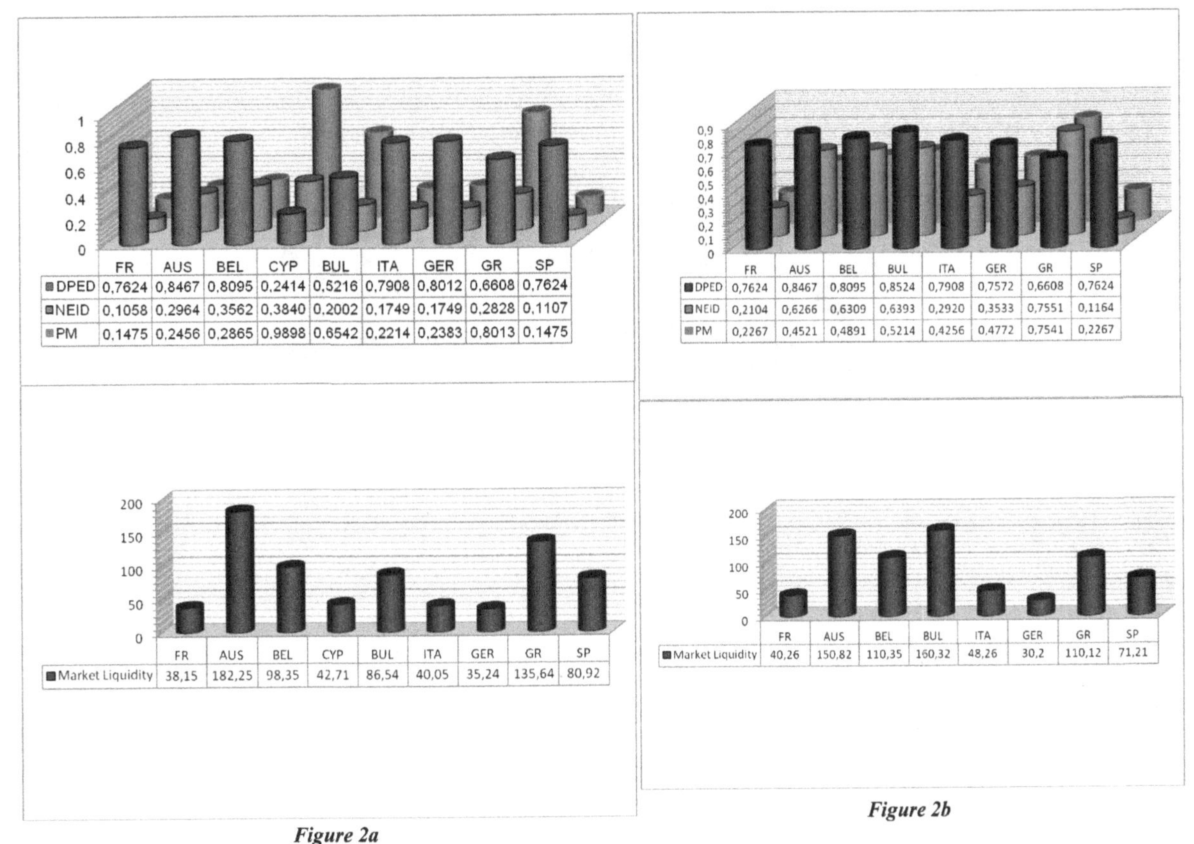

Figure 2a

Figure 2b

As it can be observed from figure 2a, in terms of oil the less dependant country in terms of oil is France whilst the most dependant one is Cyprus. Accordingly, at figure 2b the less dependant country in natural gas is Spain and the most dependant, is Greece.

The Market Liquidity indicator is derived by assuming an exponential function that adjusts the market concentration for risk due to limited market liquidity as Gupta (2007) observes. It is measured as the percentage of world oil imports relative to the oil imports of the case study countries.

A higher value of Market Liquidity indicates a higher potential for a country to switch between different oil suppliers and therefore to increase its energy sources diversification and consequently its energy supply security.

Among the best performers in terms of market liquidity are Austria and Greece. On the other hand Germany and Cyprus are among the worst performers in terms of market liquidity.

Finally, the Political Measure indicator is calculated using the Herfindahl-Hirschman index and measures the exposure of the countries' economies in long term geopolitical risks of oil and natural gas supply. The greater the value of Political Measure indicator the larger the political risk that the sample countries are facing. Particularly, Cyprus is the most vulnerable country in terms of geopolitical risk as it is mainly oil dependant and imports from politically difficult and unstable OPEC countries.

Figure 3. Oil and Natural Gas Market Risk Indicators

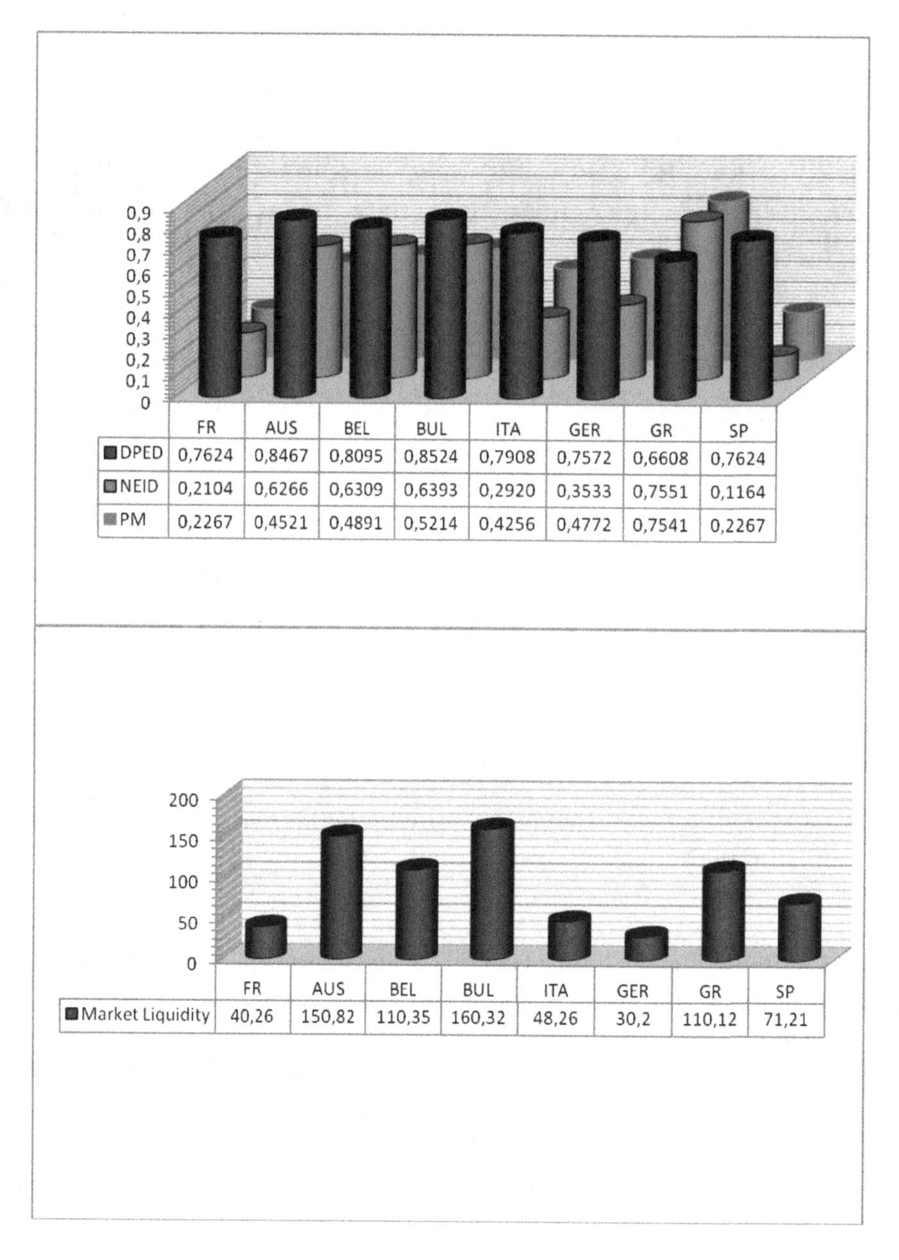

Market Risk Indicators

As it was previously mentioned the market risk that the countries are facing, is an important factor which affects the security of energy supply for a particular country or region. The macroeconomic effects that steam from the high dependency of the economies in oil and natural gas can disrupt the energy balance of a particular country.

Particularly, the high oil vulnerability that a country exhibits can lead to the increase of inflation due to the higher oil prices. Additionally, the high natural gas vulnerability can create economic and

Figure 4. Oil Vulnerability Index

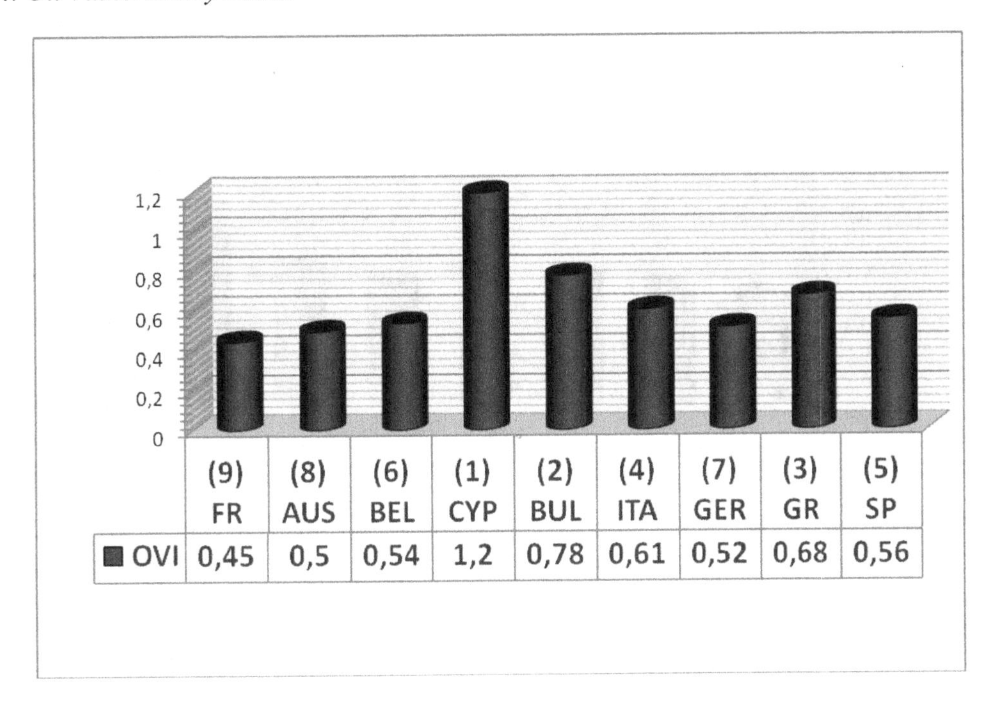

political problems due to the fact that the natural gas production and distribution is concentrated in a few countries.

The two indicators that were used for the calculation of market risk, measure the oil and natural gas consumption of a country in tones relative to its GDP and the percentage of oil and natural gas consumption to its total Primary Energy Consumption. The P.E.S. consumed in an economy to its GDP measures the impact that an increase in oil and natural gas prices will have to the country's economy, whilst the P.E.S consumption in total primary energy consumption indicator measures the exposure of a country's economy to the degree of dependence on imported oil and natural gas (Figure 3).

Aggregated Indices

Having ranked the countries on the basis of individual Indicators, we now proceed to the display of the final values of the O.V.I. and N.G.V.I. for all the case study countries. Figure 4, displays the Oil Vulnerability Index of all the nine countries and figure 5, exhibits the Natural Gas Vulnerability Index.

The country that is ranked in the first position is the most vulnerable one, while the one who is ranked in the 9th position is the least vulnerable.

Sub-Module 1: Oil Vulnerability Index

As it can be observed, the most vulnerable country in terms of oil is Cyprus. This result can be explained by the fact that Cyprus is highly dependent and thus vulnerable to oil, as its energy mix for the year 2006 is constituted in a 94% percentage from oil and only a 2% and 4% is solid fuels and renewable accordingly. Moreover, Cyprus imports are mainly oil and its origins were Russia and Syrian Arab

Figure 5. Natural Gas Vulnerability Index

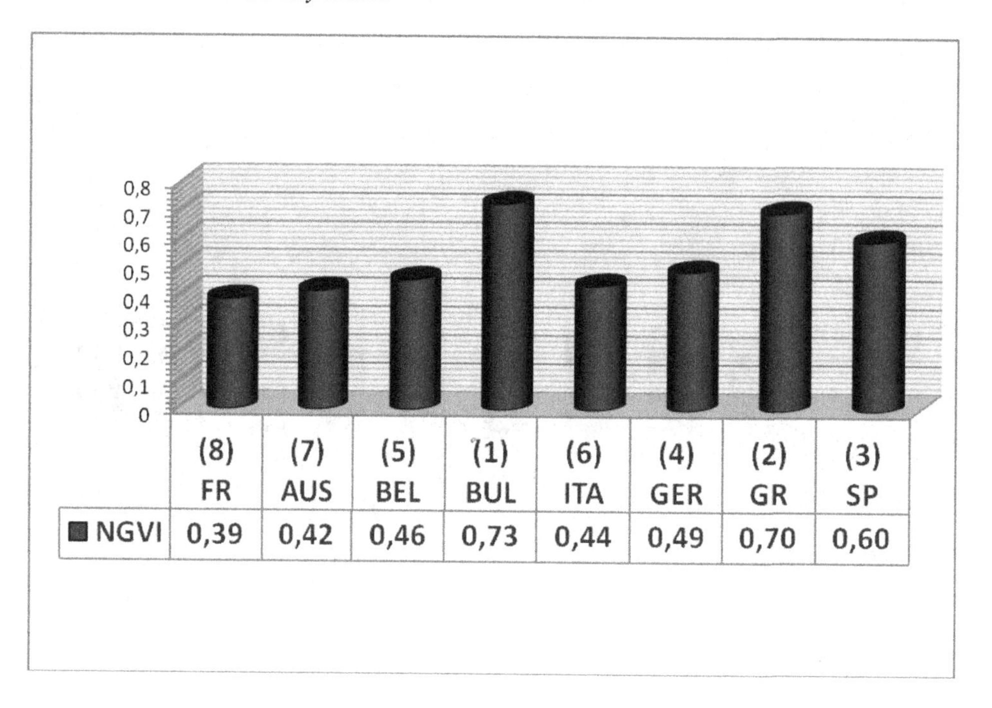

Republic. France ranked 9th in terms of oil vulnerability and is the least dependant country in oil. The energy mix of France, which is constituted from 40% nuclear, 33% oil, 14% natural gas, 5% solid fuels 6% renewable and the rest 2% from other fuels explains why it is considered to be one of the more secure countries in terms of energy supply. Additionally, the fact that France is not dependant in oil is mirrored in its national accounts as the ratio oil consumed to France GDP is very low (0,068). The rest countries of our sample can be separated in two groups: Those that have a high vulnerability to oil and those that are not so dependant to oil. In the first group we find Greece and Bulgaria and in the second Spain, Italy, Austria, Germany and Belgium. From these intercountry differences, with respect to the OVI, certain regional patterns can be identified. The central Europe countries are exhibiting similarities in the vulnerability to oil. This observation can lead to the conclusion that a similar energy strategy is adopted from countries that operate in the same geographical area.

Sub-Module 2: Natural Gas Vulnerability Index

Similar conclusions are obtained from the results of the Natural Gas Vulnerability Index. Bulgaria is the most vulnerable country in terms of natural gas and France is again the less vulnerable one. At this point it must be stated that Cyprus is not included in the N.G.V.I., as Natural Gas is not part of its energy mix.

The results that are obtained from the N.G.V.I. are in convergence with the results that the O.V.I. provides us with. This can be explained by the fact that for the synthesis of the two indices we used the same indicators and therefore the same concepts of security of supply were examined.

CONCLUSION

The E.S.R.A.S brings a significant potential to be elevated at the level of a facilitative Decision Support Systems on critical issues of energy security of supply. However, its modules and sub-modules should be carefully tested before proceeding with their results verification. More specifically, the R.D.M. module could enable us to compare energy policies in terms of various robustness/discount rate areas in which the energy system can be found and in this regard facilitate the long-term energy decisions. The E.S.I.C module assessed two composite vulnerability indices in order to measure the vulnerability that the countries under examination exhibit relative to Oil and Natural Gas. Although, despite the useful results that were obtained from the application of the two indices, the analysis that was performed was static as the indices were synthesized for the year 2006. It is in the authors' plans to develop indices that will operate in a more dynamic environment and to incorporate more indicators capturing all the essences of vulnerability and security of supply. Finally, it should be considered that the results of traditional risk management techniques in the energy sector may be proved not suitable for constructing and quantifying an Energy Security Index which will incorporate all risk parameters that endanger the uninterrupted flow of energy supplies among the numerous energy routes from their destination to European countries. Following an extensive desktop research, the authors are currently considering on financial techniques and concepts that may provide appropriate methods for constructing such an Index. The E.S.R.A.S architecture enables the almost unlimited horizontal and vertical expandability that will allow the authors team to further enhance it with additional modules and sub-modules towards its elevation to an integrated Decision Support System on energy systems analysis and policy advice.

REFERENCES

Basu, S. (2002). Does Governance Matter? Some Evidence from Indian States. *VIIth Spring Meeting of Young Economists,* Paris.

Ben-Haim, Y. (2006). *Info-Gap Theory: Decisions Under Severe Uncertainty,* (2nd ed.). Academic Press: London.

British Petroleum (BP). (2006) *Statistical Review.*

Chevalier, M., (2005). Security of energy supply for the European Union. *International Journal of European Sustainable Energy Market.*

Costantini, V., Gracceva, F., Markandya, A., & Vicini, G. (2007). Security of energy supply. Comparing scenarios from European perspective. *Energy Policy, 35,* 210–226. doi:10.1016/j.enpol.2005.11.002

Fox, D., Ben-Haim, Y., Hayes, K., McCarthy, M., Wintle, B., & Dunstan, P. (2007). An info-gap approach to power and sample size calculations. *Environmetrics, 18,* 189–203. doi:10.1002/env.811

Gnansounou, E. (2008). Assessing the energy vulnerability: Case of industrialised countries. *Energy Policy, 36*(10), 3734–3744. doi:10.1016/j.enpol.2008.07.004

Groves, D. (2005). *New Methods for Identifying Robust Long-Term Water Resources Management Strategies for California.* Ph.D. dissertation, Pardee Rand Graduate School, Santa Monica, CA.

Grubb, M., Butler, L., & Twomey, P. (2006). Diversity and security in UK electricity generation: The influence of low-carbon objectives. *Energy Policy, 34*(18), 4050–4062. doi:10.1016/j.enpol.2005.09.004

Gupta E. (2007). *Assessing the relative geopolitical risk of oil importing countries* [Working paper].

Gupta, E. (2008). Oil vulnerability index of oil-importing countries. *Energy Policy, 36*, 1195–1211. doi:10.1016/j.enpol.2007.11.011

IEA (International Energy Agency). (2006a). *Energy Balances of Non- OECD Countries 2006*. Paris: IEA.

IEA (International Energy Agency). (2006b). *Energy Balances of OECD Countries 2006*. Paris: IEA.

IEA (International Energy Agency). (2006c). *World Energy Outlook 2006*. Paris: IEA.

Jaffe, A., & Wilson, W. (2005). *Energy Security: Oil-geopolitical and Strategic Implications for China and the United States*. Houston, TX: James A Baker Institute for Public Policy, Rice University.

Jansen, J. C., van Arkel, W.G., Boot, M.G. (2004). *Designing indicators of long term energy supply security*, (ECN-C-04-007).

Kouvelis, P., & Yu, G. (1997). *Robust Discrete Optimization and Its Applications*. Dordrecht, the Netherlands: Kluwer Academic Publishers.

Krokhmal, P. (2003). *Risk Management Techniques for Decision making in Highly Uncertain Environments*. Phd dissertation, University of Florida, Gainesville, FL.

Krokhmal P., Murphey R., Pardalos P., Uryasev S., Zrazhevski G.(2004). *Robust Decision Making: Addressing Uncertainties in Distributions*. [DTIC, Final report].

Lempert, R. J., & Bonomo, J. (1998). *New Methods for Robust Science and Technology Planning*. Santa Monica, CA: RAND.

Lempert, R. J., Groves, D. G., Popper, S. W., & Bankes, S. C. (2005). (Submitted to). A General [Management Sciences.]. *Analytic Method for Generating Robust Strategies and Narrative Scenarios*.

Lempert, R. J., Popper, S. W., & Bankes, S. C. (2003). *Shaping the Next One Hundred Years: New methods for quantitative, long-term policy analysis*. Santa Monica, CA: RAND.

Lempert, R. J., & Schlesinger, M. E. (2000a). Robust Strategies for Abating Climate Change. *Climatic Change, 45*(3/4), 387–401. doi:10.1023/A:1005698407365

Lempert, R. J., Schlesinger, M. E., Bankes, S. C., & Andronova, N. G. (2000b). The Impact of Variability on Near-Term Climate-Change Policy Choices. *Climatic Change, 45*(1), 129–161. doi:10.1023/A:1005697118423

Rahman, T., Mittelhammer, C., & Wandschneider, P. (2005). *Measuring the Quality of Life Across Countries: a sensitivity analysis of well-being indices*. United Nations University and WIDER (World Institute For development Economic Research), USA [Research Paper No. 2005/06].

Savage, L. J. (1954). *The Foundation of Statistics*. New York: Dover Publications.

Schaepers, M., Seebregts, A., de Jong, J., Maters, H. (2007). *EU Standards for Energy Security of Supply*, (ECN-E-07-004/CIEP).

Shukla, R., & Kakar, P. (2006). *Role of science and technology, higher education and research in regional socio-economic development*. Working Paper, NCAER.

Smith, L. (2006). *A tutorial on Principal Components Analysis*.

Sterling. (1999). *On economic and analysis of diversity.* SPRU Electronic Working Paper Series Paper No. 28.

Suehiro, S. (2007). *Energy intensity of GDP as Index of Energy Conservation*. Problems in International Comparison of Energy Intensity of GDP and Estimate using Sector-Based Approach. Japan: Institute of Energy Economics, Japan, IEEJ.

World Energy Council. (2008). *Europe's vulnerability to energy crisis: Executive summary*.

FURTHER READING

EC. (2001a). *Towards a European Strategy for the Security of Energy Supply*. European Communities.

IAEA (International Atomic Energy Agency). (2005). *Energy Indicators for Sustainable Development: Guidelines and methodologies*. Vienna, Austria: United Nations Department of Economic and Social Affairs. Details available at /http://www-ub.iaea.org/MTCD/publications/PDF/ Pub1222_web.pdfS, last accessed on 20 October 2007.

IEM (International Energy Markets). (2005). *Security of energy supply: comparing scenarios from a European perspective*. Social Science Research Network Electronic Paper Collection. Details available /http://ssrn.com/abstract=758225S, last accessed on 10 October 2007.

International Country risk guide. (2004–05). http://www.prsgroup.com/ FreeSamplePage.aspx.

Chapter 13
Formulating National Action Plans for Energy Business Environment:
An Intelligent Information System

Konstantinos D. Patlitzianas
REMACO S.A. – Research Department, Greece.

ABSTRACT

The penetration of the Renewable Energy Sources (RES) and the development of the Energy Efficiency (EE) is related to the synthesis of an appropriate action plan by each state for its energy business environment (companies such as "clean" energy producers, energy services companies etc.). The aim of this chapter is to present an information intelligent system which consists of an expert subsystem, as well as a Multi Criteria subsystem. The system supports the state towards the formulation of a modern business environment, since it incorporates the increasing needs for energy reform, successful energy planning, rational use of energy as well as climate change. The system was successfully applied to the thirteen "new" member states of the EU.

INTRODUCTION

The Brussels European Council of 8 / 9 March 2007 adopted an Action Plan for energy market 2007–2009 and committed the EU to achieving at least a 20% reduction in greenhouse gas emissions by 2020 compared to 1990. The European Council also endorsed a binding target of a 20% share of Renewable Energy Sources (RES) in overall EU energy consumption by 2020, supplemented by a binding minimum target of 10% for the share of biofuels in petrol and diesel consumption for transport. Furthermore, the European Council stressed the need to increase Energy Efficiency (EE) in the EU so as to achieve the objective of saving 20% of energy consumption compared to projections for 2020. In particular, the European Commission has made a first assessment of National Energy Efficiency Action Plans (NEEAP), which Member States were required to submit by 30 June 2007. The Plans present national strategies

DOI: 10.4018/978-1-60566-737-9.ch013

on how Member States intend to achieve their adopted energy savings target by 2016. As a result, the European Community has undertaken an ambitious program aimed at improving the sustainability of energy use across Europe (Wu, 2007).

According to the basic literature of the economy's science, Chandler (1962) outlines that the business environment (operational environment of companies) determines substantially the main long-lasting objectives and aims of each company, fires a line of action and indicates the necessary means for the accomplishment of these objectives. Johnson and Scholes (1999) note that the environment directs decisively the activities of a company in the long run. In addition to this, Ansoff (1985) supports that the existence of the companies' operational environment is the base of creation of common lines between the activities of a company.

In the above context, one of the most important parameters for the RES penetration is the enhancement of the producers. These producers can be either companies deriving from utilities producing energy from conventional sources that have decided to be activated in the field of RES or Independent Power Producers (IPPs) (Patlitzianas, Ntotas, Doukas, & Psarras, 2007). The positive impacts of an increasing share of renewable energy on the mitigation of climate change as well as on the decrease of the dependency of energy imports are indisputable.

Moreover, Energy Service Companies (ESCOs) have been developed and their role is crucial for the promotion of Energy Efficiency (EE) in demand side (Patlitzianas, Doukas, Psarras, 2006). The success of the above energy companies is based on the formulation of a modern environment especially in each Europe Union (EU) member state. As a result, each member-state needs to formulate an up-to-date energy companies's environment, which has to be enhanced, giving thus the opportunity to more companies in these member-states to be properly activated.

Nowadays, sustainable energy is about delivering affordable energy with reduced environmental impacts in ways that are financially viable. This helps to tackle fuel poverty, contributes to fighting climate change and boosts economic benefits to communities and businesses alike. Traditionally, energy companies have prioritised financial benefits over other objectives. Their need to deliver shareholder value obliges them to require financial returns that marginalise technical options which could provide lower carbon impacts and better energy services to end users. Typically, investment decisions are taken remote from the communities where delivery occurs.

Based on the international literature, a large body of scientific papers examines the external factors of the energy companies in terms of policies, regulations and financing support schemes of these states. However, there are no studies investigating the operational environment of energy companies in an integrated way.

On the other hand, the use of the expert systems, as well as the use of the Multi Criteria Decision Making (MCDM) systems, can help states assess and evaluate the companies' operational environment. In particular, during the last years the expert systems are considered to be programs with a wide knowledge base in a limited space which use complex knowledge for the execution of works that an expert human could do. An expert system is the incorporation in the computer of a component based on knowledge, so that the system can be able to provide a logical advice regarding a processing function or to have the ability of decision making. A conventional system uses mathematical models for the simulation of the problem, while the expert system simulates the human arguing regarding the problem (Welbank, 1985). Furthermore, the purpose of MCDM systems is to correlate efficiently the various characteristics of any given problem (Prastakos, 2008) and as a result to demonstrate the best possible solution to any problem (Greening, & Bernow, 2004).

The use of well organized intelligent systems can help states assess and evaluate the energy business environment (Meyer, 2003). In this context, the aim of the proposed chapter will be to present an intelligent information system, which consists of an expert unit as well as a decision making unit. It is noted that the system incorporates the increasing needs for effective use of Renewable Energy Sources, successful energy planning, energy efficiency as well as climate change (Pablo, Hernandez, Gual, 2005). This intelligent system is based on the philosophy of expert and MCDM systems. It was applied in order to support the decisions towards the formulation of the operational environment of the energy companies in the thirteen (13) new member states of the EU.

The paper is structured in fifth sections. Apart from the introduction, the second part presents a brief literature review for the related systems aiming at supporting the innovation of the current system. The third part provides a description of the adopted methodology and the fourth part presents the developed information system and its application. The last section presents the conclusions, which summarize the main points that have been drawn up in this paper.

A BRIEF REVIEW OF RELATED SYSTEMS

Based on the literature survey, there exist a number of energy planning systems that are related to the current problem (Kagiannas, et al., 2006; Kavrakoglou, 1987; Jebaraj, 2006). In particular, some of the requirements of current environment's analysis could be supported through the supply systems (EFOM, ELFIN, MARKAL, MESSAGE, TESOM, UPLAN-E and WASP), demand systems (IIASA, COMMEND, ENPEP, MAED, HELM, MED-PRO) (Weyant, 1990) or the system of integrated planning of resources (IRP-MANAGER, MARKAL) (D'Sa, 2005; Chen, 2005).

On a global level integrated energy system analysis built in a modular structure has become the focus of developers of large energy system models in recent years to accommodate the needs of policy makers in terms of market liberalisation, sustainable development and climate mitigation analysis. As regards the origin energy models, five main categories can be distinguished (Kagiannas et al, 2006):

- Energy Information Administration (EIA) models, which the primary model used in policy decision-making is the NEMS. This is a general equilibrium model of the interactions between the U.S. energy markets and the economy.
- The European Commission (EC) (DG-TREN ex-DG-XVII) over the past three decades has sponsored the development of numerous models. The primary models in use are PRIMES and POLES, both of which have evolved through European Commission sponsored programmes. Other EU models are GEM-E3, HERMES, QUEST, MIDAS, MEDEE and EFOM.
- The International Atomic Energy Agency (IAEA) model, which has come to the forefront for energy analysis and planning with ENPEP that has been distributed free of charge to more than 80 countries. Included in the latest version of ENPEP as a module, is a windows based version of WASP. Other IAEA models are GTMAx, MAED and EMCAS.
- The International Energy Agency (IEA) model is MARKAL. Over 70 teams in more than 35 counties around the world make use of the MARKAL family of energy/economy/environment models. The primary objective of developing MARKAL was to assess the long-term role of new technologies in the energy systems of the 17 IEA member countries.

- More than 200 government agencies, NGOs and academic organisations worldwide use LEAP for a variety of tasks including, energy forecasting, greenhouse mitigation analysis, and integrated resource planning. Other general model is MESAP by the university of Stuttgart.

All the above energy modelling approaches are similar in that they are essentially quantitative and use mathematical formulations and equations, in which the parameters can be econometrically estimated. Each energy model has advantages and disadvantages that users have to trade off. The energy models in use globally are large sophisticated models that require, in order to be fully utilised, links to large databases and some times to other models to provide exogenous variables.

As a result, there are no decision systems supporting the operational environment of energy companies in an integrated way.

Intelligent Systems

The expert systems are used for at least 15 years in the energy market. To the best of our knowledge, one of the first expert systems is a system (ALFA) of short-term forecast of electrical charge that was materialised by Jabbour et al. (1998). Since then, many forecast systems were presented, while the most important of them were reported in the works of Hung et al. (1997) and include the following: SOLEXPERT, ALEX, PANexpert, TRANSEPT and CISELEC.

Furthermore, many systems for short-term forecast load exist, as reported in the review of Metaxiotis et al. (2003). Other models, as described by Markovic, & Fraissler (1997) and by Kandil (2003), use the expert system for short-term electrical charge forecast with parallel control of the demand and are based on historical data of the last 5 to 10 years. The system of David & Zhao (1989) simulates the extension of electrical installations via a completed expert system which makes use of dynamic planning. Another similar expert system is the NETMAT which was developed by Zitouni & Irving (1999), while COLOMBO (Melli & Sciuba, 1999) is a system which proposes the planning of electrical generation, as well as the planning of cogeneration plants. Additionally, expert systems exist for the short-term engagement of units (Ouyang & Shahidehpour, 1990; Li, Shahidehpour, Wang, 1993) as well as for various other applications of energy and climate policy (Humphreys, McIvor, & Huang, 2002; Tsamboulas, & Mikroudis, 2005; Huang et al 2005).

Furthermore, expert systems are used mainly for the registration and analysis of errors in stations or substations (Kezunovic, 2003; Sherwali, & Crossley, 1994; Yongli et al., 1994; Minakawa et al., 1995), in situations of network overload, in the regulation of relays so that they function more precisely and efficiently (Lee, Yoon, Yoon, & Jang, 1990), as well as in an alarm process (Eickhoff et al., 1992; Hasan et al., 1997) for the protection of the energy system. Another expert system is SOCRATES (Vale et al., 1998), which is used for the confrontation of extraordinary incidents in the networks transmission and the distribution of electricity, while a study by Khosla and Dillon (1994) constitutes an approach with a combination of an expert system and some networks.

Moreover, the expert systems have penetrated in the restoration of the network after power-offs (Adibi, Kafka, & Milanicz, 1994). The expert systems are mainly used for data collection and decision making after the collapse of the network, but also for the simulation in computer of the situation at which the system collapses Most of them are connected "on-line", as for example the system proposed by Brunner, Nejdl, Schwarzjirg, & Sturm (1993), in order to provide help in the operators. Also, the proposal of Nagata et al. (1995) is very interesting, because it examines a model, which, combining expert systems with mathematic planning, analyzes the cost involved in the restoration of the operation network. Another expert system in the energy sector is SPARSE (Zita, Vale, Santos, & Ramos, 1995), which is used in the Portuguese distribution network. The application of expert systems in the restoration of networks occupied also Park and Lee (1995) who tested their system in a local network. In addition, the desirable regulations of voltage are achieved automatically via expert systems (Matsuda et al., 1990; Azzam, & Nour, 1995; Chokri, Rabei, & Dai, 1996; Le, Negnevitsky, & Piekutowski, 1995; Singh, Raju, & Gupta, 1993). Lastly, important applications of expert systems dealt with the safety of RES production units (Christie, Talukdar, Nixon, 1990; Chang, & Chung, 1990; Fouad, Vekataraman, & Davis, 1990), while various applications involve the energy efficiency (Patlitzianas et al. 2005; Jaber, Mamlook, & Awad, 2005; Doukas, Patlitzianas, Iatropoulos, & Psarras, 2006; Hung, Batanovand, & Lefevre, 1997).

Turton (2008) described the development of the energy and climate policy and scenario Evaluation (ECLIPSE) model—a flexible integrated assessment tool for energy and climate change policy and scenario assessment. Druckman et al. (2008) presented a methodology that can be used in various ways to inform policy-making. In addition, Koumbraroglou et al. (2008) presented a policy planning model that integrates learning curve information on renewable power generation technologies into a dynamic programming formulation featuring real options analysis. Cosmi et al (2008) described a new model for the Italian energy system implemented with a common effort in the framework of an integrated project under the Sixth Framework Programme. Mahadevan et al. (2007) presented the correction energy policy model using data on 20 net energy importers and exporters from 1971 to 2002. Iniyan et al. (2007) described three models have been projected namely Modified Econometric Mathematical (MEM) model, Mathematical Programming Energy-Economy-Environment (MPEEE) model, and Optimal Renewable Energy Mathematical (OREM) model.

Based on the above, the expert systems have many important applications in various issues regarding the operation of the energy companies and as a result they are intensely used in specialised technically energy problems. However, their use in problems of knowledge representation for the support of decision making for the formulation of a modern energy companies' environment is inexistent.

MCDM Systems

The most important research MCDM methodologies and systems that are related to the parts of the current problem are presented as follows.

Siskos & Hubert (1983) presented an evaluation of RES, while Mirasgedis & Diakoulaki (1997) evaluated the environmental and social consequences, as well as the consequences on the human health, which are calculated for a number of energy production units when alternative fuels are used. Marks (1997) dealt with the energy conservation and in particular with the minimization of the annual expenditures for heat by using the CAMOS program.

Pokharel & Chandrashekar (1998) proposed a new way of approaching the analysis of different energy alternatives in isolated regions, as far as it concerns the coverage of energy demand, with the use of multi-criteria systems. Gandibleux (1998) presented an interactive multi-criteria procedure (CASTART) for the selection of alternative ways of energy production.

Goumas et al. (1999) presented an application of some decision making systems that have been developed in operational research for the best utilization of geothermal sources. Gupta et al. (2003) searched the risk evaluation procedure for companies that apply the venture capital mechanisms. Beccali et al. (2003) presented an application for the promotion of RES technologies in local level. Nigim et.

Table 1. The actions

D_1 - Political Dimension	D_2 - Financial Dimension	D_3 - Social Dimension	D_4 - Research and Technology Dimension
$A_{1.1}$ - Supporting Policy Programmes of Energy Production from RES	$A_{2.1}$ – Investment Support of RES	$A_{3.1}$ – Supporting Employment for RES-EE	$A_{4.1}$ - Supporting R&D on Technologies of Energy Production by RES
$A_{1.2}$ - Verification System for ESCOs	$A_{2.2}$ – Investment Support of Energy Management	$A_{3.2}$ – Supporting Social Acceptance for the RES Projects	$A_{4.2}$ - Supporting R&D on EE's Technologies
$A_{1.3}$–Standardization of Energy Services Contracts	$A_{2.3}$ – Investment Support of EE	$A_{3.3}$ – Supporting Education for RES – EE	$A_{4.3}$ - Supporting Best Practices and Technology
$A_{1.4}$ - Supporting Policy Programmes of EE Promotion	$A_{2.4}$ - Promotion of New Financing Sources for RES – EE Projects	$A_{3.4}$ –Developing Energy Companies in the Region	
$A_{1.5}$ - Enhancement of Energy Cooperation			

al. (2004) presented a classification tool as well as an interactive tool for the rural retention (SIMUS). Koroneos et al. (2004) applied a PYA methology in the island of Lesvos, in Greece, where there is adequate amount of RES in order to cover part of the island's energy needs. Giannantoni et al. (2005) proposed a repeatable system for the evaluation and improvement of the energy systems' design; this work deals with the advantages and disadvantages of the suggested solutions. Diakoulaki et al. (2006) examined four independend scenarios for the expansion of three Greek energy system.

Cavallaro & Ciraolo (2005) presented a preliminary evaluation regarding the possibility of setting some wind generators in the island of Salina (part of the Aeolian islands in Italy). Madlener & Stagl (2005) proposed a system for the design of tools for RES policy, taking into consideration the new parameters of energy market. Kaklauskas et al. (2006) presented a system of complex multi-criteria evaluation (COPRAS) for the energy conservation in buildings. Stagl (2006) presented a work that examines a case of decision making regarding some choices of energy offer in the means of rapid development where environmental, social and financial goals are taken into consideration.

METHODOLOGY

The methodology incorporates the following five steps.

Step A: Identification

This component concerns the identification, based on the experience (mainly in the EU), of sixteen actions. Firstly, the dimensions (Di) of the environment can be categorized in four dimensions - the political, financial and social dimension, as well as the dimension of research – technology (i=1,2,3 and 4) - taking into consideration the literature that is related to the companies's operational environment and its strategy (Coulter, 2002). Based on the literature review, the necessary actions towards the formulation of a modern energy companies' operational environment (Aij) are categorized as they have been presented by Patlitziana, & Psarras (2006) in the following *Table 1.*

Step B: Synthesis

The second component concerns the synthesis of the energy companies' operational environment, via the development of a group of appropriate indicators as they have been recently presented by Patlitziana, et al. (2008). In particular:

- Sixteen (16) Basic Indicators *(B_{ij}, i- 1,2,3 and 4, j=1,2, ...,a)* present the most essential information regarding the diagnosis of the country's performance, in terms of it's energy companies' operational environment. The basic indicator is the key means of decision making for the necessity of taking intervention measures or not.
- Some other indicators create a new group that includes the Secondary Indicators S_k, *(k=1,2, ...,m)*. The secondary indicators act accessory to the estimation of the weaknesses of the energy companies' operational environment. These indicators are focused on specified issues of the weaknesses of the energy companies' operational environment and describe specific activities for selected aspects of the sectors they examine. The selected secondary indicators are fourty (40).
- A third group of indicators is created, representing the effects that the "New Parameters" of the companies' market involve in the decision making regarding the formulation of their operational environment N_l, *(l=1,2, ...,n)*. The new parameters' selected indicators are twenty-one (21).

The selected indicators are based on the literature survey and they consist apart from the technology and techno-physical indicators of indicators, such as Research & Development (R&D) expenditures and some socio-economical indicators (e.g. employment, turnover)

Step C: Estimation

The third component concerns the estimation of the necessity for each Action (Aij) of the energy companies' operational environment. Concretely, the Basic Indicator is related to a group of Secondary Indicators. Moreover, a second group of indicators, reflecting the impact of "new parameters", is related to every Basic Indicator. As a result, there are selected:

- The group of the correlated indicators of "New Parameters" (N_{ilx}, *x=1,2,3, ...,b*).
- The group of the correlated Secondary Indicators (S_{ijy}, *y=1,2,3, ...,c*).

In this way, a "group of decision indicators" is created, the price control of which portrays the estimation of the "existence of the necessity or not" for improving the companies' operational environment,

Step D: Choice

After the estimation of the action's necessity, the model investigates the intervention choices, based on the evolution indicator's DBij values. This indicator illustrates the evolution of the Basic Indicator's performance during the past year.

The value of the above mentioned indicator is estimated according to appropriate thresholds and the existence or not of appropriate measures in the past year is examined. In this context, the continuation of the existing measures (I), their modification (II) or the formulation of new measures (III) is proposed.

Table 2. The criteria

Category	*Priority Criteria*
Basic Needs	C1: Contribution to the increase of the RES proportion
	C2: Contribution to the increase of the EE
	C3: Contribution to the security of supply
	C4: Contribution to the sustainable development
New Parameters	C5: Progress regarding the liberalization of the energy market
	C6: Contribution to the reduction of the greenhouse gases

Step E: Order

The last component receives as input the results of the previous components, in order to evaluate the direct actions that have to be implemented for the development of the energy companies' operational environment in each country, and involves a methodology of quantifying multiple qualitative judgments based on the ELECTRE III method (Keeny & Raiffa, 1993). The six criteria are selected so as to incorporate all the needs of the companies' operational environment, as well as the emerging needs and opportunities of the "new parameters" which determine the final decision.

In addition to this, the member-states' performance to each one of the criteria is based on a 1-5 order qualitative scale, with "1" illustrating an insignificant progress of the country regarding the particular criterion, "2" a low, "3" a moderate, "4" a high and "5" a very high progress of the member-state regarding the particular criterion. The criteria are presented in *Table 2*.

The weights of the first four criteria of each dimension were defined to be "0.200", while the last two criteria of each dimension, which express the impact of the "new parameters" (liberalization of energy market and climate change) in the final result, were defined to be "0.100". The weights express the view of the "decision maker", and as a consequence the results range between subjectivity and objectivity.

The methodology procedure is described in the following *Figure 1*.

INFORMATION INTELLIGENT SYSTEM

The methodology has been incorporated in the information system by the name of INIS-EBE (Intelligent Information System for Energy Business Environment). The following *Figure 2* plots the architecture of the system that has been created.

As it is shown in Figure 2, the system consists of two (2) subsystems as follows.

Expert Subsystem

The subsystem that is responsible for the estimation of the necessity for action and the choice of the way that an action should be made, is named "Expert Subystem for Energy Business Environment" (ES-EBE)".

This subsystem has been developed with XpertRule Knowledge Builder, a product of Attar Software Limited, and incorporates the expert system for the evaluation of the situation and the selection of the way that an action should be made in the energy companies' operational environment. The subsystem uses the values of certain indicators and consists of the typical and essential parts of every expert system. In particular:

- *Data base:* The data base consists of objects, attributes and strings. The larger amount of data consists of (a) the gathered information about the actions that should be made in order to formulate a modern operational environment for the energy companies, (b) some extra information

Figure 1. The methodology procedure

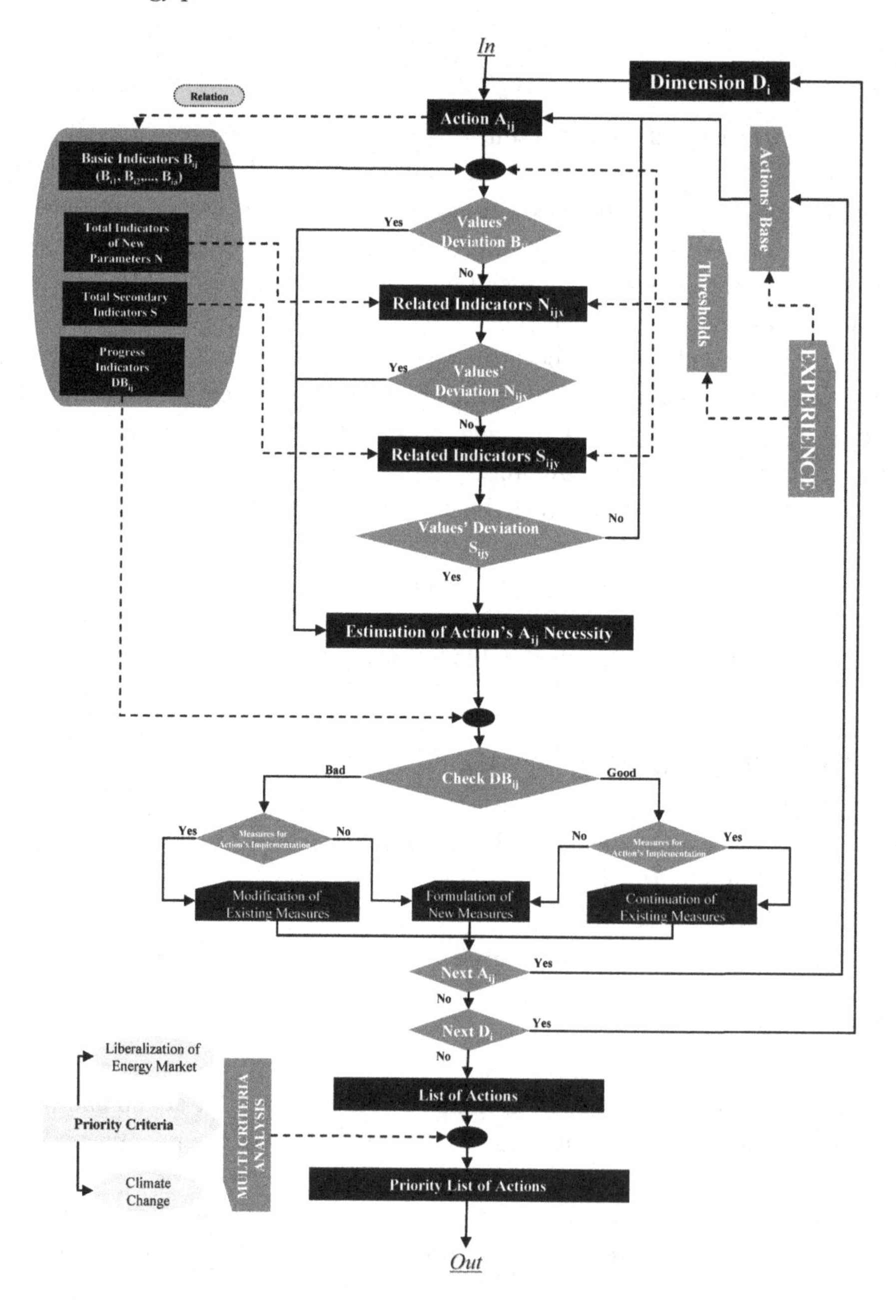

Figure 2. The architecture of the information intelligent system

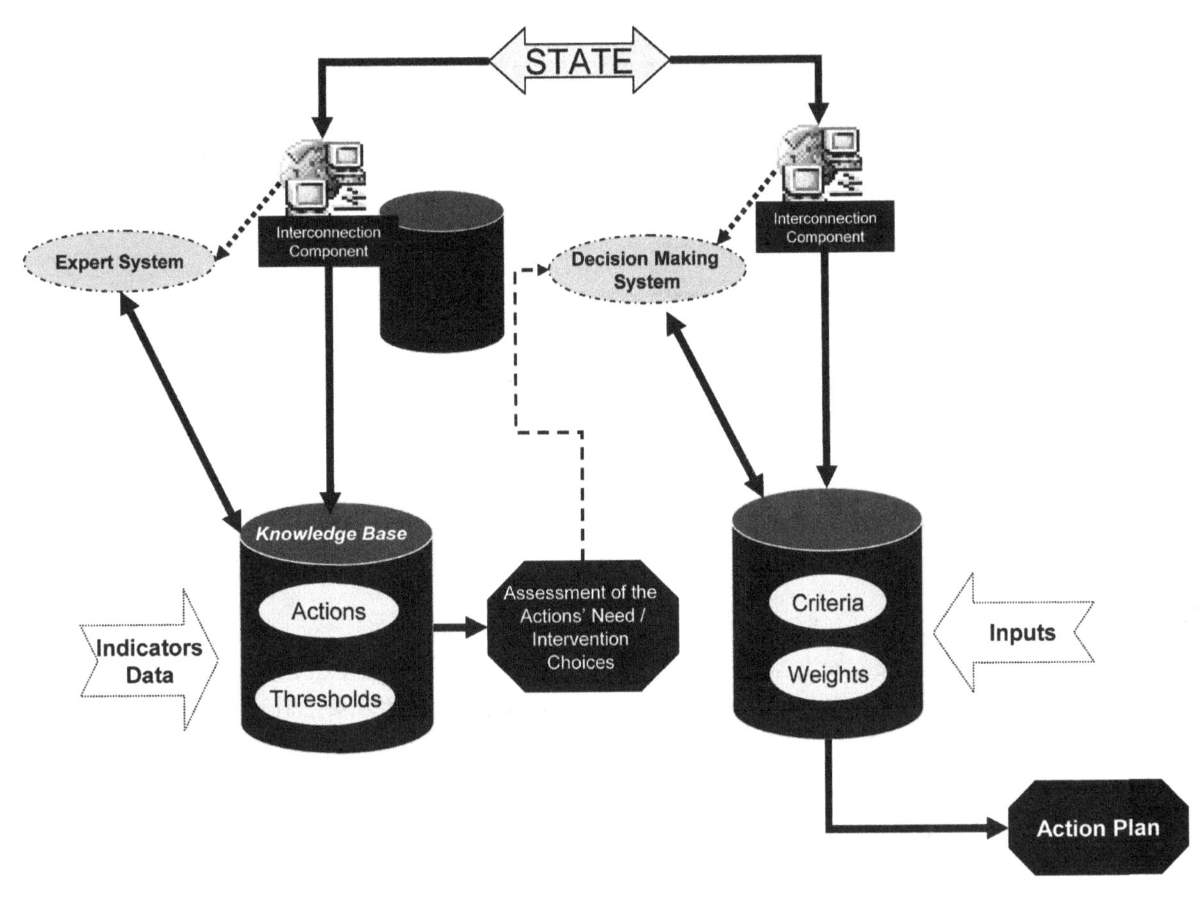

that are necessary for the calculation of the indicators which are based on annual energy data and the "library" of the estimation conclusions. The data base also stores the list of rules that are examined and applied, and therefore the program can provide the user anytime with the proper justification.

- *Knowledge base:* It consists of all the rules which present the knowledge that has been modeled. Those rules evade from the analysis of the procedures of the 3rd and 4th component of the suggested methodology. Totally, the knowledge base consists of approximately 350 rules, which have been developed using rule sets («if … then …»).
- *Interconnection component:* Between the program and the user it consists of a number of dialogs, which are categorized in: (a) forms of data input and (b) forms of results presentation. The

Figure 3. Indicative forms of expert system's inputs

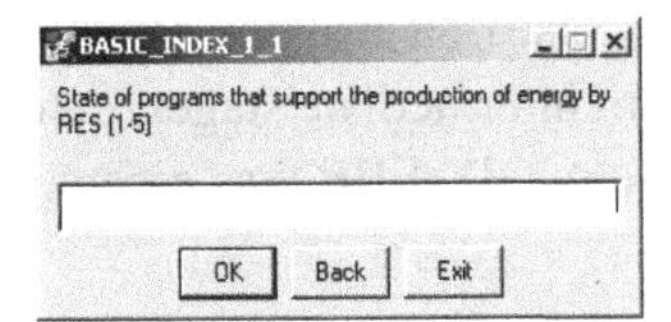

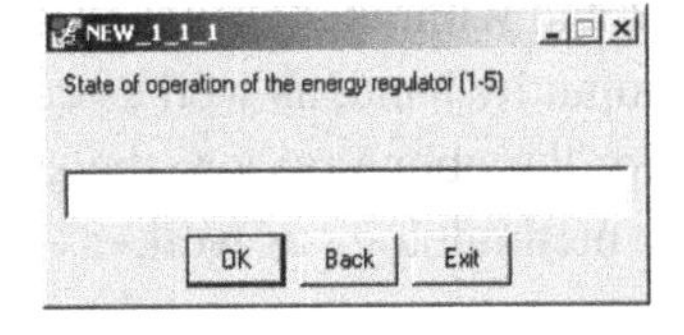

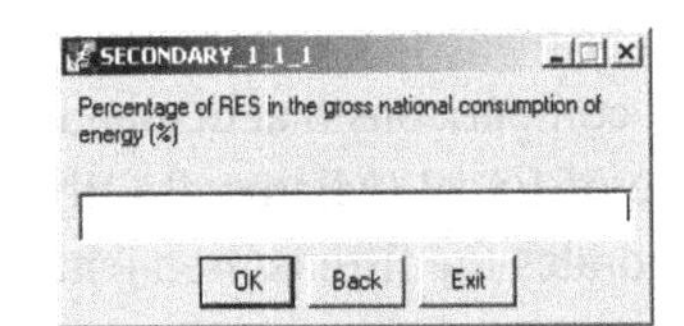

Figure 4. Indicative form of expert system's outputs

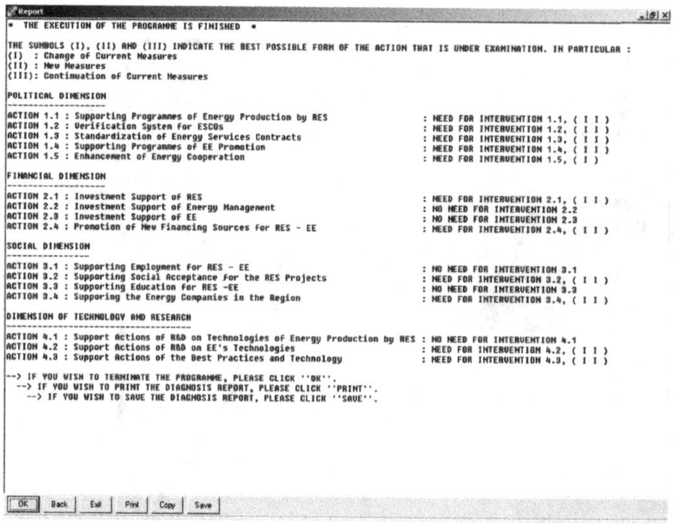

```
Report
*  THE EXECUTION OF THE PROGRAMME IS FINISHED  *

THE SUMBOLS (I), (II) AND (III) INDICATE THE BEST POSSIBLE FORM OF THE ACTION THAT IS UNDER EXAMINATION. IN PARTICULAR :
(I)  : Change of Current Measures
(II) : New Measures
(III): Continuation of Current Measures

POLITICAL DIMENSION
-------------------
ACTION 1.1 : Supporting Programmes of Energy Production by RES          : NEED FOR INTERVENTION 1.1, ( I I )
ACTION 1.2 : Verification System for ESCOs                              : NEED FOR INTERVENTION 1.2, ( I I )
ACTION 1.3 : Standardization of Energy Services Contracts               : NEED FOR INTERVENTION 1.3, ( I I )
ACTION 1.4 : Supporting Programmes of EE Promotion                      : NEED FOR INTERVENTION 1.4, ( I I )
ACTION 1.5 : Enhancement of Energy Cooperation                          : NEED FOR INTERVENTION 1.5, ( I )

FINANCIAL DIMENSION
-------------------
ACTION 2.1 : Investment Support of RES                                  : NEED FOR INTERVENTION 2.1, ( I I )
ACTION 2.2 : Investment Support of Energy Management                    : NO NEED FOR INTERVENTION 2.2
ACTION 2.3 : Investment Support of EE                                   : NO NEED FOR INTERVENTION 2.3
ACTION 2.4 : Promotion of New Financing Sources for RES - EE            : NEED FOR INTERVENTION 2.4, ( I I )

SOCIAL DIMENSION
----------------
ACTION 3.1 : Supporting Employment for RES - EE                         : NO NEED FOR INTERVENTION 3.1
ACTION 3.2 : Supporting Social Acceptance for the RES Projects          : NEED FOR INTERVENTION 3.2, ( I I )
ACTION 3.3 : Supporting Education for RES -EE                           : NO NEED FOR INTERVENTION 3.3
ACTION 3.4 : Supporing the Energy Companies in the Region               : NEED FOR INTERVENTION 3.4, ( I I )

DIMENSION OF TECHNOLOGY AND RESEARCH
------------------------------------
ACTION 4.1 : Support Actions of R&D on Technologies of Energy Production by RES : NO NEED FOR INTERVENTION 4.1
ACTION 4.2 : Support Actions of R&D on EE's Technologies                : NEED FOR INTERVENTION 4.2, ( I I )
ACTION 4.3 : Support Actions of the Best Practices and Technology       : NEED FOR INTERVENTION 4.3, ( I I )

--> IF YOU WISH TO TERMINATE THE PROGRAMME, PLEASE CLICK ''OK''.
  --> IF YOU WISH TO PRINT THE DIAGNOSIS REPORT, PLEASE CLICK ''PRINT''.
    --> IF YOU WISH TO SAVE THE DIAGNOSIS REPORT, PLEASE CLICK ''SAVE''.

[OK] [Back] [Exit] [Print] [Copy] [Save]
```

operation of the subsystem has the following stages:

- Introduction: The user is welcomed and introduced to the library of actions.
- Data Input: The program presents the forms that concern the selection of the operational environment for a certain country and for a certain year. Appropriate forms ask the user to insert the values of certain energy indicators in each one of the four dimensions of the energy companies' operational environment (Streimikiene et al, 2008). In particular, the program introduces the user to the examination of the political dimension. Next, the user is introduced to the examination of the first action of the political dimension through the incorporation of the values of the indicators that the program asks (*Figure 3)* until the evaluation of the necessity for intervention "ACTION1.1" is made.
- Output: The main results of the subsystem are shown on the final form for the evaluation of the necessity in each dimension of the energy companies' operational environment, as well as the way that each action should be made (I, II and III), as it appears in *Figure 4.*

Multicriteria Subsystem

The subsystem that is responsible for the evaluation and the order of the proposed actions is named "Multicriteria Information Subsystem for Energy Business Environment (MIS-EBE)". This subsystem has been developed with Java and incorporates the methods of ELECTRE III customized according to the specific problem for ranking the actions in energy companies' operational environment.

Due to the large amount of calculations and in particular due to the large amount of the records' production that evade from the binary comparisons in ELECTRE III, it is obvious that those calculations cannot be made in a short period of time and without mistakes without the use of a proper software. The binary comparisons that demand data input by hand, as well as control in intermediate stages, deterred the use of Excel (Microsoft Office). For the above reason, the subsystem MIS-EBE was created and incorporates the fifth component of the methodology as well.

Figure 5. Indicative form of decision making system's inputs

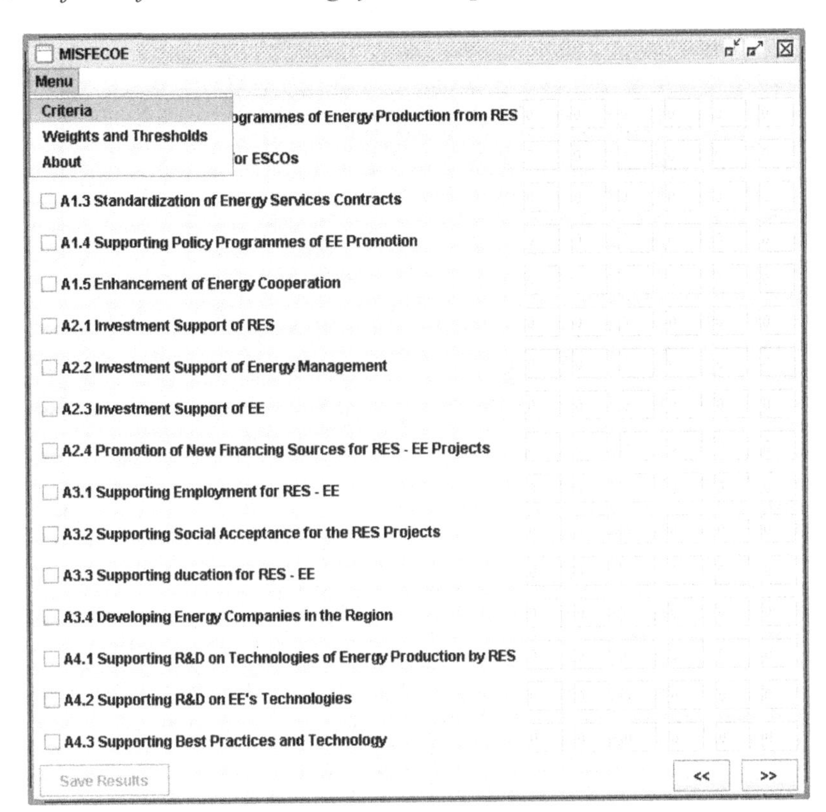

This subsystem was developed with Java, so that it can give the user the opportunity of dynamic variation of certain amounts, such as the number of actions or criteria, which is very essential in case of equality in marks of some alternatives as it presented in the *Figure 5*. In case that some alternatives get the same marks the system must be applied again.

In particular, the subsystem starts and has as defaults elements the criteria and the actions for the formulation of a modern operational environment for the energy companies. A very important advantage of the subsystem is the functionality and the easy of change of all the parameters. Through ".txt" - type files the user has the chance to save (or even to insert) the actions, the criteria, the assignment of performances, as well as the weights and thresholds. The dynamic form of the subsystem makes possible its interconnection with the 1st subsystem.

The subsystem is also user-friendly since it provides the user the opportunity to change all the related parameters. In particular, the subsystem gives the user the opportunity to examine two additional cases that will represent the effect of the "new parameters" of the energy market in the final decision.

In particular, initially is chosen the case of increased activation for the energy liberalization and the climate change, and then the reverse case. The results of the subsystem are stored in a proper file and are presented in descending order. In another «txt» file the relevant assignment of performances is stored. In case of equality in marks, the user is asked to run the program again only with the alternatives that get the same marks.

In addition, for accuracy control reasons of the application, the program creates a number of files that include the agreement, disagreement and reliability matrices, as well as the graduation of the alternatives

Figure 6. Indicative form of decision making system's outputs

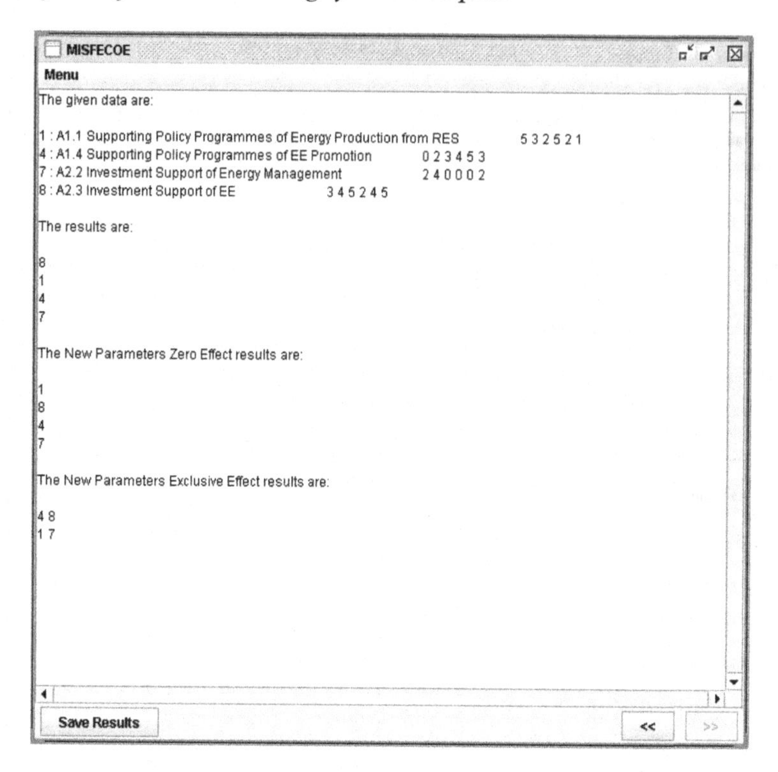

in order of their input in the program and the matrix that leads in the classification of the alternatives.

Therefore, the subsystem presents in the final form (*Figure 6*) the classification for the cases (all criteria, only basic criteria, only criteria of new parameters).

Through the settings menu, the user can see at anytime the definitions of the actions' classification criteria through the choice "criteria legend" as it has been described above. In addition, the user can change the weights and thresholds that formulate the initial stage of the criteria. By choosing "change of thresholds and weights" the user is introduced to a form for changing them.

Application

The information system was successfully applied in the ten "new" member states (Cyprus-CY, Czech Republic-CZ, Estonia-EE, Hungary-HU, Lithuania-LT, Latvia-LV, Malta-MT, Poland-PL, Slovenia-SL, Slovakia-SK), which joined the European Union (EU) in 2004, in Bulgaria-BG and Romania-RO, that joined the European Union in 2007, as well as in Croatia-HR that is joined with EC. The inputs are based to a large extent on the context of the FP6 project: "Scientific Reference System on New Energy Technologies, Energy End-use Efficiency and Energy RTD" (Final Report, 2008).

In particular, the outputs of qualitative indicators as well as the performance of criteria are based on the conclusions derived from the core-group meetings of the "Scientific Technical Reference System on Renewable Energy and Energy End-Use Efficiency" (2004-2008).

The results of the proposed model's application in the above states are presented in the following Table 3. In particular, the actions examined are ranked, from the most necessary ones to those of lowest

Table 3. Results - The action plan

State	Case A - All the Criteria	Case B- Only the Basic Criteria
EE	A2.1(II) > A1.1(III) > A2.3(II) > A3.4(II) > 2.4(II) > A2.2(II) > A4.2(II)	A1.1(III) > A2.3(II) > A2.1(II) > A3.4(II) > A2.2(II) > 2.4(II) > A4.3(II) > A4.2(II)
CY	A1.1(III)>A2.1(II)>A2.3(II)>A2.4(II)>A1.3(II)>A3.4(II) > A1.5(II) > A1.2(II) > A4.2(II) > A4.1(II)	A1.1(III)>A2.1(II)>A1.2(II)>A1.3(II)>A3.4(II)>A2.3(II) > A2.4(II) > A1.5(II) > A4.2(II) > A4.1(II)
LT	A1.5(II)>A1.2(II)>A2.3(II)>A2.2(II)>A2.4(II)>A3.4(II) > A3.3(I) > A3.2(II) > A4.2(I) > A3.1(II) > A4.1(I)	A1.2(II) > A2.3(II) > A3.4(II)> A1.5(II) > A2.2(II) > A3.3(I) > A2.4(II) > A3.1(II) > A3.2(II) > A4.1(I) > A4.2(I)
LV	A2.1(II)>A1.5(II)>A2.3(II)>A3.4(II)>A2.4(II)>A3.3(II) > A3.1(II) > A4.1(II) > A4.2(II)	A2.1(II)>A2.3(II)>A3.4(II)>A1.5(II)>A2.4(II)>A3.3(II) > A3.1(II) > A4.1(II) > A4.2(II)
MT	A1.1(II)>A2.1(II)>A1.4(II)>A2.4(II)>A1.2(II)>A1.3(II) > A3.4(II) > A1.5(I) > A4.3(II) > A4.2(II) > A3.2(II)	A1.1(II)>A2.1(II)>A1.4(II)>A1.2(II)>A1.3(II)>A3.4(II) > A1.5(I) > A2.4(II) > A3.2(II)> A4.3(II) > A4.2(II)
HU	A2.1(II)>A2.3(II)>A3.4(II)>A2.4(II)>A3.3(II)>A3.1(II) > A4.1(II)	A2.1(II)>A2.3(II)>A3.4(II)>A2.4(II)>A3.3(II)>A3.1(II) > A4.1(II)
PL	A2.1(II)>A1.1(III)>A1.4(II)>A3.4(II)>A4.3(II)>A4.2(II) > A4.1(II)	A1.1(III)>A2.1(II)>A1.4(II)>A3.4(II)>A4.3(II)>A4.2(II) > A4.1(II)
SK	A1.1(I) > A2.1(II) > A1.4(II) > A2.4(II) > A1.2(II) > A1.5(II) > A1.3(II) > A3.4(II) > A3.3(I) > A4.2(II) > A3.1(II) > A3.2(II)	A1.1(I) > A1.4(II) > A2.1(II) > A1.2(II) > A1.3(II) > A1.5(II) > A2.4(II) > A3.4(II) > A3.3(I) > A3.1(II) > A3.2(II) > A4.2(II)
SL	A2.1(II) > A1.4(II) > A2.3(I) > A1.3(II) > A2.2(II) > A2.4(II) > A3.3(II) > A3.4(II) > A4.2(II) > A3.2(II) > A3.1(II) > A4.1(II)	A2.1(II) > A1.4(II) > A1.3(II) > A2.3(I) > A3.4(II) > A3.3(II) > A2.2(II) > A2.4(II) > A3.1(II) > A3.2(II) > A4.1(II) > A4.2(II)
CZ	A1.1(III)>A1.4(II)>A2.1(II)>A2.2(II)>A2.3(II)>A1.2(II) > A1.3(II) > A3.4(II) > A1.5(II)	A1.4(II)>A1.2(II)>A1.3(II)>A1.1(III)>A2.1(II)>A2.2(II) > A2.3(II) > A3.4(II) > A1.5(II)
BG	A1.1(II)>A2.1(II)>A2.3(II)>A2.2(II)>A2.4(II)>A3.4(II) > A1.2(II) > A1.3(II) > A4.1(II)	A1.1(II) > A1.2(II) > A1.3(II) > A2.1(II) > A2.3(II) A3.4(II) > A2.2(II) > > A2.4(II) > A4.1(II)
RO	A1.4(I) > A1.2(II) > A1.5(II) > A1.3(II) > A2.3(II) > A2.2(II) > A2.4(II) > A3.4(II) > A3.3(I) > A3.2(II) > A3.1(II) > A4.1(II)	A1.4(I) > A1.2(II) > A1.3(II) > A1.5(II) > A2.3(II) > A3.4(II) > A3.3(I) > A2.2(II) > A2.4(II) > A3.1(II) > A3.2(II) > A4.1(II)
HR	A1.1(II) > A2.1(II) > A1.4(II) > A2.3(II) > A2.2(II) > A1.5(II) >A1.2(II)>A1.3(II)>A2.4(II)>A3.4(II)>A3.3(II)>A4.3(II) > A4.2(II) > A3.2(II) > A3.1(II) > A4.1(II)	A1.1(II) > A1.4(II) > A2.1(II) > A1.2(II) > A1.3(II) > A1.5(II) >A2.3(II)>A3.4(II)>A2.2(II)>A2.4(II)>A3.3(II)>A4.3(II) > A3.2(II) > A3.1(II) > A4.2(II) > A4.1(II)

priority for two cases (all criteria, only basic criteria).

Some critical points are discussed as follows:

- The above member states have made considerable strides towards becoming fully competitive market economies not least through the effort to align their policies with the EU. As regards RES, however, further progress largely depends on whether procedures for adopting and implementing legislation that explicitly supports RES and sustainable development, can be speeded up.
- Despite some present difficulties there are reasons to be optimistic concerning future development of RES in these countries. Beside the described success conditions, the Kyoto mechanisms and the phasing out of nuclear power in some countries the natural conditions are in most cases very good. Whereas hydropower has been already exploited to a great extent in many countries and there is often a tradition of using biomass for heating purposes, all the other RES are still at the beginning of their development.
- Therefore, on and offshore wind energy potential is quite high in the ten coast countries, and even

the landlocked Czech Republic, Hungary, and Slovakia have considerable possibilities. Hungary has the most significant geothermal potential among the Eastern European countries. Poland also belongs to the countries with good geothermal possibilities. There are possibilities for further growth in the biomass sector in nations with a large forest and agricultural sector such as Bulgaria, the Czech Republic, Hungary, Lithuania, Poland, and Romania. Whereas solar thermal shows considerable growth in Poland, for instance, photovoltaic is seen as an expensive luxury in Eastern Europe. Most of the PV-installations are stand-alone applications, e.g. in the agricultural sector.

- Another success condition is the existence of domestic manufacturers producing RES equipment. Whereas wind turbines and photovoltaic-cells mainly have to be imported from Western countries, there are solar collectors manufactured in Hungary, Slovakia and Poland, for example; boilers and boiler equipment as well as hydro turbines are manufactured e.g. in Latvia and Slovenia.

CONCLUSION

Summarizing the above, the information intelligent system has the following advantages:

- Possibility to export a suitable actions list that contains and classifies all necessary actions,in exceptionally less time than the time needed without the use of an information system that incorporates the methodology.
- Possibility to control the flexibility of results, thus offering an additional list of actions by the information system without taking into consideration the new parameters (null effect of criteria of new parameters). Based on the analysis of the results' flexibility, the variations of the places of each action are rather small and insignificant in Case A. This is an expected result because this particular study is focused on countries that have not yet incorporated totally the "new parameters" in their internal energy market.
- Possibility to control the thresholds that were used (in quantitative/ qualitative indicators and in criteria) for exporting the results, fact that can essentially contribute to the control and correction of errors and false choices.

Therefore, the information decision support systems, such as the one presented in this paper, are necessary in order to identify, diagnose, and classify the appropriate actions in a consistent way, to assist policy making and to formulate a modern energy companies' operational environment. In addition, the model's procedure assisted the specific decision making problem and the outcome might have been quite different if different indicators and criteria had been chosen. However, the information system's conceptual can provide a sufficient framework for supporting decisions for other problems.

ACKNOWLEDGMENT

The data of the application was based on research of the "Scientific Reference System on New Energy Technologies, Energy End-use Efficiency and Energy RTD" (project number: SSP6-CT2004-006631) project of the FP6 Programme managed by the European Commission (2005-2008). The content of the paper is the sole responsibility of its authors and does not necessarily reflect the views of the EC.

REFERENCES

Adibi, M. M., Kafka, R. J., & Milanicz, D. P. (1994). Expert system requirements for power system restoration. *IEEE Transactions on Power Systems*, *9*(3), 1592–1600. doi:10.1109/59.336099

Ansoff, I. (1985). *The Concept of Corporate Strategy.* London: Penguin. London.

Azzam, M., & Nour, M. A. S. (1995). Expert system for voltage control of a large-scale power system, Modelling, Measurement Control A: General Physics, Electronics. *Electrical Engineering*, *63*(1-3), 15–24.

Beccali, M., Cellura, M., & Mistretta, M. (2003). Decision-making in energy planning. Application of the Electre method at regional level for the diffusion of renewable energy technology. *Renewable Energy*, *28*(13), 2063–2087. doi:10.1016/S0960-1481(03)00102-2

Brunner, T., Nejdl, W., Schwarzjirg, H., & Sturm, M. (1993). On-line expert system for power system diagnosis and restoration. *Intelligent Systems Engineering*, *2*(1), 5–24.

Cavallaro, F., & Ciraolo, L. (2005). A multicriteria approach to evaluate wind energy plants on an Italian island. *Energy Policy*, *33*(2), 235–244. doi:10.1016/S0301-4215(03)00228-3

Chandler, A. (1962). *A Strategy and Structure: Chapters in the History of the American Industrial Enterprise.* New York: Massachusetts Institute of Technology, New York.

Chang, C. S., & Chung, T. S. (1990). Expert system for on-line security – Economic load allocation on distribution systems. *IEEE Transactions on Power Delivery*, *5*(1), 467–473. doi:10.1109/61.107314

Chen, W. (2005). The costs of mitigating carbon emissions in China: findings from China MARKAL-MACRO modeling. *Energy Policy*, *33*(7), 885–896. doi:10.1016/j.enpol.2003.10.012

Chokri, B., Rabei, M., & Dai, D. X. (1996). *An integrated power system global controller using expert system.* Calgary, Canada: Canadian Conference on Electrical and Computer Engineering.

Christie, R. D., Talukdar, S. N., & Nixon, J. C. (1990). A hybrid expert system for security assessment. *IEEE Transactions on Power Systems*, *5*(4), 1503–1509. doi:10.1109/59.99405

Cosmi, C., Di Leo, S., Loperte, S., Macchiato, M., Pietrapertosa, F., Salvia, M., & Cuomo, V. (2008). *A model for representing the Italian energy system: The NEEDS-TIMES experience.*

Coulter, M. (2002). *Strategic Management in Action.* New York: NJ, PPrentice Hall.

D'Sa, A. (2005). Integrated resource planning (IRP) and power sector reform in developing countries. *Energy Policy*, *33*(10), 1271–1285. doi:10.1016/j.enpol.2003.12.003

David, AK., & Zhao, R. (1989). Integrating expert systems with dynamic programming in generation expansion planning. *IEEE Transactions on Power Systems, 4*(3):), 1095-1101.

Diakoulaki, D., & Karangelis, F. (2007). Multi-criteria decision analysis and cost–benefit analysis of alternative scenarios for the power generation sector in Greece. *Renewable & Sustainable Energy Reviews*, *11*(4), 716–727. doi:10.1016/j.rser.2005.06.007

Doukas, H., Patlitzianas, K. D., Iatropoulos, K., & Psarras, J. (2006). Intelligent Building Energy Management System Using Rule Sets. *Building and Environment, 42*(10), 3562–3569. doi:10.1016/j.buildenv.2006.10.024

Druckman, A., & Jackson, T. (2008). Household energy consumption in the UK: A highly geographically and socio-economically disaggregated model. *Energy Policy, 36*(8), 3177–3192. doi:10.1016/j.enpol.2008.03.021

Eickhoff, F., Handschin, E., & Hoffmann, W. (1992). Knowledge based alarm handling and fault location in distribution networks. *IEEE Transactions on Power Systems, 7*(2), 770–776. doi:10.1109/59.141784

European Commision – JRC. (2004-2008). *Minutes of the Core Group Meetings of the Scientific Technical Reference System*. Milan-Brussels-Vienna-Denmark: Author.

European Commission. -DG-TREN. (2008). *SRS NET & EEE: Scientific Reference System on New Energy Technologies, Energy End-Use Efficiency and Energy RTD.* Final Report ((Project Ref: 006631) – Final Report. Brussels, Belgium: Author.

Fouad, A. A., Vekataraman, S., & Davis, J. A. (1991). An expert system for security trend analysis of a stability-limited power system. *IEEE Transactions on Power Systems, 6*(3), 1077–1084. doi:10.1109/59.119249

Gandibleux, X. (1998). Interactive multicriteria procedure exploiting a knowledge-based module to select electricity production alternatives: The CASTART system. *European Journal of Operational Research, 113*(2), 355–373. doi:10.1016/S0377-2217(98)00221-5

Giannantoni, C., Lazzaretto, A., Macor, A., Mirandola, A., Stoppato, A., Tonon, S., & Ulgiati, S. (2005). Multicriteria approach for the improvement of energy systems design. *Energy, 30*(10), 1989–2016. doi:10.1016/j.energy.2004.11.003

Gonzalez, P., Hernandez, F., & Gual, M. (2005). The implications of the Kyoto project mechanisms for the deployment of renewable electricity in Europe. *Energy Policy, 33*, 2010–2022. doi:10.1016/j.enpol.2004.03.022

Goumas, M. G., Lygerou, V. A., & Papayannakis, L. E. (1999). Computational methods for planning and evaluating geothermal energy projects. *Energy Policy, 27*(3), 147–154. doi:10.1016/S0301-4215(99)00007-5

Greening, L. A., & Bernow, S. (2004). Design of coordinated energy and environmental policies: use of multi-criteria decision-making. *Energy Policy, 32*(6), 721–735. doi:10.1016/j.enpol.2003.08.017

Gupta, J. P., Chevalier, A., & Dutta, S. (2003). Multicriteria model for risk evaluation for venture capital firms in an emerging market context. *European Journal of Operational Research*, 1–18.

Hasan, K., Ramsay, B., & Moyes, I. (1995). Object oriented expert systems for real-time power system alarm processing, Part I, Selection of a toolkit; Part II, Application of a toolkit. *Electrical Power Systems Research, 30*(1), 69-75/77-82.

Huang, Y. F., Huang, G. H., Hu, Z. Y., Maqsood, I., & Chakma, A. (2005). Development of an expert system for tackling the public's perception to climate-change impacts on petroleum industry. *Expert Systems with Applications*, *29*(4), 817–829. doi:10.1016/j.eswa.2005.06.020

Humphreys, P., McIvor, R., & Huang, G. (2002). An expert system for evaluating the make or buy decision. *Computers & Industrial Engineering*, *42*, 567–585. doi:10.1016/S0360-8352(02)00052-9

Hung, C. Q., Batanovand, D. N., & Lefevre, T. (1997). *KBS and macro-level systems: support of energy demand forecasting*. Thailand: Asian Institute of Technology.

Iniyan, S., Suganthi, L., & Samuel, A. (2006). Energy models for commercial energy prediction and substitution of renewable energy sources. *Energy Policy*, *34*(17), 2640–2653. doi:10.1016/j.enpol.2004.11.017

Jabbour, K., Riveros, J. F. V., Landsbergen, D., & Meyer, W. (1988). ALFA: automated load forecasting assistant. *IEEE Transactions on Power Systems*, *3*(3), 908–914. doi:10.1109/59.14540

Jaber, J. O., Mamlook, R., & Awad, W. (2005). Evaluation of energy conservation programs in residential sector using fuzzy logic methodology. *Energy Policy*, *33*, 1329–1338. doi:10.1016/j.enpol.2003.12.009

Jebaraj, S., & Iniyan, S. (2006). A review of energy models. *Renewable & Sustainable Energy Reviews*, *10*(4), 281–311. doi:10.1016/j.rser.2004.09.004

Johnson, G., & Scholes, K. (1999). *Exploring Corporate Strategy: Text and Cases* (5th ed.). London: Prentice Hall Europe. 5th ed. London.

Kagiannas, A., Patlitzianas, K., Metaxiotis, K., Askounis, D., & Psarras, J. (2006). Energy Models in the Mediterranean Countries: A survey towards a common strategy. *International Journal of Power and Energy Systems*, *26*(3), 260–268. doi:10.2316/Journal.203.2006.3.203-3526

Kaklauskas, A., Zavadskas, E. K., Raslanas, S., Ginevicius, R., Komka, A., & Malinauskas, P. (2006). Selection of low-e windows in retrofit of public buildings by applying multiple criteria method COPRAS: A Lithuanian case. *Energy and Building*, *38*(5), 454–462. doi:10.1016/j.enbuild.2005.08.005

Kandil, M. S. (2001). The implementation of long-term forecasting strategies using a knowledge-based expert system: part II. *Electric Power Systems Research*, *58*(1), 19–25. doi:10.1016/S0378-7796(01)00098-0

Kavrakoglou, I. (1987). Energy models. *European Journal of Operational Research*, *16*(2), 2231–2238.

Keeny, R. L., & Raiffa, H. (1993). *Decision Making with Multiple Objectives: Preferences and Value Tradeoffs*. Cambridge, UK: Cambridge University Press 1993. Cambridge, UK.

Kezunovic, M., Fromen, C. W., & Sevcik, D. R. (1993). Expert system for transmission substation event analysis. *IEEE Transactions on Power Delivery*, *8*(4), 1942–1949. doi:10.1109/61.248306

Khosla, R., & Dillon, T. S. (1994). Neuro-expert system approach to power system problems. *International Journal of Engineering Intelligent Systems for Electrical Engineering and Communications*, *2*(1), 71–78.

Koroneos, C., Michailidis, M., & Moussiopoulos, N. (2004). Multi-objective optimization in energy systems: the case study of Lesvos Island, Greece. *Renewable & Sustainable Energy Reviews*, *8*(1), 91–100. doi:10.1016/j.rser.2003.08.001

Kumbaroğlu, G., Madlener, R., & Demirel, M. (2008). A real options evaluation model for the diffusion prospects of new renewable power generation technologies. *Energy Economics*, *30*(4), 1882–1908. doi:10.1016/j.eneco.2006.10.009

Kumbaroğlu, G., Madlener, R., & Demirel, M. (2008). A real options evaluation model for the diffusion prospects of new renewable power generation technologies.

Le, T. L., Negnevitsky, M., & Piekutowski, M. (1995). Expert system application for voltage control and VAR compensation. *International Journal of Engineering Intelligent Systems for Electrical Engineering and Communications*, *3*(2), 79–85.

Lee, S. J., Yoon, S. H., Yoon, M. C., & Jang, J. K. (1990). Expert system for protective relay setting of transmission systems. *IEEE Transactions on Power Delivery*, *5*(2), 1202–1208. doi:10.1109/61.53142

Li, S., Shahidehpour, S. M., & Wang, C. (1993). Promoting the application of expert systems in short-term unit commitment. *IEEE Transactions on Power Systems*, *8*(1), 286–292. doi:10.1109/59.221229

Madlener, R., & Stagl, S. (2005). Sustainability-guided promotion of renewable electricity generation. *Ecological Economics*, *53*(2), 147–167. doi:10.1016/j.ecolecon.2004.12.016

Mahadevan, R., & Asafu-Adjaye, J. (2007). Energy consumption, economic growth and prices: A reassessment using panel VECM for developed and developing countries. *Energy Policy*, *35*(4), 2481–2490. doi:10.1016/j.enpol.2006.08.019

Markovic, M., & Fraissler, W. (1993). Short-term load forecast by plausibility checking of announced demand: an expert-system approach. *European Transactions on Electrical Power Engineering*, *3*(5), 353–358.

Marks, W. (1997). Multicriteria optimisation of shape of energy-saving buildings. *Building and Environment*, *32*(4), 331–339. doi:10.1016/S0360-1323(96)00065-0

Matsuda, S., Ogi, H., Nishimura, K., Okataku, Y., & Tamura, S. (1990). Power system voltage control by distributed expert systems. *IEEE Transactions on Industrial Electronics*, *37*(3), 236–240. doi:10.1109/41.55163

Melli, R., & Sciuba, E. (1997). A prototype expert system for the conceptual synthesis of thermal processes. *Energy Conversion and Management*, *38*(15-17), 1737–1749. doi:10.1016/S0196-8904(96)00186-0

Metaxiotis, K., Kagiannas, A., Askounis, D., & Psarras, J. (2003). Artificial intelligence in short term electric load forecasting: a state-of-the-art survey for the researcher. *Energy Conversion and Management*, *44*(9), 1525–1534. doi:10.1016/S0196-8904(02)00148-6

Meyer, N. I. (2003). European schemes for promoting renewables in liberalized markets. *Energy Policy*, *31*, 665–676. doi:10.1016/S0301-4215(02)00151-9

Minakawa, T., Ichikawa, Y., Kunugi, M., Shimada, K., Wada, N., & Utsunomiya, M. (1995). Development and implementation of a power system fault diagnosis expert system. *IEEE Transactions on Power Systems, 10*(2), 932–939. doi:10.1109/59.387936

Mirasgedis, S., & Diakoulaki, D. (1997). Multicriteria analysis vs. externalities assessment for the comparative evaluation of electricity generation systems. *European Journal of Operational Research, 102*, 364–379. doi:10.1016/S0377-2217(97)00115-X

Nagata, T., Sasaki, H., & Yokoyama, R. (1995). Power system restoration by joint usage of expert system and mathematical programming approach. *IEEE Transactions on Power Systems, 10*(3), 1473–1479. doi:10.1109/59.466501

Nigim, K., Munier, N., & Green, J. (2004). Pre-feasibility MCDM tools to aid communities in prioritizing local viable renewable energy sources. *Renewable Energy, 29*(11), 1775–1791. doi:10.1016/j.renene.2004.02.012

Ouyang, Z., & Shahidehpour, S. M. (1990). Short-term unit commitment expert system. *Electric Power Systems Research, 20*(1), 1–13. doi:10.1016/0378-7796(90)90020-4

Park, Y. M., & Lee, K. H. (1995). Application of expert system to power system restoration in local control center. *International Journal of Electrical Power & Energy Systems, 17*(6), 407–415. doi:10.1016/0142-0615(94)00011-5

Patlitzianas, K., Doukas, H., & Psarras, J. (2006). Designing an Appropriate ESCOs' Environment in the Mediterranean. *Management of Environmental Quality, 17*(5), 538–554. doi:10.1108/14777830610684512

Patlitzianas, K., Ntotas, K., Doukas, H., & Psarras, J. (2007). Assessing the Renewable Energy Producers' Environment in the EU Accession Member States. *Energy Conservation and Management, 48*(3), 890–897. doi:10.1016/j.enconman.2006.08.014

Patlitzianas, K., Papadopoulou, A., Flamos, A., & Psarras, J. (2005). CMEM: The Computerized Model for Intelligent Energy Management. *International Journal of Computer Applications in Technology, 22*(2/3), 120–129. doi:10.1504/IJCAT.2005.006943

Patlitzianas, K. D., Doukas, H., & Psarras, J. (2008). Sustainable energy policy indicators: Review and recommendations. *Renewable Energy, 33*(5), 966–973. doi:10.1016/j.renene.2007.05.003

Patlitzianas, K. D., & Psarras, J. (2006). Formulating a Modern Energy Companies' Environment in the EU Accession Member States through a Decision Support Methodology. *Energy Policy, 35*(4), 2231–2238. doi:10.1016/j.enpol.2006.07.010

Pokharel, S., & Chandrashekar, M. (1998). A Multi-objective approach to rural energy policy analysis. *Energy, 23*(4), 325–336. doi:10.1016/S0360-5442(97)00103-5

Prastakos, G. (2008). *Management Science: Decision making in the information society (Vol 2)*. Athens: Stamoulis Publications.

Sherwali, H., & Crossley, P. (1994). Expert system for fault location on a transmission network. *Proceedings 29th Universities Power Engineering Conference – Part,* (pp. 751-754).

Singh, S. P., Raju, G. S., & Gupta, A. K. (1993). Sensitivity based expert system for voltage control in power system. *International Journal of Electrical Power & Energy Systems, 15*(3), 131–136. doi:10.1016/0142-0615(93)90027-K

Siskos, J., & Hubert, P. (1983). Multi-criteria analysis of the impacts of energy alternatives: A survey and a new comparative approach. *European Journal of Operational Research, 13*, 278–299. doi:10.1016/0377-2217(83)90057-7

Stagl, S. (2006). Multicriteria evaluation and public participation: the case of UK energy policy. *Land Use Policy Ecological Economics, 23*(1), 53–62. doi:10.1016/j.landusepol.2004.08.007

Streimikiene, D., & Šivickas, G. (2008). The EU sustainable energy policy indicators framework. *Environment International.*

Tsamboulas, D. A., & Mikroudis, G. K. (2005). *TRANS-POL: A mediator between transportation models and decision makers' policies.* Decision Support Systems.

Turton, H. (2008). (in press). ECLIPSE: An integrated energy-economy model for climate policy and scenario analysis. *Energy.*

Vale, Z., Ramos, C., Silva, A., Faria, L., Santosm, J., Fernandes, M., et al. (1998). SOCRATES – An Integrated Intelligent System for Power System Control Center Operator Assistance and Training. *International Conference on Artificial Intelligence and Soft Computing (IASTED),* Cancun, Mexico.

Welbank, M. (1985). *A review of knowledge acquisition techniques for expert systems.* Ipswich, UK: British Telecommunications Research Laboratories Technical Report. Ipswich.

Weyant, J. P. (1990). Policy modeling: an overview. *Energy, 15*, 203–206. doi:10.1016/0360-5442(90)90083-E

Wu, L.-M., Chen, B.-S., Bor, Y.-C., & Wu, Y.-C. (2007). Structure model of energy efficiency indicators and applications. *Energy Policy, 35*(7), 3768–3777. doi:10.1016/j.enpol.2007.01.007

Yongli, Z., Yang, Y. H., Hoggm, B. W., Zhang, W. Q., & Gao, S. (1994). Expert system for power systems fault analysis. *IEEE Transactions on Power Systems, 9*(1), 503–509. doi:10.1109/59.317573

Zita, A. Vale, Santos, J., & Ramos, C. (1995). SPARSE - A Prolog Based Application for the Portuguese Transmission Network: Verification and Validation. London, UK: PAP'97 – Practical Application in Prolog.

Zitouni, S., & Irving, MR. (1999). NETMAT: A knowledge-based grid system analysis tool.

Chapter 14
Information Technology in Power System Planning and Operation under Deregulated Markets:
Case Studies and Lessons Learnt

Fawwaz Elkarmi
Amman University, Jordan

ABSTRACT

Power systems have grown recently in size and complexity to unprecedented levels. This means that planning and operation of power systems can not be made possible without the aid of information technology tools and instruments. Even small systems need such aid because of the complexity factor. On the other hand, new trends have recently emerged to solve the problems arising from increased size of power systems. These trends are related to the market structure, legal, and business issues. Other trends also cover technological developments, and environmental issues. Moreover, power systems have special characteristics and features that are not duplicated in other infrastructures. All these issues confirm the need for special information technology tools and instruments which aid in planning and operation of power systems.

INTRODUCTION

This chapter aims at introducing the new emerging trends and critical factors which have shaped and continue to influence decisions of power system planners and operators. Some of these are technical issues while others are economic and financial. Legal issues are also of great concern in modern power system business. Moreover, some issues are related to the institutional setup of the electricity companies, the regulatory bodies and government. The interaction of the four thrusts; namely, technical, economic/financial, legal, and institutional introduce another layer of complexity in the planning and operation of power systems.

DOI: 10.4018/978-1-60566-737-9.ch014

This chapter will hopefully shed some light on all these issues and how they affect the processes of decision making and conducting business in the electricity sector. It will introduce new terminologies, discuss new procedures and tools, and present the philosophy underlying the changes and trends which lie ahead.

There are so many issues that nowadays influence and shape the functions of a power system. These issues have emerged from the trends which have evolved due to the deregulation strategies adopted by almost all power utilities. The model of a comprehensive monopolistic utility is almost a fact of the past; although in few countries it is still in place. Presently there is the vertical model which distributes the country into geographical regions and gives a concession to one utility to serve one particular region or area. Then there is the functional model which separates generation from transmission from distribution and gives each one to one or more companies. In between the two models there are several variants. These variants depend on the ownership of the power utilities. For example generation and distribution are privatized while transmission is kept as a government entity. In others generation is kept with government while transmission and distribution are privatized.

In almost all cases, government is moving away from controlling the power system and more into regulatory roles and duties. These regulatory roles include tariff setting, licensing, power quality issues, and more. The relationship between the regulatory body, which does not by default represent government per se, and the power companies could be a complex one as the criteria used involves customer satisfaction, companies' profitability in addition to quality of the power delivered.

In certain aspects these ownership models have created competition and a drive for better quality. Therefore, new technologies and procedures have been tried and put into use. Moreover, customer satisfaction has become an important factor in the electricity business to the extent that in certain cases customers dictate their preferences as to green power over other environmentally polluting sources.

All these new trends have created new functions and duties for the power system planners and operators. On one hand better tools have been developed to improve the planning aspect of the power system including peak load and energy forecasting, risk assessment and reliability enhancement, integrated resource planning, and future expansion and investment planning. On the other hand other tools have been developed for the proper and cost-effective operation of the power system. These include: contingency analysis, economic dispatch with provision for tie lines control and power exchange, demand side management, reliability and availability monitoring, optimum power flows and loss reduction, interruption management and power restoration, and billing and payments follow up.

The interconnection with other systems also imposes other criteria and therefore requires new tools and methodologies for better operation and control. Moreover, the economic and financial aspects of power system planning and operation have taken more roles. For example, the issues of power purchase agreements (PPA), independent power producers (IPP), take or pay options, public private participation (PPP), etc. are important issues that take due place in power system decision making. The same also applies to the legal issues such as the various agreement forms which are needed to forge the business relationship between the parties. Finally the institutional aspects which are a direct reflection of the ownership model adopted play an important role in dictating the functions of the various entities. On top of all the issues mentioned the environment dimension is strongly present. It has technical, economic, financial, legal and institutional complications which must be properly addressed by the electricity sector entities.

In face of these new trends and technologies the educational institutions responsible for graduating engineers and other professionals who will work in entities of the electricity sector must adapt their curricula to cover such changes. Moreover, extended training is needed to equip electricity sector employees with the necessary skills mandated by the introduction of the new trends. This means that the understanding of economic and financial, legal, and institutional (or administrative) issues must be guaranteed for future engineers in order to be able to be productive in the electricity sector.

This chapter is organized into four sections in addition to this introduction. Section II discusses the new trends in the electricity supply industry including the technical and technological trends; regulatory and legal status trends; economic and financial tends; and environmental trends. Section III deals with the studies and analyses needed for the proper planning and operation of the power system. Finally section IV sheds some light on other considerations and interrelated trends which play an important role on the future of the power sector reform and development.

NEW TRENDS IN THE ELECTRICITY SUPPLY INDUSTRY

The electricity supply industry has faced and is still facing challenges and threats from emerging new trends. As a matter of fact the structure and shape of the electricity supply industry have been greatly affected and changed as a result of these trends. These trends cover a wide spectrum of areas and they are inter-disciplinary and interrelated. Therefore, functions and responsibilities of the various entities must be such as to be able to cope with this fact. Moreover, engineers currently working in the electricity supply industry are quite different from the older ones who used to work in the recent past. Although the basic education is almost the same, newer curricula have introduced newer sciences and tools, which were not present in the past. The new trends in the electricity supply industry are as follows.

Technical and Technological Trends

It is a known fact that the majority of equipment and hardware that make up the generation, transmission and distribution components of electrical power systems has not changed since their "invention" about 100 years ago (Schavemaker & Sluis, 2008). In essence any changes which might have taken place can be categorized as "evolution" rather than "revolution" (Schavemaker & Sluis, 2008). These changes did not alter the basic designs and functions of equipment but rather came to improve safety, reliability, performance, cost, and other operational factors. In essence there are no breakthroughs in the electricity supply industry as in other industries such as the computer, communications, automotive, aviation or space industries. This is probably attributed to the following factors:

- High complexity of the power system
- High capital cost of equipment
- Relatively long life of equipment

An analysis of the new technical and technological trends in the electricity supply industry reveals the following major achievements and modifications:

A) Generation
  - Renewable energy and green technologies
  - Higher efficiency plants such as combined cycle and mega plants

- Nuclear power plants
- Life-extension projects for existing plants
- Retrofit projects to improve efficiency, reduce pollution, or reduce cost of existing plants
- Energy storage

B) Transmission
- Higher voltage transmission lines
- Flexible AC Transmission Systems (FACTS)
- Bundling of conductors
- Super conductors
- HVDC transmission
- Submarine cables
- Reactive power compensation and voltage control
- Increased power transfer and stability through series capacitors
- Autotransformers and OLTC
- GIS switchgear
- Digital protective schemes

C) Distribution and Loads
- Compact substations
- More dependence on underground cables
- Autoreclosers
- Power factor improvement
- Demand side management
- Energy efficient appliances
- Computerized meter reading and billing
- Decentralized generation at consumer or group of consumers levels

D) Supporting systems
- SCADA
- Power line carrier communication
- Fiber optical communication
- Interconnection with other systems
- Live line maintenance and live line washing
- State estimation tools
- Monitoring and diagnosis tools
- Analyses software tools
- Power electronics

Regulatory and Legal Status Trends

This set of new trends deals with the structure and composition of the electricity supply industry. In the past one utility was entrusted with generation, transmission, and distribution of electricity to all consumers. Moreover, this same utility used to study and decide on the tariffs to be used in billing all consumers. This utility enjoyed monopoly in the geographic area granted to it by government. In essence this utility was a government body which operated on the basis of a law.

This has changed in almost all countries of the world, except for very few developing or centralized government countries. In still other countries the older "setup" still exists because these countries provide a huge subsidy through the electricity tariff to consumers and they want to keep this privilege.

The reform of the electricity supply industry was spearheaded by a deregulation trend. The deregulation trend started very early in certain countries and late in others. The essence of the deregulation trend was to reform the electricity market from one regulated by government to another that works on market forces. This means that newer players must come in and replace the government, or utility representing government. This was referred to as the privatization trend. In reality both deregulation and privatization trends are one and the same, since if we are to relief government of its duties as electricity supplier naturally there must be some body to take up this responsibility.

There are several motives for the electricity supply industry reform or restructuring (Rothwell & Gomez, 2003, p.3). One such motive is the fact that the competitive global economy requires electricity cost reduction; another is that government cannot respond to economic and technological changes, let alone bear the financial burden of power system investments. Finally information technologies and communications systems make possible the exchange of large volume of data needed to manage the electricity markets. These factors have helped in shaping the evolution of power sector reform or restructuring.

The philosophy of reform varied from a vertical model to a horizontal model and an in-between model. The vertical model is based on separating the country into geographical areas and giving the complete responsibility of electricity supply to one company (covering the functions of generation, transmission, and distribution). This means that each area will be served by one electricity company which will operate on commercial basis. The government's role, in this case, will be in licensing such companies and in setting the electricity tariff.

The horizontal model, on the other hand, is based on separating the generation, transmission, and distribution functions in the whole country or in certain geographical areas. Each function is then given to one company or more depending on the market size and economics of electricity supply. Naturally this should be done through open bidding and the best offer wins. This model guarantees some sort of competition among the generation companies whereas the transmission and distribution companies have little space for competition as they are restricted to certain geographical areas. In most countries the transmission is kept with one company because it involves huge investments and does not have very many alternatives.

Therefore, the horizontal model entails having several generation and distribution companies while having only one transmission company. Usually the transmission company is called transmission system operator (TSO) or independent system operator (ISO). In addition to purchasing electricity from generation companies the TSO or ISO also is responsible for selling electricity to distribution companies, and monitoring and controlling the transmission system as well as administering the power purchase over the interconnection lines with other countries. In this model as in the previously discussed vertical model the role of government is reserved for licensing electricity companies and setting the tariff.

Many countries have undergone the transformation in the electricity supply industry during the 1990s and others are still on their way to do so. In the European Union, for example, the European Parliament issued the Internal Electricity Market Directive 96/92/EC in 1996. This directive sets the goals for opening up the electricity markets in the member states (Rothwell & Gomez, 2003).

The following discussion (shown in Box 1) describes the electricity supply industry structure in Jordan as a model for the new trends in the regulatory and legal status trends.

The case of Jordan as depicted in Box 1 above is a typical electricity supply structure recommended for all small electricity market countries. The set up is clearly dependent on opening up the genera-

Figure 1. Current trading arrangements in Jordan

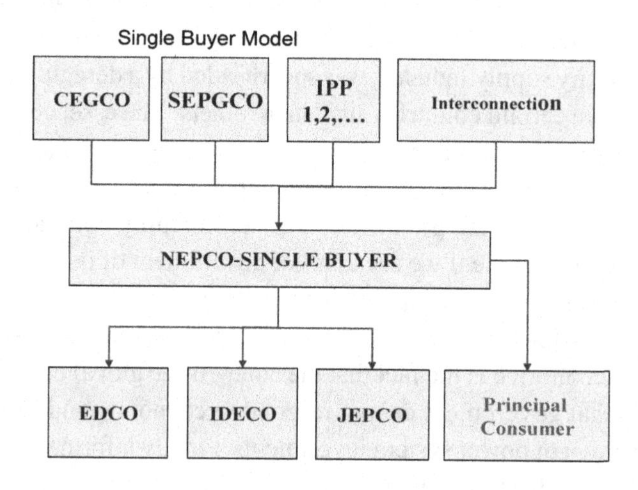

Figure 2.

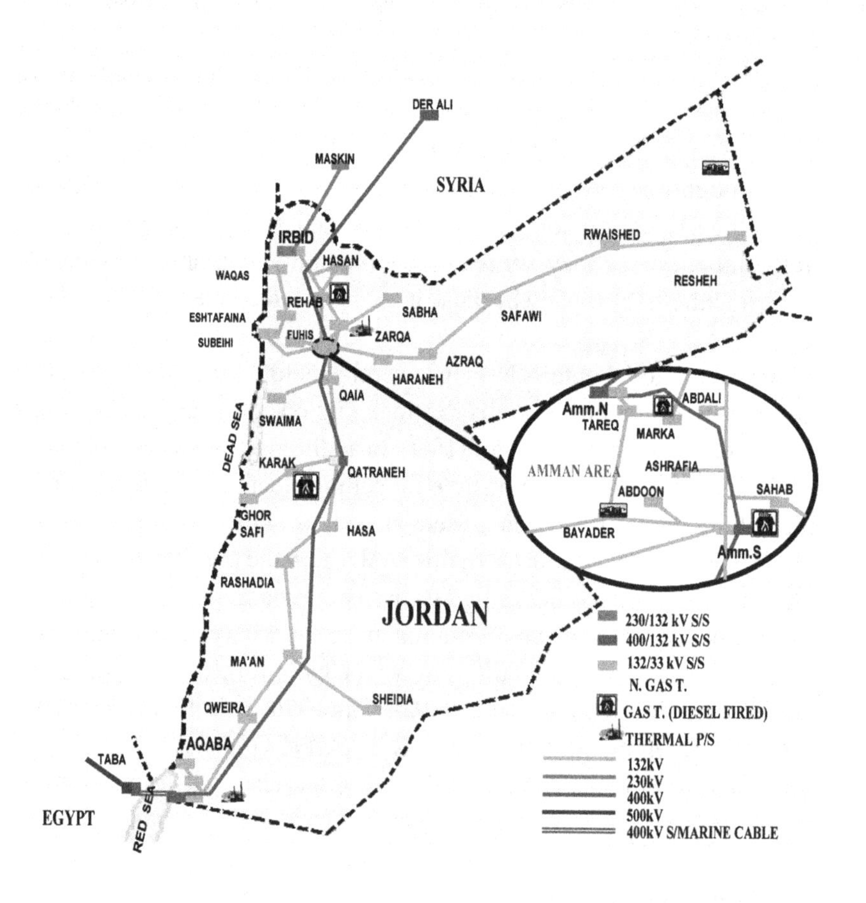

Table 1. Private, public interest in electricity markets

Sub-sector	Private sector interest	Government approach
Generation	High/investment climate must be amenable to private sector participation in energy	Highly receptive/public support to cover policy, and regulatory risks
Transmission and system operation	Moderate/with right structure there might be significant interest in management contracts	Moderately receptive/transmission business may be complicated by single buyer responsibilities
Distribution and access	Low/key challenge is how to package government reform initiatives	Moderately receptive/reluctant to push pricing reform too fast
Regulatory and market framework	Fair/predictable and transparent regulatory framework is critical	Some reluctance regarding regulatory independence
Rural electrification and access	Moderate to low	Moderately to highly receptive/limited potential of pure public approaches to the access problem
Renewable energy	Moderate to high/subsidy will almost always be necessary	Highly receptive
Environment	Moderate to high/subsidy may be applicable in some scenarios	Moderately receptive/coordination between energy and environment agencies is critical
Regional integration	Low to moderate/cross-border guarantee likely to be essential	Moderately receptive/ absence of functioning internal and/or regional markets makes it difficult

tion and distribution components for private investors. There are three privately owned distribution companies in Jordan. On the other hand there are three privately owned generation companies and one government owned generation company. This latter will be subject to privatization soon. Finally the transmission company is fully owned by government. This is called the "single buyer" scheme whereas this transmission company (NEPCO) is responsible for the purchase of power from producers and sells it to the distribution companies; each with a geographical area to serve. NEPCO is also responsible for the purchase of natural gas needed for the generation of power on behalf of the government. It is also responsible for power exchange with Egypt, Syria and The Palestinian Authority.

The electricity trading model adopted in Jordan is as depicted in figure 1 (Maabrah, 2008).

The generation and transmission systems of Jordan are shown on Figure 2 (Maabrah, 2008, p.15) to provide an insight on the electricity supply system in Jordan in order to complete the picture.

One of the most recent market structures is what is called public private partnership (PPP). According to the World Bank (World Bank, 2004, p.13-16), Table 1 lists the degree of attractiveness of each electricity supply industry sub-sector for the private and public groups:

It is clear from the table that the generation, renewable energy, and environment are the three most attractive sub-sectors in the electricity supply industry for the private sector. This is also matched by similar enthusiasm from governments, to some extent, for the same sub-sectors. The other sub-sectors are either not ready presently or require more regulatory modifications and liberalization to become more attractive for private sector participation.

The following discussion compares the electricity market structure in California, Norway, Spain and Argentina with the situation in Jordan in order to prove that deregulation works in any country in the world (Rothwell & Gomez, 2003). Table 2 lists the electricity restructuring reforms in the four case studies mentioned above.

Table 2. Electricity Restructuring Reforms in California, Norway, Spain and Argentina (Rothwell & Gomez, 2003)

Issue	California	Norway	Spain	Argentina
Conditions driving restructuring	• State's electricity prices higher than national average • Inefficient centralized regulation	• Promote efficiency in investments and reduction of regional price differences • Avoid cross-subsidy among consumers • Create cost reduction incentives	• Lower electricity prices • Eliminate subsidies to the coal industry	• Power shortages due to lack of investments and generation unavailability • Highly inefficient public owned sector • Need for new investment
Structural Changes	• Functional separation of generation, transmission and distribution • Recovery of standard costs • Generation divestiture	• Accounting separation of regulated and competitive functions • Transmission grid as a new company • No privatization of publicly owned sector	• Legal separation of regulated and competitive functions • Privatization of standard costs	• Vertical (G, T, D) and horizontal disintegration • Privatization of federal and provincial companies
Wholesale market	• Centralized and physical bilateral trades • Several transmission owners	• Centralized and physical bilateral trades • Trading in the Nordic Pool	• Centralized and physical bilateral trades	• Mandatory pool with financial and bilateral contracts
Retail Competition and customer choice	• All customers • Metering and billing competition	• All customers	• Gradual implementation in a 5-year period	• Large users now and small customers in the future.

Economic and Financial Trends

On top of the trends discussed above there are new economic and financial trends that have a great influence on the electricity supply industry. Naturally these trends go hand-in-hand with the regulatory and legal status trends. The electricity "market" is similar to other markets in the sense that there are sellers, buyers and commodity. However, the main difference with other markets is that electricity as a commodity cannot be stored and needs to be consumed immediately (Schavemaker & Sluis, 2008). In other words electricity is a perfect just-in-time (JIT) model as it is delivered to the consumer the second he/she switches an electrical load on instantly. Delay is intolerable as in other markets and it is a curtailment or interruption of service rather than delay. In reality this is one major indicator of electricity supply performance measured in hours of interruption per consumer per a time period. If a consumer had a choice between two supply companies naturally he/she will elect the one with lower interruption periods/frequency.

Other factors also complicate the issue of considering electricity as product or commodity. The first is that electricity is not labeled or tagged as to the generating plant or transmission line that delivered it to the consumer. Electricity network is common to all generating plants and transmission lines delivering to all consumers that happen to request electricity in a certain moment in time. Furthermore, power flows cannot be controlled by economic or financial means; i.e. power cannot be delivered to a certain consumer from a certain generating plant and through a certain transmission route just on the basis of cost or preferred financial terms (Schavemaker & Sluis, 2008).

One of the major obstacles for power sector reform is the volume of capital investment needed. IEA estimated that developing countries will have to invest annually around US$ 120 billion over the period

2001-2010 in order to keep pace with growing demand (International Energy Agency, 2003, p.364). Most of this investment must come from private investors, since the one of the reasons for the reform was that governments in developing countries cannot continue the investment trend of past years. According to World Bank reports (World Bank, 2004, p.2) a substantial amount of this investment indeed came from private investors. This trend has peaked in 1997 with US$ 50 billion; however, investment has declined to less than 20% of that peak investment year in recent years.

The introduction of deregulation and restructuring is, therefore, motivated by the desire of governments to involve private sector participation in the financing of the electricity sector. This would diversify markets in order to hopefully increase efficiency and reduce prices. Deregulated markets also could attract foreign capital and neighboring countries resulting in more financing and lower production costs (Rothwell & Gomez, 2003, p.3).

Deregulation will encourage private investment needed for system expansion and operation and at the same time provide investors with acceptable rate of return on their investments. The creation of competitive electricity markets, however, involves lowering of market power that could, otherwise, be exerted by the former monopolies (Rothwell & Gomez, 2003, p.3). This means that the regulator has to play a crucial and sensitive role.

In general electricity supply market can take one of four models (Weedy & Cory, 1998, p.502-505). Model 1 is the monopolistic or vertically integrated model whereby generation, transmission, and distribution are all grouped in one entity, in most cases government owned. Model 2 is the purchasing agency or single buyer model whereby this entity purchases power from generation companies (and interconnections) and sells it to distribution companies. This model is very common in many developing countries and Jordan is one such example. Model 3 is called wholesale competition. In this model, distribution companies purchase power directly from the generation companies on competitive basis. Finally model 4 is called retail competition whereby all consumers can choose to purchase from any supplier and pay fees for using the transmission and distribution networks owned by other companies.

Each of these models has certain economic advantages or constraints. Therefore, the selection of one model over the other will have to have some economic preference, in addition to the other factors. As the choice of electricity supply market model is a national concern, consumers will have to make the best of the available options to make economic gains. This is true for models 3 and 4 but not for models 1 and 2.

The possibility of interconnecting with other power systems adds more complexity to the new economic and financial trends. Newer terminologies have been added to the glossary of economic and financial electricity terms. The cross-border exchange of electricity can take one of the following types of services (World Bank, 2003, p.8-9):

- Reserve (generation) capacity supply service including emergency supply at cost for a limited period; scheduled outages of plants to be covered from another utility; a proportion of spinning reserve capacity.
- Economy energy exchange services which are mutually beneficial depending on the short-run marginal costs (SRMC) of both utilities at certain times.
- Firm energy supply services which can delay the need for additional capacity in one utility if that capacity is available for the long term from the other utility. The basis for calculating the economics of this type is the long-run marginal cost (LRMC).

- Energy banking and interchange services which apply for a season or months rather than longer term like the previous type of services. This category of services is applicable for systems with quite different generation mix; i.e. mainly thermal in one utility and mainly hydro in the other.
- Power wheeling services whereby the situation involves three utilities. Two exchanging energy and power and one in between geographically. Wheeling charges include a contribution to fixed cost in the form of connection charge and a losses charge.

Environmental Trends

It is well known that burning fossil fuels is one major source of greenhouse gas emissions. The major greenhouse gas associated with burning fossil fuels is carbon dioxide. Therefore, power plants burning fossil fuels are responsible for the production of CO_2. As a matter of fact the increased burning of fossil fuels to meet the growing global demand could lead to a serious environmental problem. Carbon emissions grew much faster in developing countries than developed countries over the past several decades. The fact of the matter is that all countries of the world are partners in keeping the global environment clean and the pace of pollution must be curbed if not reversed. The solution to this problem takes several forms and shapes. The first came from the Kyoto Protocol which is an agreement made under the United Nations Framework Convention on Climate Change (UNFCCC). In essence this protocol obligates countries that ratified it to reduce their emissions of carbon dioxide and five other greenhouse gases or engage in emissions trading if they cannot reduce these emissions. The overall objective of the protocol is to limit greenhouse gases in a collective fashion, since environmental pollution is a global concern (United Nations, 1998, p.2).

As a result, many countries have engaged themselves with what is known as carbon trading as a result of carbon credits. That is to say that a country that does not exceed its quota of carbon emissions is allowed to trade this "emission privilege" to another country which has exceeded its quota and cannot reduce its emissions. The carbon credit for a certain country can come from one of the following sources (United Nations, 1998, p.7,15):

- Carbon emission reduction programs implemented in the country involving some existing plants.
- New renewable energy plants developed in the country.
- Emissions allowances originally allocated or auctioned through previous trading mechanisms.

These carbon credits, if approved through one of the UNFCCC's approved mechanisms, are then termed certified emissions reductions (CERs). The Kyoto Protocol provides for three mechanisms that enable developed countries to acquire CERs as follows:

- Under Joint Implementation (JI) a developed country with relatively high costs of domestic greenhouse reduction would set up a project in another developed country.
- Under the Clean Development Mechanism (CDM) a developed country can sponsor a greenhouse gas reduction project in a developing country where the cost of greenhouse gas reduction project activities is usually much lower, but the atmospheric effect is globally equivalent. The developed country would be given credits for meeting emission reduction targets, while the developing country would receive the capital investment and clean technology or beneficial change in land use.

- Under International Emissions Trading (IET) countries can trade in the international carbon credit market to cover their shortfall in allowances. Countries with surplus credits can sell them to countries with capped emissions under the Kyoto Protocol. (United Nations, 1998, p.11)

Other pollution abatement measures include national limitations on the production of greenhouse gases, introduction of environmentally benign energy production technologies and curbing the levels of consumption of energy. To some extent the quest for a clean environment is strongly linked to renewable energy and energy efficiency programs and projects.

The environmental awareness has also created renewed interest in renewable energy. Some of these renewable energy projects classify as a JI or a CDM. Therefore, research and development in renewable energy sources witnessed a surge in last twenty years, driven by fossil fuel escalating prices (over US$100 per barrel for crude oil), and the concern for a clean environment. As a result we see today renewable power plants spreading in many countries. Wind energy is considered the most promising at the present time, although other technologies such as photovoltaic solar, solar thermal, geothermal, fuel cells, micro combined heat and power (MCHP), and others are also progressing very well.

Another issue of paramount importance is the group of projects in the domain of energy efficiency. This category of projects focuses on increasing the efficiency of existing plants or appliances. The end result is a net reduction of emissions and reduction of electricity bills. These projects cover the areas of energy conservation, demand side management, loss reduction, and plant rehabilitation. All these activities will one way or another lead to increased energy efficiency.

According to a World Bank report (World Bank,2007, p.37-38) a total of US$1.43 billion supported 63 renewable and energy efficiency projects in 32 countries in 2007 alone. Moreover, the average share of renewable and energy efficiency projects of the total energy projects has doubled from 13% in the period 1990-1994 to 25% in the period 2005-2007. This figure reached 40% in 2007.

POWER SYSTEM STUDIES

The task of running a power system is really very huge and it is not practical to do so without the aid of various software tools. Some aspects of the operation are also run automatically; thus computers are responsible for some of the action, with human interference only in emergency cases. Broadly speaking the following listing is an attempt to list all studies, which are used in power system planning, operation, and control (Elkarmi & Abu Shikhah, 2007, p.1). This discussion of types of power system studies and analyses is useful in the identification of IT models and systems. Some applications are already available commercially, while others are either under development or still need to be developed.

Planning Criteria

Planning criteria are actually a set of parameters, which are determined through analysis and benchmarking. These parameters need verification and updating every now and then. This requirement comes from changes which take place in the power system or surrounding environment. The parameters are used as input in other studies and analyses. Therefore, they will have an impact on the power system planning, operation and control.

The parameters include the following:

- Loss of load probability (LOLP).
- Energy not served (ENS) and its cost.
- Contingency criterion, i.e. what type of incidents the system should withstand without collapsing. It is designated as N-1 or N-2 etc. It is used for transmission as well generation sub-systems.
- Permissible loading of transformers, transmission lines, or other equipment such as reactors or capacitors.
- Reserve margin, which is the percentage of excess generation required to sustain the system intact in cases of contingencies. This percentage varies from an isolated system to a strongly interconnected system. It simply means that an isolated system will require certain percentage of generating plant in excess of expected load more than an interconnected system.
- Permissible voltage and frequency variations. These are actually power quality criteria, but they are used in determining the level with which power system operators can tolerate contingencies. As a matter of fact, any excursions on these limits are for very short durations and are, thus for operational purposes only. However, in certain instances they are used for some exceptional cases for longer durations.
- Economic parameters such as discount rates, cost of energy not served, inflation rates, financing rates, etc.

Load Forecasting

Short-Term for Operational Purposes

This is load forecast for several hours to several days ahead. This forecast is used for economic dispatch of power plants, scheduling power exchange with other systems, and optimizing the operation of the transmission network to minimize losses. There are several short term forecast algorithms and commercial computer packages. Some power companies have developed their own algorithms based on experience and deterministic parameters. Some use more sophisticated commercial or developed in-house software. The methods used are grouped as follows:

- Deterministic time series analysis or regression analysis (curve fitting)
- Probabilistic time series analysis
- Moving average methods
- Expert systems
- Artificial neural networks
- Fuzzy logic.

The first three methods are called conventional methods, while the last three are called intelligent-based methods. The intelligent-based load forecasting techniques, such as *expert systems*, *artificial neural networks*, and *fuzzy logic*, have been developed recently, showing encouraging results. Among them, artificial neural networks methods are particularly attractive. They have been applied to many time-series modeling and forecasting problems for their flexible mathematical structure which is capable of identifying the *complex* and *non-linear* relationships between inputs and outputs. The success of neural networks modeling mostly depends on selecting suitable model and the method used to fit the model and compute prediction. Most of these applications assume predefined network architecture, including connectivity and node transfer functions, and use a training algorithm, such as *back-propagation* (BP) to learn weights and biases (El-Telbany & Elkarmi, 2008, p.425-426). The Particle Swarm Optimization (PSO) algorithm is a new adaptive algorithm based on a social-psychological metaphor that may be used to find optimal (or near optimal) solutions to numerical and qualitative problems. Most particle swarms are based on two socio-metric principles. Particles fly through the solution space and are influenced by both the best particle in the particle population and the best solution that a current particle has discovered so far. The best particle in the population is typically denoted by (global best), while the best position that has been visited by the current particle is donated by (local best). The (global best) individual conceptually connects all members of the population to one another. That is, each particle is influenced by the very best performance of any member in the entire population. The (local best) individual is conceptually seen as the ability for particles to remember past personal success. The particle swarm optimization makes use of a velocity vector to update the current position of each particle in the swarm. The position of each particle is updated based on the social behavior that a population of individuals adapts to its environment by returning to promising regions that were previously discovered (El-Telbany & Elkarmi, 2008, p.426).

Medium-Term for Maintenance Planning

As the name implies the purpose of medium term load forecasting is for maintenance planning. The forecast period is from several weeks to 12 months ahead. This type of forecast depends mainly on growth factors, i.e. factors that influence demand such as main events, addition of new loads, demand patterns of large facilities, and maintenance requirements of large consumers. This type of forecast is not concerned with hourly loads like short term forecast, but rather predicts the peak load of days or for the weeks ahead. With this information it can be decided to whether take certain facilities/plants for maintenance or not during a given period of time. The methods used for this type of forecast are similar to the short term forecast except that there is less need for accuracy and simpler methods might do.

Long-Term for Long System Expansion Planning

Again as the name implies this type of forecast is used to plan the expansion of the power system, i.e. what type of generation or transmission plant(s) are needed, when, where, and what size. Usually generation system planning is done separately from transmission system planning. The study period of this forecast is from 2 years to 15-20 years ahead. The output of this forecast is usually the peak load and annual energy requirement of the system. That is to say that the peak load and energy requirement for the coming years of the study period are determined by the forecasting method. Usually econometric or regression analysis methods are widely used in this type of forecast. However, end-use and expert system methods are also used.

Energy Management

Load Research

Consumers of electricity vary in the amount of electrical energy and electrical demand drawn from the network. That is why electrical power companies classify such consumers, or customers, into several classes. This classification is based on characteristics of demand; behavioral issues; and other considerations related to location, climate, and status to name a few. The quest and accumulation of this customer-related information is called load research. With load research the electricity company will be aware and hopefully knowledgeable in the composition and trends of demand of consumers. This essential knowledge will be the basis for pricing electricity properly, and fairly. Moreover, the electricity companies can predict future demand based on such information. Therefore, new expansions, enforcements, or extensions will be affected to cater for any future demand. On the other hand proper operation and control of the power system requires that all the information available on consumers and consumption patterns be on hand. With the aid of this information the electricity companies can minimize production cost, plan maintenance schedules, and control the quality of power delivered to all consumers. This would be reflected in leaner electricity tariff and consequently in affordable bills (Elkarmi, 2008, p.1758).

Demand Side Management

Demand side management (DSM) is a set of functions and activities to effect change on consumption patterns. Demand side management is the act of attempting to manage demand of consumers to achieve energy efficiency. This is usually done through incentives and regulations. Consumers' conviction and belief in the benefits of DSM programs is essential for the success of such programs. This comes from awareness and education. The power company should determine which program(s) is (are) more suitable than others. It is not only a technical study to determine the technical feasibility of the DSM programs, but also an economic analysis to confirm if it is economically feasible as well. DSM programs include targets such as: peak shaving, peak shifting, valley filling, demand growth, and many more.

Integrated Resource Planning Studies

This type of studies is a combination of the demand side as well the supply side. In other words these studies concern the system expansion with both demand and supply sides together. If for example the system will need x MW in a certain year in the future the IRP study will first attempt to ascertain if it can be fully or partially met from DSM programs. If not then alternative supply sources will be studied to meet the demand.

Energy Efficiency and Conservation

Energy efficiency and conservation studies are similar to DSM but are related to methods and means to reduce energy consumption rather than demand reduction.

The type of appliances used and their consumption trends are in the center of these studies.

Load Flow

Load flow (or power-flow) study is a backbone tool used in different power system operational studies. Load flow involves applying numerical analysis techniques to a power system. This type of study results in solving for different system parameters based on a given snapshot of that system under certain operational conditions. By solving a system at specified conditions the base case for such system is obtained, which can be further used to conduct other studies such as fault analyses, stability analysis, economic analysis. By implementing linear programming methods the optimal power flow resulting in the lowest cost per MW generated. It should be seen that a power flow study usually uses simplified notation such as a one-line diagram and per-unit system, and focuses on various forms of AC power (i.e.: reactive, real, and apparent) rather than voltage and current. There exist a number of software implementations of power flow studies. The great importance of power flow or load-flow studies is in the planning the future expansion of power systems as well as in determining the best operation of existing systems (Saadat, 1999, p.189).

Short Circuit Analysis (Fault Levels)

This is done on as need basis, i.e. when new substations, transmission lines or transformers are added to the network. The new additions usually will either add a new feed source to a certain point or add impedance at a certain point in the network. Naturally these cases will increase the short circuit level. This fault level, of all important points in the transmission network, is very important for the design and implementation of the protection systems and schemes.

Loss Reduction Studies

As the transmission network evolves and is expanded by adding new substations and transmission lines the impedances change and new flow routes are formed, thus resulting in a different loss situation. The continuous monitoring of the levels of losses and consequent analyses of various alternatives for network expansion constitute the core of loss reduction studies. Moreover, in certain instances the re-wiring of certain lines for the purpose of increasing current carrying capabilities also serve the purpose of loss reduction. The use of suitable cooling for critical cables also helps in reducing losses. Finally the possibility of using super conductors will reduce losses tremendously.

Reliability/Availability Assessment

Electricity is one of the most efficient forms of energy. It is further clean and can economically reach any consumer. Electricity is sometimes termed "energy by wire". The production, transmission and distribution of electricity must conform to certain standards and best practices. Reliability and continuity of supply are two important performance indicators for electric system's operation. A third indicator is cost of production which must reflect the true cost of production, transmission and distribution. At the same time this cost must be affordable to consumers; especially consumers that are economically active. If cost of electricity to consumers is much less than the true cost (heavily subsidized tariff) then wasteful consumption takes place. On the other hand if it is more than the true cost (excessive margins or profits) or if it is beyond the paying capability of consumers then electricity use will be curtailed in a manner to harm consumer economic activities. Reliability indexes include the following:

- SAIFI (System Average Interruption Frequency Index) unit: int./cust./yr.
- SAIDI (System Average Interruption Duration Index) unit: hr./cust./yr.
- CAIDI (Customer Average Interruption Duration Index) unit: hr/int.
- AENS (Average Energy Not Served) unit: kwh/cust./yr.

Performance Assessment and Quality of Supply

This includes harmonics, voltage, and frequency excursions. In other words the electricity delivered to consumers must conform to certain acceptable limits. Variation within these limitations is considered acceptable and tolerable. Performance also covers interruptions of electricity supply both in number and duration. If interruptions exceed an acceptable average number or duration then electricity service can be considered low quality.

Reactive Power Compensation

Rising energy costs, increased transmission distances and use of large generating machines are resulting in increased demands for reliable and more economic operation of transmission and distribution systems. However, even in the most efficient power transmission and distribution systems, some transmission losses are inevitable. In weak networks these losses can be considerable. Efficient power transfer from source to consumer with high availability is an important link in the energy chain.

But how can we satisfy the growth in power consumption and demands for increased capacity and power quality on transmission and distribution lines? Reactive Power Compensation is a well established technology. By reducing the negative impact of reactive power in high voltage AC networks, the capacity of transmission systems can be dramatically increased. This means greater reliability and better use of existing lines.

Economic Dispatch and Unit Commitment

In a power system there are several generating plants each with different size, technology, and cost of production. These plants are located in various geographical places with varying distances from load centers. Therefore, the cost of production and delivering electricity from any of the generating plants will be different from other generating plants. The objective is to find the real and reactive power scheduling of each power plant such as to minimize the overall operating and transmission cost. Plants or units are usually committed or ranked in a merit order according to production costs. System operators select plants or units to produce scheduled power as per this merit order.

System Expansion Studies

This type of studies is very important for the future of the power system. Such studies determine when new generation and/or transmission units/plants/facilities will be needed. The size of units, type, data needed, and location are output from system expansion studies. Usually generation system planning is done first. Based on the results of the generation system expansion transmission system planning is done.

A great deal of information obtained from other studies is used as input to system expansion studies. This includes: planning criteria, load forecast, load flow, load research, DSM, and energy efficiency.

The output of system expansion studies includes the units to be added each year, the capital and operational costs associated with this optimal plan, and other performance indices. It also includes an investment plan to cover the capital costs required for the expansion.

Power System Stability

Present power systems are interconnected leading to a highly non-linear dynamic behavior, which must be understood either at the smooth operation times (steady- state) or under disturbance conditions. The specialist in this field must have good profile in this field that covers the following areas (Asare, Diez, Galli, O'Neill-Carillo, Roberston & Zhao, 1994, p.10):

- The basic concepts and definitions of angular stability (transient, small signal), and voltage stability, bifurcation (saddle-node, and Hopf bifurcation and chaos).
- Dynamic modeling of power system components covering generators (non-linear and linear models using d-q transformation, power capability curve), excitation system (IEEE Standard Models), turbine and speed governing system, loads (induction motors and composite loads), and Flexible AC Transmission System (FACTS) Devices.
- Transient stability analysis starting with single machine - infinite bus system and equal area criterion. multi-machine stability, network reduction and numerical integration methods, and methods of improvement must be mastered.
- Small signal stability analysis including eigenvalue and participation factor analysis, single machine - infinite bus and multi-machine simulation, effect of excitation system and AVR, and improvement of damping - power system stabilizer and SVS supplementary controls.
- Sub synchronous oscillations where sub synchronous resonance (SSR) phenomenon, and counter measures to SSR problems are considered.
- Voltage stability which deals with P-V and Q-V curves, impact of load and tap-changer dynamics, static analysis, sensitivity and continuation methods, proximity indices, and methods to enhance stability margin.

Power System Protection Studies

The protection scheme of the power system is very complicated and is heavily dependent on digital and logical elements. As the transmission network evolves the protection system must be updated to cover the additions. Since protection coordination is one essential principle in the design and operation of protection systems, it needs to be checked and modified whenever new additions or expansions take place. Moreover, protection system settings need continuous revision to guarantee swift, correct operation and clearing of faults. Therefore, this important aspect of the power system deserves to be continually studied and corrective actions taken to guarantee that the protection system will function properly when needed. As it is usually said, the protection system is occasionally put to test, but when it is, it should act error-free as the consequences can be devastating.

Major Incidents/Faults Analysis

Every power system is subject to major disturbances which could lead to a complete shutdown or "blackout". Some such events are readily and logically explainable, while others are more complicated; thus requiring thorough investigations, testing, and assessment. In certain cases there might be different viewpoints. Simulation techniques are used to attempt to mimic or repeat what has happened in reality. If the simulation results, using complicated computer programs, are closer to reality then the thesis of this simulation is closer to reality. Otherwise other viewpoints may be investigated. Sometimes testing is the only way to find out what has really happened.

OTHER CONSIDERATIONS

In spite of the tremendous development in the reform of the power sector or electricity supply industry and the evolvement of private sector participation, some difficulties still shadow the reform process. This stems mainly from the particularities of the power sector. The following is some account of these particularities, which could pose some obstacles and hindrance to the reform process. This analysis is useful in identifying any possible intervention or introduction of IT models which could remove or ease these constraints:

1. Transmission systems in particular and, to some extent, distribution systems are better-off if they are monopolies because of the high investment involved and the difficulty in operating different systems by more than one company. Therefore, there is no chance for horizontal unbundling in these two functions. However, geographical basis may be used to create competition by allocating each geographical area to one company. This applies more readily to distribution rather than transmission again due to the high per unit investment needed for transmission of power. This is somewhat conflicting with the privatization trend.
2. Renewable energy (solar and wind in particular) has the following unique characteristics:
    - Small power output compared to conventional technologies, thus losing the economy of scale cost advantage,
    - usually connected to the distribution network ; as a result of small power output,
    - since renewable energy sources are connected to the distribution networks or to consumer premises directly, they compete with the distribution companies in their areas,
    - in certain cases there is a need to arrange for import/export metering between renewable sources and distribution companies to financially manage the power exchange,
    - renewable energy sources, especially wind, are site-specific in the sense that each site will have to be assessed as to the energy potential. Usually countries produce wind and solar maps showing energy potentials,
    - renewable energy sources are superior to conventional sources in the environmental aspect, therefore, special cost advantage for environmental protection should be added to renewable energy sources, but this is not always done,
    - renewable energy sources are, in most cases intermittent and depend on uncontrollable factors such as weather, therefore, they need to be augmented with energy storage or be part of a hybrid system or a power exchange scheme with the distribution network,

- some renewable energy sources are still in the R&D stage and the cost of production is still very uncompetitive to conventional sources,
- many renewable resources are abundant in developing countries, while the rich developed countries are not endowed with these sources.

3. The trend in the past was to build large centralized power plants connected to bulk supply points through super transmission grids, In certain countries this trend is being reversed in favor of decentralized small local plants may be as partial solution. This is driven by the availability of renewable energy sources at the locations designated for decentralized plants or the remoteness of certain loads from the network. Decentralized generation, nevertheless, suffers from some operational difficulties such as voltage control, reliability, and stability problems. On the other hand large centralized and super transmission grids model also has operational difficulties of its own, and some argue that "small is beautiful". It is true that operating a decentralized system is easier than operating a centralized system, but reliability and continuity of supply of the decentralized system negates its advantages. Moreover, decentralized systems might not be capable of supplying a large industrial or commercial complex. The optimum set up would be to have both systems in place. The centralized system to supply large industrial or commercial loads, and decentralized systems for remote local loads.
4. If in the quest for more reliability and continuity of supply more than one decentralized system is interconnected this would complicate the operation of the overall system and would bring us back to where we were.
5. Another operational difficulty of decentralized generation is that it does not have a rotating mass as opposed to centralized generation; therefore, there is no stored energy in the form of spinning reserve for frequency control. Moreover, the little rotating mass of decentralized generation units, especially wind turbines is separated from the grid by power-electronic interfaces which decouple the speed of the rotor from the frequency of the grid. This problem may be alleviated through the use of energy storage units as part of the decentralized systems. These units can help in cases of supply-demand mismatch and the associated frequency deviations (Schavemaker & Sluis, 2008, p.225).
6. The electricity supply industry reform has succeeded in encouraging private sector participation and there are many success stories to stand witness to this trend. However, issues such as energy efficiency with its broad coverage of demand side management, and energy conservation programs have been negatively affected. These programs are cost-effective at the national level. In other words if the societal cost is taken into consideration almost all programs are feasible, however, at the company level there is great doubt that they are. These programs used to enjoy a great deal of government support in the form of subsidy which could explain a large part of their feasibility. With subsidy removed, these programs do not stand a chance of being adopted by the consumers. The only remaining hope for such programs is the voluntary willingness of consumers out of protecting the environment or other patriotic feelings.
7. Pricing of energy forms including electricity should be done on a comprehensive basis. In other words it is not advisable to change the tariff for petroleum products in isolation from electricity tariff. This is because of the switching action between energy forms. Consumers will shift from one energy form to another, provided this is technically possible, once the price structure of these forms suffers from deformations. This is especially true for heating purposes, for example. Once diesel oil prices have increased much more than electricity prices, then consumers will shift to

electricity heating, even if this required capital investment. This is exactly what happened in Jordan two years ago when the prices of diesel oil was increased to very high levels, to remove the chronic subsidy, with much less simultaneous increase of electricity tariff. The result was that most domestic, commercial and even industrial consumers shifted to electric heating, therefore, the electrical peak load also shifted, since then, to winter season instead of the historic summer time. If the diesel price was studied along with the electricity tariff and a balanced increase was adopted this would not have happened.

8. The power sector reform process is not free from political and socio-economical forces. In certain respects governments are enthusiastic about the reform process and especially the privatization trend, which will relief governments from shouldering the investment burden. On the other hand, the socio-economic factors have a strong bearing on governments, especially in democratic countries. In general there should be an intricate balance among all key factors. In other words the reform process should be comprehensive in the sense that it should include all aspects. Moreover, the pace of the process should balance the actions and have due consideration to possible future national, regional and international events and circumstances.
9. Some might wrongly think that with privatization electricity companies will seek to maximize their profits and gains with little due attention to customer satisfaction and power quality. However, this thinking neglects the role of the regulator. One of the most important roles of the regulator is to safeguard the interests of consumers and to make sure that the reliability level of the electricity supply is within the acceptable ranges. In order to be effective certain regulators impose penalties and fines on the companies that do not meet the customer satisfaction criteria or deliver power that is inferior in terms of several reliability indexes.
10. There are great potentials for energy savings in any community, municipality or city. These energy saving potentials combine electricity and other energy forms as well as water, waste, and all other utilities in an effort to maximize the benefit. These energy saving potentials could be translated into future IT models to be developed by researchers. The energy saving opportunities can be classified as per the introduction stage as follows:
    A. Savings in the design stage of buildings and facilities:
        - Selection of the most energy efficient equipment (including central air conditioning systems, air handling units, chillers, lighting systems, etc)
        - Passive architecture designs (including natural lighting, natural air circulation and ventilation, double glazing, sun curtains, insulation materials for walls and roofs, etc.)
        - Selection of most efficient utilities (combined heat and power, district heating system, common large-size units, pressurized water systems, etc)
        - Proper design of the local electricity network (including voltage level, conductor cross sections, transformer ratings, substation arrangements, etc.)
        - Selection of efficient street lighting system.
        - Provision for mass transit systems for passengers and freight (including light rail, escalators and people movers, etc)
        - Provision for the introduction of renewable energy sources (including solar water heaters, photovoltaic electricity generators, fuel cells, etc.)
        - Provision for biogas plant for electricity generation from solid, municipal and industrial wastes.

- Selection of local raw materials, as much as applicable, to conserve energy in transporting such materials from long distances.

B. Savings in the implementation stage of buildings and facilities
- Proper installation of systems and utilities.
- Use of original equipment and parts from genuine OEMs.
- Following recommendations of OEMs.
- Selecting energy efficiency knowledge and experience as a main criterion in selecting contractors.
- Use of energy efficiency consultants in preparing tender documents including specifications, performance criteria, minimum requirements, etc.
- Short listing of appropriate certified and qualified contractors.
- Use of energy efficiency consultants to monitor progress of contracts.
- Use of contract bonus schemes for engineering value judgment of contractors.
- Third party inspection of equipment for large scale utilities.
- Performance testing prior to acceptance.
- Conformance to energy efficiency specifications guarantees.
- Coordination meetings with all contractors during implementation to monitor energy efficiency issues.

C. Savings at a later stage as retrofit for existing buildings and facilities and promotion of energy efficiency programs
- Conducting life-cycle cost replacement feasibility studies for non-efficient equipment.
- Replacement of non-efficient equipment.
- Introducing Demand Side Management (DSM) studies to determine cost-effectiveness of certain programs for implementation.
- Introducing power factor improvement measures.
- Introduce electricity loss reduction programs.
- Introduce harmonics reduction programs.
- Increase electrical system reliability to increase customer satisfaction and therefore, willingness to pay for electricity.
- Designing certain incentive schemes for energy efficiency programs.
- Conducting energy audits for industries, commercial buildings, office buildings, recreational facilities, etc.)
- Introducing energy efficiency programs as a result of the energy audits.
- Provision of financing schemes for the replacement process.
- Introducing Energy Service Companies (ESCOs) to provide for energy efficiency programs for consumers.
- Negotiating with the power, water, gas, municipal waste utilities to have long term supply contracts to benefit from reduced rates.
- Negotiating with the power utility/company to have an agreement to supply power to the network for the same rate as power purchased.
- Negotiating with the power utility/company to have interruptible load agreements for certain non-critical loads to reduce electricity bills.
- Introduce local distribution tariffs for electricity, fuels, gas, water, municipal waste and all other utilities.

- Conduct awareness campaigns for energy efficiency and appropriate programs.
- Conduct training programs for energy staff of large commercial complexes, industries, high-rise office buildings, facilities, etc.
- Promote the introduction of energy manger posts at large commercial complexes, industries, high-rise office buildings, facilities, etc.
- Promote the concept of green energy for businesses that use or purchase power from green energy sources.
- Introduce energy efficiency labels for equipment as a mandatory requirement for manufacturing, importing or selling any equipment.
- Introduce energy efficiency standards and codes for building insulation.
- Introduce energy efficiency awards and prizes.

REFERENCES

Asare, P., Diez, T., Galli, A., O'Neill-Carillo, E., Roberston, J., & Zhao, R. (1994).*An Overview of Flexible AC Transmission Sytems*. e-Pubs, Purdue University Library. Retrieved from http://docs.lib.purdue.edu/ecetr/205

El-Telbany, M., & Elkarmi, F. (2008). Short-Term Forecasting of Jordanian Demand Using Particle Swarm Optimization. *Electric Power Systems Research Journal*, *78*(3), 425–433. doi:10.1016/j.epsr.2007.03.011

Elkarmi, F. (2008). Load Research as a Tool in Electric Power System Planning, Operation and control: the Case of Jordan. *Energy Policy*, *36*(5), 1757–1763. doi:10.1016/j.enpol.2008.01.033

Elkarmi, F., & Abu Shikheh, N. (2007, September 3-5). *Power System Analysis and Studies: Cooperation between Academia and Electricity Companies*. Paper presented at CIGRE Conference, Amman, Jordan.

International Energy Agency. (2003). *World Energy Investment Outlook, 2003 Insights*. Paris: Author.

Maabrah, G. (2008, May 26-27). Electricity Structure in Jordan and Regional Market. *Joint Arab Union of Producers, Transporters and Distributors of Electricity (AUPTDE) and MEDELEC Conference, Sharm El-Sheikh, Egypt.*

Rothwell, G., & Gomez, T. (2003). *Electricity Economics: Regulation and Deregulation (1st ed.)*. New York: IEEE-Wiley Press.

Saadat, H. (1999). *Power System Analysis*. New York: McGraw-Hill

Schavemaker, P., & Sluis, L. (2008). *Electric Power System Essentials*. New York: John Wiley & Sons.

United Nations. (1994). *Koyoto Protocol to the United Nations Framework Convention on Climate Change*.

Weedy, B. M., & Cory, B. J. (1998). *Electric Power Systems* (4th ed.). New York: John Wiley & Sons.

World Bank (2003). *Building Regional Power Tools: A Toolkit.*

World Bank (2004). *Public and Private Sector Roles in the Supply of Electricity Services,* (World Bank Paper No. 37476).

World Bank (2007). *Catalyzing Private Investment for a Low-Carbon Economy.*

Chapter 15
An Intelligent Motor-Pump System

P. Giridhar Kini
Manipal University, India

Ramesh C. Bansal
The University of Queensland, Australia

ABSTRACT

Process industries are energy intensive in nature and are one of the largest consumers of electrical energy that is commercially generated for utilization. Motor driven systems consume more than two-thirds of the total energy consumed by the industrial sector; among which, centrifugal pumps are the most widely used equipment mainly for the purpose of fluid transportation. The efficiency of pumping units is around 40 to 50%, hence they offer tremendous opportunities of not only improving the efficiency of the process, but also ensure effective energy utilisation and management. With the increasing use of power electronics equipment, power quality (PQ) has become a very serious issue of consideration. On account of the random switching of single-phase loads in addition to time varying operations of industrial loads, PQ problem of voltage variation and unbalance is inevitable across three-phase systems. Application of varying or unbalanced voltages across the three-phase motor terminals results in performance variations leading to inefficient operation. For the purpose of study, the performance of a motor-pump system can be separately analyzed from the motor and pump points of view. The motor efficiency may vary in a very narrow band, pump efficiency depends upon the system head and flow rate but the system efficiency is a combination of the two; hence, necessary to analyze separately. As centrifugal pumps are classified under variable torque-variable speed load category, variation on the input side has a significant effect on the output side. Therefore the system efficiency now becomes an important index for ensuring efficient energy utilisation and efficiency. The main objective of the chapter is to put forward a methodology to analyze the working performance of a three-phase induction motor driven centrifugal pump under conditions of voltage and load variations by, defining additional factors for correct interpretation about the nature and extent of voltage unbalance that can exist in a power system network; define induction motor derating factors for safe and efficient operation based on operational requirements and devise energy management strategies for efficient utilization of electrical energy by the motor-pump system considering the voltage and load conditions.

DOI: 10.4018/978-1-60566-737-9.ch015

INTRODUCTION

The industrial sector is the largest consumer of all the electrical energy that is commercially generated for utilization with process industries like fertilizer, cement, sugar, textile, aluminum, paper etc; being the major consumers (Palanichamy et al., 2001). With the conventional fuel supplies on the verge of becoming scarce and increasingly expensive, and initial investment for harnessing energy from renewable sources being too high; the concept of energy auditing and energy conservation / efficiency practices have gained significant importance. The process of energy management embodies engineering, design, applications, utilization, and to some extent the operation and maintenance of electric power systems to provide the optimal use of electrical energy (IEEE Std. 739-1995; Lee and Kenarangui, 2002). Energy audit programmes are inexpensive investments, as compared to the cost of energy utilization, and are an important tool in analyzing and controlling the demand-supply situation. On successful completion of energy audit, a list of number of energy conservation proposals can be prepared, which can then be prioritized for energy management.

As per IEEE Std. 1159-1995, power quality (PQ) refers to a wide variety of electromagnetic phenomena that characterize the voltage and current at a given time and at a given location on the power system (IEEE Std 1159-1995). As per International Electrotechnical Commission (IEC), "Electromagnetic compatibility is the ability of an equipment or system to function satisfactorily in its electromagnetic environment without introducing intolerable electromagnetic disturbances to anything in that environment" (IEC 61000-1-1). Power quality disturbances are the result of various events that are internal and external to industrial utilities. Because of interconnection of grid network, internal PQ problems of one utility become external PQ problems for the other. With increasing stress towards improving energy efficiency, industrial utilities adopt energy saving technologies in the form of adjustable speed drives (ASDs), which are one of the main contributors towards PQ degradation. With the widespread use of non-linear loads, time varying single and three-phase loads, problem of controlling PQ is becoming a relevant issue (Aquila et al., 2004). The main concern of all industrial utilities is with regard to voltage and current quality. As voltage and current are closely related, deviation of any of them from the ideal may cause the other to deviate from the ideal case (Bollen, 2000). PQ variations are generally divided into four basic categories while the various kinds of disturbances that can possibly occur in a power system networks are impulsive transients, oscillatory transients, sag, swell, sustained interruption, undervoltages, overvoltages, voltage unbalance, dc offset, harmonics, interharmonics, notching, noise, voltage fluctuation, frequency fluctuations (IEEE Std 1159-1995; Dugan et al., 1996; Oliver et al., 2002). The sources that lead to PQ problems are quite complex in nature and sometimes difficult to detect; but most of the PQ problems are caused by factors that are beyond the control of the utilities; hence they can never be totally eliminated.

Motor driven systems consume about two-thirds of the total energy consumed by the industrial sector; among which, centrifugal pumps are the most commonly used equipment, and are the largest consumers of electrical energy (Mircevski et al., 1998). Pumping systems are the major energy consumers in the industrial sector while in many process plants; pumping systems are estimated to account for 40 to 60% of total plant energy consumption. With the efficiency of pumping units around 40 to 50%, they offer tremendous opportunities of not only improving the efficiency of the process, but also ensure effective energy utilisation and management.

METHODOLOGY OF ANALYSIS

The methodology for steady state performance analysis of a three-phase induction motor driven pump system subjected to voltage variation and unbalance is studied by splitting up into 3 parts: input side, process side and output side.

The input side involves the analyses of the three-phase voltages applied to the motor-pump system. The process side involves the steady state three-phase induction motor equivalent circuit analysis. The output side involves the pump side analysis.

Input Side Analysis

Voltage variation and unbalance is one of the commonly occurring PQ problem faced by the industrial sector. In simple terms, voltage unbalance is the deviation of the three-phase voltage magnitudes away from the normal / rated values. Voltage unbalance has been interpreted and mathematically expressed in a number of ways (Jeong, 2002; Lee, 1999; Pillay and Manyage, 2001; Wang, 2001; Bollen, 2002; Pillay et al., 2002; Faiz et al., 2004; Ghijselen and Van den Bossche, 2005).

The voltage unbalance factor (VUF) as expressed by IEC is

$$\% \, VUF = \frac{V_N}{V_P} \times 100 \, \% \qquad (1)$$

V_N and V_P are magnitudes of negative and positive sequence voltage components respectively.

Equation (1) gives a better understanding of voltage unbalance as it takes into account the resulting sequence components by applying the symmetrical component technique, but for a particular VUF, there will be various combinations of positive and negative sequence voltages (Wang, 2001; Pillay et al., 2002). These infinite combinations will lead to various interpretations and computational errors. The infinite combinations can be narrowed to a unique case by considering the complex nature of the sequence components i.e.; complex voltage unbalance factor (CVUF) (Pillay et al., 2002; Faiz et al., 2004), defined as the ratio of negative sequence voltage phasor to the positive sequence voltage phasor, and can be expressed as

$$\text{CVUF} = K_V \angle \theta_V \qquad (2)$$

where $K_V = \left| \frac{V_N}{V_P} \right|$ is magnitude of CVUF and $\theta_V = (\theta_N - \theta_P)$ is phase angle by which the negative sequence component leads the positive sequence component.

Fortescue (1918) has proved that an unbalanced system of 'n' phasors can be resolved into 'n' systems of balanced phasors called the symmetrical components of original phasors. For most motor driven systems, it is quite difficult to measure the individual phase voltages, as the neutral point is not externally available. Since all calculations are computed on a per-phase basis, for computing the correct CVUF, there must exist a definite relationship between the magnitudes of the CVUF computed using line and phase voltage values. Industrial utilities have to pay for both active power and reactive power

Figure 1. Line and phase voltage triangle

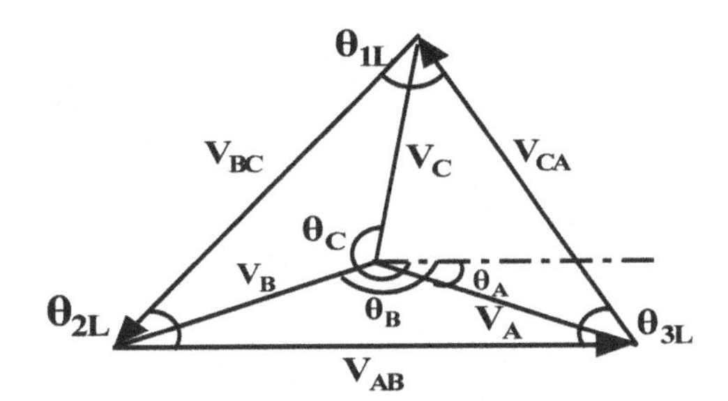

consumption, and therefore inaccurate or approximate estimation or measurements will be reflected in all operating parameters leading to inaccurate result analysis. To avoid assumptions or approximations as far as possible, proper methodology must be followed for appropriate definition and understanding of VUF.

As shown in Figure 1, if V_{AB}, V_{BC} and V_{CA} are the measured line voltage magnitudes with V_{AB} as the reference phasor, since the line voltage phasors form a close triangle, the phase angle displacement with respect to each other can be obtained using the basic trigonometric property of triangles (Jeong, 2002; Kim et al., 2005). The line voltages can now be expressed in phasor form as $V_{AB}\angle 0$, $V_{BC}\angle\theta_{BC}$ and $V_{CA}\angle\theta_{CA}$ with $\theta_{BC} = (\theta_{2L} - 180)$ and $\theta_{CA} = (180 - \theta_{3L})$.

The sequence components using the line voltage combinations can be represented as below

$$\begin{bmatrix} V_{PL}\angle,_{PL} \\ V_{NL}\angle,_{NL} \end{bmatrix} = \frac{1}{3}\begin{bmatrix} 1 & a & a^2 \\ 1 & a^2 & a \end{bmatrix}\begin{bmatrix} V_{AB}\angle 0 \\ V_{BC}\angle,_{BC} \\ V_{CA}\angle,_{CA} \end{bmatrix} \quad (3)$$

Continuing with the theory of triangles, the phase voltages can now be expressed in phasor form as $V_A\angle\theta_A$, $V_B\angle\theta_B$ and $V_C\angle\theta_C$ with $\theta_A = (90 - \theta_1)$, $\theta_B = (\theta_A - \theta_2)$ and $\theta_C = (\theta_A - \theta_2 - \theta_3)$.

The sequence components using the phase voltage combinations can be represented as below

$$\begin{bmatrix} V_{PP}\angle,_{PP} \\ V_{NP}\angle,_{NP} \end{bmatrix} = \frac{1}{3}\begin{bmatrix} 1 & a & a^2 \\ 1 & a^2 & a \end{bmatrix}\begin{bmatrix} V_A\angle,_A \\ V_B\angle,_B \\ V_C\angle,_C \end{bmatrix} \quad (4)$$

With regard to three-phase, three-wire system, measuring line-to-line voltages is an easier proposition, but since all computations are carried out on a per phase basis, the line voltages now need to be converted to their phase equivalents considering both magnitude and their phase displacements.

If line voltage values are used, equation (3) can be rewritten as

$$CVUF_L = K_{VL}\angle\theta_{VL} \quad (5)$$

Figure 2. NEMA derating curve

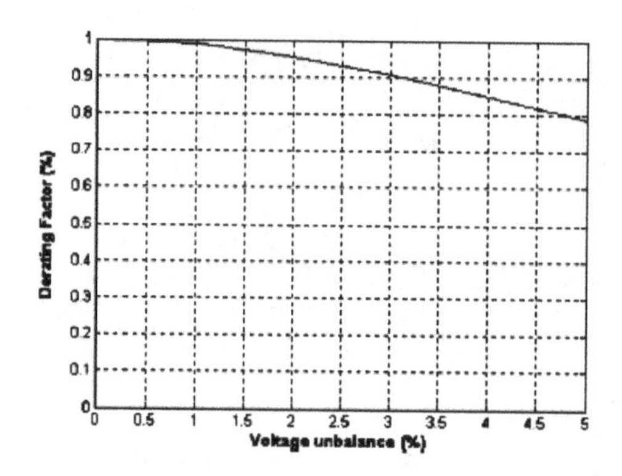

Table 1. Line voltage magnitudes corresponding to 3% voltage unbalance

Set	V_{AB}	V_{BC}	V_{CA}	V_{PL}	Nature
1	420.00	417.00	400.00	412.24	$V_{PL} > V_{RL}$
2	419.50	415.00	399.00	411.07	$V_{PL} > V_{RL}$
3	399.40	384.40	380.40	387.98	$V_{PL} < V_{RL}$
4	405.00	390.50	385.40	393.54	$V_{PL} < V_{RL}$
5	410.55	400.00	389.75	400.00	$V_{PL} = V_{RL}$

$K_{VL} = \left|\frac{V_{NL}}{V_{PL}}\right|$ is magnitude of $CVUF_L$ and $\theta_{VL} = (\theta_{NL} - \theta_{PL})$ is phase angle by which the negative sequence component leads the positive sequence component.

If the phase voltage values are used, equation (4) can be rewritten as

$$CVUF_P = K_{VP}\angle\theta_{VP} \tag{6}$$

$K_{VP} = \left|\frac{V_{NP}}{V_{PP}}\right|$ is magnitude of $CVUF_P$ and $\theta_{VP} = (\theta_{NP} - \theta_{PP})$ is phase angle by which the negative sequence component leads the positive sequence component.

As a matter of check, it should be verified that

Table 2. Phase voltage magnitudes corresponding to 3% voltage unbalance

Set	V_A	V_B	V_C	V_{PP}	Nature
1	233.95	248.74	231.37	238.01	$V_{PP} > V_{RP}$
2	233.99	247.92	230.12	237.33	$V_{PP} > V_{RP}$
3	227.15	230.64	214.23	223.99	$V_{PP} < V_{RP}$
4	229.90	234.35	217.41	227.21	$V_{PP} < V_{RP}$
5	231.00	239.94	221.92	230.94	$V_{PP} = V_{RP}$

$$CVUF_L = CVUF_P \angle -60 \tag{7}$$

Thus the magnitude of CVUF determined using the line voltages and the phase voltages must be the same for correct estimation of the voltage unbalance that exists in the system.

Nature and Extent of Voltage Unbalance (IEEE Std. 1346-1998)

Let the rated three-phase line voltage (V_{RL}) be 400 V, rated phase voltage (V_{RP}) be 230.94 V i.e. $V_{RL} = \sqrt{3}\, V_{RP}$ and consider the NEMA derating curve for a three-phase induction motor operated under voltage unbalance conditions as shown in Figure 2. As an example, consider the case of 3% voltage unbalance; the question now arises as to what is the combination of voltages that corresponds to this 3% voltage unbalance. A sample set of magnitudes of line voltage combinations that correspond to 3% voltage unbalance are presented in Table 1.

Using the theory of triangles, the respective phase voltage magnitudes are obtained and are shown in Table 2. It is common knowledge that voltage variation and unbalance are dynamically occurring events; hence individual line or phase voltage magnitudes can be greater or lesser than the rated value.

For the present example, it is interesting to observe the line and phase voltage magnitudes in Tables 2 and 3, and it can be seen that

For sets 1 and 2, positive sequence voltage V_P is greater than rated voltage V_R
For sets 3 and 4, positive sequence voltage V_P is less than the rated voltage V_R
For set 5, positive sequence voltage V_P is approximately equal to the rated voltage V_R

The above results highlight the fact that for a certain voltage unbalance that can exist in a power system network; there can be overvoltages or undervoltages either in line or phase magnitudes. In order to segregate the various sets of voltages for further clarity, positive sequence voltage component is made use of as a reference quantity.

As per IEEE Standards (IEEE Std 1159-1995; IEEE Std. 1346-1998), the term, overvoltage is used to describe a specific type of long duration variation and refers to a measured voltage having a value greater than the nominal voltage for a period of time greater than one minute. Similarly, the term, undervoltage is used to describe a specific type of long duration variation and refers to a measured voltage having a value lesser than the nominal voltage for a period of time greater than one minute.

Further analysis now leads to the following observations.

In case of set 1, all the individual voltage magnitudes (line and phase) are greater than or equal to the rated voltage (line and phase) respectively. This can be termed as over voltage unbalance, defined as an unbalance factor (K_{OU}) wherein the positive sequence voltage component is greater than the rated voltage. In addition for clarity, individual line voltages (V_{AB}, V_{BC} and V_{CA}) along with its positive sequence voltage (V_{PL}) should be greater than rated value (V_{RL}) and individual phase voltages (V_A, V_B and V_C) along with its positive sequence voltage (V_{PP}) should be greater than rated voltage (V_{RP}).

$$\% K_{OU} = \left|\frac{V_{PL}}{V_{RL}}\right| \times 100\,\% \quad or \quad \% K_{OU} = \left|\frac{V_{PP}}{V_{RP}}\right| \times 100\,\% \tag{8}$$

The overvoltage unbalance case is referred to as unbalanced overvoltage (UBOV).

In case of set 2, all individual voltage magnitudes (line and phase) may or may not be greater than the rated value (line and phase) respectively. This can be termed as mixed over voltage unbalance defined as an unbalance factor (K_{MOU}) wherein the positive sequence voltage component is greater than the rated voltage but the individual voltage (line or phase) combinations may be greater or lesser than the rated voltage. In addition for clarity, individual line voltages along with its positive sequence voltage may or may not be greater than rated voltage and individual phase voltages along with its positive sequence voltage may or may not be greater than rated voltage.

$$\% K_{MOU} = \left| \frac{V_{PL}}{V_{RL}} \right| \times 100\% \quad or \quad \% K_{MOU} = \left| \frac{V_{PP}}{V_{RP}} \right| \times 100\% \qquad (9)$$

In case of set 3, all individual voltage magnitudes (line and phase) are lesser than the rated value (line and phase) respectively. This can be termed as under voltage unbalance defined as an unbalance factor (K_{UU}) wherein the positive sequence voltage component is lesser than the rated voltage. In addition for clarity, individual line voltages along with the positive sequence voltage is lesser than the rated voltage, the individual phase voltages along with the positive sequence voltage is also lesser than the rated voltage.

$$\% K_{UU} = \left| \frac{V_{PL}}{V_{RL}} \right| \times 100\% \quad or \quad \% K_{UU} = \left| \frac{V_{PP}}{V_{RP}} \right| \times 100\% \qquad (10)$$

The undervoltage unbalance case is referred to as unbalanced undervoltage (UBUV).

In case of set 4, all individual voltage magnitudes (line and phase) may or may not be greater than the rated value (line and phase) voltage respectively. This can be termed as mixed over voltage unbalance defined as an unbalance factor (K_{MUU}) wherein the positive sequence voltage component is lesser than the rated voltage but the individual voltage (line or phase) combinations can be greater or lesser than the rated voltage. In addition for clarity, individual line voltages along with their positive sequence voltage may or may not be greater than rated voltage Additionally individual phase voltages along with its positive sequence voltage may or may not be greater than rated voltage.

$$\% K_{MUU} = \left| \frac{V_{PL}}{V_{RL}} \right| \times 100\% \quad or \quad \% K_{MUU} = \left| \frac{V_{PP}}{V_{RP}} \right| \times 100\% \qquad (11)$$

For understanding set 5, voltage unbalance definition in IEEE Standard (IEEE Std 1159-1995) is used, which states that; "Voltage imbalance (or unbalance) is defined as the ratio of the negative or zero sequence component to the positive sequence component". This definition does not make any distinction with regard to the nature of unbalance, but as in Tables 1 and 2 for set 5 data, there are possibilities of positive sequence voltage being equal to the rated voltage.

Having already classified the voltage unbalance into various categories above, it can be said that this case can now be considered to be in line with the IEEE Standard (IEEE Std 1159-1995) definition of VUF. Thus definition of voltage unbalance factor (K_V) can be further refined as a ratio of negative sequence to positive sequence voltage, exception being that the positive sequence voltage (line or phase) equals the rated voltage (line or phase).

$$\% K_{VL} = \left|\frac{V_{NL}}{V_{PL}}\right| \times 100\ \% \quad or \quad \% K_{VP} = \left|\frac{V_{NP}}{V_{PP}}\right| \times 100\ \% \tag{12}$$

This voltage unbalance case is referred to as unbalanced equal voltage (UBEV).

In the definitions for eqns. (8) to (12), the complex nature of sequence components is made use of and since it is only a factor of reference, only the magnitude is considered for evaluation. The summary of above discussion as applied to voltage magnitudes in Tables 1 and 2, is shown in Table 3.

For the balanced voltage case, there will not be any negative sequence component and with the possibility of individual voltage (line or phase) magnitudes being greater or lesser than the rated value, two additional factors K_{OB} (overvoltage balance) and K_{UB} (undervoltage balance) can now be expressed for the purpose of clarity.

Overvoltage balance can be defined as a factor (K_{OB}) wherein the positive sequence voltage magnitude (line or phase) is greater than the rated voltage (line or phase).

$$\% K_{OB} = \left|\frac{V_{PL}}{V_{RL}}\right| \times 100\ \% \quad or \quad \% K_{OB} = \left|\frac{V_{PP}}{V_{RP}}\right| \times 100\ \% \tag{13}$$

This voltage balance case is referred to as balanced over voltage (BOV) case in the thesis hereon.

Undervoltage balance (K_{UB}) can be defined as a factor wherein the positive sequence voltage magnitude (line or phase) is lesser than the rated voltage (line or phase).

Table 3. Interpretation of voltage combinations based on positive sequence voltage

Set	V_{AB}	V_{BC}	V_{CA}	V_{PL}	Nature	Extent of unbalance in %
1	420.00	417.00	400.00	412.24	Over voltage	$K_{OU} = 3.06$
2	419.50	415.00	399.00	411.07	Mixed over voltage	$K_{MOU} = 2.97$
3	399.40	384.40	380.40	387.98	Under voltage	$K_{UU} = 3.00$
4	405.00	390.50	385.40	393.54	Mixed under voltage	$K_{MUU} = 1.61$
5	410.55	400.00	389.75	400.00	Voltage unbalance	$K_{V} = 3.00$

Table 4. Line voltage magnitudes corresponding to 3% balance

Set	V_{AB}	V_{BC}	V_{CA}	V_{PL}	Nature
1	412	412	412	412	$V_{PL} > V_{RL}$
2	388	388	388	388	$V_{PL} < V_{RL}$

$$\% K_{UB} = \left| \frac{V_{PL}}{V_{RL}} \right| \times 100\,\% \quad or \quad \% K_{UB} = \left| \frac{V_{PP}}{V_{RP}} \right| \times 100\,\% \tag{14}$$

This voltage balance case is referred to as balanced under voltage (BUV). The above two factors are summarized in the Table 4.

In summary, whatever may be the magnitudes of line voltage, for the voltage unbalance case, once Eqn. (7) is satisfied and making use of Eqns (8) to (12); and for voltage balance case, making use of Eqns (13) to (14); the nature and extent of voltage variation and / unbalance can be clearly identified and defined; the subsequent result analysis will therefore be more logical and correct in nature.

As per IEEE Std. definitions of overvoltage and undervoltage, the reference of over or under is made with respect to the nominal value, and the nominal value is defined as a value that is assigned to a circuit or system for the purpose of conveniently designating its voltage class (208 / 120, 480 / 277, 600).

Here too, the true interpretation of nominal voltage needs some clarification. Generally, the three-phase line-to-line power system voltage is 400 V, but motor companies manufacture three-phase induction motors with name plate voltage ratings of 380 V, 400 V and 415 V. If a 400 V motor is connected to a power system of 400 V, then there is no confusion about the value of the nominal voltage. The confusion arises when a 380 V or 415 V motor is connected across a power system of 400 V, or a 380 V or 400 V motor is connected across a power system of 415 V. In such a situation, the question arises as to what is the value of the nominal voltage.

Consider a simple example of two motors of rated name plate voltage being 380 V and 415 V respectively, with the rated power system line-to-line voltage being 400 V. Now assume the positive sequence voltage to be 390 V with some degree of unbalance. As per the definitions of overvoltage and undervoltage unbalance as defined above, it will be treated as K_{UU} or K_{MUU} to be 2.5%

- For the 380 V motor, if the motor name plate voltage itself is to be considered as the rated value, then it will be treated as K_{OU} or K_{MOU} to be 2.63%.
- For the 415 V motor, if the motor name plate voltage itself is to be considered as the rated value, then it will be treated as K_{UU} or K_{MUU} to be 6.02%.

So now question arises that what is the true value of the rated voltage for reference in Eqns. (8) to (11), (13), (14)? A possible solution to this problem is to take the motor name plate voltage itself as the rated voltage. This is because, for obtaining the parameters of the induction motor equivalent circuit, the no load and blocked rotor tests are done using the name plate voltage and current values. Hence, motor name plate voltage should be taken as the reference rated voltage.

Use of Upper Limit (IEEE Std. 1346-1998)

Having reduced the ambiguity to a considerable extent, it is still possible to show that a particular complex voltage unbalance factor can be obtained using multiple sets of voltage combinations. To highlight the harmful effects of voltage variation and unbalance on any equipment, it would be quite prudent to consider the highest possible K_{OU}, K_{UU}, K_{MOU} or K_{MUU} factors together with complex voltage unbalance factor K_V. To limit impractical voltage combinations there must now be an upper limit such that when in conjunction with the additional factors defined in Section Nature and Extent of Voltage Unbalance, the understanding of the extent of voltage unbalance would be more complete.

Consider the example of three-phase induction motor subjected to voltage unbalance wherein the present NEMA guideline does not recommend the motor operation for any length of time, when the voltage unbalance is 5%. This is because when the voltage unbalance reaches and beyond 5%, temperature rise is so fast that damage may be imminent (Pillay et al., 2002). Keeping the above discussions in mind, the question now arises as to what is the true meaning of this 5% voltage unbalance, is it K_{OU}, K_{MOU}, K_{UU}, K_{MUU} or is it still complex voltage unbalance factor K_V? The logical reason would be to take K_V = 5%. If that is so, then what is the extent of K_{OU}, K_{UU}, K_{MOU} or K_{MUU} (which ever is applicable)? In this regard, if the generally followed NEMA standard of maximum 5% is taken as reference factor for

Table 5(a). Balanced line voltage combinations for 400 V reference

Set	Nature	% K	V_{AB}	V_{BC}	V_{CA}
1	$V_{PL} > V_{RL}$ (BOV)	1	404	404	404
		2	408	408	408
		3	412	412	412
		4	416	416	416
		5	420	420	420
2	$V_{PL} < V_{RL}$ (BUV)	1	396	396	396
		2	392	392	392
		3	388	388	388
		4	384	384	384
		5	380	380	380

Table 5(b). Unbalanced line voltage combinations for 400 V reference

Set	Nature	% K_V	V_{AB}	V_{BC}	V_{CA}	V_{PL}
1	V_{PL} = 1.05 V_{RL} (UBOV)	1	424.00	419.30	416.85	420.04
		2	426.35	421.75	412.05	420.01
		3	427.35	425.55	407.45	420.02
		4	432.40	424.25	403.95	420.03
		5	434.00	427.50	400.00	420.25
2	V_{PL} = 0.95 V_{RL} (UBUV)	1	382.60	381.10	376.25	379.97
		2	387.60	376.30	376.15	379.98
		3	391.25	372.85	376.10	379.98
		4	395.00	375.05	370.40	379.99
		5	398.85	373.20	368.65	379.99
3	$V_{PL} = V_{RL}$ (UBEV)	1	403.50	400.00	396.55	400.00
		2	407.05	400.00	393.15	400.00
		3	410.55	400.00	389.75	400.00
		4	414.10	400.00	386.40	400.00
		5	417.70	400.00	383.10	400.00

defining the upper limit i.e. $K_{OU;}$ $K_{UU,}$ $K_{MOU,}$ and K_{MUU} are considered as 5%, then the upper limit for a particular voltage unbalance condition can be correctly defined i.e. K_V = 5% with K_{OU} = 5% or K_{MOU} = 5% can be taken to the upper limit for overvoltage unbalance while K_V = 5% with K_{UU} = 5% or K_{MUU} = 5% can be taken to the upper limit for undervoltage unbalance. By doing so, all possible unrealistic values can be totally eliminated from being considered for analysis to highlight the harmful effects of voltage unbalance. Table 5 (a) shows the balanced line voltage combinations for voltage balance (% K) ranging from 1 to 5%, the rated line voltage used for reference being 400 V. Table 5 (b) shows the unbalanced line voltage combinations for CVUF ranging from 1 to 5% with K_{OU} and K_{UU} limit set as 5% for each case, rated line voltage used for reference being 400 V.

For sake the of simplicity in explanation and difficulty in obtaining a range of values for mixed overvoltage and undervoltage cases, factors K_{MOU} and K_{MUU} have not been considered for the result analysis.

It can thus be concluded that voltage unbalance needs to correctly classified and differentiated on the basis of the nature and extent of unbalance that exists in the system.

Process Side

In the process side analysis, performance analysis of a three-phase induction motor is considered using the steady state equivalent circuit approach. Under varying voltage and load conditions, core loss and / or, friction and windage loss have always been neglected, the reason being ease of understanding and result analysis (Lee, 1999; Wang, 2001; Pillay et al., 2002; Faiz et al., 2004; Kersting and Phillips, 1997; Wang, 2000). Though it is extremely difficult to be as exact as possible but still, it is important to consider all possible quantifiable parameters during analysis. Therefore accurate estimation of losses is extremely important else there will be significant errors in the efficiency estimation (El-Ibiary, 2003; Kini et al., 2005). The per phase equivalent circuit for steady state analysis suitably modified to take into consideration core loss and, friction and windage loss under running conditions is as shown in Figure 3 (Kothari and Nagrath, 2004).

In Figure 3, V is the applied voltage, R_1 and X_1 are stator resistance and reactance respectively, and rotor resistances, R'_2 and X'_2 are equivalent rotor resistance and reactance as referred to the stator, R_C is the core loss resistance, R_{FW} is the resistance representing the friction and windage loss, X_M is the magnetizing reactance, s is the operating slip, I_1 is the stator current, I_o is the no load current component and I'_2 is the rotor current referred to stator side.

Figure 3. Per-phase equivalent circuit with rotational loss components

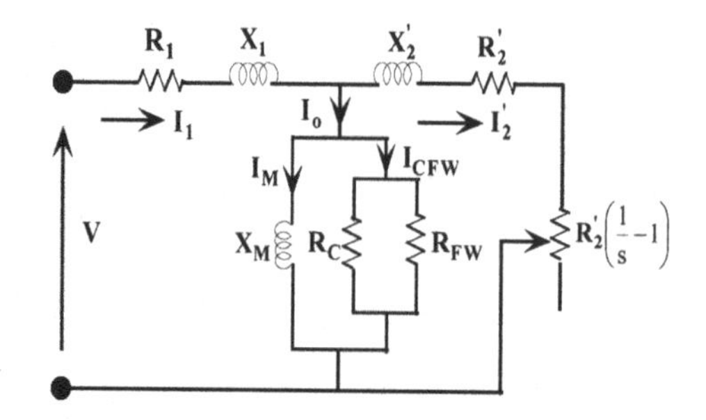

Proposed Steady State Model

For balanced voltage conditions, the equivalent circuit shown in Fig 3 is taken for analysis. Under conditions of voltage unbalance, with the application of symmetrical component technique, per phase induction motor equivalent circuit can now be split up into a positive sequence equivalent circuit and negative sequence equivalent circuit. Thus under voltage unbalance conditions, the induction motor can be considered as two separate motors in operation, one operating with a positive sequence voltage V_P at slip 's', and other operating with a negative sequence voltage V_N at slip "(2 - s)" (Pillay et al., 2002).

Let V_{AB}, V_{BC} and V_{CA} be the measured line-to-line voltage magnitudes, with V_{AB} being taken as the reference phasor. Using the procedure outlined in Section 2.1, the measured line voltage magnitudes are used to determine the phase voltages $V_A\angle\theta_A$, $V_B\angle\theta_B$ and $V_C\angle\theta_C$, positive sequence voltage $V_{PP}\angle\theta_{PP}$ and negative sequence voltage $V_{NP}\angle\theta_{NP}$.

For the positive sequence equivalent circuit,

$$I_{1P}\angle,_{1P} = \frac{V_{PP}\angle,_{PP}}{Z_P\angle\phi_P} \tag{15}$$

$$\text{Power output, } P_P = \frac{(I'_{2P})^2 R'_2 (1-s)}{s} \tag{16}$$

$$\text{Torque } T_P = \frac{(I'_{2P})^2 R'_2}{s\omega_s} \tag{17}$$

$V_{PP}\angle\theta_{PP}$ is the positive sequence voltage; $Z_P\angle\varphi_P$ is the positive sequence input impedance; $I_{1P}\angle\theta_{1P}$, $I'_{2P}\angle\theta_{2P}$, P_P, T_P are the positive sequence components of stator, rotor currents, power output and torque respectively.

For the negative sequence equivalent circuit,

$$I_{1N}\angle,_{1N} = \frac{V_{NP}\angle,_{NP}}{Z_N\angle\phi_N} \tag{18}$$

$$\text{Power output, } P_N = \frac{(I'_{2N})^2 R'_2 (s-1)}{(2-s)} \tag{19}$$

$$\text{Torque } T_N = \frac{(I'_{2N})^2 R'_2}{(2-s)\omega_s} \tag{20}$$

$V_{NP}\angle\theta_{NP}$ is the negative sequence voltage; $Z_N\angle\varphi_N$ is the negative sequence input impedance; $I_{1N}\angle\theta_{1N}$, $I'_{2N}\angle\theta_{2N}$, P_N, T_N are the negative sequence components of stator, rotor currents, power output and torque respectively

For steady state operation, motor torque (T_M) = load torque (T_L) (21)

Net motor torque $T_M = T_P + T_N$ (22)

Net power output $P_{OUT} = P_P + P_N$ (23)

Net motor power input $P_{IN} = \mathrm{Real}\,[(V_{PP}\, I^{*}_{1P} + V_{NP}\, I^{*}_{1N})]$ (24)

where '*' indicates the conjugate value.

Motor efficiency is given by $\% \cdot = \frac{P_{OUT}}{P_{IN}} x\, 100\,\%$ (25)

Output Side

The output side involves measuring the flow rate and determining the system head. To estimate the working system head, it is very much necessary that all possible loss components be taken into account for near accurate results. The performance of a pump can be graphically expressed by plotting curves of head v/s flow rate as shown in Figure 4. The pump curve is interpreted as what a pump can offer at a particular speed while a system curve is interpreted as what is desired by the system. The total system head is composed of static head, frictional head and delivery side velocity head component wherein, the static head remains constant if the water level remains constant else static head also varies for varying water levels. The frictional head is a combination of head loss in the pipe and velocity head on the delivery side. The delivery side velocity head is usually small, and the frictional head varies as the square of the flow rate. When the pump curve and system curve are superimposed upon each other, they intersect at point 'O' as in Figure 4. This point is termed as the operating point i.e.; the pump delivers whatever is demanded by the system.

On installing the motor-pump system, when the desired flow rate is not achieved, throttling on the delivery side is adopted so as to achieve the desired flow rate. The main disadvantage is the increase in frictional losses. As shown in Figure 5, on throttling from the fully open position to partly open position the pump curve shifts from speed N_1 to speed N_3, the head increases from H_1 to H_3 and the flow rate decreases from Q_1 to Q_3. Thus on throttling, the pump curve and the system curve shifts resulting in a varying operating points.

Figure 4. Pump head vs flow rate curve

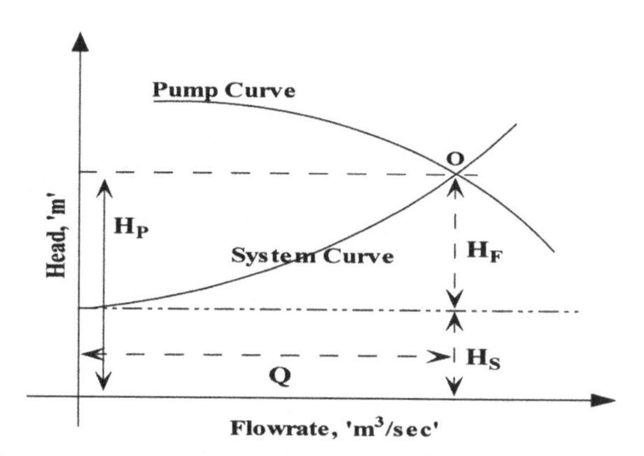

Affinity Laws

Affinity laws also termed as pump laws or fan laws, establish the mathematical relationship between the variables involved in defining the performance of a pump, thereby enabling us to predict the performance characteristics of a pump when operating at speeds other than the speed at which it has been tested. The mathematical relations are given as follows:

(a) With the impeller diameter 'D' kept constant,

$$\left.\begin{aligned} \frac{Q_1}{Q_2} &= \frac{N_1}{N_2} \\ \frac{H_1}{H_2} &= \left[\frac{N_1}{N_2}\right]^2 \\ \frac{P_1}{P_2} &= \left[\frac{N_1}{N_2}\right]^3 \end{aligned}\right] \tag{26}$$

(b) With the operating speed 'N' kept constant,

$$\left.\begin{aligned} \frac{Q_1}{Q_2} &= \frac{D_1}{D_2} \\ \frac{H_1}{H_2} &= \left[\frac{D_1}{D_2}\right]^2 \\ \frac{P_1}{P_2} &= \left[\frac{D_1}{D_2}\right]^3 \end{aligned}\right] \tag{27}$$

Figure 5. Effect of throttling on operating point

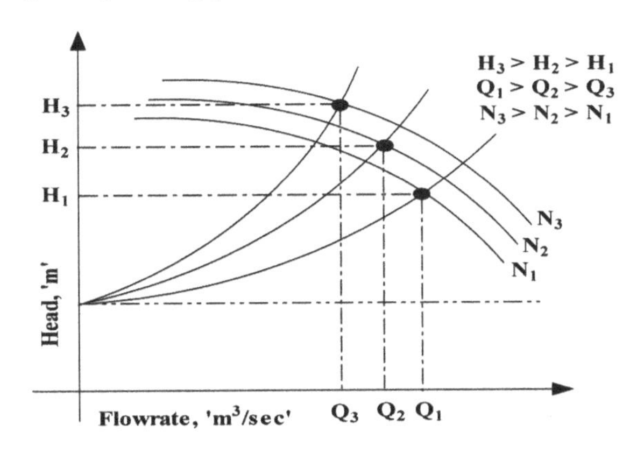

In (26) and (27), Q_1, Q_2 are flow rate; N_1, N_2 are operating speed; H_1, H_2 are total head; P_1, P_2 are power required (Sahu, 2000; Carlson, 2000). It is very important to note that the equations (26) or (27) are only useful in describing the shift that occurs with the pump curve (Rice, 1988). In actual applications, pump head and the operating speed are a function of the pump and system characteristics for which the pump system is made to work for. Hence it is necessary to be extremely careful of using the above relations to determine the operating parameters under working conditions.

Pump Parameters

The head (H_P) of the pump is the total head against which a pump has to work and is expressed as the sum of static head (H_S), frictional head losses (H_L) and delivery side velocity head. If the delivery side velocity head is relatively small, it may be neglected and the net head against which the pump has to work is given by

$$H_P = H_S + \Sigma \text{ friction loss head} \tag{28}$$

For a typical pump system as shown in Figure 6, the friction loss head include the head loss at suction entry and delivery exit, friction over the entire length of suction pipe and delivery pipe, pipe bends, sudden contraction and / or enlargement, gradual contraction and / or enlargement, obstruction in flow passage and head loss in various pipe fittings on the suction and delivery sides.

The head to be supplied for the pump is usually expressed in the form

$$H_P = H_S + K\,Q^2 \tag{29}$$

H_P is the total head against which the pump has to work in metres, H_S is the total static head in metres, K is the system frictional coefficient and Q is the flow rate in m^3/s.

It is well known that the pipe material on the inside, degrade over time owing to the impurities present in the liquid flow through the pipe system. Hence the friction factor never remains the same over a period of time. In order to account for all the uncertain factors under dynamically varying conditions, for a particular throttle setting the graphical approach is used to determine the system head H_P, the methodology of which is as follows.

Figure 6. Typical pump system layout

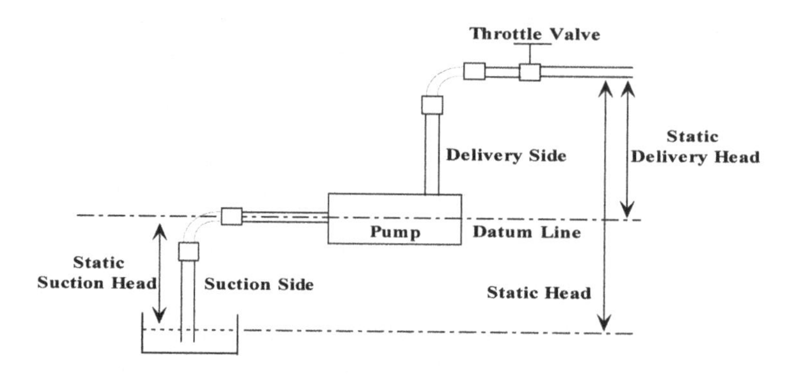

Step 1: Measure the motor operating speed in rpm.
Step 2: Plot the pump curve corresponding to this measured operating speed, using the affinity laws with reference to the head-flow rate values at rated operating speed
Step 3: Measure the flow rate in m^3/s.
Step 4: Draw a straight line perpendicular to the x-axis such that it cuts the pump curve at operating point 'O'.
Step 5: Draw a straight line from point 'O' perpendicular to the y-axis so as to cut the y-axis to obtain the pump head H_P.
Step 6: The total static head H_S is measured from the water level to the point of discharge. Using this, the unknown frictional coefficient K in eqn. (29) is now calculated.
Step 7: For a given throttle position, the system curve is now formulated as per eqn. (29).
Step 8: If the throttle setting is changed, the pump and the system curve shifts. To express the nature of the system curve for the changed throttle setting, steps 1-7 is now repeated.
Step 9: The characteristic head-flow rate curve is now graphically represented as per Figure 4.

The pump power output (P_{POUT}) is defined as the energy delivered to the fluid and it is expressed as

$$P_{POUT} = \gamma QH \qquad (30)$$

γ is the specific weight of the fluid in N/m^3, Q is the pump flow rate in m^3/s and H is the head in m to be supplied by the pump.

The pump efficiency is expressed as the ratio of pump power output to pump power input

$$\% \cdot_P = \frac{pump\ power\ output}{pump\ power\ input} \times 100\ \% \qquad (31)$$

The system efficiency is expressed as the ratio of pump power output to motor power input

$$\% \cdot_S = \frac{pump\ power\ output}{motor\ power\ input} \times 100\ \% \qquad (32)$$

ENERGY MANAGEMENT

The energy management strategy for efficient energy utilisation should take into account the present and the expected working conditions. For a motor-pump system operating under voltage variation and unbalance, the energy management strategy involves derating using the throttle valve placed on the delivery side.

Derating

Derating is defined as a reduction in the load on the motor and is usually done when the line current flowing in the lines exceeds the rated value (Faiz and Ebrahimpour, 2005). As per NEMA guidelines, operating a motor for any length of time under conditions of voltage unbalance over 5% is not recommended, as it makes the temperature to rise so fast that protection from damage becomes impractical, thus the solution is to derate the motor i.e. reduce the power output ; (Pillay et al., 2002; Kersting and Phillips, 1997; 2002; ANSI/NEMA Standards Publication no MG1-1978, 1993). In order to derate, a factor termed as a derating factor is expressed, which is defined as the ratio of motor power output under unbalanced voltage conditions to that under balanced conditions, with the line currents remaining under the rated value under both cases. It is of common knowledge that motor driven systems are never operated at full load. This is basically to incorporate a safety factor to account for unforeseen situations of varying voltage (balanced or unbalanced) and load conditions. In such situations, the definition of derating factor now needs to be slightly modified so as to correctly interpret the process of reducing the power output. The derating factor now needs to be expressed as the ratio of motor power output after derating to the motor power output before derating. Whether balanced or unbalanced voltage condition, for the motor-pump system, the condition of motor line current exceeding the rated value is frequently encountered if the dimensions of the pipe system network are not correctly evaluated and implemented. In such a scenario too, the motor has to be derated to prevent harmful temperature rise and subsequent insulation damage.

As shown in Figure 7, for a particular throttle setting, the flow rate may be on either side of the maximum pump efficiency point. On operating the throttle valve there will be a change in speed, thus displacing the pump curve and at the same time there is the variation of the frictional head, thus displacing the system curve. It is therefore extremely necessary to know what is the flow rate and where it is with respect to the maximum pump efficiency point i.e. if the flow rate point is at the extreme right hand side of the maximum pump efficiency point, then reducing the flow rate by throttling actually increases the efficiency of operation or if the flow rate point is at the extreme left hand side of the maximum pump efficiency point, then increasing the flow rate by opening the throttle valve also increases the efficiency of operation.

Figure 8 shows the variation of motor, pump and system efficiencies when moving from the fully open throttle position (maximum flow rate) to fully closed throttle position (zero flow rate). For the pump and the system as a whole, on throttling from the fully open position to the partly open position, it is seen that the pump efficiency and in turn the system efficiency, starts to increase, reaches a maximum value and decreases, with the increase in pump power output and system power output respectively. Since derating

Figure 7. Head vs flow rate characteristic

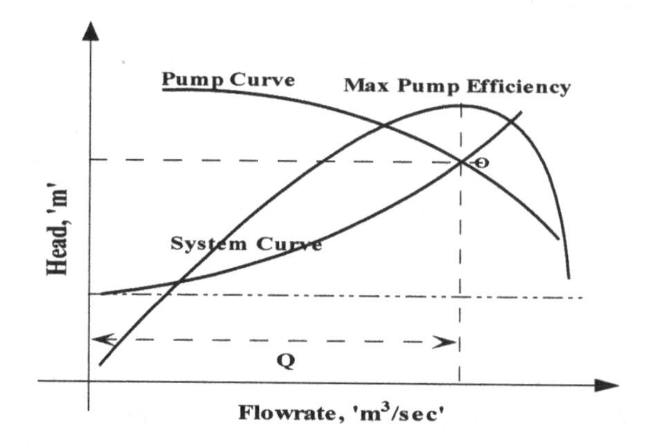

means reduction in power output, the discussion can be in terms of motor derating, pump derating and system derating separately. This is possible if we have a parameter of reference.

The usual way of motor derating is to reduce the motor power output by reducing the currents flowing through the lines to below the rated name plate value without due consideration to the flow rate being delivered. For balanced over voltage or under voltage cases, the process of derating is quite easy as all the three line current magnitudes are the same. For the unbalanced over voltage, under voltage and equal voltage cases, due care should be adopted as the magnitudes of the three line currents are not the same, also, the current magnitudes in one, two or all three of the lines are greater than the rated name plate value. It is a usual tendency that one of the three phases is taken to be the reference phase, with the assumption that the current magnitudes in the remaining two lines are always lesser than the reference phase. For varying voltage unbalance conditions, as the voltage magnitude in any of the three phases can be greater than the other by different extents, with respect to the voltage magnitudes, the current magnitudes in the three lines can also be greater than the other with varying extents. In such a scenario, it is very important to define a current magnitude reference for derating. The simplest possible option is to determine the maximum of the three line currents and take it as a reference. This helps to a great extent wherein on derating if the maximum of the three line currents is brought to the safe value, the other two line currents will automatically be lesser than the safe value. The pump and in turn the system derating can be effected by throttling the delivery side to reduce the flow rate to acceptable / desired values without giving due consideration to the line current magnitudes. For safe operation and efficient operation it is necessary that the line currents flowing in the lines be below the rated (safe) value. For the motor-pump system, it can be achieved by throttling. As shown in Figure 8, it can be seen that there are points of maximum efficiencies of motor, pump and system for a particular flow rate i.e.; between the limits of fully open and partly open position for the throttle there is a variation of pump and system efficiencies. Depending upon the system layout, the motor efficiency may vary like the pump and system efficiency curves or may continuously reduce from a maximum value on throttling from the fully open position. In other words, the curves shown in Figure 7 may be applicable for a particular voltage setting. Thus there will be different efficiencies curves for balanced and unbalanced over and under voltages.

The concept of derating i.e. reduction in the motor power output can also be extended to achieving the best possible efficiency, irrespective of balanced or unbalanced voltage conditions. For the motor-pump system, there are three efficiencies of operation; motor efficiency, pump efficiency and system

Figure 8. Efficiency vs power output graph superimposed on the head vs flow rate graph

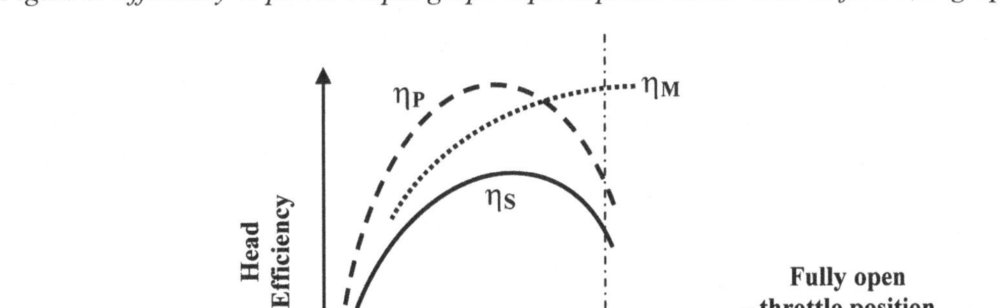

efficiency, all of them being interrelated. In such a scenario, derating of the motor-pump system also needs to be analysed from the efficiency points of view, because for energy management strategy, efficiency of operation is the importance parameter. It should be kept in mind that voltage variations are not under the control of the user while the load variations can be easily achieved / controlled by means of the throttle valve. Whenever motor-driven pump systems are operated under balanced rated voltage conditions, there will be a unique point of maximum motor, pump and system efficiencies. Thus these unique maximum points under balanced rated voltage conditions may or may not be achieved under varying voltage and load conditions. Even if they are achieved, other operating parameters will vary leading to different operating conditions. In such a scenario, in order to derate for achieving the best efficiency, instead of trying to achieve the best efficiency under balanced rated conditions, an effort must be made to achieve the maximum efficiency points for that condition itself.

For example, if M_{eff}, P_{eff} and S_{eff} are the maximum motor, pump and system efficiencies under rated balanced condition respectively; then these unique values of M_{eff}, P_{eff} and S_{eff} may not be achieved for balanced and unbalanced, over and under voltage conditions. Instead, an attempt can always be made to achieve the maximum possible efficiencies of M_{eff1}, P_{eff1} and S_{eff2} for BOV 1; M_{eff2}, P_{eff2} and S_{eff2} for BOV 2; M_{eff3}, P_{eff3} and S_{eff3} for BOV 3 and so on for rest of the cases. In this way, the operation can be at best possible maximum efficiencies under balanced and unbalanced, over and under voltage conditions for those voltage combinations by suitably varying the throttle valve.

As the pump efficiency is dependent on the motor power output, the throttle valve can be used to reduce the motor power output to make the pump part of the motor-pump system, and in turn the system to operate better. Thus the motor derating process can also be carried out to achieve the best pump efficiency for the respective voltage conditions. As the system efficiency is the product of motor and pump efficiency, the motor power output can be reduced to make the system as a whole to work better. Thus the motor derating process can also be applied to achieve the best possible system efficiency. Thus there will be different derating factors for different throttle settings based on the various balanced and unbalanced voltage conditions.

In attempting to derate the motor-pump system by trying to obtain the maximum efficiency points, it is still possible that the magnitude of one or all of the line currents may be in excess of the safe value. In order to ensure safe operation, the throttle valve must be further operated to reduce the line current(s) to below the safe value; the resulting efficiencies will then be the maximum possible efficiencies. The derating factors would be termed as the corrected derating factors that take into account both the line current magnitudes as well as the efficiencies.

The true meaning of derating factors obtained will be best understood when the concept of energy management is made use of to optimize the utilisation of energy under conditions of voltage and load variation. Thus derating factor is dependent not only on the nature and extent of voltage variations but also on the reference parameter used for derating with respect to load variation. Also, during the process of derating, it should be ensured that safe operating condition is the first priority during derating, only then the correct derating factor should be determined so as to avoid erroneous result analysis.

CONCLUSION

Voltage variation and unbalance is one of the commonly occurring power quality problem faced by the industrial sector, and with motor driven systems being one of the major energy consuming systems within plant premises, poor power quality severely affects plants operating performance in terms of energy efficiency and product quality. By making use of the complex nature of voltage unbalance, the voltage combinations that lead to the calculation of CVUF can be narrowed down to a great extent. With a sample set of voltage combinations, the ambiguity has been highlighted, and the solution obtained in the form of defining balanced and unbalanced, over and under voltages. Given the fact that voltage variation and unbalance are dynamic events, it has been shown that by comparing the deviation of positive sequence voltage with the rated voltage value, two more cases were brought forward i.e.; mixed overvoltage and mixed undervoltage. The present definition of VUF should be restricted to cases wherein the magnitude of positive sequence voltage component equals the rated voltage. By considering the IEEE definitions of over voltage and under voltage balance, interpretation is made even more clearer by taking the motor name plate voltage as reference voltage, in the event of motor name plate and power system voltages being not equal to each other. Also, to avoid impossible voltage combinations being taken into account to highlight dangerous / harmful motor operation, the concept of upper limit is incorporated into the definitions of over and under voltages, thereby lending more credibility to the voltage definitions and its combinations.

The equivalent circuit used for steady state analysis has taken into consideration the core loss; and friction and windage loss, and represented them as resistances; CVUF having both magnitude and angle is used for purposes of greater accuracy.

Derating for a motor-pump system is generally thought of whenever the line currents exceed the rated value, which usually does when there is either some amount of unbalance among the voltages or if the motor is loaded to the extent that excess current flows in the lines. Prolonged operation of any motor driven system under voltage unbalance conditions shortens the motor life in the long run, thus necessitating derating. Under conditions of excess current flow, maximum of the three line currents is taken as the reference parameter for current limiting. Unlike the previous research studies wherein the derating process was applied only for motor performance analysis with no study on the pump or the system parameters as a whole, a methodology is presented whereby the concept of derating i.e. reduction in the motor power output; is carried out to obtain the maximum pump and system efficiency. Appropriately the derating factors can be tabulated for various voltage cases in addition to the throttle settings, and in a couple of cases, corrected derating factors also need to be tabulated as at the points of occurrence of maximum motor, pump and system efficiencies, the maximum of the three line current magnitude will still in excess of the rated value, thus requiring further derating.

System efficiency being the product of motor and pump efficiency, pump efficiency is generally found to be the dominant parameter. Thus the motor-pump system needs to be derated in such a way that it operates near the maximum pump or the system efficiency point. There will now be a small compromise on flow rate as flow rate at maximum motor efficiency point will be higher that the flow rates at maximum pump efficiency and maximum system efficiency points. But as the flow rate at the maximum pump and maximum system efficiency points are relatively close to each other, the system as a whole will be performing better, as motor power input is lesser at maximum pump and system efficiency points. Just as the voltage variation and unbalance are classified into balanced rated voltage, BOV, UBOV, BUV, UBUV and UBEV cases appropriately; derating factors for all these cases will also be different. Moreover with the use of throttle enabling input power and flow rate variation at different set points on derating to obtain maximum motor, pump or system efficiency as the selection criteria, it can be seen that derating factors will be different for each of the them. This indicates that extensive performance analysis is very much necessary for varying voltage and load conditions from the derating point of view. In doing so, the derating factors obtained will now give a more accurate picture of status of the motor-pump system performance, before and after derating. By applying current limit derating whenever necessary, erroneous result analysis can be avoided to a large extent.

REFERENCES

ANSI/NEMA Standards Publication no MG1-1978 (n.d.). Washington, DC: National Electrical Manufacturers Association.

Aquila, A. D., Marinelli, M., Monopoli, V. G., & Zanchetta, P. (2004). New power quality assessment criteria for supply systems under unbalanced and non sinusoidal conditions. *IEEE Transactions on Power Delivery*, *19*(3), 1284–1290. doi:10.1109/TPWRD.2004.829928

Bollen, M. H. J. (2000). *Understanding power quality problems: voltage sags and interruptions*. New York: IEEE Press.

Bollen, M. (2002). Definitions of voltage unbalance. *IEEE Power Engineering Review*, *22*(11), 49–50. doi:10.1109/MPER.2002.1045567

Carlson, R. (2000). The correct method of calculating energy savings to justify adjustable-frequency drives on pumps. *IEEE Transactions on Industry Applications*, *36*(6), 1725–1733. doi:10.1109/28.887227

Dugan, R. C., McGranaghan, M. F., & Beaty, H. W. (1996). *Electrical power systems quality*. New York: McGraw-Hill.

El-Ibiary, Y. (2003). An Accurate low cost method for determining electric motors efficiency for the purpose of plant energy management. *IEEE Transactions on Industry Applications*, *39*(4), 1205–1210. doi:10.1109/TIA.2003.813686

Faiz, J., Ebrahimpour, H., & Pillay, P. (2004). Influence of unbalanced voltage on the steady state performance of a three phase squirrel cage induction motor. *IEEE Transactions on Energy Conversion*, *19*(4), 657–662. doi:10.1109/TEC.2004.837283

Faiz, J., & Ebrahimpour, H. (2005). Precise derating of three-phase induction motors with unbalanced voltages. In *Proceedings of Fortieth IAS Annual Meeting of Industry Applications Conf.*, Hong Kong, 1, Oct., (pp. 485-491).

Fortescue, C. L. (1918). Method of symmetrical coordinates applied to the solution of polyphase networks. *Trans. AIEE*, *37*(part II), 1027–1140.

Ghijselen, J., & Van den Bossche, A. (2005). Exact voltage unbalance assessment without phase measurements. *IEEE Transactions on Power Systems*, *20*(1), 519–520. doi:10.1109/TPWRS.2004.841145

IEEE Std. 739-1995, (n.d.). IEEE Recommended practice for energy management in industrial and commercial facilities.

IEEE Std 1159-1995 (n.d.). *IEEE recommended practice for monitoring electric power quality*.

IEEE Std. 1346-1998 (n.d.). *IEEE recommended practice for evaluating electric power system compatibility with electronic process equipment.*

IEC 61000-1-1, (n.d.). Electromagnetic compatibility (EMC) Part I: General, Section 1: Application and interpretation of fundamental definitions and terms.

Jeong, S. (2002 October). Representing line voltage unbalance. *Conference Record of 2002 IEEE Industry Applications Conf.*, Pennsylvania, USA, *3,* 1724-1732.

Kersting, W. H., & Phillips, W. H. (1997). Phase frame analysis of the effects of voltage unbalance on induction machines. *IEEE Transactions on Industry Applications, 33*(2), 415–420. doi:10.1109/28.568004

Kim, J. G., Lee, E. W., Lee, D. J., & Lee, J. H. (2005). Comparison of voltage unbalance factor by line and phase voltage. *Proc. 8th Int. Conf. on Electrical Machines and Systems (ICEMS)*, Nanjing, China, *3*, 1988-2001.

Kini, P. G., Bansal, R. C., & Aithal, R. S. (2007). A novel approach towards interpretation and application of voltage unbalance factor. *IEEE Transactions on Industrial Electronics, 54*(4), 2315–2322. doi:10.1109/TIE.2007.899935

Kini, P. G., Sreedhar, P. N., & Varmah, K. R. (2005, November). Importance of Accuracy For Steady State Performance Analysis of 3φ Induction Motor. *Proc. of IEEE Region 10 conf. of TENCON 2005,* Melbourne, Australia, (pp. 1994-1998).

Kothari, D. P., & Nagrath, I. J. (2004). *Electric Machines, 3rd Edition.* New Delhi, India: Tata McGraw Hill.

Lee, C. Y. (1999). Effects of unbalanced voltage on the operation performance of a three-phase induction motor. *IEEE Transactions on Energy Conversion, 14*(2), 202–208. doi:10.1109/60.766984

Lee, W., & Kenarangui, R. (2002). Energy management for motors, systems, and electrical equipment. *IEEE Transactions on Industry Applications, 38*(2), 602–607. doi:10.1109/28.993185

Mircevski, S., Kostic, Z. A., & Andonov, Z. (1998, May). Energy saving with pumps ac adjustable speed drives. *Proceedings of 9th Mediterranean Electrotechnical Conf. (MELECON)*, Tel-Aviv, Israel, *2*, 1224-1227.

NEMA Standards Publication no MG1-1993, Motors and Generators. (n.d.). Washington DC: National Electrical Manufacturers Association.

Oliver, J. A., Lawrence, R., & Banerjee, B. (2002). Power quality – how to specify power quality tolerant process equipment. *IEEE Industry Applications Magazine, 8*(5), 21–30. doi:10.1109/MIA.2002.1028387

Palanichamy, C., Nadarajan, C., Naveen, P., Babu, N.S., & Dhanalakshmi, (2001). Budget constrained energy conservation - an experience with a textile industry . *IEEE Trans. Energy Conversion,16*(4), 340-344. doi:10.1109/60.969473

Pillay, P., Hofmann, P., & Manyage, M. (2002). Derating of induction motors operating with a combination of unbalanced voltages and over or undervoltages. *IEEE Transactions on Energy Conversion, 17*(4), 485–491. doi:10.1109/TEC.2002.805228

Pillay, P., & Manyage, M. (2001). Definitions of voltage unbalance. *IEEE Power Engineering Review, 21*(11), 50–51. doi:10.1109/39.920965

Rice, D. E. (1988). A suggested energy-savings evaluation method for ac adjustable-speed rive applications. *IEEE Transactions on Industry Applications, 24*(6), 1107–1117. doi:10.1109/28.17486

Sahu, G. K. (2000). *Pumps*. New Delhi, India: New Age International.

Wang, Y. J. (2000, January). *An analytical study on steady state performance of an induction motor connected to unbalanced three-phase voltage*. IEEE Power Engineering Society Winter Meeting, Singapore.

Wang, Y. J. (2001). Analysis of effects of three phase voltage unbalance on induction motor with emphasis on the angle of the complex voltage unbalance factor. *IEEE Transactions on Energy Conversion, 16*(3), 270–275. doi:10.1109/60.937207

Chapter 16

Intelligent Information Systems for Strengthening the Quality of Energy Services in the EU: Case Study in the Greek Energy Sector

Alexandra G. Papadopoulou
National Technical University of Athens, Greece

Andreas Botsikas
National Technical University of Athens, Greece

Charikleia Karakosta
National Technical University of Athens, Greece

Haris Doukas
National Technical University of Athens, Greece

John Psarras
National Technical University of Athens, Greece

ABSTRACT

Nowadays, taking into consideration the prevailing situation of price fluctuations, the rapid population increase and the technology's evolution, the energy efficiency unexploited potential is considered to be extremely significant as a means of partly tackling energy dependence and climate change. This potential can be utilised through the provision of energy services, with the support of intelligent information systems. In particular, up to date several researchers, have proposed energy management tools and methodologies that provide specialized energy management services. However, the majority of the known energy tools are limited to a single equipment type, fuel, or locality. The present paper introduces an intelligent information decision support system, addressed to Energy Service Companies (ESCOs) for assessing an operational unit's (building or industrial sector) energy behaviour and suggesting the appropriate interventions. Its overall scope is to facilitate the ESCO in reaching a decision quickly and accurately, by simulating the whole unit's energy behaviour.

DOI: 10.4018/978-1-60566-737-9.ch016

INTRODUCTION

Current Status and Needs for Enhanced Energy Services

Buildings and industry have become two of the fastest growing energy consuming sectors especially in the European Union (EU) countries, following shortly behind the transport sector. According to statistic data available through Eurostat (2008), the amount of the energy consumed in the EU buildings reaches 42% of total energy consumption, around two thirds of which is used in dwellings. In the current decade, according to EC Directorate (2003), energy demand of the tertiary and residential sectors are increasing with 1,2% and 1,0% per annual respectively. As a result the European Environment Agency (2004) states that energy usage in the above sectors is responsible for approximately 50% of the union's Greenhouse Gas (GHG) emissions that contribute to climate change.

Nowadays, taking into consideration the prevailing situation of price fluctuations, the rapid population and the technology's evolution, the energy efficiency unexploited potential is considered to be extremely significant as a means of partly tackling energy dependence and climate change. Based on the European Commission's (EC) Green paper (2005), it is estimated that by 2030, on the basis of present trends, the EU will be 90% dependent on imports for its oil requirements and 80% dependent on gas. Total consumption is currently around 1.725 Mtoe. Estimations indicate that, if current trends continue, consumption will reach 1.900 Mtoe in 2020. Moreover, it is estimated that if no measures at all were taken, at this moment the EU-25 would be facing a primary energy demand of approximately 2.550 Mtoe. This clearly highlights the important results achieved till now in the energy consumption's reduction for the period after the first oil crisis through energy efficiency as well as the huge unexploited energy efficiency potential.

According to the EU Action Plan for Energy Efficiency (2006) the Commission considers the energy savings in these sectors will reach 27% and 30% respectively for the residential and commercial buildings (tertiary sector), as well as 27% for the manufacturing industry. Indeed, energy efficiency according to Lechtenböhmer (2005) and Blok (2004) could contribute to the reduction of the current energy consumption by at least 20%, which is equivalent to the savings of 60 billion euros annually. Moreover, the same studies conclude that an average EU household could save between 200 and 1.000 euros per year in a cost effective manner, depending on its energy consumption. On the other hand, Ecofys (2003) states that the energy efficiency industry's boom will result in the creation of high quality working opportunities that directly or indirectly may even reach 1 million jobs. It is indicatively mentioned by the Association for the Conservation of Energy (2000) that only in the United Kingdom, where there is a wide range of energy efficiency programmes and initiatives, the last decade more than approximately 55.000 new jobs have been created. Moreover, according to the Green Paper (2005) the R&D budget dedicated by EC for energy efficiency's promotion for the period 2007–13 through the programme 'Intelligent Energy–Europe' reaches 780 million euros, creating significant employment opportunities.

Especially for Greece, Eurostat (2008) states that the use of energy in buildings, such as public and private buildings, schools, hospitals, hotels and athletic facilities, constitutes of 30% of total national energy demand and contributes of about 40% of the national carbon dioxide emissions. Paravantis (1995) and Athanasakou (1996-97) indicate that heating and refrigeration of buildings consume the largest part of energy expended in domestic uses. As regards the industrial sector's participation in the energy consumption, according to Eurostat (2008), it seems that it covers approximately 20% of the final energy consumption. With respect to the above, evident is the need for enhanced energy services so as to tackle the increasing energy demand and exploit the energy efficiency potential.

European Efforts for Improving the Quality of Energy Services

Among the important efforts in applying energy management processes for improving the quality of energy services, the European Commission released the EC Directive 2002/91 (2002) on buildings' energy efficiency. This directive sets target to realize a savings potential of around 22% by 2010 for energy used in heating, air-conditioning, hot water and lighting through the implementation of measures such as improved standards, certification of buildings, subsidies form EU structural funds for energy efficiency improvements in public buildings etc.

Alongside, with the EC Directive 2006/32 (2006), which concerns the energy end use efficiency at the energy services, the European Commission has set a suitable environment for the systematization of the energy management procedure in energy consuming buildings such as hospitals, hotels, malls and offices, as well as small and medium sized industrial customers. In this directive's framework the EC states that all Member States must adopt and achieve an indicative target for saving energy of 9% by 2015.

The EC Directive 2004/8 (2004) promotes the cogeneration based on a useful heat demand in the internal energy market. This directive finds application field to all industrial companies that produce electricity and useful heat. Streimikiene (2008) clearly explains that the strategic goal of EU-15 is to double the share of electricity produced by combined heat and power pants (CHP) by 2010. The different mechanisms can be applied to support cogeneration at the national level, including investment aid, tax exemptions or reductions, green certificates and direct price support schemes, information disclosure, etc.

Energy Service Companies (ESCOs) are important agents to enhance energy services and promote energy efficiency improvements in this respect. Thus, one of the most important goals of the European energy policy, involves the implementation of energy efficiency measures in large scale so as to promote sustainable development in an EU level. This policy direction is shaped in order to encounter the challenges of the Kyoto Protocol's emission commitments, the security of energy supply and the increase of competitiveness, as stated in the EU Energy Policy (2007) and the EC Green Paper (2005).

The majority of the ESCOs being activated in the European Union are usually subsidiaries of construction and equipment companies or utilities. Although the European Commission has supported a number of policy initiatives to foster the ESCO industry so as to utilize the existing potential, a recent survey of ESCO businesses in EU-27, Switzerland and Norway by Bertoldi (2005) has indicated that major differences exist in their development among the various countries. With a few notable exceptions, such as Germany, Austria, Hungary, and to certain extent France, the ESCO industry is still in its infancy stage and ESCOs are struggling to get off the ground.

Especially in Greece, ESCO activities are still in an experimental phase, mainly regarding RES activities. The main obstacles for the national ESCOs market development are focused on the absence of an integrated regulatory framework supporting their establishment and operation, on the limited experience and know how of the candidate ESCOs, as well as the insufficient financing processes for the projects.

Applying Intelligent Information Systems

Energy services are considered as a significant factor to the energy savings of an operational production unit and an essential procedure for energy efficiency, environmental protection as well as for the mitigation of currency outflow from the national economy. According to Patlitzianas (2005), the methodical organization and application of energy services procedures can achieve a reduction of 5-25% in the energy costs, depending on the sort of intervention required.

Today, with the evolution of Information Technology (IT), Laudon (2000) states that intelligent information systems have been developed for the units' energy consumption, aiming to preserve the comfort conditions of buildings' occupants or the industrial production levels correspondingly, minimizing the energy consumption. In addition to this, Mařík (2008) and Pohekar (2004) highlight that the use of intelligent decision support models and information systems by energy managers for the conduction of energy audits and for the optimized operation of their units, especially concerning long term strategic proposals for retrofit measures and interventions on building facilities, are of significant importance.

Towards this direction, the role of the Building Energy Management Systems (BEMS) is known and significant, since these systems can contribute to the continuous energy management and therefore to the achievement of the possible energy and cost savings. The BEMS are generally applied to the control of active systems, i.e. Heating, Ventilation, and Air-Conditioning (HVAC) systems, while also determining their operating times. In the above efforts, the performance of the BEMS is directly related to the amount of energy consumed in the buildings and the comfort of the buildings' occupants.

In addition, BEMS are currently being developed to be applied in buildings, namely the "intelligent buildings" and a number of studies, such as Kua (2002), Al-Rabghi (2004) and Wong (2005) have been presented for modern intelligent buildings and control systems, revealing the ongoing interest of the scientific community on this topic.

Several researchers, such as Alcala (2005), Patlitzianas (2005), Doukas (2007), Clarke (2002) and Doukas (2008), have proposed energy management tools and methodologies that provide specialized energy management services. Laudon (2000) states that the evolution of the energy management systems is vastly affected by the technological progress of microprocessors. Less sophisticated systems were used in the past, served mainly as monitoring systems with negligible control functions while the degree of dependence on the human factor was very high. In addition to this, energy tools have been developed focused on limited to a single equipment type, such as studies by Guo (1993) and Mathews (2003), or on a specific type of buildings, such as tools developed by Pfafferott (2004) and Askounis (1998). Furthermore, there are a number of energy management tools that apart from assisting in conducting energy audits have a number of additional features, such as the ones presented by Farag (1999), Mills (2006) and O'Sullivan (2004), as well as more sophisticated tools, which incorporate intelligence, for example Clark (1997). Moreover, methodologies have been developed regarding the renovation of buildings in a way that will render them energy efficient. They proceed to initial energy audits for the estimation of the building's energy status and for the evaluation of different scenarios regarding their upgrade to a more energy efficient level. Such methodologies are the "Tobus", "Xenios" and "Epiqr – Investimno", which have been introduced through the Joule II, Altener and Growth programmes.

As regards the operational units of the industrial sector, they have also attracted the researchers' interest. Boyd (2008) has used the evolution of the ENERGY STAR® energy performance indicator for benchmarking industrial plant manufacturing energy use, while Larsson (2006) developed a methodology for a steel plant and Curtis (2004) has focused on the energy conservation in electric utilities. Several

Figure 1. System core idea

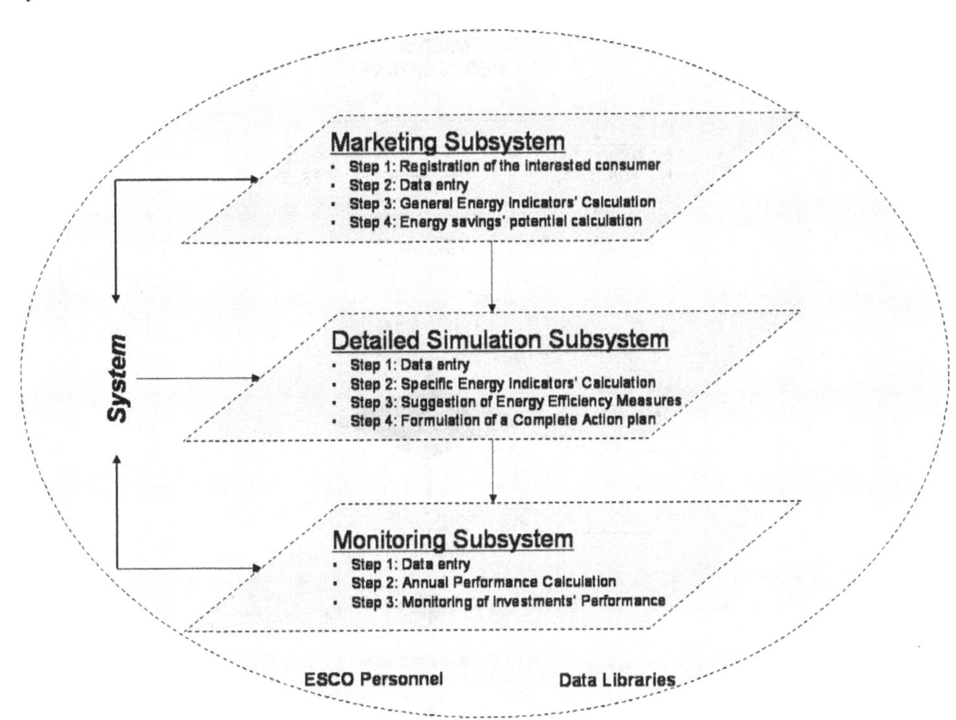

efforts have also been made under the umbrella of the Intelligent Energy for Europe Programme, where benchmarking and self assessment tools have been created for a number of industries, such as the wine-, ceramics-, textile- and plastics industry, in the framework of IEE programmes "Amethyst", "Ceramin", "EMS Textile" and "Recipe".

The rest of the manuscript focuses on the system's methodology, its application to the Greek energy sector and the conclusions drawn up from this study.

METHODOLOGY FOR ENHANCING THE QUALITY OF ENERGY SERVICES

The core idea is the development of an intelligent tool to support the provision of integrated energy services. This tool includes three fully functional, distinct but yet cooperative subsystems, and has been designed in an open way, so as to be easily updated with data regarding the specific requirements and characteristics of any operational unit, contributing thus in its appropriate "mapping". Moreover, its open architecture allows its enhancement with data for additional operational units of the wider building and industrial sector. Figure 1 presents the system's core idea.

The system's basic components include:

- Indicators' Database: It includes the standard general and specific energy intensity indicators, classified per operational unit and production stage or consumption equipment correspondingly.
- External Parameters' Database: This database includes all the information required for the calculation of the specified energy, environmental and financial parameters.

Figure 2. The marketing subsystem's procedure

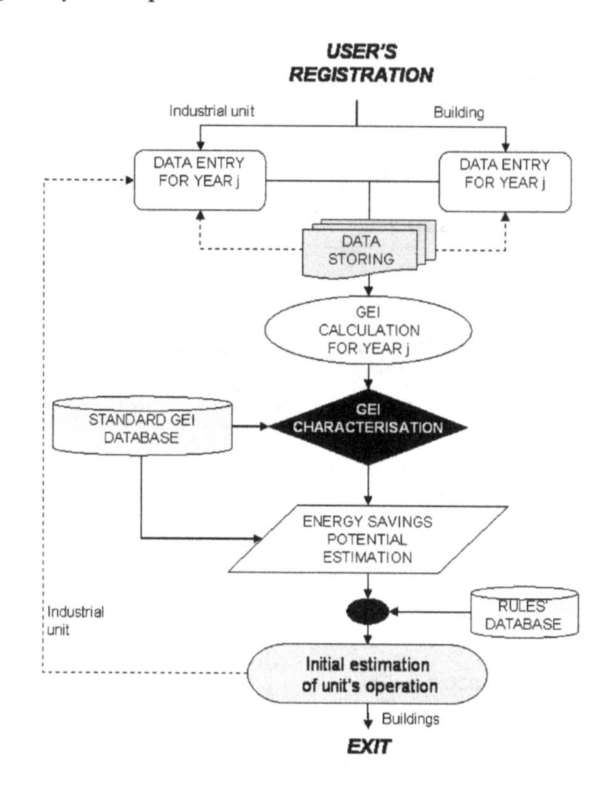

- Interventions' Database: The specific database includes all interventions suggested by the system, classified per basic equipment consumption category or production stage, as well as information regarding their implementation and operational cost and estimation on their energy savings potential.
- Rules' Database: It is only used in the marketing subsystem and it includes a total number of 135 rules, which characterize the unit's operation.

The subsystems are presented in the following paragraphs.

Marketing Subsystem

The Marketing Subsystem is a web based tool enabling a brief energy audit of the operational unit, assessing its energy operation and providing an estimation of the energy savings potential. This tool is primarily intended to attract potential customers, as it requires only the user's registration. The marketing subsystem's procedure is presented in Figure 2.

The Marketing Subsystem's procedure incorporates the following basic steps:

- *Step I: Users' Registration.* The registration of the user enables full access to the specific subsystem and the opportunity of a draft evaluation of the operational unit's status.
- *Step II: Operational Unit Choice.* Choice whether the unit simulated will be a building or an industrial one. In the latter case a subsequent option is made among the industrial units simulated

Figure 3. The detailed simulation subsystem's procedure

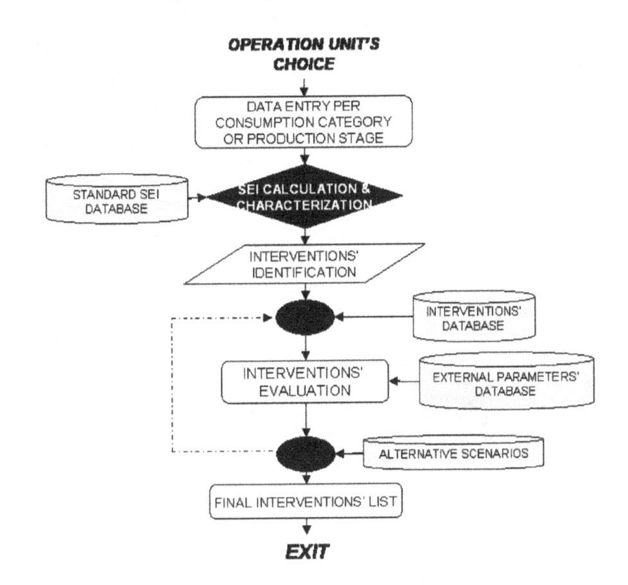

by the system.

- *Step III: Data Entry.* Data entry is realised for a three years period, including information on the consumed fuel quantity, the total surface of the building unit in square meters or the total production level of the industrial unit.
- *Step IV: General Energy Intensity Indicator's Calculation & Characterisation.* The calculation of the general energy indicator (GEI) is realized utilizing the data entered in the previous stage. Then, a comparison of the general energy indicator with standard values from the general indicators' database results to its characterization, which may vary from «good», «satisfactory», «moderate», «weak», to «very weak».
- *Step V: Estimation of Energy Savings Potential.* The estimation of the energy savings potential is realized using the standard GEIs database.
- *Step VI: Energy Status's Evaluation.* The evaluation of the unit's energy status according to the GEI is realized and estimation of the necessity for the implementation of energy saving measures takes place through the use of the rules' database. In case of an industrial unit, the procedure is repeated for the building part.

Detailed Simulation Subsystem

The Detailed Simulation Subsystem is the core of all operations executed by the system, it enables the operational unit's detailed energy audit, the suggestion of interventions where required, as well as their complete evaluation. It is of great assistance to the ESCO and its energy management personnel.

The philosophy behind the subsystems' design and development is quite common. To this end, Figure 3 presents the Detailed Simulation Subsystem's procedure.

The subsystem's procedure includes the following basic steps.

Figure 4. The monitoring subsystem's procedure

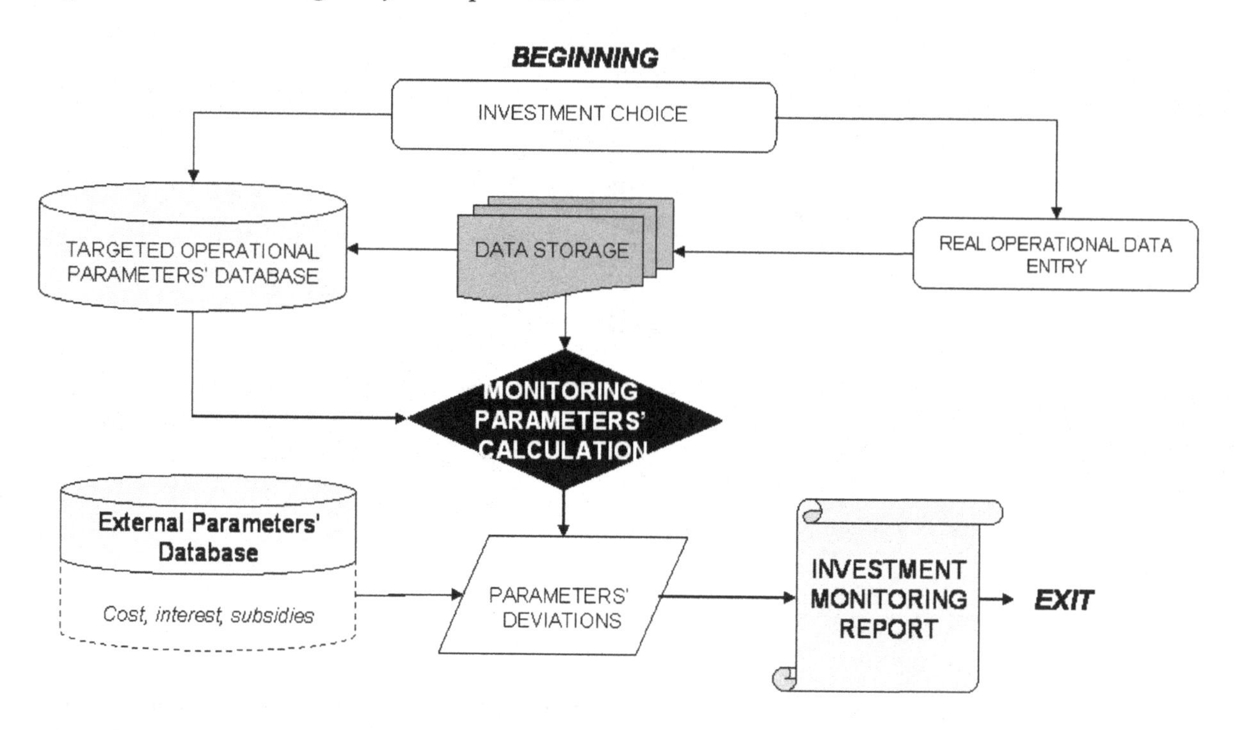

- *Step I: Operational Unit's choice.*
- *Step II: Data Entry.* Data entry takes place, depending on the type of the operational unit being simulated. In case of a building, data entry is realized for the four main consumption categories, namely lighting, air conditioning, heating and other electromechanical equipment. For an industrial unit, data is entered depending on the industry's activities.
- *Step III: Specific Energy Intensity Indicators' Calculation & Characterization.* The calculation of the specific energy intensity indicator (SEI) is realized based on the consumed energy per stage or equipment category and the produced quantity or surface. Corresponding to the GEIs case, SEIs receive a classification depending on their performance in comparison to the standard indicators.
- *Step IV: Detection of Interventions.* Depending on the specific energy intensity indicators' characterization, the equipment or production stage requiring the implementation of energy saving measures are detected.
- *Step V: Interventions' Evaluation.* All investments included in the proposed interventions' list are evaluated with the use of a multicriteria decision support method for three distinct scenarios; the baseline scenario, the increased environmental concern scenario and the instability scenario.
- *Step VI: Final Energy Investment Lists.* The list with the final energy actions proposed by the system for each one of the scenarios is formed.

Monitoring Subsystem

The Monitoring Subsystem's role is to monitor energy investments' operation and to identify any deviations from the agreed target values. It fully collaborates with the previous two subsystems, and its broad application range may include the monitoring of all type of energy investments included in its database.

The procedure for this subsystem is depicted in Figure 4.

The Monitoring Subsystem's procedure incorporates the following basic stages:

- *Step I: Choice of Energy Investment.* The user chooses the specific investment agreed with the client. During the monitoring subsystem's initiation, all relative data stored in the 2nd subsystem's database are obtained. These data, including the calculated potential energy savings, can be modified by the user, depending on the ESCO's agreement with the client, thus resulting to the new calculation of the investment's feasibility.
- *Step II: Investment Real Operational Data Entry.* At this stage, the user enters into the system the real operation data required for the calculation of the investment's monitoring parameters.
- *Step III: Calculation of Investment's Target Parameters.* The investment's target parameters are set based on the agreement between the client and the ESCO. Depending on the existence of any modifications, these data either originate from the 2nd subsystem, or they are inserted by the user. More specifically, the parameters used by the system include the energy consumption, savings and cost, the CO_2 emission reduction levels and the subsequent financial profit, as well as the investment's net present value.
- *Step IV: Calculation of Investment's Monitoring Parameters.* At this step, the investment's parameters are calculated based on the real operation data deriving after the investment's implementation. These parameters are called monitoring parameters and their calculation is supported by the external parameters' database.
- *Step V: Investment Monitoring Report:* The final methodological stage of the subsystem's operation includes the calculation of the deviations among the monitoring parameters and the targeted ones. The produced monitoring report includes detailed information on the investment's progress and the declinations encountered.

Information System Development

The system was realized with the following software tools and applications:

- Microsoft SQL Server: It is used for the development of all the databases (Standard Energy Intensity Indicators' Database, External Parameters' Database, Interventions' Database, Rules' Database), as well as the provisional storing of the data.
- Microsoft .NET 2.0: It is used as development platform of the system's graphic environment, the user interacts with.

MS SQL Server's architecture makes it an ideal tool for the development of complex applications that require access to a large amount of data. Moreover, it can fully support Web portals and e-marketplaces of high requirements.

With respect to the above, the selected development platform provides scalability, reliability and compatibility with most desktop personal computers. In this context, it was assured that the presented model can be readable and accessible to everyone from novice programmers to advanced system architects.

PILOT APPLICATION IN THE GREEK ENERGY SECTOR

The more and more observed phenomenon of combustible fuels' prices galloping course, in comparison to the rapid utilization of the fuel reserves causing their exhaustion have contributed to the search of a solution. The provision of energy services is one of the solutions offered today, although in Greece ESCOs are currently at their infancy.

In this context, the Hellenic General Secretariat for Research and Technology (GSRT) has funded the project "Decision Support System for the Development and Provision of Energy Services", where the authors' team participated, aiming to develop an intelligent decision support system, that can significantly support the developing ESCOs in the energy services' provision, tackling the lack of know how and required expertise to a certain extent.

The system was tested in real life case studies, so as to extract useful conclusions regarding its operation.

Among the pilot runs made, the marketing subsystem was used to evaluate the overall energy operation of an office building and a small hotel. Based on the energy consumption and surface data entry for the most recent triennium, the monitoring subsystem calculated the general energy indicators and evaluated the overall energy operation for these establishments.

During the pilot appraisal the monitoring subsystem presented normal operation and satisfactory intercommunication within the subsystem, while it reacted immediately to the user's commands. Moreover, the interface was considered by the systems' users as simple and easy to operate, without requiring a long period of time for the user to get acquainted with it. As potential actions for the subsystem's future optimization were identified the addition of more industrial or building operational units that can be simulated.

Afterwards, the detailed simulation subsystem's evaluation took place for these two establishments. The procedure regarding the office building and hotel includes the following steps: As a first step, all the required information regarding the unit's building envelope was inserted, such as data regarding the wall, floor and roof material, the type and number of windows etc. Next step was the insertion of all required energy data per type of energy consuming equipment, so as to realize a true evaluation of the unit's operation and the necessary interventions required.

The subsystem's results for the office building can be summarized as follows:

- Although the unit's operation in each one of the 4 major equipment consumption categories was satisfactory according to the specific energy indicators, additional low cost measures would improve its performance.
- The evaluation of all proposed measures resulted in the suggestion of the following options as the most suitable ones: Installation of PV systems; Better maintenance of the air-conditioning units; More frequent maintenance of the boiler; Installation of solar protecting membranes; Installation of a lighting time switch.

As regards the subsystem's results for the small hotel establishment, these are summarized below:

- The energy indicator for the hotel's lighting consumption was characterized as weak, due to their operation in a 24-hour basis, the existence of incandescence lamps, as well as the unreasonable capacity of the existing lamps in the hotel public areas.

- As regards water and space heating, the indicator was moderate. This is mainly due to the inexistence of a central boiler control system, to the old water heating network, which presents energy losses, and the insufficient insulation of the building shell.
- The indicator for the rest of the electromechanical equipment was weak, mainly because of the large energy consumption for cooking and the elevators.
- Air conditioning is another important consumption category of this hotel establishment, which has been characterized by the system as satisfactory.
- The above indicators highlight the need to take immediate action with low or high cost measures where required. From all the measures suggested by the system, their evaluation resulted in the suggestion of the following, as the most promising ones:
    - Installation of new insulation in the hotel roof.
    - Replacement of the incandescence lamps with fluorescent ones.
    - Use of natural gas instead of diesel oil, modifications in the central heating system.
    - Installation of a BEMS.

The pilot appraisal of the detailed simulation subsystem resulted in the following improvement actions:

- Detailed instructions and help were added to the system, so as to make it more user friendly;
- The interventions' database was enriched with more data.

In general, the subsystem's operation presented no significant operational problems.

The monitoring subsystem was in general tested with data from interventions coming from past investments and data from literature case studies, as due to the specified project period there was no time to test it in real time conditions. Its pilot application was realised with data from a textile company, where the subsystem's results showed less expected income by approximately 3,6%, coinciding with the investments' outcome. The monitoring subsystem's operation presented minor operational problems that were restored.

It should be noted that the system in its current form contains databases updated with data and practices from the Greek reality. A potential expansion of its use outside the national borders would require additional data on the country's energy market.

CONCLUSION

The use of intelligent energy management systems for energy auditing of operational units has been very popular within the past years, and they continue to expand their application field. These systems are able to evaluate a specific unit's energy operation and to propose energy interventions or to apply predefined measures so as to reduce energy consumption. In this effort, significant is the role of the decision support systems, which can contribute to the proper evaluation of the operational units' performance, the suggestion of energy saving measures and their evaluation, as well as the monitoring of the implemented energy investments.

In the above context, the presented intelligent system can significantly assist the energy manager in receiving their decisions quickly and accurately, as it enables complete evaluation of any building or industrial units' operation, using standard general and specific energy indicators from the international literature and the ESCO's experience. In addition, it suggests a complete set of interventions per production stage or equipment category, depending on the type of unit simulated. Finally, the system performs detailed monitoring of the investment's performance.

Based on the results of its pilot application, it can be considered that its operation was satisfactory. In addition, its interface was characterized as very user friendly and facilitative, based on the users' comments. Moreover, its open architecture allows easy and continuous updates in the directions indicated within the paper. Therefore, the model's design allows its application to a large group of different types of building and industrial units.

ACKNOWLEDGMENT

This paper was based on research conducted within the "Decision Support System for the Development and Provision of Energy Services" project of the Hellenic General Secretariat for Research and Technology (GSRT). The content of the paper is the sole responsibility of its authors and does not necessarily reflect the views of the GSRT.

REFERENCES

Al-Rabghi, O. M., & Akyurt, M. M. (2004). A survey of energy efficient strategies for effective air conditioning. *Energy Conversion and Management, 45*, 1643–1654. doi:10.1016/j.enconman.2003.10.004

Alcala, R., Casillas, J., Cordon, O., Gonzalez, A., & Herrera, F. (2005). A genetic rule weighting and selection process for fuzzy control of heating, ventilating and air conditioning systems. *Artificial Intelligence, 18*, 279–296.

Askounis, D. T., & Psarras, J. (1998). Information system for monitoring and targeting (M&T) of energy consumption in breweries. *Energy, 23*(5), 413–419. doi:10.1016/S0360-5442(97)00107-2

Association for the Conservation of Energy (September 2000). *Energy efficiency and jobs: UK issues and case studies, A report to the Energy Saving Trust.* EC SAVE Project "National and Local Employment Impacts of Energy Efficiency Investment Programmes". London: Association for the Conservation of Energy.

Athanasakou, E. (1996–1997, December–January). Solar passive systems and technologies [in Greek] [in Greek]. *Energy, 24*, 55–66.

Bertoldi, P., & Rezessy, S. (2005). *Energy Service Companies in Europe, Status Report 2005.* Institute for Environment and Sustainability, Renewable Energies Unit, Office for Official Publication of the European Communities.

Blok, K. (2004). Improving energy efficiency by 5% and more per year. *Journal of Industrial Ecology, 8*(4), 87–99. doi:10.1162/1088198043630478

Boyd, G., Dutrow, E., & Tunnessen, W. (2008). The evolution of the ENERGY STAR® energy performance indicator for benchmarking industrial plant manufacturing energy use. *Journal of Cleaner Production, 16*(6), 709–715. doi:10.1016/j.jclepro.2007.02.024

Clark, G., & Mehta, P. (1997). Artificial intelligence and networking in integrated building management systems. *Automation in Construction, 6*, 481–498. doi:10.1016/S0926-5805(97)00026-5

Clarke, J. A., Cockroft, J., Conner, S., Hand, J. W., Kelly, N. J., & Moore, R. (2002). Simulation - assisted control in building energy management systems. *Energy and Building, 34*, 933–940. doi:10.1016/S0378-7788(02)00068-3

EU Commission (10 January 2007). *An Energy Policy For Europe, COM(2007) 1 final.*

EUCommission (11 February 2004). Directive 2004/8/EC of the European Parliament and of the Council on the promotion of cogeneration based on a useful heat demand in the internal energy market and amending Directive 92/42/EEC.

Curtis, M., & Khare, A. (2004). Energy conservation in electric utilities: an opportunity for restorative economics at SaskPower. *Technovation, 24*, 395–402. doi:10.1016/S0166-4972(02)00116-5

Doukas, H., Nychtis, C., & Psarras, J. (2008). (in press). Assessing energy-saving measures in buildings through an intelligent decision support model. [Corrected Proof.]. *Building and Environment.*

Doukas, H., Patlitzianas, K. D., Iatropoulos, K., & Psarras, J. (2007). Intelligent building energy management system using rule sets. *Building and Environment, 42*, 3562–3569. doi:10.1016/j.buildenv.2006.10.024

Ecofys (2003). *Rat für Nachhaltige Entwicklung.* Berlin: German Council for Sustainable Development.

European Commission. (2002). *EC Of The European Parliament And Of The Council. Directive 2002/91 on the energy performance of buildings.* EC- Official Journal L 001. Brussels: Author.

European Commission (2005). *Green paper "Doing more with less"*. Luxemburg: European Communities.

European Commission. (2006). *EC Of The European Parliament And Of The Council. Directive 2006/32 EC on energy end-use efficiency and energy services and repealing Council Directive 93/76/EEC.* Official Journal L 114. Brussels: Author.

European Commission (19 October 2006). *Action Plan for Energy Efficiency: Realizing the Potential.* COM (2006)545 final.

European Commission Directorate — General for Energy and Transport. (2003). *European Energy and Transport Trends to 2030.* Brussels: Author.

European Environment Agency (EEA). (2004).Annual European Community greenhouse gas inventory 1990-2002 EEA report.

Eurostat (2008). *Energy Balance Sheets. Data 2006.* Luxemburg: Eurostat.

Farag, A. S., Mousab, A. E., Chengc, T. C., & Beshir, M. (1999). Cost effective utilities energy plans optimization and Management. *Energy Conversion and Management, 40*, 527–543. doi:10.1016/S0196-8904(98)00120-4

Guo, B., Belcher, C., & Roddis, W. M. K. (1993). RetroLite: an artificial intelligence tool for lighting energy-efficiency Upgrade. *Energy and Building, 20*, 115–120. doi:10.1016/0378-7788(93)90002-C

Kua, H. W., & Lee, S. E. (2002). Demonstration intelligent building—a methodology for the promotion of total sustainability in the built environment. *Building and Environment, 37*, 231–240. doi:10.1016/S0360-1323(01)00002-6

Larsson, M., Wang, C., & Dahl, J. (2006). Development of a method for analysing energy, environmental and economic efficiency for an integrated steel plant. *Applied Thermal Engineering, 26*(13), 1353–1361. doi:10.1016/j.applthermaleng.2005.05.025

Laudon, K., & Laudon, J. (2000). *Management information systems, Organization and technology in the networked enterprise* (6th Ed.). Englewood Cliffs, NJ: Prentice-Hall.

Lechtenböhmer, S., & Thomas, S. (2005). *The mid-term potential for demand-side energy efficiency in the EU.* Wuppertal Institute for Climate, Environment, Energy.

Mařík, K., Schindler, Z., & Stluka, P. (2008). Decision support tools for advanced energy management. *Energy, 33*(6), 858–873. doi:10.1016/j.energy.2007.12.004

Mathews, E. H., & Botha, C. P. (2003). Improved thermal building management with the aid of integrated dynamic HVAC simulation. *Building and Environment, 38*, 1423–1429. doi:10.1016/S0360-1323(03)00070-2

Mills, E., Kromerb, S., Weissc, G., & Mathewd, P. A. (2006). From volatility to value: analysing and managing financial and performance risk in energy savings projects. *Energy Policy, 34*, 188–199. doi:10.1016/j.enpol.2004.08.042

O'Sullivan, D. T. J., Keane, M. M., Kelliher, D., & Hitchcock, R. J. (2004). Improving building operation by tracking performance metrics throughout the building lifecycle (BLC). *Energy and Building, 36*, 1075–1090. doi:10.1016/j.enbuild.2004.03.003

Paravantis, J. (1995, November). Energy savings in buildings: heating and passive cooling [in Greek] [in Greek]. *Energy, 17*, 49–59.

Patlitzianas, K. D., Papadopoulou, A., Flamos, A., & Psarras, J. (2005). CMIEM: the computerised model for intelligent energy management. *International Journal of Computer Applications in Technology, 22*(2/3), 120–129. doi:10.1504/IJCAT.2005.006943

Pfafferott, J., Herkel, S., & Wambsgan, M. (2004). Design, monitoring and evaluation of a low energy office building with passive cooling by night ventilation. *Energy and Building, 36*, 455–465. doi:10.1016/j.enbuild.2004.01.041

Pohekar, S. D., & Ramachandran, M. (2004). Application of multi-criteria decision making to sustainable energy planning – A review. *Renewable & Sustainable Energy Reviews, 8*, 365–381. doi:10.1016/j.rser.2003.12.007

Streimikiene, D., Šivickas, G. (2008). The EU sustainable energy policy indicators framework. *Environment International*, in press.

Wong, J. K. W., Li, H., & Wang, S. W. (2005). Intelligent building research: a review. *Automation in Construction, 14*, 143–159. doi:10.1016/j.autcon.2004.06.001

Compilation of References

Aageng, Q., Zongxiang, L., & Yong, M. (2006). Research on isolated diesel-wind power system equipped with doubly fed induction generator, *Proc. 8th IEE International Conf. AC and DC Power Transmission*, March, (pp. 246 – 250).

Abido, M. A. (2003). A novel multiobjective evolutionary algorithm for environmental/economic power dispatch. *Electric Power Systems Research*, *65*, 71–81. doi:10.1016/S0378-7796(02)00221-3

Abido, M. A. (2003). Environmental/Economic power dispatch using multiobjective evolutionary algorithms. *IEEE Transactions on Power Systems*, *18*, 1529–1537. doi:10.1109/TPWRS.2003.818693

Abido, M. A. (2006). Multiobjective evolutionary algorithms for electric power dispatch problem. *IEEE Transactions on Evolutionary Computation*, *10*, 315–329. doi:10.1109/TEVC.2005.857073

Abido, M. A. (2007, July). Two-level of nondominated solutions approach to multiobjective particle swarm optimization. *Proceedings of the 2007 Genetic and Evolutionary Computation Conference GECCO'2007*, (pp. 726-733).

Achayuthakan, C., & Srivastava, S. C. (1998). A genetic algorithm based economic load dispatch solution for eastern region of EGAT system having combined cycle and cogeneration plants. *Proceedings of International Conference on Energy Management and Power Delivery* (pp. 165-170).

Adapa, R. (1994). Expert system applications in power system planning and operations. *IEEE Power Engineering Review*, *14*(2), 15–18. doi:10.1109/MPER.1994.262142

Adibi, M. M., Kafka, R. J., & Milanicz, D. P. (1994). Expert system requirements for power system restoration. *IEEE Transactions on Power Systems*, *9*(3), 1592–1600. doi:10.1109/59.336099

AFAQ. *PAS 55 Assessment: Ensuring Effective Asset Management.* Retrieved November 6, 2008, from http://www.afnor.co.uk/webai/uk/Site.nsf/0/0B6DE01A86A0C28BC125735400557A6A/$file/PAS55.pdf

Afgan, N. H., & Carvalho, M. G. (2008). Sustainability assessment of a hybrid energy system. *Energy Policy*, *36*(8), 2903–2910. doi:10.1016/j.enpol.2008.03.040

Afgan, N. H., Gobaisi, D. A., Carvalho, M. G., & Cumo, M. (1998). Sustainable energy development. *Renewable & Sustainable Energy Reviews*, *2*, 235–286. doi:10.1016/S1364-0321(98)00002-1

Afgan, N. H., Pilavachi, P. A., & Carvalho, M. G. (2007). Multi-criteria evaluation of natural gas resources. *Energy Policy*, *35*(1), 704–713. doi:10.1016/j.enpol.2006.01.015

Agrawal, J. P. (2001). *Power electronic systems: theory and design.* Upper Saddle River, N.J.: Prentice Hall.

Aguiar, A. B. M., Pinto, J. O. P., Andrea, C. Q., & Noguerira, L. A. (2007). Modeling and Simulation of Natural Gas Microturbine Application for Residential Complex Aiming Technical and Economical Viability Analysis. *2007 IEEE Canada Electrical Power Conference* (pp. 376-381).

Aigner, D. J., Sorooshian, C., & Kerwin, P. (1984). Conditional demand analysis for estimating residential end-use load profiles. *The Energy Journal (Cambridge, Mass.)*, *5*(3), 81–97.

Akagi, H., & Sato, H. (1999). Control and performance of a flywheel energy storage system based on a doubly fed induction generator-motor for powering conditions, *Proc. 30th Annual IEEE Power Electronics Specialists Conf.*, (p. 1), June – July, 32 – 39.

Aki, H., Murata, A., Yamamoto, S., Kondoh, J., Maeda, T., Yamaguchi, H., & Ishii, I. (2005). Penetration of residential fuel cells and CO_2 mitigation—case studies in Japan by multi-objective models. *International Journal of Hydrogen Energy*, *30*, 943–952. doi:10.1016/j.ijhydene.2004.11.009

Aki, H., Oyama, T., & Tsuji, K. (2000). Analysis on the energy pricing and its reduction effect on environmental impact in urban area including DHC. *2000 IEEE Power Engineering Society Winter Meeting* (pp. 1097-1102).

Aki, H., Oyama, T., & Tsuji, K. (2003). Analysis of energy pricing in urban energy service systems considering a multiobjective problem of environmental and economic impact. *IEEE Transactions on Power Systems*, *18*(4), 1275–1282. doi:10.1109/TPWRS.2003.818599

Aki, H., Oyama, T., & Tsuji, K. (2006). Analysis of energy service systems in urban areas and their CO2 mitigations and economic impacts. *Applied Energy*, *83*, 1076–1088. doi:10.1016/j.apenergy.2005.11.003

Alanne, K., & Saari, A. (2004). Sustainable small-scale CHP technologies for buildings: the basis for multi-perspective decision-making. *Renewable & Sustainable Energy Reviews*, *8*, 403–431. doi:10.1016/j.rser.2003.12.005

Alanne, K., Salo, A., Saari, A., & Gustafsson, S. I. (2007). Multi-criteria evaluation of residential energy supply systems. *Energy and Building*, *39*, 1218–1226. doi:10.1016/j.enbuild.2007.01.009

Alcala, R., Casillas, J., Cordon, O., Gonzalez, A., & Herrera, F. (2005). A genetic rule weighting and selection process for fuzzy control of heating, ventilating and air conditioning systems. *Artificial Intelligence*, *18*, 279–296.

Alfonso, D., Perez-Navarro, A., Encinas, N., Alvarez, C., Rodrıguez, J., & Alcazar, M. (2007). Methodology for ranking customer segments by their suitability for distributed energy resources applications. *Energy Conversion and Management*, *48*(5), 1615–1623. doi:10.1016/j.enconman.2006.11.006

AlFuhaid, A. S., El-Sayed, M. A., & Mahmoud, M. S. (1997). Cascaded artificial neural networks for short-term load forecasting. *IEEE Transactions on Power Systems*, *12*(4), 1524–1529. doi:10.1109/59.627852

Algie, C., & Wong, K. O. (2004). A test system for combined heat and power economic dispatch problems. *2004 IEEE International Conference on Electric Utility Deregulation, Restructuring and Power Technologies* (vol.1, pp. 96-101).

Al-Mansour, F., & Kozuh, M. (2007). Risk analysis for CHP decision making within the conditions of an open electricity market. *Energy*, *32*, 1905–1916. doi:10.1016/j.energy.2007.03.009

Al-Rabghi, O. M., & Akyurt, M. M. (2004). A survey of energy efficient strategies for effective air conditioning. *Energy Conversion and Management*, *45*, 1643–1654. doi:10.1016/j.enconman.2003.10.004

Al-Rashdan, D., Al-Kloub, B., Dean, A., & Al-Shemmeri, T. (1999). Environmental impact assessment and ranking the environmental projects in Jordan. *European Journal of Operational Research*, *118*(1), 30–45. doi:10.1016/S0377-2217(97)00079-9

Al-Rashidi, M. R., & El-Hawary, M. E. (2007). Hybrid particle swarm optimization approach for solving the discrete OPF problem considering the valve loading effects. *IEEE Transactions on Power Systems*, *22*(4), 2030–2038. doi:10.1109/TPWRS.2007.907375

Al-Shehri, A. (1999). Artificial neural network for forecasting residential electrical energy. *International Journal of Energy Research*, *23*(8), 649–661. doi:10.1002/(SICI)1099-114X(19990625)23:8<649::AID-ER490>3.0.CO;2-T

Amann, M., Bertok, I., Cofala, J., Gyarfas, F., Heyes, C., Klimont, Z., & Schöpp, W. (1996). *Cost-effective*

Control of Acidification and Ground-Level Ozone. International Institute for Applied Systems Analysis, Laxenburg, Austria.

American Public Power Association. (2005). *Work Force Planning for Public Power Utilities: Ensuring Resources to Meet Projected Needs.* Retrieved September 3, 2008, from: http://www.appanet.org/files/pdfs/workforceplanningforpublicpowerutilities.pdf

Anagnostopoulos, K., Flamos, A., Kagiannas, A., & Psarras, J. (2004). The impact of clean development mechanism in achieving sustainable development. *International Journal of Environment and Pollution, 21*(1), 1–23.

Andersen, A. N., & Lund, H. (2007). New CHP partnerships offering balancing of fluctuating renewable ectricity productions. *Journal of Cleaner Production, 15*, 288–293. doi:10.1016/j.jclepro.2005.08.017

Andersen, C. A., Christensen, M. O., & Korsgaard, A. R. (2002). *Design and Control of Fuel Cell System for Transport Application* (Tech. Rep.). Aalborg University, Project Group EMSD 10 - 11A, Denmark.

ANSI/NEMA Standards Publication no MG1-1978 (n.d.). Washington, DC: National Electrical Manufacturers Association.

Ansoff, I. (1985). *The Concept of Corporate Strategy.* London: Penguin. London.

Anstett, M., & Kreider, J. F. (1993). Application of neural networking models to predict energy use. *ASHRAE Transactions, 99*(1), 505–517.

Aquila, A. D., Marinelli, M., Monopoli, V. G., & Zanchetta, P. (2004). New power quality assessment criteria for supply systems under unbalanced and non sinusoidal conditions. *IEEE Transactions on Power Delivery, 19*(3), 1284–1290. doi:10.1109/TPWRD.2004.829928

Arcuri, P., Florio, G., & Fragiacomo, P. (2007). A mixed integer programming model for optimal design of trigeneration in a hospital complex. *Energy, 32*, 1430–1447. doi:10.1016/j.energy.2006.10.023

Arivalagan, A., Raghavendra, B. G., & Rao, A. R. K. (1995). Integrated energy optimization model for a cogeneration based energy supply system in the process industry. *Electrical Power & Energy Systems, 17*, 227–233. doi:10.1016/0142-0615(95)00037-Q

Arnalte, S., Burgos, J. C., & Amenedo, J. L. R. (2002). Direct torque control of a doubly-fed induction generator for variable speed wind turbines. *Electric Power Components and Systems, 30*(2), 199–216. doi:10.1080/153250002753427851

Arondel, C., & Girardin, P. (1998). *Sorting cropping systems on the basis of their impact on groundwater quality.* Cahiers du LAMSADE No. 158, Universite Paris – Dauphine.

Asano, H., Sagi, S., Imamura, E., Ito, K., & Yokoyama, R. (1992). Impacts of time-of-use rates on the optimal sizing and operation of cogeneration systems. *IEEE Transactions on Power Systems, 7*(4), 1444–1450. doi:10.1109/59.207366

Asare, P., Diez, T., Galli, A., O'Neill-Carillo, E., Roberston, J., & Zhao, R. (1994). *An Overview of Flexible AC Transmission Sytems.* e-Pubs, Purdue University Library. Retrieved from http://docs.lib.purdue.edu/ecetr/205

Ashari, M., & Nayar, C. V. (1999). An optimum dispatch strategy using set points for a photovoltaic (PV)–diesel–battery hybrid power system. *Solar Energy, 6*(1), 1–9. doi:10.1016/S0038-092X(99)00016-X

Ashok, S., & Banerjee, R. (2003). Optimal operation of industrial cogeneration for load management. *IEEE Transactions on Power Systems, 18*, 931–937. doi:10.1109/TPWRS.2003.811169

Askounis, D. T., & Psarras, J. (1998). Information system for monitoring and targeting (M&T) of energy consumption in breweries. *Energy, 23*(5), 413–419. doi:10.1016/S0360-5442(97)00107-2

Association for the Conservation of Energy (September 2000). *Energy efficiency and jobs: UK issues and case studies, A report to the Energy Saving Trust.* EC SAVE Project "National and Local Employment Impacts of

Energy Efficiency Investment Programmes". London: Association for the Conservation of Energy.

Athanasakou, E. (1996–1997, December–January). Solar passive systems and technologies [in Greek] [in Greek]. *Energy, 24*, 55–66.

Augenstein, E., Wrobel, M., Kuperjans, I., & Plessow, M. (2004). TOP-Energy computational support for energy system engineering processes. In *First international conference from scientific computing to computational engineering*, Athens.

Auld, A., Mueller, F., Smedley, K., Samuelsen, S., & Brouwer, J. (2008). Applications of one-cycle control to improve the interconnection of a solid oxide fuel cell and electric power system with a dynamic load. *Journal of Power Sources, 179*(1), 155–163. doi:10.1016/j.jpowsour.2007.12.072

Aydinalp Koksal, M., & Ugursal, V. I. (2008). Comparison of neural network, conditional demand analysis, and engineering approaches for modeling end-use energy consumption in the residential sector. *Applied Energy, 85*(4), 271–296. doi:10.1016/j.apenergy.2006.09.012

Aydinalp, M. (2002). *A new approach for modeling of residential energy consumption*. Unpublished doctoral dissertation, Dalhousie University, Halifax, Nova Scotia, Canada.

Aydinalp, M., Ugursal, V. I., & Fung, A. (2002). Modeling of the appliance, lighting, and space-cooling energy consumptions in the residential sector using neural networks. *Applied Energy, 72*(2), 87–110. doi:10.1016/S0306-2619(01)00049-6

Aydinalp, M., Ugursal, V. I., & Fung, A. (2004). Modeling of the space and domestic hot-water heating energy-consumption in the residential sector using neural networks. *Applied Energy, 79*(2), 159–178. doi:10.1016/j.apenergy.2003.12.006

Azzam, M., & Nour, M. (1996). An expert system for voltage control of a large-scale power system. *Energy Conversion and Management, 37*, 81–86. doi:10.1016/0196-8904(95)00014-5

Azzam, M., & Nour, M. A. S. (1995). Expert system for voltage control of a large-scale power system, Modelling, Measurement Control A: General Physics, Electronics. *Electrical Engineering, 63*(1-3), 15–24.

Babu, C. A., & Ashok, S. (2008). Peak Load Management in Electrolytic Process Industries. *IEEE Transactions on Power Systems, 23*(2), 399–405. doi:10.1109/TPWRS.2008.920732

Badrzadeh, B., & Salman, S. K. (2006). Performance comparison of fixed-speed and doubly fed induction generator under torsional oscillations. *Proc. 8th IEE Int. Conf. AC and DC Power Transmission*, March, (pp. 167 – 171).

Bahn, O., Barreto, L., Büeler, B., & Kypreos, S. (1997). *A multi-regional MARKAL-MACRO model to study an international market of CO2 emission permits: a detailed analysis of a burden sharing strategy among the Netherlands, Sweden and Switzerland*. PSI Technical Report, no. 97-09, Villigen, Switzerland.

Bai, H., & Zhao, B. (2006, June). A survey on application of swarm intelligence computation to electric power system. *Proceedings of the 6th World Congress on Intelligent Control and Automation*, (pp. 7587-7591).

Bakken, B. J., Skjelbred, H. I., & Wolfgang, O. (2007). eTransport: Investment planning in energy supply systems with multiple energy carriers. *Energy, 32*, 1676–1689. doi:10.1016/j.energy.2007.01.003

Balestieri, J. P., & De Borros Correia, P. (1997). Multiobjective linear model for pre-feasibility design of cogeneration systems. *Energy, 22*, 537–548. doi:10.1016/S0360-5442(96)00151-X

Balog, R. S., & Krein, P. T. (2005). Commutation technique for high- frequency link cycloconverter based on state-machine control. *IEEE Power Electronics Letters, 3*(3), 101–104. doi:10.1109/LPEL.2005.858422

Balzani, M., & Reatti, A. (2005). Neural network based model of a PV array for the optimum performance of PV system. *Research in Microelectronics and Electronics, 2*, 123–126. doi:10.1109/RME.2005.1542952

Banerjee, S., & Verghese, G. C. (2001). *Nonlinear phenomena in power electronics: attractors, bifurcations, chaos, and nonlinear control.* New York: IEEE Press.

Banerjee, S., Kastha, D., & SenGupta, S. (2002). Minimising EMI problems with chaos. *Proceedings of the International Conference on Electromagnetic Interference and Compatibility,* (pp. 162-167). Retrieved December 15, 2002, from http://ieeexplore.ieee.org

Bansal, R. C. (2005). Three-phase self-excited induction generators (SEIG): an overview. *IEEE Transactions on Energy Conversion, 20*(2), 292–299. doi:10.1109/TEC.2004.842395

Bansal, R. C., Bhatti, T. S., & Kothari, D. P. (2003). A bibliographical survey on induction generators for application of non conventional energy systems. *IEEE Transactions on Energy Conversion, 18*(3), 433–439. doi:10.1109/TEC.2003.815856

Bansal, R. C., Zobaa, A. F., & Saket, R. K. (2005). Some issues related to power generation using wind energy conversion systems: an overview. *Int. Journal of Emerging Electric Power Systems, 3*(2), 1070.

Barbounis, T. G., & Theocharis, J. B. (2007). Locally recurrent neural networks for wind speed prediction using spatial correlation. *Information Sciences, 177,* 5775–5797. doi:10.1016/j.ins.2007.05.024

Barbounis, T. G., Theocharis, J. B., Alexiadis, M. C., & Dokopoulos, P. S. (2006). Long-term wind speed and Power Forecasting Using Local Recurrent Neural Network Models. *IEEE Transactions on Energy Conversion, 21,* 273–284. doi:10.1109/TEC.2005.847954

Bardouille, P. (1998). *Towards the Operationalisation of Sustainable Energy Systems: Examination of Environmental Decision-Support Tools and Models, and Elaboration of a Conceptual Framework for Sustainable Power Sector Investment Analysis and Decision- Support.* Masters Dissertation, International Institute for Industrial Environmental Economics, Lund University, Lund, Sweden.

Barr, A., & Feigenbaum, E. (1982). *The handbook of artificial intelligence* (Vol. 1). Los Altos, CA: Kaufman Publishing Co.

Bartels, R., & Fiebig, D. G. (2000). Residential end-use electricity demand: results from a designed experiment. *The Energy Journal (Cambridge, Mass.), 21*(2), 51–81.

Bartol, K. M. (2002). *Encouraging knowledge sharing: the role of organizational reward systems.* Retrieved November 6, 2008, from http://findarticles.com/p/articles/mi_m0NXD/is_1_9/ai_n25057533

Basu, S. (2002). Does Governance Matter? Some Evidence from Indian States. *VIIth Spring Meeting of Young Economists,* Paris.

Basu, S. (2007). *Recent Trends in Fuel Cell Science and Technology.* New York: Springer.

Batarseh, I. (2003). *Power electronic circuits.* Hoboken, NJ: John Wiley & Sons Inc.

Batlle, C., Cerezo, A. D., & Ortega, R. (2004). Power flow control of a doubly-fed induction machine coupled to a flywheel. *Proc. IEEE Int Conf. – Control Applications,* 2, (pp. 1645 – 1650).

Baumhögger, F., Baumhögger, J., Baur, J.-M., Kühner, R., Schellmann, U., Schlenzig, C., et al. (1998). *Manual MESAP, Version 3.1.* Institute of Energy Economics and the Rational use of Energy (IER), Germany.

Bauwens, L., Fiebig, D., & Steel, M. (1994). Estimating end-use demand: a bayesian approach. *Journal of Business & Economic Statistics, 12*(2), 221–231. doi:10.2307/1391485

Bazaraa, M. S., & Shetty, C. M. (1993). *Nonlinear programming theory and algorithms.* New York: Wiley.

Beccali, M., Cellura, M., & Mistretta, M. (2003). Decision-making in energy planning. Application of the ELECTRE method at regional level for the diffusion of renewable energy technology. *Renewable Energy, 28*(13), 2063–2087. doi:10.1016/S0960-1481(03)00102-2

Becherif, M., Ortega, R., Mendes, E., & Lee, S. (2003). Passivity-based control of a doubly fed induction generator interconnected with an induction motor. *Proc. 42nd IEEE Conf. Decision and Control,* 6, December, (pp. 2711 – 2716).

Begic, F., & Afgan, N. H. (2007). Sustainability assessment tool for the decision making in selection of energy system—Bosnian case. *Energy, 32*, 1979–1985. doi:10.1016/j.energy.2007.02.006

Bellman, R. E., & Zadeh, L. A. (1977). *Local and fuzzy logics in modern uses of multiple-valued logic.* J. M. Dunn & G. Epstein, (Eds.), (pp. 103-165). Dordrecht, the Netherlands: Reidel.

Bender, E. A. (1996). *Mathematical methods in artificial intelligence*. Los Alamitos, CA: IEEE Computer Society Press.

Bengiamin, N. N. (1983). Operation of cogeneration plants with power purchase facilities. *IEEE Transactions on Power Appratus Systems PAS, 102*(10), 3467–3472. doi:10.1109/TPAS.1983.317845

Ben-Haim, Y. (2006). *Info-Gap Theory: Decisions Under Severe Uncertainty*, (2nd ed.). Academic Press: London.

Bergman, L., & Henrekson, M. (2003). *CGE Modeling of Environmental Policy and Resource Management.* Retrieved June, 21, 2008, from http://users.ictp.it/~eee/workshops/smr1533/Bergman%20-%20Handbook-1.doc.

Bergveld, H. J., Wanda, S. J., Kruijt, S., & Notten, P. H. L. (2002). *Battery Management Systems: Design by Modeling.* Boston: Kluwer Academic Press.

Bertoldi, P., & Rezessy, S. (2005). *Energy Service Companies in Europe, Status Report 2005.* Institute for Environment and Sustainability, Renewable Energies Unit, Office for Official Publication of the European Communities.

Bizon, N. (2004). Design Consideration for Voltage Regulator Modules used in Automotive Control. In Society of Automotive Engineers of Romania (Ed.) under SAE and FISITA patronage, *International Automotive Congress - CONAT 2004* (pp. 25-29). Brasov, Romania: Transilvania University Press.

Bizon, N. (2007). Intelligent Integrated Control of the Power Flows into an Energy Generation System. *Mediterranean Journal of Measurement and Control, 3*(3), 113–125.

Bizon, N. (in press). Chaotification of the buck converter using a modified Chua's diode. *Fuzzy Systems and AI journal - Reports and Letters.*

Bizon, N., & Oproescu, M. (2004). Hysteretic Fuzzy Control of the Power Interface Converter. In Romanian Academy, (Ed.), *Fuzzy Systems and AI journal - Reports and Letters,* (pp. 139-158). Iasi, Romania: Publishing House of the Romanian Academy.

Bizon, N., & Oproescu, M. (2005). Hysteretic Fuzzy Control of the Boost Converter. *Electronics, Computers and Artificial Intelligence: University of Pitesti Journal – ECAI'05, 5*(S2), 1-10.

Bizon, N., & Oproescu, M. (2005). Clocked hysteretic fuzzy control of the boost converter. *Electronics, Computers and Artificial Intelligence - ECAI'05: University of Pitesti Journal – Electronics and Computer Science, 5*(S2), 11-20.

Bizon, N., & Oproescu, M. (2006). Time Equivalent Systems. *University of Pitesti Journal - Electronics and Computer Science, 6*(1), 12-36.

Bizon, N., & Oproescu, M. (2006). Modeling and Control of the PEMFC Power Interface. In ICSI (Ed.), *International Conference on Progress in Cryogenics and Isotopes Separation* (pp. 155-168). Ramnicu Valcea, Romania: ICSI Press.

Bizon, N., & Oproescu, M. (2006). Control of the DC-DC Converter used into Energy Generation System. In ICSI (Ed.), *International Conference on Progress in Cryogenics and Isotopes Separation* (pp. 173-177). Ramnicu Valcea, Romania: ICSI Press.

Bizon, N., & Oproescu, M. (2007). Energy Generation System Behaviour using a Clocked Fuzzy Peak Current Control. In European Power Electronics Association & University of Aalborg (Eds.), *12th European Conference on Power Electronics and Application - EPE 2007: IEEE Catalog Number 07EX1656C* (pp. 1-8). Denmark: University of Aalborg Press.

Bizon, N., & Oproescu, M. (2007). Feed-Forward Control of the PWM Sine Inverter. In Romanian Academia (Ed.), *European Conference H2_Fuel_Cells_Millennium _Convergence* (pp. 35-45). Bucharest: IPA Publishing House.

Bizon, N., & Oproescu, M. (2007). *Power converters used in Energy Generation Systems (Convertoare de Putere utilizate in Sistemele de Generare a Energiei).* Bucharest, Romania: MatrixROM.

Bizon, N., & Oproescu, M. (in press). Instabilities Analysis of an Energy Generation System with a Fuzzy Hysteretic Control. *Fuzzy Systems and AI journal - Reports and Letters.*

Bizon, N., & Voukalis, D. C. (2007). *Fundamentals of Electromagnetic Compatibility: Theory and Practice.* Bucharest, Romania: MatrixROM Press.

Bizon, N., Laurentiu, I., Mazare, A., & Oproescu, M. (2007). Analyze of the Feed-Forward Control for a Pure Sine Inverter. In N. Bizon (Ed.), *International Conference on Electronics, Computers and Artificial Intelligence - ECAI'07* (Vol. 2, pp. 71-79). Bucharest: MatrixROM & University of Pitesti Press.

Bizon, N., Lefter, E., & Oproescu, M. (2007). Modeling and Control of the Energy Sources Power Interface for Automotive Hybrid Electrical System. In Society of Automotive Engineers – SAE of Serbia (Ed.) under FISITA patronage, *21st JUMV International Automotive Conference on Science and Motor Vehicles 2007* (pp. 46-50). Belgrade, Serbia: SAE Press.

Bizon, N., Mazare, A., Laurentiu, I., Oproescu, M., & Raducu, M. (2007). Fuzzy Control of the DC-DC Converter used as Power Interface for a Fuel Cell. In ICSI (Ed.), *International Conference on Progress in Cryogenics and Isotopes Separation* (pp. 169-172). Ramnicu Valcea, Romania: ICSI Press.

Bizon, N., Oproescu, M., & Raducu, M. (2006). Cycloconverter Operation in the Low-Cost Energy Generation Systems. *University of Pitesti Journal - Electronics and Computer Science, 6*(1), 24-36.

Bizon, N., Oproescu, M., & Raducu, M. (2008), Fuzzy bang-bang control of a switching voltage regulator. In IEEE Computer Society (Ed.), *16th IEEE International Conference on Automation, Quality and Testing, Robotics –AQTR, IEEE Catalog Number CFP08AQT-CDR, Vol. 2* (pp. 192-197). Cluj-Napoca, Romania: IEEE Press.

Bizon, N., Raducu, M., & Oproescu, M. (2008), Fuel cell current ripple minimization using a bi-buck power interface. In EPE-PEMC (Ed.), *13th International Power Electronics and Motion Control Conference, IEEE Catalog Number CFP0834A-CDR* (pp. 621-628). Poznan, Poland: IEEE Press.

Bizon, N., Sofron, E., & Oproescu, M. (2005). Some Aspects of the PEMFC – Battery Interface Simulation in Automotive Applications. *International Conference for Road Vehicles - CAR2005.* Retrieved December 11, 2005, from http://www.fisita.com/publications/papers

Bizon, N., Sofron, E., & Oproescu, M. (2006). An Investigations into the Fast- and Slow – Scale Instabilities of an Energy Generation System with a Fuzzy Hysteretic Control. In H. N. Teodorescu (Ed.) *European Conference on Intelligent Systems and Technologies - ECIT2006; Advances in Intelligent Systems and Technologies* (pp. 19-36). Iasi, Romania: Performantica.

Bizon, N., Sofron, E., & Oproescu, M. (2007). Intelligent Control of the Power Flows on an Energy Generation System. In International Institute of Informatics and Systemics – IIIS (Ed.), *Fifth Multi-Conference on Systemics, Cybernetics and Informatics – EIC2007, Vol 5* (pp. 314-319). Orlando, FL: IIIS Press.

Bizon, N., Sofron, E., Oproescu, M., & Raducu, M. (2007). Multi-stage Inverter Topologies for an Energy Generation Systems, *13th International Symposium on Modeling, Simulation and System's Identification.* Retrieved May 10, 2007, from http://www.simsis.ugal.ro/simsis13

Bizon, N., Zafiu, A., & Oproescu, M. (2005). PEMFC modeling using a genetics algorithm for parameters tuning, In Bibliotheca Publishing House (Ed.), *International Symposium on Electrical Engineering - ISEE 2005* (pp. 163-169). Targoviste, Romania: Valahia University.

Blok, K. (2004). Improving energy efficiency by 5% and more per year. *Journal of Industrial Ecology, 8*(4), 87–99. doi:10.1162/1088198043630478

Bloom, J. A. (1982). Long-range generation planning using decomposition and probabilistic simulation. *IEEE Transactions on Power Apparatus and Systems, 101*(4), 797–802. doi:10.1109/TPAS.1982.317144

Bogalecka, E., & Krzeminski, Z. (2002). Sensorless control of a double-fed machine for wind power generators. *Proc. Eur. Power Electron Conf.-Power Electron., Machines Control*, Dubrovnik and Cavtat, Slovenia.

Bojic, M., & Stojanovic, B. (1998). MILP optimization of a CHP energy system. *Energy Conversion and Management, 37*, 637–642. doi:10.1016/S0196-8904(97)00042-3

Boldea, I. (2003). Control of electric generators: a review. *IEEE 26th annual Industrial Electronics Society Conf.*, 1, (pp. 972-980).

Boldea, I. (2005). *The electric generators handbook.* New York: CRC Press.

Boldea, I. (2006). *Variable speed generators: the electrical handbook.* New York: Taylor and Francis.

Bollen, M. (2002). Definitions of voltage unbalance. *IEEE Power Engineering Review, 22*(11), 49–50. doi:10.1109/MPER.2002.1045567

Bollen, M. H. J. (2000). *Understanding power quality problems: voltage sags and interruptions.* New York: IEEE Press.

Bos, M. F. J., Beune, R. J. L., & van Amerongen, R. A. M. (1996). On the incorporation of a heat storage device in Lagrangian Relaxation based algorithms for unit commitment. *Electrical Power & Energy Systems, 18*(4), 205–214. doi:10.1016/0142-0615(95)00059-3

Boyd, G., Dutrow, E., & Tunnessen, W. (2008). The evolution of the ENERGY STAR® energy performance indicator for benchmarking industrial plant manufacturing energy use. *Journal of Cleaner Production, 16*(6), 709–715. doi:10.1016/j.jclepro.2007.02.024

Bozhko, S. V., Gimnez, R. B., Li, R., Clare, J. C., & Asher, G. M. (2007). Control of offshore DFIG-based wind farm grid with line-commutated HVDC connection. *IEEE Transactions on Energy Conversion, 22*(1), 71–78. doi:10.1109/TEC.2006.889544

Brady, F. J. (1984). A mathematical model for the doubly fed wound rotor generator - Part I. *IEEE Transactions on Power Apparatus and Systems, 103*(4), 798–802. doi:10.1109/TPAS.1984.318356

Brady, F. J. (1986). A mathematical model for the doubly fed wound rotor generator - Part II. *IEEE Transactions on Energy Conversion, 1*(2), 180–183. doi:10.1109/TEC.1986.4765718

Brekken, T., & Mohan, N. (2003). A novel doubly-fed induction wind generator control scheme for reactive power control and torque pulsation compensation under unbalanced grid voltage conditions. *Proc. IEEE 34th Annual Conf. on Power Electronics Specialist*, 2, June, (pp. 760 – 764).

Bretthauer, G., Handschin, E., & Hoffmann, W. (1992). Expert systems application to power systems – state-of-the-art and future trends. *IFAC Symposium on Control of Power Plants and Power Systems*, Munich, Germany, (pp. 463-468).

British Broadcasting Corporation. (2006). *Nuclear power plants get go-ahead.* Retrieved September 3, 2008, from http://news.bbc.co.uk/1/hi/uk_politics/5166426.stm

British Petroleum (BP). (2006) *Statistical Review.*

Brown, J. S., & Duguid, P. (2000). Balancing act: How to capture knowledge without killing it. *Harvard Business Review, 78*(3), 73–80.

Brunner, T., Nejdl, W., Schwarzjirg, H., & Sturm, M. (1993). On-line expert system for power system diagnosis and restoration. *Intelligent Systems Engineering, 2*(1), 5–24.

Buchholz, T. S., Volk, T. A., & Luzadis, V. A. (2007). A participatory systems approach to modeling social, economic, and ecological components of bioenergy. *Energy Policy, 35*(12), 6084–6094. doi:10.1016/j.enpol.2007.08.020

Buja, G. S., & Kazmierkowski, M. P. (2004). Direct torque control of PWM Inverter-fed ac motors – a survey'. *IEEE Transactions on Industrial Electronics, 51*(4), 744–757. doi:10.1109/TIE.2004.831717

Burer, M., Tanaka, K., Favrat, D., & Yamada, K. (2003). Multi-criteria optimization of a district cogeneration plant integrating a solid oxide fuel cell-gas turbine combined cycle, heat pumps and chiller. *Energy, 28*, 497–518. doi:10.1016/S0360-5442(02)00161-5

Burton, T., Sharpe, D., Jenkins, N., & Bossanyi, E. (2001). *Wind energy handbook*. Chichester, UK: John Wiley.

Cadirci, I., & Ermis, M. (1992). Double output induction generator operating at subsynchronous and supersynchronous speeds: steady state performance optimization and wind-energy recovery. *IEE Proceedings. Electric Power Applications, 139*(5), 429–442.

Cadirci, I., & Ermis, M. (1994). Commutation angle analysis of a double output induction generator operating in sub- and super- synchronous modes. *Proc. 7th IEEE Mediterranean Electrotechnical Conf., 2*, 793-796.

Calderan, R., Spiga, M., & Vestrucci, P. (1992). Energy modelling of a cogeneration system for a food industry. *Energy, 17*, 609–616. doi:10.1016/0360-5442(92)90096-I

Capros, P. (n.d.). *The PRIMES Energy System Model. Summary Description*. National Technical University of Athens - European Commission Joule-III Programme. Retrieved June, 25, 2008, from http://www.e3mlab.ntua.gr/manuals/PRIMsd.pdf.

Capros, P., Mantzos, L., Vouyoukas, L., & Petrellis, D. (1999). European Energy and CO2 Emissions Trends to 2020: PRIMES model v.2. *Bulletin of Science, Technology & Society, 19*(6), 474–492. doi:10.1177/027046769901900604

Caratozzolo, P., Fossas, E., Pedra, J., & Riera, J. (2000 October). Dynamic modeling of an isolated system with DFIG. *Proc. 8th IEEE Int. Power Electronics Congress Conf.*, (pp. 287 – 292).

Cardenas, R., Pena, R., Proboste, J., Asher, G., & Clare, J. (2005). MRAS observer for sensorless control of standalone doubly fed induction generators. *IEEE Transactions on Energy Conversion, 20*(4), 710–718. doi:10.1109/TEC.2005.847965

Cardona, E., & Piacentino, A. (2007). Optimal design of CHCP plants in the civil sector by thermoeconomics. *Applied Energy, 84*(7-8), 729–748. doi:10.1016/j.apenergy.2007.01.005

Cardona, E., Sannino, P., Piacentino, A., & Cardona, F. (2006). Energy saving in airports by trigeneration. Part II: Short and long term planning for the Malpensa 2000 CHCP plant. *Applied Thermal Engineering, 26*, 1437–1447. doi:10.1016/j.applthermaleng.2006.01.020

Carlson, R. (2000). The correct method of calculating energy savings to justify adjustable-frequency drives on pumps. *IEEE Transactions on Industry Applications, 36*(6), 1725–1733. doi:10.1109/28.887227

Carraretto, C., & Lazzaretto, A. (2004). A dynamic approach for the optimal electricity dispatch in the deregulated market. *Energy, 29*, 2273–2287. doi:10.1016/j.energy.2004.03.025

Carter, C. (2003). *The power of listening*. Presentation material. Retrieved September 15, 2008, from www.etsu.edu/edc/EDC%20Training%20Handouts/The%20Power%20of%20Listeningpowerpoint2.ppt

Casella, F., Maffezzoni, C., Piroddi, L., & Pretolani, F. (2001). Minimising production costs in generation and cogeneration plants. *Control Engineering Practice, 9*, 283–295. doi:10.1016/S0967-0661(01)00002-8

Cathey, J. J. (2001) Electric machines: analysis and design applying Matlab®, *McGraw Hill*, Singapore.

Cavallaro, F., & Ciraolo, L. (2005). A multicriteria approach to evaluate wind energy plants on an Italian island. *Energy Policy, 33*(2), 235–244. doi:10.1016/S0301-4215(03)00228-3

Caves, D. W., Herriges, J. A., Train, K. E., & Windle, R. J. (1987). A bayesian approach to combining conditional demand and engineering models of electric usage. *The*

Review of Economics and Statistics, 69(3), 438–448. doi:10.2307/1925531

CEC Commission of the European Communities. (1997). *Communication from the Commission to the Council, the European Parliament, the Economic and Social Committee and the Committee of the Regions—A Community strategy to promote combined heat and power (CHP) and to dismantle barriers to its development,* COM (1997) 514 final, Brussels.

Celli, G., Pilo, F., Pisano, G., & Soma, G. G. (2005). Optimal Participation of a Microgrid to the Energy Market with an Intelligent EMS. *International Power Engineering Conference* (pp. 663-668).

Cerri, G., Borghetti, S., & Salvini, C. (2006). Neural management for heat and power cogeneration plants. *Engineering Applications of Artificial Intelligence, 19*, 721–730. doi:10.1016/j.engappai.2006.05.013

Chadwick, B. P. (2006). Transaction costs and the clean development mechanism. *Natural Resources Forum, 30*, 256–271. doi:10.1111/j.1477-8947.2006.00126.x

Chan, J. K. W. (2006). *From Knowledge to Wisdom: Practical KM Implementation Experience in CLP Power.* Advances in Power System Control, Operation and Management.

Chan, J. K. W. (2007). *Scrutinizing the 4 Key Dimensions in CLP Power's Knowledge Management - The Knowledge Space, Process Space, Technology Space and Human Touch.* International Conference on Engineering Education, 2007. More information and proceedings available from http://icee2007.dei.uc.pt/proceedings/

Chan, J. K. W., & Leung, V. S. Y. (2006). *Integrating KM with Organization Learning - A Case Study on CLP Power.* Knowledge Management Asia Pacific Conference 2006.

Chandler, A. (1962). *A Strategy and Structure: Chapters in the History of the American Industrial Enterprise.* New York: Massachusetts Institute of Technology, New York.

Chang, C. S., & Chung, T. S. (1990). Expert system for on-line security – Economic load allocation on distribution systems. *IEEE Transactions on Power Delivery, 5*(1), 467–473. doi:10.1109/61.107314

Chang, C. S., & Fu, W. (1998). Stochastic multiobjective generation dispatch of combined heat and power systems. *IEE Proc. Generation . Transmission and Distribution, 145*, 583–591. doi:10.1049/ip-gtd:19981997

Chang, C., Wong, K., & Fan, B. (1995). Security-constrained multiobjective generation dispatch using bicriterion global optimization. *IEE Proceedings. Generation, Transmission and Distribution, 142*, 406–414. doi:10.1049/ip-gtd:19951806

Charniak, E., & McDermott, D. (1985). *Introduction to artificial intelligence.* Reading, MA: Addison-Wesley.

Chen, B. K., & Hong, C. C. (1996). Optimum Operation for a back-Pressure cogeneration system under time-of-use rates. *IEEE Transactions on Power Systems, 11*, 1074–1082. doi:10.1109/59.496197

Chen, C. S., Tzeng, Y. M., & Hwang, J. C. (1996). The application of artificial neural networks to substation load forecasting. *Electric Power Systems Research, 38*(2), 153–160. doi:10.1016/S0378-7796(96)01077-2

Chen, S., Tsay, M., & Gow, H. (2005). Scheduling of cogeneration plants considering electricity wheeling using enhanced immune algorithm. *Electrical Power & Energy System, 27*, 31–38. doi:10.1016/j.ijepes.2004.07.008

Chen, W. (2005). The costs of mitigating carbon emissions in China: findings from China MARKAL-MACRO modeling. *Energy Policy, 33*(7), 885–896. doi:10.1016/j.enpol.2003.10.012

Chen, Z. (2000). *Computational intelligence for decision support.* Boca Raton, FL: CRC Press.

Chen, Z., & Blaabjerb, F. (2006). Wind energy – the worlds fastest growing energy source. *IEEE Power Electronics Society Newsletter*, (pp. 15 – 19).

Cherni, J. A., Dyner, I., Henao, F., Jaramillo, P., Smith, R., & Font, R. O. (2007). Energy supply for sustainable rural livelihoods. A multi-criteria decision-support system. *Energy Policy, 35*(3), 1493–1504. doi:10.1016/j.enpol.2006.03.026

Chevalier, M., (2005). Security of energy supply for the European Union. *International Journal of European Sustainable Energy Market.*

Chicco, G., & Mancarella, P. (2006). From cogeneration to trigeneration: Profitable alternatives in a competitive market. *IEEE Transactions on Energy Conversion, 21,* 265–272. doi:10.1109/TEC.2005.858089

Chicco, G., & Mancarella, P. (2007). (article in press). Distributed multi-generation: A comprehensive view. *Renewable & Sustainable Energy Reviews.* doi:10.1016/j.reser.2007.11.014

Chicco, G., & Mancarella, P. (2008). (article in press). Matrix modelling of small-scale trigeneration systems and application to operational optimization. *Energy.* doi:. doi:10.1016/j.energy.2008.09.011

Choi Y. H. & Lau, T. W. K. (2003). *Knowledge Management in Transmission Networks and Equipment.* Advances in Power System Control, Operation and Management, 2003. (Conf. Publ. No. 497).

Chokri, B., Rabei, M., & Dai, D. X. (1996). *An integrated power system global controller using expert system.* Calgary, Canada: Canadian Conference on Electrical and Computer Engineering.

Chowdhury, B. H., & Chellapilla, S. (2006). Double-fed induction generator control for variable speed wind power generation. *Electric Power Systems Research, 76*(9-10), 786–800. doi:10.1016/j.epsr.2005.10.013

Christie, R. D., Talukdar, S. N., & Nixon, J. C. (1990). A hybrid expert system for security assessment. *IEEE Transactions on Power Systems, 5*(4), 1503–1509. doi:10.1109/59.99405

Clark, G., & Mehta, P. (1997). Artificial intelligence and networking in integrated building management systems. *Automation in Construction, 6,* 481–498. doi:10.1016/S0926-5805(97)00026-5

Clarke, J. A., Cockroft, J., Conner, S., Hand, J. W., Kelly, N. J., & Moore, R. (2002). Simulation - assisted control in building energy management systems. *Energy and Building, 34,* 933–940. doi:10.1016/S0378-7788(02)00068-3

CLP. (2008). *China Light and Power website.* Retrieved September 15, 2008, from https://www.clpgroup.com/Abt/Overview/Pages/default.aspx

Coello, C. A. C. (1999). A comprehensive survey of evolutionary-based multiobjective optimization techniques. *Knowledge and Information Systems, 1*(3), 269–308.

Cohen, D. A., & Krarti, M. (1995). A neural network modeling approach applied to energy conservation retrofits. *Proceedings of the Fourth International Conference on Building Simulation,* (pp. 423 – 430).

Colin, R. R., & Jonathan, E. R. (2002). *Genetic algorithms- Principles and perspectives, A guide to GA Theory.* Dordrecht, the Netherlands: Kluwer Academic Publishers.

Collison, C., & Parcell, G. (2001). *Learning to Fly: Practical Lessons from one of the World's Leading Knowledge Management Companies.* Oxford, UK: Capstone. CommmonKADS. *CommonKADS website.* Retrieved November 6, 2008, from http://www.commonkads.uva.nl/

Commission of the European Communities. (2007). *An Energy Policy for Europe COM(2007)1.* Retrieved April, 9, 2008, from http://ec.europa.eu/energy/energy_policy/doc/01_energy_policy_for_europe_en.pdf.

COMMunity for ENergy environment & Development - COMMEND (n.d.). Retrieved August, 18, 2008, from http://www.energycommunity.org/default.asp?action=71.

Community Knowledge. *Introduction to Knowledge Management.* Retrieved September 15, 2008, from http://communityknowledge.co.uk/KMIntro/part_e.html

Consultants, L. U. (n.d.). Retrieved July 25, 2008, from http://www.landuse.co.uk/Bristol/Downloads/NorthYorkshire/Explanatory_Note.pdf.

Corbo, P., Corcione, F. E., Migliardini, F., & Veneri, O. (2006). Energy management in fuel cell power trains. *En-*

ergy Conversion and Management, 47(18-19), 3255–3271. doi:10.1016/j.enconman.2006.02.025

Cormio, C., Dicorato, M., Minoia, A., & Trovato, M. (2003). A regional energy planning methodology including renewable energy sources and environmental constraints. *Renewable & Sustainable Energy Reviews*, *7*, 99–130. doi:10.1016/S1364-0321(03)00004-2

Cosmi, C., Cuomo, V., Macchiato, M., Mangiamele, L., & Salvia, M. (2000). Chapter 5.1: Basilicata Regional Environmental Plan. In R. Jank (Ed.), *Annex 33 Advanced Local Energy Planning (ALEP) – A Guidebook* (pp. 171-187). Jülich, Germany: Forschungszentrum Jülich, BEO on behalf of the International Energy Agency, Energy Conservation in Buildings and Community Systems Programme.

Cosmi, C., Di Leo, S., Loperte, S., Macchiato, M., Pietrapertosa, F., Salvia, M., & Cuomo, V. (2008). *A model for representing the Italian energy system: The NEEDS-TIMES experience*.

Costantini, V., Gracceva, F., Markandya, A., & Vicini, G. (2007). Security of energy supply. Comparing scenarios from European perspective. *Energy Policy*, *35*, 210–226. doi:10.1016/j.enpol.2005.11.002

Coulter, M. (2002). *Strategic Management in Action*. New York: NJ, PPrentice Hall.

Crompton, T. R. (2003). *Battery reference book*. Oxford, UK: Elsevier Science - Newnes.

Cuomo, V., Salvia, M., & Schlenzig, C. (2000). Appendix A.1: Modelling the Energy System. In R. Jank (Ed.), *Annex 33 Advanced Local Energy Planning (ALEP) – A Guidebook* (pp. 171-187). Jülich, Germany: Forschungszentrum Jülich, BEO on behalf of the International Energy Agency, Energy Conservation in Buildings and Community Systems Programme.

Curti, V., von Spakovsky, M. R., & Fvarat, D. (2002). An environomic approach for the modeling and optimization of a district heating network based on centralized and decentralized heat pumps, cogeneration and / or gas furnace, Part I: Methodology. *International Journal of Thermal Sciences*, *39*, 721–730. doi:10.1016/S1290-0729(00)00226-X

Curti, V., von Spakovsky, M. R., & Fvarat, D. (2002). An environomic approach for the modeling and optimization of a district heating network based on centralized and decentralized heat pumps, cogeneration and / or gas furnace, Part II: Application. *International Journal of Thermal Sciences*, *39*, 731–741. doi:10.1016/S1290-0729(00)00225-8

Curtis, M., & Khare, A. (2004). Energy conservation in electric utilities: an opportunity for restorative economics at SaskPower. *Technovation*, *24*, 395–402. doi:10.1016/S0166-4972(02)00116-5

Curtiss, P. S., Brandemuehl, M. J., & Kreider, J. F. (1995). Energy Management in Central HVAC Plants using Neural Networks. In J.S. Haberl, R.M. Nelson & C.C. Culp, (Eds.). *The use of Artificial Intelligence in Building Systems, ASHRAE*, (pp.199-216).

D'Sa, A. (2005). Integrated resource planning (IRP) and power sector reform in developing countries. *Energy Policy*, *33*(10), 1271–1285. doi:10.1016/j.enpol.2003.12.003

Dantzig, G. (1963). *Linear programming and extensions*. Princeton, NJ: Princeton University Press.

Das, D. B., & Patvardhan, C. (1998). New multi-objective stochastic search technique for economic load dispatch. *IEE Proceedings. Generation, Transmission and Distribution*, *145*, 747–752. doi:10.1049/ip-gtd:19982367

Data mining. Retrieved November 6, 2008, from http://en.wikipedia.org/wiki/Data_mining

Datta, R., & Ranganathan, V. T. (2001). A simple position sensorless algorithm for rotor side field oriented control of wound rotor induction machine. *IEEE Transactions on Industrial Electronics*, *48*(4), 786–793. doi:10.1109/41.937411

Datta, R., & Ranganathan, V. T. (2002). Variable-speed wind power generation using doubly fed wound rotor induction machine – A comparison with alternative schemes. *IEEE Transactions on Energy Conversion*, *17*(3), 414–421. doi:10.1109/TEC.2002.801993

David, A. K., & Zhao, R. (1989). Integrating expert systems with dynamic programming in generation expansion planning. *IEEE Transactions on Power Systems, 4*(3), 1095–1101. doi:10.1109/59.32604

DeBattista, H., & Mantz, R. J. (1998, September). Sliding mode control of torque ripple in wind energy conversion systems with slip power recovery. *Proc. 24th Annual Conf. IEEE Industrial Electronics Society, 2*, 651 – 656.

De Battista, H., Puleston, P. F., Mantz, R. J., & Christiansen, C. F. (2000). Sliding mode control of wind energy systems with DOIG-power efficiency and torsional dynamics optimization. *IEEE Transactions on Power Systems, 15*(2), 728–734. doi:10.1109/59.867166

De Brún, C. (2005). *Knowledge audit: conducting a knowledge audit.* Retrieved September 15, 2008, from: http://www.library.nhs.uk/KnowledgeManagement/ViewResource.aspx?resID=93807

De Paepe, M., & Mertens, D. (2007). Combined heat and power in a liberalised energy market. *Energy Conversion and Management, 48*, 2542–2555. doi:10.1016/j.enconman.2007.03.019

De, S., Kaiadi, M., Fast, M., & Assadi, M. (2007). Development of an artificial neural network model for the steam process of a coal biomass cofired combined heat and power (CHP) plant in Sweden. *Energy, 32*, 2099–2109. doi:10.1016/j.energy.2007.04.008

Dean, T., Allen, J., & Aloimonds, Y. (1995). Artificial intelligence: Theory and practice. Addison-Wesley, Reading, MA.

DECADES. Inter-agency Joint Report on Databases and Methodologies for Comparative Assessment of Different Energy Sources for Electricity Generation (1993). *Computerized Tools for Comparative Assessment of Electricity Generation Options and Strategies.* An inter-agency project of CEC, ESCAP, IAEA, IBRD, IIASA, OECD, OPEC, UNIDO, WMH, DECADES-Project, Working Paper No. 5.

DeJong, K. A. (1975). An Analysis of the behavior of a class of genetic adaptive systems. Ph.D Dissertation, University of Michigan.

DeLong, S. O., & Speegle, C. T. (2003). Applying the systems methodology in the design of the West Point cogeneration power plant. *Proceedings of the 2003 IEEE Systems and Information Engineering Design Symposium.*

den Ouden, A. C. B. (2000). PRODESign: a program for optimisation and dimensioning of energy conversion systems. In GG Hirs (ed). *Proceedings of ECOS 2000: The International Conference on Efficiency, Cost, Optimisation, Simulation and Environmental Aspects of Energy and Process Systems.* 05–07 July 2000, University of Twente, Enschede, The Netherlands. Part 3, p1691–1703. [ECN report number: ECN-RX--00-024].

Deyi, L., & Yi, D. (2007). Artificial intelligence with uncertainty. Chapman & Hall/CRC, First edition.

Dhillon, J. S., Parti, S. C., & Kothari, D. P. (1993). Stochastic economic emission load dispatch. *Electric Power Systems Research, 26*, 197–186. doi:10.1016/0378-7796(93)90011-3

Diakoulaki, D., & Karangelis, F. (2007). Multi-criteria decision analysis and cost–benefit analysis of alternative scenarios for the power generation sector in Greece. *Renewable & Sustainable Energy Reviews, 11*(4), 716–727. doi:10.1016/j.rser.2005.06.007

Diakoulaki, D., Georgiou, P., Tourkolias, C., Georgopoulou, E., Lalas, D., Mirasgedis, S., & Sarafidis, Y. (2007). A multicriteria approach to identify investment opportunities for the exploitation of the clean development mechanism. *Energy Policy, 35*(2), 1088–1099. doi:10.1016/j.enpol.2006.02.009

Diakoulaki, D., Kormentza, Y., & Hontou, V. (2000). *An MCDA approach to burden-sharing among industrial branches for combating climate change.* Paper Presented at the 51st Meeting of the European Working Group on MCDA, Madrid.

Dinca, C., Badea, A., Rousseaux, P., & Apostol, T. (2007). A multi-criteria approach to evaluate the natural gas energy systems. *Energy Policy, 35*, 5754–5765. doi:10.1016/j.enpol.2007.06.024

Dodier, R., & Henze, G. (1996). Statistical analysis of neural network as applied to building energy prediction. *Proceedings of the ASME ISEC*, San Antonio, TX, (pp. 495 – 506).

Dotzauer, E. (2001). *Energy system operation by Lagrangian Relaxation*. Ph.D. Thesis, Department of Mathematics, Linköping University, Sweden.

Dotzauer, E. (2003). Experiences in mid-term planning of district heating system. *Energy, 28*, 1545–1555. doi:10.1016/S0360-5442(03)00151-8

Dotzauer, E., & Ravn, H. F. (2000). Lagrangian relaxation based algorithms for optimal economic dispatch of cogeneration systems with a storage. In *proceedings from the second International Conference on Simulation, Gaming, Training and Business Process Reengineering in Operations* (pp. 51-55). Riga, Latvia.

Douglas, H., Pillay, P., & Barendse, P. (2005, October). The detection of interturn stator faults in doubly-fed induction generators. *Proc. 40th Industry Applications Conf.*, 2, 1097 – 1102.

Doukas, H., Karakosta, C., & Psarras, J. (in press). RES technology transfer within the new climate regime: A "helicopter" view under the CDM. *Renewable & Sustainable Energy Reviews*. doi:.doi:10.1016/j.rser.2008.05.002

Doukas, H., Nychtis, C., & Psarras, J. (2008). (in press). Assessing energy-saving measures in buildings through an intelligent decision support model. [Corrected Proof.]. *Building and Environment.*

Doukas, H., Patlitzianas, K. D., & Psarras, J. (2006). Supporting the Sustainable Electricity Technologies in Greece Using MCDM. *Resources Policy, 31*(2), 129–136. doi:10.1016/j.resourpol.2006.09.003

Doukas, H., Patlitzianas, K. D., Iatropoulos, K., & Psarras, J. (2006). Intelligent Building Energy Management System Using Rule Sets. *Building and Environment, 42*(10), 3562–3569. doi:10.1016/j.buildenv.2006.10.024

Doukas, H., Patlitzianas, K. D., Kagiannas, A. G., & Psarras, J. (2008). Energy Policy Making: An Old Concept or a Modern Challenge? *Energy Sources (Part B).*

Drid, S., Said, M., Sait, N., Makouf, A., & Tadjine, M. (2006). Doubly fed induction generator modelling and scalar controlled for supplying an isolated site. *J. Electrical Systems, 2*(2), 103–115.

Druckman, A., & Jackson, T. (2008). Household energy consumption in the UK: A highly geographically and socio-economically disaggregated model. *Energy Policy, 36*(8), 3177–3192. doi:10.1016/j.enpol.2008.03.021

Dubois, D., & Prade, H. (1980.) Fuzzy Sets Systems: Theory and applications. Academic Press, Orlando, FL.

Dufour, C., & Belanger, J. (2004). Real-time simulation of doubly fed induction generator for wind turbine applications. *IEEE 35th Annual Power Electronics Specialists Conf.*, (pp. 3597-3603).

Dugan, R. C., McGranaghan, M. F., & Beaty, H. W. (1996). *Electrical power systems quality*. New York: McGraw-Hill.

Dyner, I., & Larsen, E. R. (2001). From planning to strategy in the electricity industry. *Energy Policy, 29*, 1145–1154. doi:10.1016/S0301-4215(01)00040-4

Earth Simulator Center. Japan Agency for Marine-Earth Science and Technology, (2008). Retrieved from http://www.jamstec.go.jp/esc/index.en.html

Ecofys (2003). *Rat für Nachhaltige Entwicklung.* Berlin: German Council for Sustainable Development.

Ecoinvent Centre. (n.d.). Retrieved September, 1, 2008, from http://www.ecoinvent.org/.

Eco-smes - Services for green products. (n.d.). Retrieved September, 5, 2008, from http://ex-elca2.bologna.enea.it/cm/navContents?l=IT&navID=info&subNavID=1&pagID=7

EIA. (2008). *State and US Historical Data Overview.* Retrieved July 26, 2008, from http://www.eia.doe.gov/overview_hd.html

Eichhammer, W. (1999). Energy Efficiency technologies: the IKARUS Database. In *New Equilibria in the Energy Markets: The Role of new Regions and Areas* (pp.379-389). Cleveland, OH: International Association for Energy Economics -IAEE, Fraunhofer-ISI.

Eickhoff, F., Handschin, E., & Hoffmann, W. (1992). Knowledge based alarm handling and fault location in distribution networks. *IEEE Transactions on Power Systems, 7*(2), 770–776. doi:10.1109/59.141784

Ekanayake, J. B., Holdsworth, L., & Jenkins, N. (2003). Comparison of 5th order and 3rd order machine models for doubly fed induction generator (DFIG) wind turbines. *Electric Power Systems Research, 67*(3), 207–215. doi:10.1016/S0378-7796(03)00109-3

Ekanayake, J. B., Holdsworth, L., Guang, W. X., & Jenkins, N. (2003). Dynamic modeling of doubly fed induction generator wind turbines. *IEEE Transactions on Power Systems, 18*(2), 803–809. doi:10.1109/TPWRS.2003.811178

El Ferik, S., & Belhadj, C. A. (2004). Neural network modeling of temperature and humidity effects on residential air conditioner load. *Proceedings of the Fourth International Conference on Power and Energy Systems*, # 442-152, 557-562.

Electric Power Research Institute – EPRI. (1998). *Inventory of available methods and processes for assessing the benefits, costs, and impacts of demand-side options, Volume 3: Description and review of computer tools for integrated planning*, (TR-108506-V3, Final report). G. Heffner, EPRI California. Retrieved June, 8, 2008, from www.epriweb.com/public/TR-108506-V3.pdf.

El-Ibiary, Y. (2003). An Accurate low cost method for determining electric motors efficiency for the purpose of plant energy management. *IEEE Transactions on Industry Applications, 39*(4), 1205–1210. doi:10.1109/TIA.2003.813686

Elkarmi, F. (2008). Load Research as a Tool in Electric Power System Planning, Operation and control: the Case of Jordan. *Energy Policy, 36*(5), 1757–1763. doi:10.1016/j.enpol.2008.01.033

Elkarmi, F., & Abu Shikheh, N. (2007, September 3-5). *Power System Analysis and Studies: Cooperation between Academia and Electricity Companies.* Paper presented at CIGRE Conference, Amman, Jordan.

El-Keib, A. A., Ma, H., & Hart, J. L. (1994). Economic dispatch in view of the clean air act of 1990. *IEEE Transactions on Power Systems, 9*, 972–978. doi:10.1109/59.317648

Elman, J. L. (1990). Finding structure in time. *Cognitive Science, 14*, 179–211.

Elshatter, T. F., Elhagree, M. T., Aboueldahab, M. E., & Elkousry, A. A. (1997). Fuzzy modeling and simulation of photovoltaic system. In 14th European photovoltaic Solar Energy Conference on CD ROM.

El-Telbany, M., & Elkarmi, F. (2008). Short-Term Forecasting of Jordanian Demand Using Particle Swarm Optimization. *Electric Power Systems Research Journal, 78*(3), 425–433. doi:10.1016/j.epsr.2007.03.011

Energy Technology Systems Analysis Programme - ETSAP (n.d.). Retrieved August, 25, 2008, from http://www.etsap.org.

Engelberger, J. F. (1980). Robotics in practice. Kogan Page, London.

Ensinas, A. V., Nebra, S. A., Lozano, M. A., & Serra, L. M. (2007). Analysis of process steam demand reduction and electricity generation in sugar and ethanol production from sugarcane. *Energy Conversion and Management, 48*, 2978–2987. doi:10.1016/j.enconman.2007.06.038

Entchev, E., & Yang, L. (2007). Application of adaptive neuro-fuzzy inference system techniques and artificial neural networks to predict solid oxide fuel cell performance in residential microgeneration installation. *Journal of Power Sources, 170*(1), 122–129. doi:10.1016/j.jpowsour.2007.04.015

Entchev, E., & Yang, L. (2007). Application of adaptive neuro-fuzzy inference system techniques and artificial neural networks to predict solid oxide fuel cell performance in residential microgeneration installation. *Journal of Power Sources, 170*(1), 122–129. doi:10.1016/j.jpowsour.2007.04.015

Ente per le Nuove Tecnologie, l'Energia e l'Ambiente – ENEA. (2000). *Il processo di attuazione del Protocollo di Kyoto in Italia. Metodi, scenari e valutazione di politiche e misure.* [Carrying out the Kyoto Protocol in Italy. Methods, scenarios and valuation of action policies] *(in Italian).*

Environment Canada. (1999). *The National Climate Data and Information Archive.* Retrieved June 29, 1999, from http://climate.weatheroffice.ec.gc.ca/prods_servs/index_e.html

Eriksen, P. B. (2001). Economic and environmental dispatch of power: CHP production systems. *Electric Power Systems Research, 57,* 33–39. doi:10.1016/S0378-7796(00)00116-4

Eskandarzadeh, I., Uctug, M. Y., & Demirekler, M. (1999). Refrence frame modeling and steady state analysis of double output induction generator. *Proc. Int. Conf. Environmental Management,* Boston, (pp. 1022-1026).

Eskander, M. N., & El Hagry, M. T. (1993). Optimal performance of double output induction generator used in WECS. *The Europeans Power Electronics Association,* (pp. 276-281).

Eskander, M. N., Abd El Motaleb, M. S., & El Khashab, H. A. (1996, March 1). Optimal control of double output induction generator using optimal regulator theory. *Proc. 4th Int. Workshop on Advanced Motion Control (AMC),* (pp. 311-315).

ESRU. (2002). *The ESP-r System for Building Energy Simulation: User Guide Version 10 Series.* Energy Systems Research Unit, University of Strathclyde, Glasgow, Scotland. Retrieved Oct 31, 2008, from http://www.esru.strath.ac.uk/

EU Commission (10 January 2007). *An Energy Policy For Europe, COM(2007) 1 final.*

EUCommission (11 February 2004). Directive 2004/8/EC of the European Parliament and of the Council on the promotion of cogeneration based on a useful heat demand in the internal energy market and amending Directive 92/42/EEC.

European Commision – JRC. (2004-2008). *Minutes of the Core Group Meetings of the Scientific Technical Reference System.* Milan-Brussels-Vienna-Denmark: Author.

European Commission - EC (2005). *Synergy Project "Business Opportunities for CDM Project Development in the Mediterranean."* Final Report (project number 4.1041/D/02-003-507.21086).

European Commission - EC (2007). *Synergy Project - EUROGULF: An EU-GCC Dialogue for Energy Stability and Sustainability.* Final Report (project number: 4.1041/D/02-008-S07 21089).

European Commission – EC, (2004). *MEDA programme: Ad-hoc Groups Energy Policy, "Interconnections and Economic Analysis of the Euro-Mediterranean Energy Forum."* Final Report (ME8/B7-4100/IB/98/0480).

European Commission (19 October 2006). *Action Plan for Energy Efficiency: Realizing the Potential.* COM (2006)545 final.

European Commission (2005). *Green paper "Doing more with less".* Luxemburg: European Communities.

European Commission Directorate—General for Energy and Transport. (2003). *European Energy and Transport Trends to 2030.* Brussels: Author.

European Commission. (2002). *EC Of The European Parliament And Of The Council. Directive 2002/91 on the energy performance of buildings.* EC- Official Journal L 001. Brussels: Author.

European Commission. (2006). *EC Of The European Parliament And Of The Council. Directive 2006/32 EC on energy end-use efficiency and energy services and repealing Council Directive 93/76/EEC.* Official Journal L 114. Brussels: Author.

European Commission. -DG-TREN. (2008). *SRS NET & EEE: Scientific Reference System on New Energy Technologies, Energy End-Use Efficiency and Energy RTD.* Final Report ((Project Ref: 006631) – Final Report. Brussels, Belgium: Author.

European Environment Agency - EEA (2003). *EEA core set of indicators* (Tech. Rep.). Revised version April, 2003.

European Environment Agency - EEA. (2007). *Indicator: EN01 Energy and non energy-related greenhouse gas emissions [2007.04].* Retrieved April, 23, 2008, from http://themes.eea.europa.eu/Sectors_and_activities/energy/indicators/EN01,2007.04

European Environment Agency (EEA). (2004).Annual European Community greenhouse gas inventory 1990-2002 EEA report.

Eurostat (2008). *Energy Balance Sheets. Data 2006.* Luxemburg: Eurostat.

Fahl, U., Rühle, B., & Voss, A. (2004). *Wissenschaftliche Begleitung des Energieprogramm Sachsen, Schlussbericht.* Institute for Energy Economics and the Rationale Use of Energy (IER), University of Stuttgart. Retrieved from www.ier.uni-stuttgart.de/sachsen

Fahlman, S. E. (1988). Faster-learning variations on back-propagation: an empirical study. *Proceedings of the 1988 Connectionist Models Summer School*, (pp.38-51). Los Altos, CA: Morgan-Kaufmann.

Faille, D., Mondon, C., & Al-Nasrawi, B. (2007). mCHP Optimization by Dynamic Programming and Mixed Integer Linear Programming. *2007 International Conference on Intelligent Systems Applications to Power Systems.*

Faiz, J., & Ebrahimpour, H. (2005). Precise derating of three-phase induction motors with unbalanced voltages. In *Proceedings of Fortieth IAS Annual Meeting of Industry Applications Conf.*, Hong Kong, 1, Oct., (pp. 485-491).

Faiz, J., Ebrahimpour, H., & Pillay, P. (2004). Influence of unbalanced voltage on the steady state performance of a three phase squirrel cage induction motor. *IEEE Transactions on Energy Conversion, 19*(4), 657–662. doi:10.1109/TEC.2004.837283

Farag, A. S., Mousab, A. E., Cheng, T. C., & Beshir, M. (1999). Cost effective utilities energy plans optimization and Management. *Energy Conversion and Management, 40*, 527–543. doi:10.1016/S0196-8904(98)00120-4

Farag, A., Al-Baiyat, S., & Cheng, T. C. (1995). Economic load dispatch multiobjective optimization procedures using linear programming techniques. *IEEE Transactions on Power Systems, 10*, 731–738. doi:10.1109/59.387910

Farahbakhsh, H., Ugursal, V. I., & Fung, A. S. (1998). A Residential End-use Energy Consumption Model for Canada. *International Journal of Energy Research, 22*(13), 1133–1143. doi:10.1002/(SICI)1099-114X(19981025)22:13<1133::AID-ER434>3.0.CO;2-E

Farinelli, U., Johansson, T. B., McCormick, K., Mundaca, L., Oikonomou, V., & Örtenvikc, M. (1981). MARKAL, a linear-programming model for energy system analysis: technical description of the BNL version. *International Journal of Energy Research, 5*, 353–375. doi:10.1002/er.4440050406

Fast, M., Assadi, M., & De, S. (2009). Development and multi-utility of an ANN model for an industrial gas turbine. *Applied Energy, 86*(1), 9–17. doi:10.1016/j.apenergy.2008.03.018

Fausett, L. (1994). *Fundamentals of Neural Networks.* Englewood Cliffs, NJ: Prentice Hall.

Feigenbaum, E. A. (1982). Knowledge engineering for the 1980. Department of Computer Science, Stanford University, Stanford, CA.

Feijoo, A., Cidras, J., & Carrillo, C. (2000). A third order model for the doubly-fed induction machine. *Electric Power Systems Research, 56*(2), 121–128. doi:10.1016/S0378-7796(00)00103-6

Fenhann, J. (2008). *CDM Pipeline Overview.* UNEP Risoe Centre; 1st September 2008, Available at http://www.cd4cdm.org.

Fernandez, A., Sebastian, J., Hernando, M. M., & Martin-Ramos, J. A. (2006). Multiple output AC/DC converter with an internal DC UPS. *IEEE Transactions on Industrial Electronics, 53*(1), 296–304. doi:10.1109/TIE.2005.862220

Fernandez, L. M., Jurado, F., & Saenz, J. R. (2008). Aggregated dynamic model for wind farms with doubly fed induction generator wind turbines. *Renewable Energy, 33*(1), 129–140. doi:10.1016/j.renene.2007.01.010

Fernandez, L.M., Garcia, C.A., Jurado, F. & Saenz, J.R. (2005, May). Control system of doubly fed induction generators based wind turbines with production limits. *IEEE Int. Conf. Electric Machines and Drives*, (pp. 1936 – 1941).

Feuston, B. P., & Thurtell, J. H. (1994). Generalized nonlinear regression with ensemble of neural nets: the great energy predictor shootout. *ASHRAE Transactions*, *100*(2), 1075–1080.

Fiebig, D. G., Bartels, R., & Aigner, D. J. (1991). A random coefficient approach to the estimation of residential end-use load profiles. *Journal of Econometrics*, *50*, 297–327. doi:10.1016/0304-4076(91)90023-7

Figueira, J., Greco, S., & Ehrgott, M. (Eds.). (2005). *Multiple criteria decision analysis: state of the art surveys.* Berlin: Springer.

Fishbone, L. G., Giesen, G., & Vos, P. (1983). *User's guide for MARKAL (BNL/KFA version 2.0), IEA Technology System analysis Project. BNL 51701.* Julich, Germany: Brookhaven National Laboratory and Kernforschungsanlange.

Flamos, A., Anagnostopoulos, K., Askounis, D., & Psarras, J. (2004a). e-Serem – A Web-Based Manual For The Estimation of Emission Reductions From JI and CDM Projects. *International Journal Mitigation and Adaptation Strategies for Global Change*, *9*(2), 103–120. doi:10.1023/B:miti.0000017728.80734.fb

Flamos, A., Anagnostopoulos, K., Doukas, H., Goletsis, Y., & Psarras, J. (2004b). Application of the IDEA-AM (Integrated Development and Environmental Additionality - Assessment Methodology) to compare 12 real projects from the Mediterranean region. [ORIJ]. *Operational Research International Journal*, *4*(2), 119–145. doi:10.1007/BF02943606

Flamos, A., Doukas, H., Karakosta, C., & Psarras, J. (2007). *MCDM Approach for Assessing CDM Energy Technologies Towards Sustainable Development.* Ninth International Conference on "Energy for a Clean Environment", Clean Air 2007, Povoa de Varzim, Portugal.

Flamos, A., Doukas, H., Patlitzianas, K., & Psarras, J. (2005). CDM – PAT: A Decision Support Tool for the Pre-Assessment of CDM Projects. [IJCAT]. *International Journal of Computer Applications in Technology*, *22*(2/3). doi:10.1504/IJCAT.2005.006939

Fogel, D. B. (1995). Evolutionary computation: Toward a new philosophy of machine intelligence, IEEE Press, Chapter 3.

Fontes, G., Turpin, C., Astier, S., & Meynard, T. (2007). Interactions between fuel cells and power converters: Influence of current harmonics on a fuel cell stack. *IEEE Transactions on Power Electronics*, *22*(2), 670–678. doi:10.1109/TPEL.2006.890008

Forchetti, D., Garcia, G., & Valla, M. I. (2002, November 2). Vector control strategy for a doubly-fed stand-alone induction generator. *Proc. IEEE 28th Annual Conf. of the Industrial Electronics Society*, (pp. 991 – 995).

Forschungszentrum Jülich - KFA. (1994). IKARUS—Instrumente für Klimagas Reduktionsstrategien, Teilprojekt 4: Umwandlungssektor Strom- und Wärmeerzeugende Anlagen auf fossiler und nuklearer Grundlage. Teil 1 u. 2, Jülich.

Fortescue, C. L. (1918). Method of symmetrical coordinates applied to the solution of polyphase networks. *Trans. AIEE*, *37*(part II), 1027–1140.

Fouad, A. A., Vekataraman, S., & Davis, J. A. (1991). An expert system for security trend analysis of a stability-limited power system. *IEEE Transactions on Power Systems*, *6*(3), 1077–1084. doi:10.1109/59.119249

Fox, D., Ben-Haim, Y., Hayes, K., McCarthy, M., Wintle, B., & Dunstan, P. (2007). An info-gap approach to power and sample size calculations. *Environmetrics*, *18*, 189–203. doi:10.1002/env.811

Frangopoulos, C. A., & Dimopoulos, G. G. (2004). Effect of reliability considerations on the optimal synthesis, design and operation of a cogeneration system. *Energy*, *29*, 309–329. doi:10.1016/S0360-5442(02)00031-2

Gandibleux, X. (1998). Interactive multicriteria procedure exploiting a knowledge-based module to select electricity production alternatives: The CASTART system. *European Journal of Operational Research, 113*(2), 355–373. doi:10.1016/S0377-2217(98)00221-5

Gansky, K. (2002). *Rechargeable batteries applications handbook*. Oxford, UK: Elsevier Science - Butterworth Heinemann.

Gao, L., Jin, H., Liu, Z., & Zheng, D. (2004). Exergy analysis of coal-based polygeneration system for power and chemical production. *Energy, 29*, 2359–2371. doi:10.1016/j.energy.2004.03.046

Gardner, D. T., & Rogers, J. S. (1997). Joint planning of combined heat and power and electric power systems: An efficient model formulation. *European Journal of Operational Research, 102*(1), 58–72. doi:10.1016/S0377-2217(96)00221-4

Geidl, M., & Andersson, G. (2007). Optimal Power Flow of Multiple Energy Carriers. *IEEE Transactions on Power Systems, 22*(1), 145–155. doi:10.1109/TPWRS.2006.888988

Geldermann, J., Spengler, T., & Rentz, O. (2000). Fuzzy outranking for environmental assessment. Case study: iron and steel making industry. *Fuzzy Sets and Systems, 115*(1), 45–65. doi:10.1016/S0165-0114(99)00021-4

Gemmen, R. S. (2003). Analysis for the effect of the ripple current on fuel cell operating condition. *Journal of Fluids Engineering, 125*(3), 576–585. doi:10.1115/1.1567307

General Algebraic Modelling System - GAMS (n.d.). Retrieved August, 25, 2008, from http://www.gams.com.

George, F. L., & William, A. S. (1998). Artificial intelligence structures and strategies for complex problem solving. George Addison Wesley Longman, INC.

Georgiou, P., Tourkolias, C., & Diakoulaki, D. (2008). A roadmap for selecting host countries of wind energy projects in the framework of the clean development mechanism. *Renewable & Sustainable Energy Reviews, 12*(3), 712–731. doi:10.1016/j.rser.2006.11.001

Georgopoulou, E., Lalas, D., & Papagiannakis, L. (1997). A multicriteria decision aid approach for energy planning problems: The case of renewable energy option. *European Journal of Operational Research, 103*, 38–54. doi:10.1016/S0377-2217(96)00263-9

Georgopoulou, E., Sarafidis, Y., & Diakoulaki, D. (1998). Design and implementation of a Group DSS for sustaining renewable energies exploitation. *European Journal of Operational Research, 109*(2), 483–500. doi:10.1016/S0377-2217(98)00072-1

Georgopoulou, E., Sararidis, Y., Mirasgedis, S., Zaimi, S., & Lalas, D. P. (2003). A multiple criteria decision-aid approach in defining national priorities for greenhouse gases emissions reduction in the energy sector. *European Journal of Operational Research, 146*(1), 199–215. doi:10.1016/S0377-2217(02)00250-3

Ghijselen, J., & Van den Bossche, A. (2005). Exact voltage unbalance assessment without phase measurements. *IEEE Transactions on Power Systems, 20*(1), 519–520. doi:10.1109/TPWRS.2004.841145

Ghosn, R., Asmar, C., Pietrzak-David, M. & De Fornel, B. (2002, Aug. 26–28). A MRAS-sensorless speed control of doubly fed induction machine, *Proc. Int. Conf. Electrical Machines*, Bruges, Belgium.

Ghosn, R., Asmar, C., Pietrzak-David, M., & De Fornel, B. (2003). A MRAS Luenberger sensorless speed control of doubly fed induction machine. *Proc. Eur. Power Electron. Conf.*, Toulose, France.

Ghoudjehbaklou, H., & Puttgen, H. B. (1988). Optimum electric utility spot price determinations for small power producing facilities operating under PURPA provisions. *IEEE Transactions on Energy Conversion, 3*(3), 575–582. doi:10.1109/60.8070

Giannantoni, C., Lazzaretto, A., Macor, A., Mirandola, A., Stoppato, A., Tonon, S., & Ulgiati, S. (2005). Multicriteria approach for the improvement of energy systems design. *Energy, 30*(10), 1989–2016. doi:10.1016/j.energy.2004.11.003

Giles, C.L., Tsungnan, L., Bill, G. (1997). Horne remembering the past: The role of embedded memory in recurrent neural network architectures. IEEE Proceedings, 34-43.

Global Wind Energy Council. (2006). Latest News: Global wind energy markets continue to boom – 2006 another record year. Retrieved May 14, 2007 from http://www.gwec.net/index.php?id=30&no_cache=1&tx_ttnews[pointer]=1&tx_ttnews[tt_news]=50&tx_ttnews[backPid]=4&cHash=e42cb8b763

Gnansounou, E. (2008). Assessing the energy vulnerability: Case of industrialised countries. *Energy Policy, 36*(10), 3734–3744. doi:10.1016/j.enpol.2008.07.004

Godart, T., & Puttgen, H. (1991). A reactive path concept applied within a voltage control expert system. *IEEE Transactions on Power Systems, 6*(2), 787–793. doi:10.1109/59.76726

Goldberg, D. E. (1989). Genetic algorithms in search, optimization and machine learning. Addison-Wesley, Reading, MA.

Goldstein, G., & Tosato, G. C. (2008). *International Energy Agency Implementing Agreement for a Programme of Energy Technology Systems Analysis - Annex X: Global Energy Systems and Common Analyses.* Retrieved July, 24, 2008, from http://www.etsap.org/official.asp.

Gomez, S. A., & Amenedo, J. L. R. (2002, November 4). Grid synchronization of doubly fed induction generators using direct torque control. *Proc. IEEE 28th Annual Conf. Industrial Electronics Society,* (pp. 3338 – 3343).

Gómez-Villalva, E., & Ramos, A. (2003). Optimal Energy Management of an Industrial Consumer in Liberalized Markets. *IEEE Transactions on Power Systems, 18,* 716–723. doi:10.1109/TPWRS.2003.811197

Gonzalez Chapa, M. A., & Vega Galaz, J. R. (2004). An economic dispatch algorithm for cogeneration systems. *2004 IEEE Power Engineering Society General Meeting.*

Gonzalez, P., Hernandez, F., & Gual, M. (2005). The implications of the Kyoto project mechanisms for the deployment of renewable electricity in Europe. *Energy Policy, 33,* 2010–2022. doi:10.1016/j.enpol.2004.03.022

Goodell, M. (2002). *Trigeneration advantage.* http://www.cogeneration.net/The%20Trigeneration%20Advantage.htm.

Gorgun, H., Arcak, M., & Barbir, F. (2006). An algorithm for estimation of membrane water content in PEM fuel cells. *Journal of Power Sources, 157*(1), 389–394. doi:10.1016/j.jpowsour.2005.07.053

Goumas, M. G., Lygerou, V. A., & Papayannakis, L. E. (1999). Computational methods for planning and evaluating geothermal energy projects. *Energy Policy, 27*(3), 147–154. doi:10.1016/S0301-4215(99)00007-5

Granelli, G. P., Montagna, M., Pasini, G. L., & Marannino, P. (1992). Emission constrained dynamic dispatch. *Electric Power Systems Research, 24,* 56–64. doi:10.1016/0378-7796(92)90045-3

Greening, L. A., & Bernow, S. (2004). Design of coordinated energy and environmental policies: use of multi-criteria decision-making. *Energy Policy, 32*(6), 721–735. doi:10.1016/j.enpol.2003.08.017

Greer, L. R. (1986). Artificial intelligence and simulation: An introduction. Proceedings of the 1986 Winter Simulation Conference, Wilson, J., Henriksen, J., Roberts, S. (Eds.), 448-452.

Grohnheit, P. E. (1993). Modelling CHP within a national power system. *Energy Policy, 21,* 418–429. doi:10.1016/0301-4215(93)90282-K

Groves, D. (2005). *New Methods for Identifying Robust Long-Term Water Resources Management Strategies for California.* Ph.D. dissertation, Pardee Rand Graduate School, Santa Monica, CA.

Grubb, M., Butler, L., & Twomey, P. (2006). Diversity and security in UK electricity generation: The influence of low-carbon objectives. *Energy Policy, 34*(18), 4050–4062. doi:10.1016/j.enpol.2005.09.004

Guo, B., Belcher, C., & Roddis, W. M. K. (1993). RetroLite: an artificial intelligence tool for lighting energy-efficiency Upgrade. *Energy and Building, 20,* 115–120. doi:10.1016/0378-7788(93)90002-C

Guo, T., Henwood, M. I., & van Ooijen, M. (1996). An algorithm for combined heat and power economic dispatch. *IEEE Transactions on Power Systems, 11,* 1778–1784. doi:10.1109/59.544642

Gupta E. (2007). *Assessing the relative geopolitical risk of oil importing countries* [Working paper].

Gupta, E. (2008). Oil vulnerability index of oil-importing countries. *Energy Policy, 36*, 1195–1211. doi:10.1016/j.enpol.2007.11.011

Gupta, J. P., Chevalier, A., & Dutta, S. (2003). Multicriteria model for risk evaluation for venture capital firms in an emerging market context. *European Journal of Operational Research*, 1–18.

Gupta, M.M., Liang, J., Noriyasu, H. (2003). Static and dynamic neural networks, from fundamentals to advanced theory. Forward by Lotfi A. Zadeh, IEEE press John Wiley.

Gustafsson, S. I. (1993). Mathematical modelling of district heating and electricity loads. *Applied Energy, 46*, 149–159. doi:10.1016/0306-2619(93)90064-V

Hall, B. *Return on Investment and Multimedia Training: A Research Study.* Multimedia Training Newsletter. Retrieved November 6, 2008, from http://www.brandon-hall.com

Hansen, A. D., & Michalke, G. (2007). Fault ride-through capability of DFIG wind turbines. *Renewable Energy, 32*(9), 1594–1610. doi:10.1016/j.renene.2006.10.008

Hansen, A. D., Sorensen, P., Iov, F., & Blaabjerg, F. (2006). Centralized power control of wind farm with doubly fed induction generators. *Renewable Energy, 31*(7), 935–951. doi:10.1016/j.renene.2005.05.011

Hansen, A.D., Iov, F., Sorensen, P. & Blaabjerg, F. (2004 March 1 -7). Overall control strategy of variable speed doubly-fed induction generator wind turbine. *Nordic Wind Power Conf.*, Chalmers University of Technology, Goteborg, Sweden.

Haraldsson, K., & Wipke, K. (2004). *Evaluating PEM Fuel Cell System Models* (Tech. Rep. No. 1). Golden, CO: National Renewable Energy Laboratory, DOE Hydrogen and Fuel Cells and Infrastructure Technologies Program.

Harp, S. A., & Samad, T. (1991). Optimizing neural networks with genetic algorithms. *Proceedings of the American Power Conference*, Chicago, 1138–1143.

Hasan, K., Ramsay, B., & Moyes, I. (1995). Object oriented expert systems for real-time power system alarm processing, Part I, Selection of a toolkit; Part II, Application of a toolkit. *Electrical Power Systems Research, 30*(1), 69-75/77-82.

Haugeland, J. (1985). (Ed.) Artificial intelligence: The very idea. MIT Press, Cambridge.

Haurie, A., Loulou, R., & Savard, G. (1992). A two-player game model of power cogeneration in New England. *IEEE Transactions on Automatic Control, 37*(9), 1451–1456. doi:10.1109/9.159591

Haykin, S. (1999). Neural networks: A comprehensive foundation. New York: Macmillan; second edition.

Hazra, J., & Sinha, A. K. (2007). Congestion management using multiobjective particle swarm optimization. *IEEE Transactions on Power Systems, 22*(4), 1726–1734. doi:10.1109/TPWRS.2007.907532

Hebb, D. O. (1949). The Organization of behavior. Wiley, New York.

Heier, S., & Waddington, R. (2006). *Grid integration of wind energy conversion systems.* New York: Wiley.

Helle, L. & Nielsen, S. M. (2001). Comparison of converter efficiency in large variable speed wind turbines. *IEEE 16th Annual Applied Power Electronics Conf. and Exposition, 1*, 628-634.

Helm, D. (2002). Energy policy: security of supply, sustainability and competition. *Energy Policy, 30*, 173–184. doi:10.1016/S0301-4215(01)00141-0

Helsin, J. S., & Hobbs, B. F. (1989). A multiobjective production costing model for analyzing emission dispatching and fuel switching. *IEEE Transactions on Power Systems, 4*, 836–842. doi:10.1109/59.32569

Henning, D., Amiri, S., & Holmgren, K. (2006). Modelling and optimisation of electricity, steam and district heating production for a local Swedish utility. *European Journal of Operational Research, 175*(2), 1224–1247. doi:10.1016/j.ejor.2005.06.026

Hernádeza, J. C. N., Medinaa, A., & Juradob, F. (2007). Optimal allocation and sizing for profitability and voltage enhancement of PV systems on feeders. *Renewable Energy, 32*, 1768–1789. doi:10.1016/j.renene.2006.11.003

Herts, J., Krogh, A., & Palmer, R. G. (1991). Introduction to the theory of neural computation. Addision-Wesley, Redwood City CA, 163.

Hewitt, C., & Planner, A. (1971). Language for proving theorems in robots, Proceedings of IJCAI, 2.

Highley, D. D., & Hilmes, T. (1993). Load forecasting by ANN. *IEEE Computer Applications in Power, 6*(3), 10–15. doi:10.1109/67.222735

Hirai, S., Hiroyasu, T., Miki, M., Tanaka, Y., Shimosaka, H., Aoki, S., & Umeda, Y. (2004). Satisfactory design of cogeneration system using genetic algorithm. *2004 IEEE Conference on Cybernetics and Intelligent Systems.*

Hiremath, R. B., Shikha, S., & Ravindranath, N. H. (2007). Decentralized energy planning; modeling and application—a review. *Renewable & Sustainable Energy Reviews, 11*, 729–752. doi:10.1016/j.rser.2005.07.005

Hoffman, W., & Okafor, F. (2002). Optimal power utilization with doubly fed full controlled induction generator. *Proc. 6th IEEE Africon Conf. Africa*, 693–698.

Hofmann, W. & Thieme, A. (1998 May). Control of a doubly-fed induction generator for wind power plants. *Proc. Power Conversion and Intelligent Motion*, Nurnberg, Germany, (pp. 275-282).

Hofmann, W. (1999 Sept.). Optimal reactive power splitting in wind power plants controlled by double-fed induction generator. *Proc. IEEE Africon*, Cape Town, South Africa, (pp. 943-948).

Hofmann, W., & Okafor, F. (2001 November - December). Optimal control of doubly-fed full-controlled induction wind generator with high efficiency. *Proc. 27th Annual Conf. IEEE Industrial Electronics Society, 3*, 1213 – 1218.

Hofmann, W., Thieme, A., Dietrich, A., & Stoev, A. (1997, September). Design and control of wind power station with double feed induction generator. *Proc. Eupopean Conf. Power Electronics, Brussels, Belgium, 2*, 723–728.

Holdsworth, L., Wu, X. G., Eianayke, J. B., & Jenkins, N. (2002). Steady state and transient behavior of induction generators (including doubly fed machines) connected to distribution networks. In *Proc. IEEE Tutorial, 'Principle and modeling of distributed generators'*, July.

Holdsworth, L., Wu, X. G., Ekanayake, J. B., & Jenkins, N. (2003). Comparison of fixed speed and doubly-fed induction wind turbines during power system disturbances. *Proc. IEE – Gener. Transm. Distrib., 150*(3), 343 – 352.

Holdsworth, L., Wu, X. G., Ekanayake, J. B., & Jenkins, N. (2003). Direct solution method for initializing doubly-fed induction wind turbines in power system dynamic models. *IEEE Proc.- . Gener. Transm. Distrib., 120*(3), 334–342. doi:10.1049/ip-gtd:20030245

Holland, J. H. (1975). Adaptation in natural and artificial systems. University of Michigan Press, Ann Arbor.

Hollmann, M., & Voss, J. (2005). Modeling of decentralized energy supply structures with "system dynamics." *2005 International Conference on Future Power Systems.*

Hong, Y. Y., & Li, C. H. (2002). Genetic algorithm based economic dispatch for cogeneration units considering multiplant and multibuyer wheeling. *IEEE Transactions on Power Systems, 17*, 134–140. doi:10.1109/59.982204

Hong, Y. Y., & Li, C. Y. (2006). Back-pressure cogeneration economic dispatch for physical bilateral contract using genetic algorithms. *2006 International Conference on Probabilistic Methods Applied to Power Systems.*

Hopfensperger, B., Atkinson, D. J., & Lakin, R. A. (2000). Stator-flux oriented control of a doubly-fed induction machine without position encoder. *Proc. IEE.- Electr. Power Appl., 147*(4), July, 241–250.

Hopfield, J. J. (1982). Neural networks and to physical systems with emergent collective computational abilities. *Proceedings of the National Academy of Sciences of the United States of America, 79*, 2554–2558. doi:10.1073/pnas.79.8.2554

Houwing, M., Heijnen, P. W., & Bouwmans, I. (2006). Deciding on micro-CHP: a multi-level decision-making approach. *Proceedings of the 2006 IEEE International Conference on Networking, Sensing and Control* (pp. 302-307).

Hsiao, C., Mountain, D. C., & Illman, K. H. (1995). Bayesian integration of end-use metering and conditional demand analysis. *Journal of Business & Economic Statistics, 13*(3), 315–326. doi:10.2307/1392191

Huang, C. M., Yang, H. T., & Huang, C. L. (1997). Bi-objective power dispatch using fuzzy satisfaction-maximizing decision approach. *IEEE Transactions on Power Systems, 12*, 1715–1721. doi:10.1109/59.627881

Huang, H., Fan, Y., Qiu, R. C., & Jiang, X. D. (2006). Quasi-steady-state rotor emf-oriented vector control of doubly fed winding induction generators for wind-energy generation. *Electric Power Components and Systems, 34*(11), 1201–1211. doi:10.1080/15325000600698597

Huang, J. P., Poh, K. L., & Ang, B. W. (1995). Decision analysis in energy and environmental modelling. *Energy, 20*, 843–855. doi:10.1016/0360-5442(95)00036-G

Huang, S. H., Chen, B. K., Chu, W. C., & Lee, W. J. (2004). Optimal operation strategy for cogeneration power plants. *Conference Record of the 2004 IEEE Industry Applications Conference*.

Huang, Y. F., Huang, G. H., Hu, Z. Y., Maqsood, I., & Chakma, A. (2005). Development of an expert system for tackling the public's perception to climate-change impacts on petroleum industry. *Expert Systems with Applications, 29*(4), 817–829. doi:10.1016/j.eswa.2005.06.020

Hudson, R.M., Stadler, F. & Seehuber, M. (2003). Large developments in power electronic converters for megawatt class wind turbines employing doubly fed generators, *Proc. Int. Power Conversion, Intelligent Motion*, Nuremberg, Germany, June .

Hughes, F. M., Anaya-Lara, O., Jenkins, N., & Strbac, G. (2005). Control of DFIG-based wind generation for power network support. *IEEE Transactions on Power Systems, 20*(4), 1958–1966. doi:10.1109/TPWRS.2005.857275

Hughes, F. M., Lara, O. A., Jenkins, N., & Strbac, G. (2006). A Power system stabilizer for DFIG-based wind generation. *IEEE Transactions on Power Systems, 21*(2), 763–772. doi:10.1109/TPWRS.2006.873037

Humphreys, P., McIvor, R., & Huang, G. (2002). An expert system for evaluating the make or buy decision. *Computers & Industrial Engineering, 42*, 567–585. doi:10.1016/S0360-8352(02)00052-9

Hung, C. Q., Batanovand, D. N., & Lefevre, T. (1997). *KBS and macro-level systems: support of energy demand forecasting*. Thailand: Asian Institute of Technology.

Hung, C., Batanov, D., & Lefevre, T. (1998). KBS and macro-level systems: support of energy demand forecasting. *Computers in Industry, 37*, 87–95. doi:10.1016/S0166-3615(98)00092-X

Hwang, C. L., & Yoon, K. (1981). *Multiple Attribute Decision Making: Methods and Application*. New York: Springer.

IEA (International Energy Agency). (2006). *Energy Balances of Non- OECD Countries 2006*. Paris: IEA.

IEA (International Energy Agency). (2006). *Energy Balances of OECD Countries 2006*. Paris: IEA.

IEA (International Energy Agency). (2006). *World Energy Outlook 2006*. Paris: IEA.

IEC 61000-1-1, (n.d.). Electromagnetic compatibility (EMC) Part I: General, Section 1: Application and interpretation of fundamental definitions and terms.

IEEE Std 1159-1995 (n.d.). *IEEE recommended practice for monitoring electric power quality.*

IEEE Std. 1346-1998 (n.d.). *IEEE recommended practice for evaluating electric power system compatibility with electronic process equipment.*

IEEE Std. 739-1995, (n.d.). IEEE Recommended practice for energy management in industrial and commercial facilities.

Illerhaus, S. W., & Verstege, J. F. (1999). Optimal operation of industrial CHP-based power systems in liberalized energy markets. *IEEE Power Tech'99 Conference,* Budapest, Hungary, Aug 29-Sept 2, 1999.

Iniyan, S., Suganthi, L., & Samuel, A. (2006). Energy models for commercial energy prediction and substitution of renewable energy sources. *Energy Policy, 34*(17), 2640–2653. doi:10.1016/j.enpol.2004.11.017

Institute for Energy Engineering - Technical University of Berlin. (2003). *deeco Related models (which embed infrastructure capacity limitations).* Retrieved August, 18, 2008, from http://www.iet.tu-berlin.de/deeco/related-models.html.

Institute of Energy Economics and the Rational Use of Energy - University of Stuttgart. (n.d.) *EcoSenseWeb.* Retrieved July, 25, 2008, from http://EcoSenseWeb.ier.uni-stuttgart.de/.

International Atomic Energy Agency – IAEA. (1980). *Wien Automatic System Planning Package (WASP), A Computer Code for Power Generating System Expansion Planning.*

International Energy Agency. (2003). *World Energy Investment Outlook, 2003 Insights.* Paris: Author.

International Institute for Applied Systems Analysis – IIASA (n.d.). Retrieved May, 25, 2008, from http://www.iiasa.ac.at/Research/ENE/model/

Ioannides, M. G. (1992). Determination of frequencies in autonomous double output asynchronous generator. *IEEE Transactions on Energy Conversion, 7*(4), 747–753. doi:10.1109/60.182658

Ismail, Y., etc. (2002). An operating method for fuel savings in a stand-alone wind/diesel/battery system. *Journal of Japan Solar Energy Society, 28*(2), 31–38.

Issa, R. R. A., Flood, I., & Asmus, M. (2001). Development of a neural network to predict residential energy consumption. *Proceedings of the Sixth International Conference on the Application of Artificial Intelligence to Civil & Structural Engineering Computing,* (pp. 65-66).

Jabbour, K., Riveros, J. F. V., Landsbergen, D., & Meyer, W. (1988). ALFA: automated load forecasting assistant. *IEEE Transactions on Power Systems, 3*(3), 908–914. doi:10.1109/59.14540

Jaber, J. O., Mamlook, R., & Awad, W. (2005). Evaluation of energy conservation programs in residential sector using fuzzy logic methodology. *Energy Policy, 33,* 1329–1338. doi:10.1016/j.enpol.2003.12.009

Jaffe, A., & Wilson, W. (2005). *Energy Security: Oil-geopolitical and Strategic Implications for China and the United States.* Houston, TX: James A Baker Institute for Public Policy, Rice University.

Jank, R., Johnsson, J., Rath-Nagel, S., Steidle, T., Ryden, B., Skoldberg, H., et al. (2005). Annex 33 information based on the final reports of the project. In R. Burton (Ed.), *Technical Synthesis Report. Annexes 22 & 33. Energy Efficient Communities & Advanced Local Energy Planning (ALEP)* (pp. 15-28). Birmingham, UK: Faber Maunsell Ltd on behalf of the international Energy Agency, Energy Conservation in Building and Community Systems Programme.

Jansen, J. C., van Arkel, W.G., Boot, M.G. (2004). *Designing indicators of long term energy supply security,* (ECN-C-04-007).

Japan Solar Energy Society. (1985). *Solar Energy Utilization Handbook* (pp. 10-88).

Jebaraj, S., & Iniyan, S. (2006). A review of energy models. *Renewable & Sustainable Energy Reviews, 10*(4), 281–311. doi:10.1016/j.rser.2004.09.004

Jeong, S. (2002 October). Representing line voltage unbalance. *Conference Record of 2002 IEEE Industry Applications Conf.,* Pennsylvania, USA, *3,* 1724-1732.

Johnson, G., & Scholes, K. (1999). *Exploring Corporate Strategy: Text and Cases* (5th ed.). London: Prentice Hall Europe. 5th ed. London.

Jong-Bae, P., Ki-Song, L., Joong-Rin, S., & Kwang, Y. L. (2005). A particle swarm optimization for economic dispatch with nonsmooth cost functions. *IEEE Transactions on Power Systems, 20,* 34–42. doi:10.1109/TPWRS.2004.831275

Jordan, M. (1988). Serial order: A parallel distributed processing approach, in Elman, J.L. & Rumelhart, D.E. (Eds.) Advanced in connectionist theory: speech Erlbaum.

Kadaba, N., Nygard, K. E., & Juell, P. J. (1991). Integration of adaptive machine learning and knowledge-cased systems for routing and scheduling applications. *Expert Systems with Applications, 2*, 15–27. doi:10.1016/0957-4174(91)90131-W

Kagiannas, A. G., Askounis, D. T., & Psarras, J. (2004). Power generation planning: a survey from monopoly to competition. *Electrical Power & Energy Systems, 26*, 413–421. doi:10.1016/j.ijepes.2003.11.003

Kagiannas, A., Flamos, A., Askounis, D., & Psarras, J. (2002). Energy Policy Indicators for the Assessment of the Euro-Mediterranean Energy Cooperation. *International Journal of Energy Technology and Policy, 2*(4), 301–322. doi:10.1504/IJETP.2004.005738

Kagiannas, A., Patlitzianas, K., Metaxiotis, K., Askounis, D., & Psarras, J. (2006). Energy Models in the Mediterranean Countries: A survey towards a common strategy. *International Journal of Power and Energy Systems, 26*(3), 260–268. doi:10.2316/Journal.203.2006.3.203-3526

Kaklauskas, A., Zavadskas, E. K., Raslanas, S., Ginevicius, R., Komka, A., & Malinauskas, P. (2006). Selection of low-e windows in retrofit of public buildings by applying multiple criteria method COPRAS: A Lithuanian case. *Energy and Building, 38*(5), 454–462. doi:10.1016/j.enbuild.2005.08.005

Kalogirou, S. (1996a). Artificial neural networks for estimating the local concentration ratio of parabolic trough collectors, *Proceedings of the EuroSun'96 Conference*, Freiburg, Germany, 1, 470-475.

Kalogirou, S. (1996b). Design of a solar low temperature steam generation system, *Proceedings of the EuroSun'96 Conference*, Freiburg, Germany, 1, 224-229.

Kalogirou, S. (2000). Applications of artificial neural networks for energy systems. Special issue of *Applied Energy* journal on Energy systems: Adaptive complexity, *67*(1-2), 17-35.

Kalogirou, S. (2000). Long-term performance prediction of forced circulation solar domestic water heating systems using artificial neural networks . *Applied Energy, 66*, 63–74. doi:10.1016/S0306-2619(99)00042-2

Kalogirou, S. (2001). Artificial neural networks in renewable energy systems: A review . *Renewable & Sustainable Energy Reviews, 5*, 373–401. doi:10.1016/S1364-0321(01)00006-5

Kalogirou, S. (2001). Artificial neural networks in renewable energy systems: A review. *Renewable & Sustainable Energy Reviews, 5*(4), 373–401. doi:10.1016/S1364-0321(01)00006-5

Kalogirou, S. (2002). Design of a fuzzy single-axis sun tracking controller . *International Journal of Renewable Energy Engineering, 4*, 451–458.

Kalogirou, S. (2007). Use of genetic algorithms for the optimum selection of the fenestration openings in buildings, Proceedings of the 2nd PALENC Conference and 28th AIVC Conference on Building Low Energy Cooling and Advanced Ventilation Technologies in the 21st Century, Crete island, Greece, 483-486.

Kalogirou, S. A. (2000). Applications of artificial neural-networks for energy systems. *Applied Energy, 67*(1-2), 17–35. doi:10.1016/S0306-2619(00)00005-2

Kalogirou, S. A. (2003). Artificial intelligence for the modeling and control of combustion processes: a review. *Progress in Energy and Combustion Science, 29*, 515–566. doi:10.1016/S0360-1285(03)00058-3

Kalogirou, S. A. (2007). Artificial Intelligence in energy and renewable energy systems. Nova Publishers, Chapter 2 and Chapter 5.

Kalogirou, S. A., Neocleous, C. C., & Schizas, C. N. (1997). Heating Load Estimation Using Artificial Neural Networks. Proc. of CLIMA 2000 Conf., Brussels, Belgium.

Kalogirou, S. A., Panteliou, S., & Dentsoras, A. (1999). Modeling of solar domestic water heating systems using artificial neural networks. *Solar Energy, 65*(6), 335–342. doi:10.1016/S0038-092X(99)00013-4

Kalogirou, S., & Bojic, M. (2000). Artificial neural networks for the prediction of the energy consumption of a passive solar building. *Energy, 25*, 479–491. doi:10.1016/S0360-5442(99)00086-9

Kalogirou, S., & Panteliou, S. (2000). Thermosyphon solar domestic water heating systems long-term performance prediction using artificial neural networks. *Solar Energy, 69*, 163–174. doi:10.1016/S0038-092X(00)00058-X

Kalogirou, S., & Panteliou, S., Dentsoras, A. (1999). Artificial neural networks used for the performance prediction of a thermosyphon solar water heater. *Renewable Energy, 18*, 87–99. doi:10.1016/S0960-1481(98)00787-3

Kalogirou, S., Eftekhari, M., & Pinnock, D. (1999). Prediction of air flow in a single-sided naturally ventilated test room using artificial neural networks. Proceedings of Indoor Air'99, The 8th International Conference on Indoor Air Quality and Climate, Edinburgh, Scotland, Vol. 2, pp 975-980.

Kalogirou, S., Neocleous, C., & Schizas, C. (1996). A comparative study of methods for estimating intercept factor of parabolic trough collectors, Proceedings of the Engineering Applications of Neural Networks (EANN'96) Conference, London, UK, 5-8.

Kalogirou, S., Neocleous, C., & Schizas, C. (1998). Artificial neural networks for modelling the starting-up of a solar steam generator . *Applied Energy, 60*, 89–100. doi:10.1016/S0306-2619(98)00019-1

Kandil, M. (2001). The implementation of long-term forecasting strategies using a knowledge-based expert system: part II. *Electric Power Systems Research, 58*(1), 19–25. doi:10.1016/S0378-7796(01)00098-0

KanORS Consulting Inc. (2003a). *VEDA-FE User Guide - Version 1.5.11*. Retrieved May 25, 2006, from http://www.kanors.com/userguidefe.htm

KanORS Consulting Inc. (2003b). *VEDA4 User Guide - Version 4.3.8*. Retrieved May 25, 2006, from http://www.kanors.com/userguidebe.htm

Kar, N. C., & Jabr, H. M. (2005, September 2). A novel PI gain scheduler for a vector controlled doubly-fed wind driven induction generator. *Proc. 8th Int. Conf. Electrical Machines and Systems*, (pp. 948 – 953).

Karakosta, C., Doukas, H., & Psarras, J. (2008). A Decision Support Approach for the Sustainable Transfer of Energy Technologies under the Kyoto Protocol. *American Journal of Applied Sciences, 5*(12), 1720–1729.

Kaufmann, A., & Gupta, M. M. (1985). Introduction to fuzzy arithmetic, Theory and applications. 2nd ed., Van Nostrand Reinhold, New York (Japanese translation by Atsuka M., Ohmsha Ltd., Tokyo, 1991. Kodratoff, Y. & Michalski, R.S. (1990). Machine learning: An artificial intelligence approach. Morgan Kaufmann, Vol. 3.

Kavrakoglou, I. (1987). Energy models. *European Journal of Operational Research, 16*(2), 2231–2238.

Kawashima, M. (1994). Artificial neural network backpropagation model with three-phase annealing developed for the building energy predictor shootout. *ASHRAE Transactions, 100*(2), 1095–1103.

Keeny, R. L., & Raiffa, H. (1993). *Decision Making with Multiple Objectives: Preferences and Value Tradeoffs*. Cambridge, UK: Cambridge University Press 1993. Cambridge, UK.

Kennedy, J., & Eberhart, R. (2001) *Swarm Intelligence*. San Francisco: Morgan Kaufmann Publishers.

Kersting, W. H., & Phillips, W. H. (1997). Phase frame analysis of the effects of voltage unbalance on induction machines. *IEEE Transactions on Industry Applications, 33*(2), 415–420. doi:10.1109/28.568004

Kezunovic, M., Fromen, C. W., & Sevcik, D. R. (1993). Expert system for transmission substation event analysis. *IEEE Transactions on Power Delivery, 8*(4), 1942–1949. doi:10.1109/61.248306

Khatounian, F., Monmasson, E., Berthereau, F., Deleleau, E., & Louis, J. P. (2003, November 3). Control of a doubly fed induction generator for aircraft application. *Proc. 29th Annual Conf. IEEE Industrial Electronics Society*, (pp. 2711 – 2716).

Khosla, R., & Dillon, T. S. (1994). Neuro-expert system approach to power system problems. *International*

Journal of Engineering Intelligent Systems for Electrical Engineering and Communications, *2*(1), 71–78.

Kiartzis, S. J., Bakirtzis, A. G., & Petridis, V. (1995). Short-term forecasting using NNs. *Electric Power Systems Research*, *33*(1), 1–6. doi:10.1016/0378-7796(95)00920-D

Kieferndorf, F. D., Forster, M., & Lipo, T. A. (2004). Reduction of DC-bus capacitor ripple current with PAM/PWM converter. *IEEE Transactions on Industry Applications*, *40*(2), 607–614. doi:10.1109/TIA.2004.824495

Kim, J. G., Lee, E. W., Lee, D. J., & Lee, J. H. (2005). Comparison of voltage unbalance factor by line and phase voltage. *Proc. 8th Int. Conf. on Electrical Machines and Systems (ICEMS)*, Nanjing, China, *3*, 1988-2001.

King, R. T. F., Harry, C. S., & Deb, K. (2006, July). Stochastic evolutionary multi-objective environmental/economic dispatch. *Proceedings of the 2006 IEEE Congress on Evolutionary Computation*, (pp. 3369-3376).

Kini, P. G., Bansal, R. C., & Aithal, R. S. (2007). A novel approach towards interpretation and application of voltage unbalance factor. *IEEE Transactions on Industrial Electronics*, *54*(4), 2315–2322. doi:10.1109/TIE.2007.899935

Kini, P. G., Sreedhar, P. N., & Varmah, K. R. (2005, November). Importance of Accuracy For Steady State Performance Analysis of 3φ Induction Motor. *Proc. of IEEE Region 10 conf. of TENCON 2005*, Melbourne, Australia, (pp. 1994-1998).

Kirkman, A.-M., Aalders, E., Braine, B., & Gagnier, D. [Eds.] (2006). *Greenhouse Gas Market Report 2006: Financing Response to Climate Change: Moving to Action* (v, 136 p.). International Emission Trading Association - IETA.

Kitamura, S., Mori, K., Shindo, S., Izui, Y., & Ozaki, Y. (2005). Multiobjective energy management system using modified MOPSO. *IEEE International Conference on Systems, Man and Cybernetics*, (pp. 3497 – 3503).

Kohonen, T. (1984). Self-organization and associative memory. Springer-Verlag, New York.

Konar, A. (1999). Artificial intelligence and soft computing behavioral and cognitive modeling of the human brain. CRC Press, Chapter 1.

Kong, X. Q., Wang, R. Z., & Huang, X. H. (2007). Energy optimization model for a CCHP system with available gas turbines. *Applied Thermal Engineering*, *25*, 377–391. doi:10.1016/j.applthermaleng.2004.06.014

Konidari, P., & Mavrakis, D. (2007). A multi-criteria evaluation method for climate change mitigation policy instruments. *Energy Policy*, *35*(12), 6235–6257. doi:10.1016/j.enpol.2007.07.007

Koroneos, C., Michailidis, M., & Moussiopoulos, N. (2004). Multi-objective optimization in energy systems: the case study of Lesvos Island, Greece. *Renewable & Sustainable Energy Reviews*, *8*(1), 91–100. doi:10.1016/j.rser.2003.08.001

Kosugi, T., etc. (1998). Evaluation of output and unit cost of power generation systems utilizing solar energy under various solar radiation in the world. *Transaction of the Institute of Electrical Engineers of Japan . Publication of Power and Energy Society*, *118*(3), 246–253.

Kothari, D. P., & Nagrath, I. J. (2004). *Electric Machines, 3rd Edition.* New Delhi, India: Tata McGraw Hill.

Kouvelis, P., & Yu, G. (1997). *Robust Discrete Optimization and Its Applications.* Dordrecht, the Netherlands: Kluwer Academic Publishers.

Krarti, M., Kreider, J. F., Cohen, D., & Curtiss, P. (1998). Estimation of energy saving for building retrofits using neural networks. *Journal of Solar Energy Engineering*, *120*, 211–216. doi:10.1115/1.2888071

Kreider, J. F., & Haberl, J. S. (1994). Predicting hourly building energy use: the great energy predictor shootout-overview and discussion of results. *ASHRAE Transactions*, *100*(2), 1104–1118.

Kreider, J. F., & Wang, X. A. (1991). Artificial neural networks demonstrations for automated generation of energy use predictors for commercial buildings. *ASHRAE Transactions*, *97*(1), 775–779.

Kreider, J. F., & Wang, X. A. (1992). Improved artificial neural networks for commercial building energy use prediction. *Solar Engineering, 1*, 361–366.

Kreider, J. F., & Wang, X. A. (1995). Artificial Neural Network Demonstration for Automated Generation of Energy Use Predictors for Commercial Buildings. In Haberl, J.S., Nelson, R.M. and Culp, C.C. (Eds.). *The use of Artificial Intelligence in Building Systems*. ASHRAE, 193-198.

Krishnamoorthy, C. S., & Rajeev, S. (1996). Artificial intelligence and expert systems for engineers. CRC Press, LLC.

Krokhmal P., Murphey R., Pardalos P., Uryasev S., Zrazhevski G.(2004). *Robust Decision Making: Addressing Uncertainties in Distributions.* [DTIC, Final report].

Krokhmal, P. (2003). *Risk Management Techniques for Decision making in Highly Uncertain Environments.* Phd dissertation, University of Florida, Gainesville, FL.

Kua, H. W., & Lee, S. E. (2002). Demonstration intelligent building—a methodology for the promotion of total sustainability in the built environment. *Building and Environment, 37*, 231–240. doi:10.1016/S0360-1323(01)00002-6

Kuisma, M. (2003). Variable frequency switching in power supply EMI-control: an overview. *Aerospace and Electronic Systems Magazine, 18*(12), 18–22. doi:10.1109/MAES.2003.1259021

Kumbaroğlu, G., Madlener, R., & Demirel, M. (2008). A real options evaluation model for the diffusion prospects of new renewable power generation technologies. *Energy Economics, 30*(4), 1882–1908. doi:10.1016/j.eneco.2006.10.009

Kurzweil, R. (1990). The Age of intelligent machines. MIT Press, Cambridge.

Kusiak, A. (1987). Artificial intelligence and operations research in flexible manufacturing systems. *Information Processing, 25*(1), 2–12.

Kusiak, A. Y., Sunderesh, D., & Heragu, S. (1989). Expert systems and optimization. *IEEE Transactions on Software Engineering, 15*, 1012–1017. doi:10.1109/32.31358

Lahdelma, R. (1994). *An objected-orientd mathematical modeling system*. Ph.D. thesis, Acta, Polytechnica, Scandinavica Mathematics and Computing in Engineering Series No. 66. Systems Analysis Labortory, Helsinki University of Technology.

Lahdelma, R., & Hakonen, H. (2003). An efficient linear programming algorithm for combined heat and power production. *European Journal of Operational Research, 148*, 141–151. doi:10.1016/S0377-2217(02)00460-5

Lahdelma, R., & Makkonen, S. (1996). Interactive graphical object-oriented Energy modelling and optimisation. In *Proceedings of the International Symposium of ECOS'96, Efficiency, Cost, Optimisation, Simulation and Environmental Aspects of Energy Systems* (pp. 425-431), June 25-27, Stockholm, Sweden.

Lahdelma, R., & Rong, A. (2005). Efficient re-formulation of linear cogeneration planning models. In M.H. Hamza (Ed.) *Proceedings of the 24th IASTED International Conference Modelling, Identification, and Control* (pp. 300-305). February 16 -18, 2005, Innsbruck, Austria.

Lahdelma, R., & Salminen, P. (2001). SMAA-2: Stochastic Multicriteria Acceptability Analysis for Group Decision Making. *Operations Research, 49*(3), 444–454. doi:10.1287/opre.49.3.444.11220

Lahdelma, R., Makkonen, S., & Salminen, P. (2006). Multivariate Gaussian criteria in SMAA. *European Journal of Operational Research, 170*, 957–970. doi:10.1016/j.ejor.2004.08.022

Lahdelma, R., Makkonen, S., & Salminen, P. (2009). Two ways to handle dependent uncertainties in multi-criteria decision problems. *Omega, 37*, 79–92. doi:10.1016/j.omega.2006.08.005

Lai, L. (1998). Intelligent system applications in power engineering. John Wiley & Sons, London, UK.

Lai, L. L., Ma, J. T., & Lee, J. B. (1997). Application of genetic algorithms to multi-time interval scheduling for daily operation of a cogeneration system. *1997 International Conference on Advances in Power System Control, Operation and Management* (pp. 327-331).

Lai, L.L., Ma, J.T., & Lee, J.B. (1998). Multitime-interval scheduling for daily operation of a two cogeneration system with evolutionary programming. *Electrical power & Energy Systems, 20*, 305-311.

Lakhmi, C. J., & Martin, N. M. (1998). Fusion of neural networks, fuzzy systems and genetic algorithms: Industrial Applications. CRC Press, LLC.

Lara-Rosano, F., & Valverde, N. (1998). Knowledge-based systems for the energy conservation programs. *Expert Systems with Applications, 14*, 25–35. doi:10.1016/S0957-4174(97)00069-9

Larminie, J., & Dicks, A. (2003). *Fuel Cell Systems Explained- Second Edition*. Chichester, UK: John Wiley & Sons Ltd.

Larsen, B. M., & Nesbakken, R. (2004). Household electricity end-use consumption: results from econometric and engineering models. *Energy Economics, 25*(2), 179–200. doi:10.1016/j.eneco.2004.02.001

Larsen, H. V., Palsson, H., & Ravn, H. F. (1998). Probabilistic production simulation including combined heat and power plants. *Electric Power Systems Research, 48*, 45–56. doi:10.1016/S0378-7796(98)00080-7

Larsson, M., Wang, C., & Dahl, J. (2006). Development of a method for analysing energy, environmental and economic efficiency for an integrated steel plant. *Applied Thermal Engineering, 26*(13), 1353–1361. doi:10.1016/j.applthermaleng.2005.05.025

Lau, C. H., & Chan, F. C. (2000). *Knowledge Management for Power Systems Application*. Hong Kong Institution of Engineers Annual Symposium 2000. More information available from www.hkie.org.hk

Laudon, K., & Laudon, J. (2000). *Management information systems, Organization and technology in the networked enterprise* (6th Ed.). Englewood Cliffs, NJ: Prentice-Hall.

Laurikka, H. (2006). Option value of gasification technology within an emissions trading scheme. *Energy Policy, 34*, 3916–3928. doi:10.1016/j.enpol.2005.09.002

Le, T. L., Negnevitsky, M., & Piekutowski, M. (1995). Expert system application for voltage control and VAR compensation. *International Journal of Engineering Intelligent Systems for Electrical Engineering and Communications, 3*(2), 79–85.

Lechtenböhmer, S., & Thomas, S. (2005). *The mid-term potential for demand-side energy efficiency in the EU*. Wuppertal Institute for Climate, Environment, Energy.

Ledesma, P., & Usaola, J. (2001, September 4). Minimum voltage protection in variable speed wind farms. *Proc, IEEE Power Tech Conf.*, Porto, Portugal.

Ledesma, P., & Usaola, J. (2004). Effect of neglecting stator transients in doubly fed induction generators models. *IEEE Transactions on Energy Conversion, 19*(2), 459–461. doi:10.1109/TEC.2004.827045

Ledesma, P., & Usaola, J. (2005). Doubly fed induction generator model for transient stability analysis. *IEEE Transactions on Energy Conversion, 20*(2), 388–397. doi:10.1109/TEC.2005.845523

Lee, C. Y. (1999). Effects of unbalanced voltage on the operation performance of a three-phase induction motor. *IEEE Transactions on Energy Conversion, 14*(2), 202–208. doi:10.1109/60.766984

Lee, J. B., & Jeong, J. H. (2001). A daily optimal operational schedule for cogeneration systems in a paper mill. *2001 Power Engineering Society Summer Meeting* (vol.3, pp. 1357-1362).

Lee, J. B., Jung, C. H., & Lyu, S. H. (1999). A daily operation scheduling of cogeneration systems using fuzzy linear programming. *1999 IEEE Power Engineering Society Summer Meeting* (vol.2, pp. 983-988).

Lee, S. J., Yoon, S. H., Yoon, M. C., & Jang, J. K. (1990). Expert system for protective relay setting of transmission systems. *IEEE Transactions on Power Delivery, 5*(2), 1202–1208. doi:10.1109/61.53142

Lee, S., & Nam, K. (2003, October 3). Dynamic modeling and passivity-based control of an induction motor powered by doubly fed induction generator. *Proc. 38th IEEE Annual Meeting Conf. Industry Applications Conf.*, (pp. 1970 – 1975).

Lee, W., & Kenarangui, R. (2002). Energy management for motors, systems, and electrical equipment. *IEEE Transactions on Industry Applications*, *38*(2), 602–607. doi:10.1109/28.993185

Lehtilä, A., & Pirilä, P. (1996). Reducing energy related emissions using an energy system optimization model to support policy planning in Finland. *Energy Policy*, *24*(9), 805–919. doi:10.1016/0301-4215(96)00066-3

Lei, Y., Mullane, A., Lightbody, G., & Yacamini, R. (2005). Modeling of the wind turbine with a doubly fed induction generator for grid integration studies. *IEEE Transactions on Energy Conversion*, *20*(2), 435–441. doi:10.1109/TEC.2005.845526

Lempert, R. J., & Bonomo, J. (1998). *New Methods for Robust Science and Technology Planning.* Santa Monica, CA: RAND.

Lempert, R. J., & Schlesinger, M. E. (2000). Robust Strategies for Abating Climate Change. *Climatic Change*, *45*(3/4), 387–401. doi:10.1023/A:1005698407365

Lempert, R. J., Groves, D. G., Popper, S. W., & Bankes, S. C. (2005). (Submitted to). A General [Management Sciences.]. *Analytic Method for Generating Robust Strategies and Narrative Scenarios.*

Lempert, R. J., Popper, S. W., & Bankes, S. C. (2003). *Shaping the Next One Hundred Years: New methods for quantitative, long-term policy analysis.* Santa Monica, CA: RAND.

Lempert, R. J., Schlesinger, M. E., Bankes, S. C., & Andronova, N. G. (2000b). The Impact of Variability on Near-Term Climate-Change Policy Choices. *Climatic Change*, *45*(1), 129–161. doi:10.1023/A:1005697118423

Lenzen, M., Schaeffer, R., & Matsuhashi, R. (2007). Selecting and assessing sustainable CDM projects using multi-criteria methods. *Climate Policy*, *7*, 121–138.

Leonard, W. (2001). *Control of electric drives.* New York: Springer Verlag.

Li, H., Nalim, R., & Haldi, P. A. (2006). Thermal-economic optimization of a distributed multi-generation energy system—a case study of Beijing. *Applied Thermal Engineering*, *26*(7), 709–719. doi:10.1016/j.applthermaleng.2005.09.005

Li, H., Song, Z., Favrat, D., & Marechal, F. (2004). Green heating system: characteristics and illustration with multi-criteria optimization of an integrated energy system. *Energy*, *29*, 225–244. doi:10.1016/j.energy.2003.09.003

Li, S., Shahidehpour, S. M., & Wang, C. (1993). Promoting the application of expert systems in short-term unit commitment. *IEEE Transactions on Power Systems*, *8*(1), 286–292. doi:10.1109/59.221229

Li, Z.-Z., Qiu, S.-S., & Chen, Y.-F. (2006). Experimental Study on the Suppressing EMI Level of DC-DC Converter with Chaotic Map. *Proceedings of the Chinese society for electrical engineering, 26*(5), 76-81. Retrieved December 20, 2006, from http://ieeexplore.ieee.org

Liao, Y., Putrus, G. A., & Smith, K. S. (2003). Evaluation of the effects of rotor harmonics in a doubly-fed induction generator with harmonic induced speed ripple. *IEEE Transactions on Energy Conversion*, *18*(4), 508–515. doi:10.1109/TEC.2003.816606

Lindholm, M., & Rasmussen, T. W. (2003). Harmonic analysis of doubly fed induction generators. IEEE *Power Electronics and Drive Sys., 5th Int. Conf.*, *2*, 837–841.

Linkevics, O., & Sauhauts, A. (2005). Formulation of the objective function for economic dispatch optimisation of steam cycle CHP plants. *2005 IEEE Russia Power Tech.*

Lins, M. P. E., DaSilva, A. C. M., & Rosa, L. P. (2002). Regional variations in energy consumption of appliances: conditional demand analysis applied to Brazilian households. *Annals of Operations Research*, *117*(1), 235–246. doi:10.1023/A:1021533809914

Liposcak, M., Afgan, N. H., Duic, N., & Carvalho, M. (2006). Sustainability assessement of cogeneration sector development in Croatia. *Energy, 31*(13), 2276–2284. doi:10.1016/j.energy.2006.01.024

Liu, C., & Lai, J. S. (2007). Low Frequency Current Ripple Reduction Technique with Active Control in a Fuel Cell Power System with Inverter Load. *IEEE Transactions on Power Electronics, 22*(4), 1453–1463. doi:10.1109/TPEL.2007.900505

Liu, C., Ma, T., Liou, K., & Tsai, M. (1994). Practical use of expert systems in power systems. *International Journal of Engineering Intelligent Systems for Electrical Engineering and Communications, 2*(1), 11–22.

Liu, P., Gerogiorgis, D. I., & Pistikopoulos, E. N. (2007). Modeling and optimization of polygeneration energy systems. *Catalysis Today, 127*, 347–359. doi:10.1016/j.cattod.2007.05.024

Loulou, R., Goldstein, G., & Noble, K. (2004). *Energy Technology Systems Analysis Programme - Documentation for the MARKAL Family of Models. Part I: Standard MARKAL*. Retrieved January 14, 2006, from http://www.etsap.org/MrklDoc-I_StdMARKAL.pdf.

Loulou, R., Remme, U., Kanudia, A., Lehtila, A., & Goldstein, G. (2005). A comparison of the TIMES and MARKAL models. In *Energy Technology Systems Analysis Programme - Documentation for the TIMES Model - Part I: TIMES concepts and theory*. Retrieved January 14, 2006, from http://www.etsap.org/Docs/TIMESDoc-Intro.pdf.

Luger, G. F., & Stubblefield, W. A. (1993). Artificial intelligence: Structures and strategies for complex problem solving. Benjamin/Cummings, Menlo Park, CA.

Maabrah, G. (2008, May 26-27). Electricity Structure in Jordan and Regional Market. *Joint Arab Union of Producers, Transporters and Distributors of Electricity (AUPTDE) and MEDELEC Conference, Sharm El-Sheikh, Egypt.*

MacGregor, P. R., & Puttgen, H. B. (1991). A spot price based control mechanism for electric utility systems with small power producing facilities. *IEEE Transactions on Power Systems, 6*(2), 683–690. doi:10.1109/59.76713

Machado, R. J., & Rocha, A. F. (1992). A hybrid architecture for fuzzy connectionist expert systems. In Kandel, A. & Langholz, G. (Eds.), Hybrid architectures for intelligent systems. CRC Press, Boca Raton, FL.

Madlener, R., & Schmid, C. (2003). Combined heat and power generation in liberalized markets and a carbon-constrained world. *GAIA-Ecological Perspective in Science . Humanities and Economics, 12*, 114–120.

Madlener, R., & Stagl, S. (2005). Sustainability-guided promotion of renewable electricity generation. *Ecological Economics, 53*(2), 147–167. doi:10.1016/j.ecolecon.2004.12.016

Mahadevan, R., & Asafu-Adjaye, J. (2007). Energy consumption, economic growth and prices: A reassessment using panel VECM for developed and developing countries. *Energy Policy, 35*(4), 2481–2490. doi:10.1016/j.enpol.2006.08.019

Maidment, G. G., & Prosser, G. (2000). The use of CHP and absorption cooling in cold storage. *Applied Thermal Engineering, 20*, 1059–1073. doi:10.1016/S1359-4311(99)00055-1

Maifredi, C., Puzzi, L., & Beretta, G. P. (2000). Optimal power production scheduling in a complex cogeneration system with heat storage. *Proceedings of 35th Intersociety Energy Conversion Engineering Conference.*

Maiorano, A., Sbrizzai, R., Torell, F., & Trovato, M. (1999). Intelligent load shedding schemes for industrial customers with cogeneration facilities. *1999 IEEE Power Engineering Society (Winter Meeting).*

Mak, C. L., & Choi, Y. H. (2003). *Knowledge Management and Project Management for Transmission Network and System.* International Conference on Engineering Education, 2003. More information available from http://www.upv.es/ICEE2003/

Mäkelä, J. (2000). *Development of an energy system model of the Nordic electricity production system.* Unpublished master's thesis, Helsinki University of Technology, Finland.

Makkonen, S. (2005). *Decision modeling tools for utilities in the deregulated energy market*. Ph.D. Thesis, Research Report A93, Systems Analysis Laboratory, Helsinki University of Technology.

Makkonen, S., & Lahdelma, R. (1998). Stochastic simulation in risk analysis of energy trade. Trends in Multicriteria Decision Making: In *Proceedings of 13th International Conference on Multiple Criteria Decision Making* (pp. 146-156). Berlin: Springer.

Makkonen, S., & Lahdelma, R. (2001). Analysis of power pools in the deregulated energy market through simulation. *Decision Support Systems, 30*(3), 289–301. doi:10.1016/S0167-9236(00)00106-8

Makkonen, S., & Lahdelma, R. (2006). Non-Convex power plant modelling in energy optimisation. *European Journal of Operational Research, 171*, 1113–1126. doi:10.1016/j.ejor.2005.01.020

Makkonen, S., Lahdelma, R., Asell, A. M., & Jokinen, A. (2003). Multi-criteria decision support in the liberated energy market. *Journal of Multi-Criteria Decision Analysis, 12*(1), 27–42. doi:10.1002/mcda.341

Manhire, B., & Jenkins, T. B. (1982). A new technique for simulating the operation of multiple assigned-energy generating units suitable for use in generation system expansion planning models. *IEEE Transactions on Power Apparatus and Systems, 10*, 3861–3868. doi:10.1109/TPAS.1982.317036

Manne, A., Mendelsohn, R., & Richels, R. (1995). MERGE: a model for evaluating regional and global effects of GHG reduction policies. *Energy Policy, 23*, 7–34.

Manolas, D. A., Frangopoulos, C. A., Gialamas, T. P., & Tsahalis, D. T. (1997). Operation optimization of an industrial cogeneration system by a genetic algorithm. *Energy Conversion and Management, 38*, 1625–1636. doi:10.1016/S0196-8904(96)00203-8

Marechal, F., & Kalitventzeff, B. (1998). Process integration: Selection of the optimal utility system. *Computers & Chemical Engineering (Supplementary), 22*, S149–S156. doi:10.1016/S0098-1354(98)00049-0

Marik, K., Schindler, Z., & Stluka, P. (2008). Decision support tools for advanced energy management. *Energy, 33*(6), 858–873. doi:10.1016/j.energy.2007.12.004

Markovic, M., & Fraissler, W. (1993). Short-term load forecast by plausibility checking of announced demand: an expert-system approach. *European Transactions on Electrical Power Engineering, 3*(5), 353–358.

Markovic, M., Fraissler, W. (1993). Short-term load forecast by plausibility checking of announced demand: an expert-system approach. *European Transactions on Electrical Power Engineering / ETEP, 3*(5), 353-358.

Marks, W. (1997). Multicriteria optimisation of shape of energy-saving buildings. *Building and Environment, 32*(4), 331–339. doi:10.1016/S0360-1323(96)00065-0

Marques, J., & Pinheiro, H. (2005). Dynamic behavior of the doubly-fed induction generator in stator flux vector reference frame. *IEEE 36th Conf. Power Electronics Specialists*, June, (pp. 2104 – 2110).

Marriott, K., & Stuckey, P. J. (1998). *Programming with constraints*. Cambridge, MA: MIT Press.

Martinot, E. (2006). *Renewables: Global Status Report, 2006 Update*. REN21, retrieved from www.ren21.net

Martinsen, D., & Krey, V. (2008). Compromises in energy policy-using fuzzy optimization in an energy systems model. *Energy Policy, 36*(8), 2973–2984. doi:10.1016/j.enpol.2008.04.005

Mathews, E. H., & Botha, C. P. (2003). Improved thermal building management with the aid of integrated dynamic HVAC simulation. *Building and Environment, 38*, 1423–1429. doi:10.1016/S0360-1323(03)00070-2

Matics, J., & Krost, G. (2007). Computational Intelligence Techniques Applied to Flexible and Auto-adaptive Operation of CHP Based Home Power Supply. *2007 International Conference on Intelligent Systems Applications to Power Systems* (pp. 1-7).

Matics, J., & Krost, G. (2008). Micro combined heat and power home supply: Prospective and adaptive management achieved by computational intelligence techniques. *Applied Thermal Engineering, 28*, 2055–2061. doi:10.1016/j.applthermaleng.2008.05.002

Matsuda, S., Ogi, H., Nishimura, K., Okataku, Y., & Tamura, S. (1990). Power system voltage control by distributed expert systems. *IEEE Transactions on Industrial Electronics*, *37*(3), 236–240. doi:10.1109/41.55163

Matsuda, S., Ogi, H., Nishimura, K., Okataku, Y., & Tamura, S. (1990). Power system voltage control by distributed expert systems. *IEEE Transactions on Industrial Electronics*, *37*(3), 236–240. doi:10.1109/41.55163

Matsumoto, T., Tamaki, H., & Murao, H. (2007). Controlling residential co-generation system based on hierarchical decentralized model. *2007 IEEE Conference on Emerging Technologies& Factory Automation* (pp. 612-618).

Matysek, A., Ford, M., Jakeman, G., Gurney, A., & Fisher, B. S. (2006). *Technology: Its Role in Economic Development and Climate Change*. ABARE Research Report 06.6, prepared for Australian government Department of Industry, Tourism and Resources, Canberra, Australia.

Mavromatisl, S. P., & Kokossis, A. C. (1998). A logic based model for the analysis and optimisation of steam turbine networks. *Computers in Industry*, *36*, 165–179. doi:10.1016/S0166-3615(98)00070-0

Mavrotas, G., Demertzis, H., Meintani, A., & Diakoulaki, D. (2003). Energy planning in buildings under uncertainty in fuel costs: The case of a hotel unit in Greece. *Energy Conversion and Management*, *44*, 1303–1321. doi:10.1016/S0196-8904(02)00119-X

Mavrotas, G., Diakoulaki, D., Florios, K., & Georgiou, P. (2008). A mathematical programming framework for energy planning in services' sector buildings under uncertainty in load demand: The case of a hospital in Athens. *Energy Policy*, *36*, 2415–2429. doi:10.1016/j.enpol.2008.01.011

Maystre, L. Y., Pictet, J., & Simos, J. (1994). *Méthodes multicritères ELECTRE*. Presses Polytechniques et Universitaires. Romandes, Collection gérer l'environnement, Lausanne.

Mazumder, S. K., Burra, R. K., & Acharya, K. (2007). A Ripple-Mitigating and Energy-Efficient Fuel Cell Power-Conditioning System. *IEEE Transactions on Power Electronics*, *22*(4), 1429–1436. doi:10.1109/TPEL.2007.900598

Mazur, V. (2007). Fuzzy thermoeconomic optimization of energy-transforming systems. *Applied Energy*, *84*, 749–762. doi:10.1016/j.apenergy.2007.01.006

McCarthy, J. (1958). Programs with common sense. Symposium on Mechanization of Thought Processes. National Physical Laboratory, Teddington, England.

McCarthy, J. (1980). Circumscription - A form of non-monotonic reasoning. *Artificial Intelligence*, *13*, 27–39. doi:10.1016/0004-3702(80)90011-9

McCulloch, W. S., & Pitts, W. A. (1943). A logical calculus of the ideas imminent in nervous activity . *The Bulletin of Mathematical Biophysics*, *5*, 115–133. doi:10.1007/BF02478259

Meacham, J. R., Jabbari, F., Brouwer, J., Mauzey, J. L., & Samuelsen, G. S. (2006). Analysis of stationary fuel cell dynamic ramping capabilities and ultracapacitor energy storage using high resolution demand data. *Journal of Power Sources*, *156*(2), 472–479. doi:10.1016/j.jpowsour.2005.05.094

Medsker, L. R. (1995). Hybrid intelligent systems. Boston: Kluwer Academic Publishers.

Medsker, L. R. (1996). Microcomputer applications of hybrid intelligent systems. *Journal of Network and Computer Applications*, *19*, 213–234. doi:10.1006/jnca.1996.0015

Mei, F., & Pal, B. C. (2005, June). Modelling and small-signal analysis of a grid connected doubly fed induction generator. *IEEE Power Engineering Society General Meeting*, *3*, 2101-2108.

Meibom, P., Kiviluoma, J., Barh, R., Brand, H., Weber, C., & Larsen, H. V. (2007). Value of electric heat boiler and heat pumps for wind power integration. *Wind Energy (Chichester, England)*, *10*, 321–327. doi:10.1002/we.224

Meisingset, M., & Ohnstad, T. M. (2004). Field tests and modeling of a wind farm with doubly fed induction generators. *Proc. Nordic Wind Power Conf.*, Goteborg, Sweden, March, 1-6.

Melli, R., & Sciuba, E. (1999). A prototype expert system for the conceptual synthesis of thermal processes. *Energy Conversion and Management*, *38*(15-17), 1737–1749. doi:10.1016/S0196-8904(96)00186-0

Mellit, A. (2006). Artificial intelligence based-modeling for sizing of a stand-alone photovoltaic power system: Proposition for a new model using Neuro-Fuzzy system (ANFIS). In Proceedings of the 3rd International IEEE Conference on Intelligent Systems, UK, 1, 605-611.

Mellit, A., & Benghanem, M. Hadj, Arab, A., Guessoum, A. (2003). Modeling of sizing the photovoltaic system parameters using artificial neural network. In Procee. of IEEE, Conference on Control Application, Istanbul, 1, 353-357.

Mellit, A., & Kalogirou, S. A. (2006). Application of neural networks and genetic algorithms for predicting the optimal sizing coefficient of photovoltaic supply (PVS) systems In *Proceedings of World Renewable Energy Congress IX and Exhibition*, Florence, Italy on CD-ROM.

Mellit, A., & Kalogirou, S. A. (2008). Artificial intelligence techniques for photovoltaic applications: A review . *Progress in Energy and Combustion Science*, *34*(5), 574–632. doi:10.1016/j.pecs.2008.01.001

Mellit, A., Benghanem, M., & Bendekhis, M. (2005). Artificial neural network model for prediction solar radiation data: application for sizing stand-alone photovoltaic power system. In *proceedings of IEEE Power Engineering Society*, General Meeting 2005;1:40-44.

Mellit, A., Benghanem, M., Hadj Arab, A., Guessoum, A., & Moulai, K. (2004). Neural network adaptive wavelets for sizing of stand-Alone photovoltaic systems. Second IEEE International Conference on Intelligent Systems, 1, 365-370.

Messner, S., & Strubegger, M. (1995). *Model-Based Decision Support in Energy Planning.* WP-95-119, International Institute for Applied Systems Analysis, Laxenburg, Austria.

Messner, S., & Strubegger, M. (1995). *User's Guide for MESSAGE III.* WP-95-69. International Institute for Applied Systems Analysis, Laxenburg, Austria.

Metaxiotis, K., & Kagiannas, A. (2005). Intelligent computer applications in the energy sector: a literature review from 1990 to 2003. *International Journal of Computation Applications in Technology*, *22*, 53–64. doi:10.1504/IJCAT.2005.006936

Metaxiotis, K., Kagiannas, A., Askounis, D., & Psarras, J. (2003). Artificial intelligence in short term electric load forecasting: a state-of-the-art survey for the researcher. *Energy Conversion and Management*, *44*(9), 1525–1534. doi:10.1016/S0196-8904(02)00148-6

Meyer, N. I. (2003). European schemes for promoting renewables in liberalized markets. *Energy Policy*, *31*, 665–676. doi:10.1016/S0301-4215(02)00151-9

Michaelowa, A., & Jotzo, F. (2005). Transaction costs, institutional rigidities and the sized of the clean development mechanism. *Energy Policy*, *33*, 511–523. doi:10.1016/j.enpol.2003.08.016

Michalewicz, Z. (1992). Genetic algorithms + Data structures = Evolution programs. Springer-Verlag, Berlin.

Mihali, F., & Kos, D. (2006). Reduced Conductive EMI in Switched-Mode DC–DC Power Converters without EMI Filters: PWM Versus Randomized PWM. *IEEE Transactions on Power Electronics*, *21*(6), 1783–1794. doi:10.1109/TPEL.2006.882910

Millan, J., Campo, R. A., & Sierra, S. (1998). A modular system for decision-making support in generation expansion planning. *IEEE Transactions on Power Systems*, *13*, 667–671. doi:10.1109/59.667398

Mills, E., Kromerb, S., Weissc, G., & Mathewd, P. A. (2006). From volatility to value: analysing and managing financial and performance risk in energy savings projects. *Energy Policy*, *34*, 188–199. doi:10.1016/j.enpol.2004.08.042

Minakawa, T., Ichikawa, Y., Kunugi, M., Shimada, K., Wada, N., & Utsunomiya, M. (1995). Development and implementation of a power system fault diagnosis expert system. *IEEE Transactions on Power Systems, 10*(2), 932–939. doi:10.1109/59.387936

Minsky, M. (1975). A framework for representing knowledge. In The psychology of computer vision, Winston, P.H. (Ed.). McGraw Hill, New York.

Minsky, M., & Papert, S. (1969). Perceptrons. MIT Press, Cambridge, Massachusetts.

Mirasgedis, S., & Diakoulaki, D. (1997). Multicriteria analysis vs. externalities assessment for the comparative evaluation of electricity generation systems. *European Journal of Operational Research, 102*, 364–379. doi:10.1016/S0377-2217(97)00115-X

Mircevski, S., Kostic, Z. A., & Andonov, Z. (1998, May). Energy saving with pumps ac adjustable speed drives. *Proceedings of 9th Mediterranean Electrotechnical Conf. (MELECON)*, Tel-Aviv, Israel, *2*, 1224-1227.

Mogel, A., Krupar, J., & Schwarz, W. (2005). EMI performance of spread spectrum clock signals with respect to the IF bandwidth of the EMC standard. *Proceedings of the 2005 European Conference on Circuit Theory and Design, 1*, 169-172. Retrieved December 25, 2005, from http://ieeexplore.ieee.org

Moghavvemi, M., Yang, S. S., & Kashem, M. A. (1998). A practical neural network approach for power generation automation. *Proceedings of International Conference on Energy Management and Power Delivery, 1*, 305–310.

Mohan, N., Undeland, T. M., & Robbins, W. P. (2002). *Power Electronics: Converters, Applications, and Design*. Hoboken, NJ: John Wiley & Sons Inc.

Mohanty, B., & Panda, H. (1993). Integrated energy system for industrial complexes. Part I: A linear programming approach. *Applied Energy, 46*, 317–348. doi:10.1016/0306-2619(93)90048-T

Monitoring and Evaluation of the RES directives implementation in EU27 and policy recommendations for 2020 – RES2020 (n.d.). Retrieved June, 2, 2008, from http://www.res2020.eu/.

Montano – Guzman. L. (2000). *Une méthodologie d'aide multicritère a la décision pour la diagnostique de l'entreprise*. Paper Presented at the 51st Meeting of the European Working Group on MCDA, Madrid.

Morel, C., Bourcerie, M., & Chapeau-Blondeau, F. (2005). Improvement of power supply electromagnetic compatibility by extension of chaos anti-control. *Journal of Circuits, Systems, and Computers, 14*(4), 757–770. doi:10.1142/S0218126605002556

Morel, L., Godfroid, H., Mirzaian, A., & Kauffmann, J. M. (1998). Double-fed induction machine: converter optimization and field oriented control without position sensor. *Proc. IEE.- Electr. Power Appl., 145*(4), 360–368.

Moren, J., de Hann, S. W. H., & Bauer, P. (2003). Comparison of complete and reduced models of a wind turbine with doubly fed induction generator. *Proc. 10th Eur. Conf. Power Electronics Applications*, Toulouse, France, September.

Moreno, A., Julve, J., Silvestre, S., & Castaer, L. (2000). A fuzzy logic controller for stand alone PV systems, IEEE, 1618-1621.

Morren, J., & de Haan, S. W. H. (2005). Ridethrough of the wind turbines with doubly-fed induction generator during a voltage dip. *IEEE Transactions on Energy Conversion, 20*(2), 435–441. doi:10.1109/TEC.2005.845526

Morse, J. N. (1980). Reducing the size of nondominated set: Pruning by clustering. *Computers & Operations Research, 7*, 55–66. doi:10.1016/0305-0548(80)90014-3

Moslehi, K., Khadem, M., Bernal, R., & Hernandez, G. (1991). Optimization of multiplant cogeneration system operation including electric and steam networks. *IEEE Transactions on Power Systems, 6*, 484–490. doi:10.1109/59.76690

Mostaghim, S., & Teich, J. (2003). Strategies for finding good local guides in multiobjective particle swarm optimization (MOPSO). *Proceedings of 2003 IEEE Swarm Intelligence Symposium*, Indianapolis, (pp. 26-33).

Mousseau, V., & Slowinski, R. (1998). Inferring an ELECTRE TRI model from assignment examples. *Journal of Global Optimization*, (12): 157–174. doi:10.1023/A:1008210427517

Mousseau, V., Slowinski, R., & Zielniewicz, P. (1999). *ELECTRE TRI 2.0a – Methodological guide and user's manual*. Documents du LAMSADE No 111, University Paris – Dauphine.

Muller, S., Deicke, M., & De Doncker, R. W. (2000). Adjustable speed generators for wind turbines based on doubly-fed induction machines and 4-quadrant IGBT converters linked to the rotor. *Proc. IEEE Industry Applications Conf.*, 4, October, 2249 – 2254.

Muller, S., Deicke, M., & De Doncker, R. W. (2002). Doubly fed induction generator systems for wind turbines. *IEEE Industry Applications Magazine*, *8*(3), 26–33. doi:10.1109/2943.999610

Munda, G. (1996). *Naiade. Manual and tutorial*. In Ispra: Joint Research Centre - EC, ISPRA SITE.

Muselli, M., etc. (1999). Design of hybrid-photovoltaic power generator, with optimization of energy management. *Solar Energy*, *65*(3), 143–157. doi:10.1016/S0038-092X(98)00139-X

Nagata, T., Sasaki, H., & Yokoyama, R. (1995). Power system restoration by joint usage of expert system and mathematical programming approach. *IEEE Transactions on Power Systems*, *10*(3), 1473–1479. doi:10.1109/59.466501

Nagata, T., Sasaki, H., & Yokoyama, R. (1995). Power system restoration by joint usage of expert system and mathematical programming approach. *IEEE Transactions on Power Systems*, *10*(3), 1473–1479. doi:10.1109/59.466501

Nakra, H. L., & Dube, B. (1988). Slip power recovery induction generators for large vertical axis wind turbines. *IEEE Transactions on Energy Conversion*, *3*(4), 733–737. doi:10.1109/60.9346

Nannariello, J., & Frike, F. R. (2001). Introduction to neural network analysis and its applications to building services engineering. *Building Services Engineering Research and Technology*, *22*(1), 58–68. doi:10.1191/014362401701524127

Narita, K. (1996). *Research on unused energy of cold region cities and utilization for district heat and cooling*. Ph.D. thesis, Dep. Socio-Environmental Eng. Faculty of Eng., Hokkaido Univ. Sapporo, Japan.

National Energy Technology Laboratory, & U.S. Department of Energy (2004). *Fuel Cell Handbook*. USA DOE/NETL: EG&G Services Parsons, Inc. Science Applications International Corporation.

National Energy Technology Laboratory. (2005). *NETL published fuel cell specifications for Future Energy Challenge 2005 Competition*. Retrieved June 1, 2005, from http://www.netl.doe.gov.

NEDO Technical information data base, (2008). *Standard meteorology and solar radiation data (METPV-3)*. Retrieved from http://www.nedo.go.jp/database/index.html

NEMA Standards Publication no MG1-1993, Motors and Generators. (n.d.). Washington DC: National Electrical Manufacturers Association.

Neto, A. H., & Fiorelli, F. A. S. (2008). Comparison between detailed model simulation and artificial neural network for forecasting building energy consumption. *Energy and Buildings*.

New Energy Externalities Developments for Sustainability - NEEDS (n.d.). Retrieved May, 25, 2008, from http://www.needs-project.org.

Newell, A., & Simon, H. A. (1972). Human problem solving. Prentice-Hall, Englewood Cliffs, NJ.

Newell, A., Shaw, J. C., & Simon, H. A. (1963). Empirical explorations with the logic theory machine: A case study in heuristics. In Computers and Thought, Feigenbaum, E. A. & Feldman, J. (Eds.), McGraw Hill, New York.

Nigim, K., Munier, N., & Green, J. (2004). Pre-feasibility MCDM tools to aid communities in prioritizing local viable renewable energy sources. *Renewable Energy*, *29*(11), 1775–1791. doi:10.1016/j.renene.2004.02.012

Niiranen, J. (2004). *Voltage dip ride through of a Doubly-fed generator equipped with active crowbar*. Proc. Nordic Wind Power Conf., Chalmers University of Technology, Goteborg, Sweden, March.

Nilsson, N. (1998). Artificial intelligence: A new synthesis. Morgan Kaufmann.

Noble, K. (2005, November). *ANSWERv6 for TIMES: Status Report.* Noble-Soft Systems, Australia. ETSAP Annex X Meeting. Oxford, UK. Retrieved June, 22, 2008, from http://www.ukerc.ac.uk/Downloads/PDF/E/ETSAP_24Noble_ANSWER.pdf

Norheim, I., Uhlen, K., Tande, J.O., Toftevaag, T. & Palsson, M. (2004). *Doubly fed induction generator model for power system simulation tools.* Nordic Wind Power Conf, Chalmers University of Technology, Goteborg, Sweden, March.

O'Sullivan, D. T. J., Keane, M. M., Kelliher, D., & Hitchcock, R. J. (2004). Improving building operation by tracking performance metrics throughout the building lifecycle (BLC). *Energy and Building, 36*, 1075–1090. doi:10.1016/j.enbuild.2004.03.003

Obara, S. (2007). Improvement of power generation efficiency of an independent micro grid composed of distributed engine generators. *Transactions of the ASME . Journal of Energy Resources Technology, 129*(Issue 3), 190–199. doi:10.1115/1.2748812

Obara, S. (2007). Energy cost of an independent micro-grid with control of power output sharing of a distributed engine generator. *Journal of Thermal Science and Technology, 2*(1), 42–53.

Obara, S., & Tanno, I. (2008). Operation Prediction of a Bioethanol Solar Reforming System Using a Neural Network. *Journal of Thermal Science and Technology, 2*(2), 256–267. doi:10.1299/jtst.2.256

Obara, S., & Tanno, I. (2008). Fuel reduction effect of the solar cell and diesel engine hybrid system with a prediction algorithm of solar power generation. *Journal of Power and Energy Systems, 2*(4), 1166–1177. doi:10.1299/jpes.2.1166

OEE. (2008). *Energy Use Data Handbook Tables (Canada)*, retrieved July 26, 2008, from http://www.oee.nrcan.gc.ca/corporate/statistics/neud/dpa/handbook_totalsectors_ca.cfm?attr=0

Ogaji, S., Sampath, S., Singh, R., & Probert, D. (2002). Novel approach for improving power-plant availability using advanced engine diagnostics. *Applied Energy, 72*, 389–407. doi:10.1016/S0306-2619(02)00018-1

Oh, S. D., Lee, H. J., Jung, J. Y., & Kwak, H. Y. (2007). Optimal planning and economical evaluation of co-generation system. *Energy, 32*, 760–771. doi:10.1016/j.energy.2006.05.007

Oh, S. D., Oh, H. S., & Kwak, H. Y. (2007). Economic evaluation for adoption of cogeneration system. *Applied Energy, 84*(3), 266–278. doi:10.1016/j.apenergy.2006.08.002

Ojanen, O., Makkonen, S., & Salo, A. (2005). A multi-criteria framework for the selection of risk analysis methods at energy utilities. *International Journal of Risk Assessement and Management, 5*, 16–35. doi:10.1504/IJRAM.2005.006609

Öko-Institut. (2008). *Global Emission Model for Integrated Systems (GEMIS) - Version 4.4.* Retrieved September, 2, 2008, from http://www.oeko.de/service/gemis/en/.

Oliver, J. A., Lawrence, R., & Banerjee, B. (2002). Power quality – how to specify power quality tolerant process equipment. *IEEE Industry Applications Magazine, 8*(5), 21–30. doi:10.1109/MIA.2002.1028387

Olofsson, T., & Andersson, S. (2001). Long-term energy demand predictions based on short-term measured data. *Energy and Building, 33*(2), 85–91. doi:10.1016/S0378-7788(00)00068-2

Onovwiona, H. I., & Ugursal, V. I. (2006). Residential cogeneration systems: review of the current technology. *Renewable & Sustainable Energy Reviews, 10*, 389–431. doi:10.1016/j.rser.2004.07.005

Ortuzar, M., Dixon, J., & Moreno, J. (2003). Design, Construction and Performance of a Buck-Boost Converter for an Ultracapacitor-Based Auxiliary Energy System for Electric Vehicles. In IEEE Industrial Electronics Society (Ed.), *IECON2003* (Vol. 3, pp. 2889 – 2894), Roanoke, VA.

Ouyang, Z., & Shahidehpour, S. M. (1990). Short-term unit commitment expert system. *Electric Power Systems Research, 20*(1), 1–13. doi:10.1016/0378-7796(90)90020-4

Padhy, N. (2000). Unit commitment using hybrid models: a comparative study for dynamic programming, expert system, fuzzy system and genetic algorithms. *Electrical Power & Energy Systems, 23*, 827–836. doi:10.1016/S0142-0615(00)00090-9

Paladini, V., Donateo, T., de Risi, A., & Laforgia, D. (2007). Super-capacitors fuel-cell hybrid electric vehicle optimization and control strategy development. *Energy Conversion and Management, 48*(11), 3001–3008. doi:10.1016/j.enconman.2007.07.014

Palanichamy, C., Nadarajan, C., Naveen, P., Babu, N.S., & Dhanalakshmi, (2001). Budget constrained energy conservation - an experience with a textile industry . *IEEE Trans. Energy Conversion, 16*(4), 340-344. doi:10.1109/60.969473

Palle, B., Simoes, M. G., & Farret, F. A. (2005). Dynamic interaction of an integrated doubly-fed induction generator and a fuel cell connected to grid. *IEEE 36th Conf. Power Electronics Specialists*, June, (pp. 185 – 190).

Palsson, O. P., & Ravn, H. F. (1994). Stochastic heat storage problem -- Solved by the progressive hedging algorithm. *Energy Conversion and Management, 35*(12), 1157–1171. doi:10.1016/0196-8904(94)90019-1

Papaefthymiou, G., Schavemaker, P. H., van der Sluis, L., Kling, W. L., Kurowicka, D., & Cooke, R. M. (2006). Integration of stochastic generation in power systems. *Electrical Power & Energy Systems, 28*, 655–667. doi:10.1016/j.ijepes.2006.03.004

Papathanassiou, S. A., & Papadopoulos, M. P. (1994). *Simulation and control of a variable speed wind turbine equipped with double output induction generator.* Proc. PEMC, Warsaw, Poland.

Paravantis, J. (1995, November). Energy savings in buildings: heating and passive cooling [in Greek] [in Greek]. *Energy, 17*, 49–59.

Park, D. C., El-Sharkawi, M. A., Marks, R. J., Atlas, L. E., & Damborg, M. J. (1991). Electric load forecasting using an ANN. *IEEE Transactions on Power Systems, 6*(2), 442–449. doi:10.1109/59.76685

Park, Y. M., & Lee, K. H. (1995). Application of expert system to power system restoration in local control center. *International Journal of Electrical Power & Energy Systems, 17*(6), 407–415. doi:10.1016/0142-0615(94)00011-5

Parti, M., & Parti, C. (1980). The total and appliance-specific conditional demand for electricity in the household sector. *The Bell Journal of Economics, 11*, 309–321. doi:10.2307/3003415

Parui, S. (2003). *Bifurcation in dc-dc converters: Effects of transition from continuous conduction mode to discontinuous conduction mode and feedback loop delay.* Published doctoral dissertation, IIT Kharagpur, India.

PAS55. *PAS 55 website*. Retrieved September 15, 2008, from http://pas55.net.

Pathapati, P. R., Xue, & Tang, J. (2005). A new Dynamic Model for Predicting Transient Phenomena in a PEM Fuel Cell System. *Renewable . The Energy Journal (Cambridge, Mass.), 30*(1), 1–8.

Patin, N., Monmasson, E., & Louis, J. P. (2005). Analysis and control of a cascaded doubly-fed induction generator. *IEEE 32nd Annual Conf. Industrial Electronics Society*, Nov., (pp. 2481 – 2486).

Patlitzianas, K. D., & Psarras, J. (2006). Formulating a Modern Energy Companies' Environment in the EU Accession Member States through a Decision Support Methodology. *Energy Policy, 35*(4), 2231–2238. doi:10.1016/j.enpol.2006.07.010

Patlitzianas, K. D., Doukas, H., & Psarras, J. (2006). Enhancing Renewable Energy in the Arab States of the Gulf. *Energy Policy, 34*(18), 3719–3726. doi:10.1016/j.enpol.2005.08.018

Patlitzianas, K. D., Doukas, H., & Psarras, J. (2008). Sustainable energy policy indicators: Review and recommendations. *Renewable Energy, 33*(5), 966–973. doi:10.1016/j.renene.2007.05.003

Patlitzianas, K. D., Doukas, H., Kagiannas, A., & Askounis, D. (2006). A Reform Strategy of the Energy Sector of the 12 countries of North Africa and Eastern Mediterranean. *Energy Conversion and Management*, *47*(13-14), 1913–1926. doi:10.1016/j.enconman.2005.09.004

Patlitzianas, K. D., Ntotas, K., Doukas, H., & Psarras, J. (2007). Assessing the Renewable Energy Producers' Environment in the EU Accession Member States. *Energy Conservation and Management*, *48*(3), 890–897. doi:10.1016/j.enconman.2006.08.014

Patlitzianas, K. D., Papadopoulou, A., Flamos, A., & Psarras, J. (2005). CMIEM: the computerised model for intelligent energy management. *International Journal of Computer Applications in Technology*, *22*(2/3), 120–129. doi:10.1504/IJCAT.2005.006943

Patlitzianas, K. D., Pappa, A., & Psarras, J. (2008). An information decision support system towards the formulation of a modern energy companies' environment. *Renewable & Sustainable Energy Reviews*, *12*, 790–806. doi:10.1016/j.rser.2006.10.014

Patlitzianas, K., Doukas, H., & Psarras, J. (2006). Designing an Appropriate ESCOs' Environment in the Mediterranean. *Management of Environmental Quality*, *17*(5), 538–554. doi:10.1108/14777830610684512

Patlitzianas, K., Papadopoulou, A., Flamos, A., & Psarras, J. (2005). CMEM: The Computerized Model for Intelligent Energy Management. *International Journal of Computer Applications in Technology*, *22*(2/3), 120–129. doi:10.1504/IJCAT.2005.006943

Patlitzianas, KD., Doukas, H., & Psarras (2008), Sustainable energy policy indicators: Review and recommendations. *Renewable Energy, 33(*5), 966-973.

Pedersen, L. (2007). Use of different methodologies for thermal load and energy estimations in buildings including meteorological and sociological input parameters. *Renewable & Sustainable Energy Reviews*, *15*(5), 998–1007. doi:10.1016/j.rser.2005.08.005

Pena, R. Cardenas, R., Clare, J. & Asher, G. (2002). Control strategy of doubly fed induction generators for a wind diesel energy system. *IEEE 28th Annual Conf.*, *4*, 3297-3302.

Pena, R. S., Asher, G. M., Clare, J. C., & Cardenas, R. (1996). A constant frequency constant voltage variable speed stand alone wound rotor induction generator. *Proc. Int. Conf. Opportunities and Advances in Int. Electric Power Generation, (Conf. Publ. 419)*, March, (pp. 111 – 114).

Pena, R., Cardenas, R. J., Asher, G. M., & Clare, J. C. (2000, Oct.). Vector controlled induction machines for stand-alone wind energy applications. *Proc. IEEE Industry Applications Conf.*, *3*, 1409 – 1415.

Pena, R., Cardenas, R. J., Asher, G. M., Clare, J. C., Rodriguez, J., & Cortes, P. (2002, November). Vector control of a diesel driven doubly fed induction machine for a standalone variable speed energy system, *Proc. IEEE 28th Annual Conf. of the Industrial Electronics Society, 2*, 985 – 990.

Pena, R., Clare, J. C., & Asher, G. M. (1996). A doubly fed induction generator using back-to-back PWM converters supplying an isolated load from a variable speed wind turbine. *IEEE Proc.-. Electric Power Applications*, *143*(5), 380–387. doi:10.1049/ip-epa:19960454

Pena, R., Clare, J. C., & Asher, G. M. (1996). Doubly fed induction generator using back-to-back PWM converters and its application to variable-speed wind-energy generation. *IEE Proceedings. Electric Power Applications*, *143*(3), 231–241. doi:10.1049/ip-epa:19960288

Pena, R., Crdenas, R., Escobar, E., Clare, J., & Wheeler, P. (2007). Control system for unbalanced operation of stand-alone doubly fed induction generators. *IEEE Transactions on Energy Conversion*, *22*(2), 544–545. doi:10.1109/TEC.2007.895393

Peng, T. M., Hubele, N. F., & Karady, G. G. (1992). Advancement in the application of NN for short-term load forecasting. *IEEE Transactions on Power Systems*, *7*(1), 250–257. doi:10.1109/59.141711

Peresada, S., Tilli, A., & Tonielli, A. (1998, September). Robust active-reactive power control of a doubly fed induction generator. *Proc. 24th Annual Conf. IEEE Industrial Electronics Society, 2*, 1621 – 1625.

Perryman, R. (1995). Condition monitoring of combined heat and power systems. *IEEE Colloquium on Condition Monitoring of Electrical Machines* (pp. 1-4).

Peterson, A. (2003). *Analysis, modeling and control of doubly fed induction generators for wind turbines.* PhD Thesis, Chalmers University of Technology, Goteborg, Sweden.

Peterson, A., Lundberg, S. & Thiringer, T. (2004, March 1 – 7). *A DFIG wind turbine ride through system influence on the energy production.* Nordic Wind Power Conf., Chalmers University of Technology, Goteborg, Sweden.

Petersson, A., Thiringer, T., Harnefors, L., & Petru, T. (2005). Modeling and experimental verification of grid interaction of a DFIG wind turbine. *IEEE Transactions on Energy Conversion, 20*(4), 878–886. doi:10.1109/TEC.2005.853750

Pfafferott, J., Herkel, S., & Wambsgan, M. (2004). Design, monitoring and evaluation of a low energy office building with passive cooling by night ventilation. *Energy and Building, 36*, 455–465. doi:10.1016/j.enbuild.2004.01.041

Piacentino, A., & Cardona, F. (2008). An original multi-objective criterion for the design of small-scale polygeneration systems based on realistic operating conditions. *Applied Thermal Engineering, 28*, 2391–2404. doi:10.1016/j.applthermaleng.2008.01.017

Piacentino, A., & Cardona, F. (2008). EABOT – Energetic analysis as a basis for robust optimization of trigeneration systems by linear programming. *Energy Conversion and Management, 49*, 3006–3016. doi:10.1016/j.enconman.2008.06.015

Pietrapertosa, F., Cosmi, C., Macchiato, M., Marmo, G., & Salvia, M. (2003). Comprehensive modelling for approaching the Kyoto targets on local scale. *Renewable & Sustainable Energy Reviews, 7*(3), 249–270. doi:10.1016/S1364-0321(03)00041-8

Pietrapertosa, F., Cosmi, C., Macchiato, M., Salvia, M., & Cuomo, V. (in press). Life Cycle Assessment, ExternE and Comprehensive Analysis for an integrated evaluation of the environmental impact of anthropogenic activities. *Renewable & Sustainable Energy Reviews.*

Pilavachi, P. A., Roumpeas, C. P., Minett, S., & Afgan, N. H. (2006). Multi-criteria evaluation for CHP system options. *Energy Conversion and Management, 47*, 3519–3529. doi:10.1016/j.enconman.2006.03.004

Pillay, P., & Manyage, M. (2001). Definitions of voltage unbalance. *IEEE Power Engineering Review, 21*(11), 50–51. doi:10.1109/39.920965

Pillay, P., Hofmann, P., & Manyage, M. (2002). Derating of induction motors operating with a combination of unbalanced voltages and over or undervoltages. *IEEE Transactions on Energy Conversion, 17*(4), 485–491. doi:10.1109/TEC.2002.805228

Pischinger, S., Schonfelder, C., & Ogrzewalla, J. (2007). Analysis of dynamic requirements for fuel cell systems for vehicle applications. *Journal of Power Sources, 154*(2), 420–427. doi:10.1016/j.jpowsour.2005.10.037

Pohekar, S. D., & Ramachandran, M. (2004). Application of multi-criteria decision making to sustainable energy planning. *Renewable & Sustainable Energy Reviews, 8*, 365–381. doi:10.1016/j.rser.2003.12.007

Pokharel, S., & Chandrashekar, M. (1998). A Multi-objective approach to rural energy policy analysis. *Energy, 23*(4), 325–336. doi:10.1016/S0360-5442(97)00103-5

Poller, M. A. (2003, June 23-26). Doubly-fed induction machine model stability assessment of wind farms. *IEEE Power Tech Conf. Proc., 3,* 1-6.

Poon, P. W. Y. (2005). *Development and Application of Knowledge Management in CLP Power.* Paper presented at IEEE KM Symposium 2005.

Prastakos, G. (2008). *Management Science: Decision making in the information society (Vol 2).* Athens: Stamoulis Publications.

PRé Consultants. (n.d.) Retrieved August, 28, 2008, from http://www.pre.nl/.

Preiss, P. (2007). EcoSenseWeb V1.2. *User's manual & "Description of updated and extended draft tools for the detailed site-dependent assessment of External Costs"* [Draft version 4]. *Risk of Energy Availability: Common Corridors for Europe Supply Security – REACCESS* (n.d.). Retrieved June 7, 2008, from http://reaccess.epu.ntua.gr/.

Pribicevic, B., Krasenbrik, B., & Haubrich, H. J. (2002). Co-generation in a competitive market. *IEEE Power Engineering Society Summer Meeting* (vol.1, pp. 422-426).

Puttgen, H. B., & MacGregor, P. R. (1989). Optimum scheduling procedure for cogenerating small power producing facilities. *IEEE Transactions on Power Systems, 4*(3), 957–964. doi:10.1109/59.32585

Quang, N. P., Dittrich, J. A., & Thieme, A. (1997). Doubly fed induction machine as generator: control algorithm with decoupling of torque and power factor. *Electrical Engineering, 80*, 325–335. doi:10.1007/BF01370969

Rabelo, B., & Hofmann, W. (2001, October). Optimal active and reactive power control with the doubly fed induction generator in the MW-class wind-turbines. *Proc. 4th IEEE Int. Conf. Power Electronics and Drive Systems, 1*, 53 – 58.

Radel, U., Navarro, D., Berger, G., & Berg, S. (2001). Sensorless field-oriented control of a slipring induction generator for a 2.5 MW wind power plant from Nordex energy GMBH. *Proc. Eur. Power Electron. Conf.*, Graz, Austria.

Raducu, M., & Bizon, N. (2007). Efficient Energy Generation System using a Thiristors Inverter Topology. In N. Bizon (Ed.), *International Conference on Electronics, Computers and Artificial Intelligence - ECAI'07* (Vol. 2, pp. 67-70). Bucharest: MatrixROM & University of Pitesti Press.

Rahman, S., & Lauby, M. (1993). Identification of Potential Areas for the Use of Expert Systems in Power System Planning. *Expert Systems with Applications, 6*, 203–212. doi:10.1016/0957-4174(93)90010-4

Rahman, T., Mittelhammer, C., & Wandschneider, P. (2005). *Measuring the Quality of Life Across Countries: a sensitivity analysis of well-being indices.* United Nations University and WIDER (World Institute For development Economic Research), USA [Research Paper No. 2005/06].

Rajalakshmi, N., Pandiyan, S., & Dhathathreyan, K. S. (in press). Design and development of modular fuel cell stacks for various applications. *International Journal of Hydrogen Energy.*

Ramos, C. J., Martins, A. P., Araujo, A. S. & Carvalho, A. S. (2002). Current control in the grid connection of the double-output induction generator linked to a variable speed wind turbine. *IEEE 28th Annual Conf., 2*, 979-984.

Ramschak, E., Peinecke, V., Prenninger, P., Schaffer, T., & Hacker, V. (2006). Detection of fuel cell critical status by stack voltage analysis. *Journal of Power Sources, 157*(2), 837–840. doi:10.1016/j.jpowsour.2006.01.009

Ramtharan, G., Ekanayake, J. B., & Jenkins, N. (2007). Frequency support from doubly fed induction generator wind turbines. *IET. - . Renew. Power Gener., 1*(1), 3–9. doi:10.1049/iet-rpg:20060019

Rashid, M. H. (2003). *Power Electronics: Circuits, Devices and Applications.* Upper Saddle River, NJ: Pearson Education.

Riascos, L. A. M., Simoes, M. G., & Miyagi, P. E. (2007). A Bayesian network fault diagnostic system for proton exchange membrane fuel cells. *Journal of Power Sources, 165*(1), 267–278. doi:10.1016/j.jpowsour.2006.12.003

Rice, D. E. (1988). A suggested energy-savings evaluation method for ac adjustable-speed rive applications. *IEEE Transactions on Industry Applications, 24*(6), 1107–1117. doi:10.1109/28.17486

Rich, E., & Knight, K. (1996). Artificial intelligence. McGraw-Hill, New York.

Riedmiller, M., & Braun, H. (1993). A Direct Adaptive Method for Faster Backpropagation Learning: The RPROP Algorithm. *The Proceedings of the IEEE International Conference on Neural Networks, 1*, 586–591. doi:10.1109/ICNN.1993.298623

Rifaat, R. M. (1997). Practical considerations in applying economical dispatch models to combined cycle cogeneration plants. *IEEE WESCANEX 97 Communications, Power and Computing* (pp. 59-63).

Rifaat, R. M. (1998). Economic dispatch of combined cycle cogeneration plants with environmental constraints. *Proceedings of International Conference on Energy Management and Power Delivery, 1*, 149–153.

Roberge, M. A., Lamarche, L., Karjl, S., & Moreau, A. (1997). Model of Room Storage Heater and System Identification Using Neural Networks, Proc. of CLIMA 2000 Conf., Brussels, Belgium, 265.

Robert, F. (1995). Neural fuzzy systems. Abo Akademi University.

Roberts, P. C., McMahon, R. A., Tavner, P. J., Maciejowski, J. M., Flack, T. J., & Wang, X. (2004). Performance of rotors in a brushless doubly fed induction machine (BDFM). *Proc. 16th Int. Conf. Electrical Machines*, (pp. 450-455).

Rodriguez, J. M., Fernandez, J. L., Beatu, D., Iturbe, R., Usaola, J., Ledesma, P., & Wilhelmi, J. R. (2002). Incidence on power system dynamics of high penetration of fixed speed and doubly fed wind energy systems: study of the Spanish case. *IEEE Transactions on Power Systems, 17*(4), 1089–1095. doi:10.1109/TPWRS.2002.804971

Rolfsman, B. (2004). Combined heat-and-power plants and district heating in a deregulated electricity market. *Applied Energy, 78*, 37–52. doi:10.1016/S0306-2619(03)00098-9

Rolfsman, B. (2004). Optimal supply and demand investment in municipal energy systems. *Energy Conversion and Management, 45*, 595–611. doi:10.1016/S0196-8904(03)00174-2

Rong, A. (2006). *Cogeneration planning under the deregulated power market and emissions trading scheme.* Ph.D. Thesis, University of Turku, Turku Center for Computer Science, Turku, Finland.

Rong, A., & Lahdelma, R. (2005). Risk analysis of expansion planning of combined heat and power energy system under emissions trading scheme. In W. Tayati (eds) *Proceedings of IASTED International Conference on Energy and Power Systems* (pp. 70-75), April 18-20, 2005, Krabi, Thailand

Rong, A., & Lahdelma, R. (2005). An efficient model and specialized algorithms for cogeneration planning. In W. Tayati (Ed.) *Proceedings of the IASTED International Conference Energy and Power System* (pp. 1-7), April 18-20, 2005, Krabi, Thailand.

Rong, A., & Lahdelma, R. (2005). An efficient linear programming model and optimization algorithm for trigeneration. *Applied Energy, 82*, 40–63. doi:10.1016/j.apenergy.2004.07.013

Rong, A., & Lahdelma, R. (2007). CO_2 emissions trading planning in combined heat and power production via multi-period stochastic optimization. *European Journal of Operational Research, 176*, 1874–1895. doi:10.1016/j.ejor.2005.11.003

Rong, A., & Lahdelma, R. (2007). Efficient algorithms for combined heat and power production planning under the deregulated electricity market. *European Journal of Operational Research, 176*, 1219–1245. doi:10.1016/j.ejor.2005.09.009

Rong, A., & Lahdelma, R. (2007). An efficient envelope-based Branch and Bound algorithm for non-convex combined heat and power production planning. *European Journal of Operational Research, 183*, 412–431. doi:10.1016/j.ejor.2006.09.072

Rong, A., & Lahdelma, R. (2007). An effective heuristic for combined heat and power production planning with power ramp constraints. *Applied Energy, 84*, 307–325. doi:10.1016/j.apenergy.2006.07.005

Rong, A., Hakonen, H., & Lahdelma, R. (2006). An efficient linear model and optimisation algorithm for multi-site combined heat and power production. *European Journal of Operational Research, 168*(2), 612–632. doi:10.1016/j.ejor.2004.06.004

Rong, A., Hakonen, H., & Lahdelma, R. (2008). A variant of the dynamic programming algorithm for the unit commitment of combined heat and power systems. *Eu-*

ropean Journal of Operational Research, 190, 741–755. doi:10.1016/j.ejor.2007.06.035

Rong, A., Hakonen, H., Makkonen, S., Ojanen, O., & Lahdelma, R. (2004). CO_2 emissions trading optimization in combined heat and power generation. In P. Neittaanmäki, T. Rossi, K. Majava, O. Pironneau (eds.), *Proc. ECCOMAS 2004.*

Rong, A., Lahdelma, R., & Grunow, M. (2009). An improved unit decomitment algorithm for combined heat and power systems. *European Journal of Operational Research, 195*, 552–562. doi:10.1016/j.ejor.2008.02.010

Rong, A., Lahdelma, R., & Luh, P. (2008). Lagrangian relaxation based trigeneration planning with storages. *European Journal of Operational Research, 188*, 240–257. doi:10.1016/j.ejor.2007.04.008

Rongve, K. S., Naess, B. I., Undeland, T. M., & Gjengedal, T. (2003, June). Overview of torque control of a doubly fed induction generator. *Proc. IEEE Power Tech. Conf.*, Bologna, Italy, 3, 292 – 297.

Rooijers, F. J., & van Amerongen, R. A. M. (1994). Static economic dispatch for co-generation systems. *IEEE Transactions on Power Systems, 9*, 1392–1398. doi:10.1109/59.336125

Rothwell, G., & Gomez, T. (2003). *Electricity Economics: Regulation and Deregulation (1st ed.)*. New York: IEEE-Wiley Press.

Rubira, S. D., & McCulloch, M. D. (2000). Control comparison of doubly fed wind generators connected to the grid by asymmetric transmission lines. *IEEE Transactions on Industry Applications, 36*(4), 986–991. doi:10.1109/28.855951

Rumelhart, D. E., & McClelland, J. L. (1986). Parallel distributed processing: Explorations in the microstructure of cognition. The PDP Research Group, MIT Press/ Bradford Books, Cambridge, Massachusetts.

Rumelhart, D. E., & McClelland, J. L. (1986). *Parallel Distributed Processing, Vol 1.* Cambridge, MA: The MIT Press.

Runcos, F., Carlson, R., Oliveira, A. M., Peng, P. K., & Sadowski, N. (2004). Performance analysis of a brushless double fed cage induction generator. *Proc. Nordic Wind Power Conf.*, Goteborg, Sweden, March, 1-8.

Russel, S., & Norvig, P. (1995). Artificial intelligence: A modern approach. Prentice-Hall, Englewood Cliffs, NJ.

Saadat, H. (1999). *Power System Analysis*. New York: McGraw-Hill

Sahu, G. K. (2000). *Pumps*. New Delhi, India: New Age International.

Sakawa, M., Kato, K., & Ushiro, S. (2002). Operational planning of district heating and cooling plants through genetic algorithms for mixed 0—1 linear programming. *European Journal of Operational Research, 137*, 677–687. doi:10.1016/S0377-2217(01)00095-9

Salameh, Z. M., & Kazda, L. F. (1986). Analysis of the steady state performance of the double output induction generator. *IEEE Transactions on Energy Conversion, 1*(1), 26–32. doi:10.1109/TEC.1986.4765666

Salameh, Z. M., & Wang, S. (1987). Microprocessor control of double output induction generation. I. Inverter firing circuit. *IEEE Transactions on Energy Conversion, 2*(2), 175–181. doi:10.1109/TEC.1987.4765826

Salgado, F., & Pedrero, P. (2008). Short-term operation planning on cogeneration systems: A survey. *Electric Power Systems Research, 78*, 835–848. doi:10.1016/j.epsr.2007.06.001

SANYO Co. Ltd. (2007). Nickel-Metal Hydride Production Information, Batteries. Retrieved from http://www.sanyo.co.jp/energy/english/product/twicell_2.html

Savage, L. J. (1954). *The Foundation of Statistics*. New York: Dover Publications.

Schaepers, M., Seebregts, A., de Jong, J., Maters, H. (2007). *EU Standards for Energy Security of Supply*, (ECN-E-07-004/CIEP).

Schalkoff, J., Culberson, J., Treloar, N., & Knight, B. (1992). A world championship caliber checkers program.

Artificial Intelligence, *53*(2-3), 273–289. doi:10.1016/0004-3702(92)90074-8

Schavemaker, P., & Sluis, L. (2008). *Electric Power System Essentials*. New York: John Wiley & Sons.

SchaVer. J.D. (1994). Combinations of genetic algorithms with neural networks or fuzzy systems. In Computational Intelligence: Imitating Life, Zurada, J.M., Marks, R.J., Robinson, C.J. (Eds). New York, IEEE Press, 371–382.

Schlenzig, C. (1998). *PlaNet: Ein entscheidungs-unterstützendes System für die Energie-und Umweltplanung (in German)*. IER Universität Stuttgart, Germany. Retrieved January, 28, 2008, from http://elib.uni-stuttgart.de/opus/volltexte/2001/742/pdf/diss_cs.pdf.

Seddon, A. P., & Brereton, P. (1996). Component selection using non-monotonic reasoning . *Artificial Intelligence in Engineering*, *1*, 235–241. doi:10.1016/0954-1810(95)00034-8

Seebregts, A. J., Goldstein, G. A., & Smekens, K. (2002). Energy/environmental modelling with the MARKAL family of models. In P. Chamoni, R. Leisten, A. Martin, J. Minnemann, and H. Stadtler (Eds). *Operations Research Proceedings 2001 — Selected Papers of the International Conference on Operations Research (OR 2001)* (pp. 75–82). Duisburg, Germany: Springer.

Seeger, T., & Verstege, J. (1991). Short term scheduling in cogeneration systems. *1991 Power Industry Computer Application Conference* (pp. 106-112).

Selvakumar, A. I., & Thanushkodi, K. (2007). A new particle swarm optimization solution to nonconvex economic dispatch problems. *IEEE Transactions on Power Systems*, *22*(1), 42–51. doi:10.1109/TPWRS.2006.889132

Seman, S., Niiranen, J., & Arkkio, A. (2006). Ride-through analysis of doubly fed induction wind-power generator under unsymmetrical network disturbance. *IEEE Transactions on Power Systems*, *21*(4), 1782–1789. doi:10.1109/TPWRS.2006.882471

Seman, S., Niiranen, J., Kanerva, S., & Arkkio, A. (2004). Analysis of a 1.7 MVA doubly fed wind-power induction generator during power system disturbances. *Proc. Nordic Workshop on Power and Industrial Electronics*, Trondheim, Norway, June.

Seman, S., Niiranen, J., Kanerva, S., Arkkio, A., & Saitz, J. (2006). Performance study of doubly fed wind-power induction generator under network disturbances. *IEEE Transactions on Energy Conversion*, *21*(4), 883–890. doi:10.1109/TEC.2005.853741

Sen, Z. (1998). Fuzzy algorithm for estimation of solar irradiation from sunshine duration. *Solar Energy*, *63*, 39–49. doi:10.1016/S0038-092X(98)00043-7

Senjyua, T., Hayashia, D., Yonaa, A., Urasakia, N., & Funabashib, T. (2007). Optimal configuration of power generating systems in isolated island with renewable energy. *Renewable Energy*, *32*, 1917–1933. doi:10.1016/j.renene.2006.09.003

Seo, H., Sung, J., Oh, S. D., Oh, H. S., & Kwak, H. Y. (2008). Economic optimization of a cogeneration system for apartment houses in Korea. *Energy and Building*, *40*(6), 961–967. doi:10.1016/j.enbuild.2007.08.002

Shannon, C. E. (1950). Programming a computer for playing chess. Philosophical Magazine. *Series*, *7*(41), 256–275.

Sharan, Y., & Vaturi, A. (2007). *Interim Report on the ENTTRANS project in Israel*. ENTTRANS Final Meeting, Brussels, December.

Sheen, J. N. (2005). Fuzzy evaluation of cogeneration alternatives in a petrochemical industry. *Computers & Mathematics with Applications (Oxford, England)*, *49*(5-6), 741–755. doi:10.1016/j.camwa.2004.10.035

Shen, B., & Ooi, B. T. (2005). Novel sensorless decoupled p-q control of doubly-fed induction generator (DFIG) based on phase locking to gamma-delta frame. *IEEE 36th Conf. Power Electronics Specialists*, June, (pp. 2670 – 2675).

Shen, M., Joseph, A., Wang, J., Peng, F. Z., & Adams, D. J. (2007). Comparison of Traditional Inverters and Z-Source Inverter for Fuel Cell Vehicles. *IEEE Transactions on Power Electronics*, *22*(4), 1437–1452. doi:10.1109/TPEL.2007.900505

Sherwali, H., & Crossley, P. (1994). Expert system for fault location on a transmission network. *Proceedings 29th Universities Power Engineering Conference - Part 2,* (pp. 751-754).

Shireen, W., Kulkarni, R. A., & Arefeen, M. (2006). Analysis and minimization of input ripple current in PWM inverters for designing reliable fuel cell power systems. *Journal of Power Sources, 156*(2), 448–454. doi:10.1016/j.jpowsour.2005.06.012

Shukla, R., & Kakar, P. (2006). *Role of science and technology, higher education and research in regional socio-economic development.* Working Paper, NCAER.

Simoes, M. G., & Farret, F. A. (2004). *Renewable energy systems: design and analysis with induction generators.* New York: CRC Press.

Singh, S. P., Raju, G. S., & Gupta, A. K. (1993). Sensitivity based expert system for voltage control in power system. *International Journal of Electrical Power & Energy Systems, 15*(3), 131–136. doi:10.1016/0142-0615(93)90027-K

Siskos, J., & Hubert, P. (1983). Multi-criteria analysis of the impacts of energy alternatives: A survey and a new comparative approach. *European Journal of Operational Research, 13,* 278–299. doi:10.1016/0377-2217(83)90057-7

Skogestad, S., & Postlethwaite, I. (2005). Multivariable Feedback Control: Analysis and Design (2nd Ed.). New York: John Wiley & Sons.

Slootweg, J. G., Polinder, H., & Kling, W. L. (2001, July). Dynamic modelling of a wind turbine with doubly fed induction generator. *IEEE Power Engineering Society Summer Meeting,* Vancouver, Canada, *1,* 644 – 649.

Smith, L. (2006). *A tutorial on Principal Components Analysis.*

Smith, R. G., Mitchell, T. M., Chestek, R. A., & Buchanan, B. G. (1977). A model for learning systems. Proceedings of IJCAI, 5.

SNNS. (1998). *SNNS User Manual - Version 4.2.* Computer Architecture Department, University of Tuebingen, Germany. Retrieved Aug 1, 2008, from http://www.ra.cs.uni-tuebingen.de/downloads/SNNS/SNNSv4.2.Manual.pdf

Sødergren, C., & Ravn, H. F. (1996). A method to perform probabilistic production simulation involving combined heat and power units. *IEEE Transactions on Power Systems, 11,* 1031–1036. doi:10.1109/59.496191

Song, Y. H., Chou, C. S., & Stonham, T. J. (1999). Combined heat and power economic dispatch by improved ant colony search algorithm. *Electric Power Systems Research, 52,* 115–121. doi:10.1016/S0378-7796(99)00011-5

Spee, R., Bhowmik, S., & Eslin, J. H. R. (1995). Novel control strategies for variable speed doubly fed wind power generation systems. *Renewable Energy, 6*(8), 907–915. doi:10.1016/0960-1481(95)00096-6

Spinney, P. J., & Watkins, G. C. (1996). Monte Carlo simulation technique and electric utility resource decisions. *Energy Policy, 24*(2), 155–164. doi:10.1016/0301-4215(95)00094-1

Srinivasan, D., Chang, C. S., & Liew, A. C. (1994). Multiobjective generation schedule using fuzzy optimal search technique. *IEE Proceedings. Generation, Transmission and Distribution, 141,* 231–241.

Stagl, S. (2006). Multicriteria evaluation and public participation: the case of UK energy policy. *Land Use Policy Ecological Economics, 23*(1), 53–62. doi:10.1016/j.landusepol.2004.08.007

Stamps, A. T., & Gatzke, E. P. (2005). Dynamic Modeling of a Methanol Reformer - PEMFC for Analysis and Design. *Journal of Power Sources, 161*(1), 356–370. doi:10.1016/j.jpowsour.2006.04.080

Statistics Canada. (1993). *Microdata User's Guide- The Survey of Household Energy Use.* Ottawa, Canada.

Sterling. (1999). *On economic and analysis of diversity.* SPRU Electronic Working Paper Series Paper No. 28.

Stevens, L. (2000). Incentives for sharing. *Knowledge Management, 3*(10), 54–60.

Stevenson, J. S. (1994). Using artificial neural nets to predict building energy parameters. *ASHRAE Transactions, 100*(2), 1081–1087.

Stimming, U., de Haart, L. G. S., & Meeusinger, J. (2005). *Fuel Cell Systems: PEMFC for Mobile and SOFC for Stationary Application*. Berlin: Wiley-VCH Verlag GmbH.

Streimikiene, D., & Šivickas, G. (2008). The EU sustainable energy policy indicators framework. *Environment International*.

Strubegger, M. (2003). *CO2DB Software: Carbon Dioxide (Technology) Database. Version 3.0.* International Institute for Applied Systems Analysis, Laxenburg, Austria. Retrieved June, 22, 2008, from http://www.iiasa.ac.at/Research/ECS/docs/CO2DB_manual_v3.pdf

Su, C. H., & Chiang, C. L. (2004). An incorporated algorithm for combined heat and power economic dispatch. *Electric Power Systems Research, 67*, 187–195. doi:10.1016/j.epsr.2003.08.006

Sudhakaran, M., & Slochanal, S. M. R. (2003). Integrating genetic algorithms and tabu search for combined heat and power economic dispatch. *Conference on Convergent Technologies for the Asia-Pacific Region*.

Sudhakaran, M., Vimal Raj, P. A. D., & Palanivelu, T. G. (2007). Application of particle swarm optimization for economic load dispatch problems. *2007 International Conference on Intelligent Systems Applications to Power Systems*.

Suehiro, S. (2007). *Energy intensity of GDP as Index of Energy Conservation*. Problems in International Comparison of Energy Intensity of GDP and Estimate using Sector-Based Approach. Japan: Institute of Energy Economics, Japan, IEEJ.

Sun, T., Chen, Z. & Blaabjerg, F. (2004). *Transient analysis of grid connected wind turbines with DFIG after an external short circuit fault*. Nordic Wind Power Conf. Chalmers University of Technology, Goteborg, Sweden, March.

Sundberg, J., & Wene, C.-O. (1994). Integrated Modelling of Material Flows and Energy Systems (MIMES). *International Journal of Energy Research, 18*(3), 359. doi:10.1002/er.4440180303

Swan, L., Ugursal, V. I., & Beausoleil-Morrison, I. (2008). A new hybrid end-use energy and emissions model of the Canadian housing stock. *Proceedings of the First International Conference on Building Energy and Environment*, Dalian, China, 1992-1999.

Systems, N.-S. (n.d.). ANSWER MARKAL Energy modeling Software. Retrieved May, 25, 2008, from http://www.noblesoft.com.au/.

Talaq, J. H., El-Hawary, F., & El-Hawary, M. E. (1994). A summary of environmental/economic dispatch algorithms. *IEEE Transactions on Power Systems, 9*, 1508–1516. doi:10.1109/59.336110

Tamura, J., Sasaki, T., Ishikawa, S., & Hasegawa, J. (1989). Analysis of the steady state characteristics of doubly fed synchronous machines. *IEEE Transactions on Energy Conversion, 4*(2), 250–256. doi:10.1109/60.17919

Tapia, A., Tapia, G., Ostolaza, J. X., & Saenz, J. R. (2003). Modeling and control of a wind turbine driven doubly fed induction generator. *IEEE Transactions on Energy Conversion, 18*(2), 194–204. doi:10.1109/TEC.2003.811727

Tapia, A., Tapia, G., Ostolaza, J. X., Saenz, J. R., Criado, R., & Berasaregui, J. L. (2001). Reactive power control of a wind farm made up of doubly fed induction generators (I, II). *Proc. IEEE Power Tech Conf.*, Porto, Portugal, 4, Sept.

Tapia, G., & Tapia, A. (2005). Wind generation optimization algorithm for a doubly fed induction generator. *Proc. IEE – Gener. Transm. Distrib., 152*(2), 253-263.

The Society of Environmental Toxicology and Chemistry -SETAC (n.d.). Retrieved July, 23, 2008, from http://www.setac.org/

The Statistical Office of the European Communities – Eurostat. (n.d.). Retrieved May, 14, 2008, from http://epp.eurostat.ec.europa.eu/

Thiringer, T., Petterson, A., & Petru, T. (2003, July). Grid disturbance response of a wind turbine equipped with induction generator and doubly fed induction generator. *IEEE Power Engineering Society General Meeting*, 3, 1542 – 1547.

Thomas, M. (1994). Combined heat and power, the global solution to voltage dip, pollution, and energy efficiency. *International Conference on Power System Technology, PowerCon 2004* (vol. 2, pp.1975-1980).

Thorin, E., Brand, H., & Weber, C. (2005). Long-term optimization of cogeneration systems in a competitive market environment. *Applied Energy, 81*, 152–169. doi:10.1016/j.apenergy.2004.04.012

Thounthong, P., Rael, S., & Davat, B. (2005). Fuel Cell and Supercapacitors for Automotive Hybrid Electrical System. *ECTI Transactions on Electrical Engineering, Electronics, and Communications*, *1*(3), 20–30.

Ting, T. O., Rao, M. V. C., & Loo, C. K. (2005). A novel approach for unit commitment problem via an effective hybrid particle swarm optimization. *IEEE Transactions on Power Systems*, *21*(1), 411–418. doi:10.1109/TPWRS.2005.860907

Toffolo, A., & Lazzaretto, A. (2002). Evolutionary algorithms for multi-objective energetic and economic optimization in thermal system design. *Energy, 27*, 549–567. doi:10.1016/S0360-5442(02)00009-9

Tsakonas, A., & Dounias, G. (2002). Hybrid computational intelligence schemes in complex domains: An extended review. Vlahavas, I.P. & Spyropoulos, C.D. (Eds.). SETN 2002, LNAI, 2308, 494–511.

Tsamboulas, D. A., & Mikroudis, G. K. (2005). TRANSPOL: A mediator between transportation models and decision makers' policies. *Decision Support Systems*, *42*(2), 879–897. doi:10.1016/j.dss.2005.07.010

Tsay, M. T., Chang, C. Y., & Gow, H. J. (2004). The operational strategy of cogeneration plants in a competitive market. *2004 IEEE Region 10 Conference*.

Tsay, M. T., Lin, W. M., & Lee, J. L. (2001). Interactive best-compromise approach for operation dispatch of cogeneration systems. *IEEE proc. Generation, Transmission . Distribution*, *148*(4), 326–332. doi:10.1049/ip-gtd:20010163

Tsay, M.T. (2003). Applying the multi-objective approach for operation strategy of cogeneration systems under environmental constraints. *Electrical Power & Energy systems, 25*, 219-226.

Tsay, M.T., & Lin, W.M. (2000). Application of evolutionary programming to optimal operational strategy cogeneration system under time-of-use rates. *Electrical Power & Energy systems, 22*, 367-373.

Tsay, M.T., Lin, W.M., & Lee, J. L. (2001). Application of evolutionary programming for economical dispatch of cogeneration system under emission constraints. *Electrical Power & Energy systems, 23*, 805-812.

Tse, C. K. (2003). *Complex behaviour of switching power converters*. Boca Raton, FL: CRC Press, Taylor & Francis Group.

Tse, K. K., Ng, R. W. M., Chung, H. S. H., & Hui, S. Y. R. (2003). An evaluation of the spectral characteristics of switching converters with chaotic carrier-frequency modulation. *IEEE Transactions on Industrial Electronics*, *50*(1), 171–182. doi:10.1109/TIE.2002.807659

Tsikalakis, A. G., & Hatziargyriou, N. D. (2008). Centralized Control for Optimizing Microgrids Operation. *IEEE Transactions on Energy Conversion*, *23*(1), 241–248. doi:10.1109/TEC.2007.914686

Tsukada, T., Tamura, T., Kitagawa, S., & Fukuyama, Y. (2003). Optimal operational planning for cogeneration system using particle swarm optimization. *Proceedings of the 2003 IEEE Swarm Intelligence Symposium* (pp. 138-143).

Turban, E. (1992). Expert systems and applied artificial intelligence. New York: Macmillan Publishing Company.

Turton, H. (2008). (in press). ECLIPSE: An integrated energy-economy model for climate policy and scenario analysis. *Energy*.

Tzeng, G. H., Shiau, T. A., & Lin, C. Y. (1992). Application of multicriteria decision making to the evaluation of new energy system development in taiwan. *Energy, 17*, 983–992. doi:10.1016/0360-5442(92)90047-4

Uche, J., Serra, L., & Sanz, A. (2004). Integration of desalination with cold-heat-power production in the agro-food industry. *Desalination, 166*, 379–391. doi:10.1016/j.desal.2004.06.093

Uctug, M. Y., Eskandarzadeh, I., & Ince, H. (1994). Modeling and output power optimization of a winf turbine driven double output induction generator. *Proc. IEEE – Electric Power Applications, 141*(2), 33 – 38.

Ugursal, V. I., & Fung, A. S. (1996). Impact of appliance efficiency and fuel substitution on residential end-use energy consumption in Canada. *Energy and Building, 24*(2), 137–146. doi:10.1016/0378-7788(96)00970-X

Ummels, B. C., Gibescu, M., Pelgrum, E., Kling, W. L., & Brand, A. J. (2007). Impacts of wind power on thermal generation unit commitment and dispatch. *IEEE Transactions on Energy Conversion, 22*, 44–51. doi:10.1109/TEC.2006.889616

UNFCCC, (2006). *Synthesis Report on Technology Needs Identified by Parties not included in Annex I to the Convention.* FCCC/SBSTA/2006/INF.1

UNFCCC. (2001). *FCCC/CP/2001/13/Add.1, Decision 4/CP.7, Annex, Seventh Conference of the Parties to the UNFCCC.* Marrakech, Morocco.

United Nations. (1994). *Koyoto Protocol to the United Nations Framework Convention on Climate Change.*

University of Geneva. (1998). *AIDAIR-GENEVA: a GIS based Decision Support System for air pollution management in an urban environment.* Retrieved June, 18, 2008, from http://ecolu-info.unige.ch/recherche/eureka/

Usaola, J., Ledesma, P., Rodriguez, J. M., Fernandez, J. L., Beato, D., Iturbe, R., & Wilhelmi, J. R. (2003 July). Transient stability studies in grids with great wind power penetration: modelling issues and operation requirements. *IEEE Power Engineering Society General Meeting*, 3, 1541-1544.

Vajpai, J., Singhal, S., & Naizi, K. (2001). Expert system based on line optimal control of reactive power. Proc. All India Seminar, Power Systems: Recent Advances and Prospects in 21st Century, Jaipur, India, 194-208.

Vale, A., Santos, J., & Ramos, C. (1997). SPARSE - A Prolog Based Application for the Portuguese Transmission Network: Verification and Validation. PAP'97 – Practical Application in Prolog, London, U.K.

Vale, Z., Ramos, C., Silva, A., Faria, L., Santosm, J., Fernandes, M., et al. (1998). SOCRATES – An Integrated Intelligent System for Power System Control Center Operator Assistance and Training. *International Conference on Artificial Intelligence and Soft Computing (IASTED)*, Cancun, Mexico.

Valenciaga, F., & Puleston, P. F. (2007). Variable structure control of a wind energy conversion system based on a brushless doubly fed reluctance generator. *IEEE Transactions on Energy Conversion, 22*(2), 499–506. doi:10.1109/TEC.2006.875447

van Beeck, N. (1999). *Classification of Energy Models.* Communicated by dr.ing. W. van Groenendaal. FEW 777. Tilburg University & Eindhoven University of Technology. Retrieved June, 18, 2008, from http://arno.uvt.nl/show.cgi?fid=3901.

Van der Gaast, W., Begg, K., Flamos. A, Deng, G., Mithulananthan, N., Theuri, D. *et al* (2008). *Synthesis report on technology descriptions 'Sustainable, low carbon technologies for potential use under the CDM'*, EC FP6 ENTTRANS Deliverable 5&6. Joint Implementation Network.

Van Regemorter, D. (2005). *Impact of the liberalisation of the electricity market: an analysis with MARKAL Belgium.* Retrieved August, 25, 2008, from http://www.ukerc.ac.uk/Downloads/PDF/E/ETSAP_1Regemorter.pdf

Van Wyk, J. D., & Enslin, J. H. R. (1983). A study of a wind power converter with microcomputer based maximal power control utilising an oversynchronous electronic Scherbius cascade. *Proc of the lnt. Power Electronics Conf.*, Tokyo, Japan, I, 766-777.

Vasebi, A., Fesanghary, M., & Bathaee, S.M.T. (2007). Combined heat and power economic dispatch by harmony search algorithm. *Electrical power & Energy system, 29*, 713-719.

Venkatesh, B. N. (1995). Decision models for management of cogeneration plants. *IEEE Transactions on Power Systems, 10*, 1250–1256. doi:10.1109/59.466530

Ventosa, M., Baillo, A., Ramos, A., & Rivier, M. (2005). Electricity market modeling trends. *Energy Policy, 33*, 897–913. doi:10.1016/j.enpol.2003.10.013

Vernados, P. G., Katiniotis, I. M., & Ioannides, M. G. (2003). Development of an experimental investigation procedure on double fed electric machine-based actuator for wind power. *Sensors and Actuators. A, Physical, 106*(1-3), 302–305. doi:10.1016/S0924-4247(03)00190-0

Vicatos, M. S., & Tegopoulos, J. A. (1989). Steady state analysis of a doubly fed induction generator under synchronous operation. *IEEE Transactions on Energy Conversion, 4*(3), 495–501. doi:10.1109/60.43254

Vicatos, M. S., & Tegopoulos, J. A. (1991). Transient state analysis of a doubly fed induction generator under three phase short circuit. *IEEE Transactions on Energy Conversion, 6*(1), 62–68. doi:10.1109/60.73790

Vlachogiannis, J. G., & Lee, K. Y. (2005). Determining generator contributions to transmission system using parallel vector evaluated particle swarm optimization. *IEEE Transactions on Power Systems, 20*(4), 1765–1774. doi:10.1109/TPWRS.2005.857014

Voogt, M. Oostvoorn, Frits van, Leeuwen, M. L. Van, & Velthuijsen, J. W. (2000). Energy modelling for economies in transition. In J. W. Maxwell & Jürgen von (Eds.), *Empirical Studies of Environmental Policies in Europe*. Boston: Kluwer Academic Publishers.

Voorspools, K.R. & D'haeseleer, W.D. (2003). Long-term unit commitment optimization for large power systems: unit decommitment versus advanced priority listing. *Applied Energy, 76*, 157–167. doi:10.1016/S0306-2619(03)00057-6

WADE. (2003). *Guide to Decentralized Energy Technologies*. Edinburgh, UK: www.localpower.org.

Walker, J. (2008). *Engineering Skill Shortage – A Degree of Concern*. Paper presented at New Zealand Electricity Engineers Association (EEA), Conference proceedings, Christchurch, June 2008.

Wang, J., Jing, Y., Zhang, C., Shi, G., & Zhang, X. (2008). A fuzzy multi-criteria decision-making model for trigeneration system. *Energy Policy, 36*(10), 3823–3832. doi:10.1016/j.enpol.2008.07.002

Wang, L., & Singh, C. (2007). Environmental/economic power dispatch using a fuzzified multi-objective particle swarm optimization algorithm. *Electric Power Systems Research, 77*, 1654–1664. doi:10.1016/j.epsr.2006.11.012

Wang, L., & Singh, C. (2008). Stochastic combined heat and power dispatch based on multi-objective particle swarm optimization. *Electrical power & Energy System, 30*, 226-234.

Wang, Lf. & Singh, C. (2007) PSO-based multi-criteria optimum design of a grid-connected hybrid power system with multiple renewable sources of energy. *Proceedings of 2007 IEEE Swarm Intelligence Symposium SIS 2007.*

Wang, Lf. & Singh, C. (2007). Environmental/economic power dispatch using fuzzified multi-objective particle swarm optimization algorithm. *Electric Power Systems Research, 77*, 1654–1664. doi:10.1016/j.epsr.2006.11.012

Wang, S., Liu, C., & Luu, S. (1994). A Negotiation methodology and its application to Cogeneration Planning. *IEEE Transactions on Power Systems, 9*, 869–875. doi:10.1109/59.317661

Wang, Y. J. (2000, January). *An analytical study on steady state performance of an induction motor connected to unbalanced three-phase voltage*. IEEE Power Engineering Society Winter Meeting, Singapore.

Wang, Y. J. (2001). Analysis of effects of three phase voltage unbalance on induction motor with emphasis on the angle of the complex voltage unbalance factor. *IEEE*

Transactions on Energy Conversion, 16(3), 270–275. doi:10.1109/60.937207

Wang, Z., Zheng, D., & Jin, H. (2007). A novel polygeneration system integrating the acetylene production process and fuel cell. *International Journal of Hydrogen Energy, 32*(16), 4030–4039. doi:10.1016/j.ijhydene.2007.03.018

Washington State University - Extension Energy Program. (n.d.). *HeatMap, District Energy Analysis Software for Steam, Hot-Water, and Chilled-Water Systems.* Retrieved July, 25, 2008, from http://www.energy.wsu.edu/software/heatmap.cfm.

Weber, C., & Woll, O. (2006). Valuation of CHP power plant portfolios using recursive stochastic optimization. *9th International Conference on Probabilistic Methods Applied to Power systems,* KTH, Stockholm, Sweden, June 11-15, 2006.

Weber, C., Marechal, F., Favrat, D., & Kraines, S. (2006). Optimization of an SOFC-based decentralized polygeneration system for providing energy services in an office-building in Tokyo. *Applied Thermal Engineering, 26*, 1409–1419. doi:10.1016/j.applthermaleng.2005.05.031

Weedy, B. M., & Cory, B. J. (1998). *Electric Power Systems* (4th ed.). New York: John Wiley & Sons.

Welbank, M. (1985). *A review of knowledge acquisition techniques for expert systems.* Ipswich, UK: British Telecommunications Research Laboratories Technical Report.

Wenying, Ch. (2005). The costs of mitigating carbon emissions in China: findings from China MARKAL-MACRO modeling. *Energy Policy, 33*(7), 885–896. doi:10.1016/j.enpol.2003.10.012

Weyant, J. P. (1990). Policy modeling: an overview. *Energy, 15*, 203–206. doi:10.1016/0360-5442(90)90083-E

Wierzbicki, A., Makowski, M., & Wessels, J. (Eds.). (2000). *Model-Based Decision Support Methodology with Environmental Applications.* Dordrecht, The Netherlands: Kluwer Academic Publishers.

Williams, K. A., Keith, W. T., Marcel, M. J., Haskew, T. A., Shepard, W. S., & Todd, B. A. (2007). Experimental investigation of fuel cell dynamic response and control. *Journal of Power Sources, 163*(2), 971–985. doi:10.1016/j.jpowsour.2006.10.016

Williams, R. J., & Zipser, D. (1989). Experimental analysis of the real-time recurrent learning algorithm. *Connection Science, 1*(1), 17–11. doi:10.1080/09540098908915631

Winograd, S. (1972). Understanding natural language, Academic Press, New York.

Winston, P. H. (1994). Artificial intelligence, Addison-Wesley, 2nd ed., Reading, MA.

Wong, H., Chan, Y., & Ma, S. W. (2002). Electromagnetic interference of switching mode power regulator with chaotic frequency modulation. *Proceedings of the 23rd Int. Conference on Microelectronics - MIEL 2002*, 2, 577 – 580. Retrieved December 15, 2002, from http://ieeexplore.ieee.org.

Wong, J. K. W., Li, H., & Wang, S. W. (2005). Intelligent building research: a review. *Automation in Construction, 14*, 143–159. doi:10.1016/j.autcon.2004.06.001

Wong, K. P. (2002). Recent development and application of evolutionary optimisation techniques in power systems. *Proceedings of International Conference on Power System Technology, 1*, 1–5. doi:10.1109/ICPST.2002.1053493

Wong, K. P., & Algie, C. (2002). Evolutionary programming approach for combined heat and power dispatch. *Electric Power Systems Research, 61*, 227–232. doi:10.1016/S0378-7796(02)00028-7

Woodhouse, J. (2006). *Putting the total jigsaw puzzle together: PAS 55 standard for the integrated, optimized management of assets.* Retrieved September 15, 2008, from http://www.twpl.com/confpapers/putting_the_total_jigsaw_together.pdf.

Woojin, C., & Jo, H. W. (2006). Development of an equivalent circuit model of a fuel cell to evaluate the effects of inverter ripple current. *Journal of Power Sources, 158*(2), 1324–1332. doi:10.1016/j.jpowsour.2005.10.038

Woojin, C., Gyubum, J., Prasad, E. N., & Jo, H. W. (2004). An Experimental Evaluation of the Effects of

Ripple Current Generated by the Power Conditioning Stage on a Proton Exchange Membrane Fuel Cell Stack. *Journal of Materials Engineering and Performance*, *13*(3), 257–264. doi:10.1361/10599490419144

World Bank (2003). *Building Regional Power Tools: A Toolkit.*

World Bank (2004). *Public and Private Sector Roles in the Supply of Electricity Services,* (World Bank Paper No. 37476).

World Bank (2007). *Catalyzing Private Investment for a Low-Carbon Economy.*

World Bank. (WB), (2005). *Technical and Economic Assessment: off-grid, mini-grid and grid electrification technologies, prepared for Energy Unit, Energy and Water Department, the World Bank.* Discussion Paper, Energy Unit, Energy and Water Department.

World Coal Institute (WCI). (2004). *Clean Coal Building a Future Through Technology.* Retrieved from http://www.worldcoal.org/assets_cm/files/pdf/clean_coal_building_a_future_thro_tech.pdf

World Energy Council. (2008). *Europe's vulnerability to energy crisis: Executive summary.*

Worldbank (1991). *Assessment of personal computer models for energy planning in developing countries. Report Number 17336.* Retrieved July, 7, 2008 from http://go.worldbank.org/LCFRW3DDZ1.

Wu, D. W., & Wang, R. Z. (2006). Combined cooling, heating and power: A review. *Progress in Energy and Combustion Science*, *32*, 459–495. doi:10.1016/j.pecs.2006.02.001

Wu, L.-M., Chen, B.-S., Bor, Y.-C., & Wu, Y.-C. (2007). Structure model of energy efficiency indicators and applications. *Energy Policy*, *35*(7), 3768–3777. doi:10.1016/j.enpol.2007.01.007

Wu, T., Shieh, S., Jang, S., & Liu, C. (2005). Optimal energy management integration for a petrochemical plant under considerations of uncertain power supplies. *IEEE Transactions on Power Systems*, *20*(3), 1431–1439. doi:10.1109/TPWRS.2005.852063

Wu, Y. J., & Rosen, W. A. (1999). Assessing and optimizing the economic and environmental impacts of cogeneration/district energy systems using an energy equilibrium model. *Applied Energy*, *62*, 141–154. doi:10.1016/S0306-2619(99)00007-0

Xiang, D., Ran, L., Bumby, J. R., Tavner, P. J., & Yang, S. (2006). Coordinated control of an HVDC link and doubly fed induction generators in a large offshore wind farm. *IEEE Transactions on Power Delivery*, *21*(1), 463–471. doi:10.1109/TPWRD.2005.858785

Xiang, D., Ran, R. L., Tavner, P. J., & Yang, S. (2006). Control of a doubly fed induction generator in a wind turbine during grid fault ride-through. *IEEE Transactions on Energy Conversion*, *21*(3), 652–662. doi:10.1109/TEC.2006.875783

Xu, L., & Cartwright, P. (2006). Direct active and reactive power control of DFIG for wind energy generation. *IEEE Transactions on Energy Conversion*, *21*(3), 750–758. doi:10.1109/TEC.2006.875472

Xu, L., & Cheng, W. (1995). Torque and reactive power control of a doubly-fed induction machine by position sensorless scheme. *IEEE Transactions on Industry Applications*, *31*(3), 636–641. doi:10.1109/28.382126

Xu, L., & Wang, Y. (2007). Dynamic modeling and control of DFIG-based wind turbines under unbalanced network conditions. *IEEE Transactions on Power Systems*, *22*(1), 314–323. doi:10.1109/TPWRS.2006.889113

Xue, Y., Chang, L., & Kjær, S. B. (2004). Topologies of Single-Phase Inverter for Small Distributed Generators: an Overview. *IEEE Transactions on Power Electronics*, *19*(5), 1305–1313. doi:10.1109/TPEL.2004.833460

Yalcintas, M. (2008). Energy-savings predictions for building-equipment retrofits. *Energy and Buildings.*

Yamamoto, M., & Motoyoshi, O. (1991). Active and reactive power control for doubly-fed wound rotor induction generator. *IEEE Transactions on Power Electronics*, *6*(4), 624–629. doi:10.1109/63.97761

Yamamoto, S., etc. (2004). An operating method using prediction of photovoltaic power for a photovoltaic-diesel hybrid power generation system. *Transactions of the Institute of Electrical Engineers of Japan, B . Power and Energy, 124*(4), 521–530. doi:10.1541/ieejpes.124.521

Yang, J., Rivard, H., & Zmeureanu, R. (2005). On-line building energy prediction using adaptive artificial neural networks. *Energy and Building, 37*(12), 1250–1259. doi:10.1016/j.enbuild.2005.02.005

Yang, K. (2001). Spread-Spectrum DC-DC Converter Combats EMI. *Electronics Design*, (pp. 86-88). Retrieved October 23, 2001, from http://electronicdesign.com/Articles.

Yazhou, L., Mullane, A., Lightbody, G., & Yacamini, R. (2006). Modeling of the wind turbine with a doubly fed induction generator for grid integration studies. *IEEE Transactions on Energy Conversion, 21*(1), 257–264. doi:10.1109/TEC.2005.847958

Yikang, H., Jiabing, H., & Rende, Z. (2005). Modeling and control of wind-turbine used DFIG under network fault. *Proc. 8th Int. Conf. Electrical Machines and Systems*, 2, Sept., 986 – 991.

Yokoyama, R., Bae, S. H., Morita, T., & Sasaki, H. (1988). Multiobjective generation dispatch based on probability security criteria. *IEEE Transactions on Power Systems, 3*, 317–324. doi:10.1109/59.43217

Yongli, Z., Yang, Y. H., Hogg, B. W., Zhang, W. Q., & Gao, S. (1994). Expert system for power systems fault analysis. *IEEE Transactions on Power Systems, 9*(1), 503–509. doi:10.1109/59.317573

Ypsilantis, J., & Yee, H. (1990). Survey of expert systems for SCADA-based power applications. 4th Conference in Control Engineering, Gold Coast, Australia, 177-183.

Yu, H. H., & Jenq-Neng, H. (2001). Handbook of neural network signal processing. CRC press.

Yu, W. (1992). *ELECTRE TRI: Aspects méthodologiques et manuel d'utilisation*. Document du LAMSADE No. 74, Université Paris – Dauphine.

Zadeh, L. A. (1965). Fuzzy sets. *Information and Control, 8*, 338–353. doi:10.1016/S0019-9958(65)90241-X

Zadeh, L. A. (1973). Outline of a new approach to the analysis of complex systems and decision processes. IEEE Transactions on Syst. *Man Cybernet., 3*, 28–44.

Zadeh, L. A. (1978). Fuzzy sets a basis for a theory of possibility. *Fuzzy Sets and Systems, 1*, 3–28. doi:10.1016/0165-0114(78)90029-5

Zadeh, L.A. (1972). A Fuzzy-set-theoretic interpretation of linguistic hedges. Cyber net, 2, 4–34.

Zhan, C., & Barker, C. D. (2006). Fault ride-through capability investigation of a doubly-fed induction generator with an additional series-connected voltage source converter. *Proc. 8th IEE International Conf. AC and DC Power Transmission*, March, 79 – 84.

Zhang, L. and. Bai, Y. (2005). Genetic algorithm-trained radial basis functions neural networks for modelling photovoltaic panels. Engineering application of artificial intelligence, 18, 833-844.

Zhang, L., & Watthanasarn, C. (1998). A matrix converter excited doubly-fed induction machine as a wind power generator. *Proc. 7th Int. Conf. on (IEE Conf. Publ. 456) Power Electronics and Variable Speed Drives*, London, Sept., 532 – 537.

Zhang, L., Watthanasarn, C. & Shepherd. (1997). Application of a matrix converter for the power control of a variable–speed wind-turbine driving a doubly fed induction generator. *Proc. 23rd Int. Conf. Industrial Electronics, Control and Instrumentation*, 2, Nov., 906 – 911.

Zhi, D., & Xu, L. (2007). Direct power control of DFIG with constant switching frequency and improved transient performance. *IEEE Transactions on Energy Conversion, 22*(1), 110–118. doi:10.1109/TEC.2006.889549

Zhong, Z.-D., & Huo, H.-B. (in press). Zhu, -J., Cao, G.-Y., & Ren [Adaptive maximum power point tracking control of fuel cell power plants. *Journal of Power Sources.*]. *Y (Dayton, Ohio).*

Zhou, P., Ang, B. W., & Poh, K. L. (2006). Decision analysis in energy and environmental modeling: An update. *Energy, 31*, 2604–2622. doi:10.1016/j.energy.2005.10.023

Zimmermann, H.-J., Tselentis, G., Van Someren, M., & Dounias, G. (2001). Advances in computational intelligence and learning: Methods and applications. Kluwer Academic Publications.

Zita, A. Vale, Santos, J., & Ramos, C. (1995). *SPARSE - A Prolog Based Application for the Portuguese Transmission Network: Verification and Validation.* London: PAP'97 – Practical Application in Prolog.

Zita, A. Vale, Santos, J., & Ramos, C. (1995). SPARSE - A Prolog Based Application for the Portuguese Transmission Network: Verification and Validation. London, UK: PAP'97 – Practical Application in Prolog.

Zitouni, S., & Irving, MR. (1999). *NETMAT: A knowledge-based grid system analysis tool.*

Zitzler, E., & Thiele, L. (1998). An evolutionary algorithm for multiobjective optimization: The strength Pareto approach. *TIK-Report, 43.*

Zopounidis, C., & Dimitras, A. I. (1998). *Multicriteria Decision Aid Methods for the Prediction of Business Failure.* Dordrecht, The Netherlands: Kluwer Academic Publishers.

Zwe-Lee, G. (2003). Particle swarm optimization to solving the economic dispatch considering the generator constraints. *IEEE Transactions on Power Systems, 18,* 1187–1195. doi:10.1109/TPWRS.2003.814889

About the Contributors

Kostas Metaxiotis, (PhD), is an assistant professor at the University of Piraeus and has served as Senior Advisor to the Secretary for the Information Society in the Greek Ministry of Economy and Finance since 2004. He has wide experience in knowledge management, artificial intelligence, object-oriented knowledge modelling, inference mechanisms, e-health, e-business. Dr. Metaxiotis has published more than 70 scientific papers in various journals and conferences, such as Journal of Knowledge Management, Journal of Information and Knowledge Management, Knowledge Management & Practice, Journal of Intelligent Manufacturing, Applied Artificial Intelligence, Industrial Management & Data Systems, Journal of Computer Information Systems. Currently, he is a member of several editorial boards and reviews in many leading journals in the field. He is also a member of the Program Committee at international conferences. Since 1996 he has been participating in various European Commission (EC)-funded projects within Tacis, Phare, MEDA and IST Programmes as Senior ICT Consultant and Manager. Since 2007 he has been serving as External Evaluator of EC-funded ICT projects.

* * *

Mohammad Abido received his B.Sc. and MSc. degrees in Electrical Engineering from Menoufia University, Egypt, in 1985 and 1989 respectively and Ph. D degree from King Fahd Univ. of Petroleum and Minerals, Saudi Arabia, in 1997. Dr. Abido is currently a Professor at KFUPM. His research interests are power system planning, operation, and optimization techniques applied to power systems. Dr. Abido is the recipient of KFUPM Excellence in Research Award 2002 and 2007, the *First Prize Paper Award* of the Industrial Automation and Control Committee of the IEEE Industry Applications Society, 2003, *Abdel-Hamid Shoman Prize for Young Arab Researchers in Engineering Sciences*, 2005, and the *Best Applied Research Award* of 15th GCC-CIGRE Conference, Abu-Dhabi, UAE, 2006. Dr. Abido has published more than 100 papers in reputable journals and international conferences.

Ramesh Bansal received the M.E. degree from Delhi College of Engineering, Delhi, India, in 1996; MBA degree from Indira Gandhi National Open University, New Delhi, India, in 1997; and the Ph. D. degree from Indian Institute of Technology (IIT), Delhi, India in 2003. He is a faculty in School of Information Technology and Electrical Engineering, The University of Queensland, Australia. Earlier he was faculty in the School of Engineering and Physics, The University of the South Pacific, Fiji. He was Assistant Prof. in the Department of Electrical and Electronics Engineering, Birla Institute of Technology and Science, Pilani in 1999-2005. Dr. Bansal is an Editor of IEEE Transactions of Energy Conversion and Power Engineering Letters, IEEE Trans. Industrial Electronics, and Editorial Board member of IET, Renewable Power Generation, Electric Power components & Systems, and Energy Sources. He is a member of Board of Directors, International Energy Foundation (IEF), Alberta, Canada. He has published more than 85 papers in national/international journals and conference proceedings. His research interests include reactive power control in renewable energy systems and conventional power systems, power system optimization, analysis of induction generators, and artificial intelligence techniques applications in power systems.

Nicu Bizon received a 5-year degree in Electronic Engineering from the University of Bucharest, Romania, in 1986 and a PhD in Automatic Systems and Control from the same university, in 1996. He is currently Professor at the University of Pitesti, Romania. He has authored 4 books in Power Converters area. In addition, he has co-authored the book "Fundamentals of Electromagnetic Compatibility, Theory and Practice". His current research interests include the broad area of nonlinear systems, on both dynamics and control, and power electronics. He has authored more than 100 papers in journals and international conference proceedings.

Campbell Booth is a lecturer within the Electronic and Electrical Engineering Department at the University of Strathclyde in Glasgow. His teaching and research interests lie in the areas of: power system protection; plant condition monitoring and intelligent asset management; applications of intelligent system techniques to power system monitoring, protection and control; knowledge management and decision support systems. He has over 15 years of experience working with the Electric Power Industry and in other sectors through several university research contracts, independent consultancy engagements and through Dynamic Knowledge Corporation, a company spun-out from the University of Strathclyde, of which Dr. Booth is a co-founder.

Andreas Botsikas is an Electrical and Computer Engineer of the National Technical University of Athens, with M.Sc in Business Administration. His area of expertise includes hybrid optimization algorithms, smart routing on GIS systems, capacity planning, time scheduling, support systems and business modeling. In the abovementioned fields, he has exhibited his results in scientific conferences and international journals. His research has been awarded in many occasions and by many people including Bill Gates during the presentation of Microsoft's Innovation center in Athens 2008. He is an associate in the NTUA in the Decision Support Systems Laboratory and has been involved in research, consultancy and technical projects.

Carmelina Cosmi is researcher at IMAA-CNR since March 1998. Her current research interests' deal with energy system analysis with regard to the characterization of the impact of anthropogenic activities on environment and multi-objective data analysis. Her research activities have been carried out in the framework of the Energy Technology Systems Analysis Programme (ETSAP) of the International Energy Agency (IEA) and in several national and international projects. From 1994 she has being coordinating the research activities of the research line "Environmental modelling and planning", in the framework of the CNR Project "Sustainability of Environmental systems".

Vincenzo Cuomo, since 1993, is Director of the Institute of Methodologies for Environmental Analysis of the Italian National Research Council (CNR- IMAA). Since 1975, the main topics of his research dealt with environmental physics, being focused on: earth observation by satellite; airborne and ground based instrumentation; development of new advanced sensors for observation from space; monitoring and control of environmental processes; protection against and prevention of natural risks; energy and environmental planning. He is member of Scientific Committee of Italian Space Agency, member of Scientific Committee of CNIT (National Inter-University Consortium for Telecommunications), chair of the Italian Coordination Group of Earth Observation, coordinator of projects funded by U.E., ESA, EUMETSAT, National Research Council, Italian Minister of University and Research, Italian Space Agency, Italian Institute of Physics Matter.

Senatro Di Leo received the MSc degree in Environmental Engineering from the University of Basilicata (Potenza), Italy, in 2002. At present he is attending the PhD in "Methods and Techniques for Environmental Monitoring" at the University of Basilicata, Italy, in collaboration with IMAA-CNR. His research activities deals with comprehensive modelling of energy systems for the evaluation of strategic energy –environmental scenarios and the multivariate statistical analysis of different sets of data for identifying key variables and information indexes for the characterization of system's environmental and energy performances.

Zhao Yang Dong (M'99, SM'06) obtained his PhD from The University of Sydney, Australia in 1999. He holds academic positions at Hong Kong Polytechnic University, Hong Kong, the University of Queensland, Australia; and National University of Singapore. He also holds industrial positions with Powerlink Queensland, and Transend Networks, Tasmanis, Australia working in power system planning, stability and protection areas. His research interests include power system stability and control, power system planning, load modelling, electricity market, data mining, computational intelligence and its application in power engineering.

Haris Doukas is a Mechanical Engineer of the Aristotle University of Thessalonica, holding a PhD degree on the decision support systems for the promotion of Renewable Energy Sources under the modern environment of the energy sector's operation. His areas of expertise include energy policy and planning, decision support systems, renewable and climate policies and strategies and Kyoto GHG emissions reduction Flexible mechanisms (CDM, JI and ET). In the abovementioned fields, he has 24 papers published in international journals with reviewers and 27 announcenments in international conferences.

Fawwaz Elkarmi is an Associate Professor and Deputy Dean at Amman University/Faculty of Engineering, Jordan. He holds a Doctor of Engineering degree from Texas A&M University obtained in 1981 and currently teaches power system analysis and engineering economics. He authored and co-authored more than 25 papers in journals, magazines, and conference proceedings, in addition to more than 50 technical reports. He worked in government, and private sectors for 25years in Jordan and 9 years abroad (Kuwait and U.S.A) in education and training, electricity utility, research and development, consulting, manufacturing, and business. Dr. Elkarmi is a Senior Member of IEEE, and a Chartered Engineer with IEE. Moreover, he is a member of the Institute of Management Consultants (IMC) and holds a Certified Management Consultant (CMC) title. He also provides consultation in energy efficiency, energy management, renewable energy, business development, feasibility studies, organizational re-engineering, business turnaround, project appraisal, and investment assessment.

Alexandros Flamos is an Electrical and Computer Engineer holding a PhD in Decision Support Systems. He has an over 10 years working experience in the scientific areas of Decision Support Systems (DSS), Energy Management & Planning and their applications for analyzing energy and environmental policy, energy and environmental modeling, climate change abatement, security of energy supply and energy pricing competitiveness. He has held the position of project manager / senior energy expert in more than 15 EU funded projects (Tacis, MEDA, Phare, Synergy, FP6, FP7, etc.) related to the EU energy policy. He is currently teaching energy management and DSS at the National Technical University and he has more than 50 publications in high impact international scientific Journals and international conferences and as invited speaker in major International energy policy cooperation events (MENAREC, Euro-Asia meetings and Int. Conferences for climate change, participation in COP 6, COP 7 etc.).

Martin Grunow is professor in Operations Management at the Technical University of Denmark. Earlier, he worked at the Technical University Berlin, Germany, and at Degussa AG's R&D department. His research interests are in production and logistics management with a focus on supply chain management in the process industries. He has coauthored more than 80 publications amongst others in International Journal of Production Economics, International Journal of Production Research, European Journal of Operational Research, CIRP Annals, and OR Spectrum. For the latter journal, he also acts as an editor (since 2001).

Soteris Kalogirou received his HTI Degree in Mechanical Engineering in 1982, his M.Phil. in Mechanical Engineering from the Polytechnic of Wales in 1991 and his Ph.D. in Mechanical Engineering from the University of Glamorgan in 1995. For more than 25 years, he is actively involved in research in the area of solar energy and particularly in flat plate and concentrating collectors, solar water heating, solar steam generating systems, desalination and absorption cooling. Additionally, since 1995 he is involved in a pioneering research dealing with the use of artificial intelligence methods, like artificial neural networks, genetic algorithms and fuzzy logic, for the modelling and performance prediction of energy and solar energy systems. He has 19 book contributions and published 179 papers; 79 in international scientific journals and 100 in refereed conference proceedings. Until now, he received more than 1250 citations on this work.

Charikleia Karakosta is a Chemical Engineer of the National Technical University of Athens (NTUA, 1999-2004) with a M.Sc. in Energy Production and Management (2004-2006). She is a PhD candidate at NTUA in the Management & Decision Support Systems Laboratory, School of Electrical and Computer Engineering. Her research focuses on energy planning and modelling, decision support systems, energy management and policy, climate change and Kyoto GHG emissions reduction Flexible mechanisms (CDM, JI and ET). She has participated in several research and consultancy projects in the fields of environmental policy, climate change, management and energy modelling. She has 17 scientific journal publications and publications/announcements in international conferences.

Giridhar Kini received the B.E. degree in Electrical Engineering, in 1995 and M.Tech degree in Engineering Management, in 1998 and PhD degree in 2008 from Manipal Institute of Technology, Mangalore University. He is currently with the Department of Electrical Engineering, Manipal Institute of Technology, Manipal University. He has published many papers in journals/conferences. His research interests are in Electrical Machinery and Energy Management Systems.

Merih Koksal is a faculty member at the Environmental Engineering Department of Hacettepe University, Ankara, Turkey. She holds a BSc in Environmental Engineering and a MSc in Mechanical Engineering from Marmara University, Istanbul, Turkey, and a PhD in Mechanical Engineering from Dalhousie University, Halifax, N.S., Canada.

Steven Kong is pursuing his PhD at the University of Queensland, Australia. His research interests are power system stability, DFIG and distributed generation.

Vinod Kumar received the B.E. degree in Electrical Engineering from Maharana Pratap University of Agriculture and Technology in 2002 and M. Tech. degree in Power Electronics from VIT, Vellore, India, in 2004. He is faculty in the Department of Electrical Engineering, MPUAT, India. Currently he is pursing for his PhD. He has published many papers in national/international journals and conference proceedings. His research interests include power converters and their application in wind power generation; advanced electric drives, and power quality.

Risto Lahdelma is professor in Energy Technology for Communities at Helsinki University of Technology. From 2000 to 2009 he has been full professor of computer science at University of Turku. He is also the chairman of the Finnish Operations Research Society. His research interests include systems and operations research, energy management, intelligent systems, embedded algorithms and multi-criteria decision support. He has published papers in Operations Research, European Journal of Operational Research, Journal of Environmental Management, Environmental Management, Journal of Multi-Criteria Decision Analysis, Decision Support Systems, Applied Energy, Socio-Economic Planning Science, Forest Policy and Economics. His email address is risto.lahdelma@tkk.fi.

Simona Loperte received the MSc degree in Environmental Engineering from the University of Basilicata (Potenza), Italy, in 2003 and the Ph.D. degree in "Methods and Techniques for Environmental Monitoring" in 2007. Currently she holds a post –doc position at CNR- IMAA. Her research interests deal with LCA (Life Cycle Assessment), techno-economic and environmental characterisation of energy systems for the evaluation of their full environmental impact and the development of scenarios for air quality improvement and mitigation strategies.

Maria Macchiato is full professor of Physics at the University of Naples Federico II. Her main areas of interests concerns: integrated methodologies for environmental planning; physical methodologies for the dynamical characterization of geophysical and environmental processes. In recent years, she has been coordinating several national and international projects concerning sustainable energy systems management and strategies for local and GHG emissions reduction. On behalf of INFM-CNR she has been responsible of the activities related to environmental modeling at Department of Physics Sciences of the University Federico II, Naples.

Adel Mellit received his BElect Eng (Hons) from the University of Blida in 1997 and MSc in Electronics option Renewable Energy from the University of Sciences Technologies Algiers in 2002 and PhD in Electronics option Renewable Energy from the University of Sciences Technologies (USTHB) Algiers in 2006. His research interest includes the application of the artificial intelligence techniques in photovoltaic power supply systems, meteorological data modelling and weather forecasting. He has authored and co-authored more than 45 technical papers in the international journal and conferences. Also he is an invited reviewer for more than 40 technical papers in different journal and proceedings. Currently, he is an Assistant Professor at Jijel University, Department of Electronics/LMD.

Yateendra Mishra received B.E and M. Tech from BIT Mesra and IIT Delhi, India in 2003 and 2005 respectively. He is now pursuing his PhD at the University of Queensland, Australia. His research interests are power system stability, DFIG and distributed generation.

Shinya Obara is Professor in the Department of Electrical and Electronic Engineering, Kitami Institute of Technology, Japan. Shinya Obara obtained a Bachelor of Engineering in Nagaoka University of Technology, 1987, a Master of Engineering in Nagaoka University of Technology, 1989 and a PhD in Engineering in Hokkaido University, 2000. Research field: Optimal planning of energy system, Operation plan of fuel cell system, Radiation, Bio-ethanol reformer, Fuel cell co-generation.

Alexandra G. Papadopoulou is a Chemical Engineer of the National Technical University of Athens, with a M.Sc in Energy Production and Management. Her area of expertise includes energy efficiency and energy management procedures, energy management decision support tools, energy planning and modelling, climate change and Kyoto Greenhouse Gas emission reduction flexile mechanisms. She is a Phd candidate and associate at NTUA in the Management and Decision Support Systems Laboratory and has been involved in several research and consultancy projects in the fields of decision making and promoting energy efficiency and climate policy. In the above mentioned fields, she has 19 scientific journal publications and publications/announcements in international conferences.

Kostas Patlitzianas is Project Manager and he has wide experience in the project development, organization and management for private and public sector clients, decision making systems, energy and climate policy, corporate finance, private investments finance through support framework programmes. Before joining Remaco SA (2008), he worked as R&D manager in the Energy Department of Raycap S.A. (2006-2007). He has worked as an energy expert in the CRES (2008) and other companies/institutions, as well as in the Management and Decision Support Systems Lab of NTUA (EPU-NTUA) from 2002 to date. Dr Patlitzianas worked as chemical engineer in the L'oreal S.A. (1998-2000) and Anelor S.A. (1997). Dr. Patlitzianas has graduated from the School of Chemical Engineering of the NTUA. He holds a PhD in Decision Support Systems on Energy Market from the NTUA, a MSc in Energy Production and Management from the NTUA and he has studied in the Department of Management Science & Technology of the Athens University of Economics and Business. He has more than 50 publications in the international scientific journals and conferences such as Energy Policy, Renewable Energy, Energy Sources, Renewable & Sustainable Energy Reviews, Energy Conservation and Management, Resources Policy, International Journal of Power and Energy Systems etc.

Filomena Pietrapertosa is researcher at IMAA-CNR since June 2005. Her main activity deals with energy system analysis using the ETSAP/IEA model generators to identify sustainable development strategies, with particular reference to the investigation on the contribution of renewable and high efficient technologies at local and national scale. Recent activities include the development of a consistent analytical platform in which MARKAL model, LCA and ExternE methodologies are linked to provide a comprehensive computation of the environmental impacts of the energy systems and to internalise environmental damage costs (externalities).

John Psarras is Professor in the Department of Electrical and Computer Engineering of National Technical University of Athens (NTUA) and Head of Energy Policy Unit, holding a Ph.D. degree in Multiobjective Mathematical Programming applied to energy and environmental systems (NTUA, 1989). He has been the Project Manager or Senior Researcher in numerous EU-funded and national projects acquiring over twenty years experience in the areas of energy policy, national and regional energy planning, energy and environmental modelling, promotion of energy and environmental friendly technologies, energy management, decision support and monitoring systems.

Aiying Rong received her master degree in Industrial Engineering and Engineering Management at Hong Kong University of Science and Technology and her Ph.D. degree in Computer Science (Algorithmics) at the University of Turku, Finland. During her PhD study, she was mainly involved in research on cogeneration planning under the deregulated power market and emissions trading scheme. Currently, she is working as a postdoctoral researcher in the Department of Management Engineering at the Technical University of Denmark. Her research interests include operations, planning and scheduling of production activities in different industrial sectors such as the energy industry, the food industry, and the iron & steel industry. She has published papers in European Journal of Operational Research, International Journal of Production Research, Journal of the Operational Research Society and Applied Energy.

Christos Roupas is an Economist holding an MSc in Accounting and Financial Management. He is also a candidate for a PhD in Decision Support Systems. He has an over 3 years working experience in the scientific areas of Accounting and Financial Management, Environmental Accounting and Modelling Energy Policies for the minimization of security of supply threats. He is a researcher in the Decision Support Systems (DSS) Laboratory of the National Technical University of Athens. He has published papers in several international scientific journals and he has participated in many conferences.

Monica Salvia is researcher at IMAA-CNR since April 2001. In 2003 she was a Visiting Researcher at Centre for Social and Economic Research on the Global Environment (CSERGE, England), funded by a NATO-CNR fellowship. Her main areas of interests concern the field of modelling for waste management and energy systems. During these years, she has been taking part in several national and international research projects in the field of energy environmental planning, participating actively at the ETSAP/IEA Programme activities.

Emmanuel Samouilidis is a Professor Emeritus in the Department of Electrical and Computer Engineering of National Technical University of Athens (NTUA) and Director of the Energy Policy Unit. He has over 35 years of experience in energy policy analysis, national and regional energy planning, energy and environmental modelling, energy management, decision support and monitoring systems. He has held top government positions having served as chief advisor to many Ministers and has a complete grasp of EU-funded projects. He has worked as Project Director and Senior Energy Consultant in numerous projects funded by EC and other donors since 1985.

Index

A

adjustable speed drives 401, 421
adjustable speed generators (ASGs) 149
artificial intelligence 1, 30, 32, 33, 35, 36, 37, 38, 39
artificial neural networks 388
automotive hybrid electrical system (AHES) 42

B

back-propagation (BP) 389
battery capacity 243, 254, 256, 260, 261, 262, 263, 265, 268, 269
biofuels 273
BOV 407, 409, 418, 419
brushless doubly fed induction machines (BDFM) 151, 174
building energy management systems (BEMS) 426, 433

C

CDM-SET3 205–222
certified emission reductions (CERs) 206, 207
classical optimization 303, 307, 309
clean development mechanism (CDM) 205, 206, 207, 208, 209, 210, 211, 212, 213, 215, 216, 217, 218, 219, 220, 221
cogeneration 296, 297, 298, 299, 300, 301, 310, 311, 312, 315, 316, 317, 318, 319, 320, 321, 322, 323, 324, 325, 326, 327, 328, 329, 330, 331, 332, 333, 334
composite index 338, 346, 347
computational flow 130, 133
conditional demand analysis 180, 201, 202, 203
continuous professional development (CPD) 102

D

decentralized systems 395
decision support (DS) 97, 98, 100, 101, 102, 114, 120
decision support systems 223, 225, 237
demand side management (DSM) 390, 393, 397
direct drive synchronous generator (DDSG) 149
direct torque control (DTC) 154, 158
doubly-fed induction generators (DFIG) 147–178
dynamic energy compensator (DEC) 42, 44, 73, 75, 77, 78
dynamic programming (DP) 303, 306, 310, 334

E

economic dispatch (ED) 123
ELECTRE III 209, 210, 363, 366
electro magnetic interference (EMI) 149, 151
energy demand forecasting (EDF) 226
energy generation system (EGS) 40, 41, 42, 43, 44, 45, 53, 54, 55, 57, 58, 59, 61, 62, 63, 64, 65, 66, 68, 69, 70, 71, 72, 73, 74, 75, 78, 81, 82, 83, 87, 89, 90

Energy Information Administration (EIA) 226
energy modeling 180, 182, 183, 198
energy not served (ENS) 388
energy planning 356, 358, 371
energy service companies (ESCOs) 423, 425, 432
energy storage device (ESD) 41, 42, 43, 44, 45, 57, 59, 61, 64, 65, 66, 68, 69, 70, 72, 75
engineering model (EM) 180, 181, 182, 188, 195, 196, 197, 198, 199, 200, 201, 203
environmental/economic dispatch (EED) 123, 124, 125, 126, 138, 142
error variance 346
expert systems 1, 3, 4, 6, 8, 9, 25, 28, 30, 36, 39, 388
externalization 102

F

field oriented control (FOC) 154
fixed speed generators (FSGs) 149
flux magnitude angle control (FMAC) 154
flywheel control 157
fuzzy-based mechanism 126, 131, 132, 139
fuzzy logic 1, 4, 6, 24, 25, 27, 28, 30, 37, 388
fuzzy logic controller 40, 82, 87
fuzzy optimization 303, 317, 327

G

general nonlinear optimization (GNLO) 303, 307, 309, 311, 312, 334
generation capacity constraint 127
generation efficiency 244, 247, 264, 265, 271
generation expansion planning 226, 238
genetic algorithm 1, 26, 27, 28, 29, 30, 33, 35, 36, 37, 38, 40, 47, 48, 49, 51
geothermal energy 216
greenhouse gas 180, 205, 224, 272, 273, 292, 424, 435

H

Hankel singular values (HSV) 164
H∞ Control 147, 160
heating, ventilation, and air-conditioning (HVAC) system 426, 436
hybrid system 1, 6, 24, 27, 28, 29, 30, 394

I

independent power producers (IPP) 357, 378
integrated fuzzy control (IFC) 40, 81, 82, 83, 84
integrated resource planning (IRP) 225, 226, 238
intelligent decision support 423, 426, 432, 435
intelligent system 356, 358, 365, 370
intelligent techniques 296, 297, 298, 300, 308, 309, 312, 317
internalization 103
International Atomic Energy Agency (IAEA) 226
International Energy Agency (IEA) 226, 358

J

Joint Centre for Energy Management 184

K

Kirchhoff's voltage law 152
knowledge capture 99, 105, 106, 107, 108, 109, 114, 115, 116
knowledge discovery in databases (KDD) 111
knowledge management (KM) 97–122
Kyoto Protocol 180, 181, 200, 205, 206, 220

L

learning algorithm 188, 189, 193, 194
life cycle assessment (LCA) 282, 283, 285, 286, 287, 290
linear programming (LP) 137, 139, 303, 304, 305, 306, 307, 309, 310, 311, 312, 315, 334
load flow 391, 393
long-run marginal cost (LRMC) 385
Loss of load probability (LOLP) 388

M

Matlab 40, 42, 51, 61, 62, 73, 164
meta-heuristics 301, 303, 316, 317, 318
methane combustion 216, 218
model reference adaptive system (MRAS) 157, 168, 170
motor efficiency 400, 417, 419
multi-attribute utility theory (MAUT) 304, 313, 334
multicriteria decision making (MCDM) 357, 358, 360, 375
multi-layer perceptron (MLP) 183, 184, 185, 186, 187, 194
multiobjective optimization 123, 124, 125, 126, 128, 129, 137, 142, 143, 144
multiobjective particle swarm optimization (MOPSO) 123, 125, 126, 129, 130, 133, 134, 135, 137, 138, 139, 140, 141, 142, 143
multiobjective stochastic search technique (MOSST) 137, 139

N

Natural Gas Vulnerability Index (NGVI) 343, 351, 352
neural networks 1, 3, 6, 10, 18, 23, 24, 27, 28, 29, 30, 32, 33, 34, 35, 36, 37, 38, 39, 180, 182, 183, 200, 201, 202, 203, 204
non-linear programming (NLP) 303, 304, 306, 307, 308, 309, 310, 311, 312, 316, 334

O

Oil Vulnerability Index (OVI) 343, 351, 352
operation plan 242, 243, 256, 257, 258, 259, 261, 263, 268, 270
outranking methods 304

P

Paretooptimal 124, 125, 128, 129, 130, 142
Pareto-optimal solutions 123, 140, 142
particle swarm optimization (PSO) 389
PAS algorithm 242, 243, 248, 251, 253, 254, 255, 257, 258, 259, 260, 261, 262, 269, 270
passivity-based controller (PBC) 157
peak current controller (PCC) 40, 42, 43, 52, 53, 56, 57, 58, 59, 61, 63, 64, 65, 66, 68, 69, 70, 71, 72, 73, 74, 75, 78, 79, 80, 81, 82, 83
permanent magnet synchronous machine (PMSM) 156
photovoltaic 215
pitch control 155, 159
poly-generation 296, 297, 298, 299, 300, 301, 302, 306, 310, 311, 312, 315, 316, 317, 318
power balance constraints 127, 133
power purchase agreements (PPA) 378
power quality 400, 401, 418, 419, 420, 421
power system planning 377, 378, 387
prediction algorithm 242, 261, 262, 269, 271
principal component analysis (PCA) 338, 345
proton exchange membrane fuel cells (PEMFC) 42, 43, 44, 46, 47, 48, 49, 50, 51, 52, 53, 56, 57, 61, 62, 63, 68, 69, 77, 78, 79, 81, 82, 83, 84, 87, 88, 89, 90, 92, 93, 95
public private participation (PPP) 378, 383
pulse width modulation (PWM) 150, 158, 159, 168, 173
pump efficiency 400, 415, 416, 417, 418, 419

Q

quadratic programming (QP) 303, 312, 335

R

Reference Energy and Materials System (REMS) 276, 278, 281, 284, 285
regression analysis 388, 389
renewable energy 224, 225, 232
renewable energy sources (RES) 356, 357, 360, 361, 363, 369, 370
robust decision making (RDM) 337, 338, 339, 340
rotor side converter (RSC) 154

S

scalar control 155
security constraints 127
sensitivity analysis 275
sequential nonlinear programming (SNLP) 303, 335
sequential quadratic programming (SQP) 303, 307, 335
sliding mode control 154, 158
socialization 102
solar cell 242, 243, 244, 246, 247, 248, 249, 250, 251, 252, 253, 254, 255, 256, 257, 258, 260, 261, 262, 263, 264, 265, 266, 268, 269, 270, 271
solar lanterns 216
solar power generation 242, 246, 248, 271
solar radiation 243, 246, 247, 248, 249, 250, 251, 252, 253, 256, 257, 258, 261, 265, 267, 268, 269, 270, 271
solar tower technology 216
stochastic optimization 303, 317, 330, 333
stochastic search 124, 137, 143
strategic planning 275
strength Pareto evolutionary algorithm (SPEA) 137, 139, 140, 141
supply risk 338, 339, 343, 344, 345, 348
supply side management 226
Survey of Household Energy Use (SHEU) 192
sustainable energy technology 205, 206, 208, 211
system efficiency 400, 415, 416, 417, 418, 419

T

tailored heuristics 303, 304, 318
teaching data 243, 249, 250, 251
thin metal film (TMF) battery 41
tri-generation 296, 297, 298, 299, 300, 301, 310, 311, 312, 315, 316, 317, 318

U

Ultracapacitors 42
unbalanced equal voltage (UBEV) 407, 409, 419
unbalanced overvoltage (UBOV) 406, 409, 419
unbalanced undervoltage (UBUV) 406, 409, 419
upper limit 408, 410, 419

V

variable speed constant frequency (VSCF) 150
vector control 153, 154, 155, 156, 157, 171
voltage unbalance factor 400, 402, 406, 408, 409, 421, 422

W

wind energy 211, 216, 220
wind turbine 147, 148, 149, 150, 151, 152, 154, 155, 158, 159, 160, 162, 166, 169, 170, 172, 173, 174, 175, 176, 177